AF787179

Marine Renewable Energy

Marine Renewable Energy

Editors

Eugen Rusu
Kostas Belibassakis
George Lavidas

MDPI

Basel • Beijing • Wuhan • Barcelona • Belgrade • Novi Sad • Cluj • Manchester

Editors

Eugen Rusu
University Dunarea de Jos of Galati
Galați, Romania

Kostas Belibassakis
National Technical University of Athens
Athens, Greece

George Lavidas
Delft University of Technology (TU Delft)
Delft, The Netherlands

Editorial Office
MDPI
St. Alban-Anlage 66
4052 Basel, Switzerland

This is a reprint of articles from the Topic published online in the open access journals *Journal of Marine Science and Engineering* (ISSN 2077-1312), *Processes* (ISSN 2227-9717), and *Energies* (ISSN 1996-1073) (available at: https://www.mdpi.com/topics/marine).

For citation purposes, cite each article independently as indicated on the article page online and as indicated below:

Lastname, A.A.; Lastname, B.B. Article Title. *Journal Name* **Year**, *Volume Number*, Page Range.

ISBN 978-3-0365-9134-6 (Hbk)
ISBN 978-3-0365-9135-3 (PDF)
doi.org/10.3390/books978-3-0365-9135-3

Cover image courtesy of Eugen Rusu

© 2023 by the authors. Articles in this book are Open Access and distributed under the Creative Commons Attribution (CC BY) license. The book as a whole is distributed by MDPI under the terms and conditions of the Creative Commons Attribution-NonCommercial-NoDerivs (CC BY-NC-ND) license.

Contents

Journal of
Marine Science and Engineering

MDPI

Article

Vertical Stiffness Functions of Rigid Skirted Caissons Supporting Offshore Wind Turbines

AbdelRahman Salem [1,2], Saleh Jalbi [3,*] and Subhamoy Bhattacharya [2]

1 Arcadis Consulting, London CR0 1EA, UK; abdelrahman.salem@arcadis.com
2 Department of Civil and Environmental Engineering, University of Surrey, Guildford GU2 7XH, UK; S.Bhattacharya@surrey.ac.uk
3 Sea & Land Project Engineering, London SE1 1UN, UK
* Correspondence: saleh.jalbi@seaandland.co.uk

Abstract: Suction Bucket Jackets (SBJs) need to be fundamentally designed to avoid rocking modes of vibration about the principal axes of the set of foundations and engineered towards sway-bending modes of tower vibration. Whether or not such type of jackets exhibit rocking modes depends on the vertical stiffness of the caissons supporting them. This paper therefore derives closed form solutions for vertical stiffness in three types of ground profiles: linear, homogenous, and parabolic. The expressions are applicable to suction caissons having an aspect ratio (depth: diameter) between 0.2 and 2 (i.e., 0.2 < L/D < 2). The work is based on finite element analysis followed by non-linear regression. The derived expressions are then validated and verified using studies available in literature. Finally, an example problem is taken to demonstrate the application of the methodology whereby fundamental natural frequency of SBJ can be obtained. These formulae can be used for preliminary design and can also be used to verify rigorous finite element analysis during detailed design.

Keywords: suction caissons; vertical stiffness functions; natural frequency; jackets; offshore wind turbines

Citation: Salem, A.; Jalbi, S.; Bhattacharya, S. Vertical Stiffness Functions of Rigid Skirted Caissons Supporting Offshore Wind Turbines. *J. Mar. Sci. Eng.* **2021**, *9*, 573. https://doi.org/10.3390/jmse9060573

Academic Editor: Eugen Rusu

Received: 27 April 2021
Accepted: 18 May 2021
Published: 26 May 2021

Publisher's Note: MDPI stays neutral with regard to jurisdictional claims in published maps and institutional affiliations.

cc BY

Copyright: © 2021 by the authors. Licensee MDPI, Basel, Switzerland. This article is an open access article distributed under the terms and conditions of the Creative Commons Attribution (CC BY) license (https://creativecommons.org/licenses/by/4.0/).

1. Introduction

The European Union's (EU) strategy to fight climate change and air pollution issues has accelerated the investments into sustainable energy sources. This is essentially to meet the targets of 55% reduction in the greenhouse emissions by 2030 as well as paving the way for climate neutrality by 2050 [1]. Offshore wind in Europe in particular has witnessed a substantial growth, with the UK leading the market (10,428 MW cumulative capacity) and is expected to add 15 GW capacity in the next 5 years. Countries with large offshore wind developments also include Germany (7689 MW capacity), Belgium (2261 MW capacity), and the Netherlands (2611 MW capacity) [2]. Other global leaders include mainland China (approximately 10 GW capacity) [3]. There are new entries to the market including Taiwan (through the Formosa 1 and 2 offshore wind farms) and in the final planning stages for the East Coast of the United States.

This growing demand for renewable energy is responsible for the rapid pace of the technological developments emerging in the industry. Such developments mainly target the turbine size and installations in deeper waters. Wind Europe [2] has stated that the rated capacity of OWT has been enhanced by 102% in the last 20 years. At present, the newly-installed turbines have an average rated capacity of 7.8 MW, though most new installments have turbine capacities exceeding 10 MW. Monopiles support 81% of all installed OWT in Europe [2]. Yet, for some locations, their design poses several engineering challenges and environmental issues to satisfy the requirements needed to support larger wind turbines in deeper waters. This explains the continuous efforts to innovate in this field. Recently, the trend in the construction of new wind farms has considered OWTs supported on jackets as an attractive alternative to conventional monopiles in deeper waters.

Jackets supported on piles or caissons, illustrated in Figure 1, are suitable for water depths of 30–60 m which allow them to be potentially used for future rounds of wind energy developments [4]. Aberdeen and Borkum Riffgrund 2 offshore wind farms are recent developments utilising caisson jackets foundations to support 8 MW turbines [5]. Other on-going projects include Seagreen and Zhuanghe 2.

Figure 1. Schematic of a 3-legged jacket on suction caissons reproduced from [6], with permission from Ørsted, 2019.

There is an inherent difference in the way monopiles and jackets resist overturning moments due to lateral loads. As illustrated in Figure 2, single foundations (typically monopiles) transfer the loads via overturning moments to the surrounding soil. On the other hand, multiple foundations such as jacket on piles/caissons mainly transfer the loads through axial push-pull interaction. This obvious difference will later drive the simple mechanical modelling of the system where the foundation is replaced by equivalent springs for analysis purposes. Hence whilst the lateral stiffness plays a major role in the dynamic performance of monopile supported offshore wind turbines, the vertical stiffness plays a more detrimental role in jackets due to the propensity of rocking type of vibrations. It has shown by Jalbi and Bhattacharya that low-frequency rocking modes of jacket vibration must be avoided as it may coincide with the low frequency 1P rotor frequency and more importantly the peak wave frequency [7].

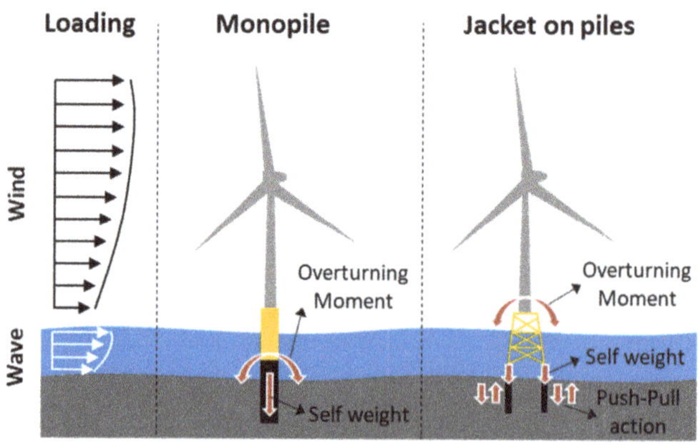

Figure 2. Load transfer in different foundation systems.

Current design aims to place the natural frequency of the bottom-fixed structures within the soft-stiff band, see Bhattacharya [4] for fundamentals of design. It is well established in literature that the natural frequency of the system is reliant on the support condition (i.e. foundation stiffness) which in turn is a function of the properties of the foundation and subsoil. Considering the dynamic sensitivity of the OWTs, modes of vibration are considered a key element in the design procedure. Similar to the load transfer process discussed above, the modes of vibration of an OWT system are primarily dependent on the foundation and superstructure stiffness [8].

Studies carried out by Bhattacharya et al. [9] showed that the first eigenfrequency of vibration for OWTs supported on multiple shallow foundations (such as jackets on three or four suction caissons) correspond to low frequency rocking modes of vibration about the principle axes. The work is based on scaled model tests on three types of foundations: monopiles, tetrapods (4-legged jacket on caissons), and asymmetric tripods (3-legged seabed frame on caissons). Rocking modes of vibration are also reported in offshore structures such as the Brent B Condeep platform, see [4].

As mentioned before, rocking modes of vibration tend to have a lower frequency and may interfere with the 1P (rotor) frequency range and wave frequency, see Figure 3 for schematics. This is particularly challenging for large turbines where the soft-stiff target frequency is shifting towards the wave frequency. For example, a typical 8 MW turbine will have a target of 0.22 Hz and a 12 MW turbine will have a target frequency in the range of 0.15 Hz. Furthermore, wave loads will have a higher energy of excitation and may impose serious fatigue damage on the structure if rocking modes are allowed. It is therefore advisable to avoid rocking modes for jackets supported on shallow foundations. In addition, for asymmetric arrangements, scaled model tests showed that they have experienced two closely-spaced natural frequencies associated with the rocking modes of vibration [10]. This corresponds to the variability of the ground reflected in the vertical stiffness of the foundation. Not only does it widen the range of frequencies that can be excited by the loading conditions but also may introduce an additional design problem such as the beating phenomenon and both can have an impact in the fatigue limit state. Moreover, through analytical methods, Jalbi et al. [11] and Jalbi and Bhattacharya [7] showed that that a jacket may be engineered towards a no-rocking solution by optimising two parameters: (a) ratio of vertical stiffness of the foundation stiffness to lateral superstructure stiffness; and (b) aspect ratio of the jacket-tower geometry. A low value of vertical foundation stiffness values together with a low aspect ratio will promote a rocking mode of vibration.

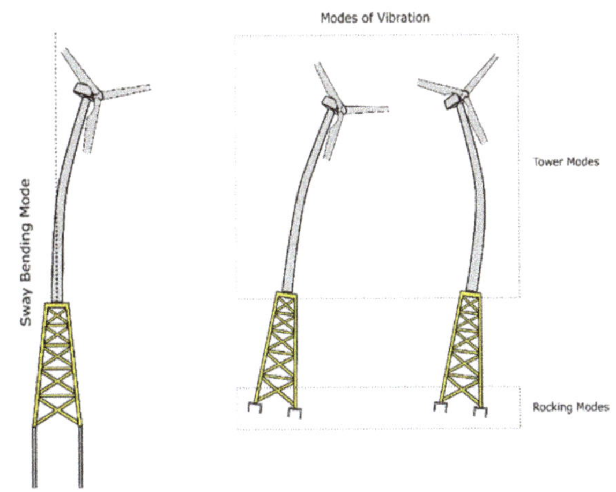

Figure 3. Vibration modes in different foundation systems.

From the discussions above, it is essential to have a method to calculate the vertical stiffness of foundations early on in the design stages of a project. Hence, given the importance of SSI on the dynamic performance, the objectives of the paper are as follows:

(1) Carry out review of the current methods in literature to predict the vertical stiffness of rigid caissons.
(2) Provide static vertical stiffness functions for caissons of aspect ratio between 0.2 and 2 i.e., $0.2 < L/D < 2$ for three types of ground: homogeneous, parabolic and linear profiles. This is based on the approach laid out in Eurocodes for ground types (see Eurocode 8 Part 5) and also a gap in the literature.
(3) Demonstrate the application of the developed methodology through a step-by-step solved example in the context of predicting the natural frequency of the system.

It should be noted that the solutions provided in this paper are intended for the concept design stage and for initial sizing of the foundation when information about the structure and the ground profile is scarce. As the design progresses from conceptual to detailed design, a higher computational complexity of the analysis is required to further optimise the foundations. This includes using refined soil constitutive models incorporated in 3D finite element analysis (FEA) packages. In addition, this would also require more input such as site-specific ground investigations and geotechnical laboratory testing.

Background Literature

Ideally, each assessment of the composite system should encompass an independent numerical analysis for the structure and foundation. For instance, the analysis of the latter would likely involve modelling the soil as continuum which is typically carried out using advanced geotechnical finite element methods. However, the limitation of the high computational cost and modelling complexities make it impractical to be utilised in preliminary design stages, yet useful in verifying the final design of the foundation. Consequently, both approaches (i.e., analytical and numerical solutions) tend to idealise the structural dynamics problem through replacing the foundation by a set of lumped springs or in the case of deep foundations distributed springs. The overall stability and foundation stiffness can be purely expressed in terms of functions that describe the force resultants and their conjugate displacements and rotations of these lumped springs. Figure 4 illustrates the breakdown of the structure-foundation problem.

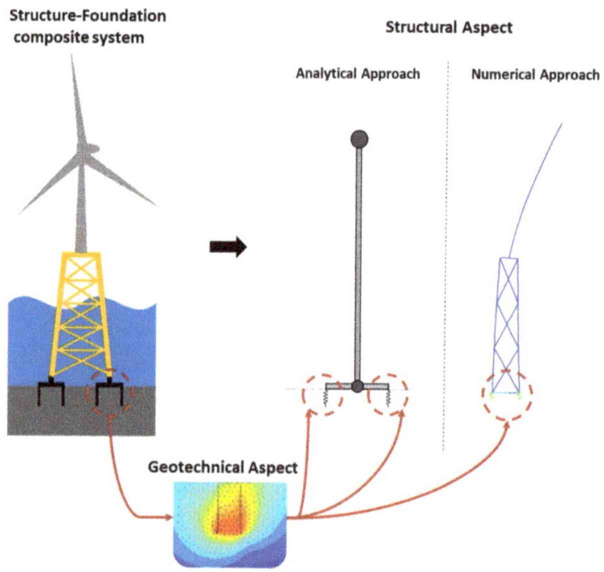

Figure 4. Breakdown of the structure-foundation problem.

The work of this paper continues the efforts of the research group which aims at providing simplified expressions for the computation of the foundation stiffness. For instance, Shadlou and Bhattacharya [12] studied the lateral dynamic stiffness of deep foundations and proposed spring stiffness functions for both rigid and flexible monopiles. In their model, the foundation is replaced by four springs; K_L (lateral spring), K_R (rocking spring), K_V (vertical spring) and K_{LR} (cross-coupling spring) to capture the degrees of freedom. This methodology was later utilised by Arany et al. [13] in designing monopile foundations. Similarly, Jalbi et al. [14] obtained static stiffness functions for the lateral stiffness terms of a rigid monopod caissons. Moreover, Jalbi and Bhattacharya [15] provided closed form solutions to calculate the natural frequency of jackets supported on multiple foundations incorporating soil-structure interaction (SSI). The foundation flexibility was represented by a set of vertical springs which emphasizes the importance of predicting the vertical stiffness of foundations. Thus, it is now essential to continue the work and obtain the vertical stiffness components of the foundations.

The assessment of static and dynamic vertical spring constants has been the subject of extensive studies in the field of machine foundations and seismic analysis. Most of the elastic solutions available in the literature provide guidance regarding surface and embedded footings. Generally, most literature reports that stiffness decreases with increasing strains and increasing forcing frequencies. However, work on the elastic and non-linear stiffness of a skirted caisson is inadequate. Bell [16] presented a comprehensive review of the existing stiffness coefficients of surface footings, whereas the effect of embedment of the circular footings was extensively discussed in Gazetas [17]. On the other hand, there is less work assessing the stiffness of suction caisson foundations. These foundations are quite similar to the embedded-type foundations with the difference of the soil mass is trapped beneath the lid and within the enclosed volume. Two extreme models could be adopted to represent the caissons: one in which the lid is treated as a rigid circular foundation on the surface while ignoring the effect of the skirts, and another in which the caisson is completely rigid. The latter was analysed by Doherty et al. [18] who provided tabulated coefficients for completely rigid caisson. The analyses incorporated practical variation in soil stiffness, embedment depth and Poisson's ratios. It also provided correction factors to account for the skirt flexibility. Skau et al. [19] focused on the effect of caisson flexibility following the observations of the extensive behaviour monitoring for Borkum Riffrung 01 -Suction Bucket Jacket (BKR01-SBJ). An elastic correction to the response of a rigid foundation response was suggested to address the foundation flexibility particularly due to the lid which appeared to significantly influence the total vertical stiffness of the system. Table 1 summarises the vertical stiffness formulae found by different researchers for different foundation types including some additional guidance found for deep foundations.

Table 1. Methods to estimate the vertical stiffness of the foundation in literature.

	Source (Year) [Reference]	Formulae and Their Applications
Surface Foundation	Lysmer (1965) [20] Spence (1968) [21]	Circular rigid footing on surface of homogenous elastic half-space: $k_v = \frac{2G_sD}{1-y_s}$ To account for the roughness of the footing base that allow full transmission of shear stress, Spence [21] proposed the following: $k_v = \frac{2G_sD\ \ln(3-4y_s)}{1-2y_s}$ The results from this analytical solution showed up to 10% increase in stiffness values at low y_s values
	Gazetas (1983) [22] DNVGL (2019) [23]	Circular footing on stratum over bedrock: $k_v = \frac{2G_sD}{1-y_s}\left(1 + 1.28\frac{D}{2H}\right)$
	Gazetas (1991) [17]	Arbitrary shaped foundation on surface of homogenous half-space: $k_v = \frac{2G_sl}{1-y_s}\left(0.73 + 1.54\chi^{0.75}\right)$ where $\chi = A_b/4l^2$, l is the base-length of the circumscribed rectangle and A_b is the area

Table 1. *Cont.*

	Source (Year) [Reference]	Formulae and Their Applications
Shallow Embedded Foundation	Gazetas (1983) [22] DNVGL (2019) [23]	Circular rigid footing embedded in homogenous stratum over bedrock; Developed for machine-type inertial loading. Range of validity: L/D<1 $k_v = \frac{2G_sD}{1-y_s}\left(1+1.28\frac{D}{2H}\right)\left(1+\frac{L}{D}\right)\left[1+\left(0.85-\frac{0.28L}{0.5D}\right)\left(\frac{L}{H-L}\right)\right]$
	Wolf (1988) [24] Wolf and Deeks (2004) [25]	General prismatic footing embedded in a linear elastic half space $k_v = \frac{G_sb}{1-y_s}\left(3.1\left(\frac{l}{b}\right)^{0.75}+1.6\right)\left(1+\left(0.25+0.25\frac{b}{l}\right)\left(\frac{e}{b}\right)^{0.8}\right)$ where 2l, 2b are the base dimensions of circumscribed rectangle and e is the embedment depth. This formulation was later simplified by Wolf and Deeks [25]; $k_v = \frac{2G_sD}{1-y_s}\left(1+1.08\left(\frac{e}{D}\right)\right)$
	Gazetas (1991) [17]	Arbitrary shaped foundation embedded in half-space $k_v = k_{v,\ surface}\left[1+\left(\frac{1}{21}\right)\left(\frac{L}{b}\right)+(1+1.3\chi)\right]\left[1+0.2\left(\frac{A_w}{A_b}\right)^{2/3}\right]$ where $k_{v,\ surface}$ is obtained using the equation provided earlier by Gazetas [17]for surface footings, b is the base-width of the circumscribed rectangle, $\chi = A_b/4l^2$ and A_W is the actual sidewall-soil contact area; for constant effective-contact height, d, along the perimeter: A_W = (d) × (perimeter). Based on Gazetas' methodology, Bordón et al. [26] developed a simplified formula for the stiffness of a rigid cylinderical foundation embedded in homogenous soil to study the group effect of multi-bucket foundations: $k_v = \frac{2G_sD\ \ln(3-4y_s)}{1-2y_s}\left[1+1.12(1-0.84y_s)\left(\frac{L}{D}\right)^{0.84}\right]$
Deep foundation	Fleming et al. (1992) [27]	Embedded piles considering shaft friction only: $k_v = \frac{2\pi LG_s}{y}$ where *St* is between 3 and 5
	Shama & El Naggar (2015) [28]	Single pile under axial load for seismic design of highway bridges: $k_v = \frac{1.25E_pA}{L}$

N.B: K_V is vertical stiffness of the foundation, G_s is shear modulus of the soil, y_s is Poisson's ratio of the medium, D is diameter, L is embedment depth, H is thickness of the soil layer and E_p is modulus of elasticity of pile material.

From the table above, it is evident that the available methodologies are limited either by the shape of the footing and the idealised soil profiles which do not reflect the actual heterogeneity in the soil. This paper aims to tackle one aspect of that where solutions are provided for rigid caissons through numerical modelling. The solutions provide the vertical stiffness K_v are for homogeneous, parabolic, and linear ground profiles.

2. Numerical Modelling

Finite element method using Plaxis 3D (continuum approach) was utilised to model the soil-structure interaction. The size of the soil contour is specified such that any stress increase on the boundary is absorbed without rebounding and disturbing the model results. Suryasentana et al. [29] presented a mesh domain of 80D (D is the diameter of the suction caisson) for both diameter and depth to analyse vertically loaded foundations, while Latini et al. [30] used 100D and 30D for the diameter and depth respectively. Moreover, Sloan [31] adopted 5D for the mesh dimensions when analysing vertically loaded rigid circular footing. This clearly shows the wide range of possibilities to eliminate the boundary effects. Considering the scope of this analysis, all the models have been set up with an extent of the soil domain 10D and depth of 15D; as shown in Figure 5 (which has been obtained through trial and error). Despite the symmetry of the problem, a full model was adopted as to avoid a rotation of the caisson if the point load was placed at the centre.

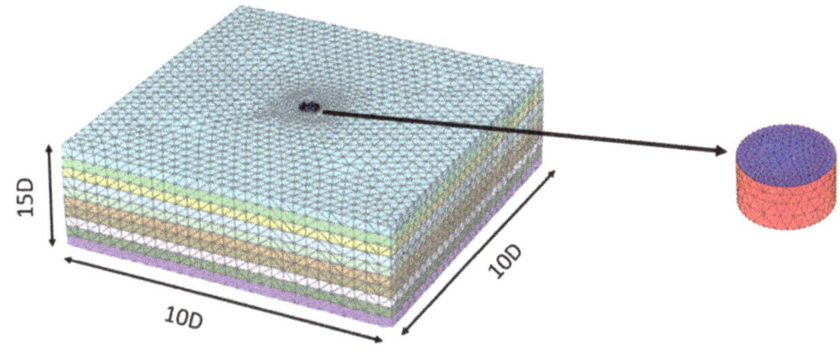

Figure 5. Mesh dimensions and shape of caissons.

Generally, the stiffness of the foundation dictating the dynamic stability of the system, is characterised by a non-linear nature. It is dependent on the strain levels; generated from the load cycles due to the soil-structure interaction, as well as the forcing frequency (expressed in terms of static and dynamic stiffness) [32]. Since the natural frequency is associated with relatively small amplitude of vibrations (linear range), the initial foundation stiffness would be sufficient for this purpose [13]. Similarly, OWTs are considered a very low frequency application compared to seismic actions based on design charts provided in [12,20,33–36]. Hence, the effect of the forcing frequency on the stiffness values can be ignored and the static stiffness value can effectively be adopted. Based on the above justifications, the soil is modelled as a linear elastic material. This model is based on Hooke's law of isotropic elasticity and requires the identification of two basic elastic parameters; Young's modulus (E) and Poisson's ratio (v_s) which can be determined using conventional site investigation techniques [37].

To encompass the realistic variation of the shear modulus of the soil with depth, three idealisations of the ground profile were adopted:

- *Homogenous soil*: typical for overconsolidated clays that show constant variation in stiffness with depth and can be easily defined on Plaxis 3D;
- *Linear inhomogeneity*: common for normally consolidated clay, sometimes referred to as Gibson soil where the stiffness increases linearly with depth [18]; and
- *Parabolic inhomogeneity*: is an intermediate condition which is typical for sandy soils [12]. Unlike the other two ground profiles, the soil stratum has to be discretised into multiple layers on Plaxis 3D to incorporate the parabolic variation of the soil stiffness. 10 distinct layers of 0.1H thickness each (H is the depth of the soil stratum equivalent to 15D) were used to model the soil stratum. For each layer, an initial stiffness and linear slope were used to mimic the parabolic behaviour.

Figure 6 illustrates the three ground profiles described above where the values of Young's modulus for the different ground profile intersect at one diameter depth. For all the cases, Poisson' ratio was assumed to be uniform within each model and the soil density was set to a constant value of 18 kN/m^3.

The structure forming the foundation, which consists of the skirt and the lid, has been treated under the assumption of being rigid for all the models. In other words, the response of the foundation system due to the applied loads is solely due to the deformations in the soil (no structural deformations of the caisson lid and skirt). This also includes the soil enclosed within the skirt. This assumption is considered valid considering the low aspect ratio of the caissons modelled and the high flexural and shear stiffness of the steel compared to those of the soil. Doherty et al. [18] investigated the effect of the skirt flexibility on the response of a caisson foundation. Results showed that the vertical stiffness values of both cases are almost similar for low aspect ratios of the caisson which reinforces and validates the previous assumption. In addition, the bucket lid can in-reality deflect and alter the foundation stiffness where the recent in-situ observations by Shonberg et al. [38] confirmed that the suction bucket's structural elements stiffness has an effect on the performance and can be idealised as pair of vertical springs in series, such that $\frac{1}{K_{total}} = \frac{1}{K_{lid}} + \frac{1}{K_{soil-skirt}}$. However, for simplicity, a lumped vertical spring compiling both elements is adopted for this study rather than treating them separately.

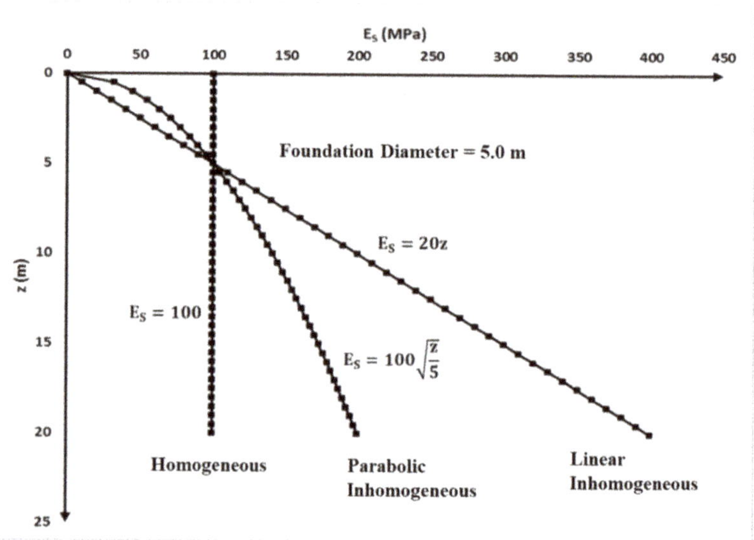

Figure 6. Variation in stiffness with depth.

Furthermore, the push-pull nature for OWT supported on jackets requires the estimation of the stiffness in both tension and compression, yet this study assumes that the vertical stiffness is the same for both tension and compression as the intention is to use these values for low amplitude vibrations. In reality, the computed compressive stiffness should be higher due to the additional contribution of the bearing below the lid. Other factors should further be investigated involving the impact of the grouted connections and its imperfections, interaction between the adjacent jacketed caissons and imperfect contact at the interface of the soil and foundation. All these factors form a strong basis for the continuation of the work produced in this study. Nevertheless, it may be reminded that the solutions provided in this paper are intended for the concept design stage and for initial sizing of the foundation when information about the structure and the ground profile is limited.

Methodology Verification and Comparison of Results

Normalized values for K_V are plotted against L/D for 0.2 < L/D < 2 considering the three different ground profiles. The results are then compared to the solutions provided by Wolf and Deeks [25] and DNVGL [23] which are in turn based on the work of Gazetas [22]. Figure 7 shows the stiffness coefficients plotted for homogeneous profile at two Poisson's ratios (0.2 and 0.499). All the simulated cases are summarised in Appendix A.

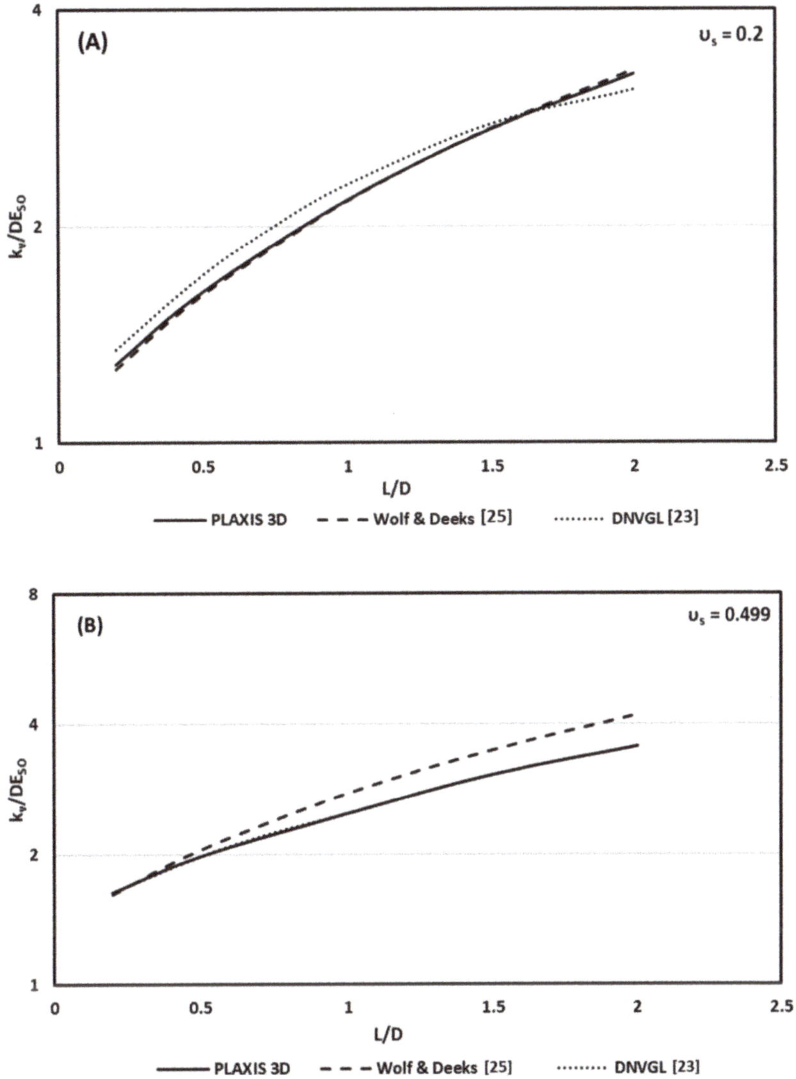

Figure 7. Vertical static stiffness functions for rigid caissons in homogeneous ground profiles at two Poisson's ratios (**A**) $\upsilon_s = 0.2$, and (**B**) $\upsilon_s = 0.499$.

The numerical model compares well with the formulations provided in the literature which justifies the method of extraction, the mesh used, the extent of the boundary conditions, and the rigid body assumption applied in the finite element model. Any discrepancy can be justified in terms of the foundation geometry implemented; where both Wolf and Deeks [25] and DNVGL [23] used a solid embedded foundation rather than a skirted caisson with soil mass enclosed within it. Another reason could be the effect of the Poisson ratio.

The impact of Poisson's ratio on the computed stiffness coefficients was evident from the FE analysis. Therefore, normalised values for the vertical foundation stiffness coefficients in a homogeneous stratum were plotted against υ_s, as shown in Figure 8. The results were normalised against their respective values at $\upsilon_s = 0.1$. The aspect ratio (L/D) also appears to influence the effect of Poisson's ratio as also shown in Figure 8. For all L/D cases, the stiffness decreases with increasing υ_s until υ_s = 0.4 and then slightly increases. Both Gazetas [22] and Wolf and Deeks [25] considered the effect of the soil's Poisson's ratio on the stiffness coefficients and incorporated it in their proposed impedance functions. Their proposed functions show similarity with the trend predicted by this study, yet, the impact of the Poisson's ratio is observed to be slightly lower than the abovementioned literature. Accordingly, a correction factor $f(\upsilon_s)$, function of both υ_s and L/D, will be developed in the subsequent section.

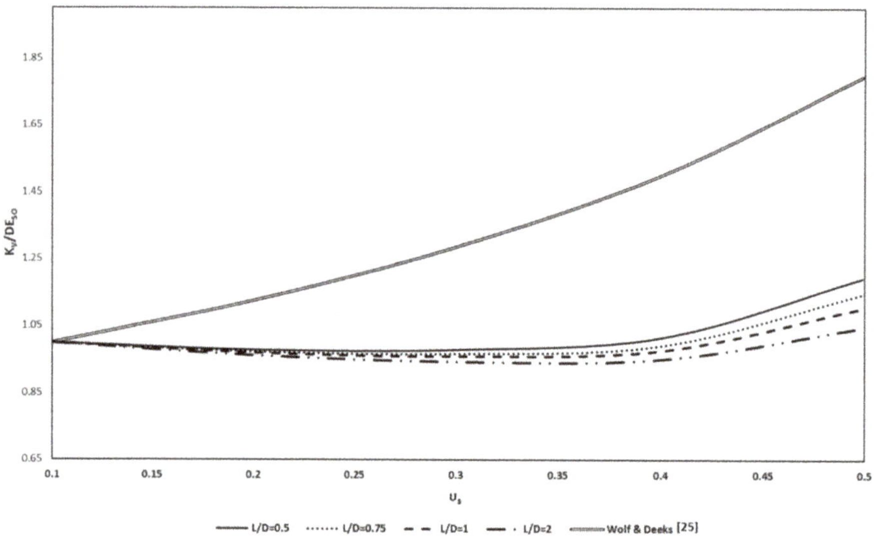

Figure 8. Variation of vertical stiffness (K_v) with Poisson's ratio (υ_s) and aspect ratio (L/D).

3. Development of the Static Stiffness Functions and Correction Factors

Functions for the vertical stiffness were developed using non-linear regression analysis on the normalised set of stiffness coefficients. A diameter of 5m was adopted and the soil stiffness at 1m depth was set as 100 (MPa), see Figure 5. The study then plots the change of the vertical stiffness with increasing L (L values include 2.5, 5, 7.5 and 10 m) and υ_s (0.1 to 0.49).

First, a new correction factor due to Poisson's ratio, $f(\upsilon_s)$, will be presented. The reason for this correction is to clarify how the Poisson's ratio modifies the vertical stiffness for different ranges of caisson dimensions. Thus, Figure 8 shows that $f(\upsilon_s)$ is not only a function the Poisson's ratio itself but also a function of L/D. This has not been extensively discussed in literature where the correction factor was only a function of the Poisson's ratio itself. Consequently, revised $f(\upsilon_s)$ functions dependent on both L/D and υ_s are suggested herein.

From Figure 8, the best fit curve for the Poisson's ratio correction was a cubic function in the form of $f(\upsilon_s) = a_0\upsilon_s^3 - a_1\upsilon_s^2 + a_2\upsilon_s + a_3$. The values of the coefficients a_0, a_1, a_2, and a_3 were recorded for L/D = 0.5, 0.75, 1,1.5, 2 normalized at L/D = 0.5. The coefficients for L/D = 0.5 are a_0 = 10.028, a_1 = -5.8814, a_2 = 0.9092, and a_3 = 0.96. Figure 9 shows the normalized values for a_0, a_1, a_2, and a_3, thus simulating the dependency of $f(\upsilon_s)$ on L/D. From the figure, it is evident that a_0 and a_1 follow a similar trend and where given the same logarithmic function whilst a_2 followed a different trend and was given another logarithmic function. Finally, a constant value of 1 was given for a_3.

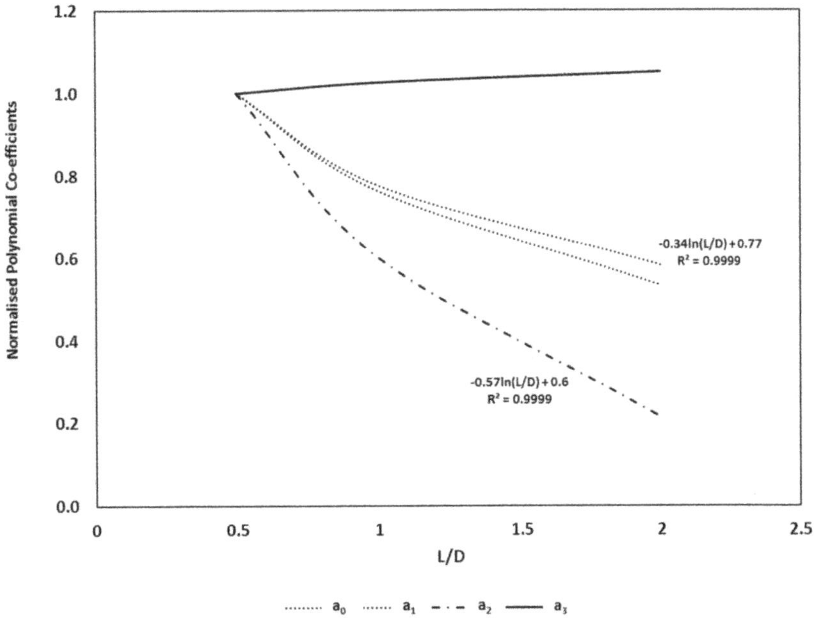

Figure 9. Variation of the polynomial coefficients with L/D.

As a result, from the analysis above, the Poisson's ratio correction may be summarized using Equation (1):

$$f(\upsilon_s) = \left[\left(10\upsilon_s^3 - 5.88\upsilon_s^2\right)\left(-0.34\ln\frac{L}{D} + 0.77\right)\right] + 0.91\upsilon_s\left(-0.57\ln\frac{L}{D} + 0.6\right) + 1 \quad (1)$$

It may be noted that the same methodology was repeated in parabolic and linear inhomogeneous ground profiles where the $f(\upsilon_s)$ obtained was in close proximity to the one shown in Equation (1). Similarly, a few trials on higher aspect ratios (L/D > 2) showed no noticeable change in the formulations:

Subsequently, the normalized values of $\frac{K_V}{DE_{SO}f(\upsilon_s)}$ were then computed and plotted against L/D for all ground profiles. Therefore, a best fit curve was applied in the form of a power function. Even though previous literature did not specifically use power functions for static vertical stiffness functions of shallow caissons, it still performs accurately for the rigid caissons as the R^2 values shown show a good correlation as shown in Figure 10 and the solutions are summarized in Table 2.

$$f(\upsilon_s) = \left[\left(10\upsilon_s^3 - 5.88\upsilon_s^2\right)\left(-0.34\ln\frac{L}{D} + 0.77\right)\right] + 0.91\upsilon_s\left(-0.57\ln\frac{L}{D} + 0.6\right) + 1$$

Table 2. Vertical stiffness for shallow skirted foundations exhibiting rigid behaviour.

Ground profile	$\frac{K_V}{DE_{SO}f(\upsilon_s)}$
Homogeneous	$2.31\left(\frac{L}{D}\right)^{0.52}$
Parabolic	$2.16\left(\frac{L}{D}\right)^{0.96}$
Linear	$2.37\left(\frac{L}{D}\right)^{1.28}$

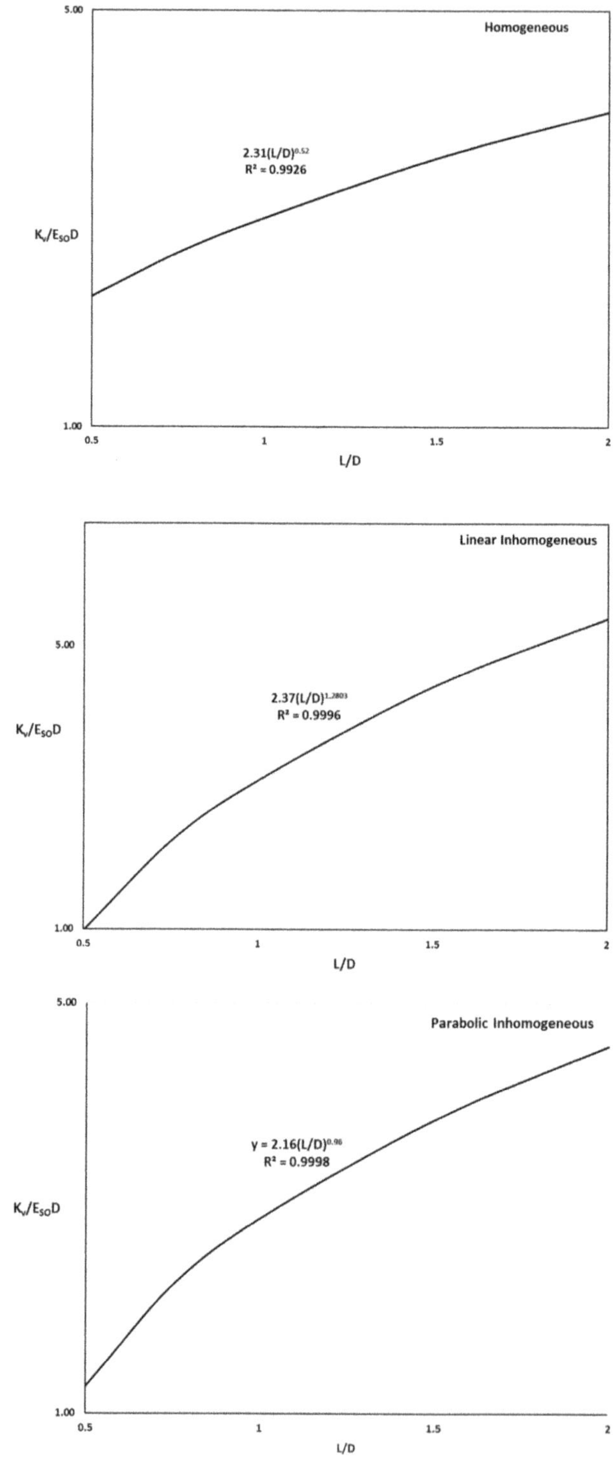

Figure 10. Best fit curves for the vertical stiffness functions.

4. Discussion and Validation of the Results

The functions provided in Table 2 were used to calculate K_V for the simulated cases and the highest recorded percentage was found to be below 10% which is considered of low practical significance. The results are also checked against the coefficients provided by [18] and summarized in Table 3. As shown in the table, the results generally show a good match with the obtained results while being applicable for a wide range of Poisson's ratio. Slight discrepancies exist at the higher Poisson ratio, and this is explained by the difference in the Poisson's ratio correction between the proposed method and existing literature.

Table 3. Comparison of vertical stiffness at $\upsilon_s = 0.2$ and $\upsilon_s = 0.499$. Data from Doherty et al. (2005) [18].

Case		$\frac{K_V}{DE_{SO}}$			$\frac{K_V}{DE_{SO}}$	
		Doherty, et al., 2005 [18]	Proposed method		Doherty et al., 2005 [18]	Proposed method
L/D = 0.5 Homogeneous	$\upsilon_s = 0.2$	1.61	1.65	$\upsilon_s = 0.499$	1.81	1.98
L/D = 0.5 Linear		1.38	1.00		2.10	1.199
L/D = 2 Homogeneous		3.29	3.146		2.86	3.21
L/D = 2 Linear		6.72	5.47		7.60	5.58

5. Application of the Methodology

In order to demonstrate the application of the proposed vertical stiffness functions, a solved example in the context of the prediction of the natural frequency of an OWT system is presented in this section. A 5 MW turbine supported on a symmetrical four-legged jacket was considered for design in deep waters, as shown in Figure 11. Details about the turbine specification and an approximate jacket dimensions were found in Jonkman et al. [39] and Alati et al. [40], respectively, and summarised in Table 4.

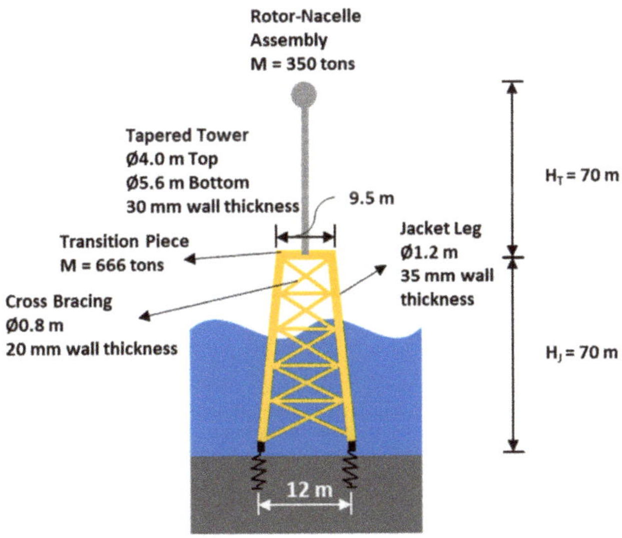

Figure 11. Example problem.

Table 4. Main input parameters of the example (Data from Jonkman et al. [39]; Alati et al. [40]).

Parameter	Value	Unit
Height of the jacket (h_J)	70	m
Jacket bottom width (L_{bottom})	12	m
Jacket top width (L_{top})	9.5	m
Area of jacket leg (A_C)	0.1281	m^2
Distributed mass of the jacket including diagonals (m_J)	8150	kg/m
Tower height (h_T)	70	m
Bottom diameter of the tower (D_{bottom})	5.6	m
Top diameter of the tower (D_{top})	4.0	m
Distributed mass of the tower (m_T)	3730	kg/m
Mass of Rotor-Nacelle Assembly (M_{RNA})	350	tons
Mass of transition piece (M_{TP})	666	tons

A homogeneous stratum over bedrock is assumed with all the foundation dimensions and soil properties summarised in Table 5.

Table 5. Foundation details for the example problem.

Parameter	Value	Unit
Foundation depth (L)	4	m
Foundation diameter (D)	4	m
Depth to bed rock (H)	50	m
Soil Young's modulus (Es)	40	MPa
Soil Poisson's ratio (vs)	0.28	Non-dimensional

Using the equations provided in Table 2, a preliminary estimate of the vertical stiffness for a rigid caisson foundation (with L/D = 1) is obtained as shown below:

$$\begin{aligned} f(\upsilon_s) &= \left[\left(10\upsilon_s^3-5.88\upsilon_s^2\right)\left(-0.34\ln\tfrac{L}{D}+0.77\right)\right]+0.91\upsilon_s\left(-0.57\ln\tfrac{L}{D}+0.6\right)+1 \\ &= \left[\left(10(0.28)^3-5.88(0.28)^2\right)(-0.34\ln(1)+0.77)\right]+0.91(0.28)(-0.57\ln(1)+0.6)+1 = 0.97 \end{aligned}$$

$$K_V = 2.31\left(\tfrac{L}{D}\right)^{0.52} DE_S f(\upsilon_s) = 2.31(1)^{0.52}(4)(40)(0.97) = 0.36\ \tfrac{GN}{m}$$

The target natural frequency of the system for a soft-stiff design should ultimately be between 0.2 and 0.35 Hz to avoid 1P/3P frequencies. Following the methodology suggested by Jalbi and Bhattacharya [15], a closed-form solution of the first natural frequency of the system which considers the soil-structure interaction can be obtained as follows:

$$f_0 = C_J \times f_{fb}$$

where f_{fb} is the fixed base natural frequency and C_J is the foundation flexibility parameter that is dependent on the vertical stiffness of the springs (K_V).

For the purpose of this explanatory example, the computed fixed base natural frequency is 0.303 Hz and the C_J is equal to 0.77. Hence, the first natural frequency of the system (with the SSI effect) is calculated to be $f_0 = C_J \times f_{fb} = 0.77 \times 0.303 = 0.23$ Hz, refer to Appendix A for the detailed calculation of the example. Although, the estimated value of the natural frequency falls within the targeted range, it is still in close proximity to the 1P frequency.

For further detailed analysis group effects can also be incorporated and readers are referred to Bordón et al. [26] which suggests formulations on how to incorporate foundation group effects correction factors which are dependent on the caisson aspect ratio, diameter and spacing. These formulations are also presented in the Appendix A. This will further lower the natural frequency and the design may have to be refined in order to allow for the additional 10% safety margin as per the DNVGL [23] recommendations. Subsequently, the FLS and ULS criteria should also be checked.

6. Conclusions

Offshore wind turbines supported on multiple shallow foundations can exhibit undesirable rocking modes of vibration strongly dependent on the vertical stiffness of the foundation. In this paper, numerical analysis is carried out to explore the axial behaviour of skirted caisson foundations for three idealised ground profiles (homogeneous, linear inhomogeneous and parabolic inhomogeneous) where the soil stiffness varies with depth. A set of static vertical stiffness functions and correction factors due to Poisson's ratio are derived. The results are validated against numerical and analytical solutions found in the literature. These formulations can be utilised for optimising the caisson's dimensions for feasibility studies purposes and preliminary design stages of a project. An explanatory example is provided to illustrate the usage of the derived formulation whereby natural frequency of a typical jacket supported OWT structure is calculated.

Author Contributions: Conceptualization, A.S., S.J. and S.B.; methodology, A.S. and S.J.; software, A.S.; validation, A.S. and S.J.; formal analysis, A.S. and S.J.; investigation, A.S. and S.J.; resources, S.B.; data curation, A.S.; writing—original draft preparation, A.S. and S.J.; writing—review and editing, S.B.; visualization, A.S.; supervision, S.B.; project administration, S.B.; funding acquisition, A.S. and S.J. All authors have read and agreed to the published version of the manuscript.

Funding: This research received no external funding.

Institutional Review Board Statement: Not applicable.

Informed Consent Statement: Not applicable.

Data Availability Statement: Not applicable.

Conflicts of Interest: The authors declare no conflict of interest.

Nomenclature

L	Foundation Depth
D	Foundation Diameter
R	Foundation Radius
Pile	Foundation with L/D > 2
Caisson	Foundation with 0.2 < L/D < 2
E_{SO}	initial soil Young's modulus at 1D depth
E_S	Vertical distribution of soil's Young's modulus
f_{FB}	Fixed base (cantilever) natural frequency
C_J	Foundation flexibility parameter
m_{RNA}	Mass of Rotor Nacelle assembly
m_T	Mass of tower
D_{Bottom}	Tower bottom diameter
D_{Top}	Tower top diameter
υ_s	Soil Poisson's ratio
K_V	Vertical stiffness of the foundation

Appendix A.

Appendix A.1. Summary of the Analysis Performed

Table A1. Summary of the Analysis Performed.

Ground Profiles	E_{SO} (MPa)	L/D (D = 5 m)	υ_s
Homogeneous Parabolic Inhomogeneous Linear Inhomogeneous	100	0.2, 0.5, 0.75, 1, 1.5, 2	0.1, 0.2, 0.3, 0.4, 0.499

Appendix A.2. Obtaining the Natural Frequency

Step 1: Calculate the fixed base natural frequency:

$$f_{fb} = \frac{1}{2\pi}\sqrt{\frac{3EI_{T-J}}{(0.243m_{eq}h_{total} + M_{RNA})(h_{total})^3}} = \frac{1}{2\pi}\sqrt{\frac{3 \times 1.635 \times 10^{12}}{[0.243(4200 \times 140) + 350000](140)^3}} = 0.303\text{Hz}$$

This is also presentative of the natural frequency if the jacket is supported on deep embedded piles

Step 2: Calculate C_J for the stiffness of the springs:

Using the equations provided in Table 3, a preliminary estimate of the vertical stiffness for a rigid caisson foundation (with $L/D = 1$) is obtained as shown below;

$$\begin{aligned} f(\upsilon_s) &= \left[\left(10\upsilon_s^3 - 5.88\upsilon_s^2\right)\left(-0.34\ln\tfrac{L}{D} + 0.77\right)\right] + 0.91\upsilon_s\left(-0.57\ln\tfrac{L}{D} + 0.6\right) + 1 \\ &= \left[\left(10(0.28)^3 - 5.88(0.28)^2\right)(-0.34\ln(1) + 0.77)\right] + 0.91(0.28)(-0.57\ln(1) + 0.6) + 1 = 0.97 \end{aligned}$$

$$K_V = 2.31\left(\tfrac{L}{D}\right)^{0.52} DE_S f(\upsilon_s) = 2.31(1)^{0.52}(4)(40)(0.97) = 0.36\ \tfrac{GN}{m}$$

$$K_{V1,2} = 0.36 \times 2 = 0.72\ \frac{GN}{m}$$

$$\text{For } \alpha = 1,\ K_R = K_{V1,2}L^2\left[\frac{\alpha}{1+\alpha}\right] = 0.72 \times 12^2 \times \left[\frac{1}{1+1}\right] = 52\ \text{GNm}$$

$$\tau = \frac{K_R h_{total}}{EI_{T-J}} = \frac{52 \times 10^9 \times 140}{1.635 \times 10^{12}} = 4.45$$

$$C_J = \sqrt{\frac{\tau}{\tau+3}} = \sqrt{\frac{4.45}{4.45+3}} = 0.77$$

$$f_0 = C_J \times f_{fb} = 0.77 \times 0.303 = 0.23\text{Hz}$$

Readers are refered to Jalbi and Bhattacharya [15] for the step-by-step derivation of EI_{T-J} and K_R. In essence, the vertical spring stiffness was factored by 2 as the method converts the 3D representation of the system into 2D. Therefore, each spring in Figure 11 is representative of two caisson foundations.

Additional Step: Bordón et al. [26] propose the following method to calculate group effects correction factors:

$$\text{Group Effect Factor} = \frac{1}{1 + 0.06(1 + 3.08N)\left[\left(1 + 1.2(L/D)^{0.53}\right)/(s/D)\right]}$$

where N is the number of foundations and s is the spacing. For the solved example, this results in reducing the vertical stiffness of the foundations by a factor of 0.63. This further reduces the natural frequency to 0.21 Hz after the repition of the calculation above.

References

1. European Commission. 2030 Climate & Energy Framework. 2020. Available online: https://ec.europa.eu/clima/policies/strategies/2030_en#tab-0-0 (accessed on 26 May 2021).
2. Wind Europe. *The European Offshore Wind Industry. Key Trends and Statistics 2019*; windeurope.org: Brussels, Belgium, 2019. Available online: https://windeurope.org/wp-content/uploads/files/about-wind/statistics/WindEurope-Annual-Offshore-Statistics-2019.pdf (accessed on 24 May 2021).
3. Global Wind Energy Council. *Global Wind Report*; Global Wind Energy Council: Brussels, Belgium, 2021.
4. Bhattacharya, S. *Design of Foundations for Offshore Wind Turbines*; John Wiley & Sons Ltd.: Chichester, UK, 2019.
5. 4COffshore Limited. Global Offshore Wind Farms Database. 2021. Available online: http://www.4coffshore.com/windfarms/ (accessed on 15 May 2021).
6. Orsted. Our Experience with Suction Bucket Jacket Foundations. 2019. Available online: https://orsted.com/en/our-business/offshore-wind/wind-technology/suction-bucket-jacket-foundations (accessed on 26 May 2021).
7. Jalbi, S.; Bhattacharya, S. Minimum foundation size and spacing for jacket supported offshore wind turbines considering dynamic design criteria. *Soil Dyn. Earthq. Eng.* **2019**, *123*, 193–204. [CrossRef]
8. Bhattacharya, S. Civil Engineering Aspects of a Wind Farm and Wind Turbine Structures. In *Wind Energy Engineering: A Handbook for Onshore and Offshore Wind Turbines*; Elsevier: London, UK, 2017.
9. Bhattacharya, S.; Nikitas, N.; Garnsey, J.; Alexander, N.; Cox, J.; Lombardi, D.; Wood, D.M.; Nash, D. Observed dynamic soil–structure interaction in scale testing of offshore wind turbine foundations. *Soil Dyn. Earthq. Eng.* **2013**, *54*, 47–60. [CrossRef]
10. Bhattacharya, S.; Cox, J.A.; Lombardi, D.; Wood, D. Dynamics of offshore wind turbines supported on two foundations. *Geotech. Eng.* **2011**, *166*, 159–169. [CrossRef]
11. Jalbi, S.; Nikitas, G.; Bhattacharya, S.; Alexander, N. Dynamic design considerations for offshore wind turbine jackets supported on multiple foundations. *Mar. Struct.* **2019**, *67*, 102631. [CrossRef]
12. Shadlou, M.; Bhattacharya, S. Dynamic stiffness of monopiles supporting offshore wind turbine generators. *Soil Dyn. Earthq. Eng.* **2016**, *88*, 15–32. [CrossRef]
13. Arany, L.; Bhattacharya, S.; Macdonald, J.; Hogan, S.J. Design of monopiles for offshore wind turbines in 10 steps. *Soil Dyn. Earthq. Eng.* **2017**, *92*, 126–152. [CrossRef]
14. Jalbi, S.; Shadlou, M.; Bhattacharya, S. Impedance functions for rigid skirted caissons supporting offshore wind turbines. *Ocean Eng.* **2018**, *150*, 21–35. [CrossRef]
15. Jalbi, S.; Bhattacharya, S. Closed form solution for the first natural frequency of offshore wind turbine jackets supported on multiple foundations incorporating soil-structure interaction. *Soil Dyn. Earthq. Eng.* **2018**, *113*, 593–613. [CrossRef]
16. Bell, R.W. The analysis of offshore foundations subject to combined loading. In *Degree of Master of Science*; University of Oxford: Oxford, UK, 1991.
17. Gazetas, G. Formulas and Charts for Impedances of Surface and Embedded Foundations. *J. Geotech. Eng.* **1991**, *117*, 1363–1381. [CrossRef]
18. Doherty, J.; Houlsby, G.; Deeks, A. Stiffness of flexible caisson foundations embedded in nonhomogeneous elastic soil. *J. Geotech. Geoenvironmental Eng.* **2005**, *131*, 1498–1508. [CrossRef]
19. Skau, K.S.; Jostad, H.P.; Eiksund, G.; Sturum, H. Modelling of soil-structure-interaction for flexible caissons for offshore wind turbines. *Ocean Eng.* **2019**, *171*, 273–285. [CrossRef]
20. Lysmer, J. Vertical Motions of Rigid Footings. Ph.D. Thesis, University of Michigan, Ann Arbor, MI, USA, 1965.
21. Spence, D.A. Self similar solutions to adhesive contact problems with incremental loading. *Proc. R. Soc. Lond. Ser. A* **1968**, *305*, 55–80.
22. Gazetas, G. Analysis of machine foundation vibrations: State of the art. *Int. J. Soil Dyn. Earthq. Eng.* **1983**, *2*, 2–42. [CrossRef]
23. DNVGL. *DNVGL-RP-C212 Offshore Soil Mechanics and Geotechnical Engineering*; DNVGL AS: Oslo, Norway, 2019.
24. Wolf, J.P. *Soil Structure Interaction Analysis in Time Domain*; Electrowatt Engineering Services Ltd.: Zurich, Switzerland, 1988.
25. Wolf, J.P.; Deeks, A.J. *Foundation Vibration Analysis: A Strength-of-Materials Approach*; Elsevier: Oxford, UK, 2004.
26. Bordón, J.D.R.; Aznarez, J.J.; Padron, L.A.; Maeso, O.; Bhattacharya, S. Closed-form stiffnesses of multi-bucket foundations for OWT including group effect correction factors. *Mar. Struct.* **2019**, *65*, 326–342. [CrossRef]
27. Fleming, W.G.; Weltman, A.J.; Randolph, M.F.; Elson, W.K. *Piling Engineering*; Blackie and Son: London, UK, 1992.
28. Shama, A.A.; El Naggar, H. Bridge Foundations. *Encycl. Earthq. Eng.* **2015**, *1*, 298–317.
29. Suryasentana, S.; Byrne, B.; Burd, H.; Shonberg, A. Simplified model for stiffness of suction caisson foundations under 6 dof loading. In Proceedings of the SUT OSIG 8th International Conference, London, UK, 12–14 September 2017.
30. Latini, C.; Cisternino, M.; Zania, V. Vertical dynamic stiffness of offshore foundations. In Proceedings of the 26th International Ocean and Polar Engineering Conference, Rhodes, Greece, 26 June–1 July 2016; pp. 775–780.
31. Sloan, S.W. Numerical Analysis of Incompressible and Plastic Solids Using Finite Elements. Ph.D. Thesis, Cambridge University, Cambridge, UK, 1981.
32. Bhattacharya, S.; Wang, L.; Liu, J.; Hong, Y. Civil engineering challenges associated with design of offshore wind turbines with special reference to China. In *Wind Energy Engineering: A Handbook for Onshore and Offshore Wind Turbines*; Elsevier: London, UK, 2017.
33. Latini, C.; Zania, V. Vertical dynamic impedance of suction caissons. *Soils Found.* **2019**, *59*, 1113–1127. [CrossRef]

34. Latini, C.; Zania, V. Dynamic lateral response of suction caissons. *Soil Dyn. Earthq. Eng.* **2017**, *100*, 59–71. [CrossRef]
35. Anoyatis, G.; Mylonakis, G.; Lemnitzer, A. Soil reaction to lateral harmonic pile motion. *Soil Dyn. Earthq. Eng.* **2016**, *87*, 164–179. [CrossRef]
36. Lysmer, J.; Udaka, T.; Seed, H.B.; Hwang, R. *FLUSH-A Computer Program for Complex Response Analysis of Soft-Structure Systems*; University of California: Berkeley, CA, USA, 1974.
37. Plaxis 3D. *Materials Models Manual*; Bentley Systems International Limited: Dublin, Ireland, 2018.
38. Shonberg, A.; Harte, M.; Aghakouchak, A.; Brown, C.S.D.; Andrade, M.P.; Lingaard, M.A. Suction bucket jackets for offshore wind turbines: Applications from in-situ observations. In Proceedings of the TC-209 Workshop—19th ICSMGE, Seoul, Korea, 20 September 2017.
39. Jonkman, J.; Buttefield, S.; Musial, W.; Scott, G. *Definition of a 5-MW Reference Wind Turbine for Offshore System Development*; National Renewable Energy Laboratory (NREL): Golden, CO, USA, 2009.
40. Alati, N.; Failla, G.; Arena, F. Seismic analysis of offshore wind turbines on bottom-fixed support structures. *Philos. Trans. R. Soc. A* **2015**, *373*, 20140086. [CrossRef] [PubMed]

Journal of
Marine Science and Engineering

MDPI

Article

Dynamic Analysis on Pile Group Supported Offshore Wind Turbine under Wind and Wave Load

Yusha Shi [1], Wenjuan Yao [2,*] and Guoliang Yu [3,*]

1 School of Mechanics and Engineering Science, Shanghai University, Shanghai 200444, China; shiyushawork@gmail.com
2 Shanghai Institute of Applied Mathematics and Mechanics, Shanghai University, Shanghai 200444, China
3 SKLOE, CISSE, School of Naval Architecture, Ocean & Civil Engineering, Shanghai Jiao Tong University, Shanghai 200240, China
* Correspondence: wjyao@shu.edu.cn (W.Y.); yugl@sjtu.edu.cn (G.Y.)

Abstract: With growing demand for renewable wind energy, the number of offshore wind turbines increased rapidly in recent years. This paper uses the improved Tajimi model and transfer matrix method to analyze the dynamic response of pile group supported offshore wind turbine under wind and wave load. The vibration equation of the structure is established by tower discretization. The calculation result is compared with the numerical simulation result. The horizontal displacement of the structure under loads with different frequencies can be obtained. The wind speed and the foundation impedance are found important to the structure displacement. The pile–pile interaction factor depends on the pile spacing, the pile embedment ratio, and the incidence angle.

Keywords: offshore wind turbine; pile–pile interaction factor; wind load; wave load; transfer matrix method

Citation: Shi, Y.; Yao, W.; Yu, G. Dynamic Analysis on Pile Group Supported Offshore Wind Turbine under Wind and Wave Load. *J. Mar. Sci. Eng.* **2022**, *10*, 1024. https://doi.org/10.3390/jmse10081024

Academic Editors: Eugen Rusu, Kostas Belibassakis and George Lavidas

Received: 6 July 2022
Accepted: 23 July 2022
Published: 26 July 2022

Publisher's Note: MDPI stays neutral with regard to jurisdictional claims in published maps and institutional affiliations.

Copyright: © 2022 by the authors. Licensee MDPI, Basel, Switzerland. This article is an open access article distributed under the terms and conditions of the Creative Commons Attribution (CC BY) license (https://creativecommons.org/licenses/by/4.0/).

1. Introduction

New energy development and utilization have become the main focus of all human beings with the consumption of traditional energy. As one kind of renewable and clean energy, wind energy is developing rapidly. With abundant energy reserves and the farmland-free characteristic, offshore wind energy source has better development prospect rather than onshore wind energy. Offshore wind turbine (OWT) developed rapidly in China. From 2014–2018, the installed offshore wind turbine in China increased average of 60.5% per year.

OWTs are mainly under wind load and wave load [1], sometimes even earthquake load [2]. The structure of OWT can be divided into two parts, the tower above the seawater and the foundation partially submerged in the seawater, and partially embedded in the soil. The tower is under wind load, whereas the foundation is under wave load and soil reaction [3]. OWT is a slender structure, which means its dynamic response will be significantly influenced by the load frequency. In OWT design, it is needed to calculate the OWT structure natural frequency to prevent it falls within the frequency ranges of main loads [4]. As shown in Figure 1, the peak frequency of the wind load is usually within 0.01 Hz. The peak frequency of the wave load is 0.08 Hz~0.2 Hz. Definitions of 1P frequency and 3P frequency are presented below:

(1) The barycenter of blades and the turbine might slightly shift from the rotating shaft, which will produce eccentric force, and its frequency is equal to the frequency of the turbine rotation frequency. This frequency is called the 1P frequency [5–7]. Since an OWT has different rotation speeds, the 1P frequency is not a single frequency but a frequency range, which is related to the highest and lowest value of the rotation speed.
(2) Rotating blades will produce the air turbulence load. Once a blade runs across a certain location, a turbulence load will be created. This load is usually called 2P or 3P

load. Most OWTs are three-blade structures, and the model used in this paper is the same, so it is a 3P frequency.

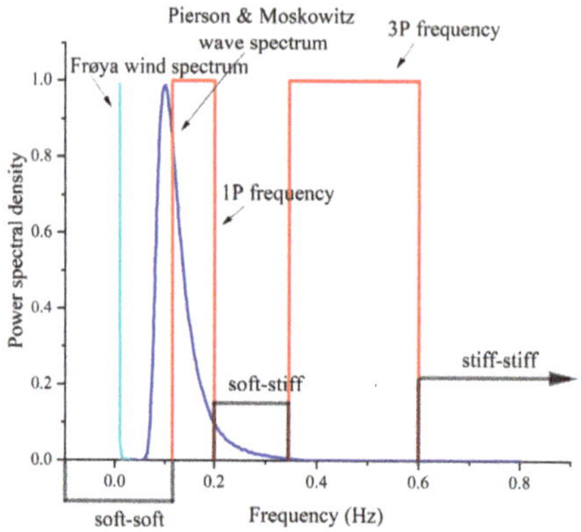

Figure 1. Frequency spectrum of the dynamic loads [4].

As shown in Figure 1, in OWT design, three different design methods are considered, which are "soft-soft", "soft-stiff", and "stiff-stiff" methods. Most installed OWTs adopted "soft-stiff" design method. Therefore, it is needed to analyze the dynamic response of OWT under different frequencies.

Previous research mainly focused on the study of monopile-supported OWTs. The numerical method is used to analyze the overall dynamic structure response, including the finite element method [8] (FEM) and boundary element method (BEM). Kjørlaug [9] used SAP2000 to analyze the acceleration and structure natural frequency of OWT under lateral and vertical earthquake loads. Corciulo [10] used the OpenSees simulation platform to investigate the dynamic response of OWT under wind and wave load. Zuo [11] used ABAQUS to establish the model of OWT including blades and analyzed the dynamic response of OWT under operating and steady conditions. Galvin [12] used the FEM-BEM method and analyzed the dynamic response of OWT under earthquake load.

Another analysis method is the analytical method. The OWT structure can be divided into the superstructure and the foundation to investigate its dynamic response [13]. As for the foundation, the P-y curve method is used in early research [14,15]. This method is still widely used [6,16]. Andersen [17] simplified the pile–soil interaction as the equivalent coupled spring model and obtained the structure natural frequency. Adhikari and Bhattacharya [18,19] established the foundation model with elastic supports based on the Euler–Bernoulli beam, used horizontal and rotation springs to simulate the foundation reaction, and validated the result with the experimental result [20].

The foundation of OWT is partially embedded in the soil, and the dynamic equation of different pile parts is different. By using the transfer matrix method [21–23], the dynamic response of different pile parts can be connected. Wang [24] analyzed the onshore wind turbine structure natural frequency using the transfer matrix method. Huang [25] analyzed the dynamic response of the pile group supported OWT using the transfer matrix method.

This paper used the Morison equation to calculate the wave load applied to the pile, and calculates the pile–soil interaction using an improved Tajimi soil model [26,27]. By using the transfer matrix method, the dynamic response of the pile group embedded in the soil and submerged in the seawater are connected, and the overall pile group impedance is obtained. The stable forced vibration equation of the multiple-degree-of-freedom OWT system is established by discretizing the tower into multiple segments. By substituting the pile group impedance into the equation, the dynamic response of OWT under different load frequencies is obtained. For the pile group, the pile–pile interaction factor is calculated, which considers the influence of the passive pile on the active pile.

The calculated result is compared with the FEM result to validate the correctness of the proposed calculation method.

2. Proposed Calculation Method

2.1. Model Establishment

As shown in Figure 2, the pile group supported 3.6 MW offshore wind turbine can be divided into two parts. The first part is the pile group foundation, which is partially embedded in the soil and partially submerged in the seawater. The second part is the superstructure, which is under distributed and thrust wind load.

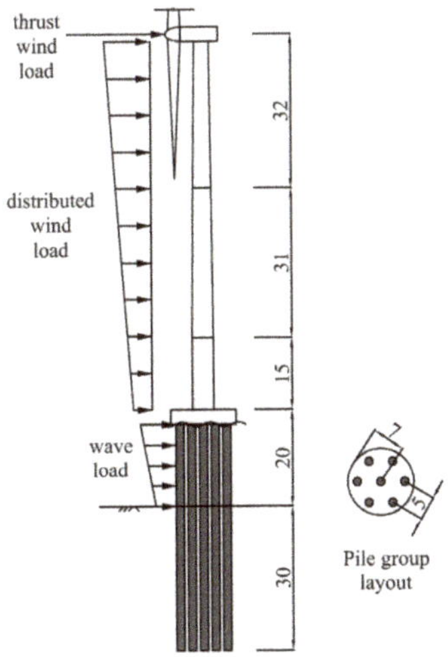

Figure 2. Pile group supported 3.6 MW offshore wind turbine.

2.2. Pile Submerged in the Seawater

The pile diameter of the pile group is relatively small compared with the wavelength ($D/L_{wl} \leq 0.2$). Therefore, the wave load applied to the pile can be calculated by the Morison equation. For linear waves, at height z of the cylinder, the horizontal wave load $q_G(z)$ can be calculated according to Equation (1) [28]:

$$\begin{gathered} q_G(z) = \frac{1}{2} C_D \rho_w D \left(\frac{\pi H_w}{T} \frac{\cosh(KZ_w)}{\sinh(KL_w)} \right)^2 \cos\theta |\cos\theta| dz + \\ C_M \frac{\pi D^2}{4} \frac{2\pi^2 H_w}{T^2} \frac{\cosh(KZ_w)}{\sinh(KL_w)} \sin\theta dz \end{gathered} \tag{1}$$

where C_D is the drag coefficient, C_M is the inertia coefficient, $\theta = -\omega t$, ρ_w is the density of the seawater, D is the pile diameter, H_w is the wave height, T is the wave period, and $K = \frac{2\pi}{L_{wl}}$, where L_{wl} is the wavelength. This paper mainly discusses the dynamic response of the pile under different frequencies; therefore, it is needed to calculate the maximum horizontal wave load applied to the whole pile body. The wave load applied to the whole pile body can be calculated according to Equation (2) [28]:

$$\begin{gathered} F_T = \int_0^{L_w} q_G(z) = F_{HD} \cos\theta |\cos\theta| + F_{HI} \sin\theta \\ F_{HD} = C_D \frac{\gamma D {H_w}^2}{2} K_1, F_{HI} = C_M \frac{\gamma \pi D^2 H_w}{8} K_2 \end{gathered} \tag{2}$$

where $K_1 = \frac{2KL_w + \sinh 2KL_w}{8 \sinh 2KL_w}$ and $K_2 = \tanh KL_w$. As shown in Equation (2), the value of θ to determine the maximum of F_{HD} and F_{HI} is not the same. Therefore, it is needed to determine the value of θ when the total wave force is the largest. According to the calculation, the maximum total wave force depends on the value of F_{HD} and F_{HI}.

(1) When $F_{HI} \geq 2F_{HD}$, the maximum total wave force happens when $\theta = \pi/2$, and the maximum total wave force $F_{T\max}$ equals F_{HI}.
(2) When $F_{HI} < 2F_{HD}$, the maximum total wave force happens when $\theta = \arcsin(F_{HI}/F_{HD})$, $F_{\max} = F_{HD}\left(1 + \frac{1}{4}\left(\frac{F_{HI}}{F_{HD}}\right)^2\right)$.

In engineering, $F_{HI} < 2F_{HD}$ is rarely seen. Therefore, in this paper, we consider $F_{HI} \geq 2F_{HD}$, which means $F_{T\max} = F_{HI}$. Then, we can calculate the wave load applied to the cylinder at height z when the maximum total wave load happens [28]:

$$q_G(z) = C_M \frac{\pi D^2}{4} \frac{2\pi^2 H_w}{T^2} \frac{\cosh(KZ_w)}{\sinh(KL_w)} \tag{3}$$

Equation (3) is used for single piles in the seawater. However, the wave load applied to the pile group can be complex due to the pile–pile interaction and pile–cap interaction. According to the research [29], considering the pile–pile interaction, this paper considers the wave load applied to each pile within the pile group as 0.8 times the wave load applied to the single pile, and Equation (3) can be modified as:

$$q_G(z) = 0.8 C_M \frac{\pi D^2}{4} \frac{2\pi^2 H_w}{T^2} \frac{\cosh(KZ_w)}{\sinh(KL_w)} \tag{4}$$

Based on the water–soil interface, the local coordinate system is established, and the z axis is towards the seabed. The pile submerged in the seawater can be divided into n parts, and the length of each part is h_i. According to dynamic equilibrium conditions, the dynamic equation of each pile section can be obtained:

$$\frac{d^4 w_i^{a1}}{dz^4} + 4\gamma^4 w_i^{a1}(z) = \frac{q_G(z)}{E_p I_p} \tag{5}$$

where w is the lateral displacement of the pile section, superscript $a1$ denotes the first part of the active pile, and $\gamma = \left(\frac{-m_p \omega^2}{4E_p I_p}\right)^{1/4}$, where ω is the load frequency, m_p is the unit mass of the pile body, E_p is the modulus of elasticity of the pile, and I_p is the cross-section inertia moment of the pile. The solution of Equation (5) is the superposition of general solution and particular solution, as shown in Equation (6):

$$w_i^{a1}(z) = e^{(-1+i)\gamma z} A_{1i}^{a1} + e^{(-1-i)\gamma z} A_{2i}^{a1} + e^{(1-i)\gamma z} A_{3i}^{a1} + e^{(1+i)\gamma z} A_{4i}^{a1} + q'_G(z) \tag{6}$$

where $q'_G = a_p q_G, a_p = \frac{1}{E_p I_p (K^4 + 4\gamma^4)}$, A_{1i}^{a1}, A_{2i}^{a1}, A_{3i}^{a1}, A_{4i}^{a1} are undetermined coefficients, which can be obtained according to boundary conditions.

According to differential relations, we can obtain the lateral displacement w, rotation angle θ, bending moment M, and shearing force Q of the pile section. For simplicity, we define $M = \overline{M}/E_p I_p$ and $Q = \overline{Q}/E_p I_p$ to represent the equivalent bending moment and shearing force. Written in matrix form:

$$\begin{Bmatrix} w_i^{a1} \\ \phi_i^{a1} \\ M_i^{a1} \\ Q_i^{a1} \\ 1 \end{Bmatrix} = \left\{ t_i^{a1} \right\} \begin{Bmatrix} A_{1i}^{a1} \\ A_{2i}^{a1} \\ A_{3i}^{a1} \\ A_{4i}^{a1} \\ 1 \end{Bmatrix} \tag{7}$$

where the matrix $\left\{ t_i^{a1} \right\}$ is shown in Appendix A.

Equation (7) can be used to calculate each pile section of the pile body. The transfer matrix between the pile top and pile tip of each pile section can then be obtained, as shown in Equation (8):

$$\begin{Bmatrix} w^{a1}(h_i) \\ \phi^{a1}(h_i) \\ M^{a1}(h_i) \\ Q^{a1}(h_i) \\ 1 \end{Bmatrix} = \left\{t_i^{a1}(h_i)\right\}\left\{t_i^{a1}(0)\right\}^{-1} \begin{Bmatrix} w^{a1}(0) \\ \phi^{a1}(0) \\ M^{a1}(0) \\ Q^{a1}(0) \\ 1 \end{Bmatrix} \quad (8)$$

Let $\{T_i^{a1}\} = \{t_i^{a1}(h_i)\}\{t_i^{a1}(0)\}^{-1}$ and $T^{a1} = T_n^{a1} \cdots T_i^{a1} \cdots T_2^{a1} T_1^{a1}$; the transfer matrix between the pile tip (water–soil interface) and the pile top (water–air interface) can then be obtained, as shown in Equation (9):

$$\begin{Bmatrix} w^{a1}(L_w) \\ \phi^{a1}(L_w) \\ M^{a1}(L_w) \\ Q^{a1}(L_w) \\ 1 \end{Bmatrix} = \{T^{a1}\} \begin{Bmatrix} w^{a1}(0) \\ \phi^{a1}(0) \\ M^{a1}(0) \\ Q^{a1}(0) \\ 1 \end{Bmatrix} \quad (9)$$

Thus, we obtain the transfer matrix of the active pile submerged in the seawater. The solution can also be used for the passive pile and the active pile under the influence of the passive pile. Let $\{T^{p1}\} = \{T^{a1}\} = \{\overline{T}^{a1}\}$, written in matrix form:

$$\begin{gathered} \begin{Bmatrix} w^{p1}(L_w) \\ \phi^{p1}(L_w) \\ M^{p1}(L_w) \\ Q^{p1}(L_w) \\ 1 \end{Bmatrix} = \{T^{p1}\} \begin{Bmatrix} w^{p1}(0) \\ \phi^{p1}(0) \\ M^{p1}(0) \\ Q^{p1}(0) \\ 1 \end{Bmatrix} \\ \begin{Bmatrix} \overline{w}^{a1}(L_w) \\ \overline{\phi}^{a1}(L_w) \\ \overline{M}^{a1}(L_w) \\ \overline{Q}^{a1}(L_w) \\ 1 \end{Bmatrix} = \{\overline{T}^{a1}\} \begin{Bmatrix} \overline{w}^{a1}(0) \\ \overline{\phi}^{a1}(0) \\ \overline{M}^{a1}(0) \\ \overline{Q}^{a1}(0) \\ 1 \end{Bmatrix} \end{gathered} \quad (10)$$

where w^{p1} is the lateral displacement of the passive pile and $\overline{w}^{a1}$ is the lateral displacement of the active pile under the influence of the passive pile.

2.3. Pile Embedded in the Soil

2.3.1. Pile–Soil Interaction

The pile embedded in the soil is under lateral reaction of the soil. To investigate the influence of the soil cut-off frequency, this paper adopts the soil model proposed by Anoyatis [26]. This model uses the improved Tajimi soil model, considers the soil as a three-dimensional continuum, and considers the influence of soil vertical displacement on the lateral soil displacement. The proposed dynamic Winkler modulus can reflect the vibration of spring coefficient and damping coefficient around the soil cut-off frequency. The proposed calculation method is shown in the equation below:

$$\begin{gathered} k^* = \pi G_s^* s\left[s + 4\frac{Y_1'(s)}{Y_0'(s)}\right] \\ s = \frac{1}{2(\eta_s)\chi}\sqrt{a_{cutoff}^2 - \frac{a_0^2}{1+2i\beta_s}} \end{gathered} \quad (11)$$

where $G_s^* = G_s(1+2i\beta_s)$, β_s is the soil material damping, valued 0.05 in this paper, $Y'_1(s)$ and $Y'_0(s)$ are the first and zero order modified Bessel function of the second kind, $\eta_s = \sqrt{2-\nu_s/1-\nu_s}$, where υ_s is Poisson's ratio, $a_{cutoff} = (\pi/2)(L/D)^{-1}$, where L is the pile length, and χ is the dimensional coefficient, valued according to Poisson's ratio, valued 3 in this paper.

As shown in Figure 3, the dynamic response of the pile embedded in the soil can be divided into four steps:

(1) The horizonal displacement of the active pile under the pile top dynamic load;
(2) The passive pile is influenced by the incident wave from the active pile displacement after the soil attenuation;
(3) The horizontal displacement of the passive pile;
(4) The active pile is influenced by the secondary wave from the passive pile displacement after soil attenuation.

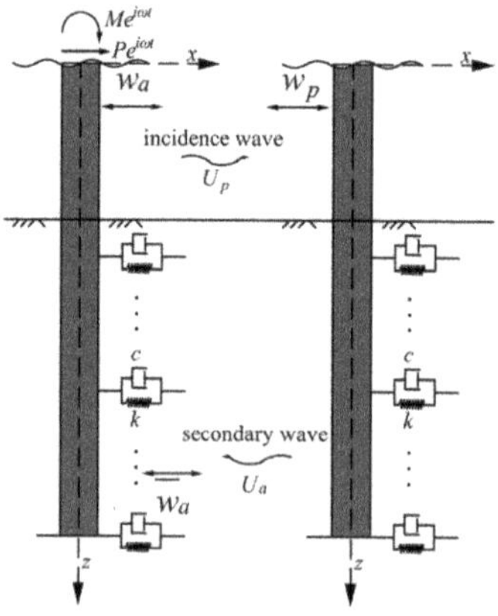

Figure 3. Pile–pile interaction.

2.3.2. Active Pile Displacement

The dynamic equation of the active pile under horizontal dynamic load can be written as:

$$\frac{d^4 w_i^{a2}}{dz^4} + 4\gamma^4 w_i^{a2}(z) = 0 \tag{12}$$

where $\gamma = \left(\frac{k+i\omega c - m_p\omega^2}{4E_pI_p}\right)^{1/4}$, $k = \mathrm{Re}(k^*)$, $wc = \mathrm{IM}(k^*)$, and w_i^{a2} means the lateral displacement of the active pile embedded in the soil. The solution of Equation (12) is written below:

$$w_i^{a2}(z) = e^{(-1+i)\gamma z}A_{a2} + e^{(-1-i)\gamma z}B_{a2} + e^{(1-i)\gamma z}C_{a2} + e^{(1+i)\gamma z}D_{a2} \tag{13}$$

where A_{a2}, B_{a2}, C_{a2}, D_{a2} are undetermined coefficients, which can be determined according to boundary conditions.

Similarly, according to differential relations, we can obtain the lateral displacement, rotation angle, bending moment, and shearing force of the active pile embedded in the soil. Written in matrix form:

$$\begin{Bmatrix} w_i^{a2} \\ \phi_i^{a2} \\ M_i^{a2} \\ Q_i^{a2} \\ 1 \end{Bmatrix} = \left\{ t_i^{a2} \right\} \begin{Bmatrix} A_{a2} \\ B_{a2} \\ C_{a2} \\ D_{a2} \\ 1 \end{Bmatrix} \tag{14}$$

where the matrix $\{t_i^{a2}\}$ is shown in Appendix A. The transfer matrix between the pile top and pile tip of each pile section is shown in Equation (15):

$$\begin{Bmatrix} w_i^{a2}(h_i) \\ \phi_i^{a2}(h_i) \\ M_i^{a2}(h_i) \\ Q_i^{a2}(h_i) \\ 1 \end{Bmatrix} = \left\{t_i^{a2}(h_i)\right\}\left\{t_i^{a2}(0)\right\}^{-1} \begin{Bmatrix} w_i^{a2}(0) \\ w_i^{a2}(0) \\ M_i^{a2}(0) \\ Q_i^{a2}(0) \\ 1 \end{Bmatrix} \tag{15}$$

Let $\{T_i^{a2}\} = \left\{t_i^{a2}(h_i)\right\}\left\{t_i^{a2}(0)\right\}^{-1}$ and $T^{a2} = T_n^{a2} \cdots T_i^{a2} \cdots T_2^{a2} T_1^{a2}$; the transfer matrix between the active pile tip and top (soil-water surface) can be obtained:

$$\begin{Bmatrix} w^{a2}(H) \\ \phi^{a2}(H) \\ M^{a2}(H) \\ Q^{a2}(H) \\ 1 \end{Bmatrix} = \{T^{a2}\} \begin{Bmatrix} w^{a2}(0) \\ \phi^{a2}(0) \\ M^{a2}(0) \\ Q^{a2}(0) \\ 1 \end{Bmatrix} \tag{16}$$

2.3.3. Soil Attenuation Function and Soil Displacement

For the soil S m away from the active pile, its attenuation function and soil displacement can be calculated according to the following equation:

$$U_p(S,z) = \psi w^{a2}(z) = \left[\psi(S,0^\circ)\cos\theta^2 + \psi(S,90^\circ)\sin\theta^2\right] w^{a2}(z) \tag{17}$$

where

$$\psi(S,0^\circ) = \sqrt{\frac{R}{S}} \exp\left(\frac{-\beta\omega(R-S)}{V_{La}}\right) \exp\left(\frac{-i\omega(R-S)}{V_{La}}\right) \tag{18}$$

$$\psi(S,90^\circ) = \sqrt{\frac{R}{S}} \exp\left(\frac{-\beta\omega(R-S)}{V_s}\right) \exp\left(\frac{-i\omega(R-S)}{V_s}\right) \tag{19}$$

where R is the radius of the pile, S is the distance between the active pile and passive pile, β is the soil damping, V_{La} and V_s are the shear wave velocity and Lysmer's simulation velocity [30], $V_{La} = \frac{3.4}{\pi(1-v)} V_s$, and θ is the angle of incidence, which will influence the pile–pile interaction factor, as shown in Figure 4.

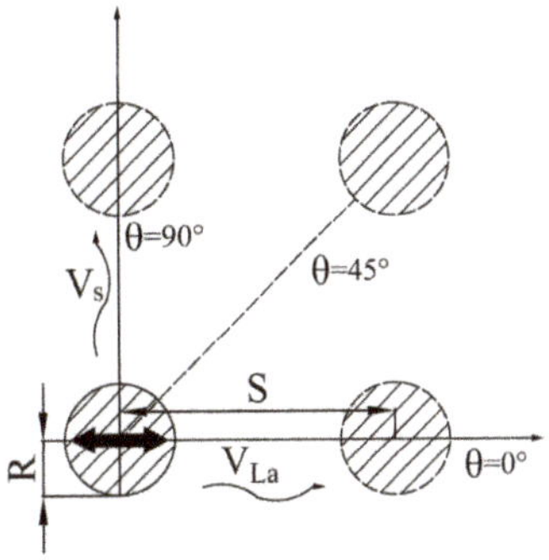

Figure 4. Pile–pile interaction.

The soil displacement around the passive pile after the soil attenuation can then be obtained. In pile groups, θ can be different between different piles, which will be considered when calculating the overall impedance of the pile group.

2.3.4. Passive Pile Displacement

The displacement of the passive pile due to the soil displacement can be calculated according to the following equation:

$$\frac{d^4 w_i^{p2}}{dz^4} + 4\gamma^4 w_i^{p2}(z) - \frac{(k+i\omega c)}{E_p I_p} U_p(S,z) = 0 \tag{20}$$

Substitute Equation (17) into Equation (20), the solution can be obtained:

$$\begin{aligned} w_i^{p2}(z) = \; & e^{(-1+i)\gamma z}A_{p2} + e^{(-1-i)\gamma z}B_{p2} + e^{(1-i)\gamma z}C_{p2} + e^{(1+i)\gamma z}D_{p2} + \\ & \tfrac{(k+i\omega c)\psi z}{E_p I_p}[(1-i)e^{(-1+i)\gamma z}A_{a2} + (1+i)e^{(-1-i)\gamma z}B_{a2} + \\ & (-1+i)e^{(1-i)\gamma z}C_{a2} + (-1-i)e^{(1+i)\gamma z}D_{a2}] \end{aligned} \tag{21}$$

where $A_{p2}, B_{p2}, C_{p2}, D_{p2}$ are undetermined coefficients.

Written in matrix form:

$$\begin{Bmatrix} w_i^{p2} \\ \phi_i^{p2} \\ M_i^{p2} \\ Q_i^{p2} \\ 1 \end{Bmatrix} = \left\{t_i^{a2}\right\} \begin{Bmatrix} A_{p2} \\ B_{p2} \\ C_{p2} \\ D_{p2} \\ 1 \end{Bmatrix} + \left\{t_i^{p2}\right\} \begin{Bmatrix} A_{a2} \\ B_{a2} \\ C_{a2} \\ D_{a2} \\ 1 \end{Bmatrix} \tag{22}$$

where the matrix $\left\{t_i^{p2}\right\}$ is shown in Appendix A. The transfer matrix between passive pile segment tip and top can be obtained after the following calculation:

$$\begin{Bmatrix} w_i^{p2}(h_i) \\ \phi_i^{p2}(h_i) \\ M_i^{p2}(h_i) \\ Q_i^{p2}(h_i) \\ 1 \end{Bmatrix} = T_{i1}^{p2} \begin{Bmatrix} w_i^{p2}(0) \\ \phi_i^{p2}(0) \\ M_i^{p2}(0) \\ Q_i^{p2}(0) \\ 1 \end{Bmatrix} + T_{i2}^{p2} \begin{Bmatrix} w_i^{a2}(0) \\ w_i^{a2}(0) \\ M_i^{a2}(0) \\ Q_i^{a2}(0) \\ 1 \end{Bmatrix} \tag{23}$$

where

$$\begin{aligned} T_{i1}^{p2} &= t_i^{a2}(h_i)\left(t_i^{a2}(0)\right)^{-1} \\ T_{i2}^{p2} &= \left(t_i^{p2}(h_i) - t_i^{a2}(h_i)\left(t_i^{a2}(0)\right)^{-1} t_i^{p2}(0)\right)\left(t_i^{a2}(0)\right)^{-1} \end{aligned} \tag{24}$$

The transfer matrix between the passive pile tip and top (soil-water surface) can then be obtained:

$$\begin{Bmatrix} w^{p2}(H) \\ \phi^{p2}(H) \\ M^{p2}(H) \\ Q^{p2}(H) \\ 1 \end{Bmatrix} = T_1^{p2} \begin{Bmatrix} w^{p2}(0) \\ \phi^{p2}(0) \\ M^{p2}(0) \\ Q^{p2}(0) \\ 1 \end{Bmatrix} + T_2^{p2} \begin{Bmatrix} w^{a2}(0) \\ w^{a2}(0) \\ M^{a2}(0) \\ Q^{a2}(0) \\ 1 \end{Bmatrix} \tag{25}$$

where $T_1^{p2} = T_{n1}^{p2} \cdots T_{i1}^{p2} \cdots T_{21}^{p2} T_{11}^{p2}$, $T_2^{p2} = T_{n2}^{p2} \cdots T_{i2}^{p2} \cdots T_{22}^{p2} T_{12}^{p2}$.

2.3.5. Active Pile Displacement Due to Secondary Wave

The passive pile displacement will produce secondary wave, which will influence the active pile displacement. The soil displacement around the active pile can be calculated according to the following equation:

$$U_a(S,z) = \psi w^{p2}(z) = \left[\psi(S,0^{\circ})\cos\theta^2 + \psi(S,90^{\circ})\sin\theta^2\right] w^{p2}(z) \tag{26}$$

The dynamic equation can be written as:

$$\frac{d^4\overline{w}_i^{a2}}{dz^4} + 4\gamma^4\overline{w}_i^{a2}(z) - \frac{(k+i\omega c)}{E_pI_p}U_a(S,z) = 0 \tag{27}$$

where $\overline{w}^{a2}$ is the lateral displacement of the active pile under the influence of the passive pile. The solution to Equation (27) is:

$$\begin{aligned}\overline{w}_i^{a2}(z) = \; & e^{(-1+i)\gamma z}\overline{A}_{a2} + e^{(-1-i)\gamma z}\overline{B}_{a2} + e^{(1-i)\gamma z}\overline{C}_{a2} + e^{(1+i)\gamma z}\overline{D}_{a2} + \\ & \tfrac{(k+i\omega c)\psi z}{16E_pI_p\gamma^3}[(1-i)e^{(-1+i)\gamma z}A_{p2} + (1+i)e^{(-1-i)\gamma z}B_{p2} + \\ & (-1+i)e^{(1-i)\gamma z}C_{p2} + (-1-i)e^{(1+i)\gamma z}D_{p2}] + \\ & \tfrac{(k+i\omega c)\psi z^2}{512E_pI_p\gamma^6}[(1-i)^2e^{(-1+i)\gamma z}A_{a2} + (1+i)^2e^{(-1-i)\gamma z}B_{a2} + \\ & (-1+i)^2e^{(1-i)\gamma z}C_{a2} + (-1-i)^2e^{(1+i)\gamma z}D_{a2}]\end{aligned} \tag{28}$$

where $\overline{A}_{a2}, \overline{B}_{a2}, \overline{C}_{a2}, \overline{D}_{a2}$ are undetermined coefficients.

Written in matrix form:

$$\begin{aligned}\left\{\begin{array}{c}\overline{w}_i^{a2}\\ \overline{\phi}_i^{a2}\\ \overline{M}_i^{a2}\\ \overline{Q}_i^{a2}\\ 1\end{array}\right\} = \left\{t_i^{a2}\right\}\left\{\begin{array}{c}\overline{A}_{a2}\\ \overline{B}_{a2}\\ \overline{C}_{a2}\\ \overline{D}_{a2}\\ 1\end{array}\right\} + \tfrac{1}{16\gamma^3}\left\{t_i^{p2}\right\}\left\{\begin{array}{c}A_{p2}\\ B_{p2}\\ C_{p2}\\ D_{p2}\\ 1\end{array}\right\} + \\ \left(\left\{\overline{t}_i^{a2}\right\} + \tfrac{3i}{1024\gamma^7}\left\{\begin{array}{cccc}(1-i)^2 & 0 & 0 & 0\\ 0 & (1+i)^2 & 0 & 0\\ 0 & 0 & (-1+i)^2 & 0\\ 0 & 0 & 0 & (-1-i)^2\end{array}\right\}\left\{t_i^{p2}\right\}\right)\left\{\begin{array}{c}A_{a2}\\ B_{a2}\\ C_{a2}\\ D_{a2}\\ 1\end{array}\right\}\end{aligned} \tag{29}$$

where the matrix $\left\{\overline{t}_i^{a2}\right\}$ is shown in Appendix A.

Let:

$$\begin{aligned}&\left\{t_i^1\right\} = \left\{t_i^{a2}\right\}\\ &\left\{t_i^2\right\} = \tfrac{1}{16\gamma^3}\left\{t_i^{p2}\right\}\\ &\left\{t_i^3\right\} = \left\{\overline{t}_i^{a2}\right\} + \tfrac{3i}{1024\gamma^7}\left\{\begin{array}{cccc}(1-i)^2 & 0 & 0 & 0\\ 0 & (1+i)^2 & 0 & 0\\ 0 & 0 & (-1+i)^2 & 0\\ 0 & 0 & 0 & (-1-i)^2\end{array}\right\}\left\{t_i^{p2}\right\}\end{aligned} \tag{30}$$

Equation (29) can be simplified:

$$\left\{\begin{array}{c}\overline{w}_i^{a2}\\ \overline{\phi}_i^{a2}\\ \overline{M}_i^{a2}\\ \overline{Q}_i^{a2}\\ 1\end{array}\right\} = \left\{t_i^1\right\}\left\{\begin{array}{c}\overline{A}_{a2}\\ \overline{B}_{a2}\\ \overline{C}_{a2}\\ \overline{D}_{a2}\\ 1\end{array}\right\} + \left\{t_i^2\right\}\left\{\begin{array}{c}A_{p2}\\ B_{p2}\\ C_{p2}\\ D_{p2}\\ 1\end{array}\right\} + \left\{t_i^3\right\}\left\{\begin{array}{c}A_{a2}\\ B_{a2}\\ C_{a2}\\ D_{a2}\\ 1\end{array}\right\} \tag{31}$$

The transfer matrix between active pile segment tip and top can then be obtained:

$$\begin{Bmatrix} \overline{w}_i^{a2}(h_i) \\ \overline{\phi}_i^{a2}(h_i) \\ \overline{M}_i^{a2}(h_i) \\ \overline{Q}_i^{a2}(h_i) \\ 1 \end{Bmatrix} = T_{i1} \begin{Bmatrix} \overline{w}_i^{a2}(0) \\ \overline{\phi}_i^{a2}(0) \\ \overline{M}_i^{a2}(0) \\ \overline{Q}_i^{a2}(0) \\ 1 \end{Bmatrix} + T_{i2} \begin{Bmatrix} w_i^{p2}(0) \\ w_i^{p2}(0) \\ M_i^{p2}(0) \\ Q_i^{p2}(0) \\ 1 \end{Bmatrix} + T_{i3} \begin{Bmatrix} w_i^{a2}(0) \\ w_i^{a2}(0) \\ M_i^{a2}(0) \\ Q_i^{a2}(0) \\ 1 \end{Bmatrix} \tag{32}$$

where

$$\begin{aligned} T_{i1} = {} & t_i^1(h_i)\left(t_i^1(0)\right)^{-1} \\ T_{i2} = {} & \left(t_i^2(h_i) - t_i^1(h_i)\left(t_i^1(0)\right)^{-1} t_i^2(0)\right)\left(t_i^{a2}(0)\right)^{-1} \\ T_{i3} = {} & -\left(t_i^2(h_i) - t_i^1(h_i)\left(t_i^1(0)\right)^{-1} t_i^2(0)\right)\left(t_i^{a2}(0)\right)^{-1} t_i^{p2}(0)\left(t_i^{a2}(0)\right)^{-1} + \\ & \left(t_i^3(h) - t_i^1(h)\left(t_i^1(0)\right)^{-1} t_i^3(0)\right)\left(t_i^{a2}(0)\right)^{-1} \end{aligned} \tag{33}$$

The transfer matrix between the active pile tip and top (soil–water surface) due to the secondary wave can then be obtained:

$$\begin{Bmatrix} \overline{w}^{a2}(H) \\ \overline{\phi}^{a2}(H) \\ \overline{M}^{a2}(H) \\ \overline{Q}^{a2}(H) \\ 1 \end{Bmatrix} = T_1 \begin{Bmatrix} \overline{w}^{a2}(0) \\ \overline{\phi}^{a2}(0) \\ \overline{M}^{a2}(0) \\ \overline{Q}^{a2}(0) \\ 1 \end{Bmatrix} + T_2 \begin{Bmatrix} w^{p2}(0) \\ w^{p2}(0) \\ M^{p2}(0) \\ Q^{p2}(0) \\ 1 \end{Bmatrix} + T_3 \begin{Bmatrix} w^{a2}(0) \\ w^{a2}(0) \\ M^{a2}(0) \\ Q^{a2}(0) \\ 1 \end{Bmatrix} \tag{34}$$

where $T_1 = T_{n1} \cdots T_{i1} \cdots T_{21} T_{11}$, $T_2 = T_{n2} \cdots T_{i2} \cdots T_{22} T_{12}$, $T_3 = T_{n3} \cdots T_{i3} \cdots T_{23} T_{13}$.

2.4. Overall Dynamic Response of the Pile Group

The transfer matrix of pile submerged in the seawater and pile embedded in the soil are obtained in Sections 2.3 and 2.4. According to the continuity condition at the water–soil interface, the dynamic response of these two pile parts can be connected, and the overall dynamic response of the pile can be obtained.

2.4.1. Active Pile

For the active pile, the continuity condition is:

$$\begin{Bmatrix} w^{a2}(0) \\ \phi^{a2}(0) \\ M^{a2}(0) \\ Q^{a2}(0) \\ 1 \end{Bmatrix} = \begin{Bmatrix} w^{a1}(L_w) \\ \phi^{a1}(L_w) \\ M^{a1}(L_w) \\ Q^{a1}(L_w) \\ 1 \end{Bmatrix} \tag{35}$$

Substituting Equations (10) and (35) into Equation (16), the overall transfer matrix can be obtained:

$$\begin{Bmatrix} w^a(L) \\ \phi^a(L) \\ M^a(L) \\ Q^a(L) \\ 1 \end{Bmatrix} = \{T^{a2}\}\{T^{a1}\} \begin{Bmatrix} w^a(0) \\ \phi^a(0) \\ M^a(0) \\ Q^a(0) \\ 1 \end{Bmatrix} \tag{36}$$

Substituting boundary conditions:

$$\begin{aligned} M_{z=0} &= \left.\frac{d^2w(z)}{dz^2}\right|_{z=0} = 0 \\ Q_{z=0} &= \left.\frac{d^3w(z)}{dz^3}\right|_{z=0} = 0 \\ w_{z=L} &= w(z)|_{z=L} = 0 \\ \theta_{z=L} &= \left.\frac{dw(z)}{dz}\right|_{z=L} = 0 \end{aligned} \tag{37}$$

Equation (36) can then be solved, and the displacement, rotation angle, bending moment and the shearing force of the active pile top can be obtained; the value of single pile swaying impedance K^*_{hh}, rocking impedance K^*_{rr}, swaying-rocking impedance K^*_{hr}, and rocking-swaying impedance K^*_{rh} can then be obtained.

2.4.2. Passive Pile

For the passive pile, the continuity condition is:

$$\begin{Bmatrix} w^{p2}(0) \\ \phi^{p2}(0) \\ M^{p2}(0) \\ Q^{p2}(0) \\ 1 \end{Bmatrix} = \begin{Bmatrix} w^{p1}(L_w) \\ \phi^{p1}(L_w) \\ M^{p1}(L_w) \\ Q^{p1}(L_w) \\ 1 \end{Bmatrix} \tag{38}$$

Substituting Equations (10), (11) and (38) into Equation (25), the overall transfer matrix can be obtained:

$$\begin{Bmatrix} w^{p}(L) \\ \phi^{p}(L) \\ M^{p}(L) \\ Q^{p}(L) \\ 1 \end{Bmatrix} = \left\{T_1^{p2}\right\}\left\{T^{p1}\right\} \begin{Bmatrix} w^{p}(0) \\ \phi^{p}(0) \\ M^{p}(0) \\ Q^{p}(0) \\ 1 \end{Bmatrix} + \left\{T_2^{p2}\right\}\left\{T^{a1}\right\} \begin{Bmatrix} w^{a}(0) \\ \phi^{a}(0) \\ M^{a}(0) \\ Q^{a}(0) \\ 1 \end{Bmatrix} \tag{39}$$

Substituting the solution to Equation (36) and boundary conditions (37), the equation can then be solved, and the displacement, rotation angle, bending moment, and the shearing force of the passive pile top can be obtained.

2.4.3. Active Pile under Secondary Wave

For the active pile under secondary wave:

$$\begin{Bmatrix} \overline{w}^{a2}(0) \\ \overline{\phi}^{a2}(0) \\ \overline{M}^{a2}(0) \\ \overline{Q}^{a2}(0) \\ 1 \end{Bmatrix} = \begin{Bmatrix} \overline{w}^{a1}(L_w) \\ \overline{\phi}^{a1}(L_w) \\ \overline{M}^{a1}(L_w) \\ \overline{Q}^{a1}(L_w) \\ 1 \end{Bmatrix} \tag{40}$$

Substituting Equations (10), (11) and (40) into Equation (34), the overall transfer matrix can be obtained:

$$\begin{Bmatrix} \overline{w}^{a}(L) \\ \overline{\phi}^{a}(L) \\ \overline{M}^{a}(L) \\ \overline{Q}^{a}(L) \\ 1 \end{Bmatrix} = T_1\{\overline{T}^{a1}\} \begin{Bmatrix} \overline{w}^{a}(0) \\ \overline{\phi}^{a}(0) \\ \overline{M}^{a}(0) \\ \overline{Q}^{a}(0) \\ 1 \end{Bmatrix} + T_2\{T^{p1}\} \begin{Bmatrix} w^{p}(0) \\ \phi^{p}(0) \\ M^{p}(0) \\ Q^{p}(0) \\ 1 \end{Bmatrix} + T_3\{T^{a1}\} \begin{Bmatrix} w^{a}(0) \\ \phi^{a}(0) \\ M^{a}(0) \\ Q^{a}(0) \\ 1 \end{Bmatrix} \tag{41}$$

Substituting the solutions to Equations (36) and (39), and boundary conditions (37), the equation can be solved. The displacement, rotation angle, bending moment, and the shearing force of the active pile top under the secondary wave can be obtained.

2.4.4. Pile–Pile Interaction and Pile Group Dynamic Response

The pile–pile interaction factor is defined as [31]:

$$\alpha^G = \frac{w^p}{w^a} \tag{42}$$

where w_a is the displacement of the active pile and w_p is the displacement of the passive pile. To reflect the influence of the passive pile to the active pile, the definition of the pile–pile interaction factor is modified as [32]:

$$\begin{aligned} \alpha^G &= \frac{w^p}{w^a - \overline{w}^a} = \frac{w^p}{(1-\kappa)w^a} \\ \kappa &= \frac{\overline{w}^a}{w^a} \end{aligned} \tag{43}$$

where $\overline{w}_a$ is the displacement of active pile under secondary wave.

Suppose the number of piles of the pile group is n, the cap is rigid, and the mass of the cap is ignored. Under the lateral harmonic load $Pe^{i\omega t}$, the lateral displacement of the pile group w^G can be considered equal to the displacement of each pile w^i, which means $w^1 = w^i = w^G$. In the pile group, each pile plays the role of both active pile and passive pile, so the single pile displacement can be calculated:

$$w^i = \left(1 - \sum_{j=1, j\neq i}^{n} \kappa_{ij}\right) w^i + \sum_{j=1, j\neq i}^{n} \alpha_{ij}^G w^j \tag{44}$$

The dynamic equation of the pile group can be written in matrix form:

$$\frac{1}{K_{hh}^*}\begin{bmatrix} 1 - \sum\limits_{j=1,j\neq 1}^{n} \kappa_{1j} & \alpha_{12}^G & \cdots & \alpha_{1n}^G \\ \alpha_{21}^G & 1 - \sum\limits_{j=1,j\neq 2}^{n} \kappa_{2j} & \cdots & \alpha_{2n}^G \\ \vdots & \vdots & \vdots & \vdots \\ \alpha_{n1}^G & \alpha_{n2}^G & \cdots & 1 - \sum\limits_{j=1,j\neq n}^{n} \kappa_{nj} \end{bmatrix} \begin{Bmatrix} P_1 \\ P_2 \\ \vdots \\ P_n \end{Bmatrix} = \begin{Bmatrix} w^1 \\ w^2 \\ \vdots \\ w^n \end{Bmatrix} \tag{45}$$

where K_{hh}^* is the single pile impedance, which can be obtained by solving the Equation (36), P_j is the load distributed to pile j, and a_{ij}^G is the interaction factor between pile i and j, $i \neq j$. The load applied to the pile group is the sum of the load applied to each pile:

$$P = \sum_{j=1}^{n} P_j \tag{46}$$

Then, Equation (45) can be solved, and the horizontal pile group impedance can be calculated:

$$K_G^* = P/w_G = K_G + iC_G \tag{47}$$

According to the previous research [31], the swaying interaction factor, the sway-rocking interaction factor, and the rock-swaying interaction factor between piles can be ignored. Therefore, K_{rr}^*, K_{hr}^*, K_{rh}^* are the sum of single pile calculation results. The pile group impedance matrix can then be obtained:

$$K^* = \begin{Bmatrix} K_G^* & K_{hr}^* \\ K_{rh}^* & K_{rr}^* \end{Bmatrix} \tag{48}$$

2.5. Superstructure Dynamic Response

2.5.1. Distributed Wind Load

The distributed wind load $F_d(z)$ can be calculated according to the equation below [33]:

$$F_d(z) = 1/2\rho_a C_D D_T(z) V^2(z) \tag{49}$$

where ρ_d is the density of the air, valued 1.225 kg/m^3 in this paper, C_D is the drag coefficient, valued according to the Reynolds number and structure surface roughness, i.e., 1.2 in this paper, V (m/s) is the average wind speed, and D_T (m) is the tower diameter, the value of which changes with increasing tower height.

The wind profile [34] can be calculated according to the following equation:

$$V(z) = V_a\left(\frac{z}{H_a}\right)^{\alpha_w} \tag{50}$$

where V_a is the wind speed at height H_a and α_w is the power law coefficient, valued 0.12 in open seas with waves.

2.5.2. Thrust Wind Load

The wind load applied to the blades and turbine will produce lateral load, which can be considered as the concentrated load F_b, calculated according to Equation (51) [35]:

$$F_b = 1/2\rho_a \pi R_T^2 V_T^2 C_T(\lambda_s) \tag{51}$$

where V_T is the wind speed at the tower top, R_T is the blade radius, and C_T is the thrust coefficient, which is related to the tip speed ratio λ_s, as shown in Figure 5.

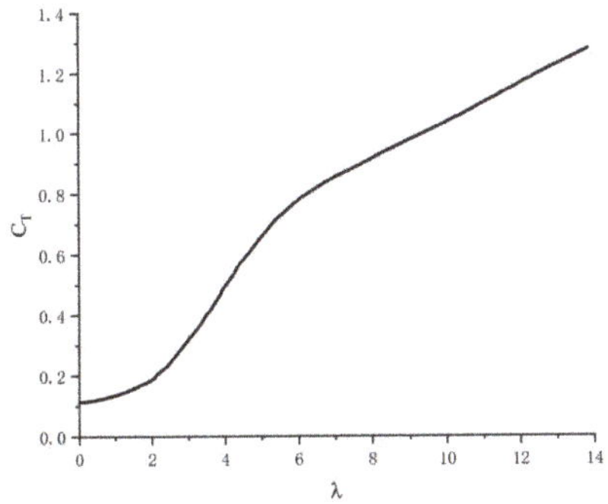

Figure 5. Thrust coefficient C_T-tip speed ratio λ_s curve [35].

2.5.3. Tower Dynamic Response

The tower can be divided into n segments, numbered 1~n from the bottom to the top. As shown in Figure 6, for the i ($i = 1, 2 \cdots n$)th segment, the number of its bottom point is i, and the number of its top point is $i + 1$. Consider each segment as equal cross-section beam, where H_i is the length of the ith segment. H_i is valued differently between segments to account for the varied cross-section geometry. The mass of each segment is concentrated to the bottom point, and the 2×2 mass matrix of each segment can be obtained, including the mass and moment of inertia. The mass of blades is considered as the concentrated mass point m_{n+1} at the tower top. Similarly, the distributed loads applied to each segment are concentrated to the bottom point, and the load matrix of each segment can be obtained, including the force and bending moment. The load matrix at the tower top includes the thrust wind force.

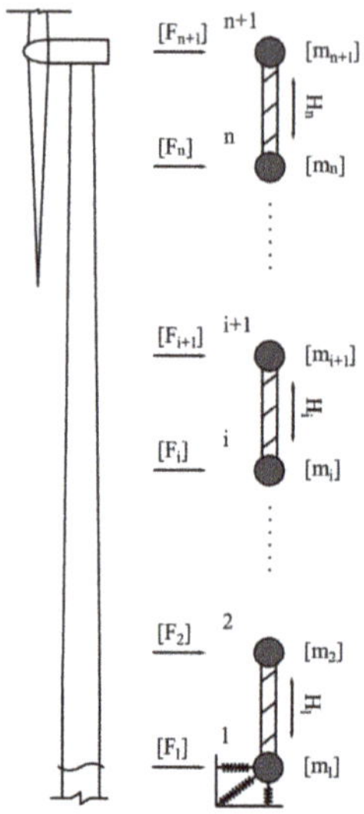

Figure 6. Tower discretization.

The swaying-rocking equation of this system with 2(n+1) degrees of freedom can be written in matrix form [25]:

$$\begin{bmatrix} \overline{K}_{11} & \overline{K}_{12} & & & & & \\ \overline{K}_{21} & \overline{K}_{22} & \overline{K}_{23} & & & & \\ \ddots & \ddots & \ddots & & & & \\ & \overline{K}_{i,j-1} & \overline{K}_{i,j} & \overline{K}_{i,j+1} & & & \\ & & \ddots & \ddots & \ddots & & \\ & & & \overline{K}_{n,n-1} & \overline{K}_{n,n} & \overline{K}_{n,n+1} \\ & & & & \overline{K}_{n+1,n} & \overline{K}_{n+1,n+1} \end{bmatrix} \begin{Bmatrix} u_1 \\ u_2 \\ \vdots \\ u_i \\ \vdots \\ u_n \\ u_{n+1} \end{Bmatrix} = \begin{Bmatrix} F_1 \\ F_2 \\ \vdots \\ F_i \\ \vdots \\ F_n \\ F_{n+1} \end{Bmatrix} \tag{52}$$

where u_i is the point displacement matrix, including the horizontal displacement and rotation angle. The stiffness in Equation (52) can be calculated according to equations below.

For the first line:

$$\overline{K}_{11} = K_{11} + K^* - \omega^2 m_1 \tag{53}$$

$$\overline{K}_{12} = K_{12} \tag{54}$$

For the second to nth line:

$$\overline{K}_{i,i-1} = K_{i,i-1} \tag{55}$$

$$\overline{K}_{i,i} = K_{i,i} - \omega^2 m_i \tag{56}$$

$$\overline{K}_{i,i+1} = K_{i,i+1} \tag{57}$$

For $n+1$th line:

$$\overline{K}_{n+1,n} = K_{n+1,n} \tag{58}$$

$$\overline{K}_{n+1,n+1} = K_{n+1,n+1} - \omega^2 m_{n+1} \tag{59}$$

where

$$K_{i,i-1} = \begin{bmatrix} -12E_pI_{i-1}/H_{i-1}^3 & -6E_pI_{i-1}/H_{i-1}^2 \\ 6E_pI_{i-1}/H_{i-1}^2 & 2E_pI_{i-1}/H_{i-1} \end{bmatrix} \tag{60}$$

$$K_{i,i} = \begin{bmatrix} 12E_pI_i/H_i^3 & 6E_pI_i/H_i^2 \\ 6E_pI_i/H_i^2 & 4E_pI_i/H_i \end{bmatrix} + \begin{bmatrix} 12E_pI_{i-1}/H_{i-1}^3 & -6E_pI_{i-1}/H_{i-1}^2 \\ -6E_pI_{i-1}/H_{i-1}^2 & 4E_pI_{i-1}/H_{i-1} \end{bmatrix} \tag{61}$$

$$K_{i,j+1} = \begin{bmatrix} -12E_p I_i/H_i^3 & 6E_p I_i/H_i^2 \\ -6E_p I_i/H_i^2 & 2E_p I_i/H_i \end{bmatrix} \quad (62)$$

When $i = 1$, Equation (60) is invalid, and the second part of Equation (61) should be removed. When $i = n + 1$, Equation (57) is invalid, and the first part of Equation (61) should be removed. Substitute the foundation impedance matrix K^* into Equation (53), the equation can then be solved, and the displacement of the tower can be obtained.

3. Validation

The calculation result is compared with the FEM result to validate the correctness of the proposed calculation model. ABAQUS simulation software is used in this paper. A pile group supported OWT [25] is used for validation. The total mass of blades and turbine is 177.1 ton, and the tower is divided into three segments, as shown in Table 1. The pile group consists of seven piles, as shown in Figure 2. The pile diameter is 1.7 m and the pile wall thickness is 30 mm. The elastic modulus of the steel is 210 GPa and the density of the steel is 7800 kg/m^3. The pile length embedded in the soil is 30 m nad the pile length submerged in the seawater is 20 m. The elastic modulus of the soil is 40 MPa, the Poisson's ratio of the soil is 0.3, and the density of the soil is 1800 kg/m^3. As shown in Figure 7, the offshore wind turbine model is established.

Table 1. Tower parameters.

Tower Parts	Length (m)	Bottom Diameter	Top Diameter	Tower Wall Thickness
Upper segment	32	3.9	3.1	50
Middle segment	31	4.5	3.9	50
Bottom segment	15	4.5	4.5	50

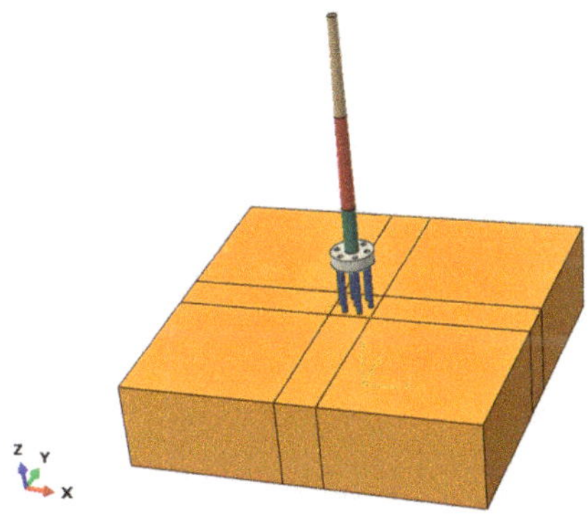

Figure 7. FEM numerical simulation model.

Since the overall displacement of the offshore wind turbine is relatively small, linear modal analysis is used to calculate its structure natural frequency. The deformation of soil foundation is also small during the dynamic analysis; therefore, the change of soil foundation stiffness is not considered, and the small strain linear elastic model is used for the soil.

For the tower, the eight-node S8R shell element is used. The tower is divided into three parts, as shown in Table 1. For each part, the tower diameter changes linearly with increasing height. To prevent the separation between tower parts, a bonding constraint is added between the interfaces, including the interface between the bottom tower part and the cap. The blades are simplified as the concentrated force applied to the tower top. The pile foundation is modeled as solid element to better simulate the pile–soil interaction. For the side face of the soil, the lateral displacement is constrained; for the bottom side of the soil, displacements of all directions are constrained. The pile–soil interaction is set as small sliding, penalty contact, while the coefficient is set as 0.4.

The dead weight is applied to the model according to parameters presented in Table 1, and the wave load is also applied to the structure. Here, we use FORTRAN to write a subroutine to accurately input wave load according to Equation (4). The offshore wind turbine is fine meshed. As for the soil, the soil around the pile is refined, as shown in Figure 8.

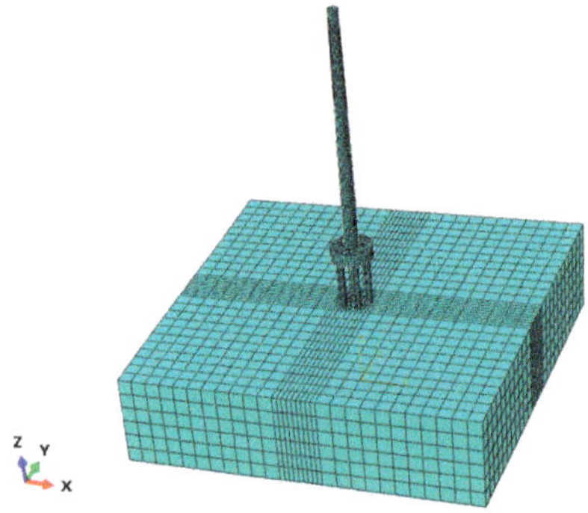

Figure 8. Meshing of the model.

After establishing the FEM model, the modal analysis is performed, and the first 10 structure natural frequencies are obtained. Here, we focus on the first lateral structure natural frequency. Then, we applied a 10 kN horizontal harmonic thrust wind load to the tower top, and the tower top displacement and load frequency curve can be obtained.

As shown in Figure 9, the FEM result and the calculating result are in good agreement, with only some differences being observed for the maximum tower displacement, which validates the correctness of the calculating result. The structure natural frequency can also be obtained from Figure 9, which falls within the "soft-stiff" design frequency range.

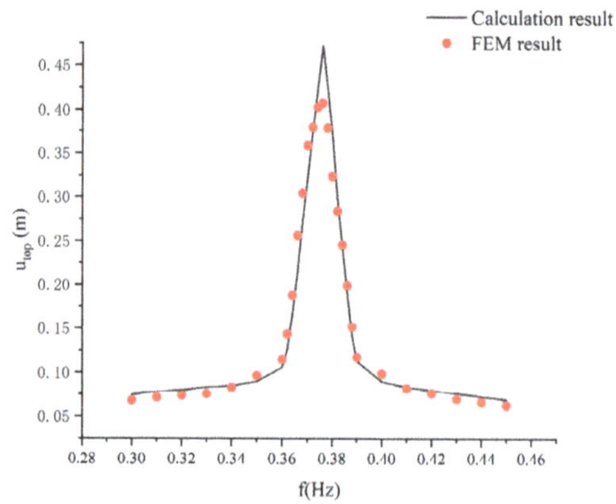

Figure 9. Comparison between the calculation result and FEM result.

4. Parametric Analysis

4.1. Tower Displacement

The tower displacement under different wind speeds is studied. The wind speed at sea level is 4 m/s, 8 m/s, and 12 m/s, respectively. The percentage of the thrust wind load is 96.31%, and the percentage of the distributed wind load is 3.69%. As shown in Figure 10, the tower displacement is largely influenced by the wind speed. When the wind speed is 4 m/s, the increasing rate of tower displacement is relatively small. When the wind speed is 8 m/s, the increasing rate rises to a certain degree. When the wind speed is 12 m/s, the increasing rate is significantly larger than the increasing rate when the wind speed is 4 m/s.

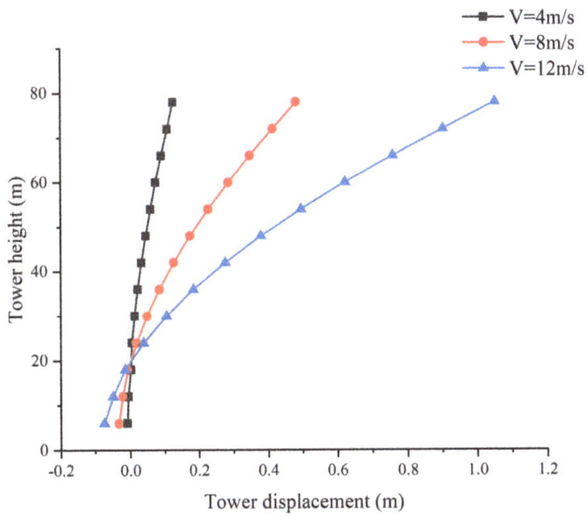

Figure 10. The influence of the wind speed on the tower displacement.

The influence of the foundation impedance on the tower displacement is also studied. Define $\alpha_K = K'^*/K^*$, where K'^* is the modified foundation impedance matrix and α_K is valued from 0.1~2 to analyze the influence of the foundation impedance on the maximum tower displacement. As shown in Figure 11, when the foundation impedance is relatively small, the tower displacement is large. With increasing foundation impedance, the tower displacement decreases rapidly. Additionally, with increasing foundation impedance, the tower displacement decreasing speed slows. When the foundation impedance reaches 2.0 its original impedance, the tower displacement remains almost unchanged.

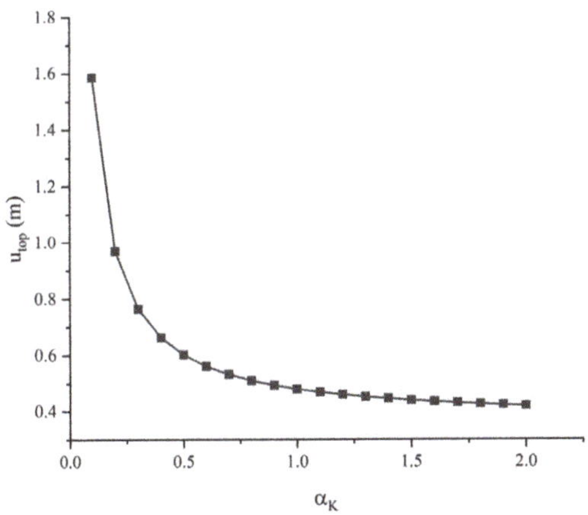

Figure 11. The influence of the foundation impedance on the tower displacement.

4.2. Pile–Pile Interaction Factor

The regressive solution (the pile is considered completely embedded in the soil) of the proposed calculation model is compared with the solution of Kaynia [36]. In this section, parameters below are used for calculation: the ratio of elastic modulus of the pile to elastic modulus of the soil $E_p/E_s = 1000$, the ratio of the pile length to the pile diameter

$L/D = 20$, and Poisson's ratio is 0.4. The ratio of the density of the pile to the density of the soil $\rho_p/\rho_s = 1.3$, while the damping ratio $\beta_s = 0.05$. The pile spacing $S = 2D$, the pile embedment ratio $L_w/L = 0.3$. The dimensionless frequency $a_0 = \omega D/V_s$ is used for analysis. Here, we consider the influence of three parameters on the pile–pile interaction factor: pile spacing, pile embedment ratio, and the angle of incidence θ.

As shown in Figure 12, when the pile spacing $S/D = 10$, the calculating result of this paper and the result from the literature are in good agreement, which validates the correctness of the calculation method. When the pile spacing $S/D = 5$, some differences can be observed compared with the results from the literature. Furthermore, when $S/D = 2$, the difference becomes more significant, especially in low frequencies. This is because when the pile spacing decreases, the influence of the passive pile on the active pile is more significant, which changes the horizontal displacement of the active pile. This analysis demonstrates that when the pile spacing is small, the influence of the passive pile on the active pile is important and cannot be ignored.

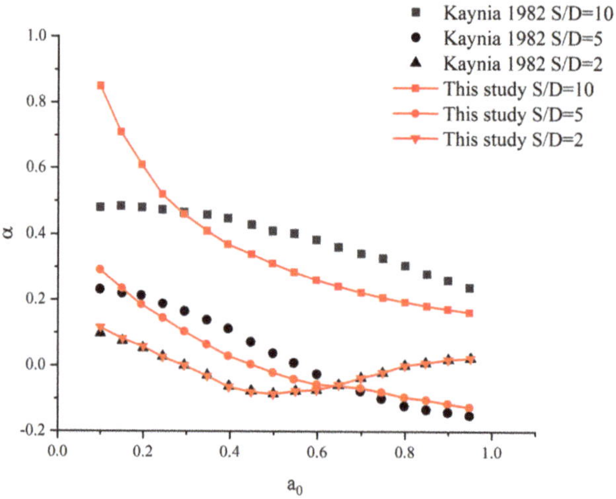

Figure 12. Pile–pile interaction under different pile spacings.

The influence of the pile embedded ratio is studied. As shown in Figure 13, when the pile is completely embedded, the pile–pile interaction factor decreases with increasing load frequency, and the value remains positive. When $L_w/L = 0.1$, the pile–pile interaction factor α at all frequencies largely decreases. When $L_w/L = 0.2$, α changes obviously. With increasing load frequency, α decreases rapidly when closing to a certain frequency and rises after reaching the lowest point. When $L_w/L = 0.3$, the curve moves from a mainly middle-high frequency to a middle-low frequency. This analysis demonstrates that α is sensitive to the pile embedment ratio. When the pile embedment ratio is relatively small, the change of α with increasing frequency is simple; when the pile embedment ratio is relatively large, the change of α becomes complex.

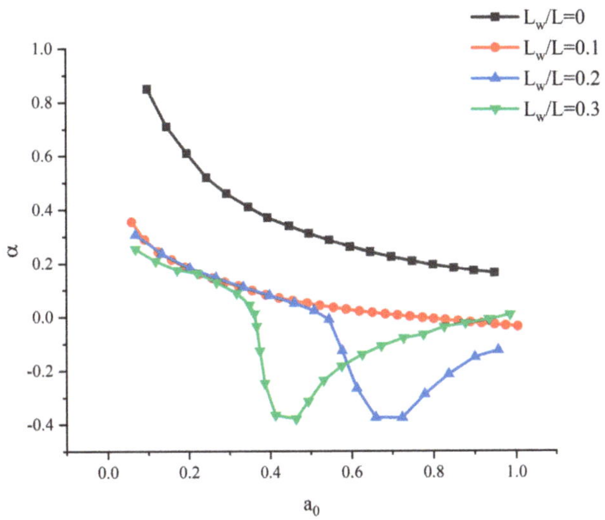

Figure 13. Pile–pile interaction under different pile embedded ratios.

The influence of different incidences angle θ is studied, i.e., 0°, 45°, and 90°, respectively. As shown in Figure 14, when $\theta = 0^{\circ}$, the curve fluctuates significantly, the minimum value of α is the smallest among three angles, and the maximum value of α is the largest among three angles. When $\theta = 90^{\circ}$, the value of α is slightly smaller than that of $\theta = 0^{\circ}$. When $\theta = 45^{\circ}$, the change of α is moderate, and no significant fluctuation is observed.

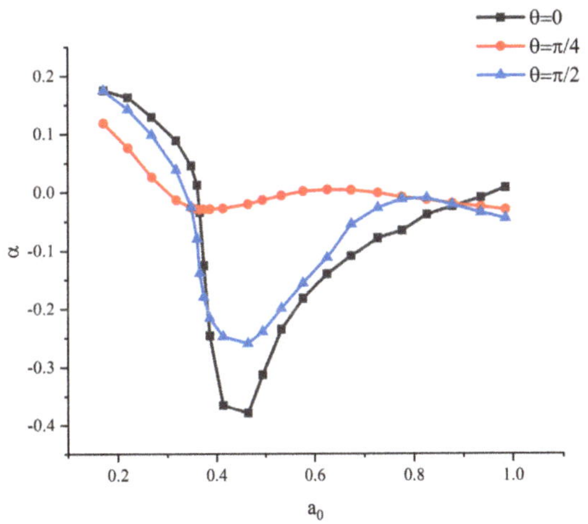

Figure 14. Pile–pile interaction under different incidence angles.

4.3. *Dynamic Foundation Impedance*

As shown in Figure 15, with increasing load frequency, the change of horizonal foundation impedance is not obvious. When the load frequency is around soil cut-off frequency, the horizontal foundation impedance slightly decreases. The foundation impedance is very sensitive to the pile length submerged in the seawater. With an increasing pile embedment ratio, the foundation impedance decreases significantly.

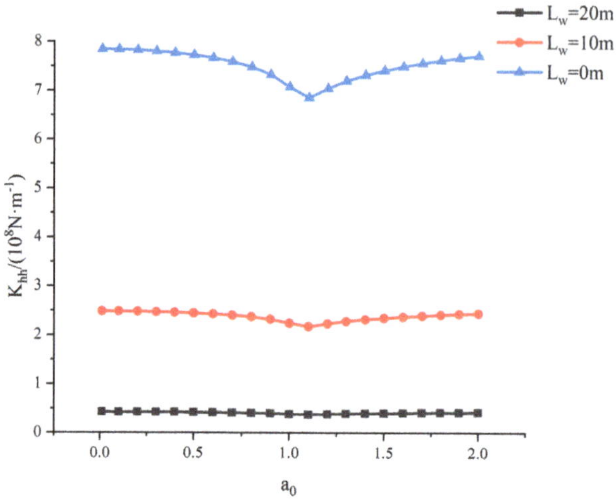

Figure 15. The influence of the pile length in seawater on the foundation impedance.

Meanwhile, the wave load can influence the foundation impedance to a certain degree. As shown in Figure 16, when the wave load is relatively small (wavelength $L_{wl} = 75$ m, wave height $H_w = 4$ m), the influence of wave load on the foundation impedance is not significant. When the wave load is relatively large (wavelength $L_{wl} = 120$ m, wave height $H_w = 8$ m), the influence of wave load on the foundation impedance is significant.

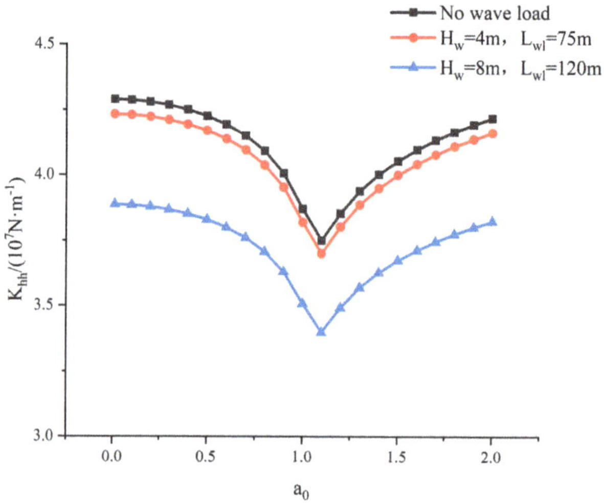

Figure 16. The influence of the wave load on the foundation impedance.

5. Conclusions

This paper establishes the model of pile group supported OWT under wind and wave load; the main findings are presented below:

(1) With increasing wind speed, the tower displacement increases significantly. The influence of foundation impedance on the tower displacement is more significant when the foundation impedance is relatively small.

(2) The pile–pile interaction factor depends largely on the pile spacing. When the pile spacing is large, the influence of the passive pile on the active pile can be ignored; when the pile spacing is small, a secondary wave effect should be considered for the pile–pile interaction factor.

(3) When the pile embedded ratio is large, the pile–pile interaction is more obvious. When the incidence angle is 45°, the pile–pile interaction is less significant compared with that of 0° and 90°.
(4) The foundation impedance decreases significantly with an increasing pile embedment ratio.
(5) The influence of wave load on the foundation impedance is more obvious when the wave load is large.

Author Contributions: Conceptualization, W.Y.; Data curation, Y.S.; Formal analysis, Y.S.; Funding acquisition, W.Y.; Investigation, G.Y.; Methodology, Y.S.; Project administration, G.Y.; Resources, G.Y.; Software, Y.S.; Supervision, W.Y.; Validation, Y.S.; Visualization, Y.S.; Writing—original draft, Y.S.; Writing—review and editing, Y.S. and W.Y. All authors have read and agreed to the published version of the manuscript.

Funding: This study was financially supported by the Key Projects of the National Natural Science Foundation of China (Grant No. 11932010).

Institutional Review Board Statement: Not applicable.

Informed Consent Statement: Not applicable.

Data Availability Statement: Not applicable.

Conflicts of Interest: The authors declare no conflict of interest.

Appendix A

$$\{t_i^{a1}\} = \begin{Bmatrix} e^{(-1+i)\gamma z} & e^{(-1-i)\gamma z} & e^{(1-i)\gamma z} & e^{(1+i)\gamma z} & q'_G(z) \\ (-1+i)\gamma e^{(-1+i)\gamma z} & (-1-i)\gamma e^{(-1-i)\gamma z} & (1-i)\gamma e^{(1-i)\gamma z} & (1+i)\gamma e^{(1+i)\gamma z} & K\frac{\sinh(Kz)}{\cosh(Kz)}q'_G(z) \\ (-1+i)^2\gamma^2 e^{(-1+i)\gamma z} & (-1-i)^2\gamma^2 e^{(-1-i)\gamma z} & (1-i)^2\gamma^2 e^{(1-i)\gamma z} & (1+i)^2\gamma^2 e^{(1+i)\gamma z} & K^2 q'_G(z) \\ (-1+i)^3\gamma^3 e^{(-1+i)\gamma z} & (-1-i)^3\gamma^3 e^{(-1-i)\gamma z} & (1-i)^3\gamma^3 e^{(1-i)\gamma z} & (1+i)^3\gamma^3 e^{(1+i)\gamma z} & K^3\frac{\sinh(Kz)}{\cosh(Kz)}q'_G(z) \\ 0 & 0 & 0 & 0 & 1 \end{Bmatrix}$$

$$\{t_i^{a2}\} = \begin{Bmatrix} e^{(-1+i)\gamma z} & e^{(-1-i)\gamma z} & e^{(1-i)\gamma z} & e^{(1+i)\gamma z} & 0 \\ (-1+i)\gamma e^{(-1+i)\gamma z} & (-1-i)\gamma e^{(-1-i)\gamma z} & (1-i)\gamma e^{(1-i)\gamma z} & (1+i)\gamma e^{(1+i)\gamma z} & 0 \\ (-1+i)^2\gamma^2 e^{(-1+i)\gamma z} & (-1-i)^2\gamma^2 e^{(-1-i)\gamma z} & (1-i)^2\gamma^2 e^{(1-i)\gamma z} & (1+i)^2\gamma^2 e^{(1+i)\gamma z} & 0 \\ (-1+i)^3\gamma^3 e^{(-1+i)\gamma z} & (-1-i)^3\gamma^3 e^{(-1-i)\gamma z} & (1-i)^3\gamma^3 e^{(1-i)\gamma z} & (1+i)^3\gamma^3 e^{(1+i)\gamma z} & 0 \\ 0 & 0 & 0 & 0 & 1 \end{Bmatrix}$$

$$\{t_i^{p2}\} = \frac{(k+i\omega c)\psi}{E_p I_p}\begin{Bmatrix} z(1-i)e^{(-1+i)\gamma z} & z(1+i)e^{(-1-i)\gamma z} \\ (1-i)e^{(-1+i)\gamma z} + z(2i\gamma)e^{(-1+i)\gamma z} & (1+i)e^{(-1-i)\gamma z} + z(-2i\gamma)e^{(-1-i)\gamma z} \\ (4i\gamma)e^{(-1+i)\gamma z} - 2z(i+1)\gamma^2 e^{(-1+i)\gamma z} & (-4i\gamma)e^{(-1-i)\gamma z} + 2z(i-1)\gamma^2 e^{(-1-i)\gamma z} \\ -6(i+1)\gamma^2 e^{(-1+i)\gamma z} + 4z\gamma^3 e^{(-1+i)\gamma z} & 6(i-1)\gamma^2 e^{(-1-i)\gamma z} + 4z\gamma^3 e^{(-1-i)\gamma z} \\ 0 & 0 \end{Bmatrix.$$
$$\begin{matrix} z(-1+i)e^{(1-i)\gamma z} & z(-1-i)e^{(1+i)\gamma z} & 0 \\ (-1+i)e^{(1-i)\gamma z} + z(2i\gamma)e^{(1-i)\gamma z} & (-1-i)e^{(1+i)\gamma z} + z(-2i\gamma)e^{(1+i)\gamma z} & 0 \\ (4i\gamma)e^{(1-i)\gamma z} + 2z(i+1)\gamma^2 e^{(1-i)\gamma z} & (-4i\gamma)e^{(1+i)\gamma z} - 2(i-1)\gamma^2 z e^{(1+i)\gamma z} & 0 \\ 6(i+1)\gamma^2 e^{(1-i)\gamma z} + 4z\gamma^3 e^{(1-i)\gamma z} & -6(i-1)\gamma^2 e^{(1+i)\gamma z} + 4\gamma^3 z e^{(1+i)\gamma z} & 0 \\ 0 & 0 & 1 \end{matrix}\Bigg\}$$

$$\left\{\bar{t}_i^{a2}\right\} = \frac{(k+i\omega c)\psi}{512E_pI_p\gamma^6}\left\{\begin{array}{rr}
-2iz^2e^{(-1+i)\gamma z} & \\
-4ize^{(-1+i)\gamma z}+2(i+1)\gamma z^2e^{(-1+i)\gamma z} & \\
-4ie^{(-1+i)\gamma z}+8(i+1)\gamma ze^{(-1+i)\gamma z}-4\gamma^2z^2e^{(-1+i)\gamma z} & \\
12(1+i)\gamma e^{(-1+i)\gamma z}-24\gamma^2ze^{(-1+i)\gamma z}+4(1-i)\gamma^3z^2e^{(-1+i)\gamma z} & \\
0 & \\
2iz^2e^{(-1-i)\gamma z} & \\
4ize^{(-1-i)\gamma z}+2(1-i)\gamma z^2e^{(-1-i)\gamma z} & \\
4ie^{(-1-i)\gamma z}+8(1-i)\gamma ze^{(-1-i)\gamma z}-4\gamma^2z^2e^{(-1-i)\gamma z} & \\
12(1-i)\gamma e^{(-1-i)\gamma z}-24\gamma^2ze^{(-1-i)\gamma z}+4(1+i)\gamma^3z^2e^{(-1-i)\gamma z} & \\
0 & \\
-2iz^2e^{(1-i)\gamma z} & \\
-4ize^{(1-i)\gamma z}-2(i+1)\gamma z^2e^{(1-i)\gamma z} & \\
-4ie^{(1-i)\gamma z}-8(i+1)\gamma ze^{(1-i)\gamma z}-4\gamma^2z^2e^{(1-i)\gamma z} & \\
-12(1+i)\gamma e^{(1-i)\gamma z}-24\gamma^2ze^{(1-i)\gamma z}-4(1-i)\gamma^3z^2e^{(1-i)\gamma z} & \\
0 & \\
2iz^2e^{(1+i)\gamma z} & 0 \\
-4ize^{(1+i)\gamma z}+2(i-1)\gamma z^2e^{(1+i)\gamma z} & 0 \\
-4ie^{(1+i)\gamma z}+8(1-i)\gamma ze^{(1+i)\gamma z}+4\gamma^2z^2e^{(1+i)\gamma z} & 0 \\
12(1-i)\gamma e^{(1+i)\gamma z}+24\gamma^2ze^{(1+i)\gamma z}+4(1+i)\gamma^3z^2e^{(1+i)\gamma z} & 0 \\
0 & 1
\end{array}\right\}$$

References

1. Feyzollahzadeh, M.; Mahmoodi, M.; Yadavar-Nikravesh, S.; Jamali, J. Wind load response of offshore wind turbine tow-ers with fixed monopile platform. *J. Wind Eng. Ind. Aerodyn.* **2016**, *158*, 122–138. [CrossRef]
2. Zheng, X.Y.; Li, H.; Rong, W.; Li, W. Joint earthquake and wave action on the monopile wind turbine foundation: An experimental study. *Mar. Struct.* **2015**, *44*, 125–141. [CrossRef]
3. Gupta, B.K.; Basu, D. Offshore wind turbine monopile foundations: Design perspectives. *Ocean Eng.* **2020**, *213*, 107514. [CrossRef]
4. Bhattacharya, S. Challenges in Design of Foundations for Offshore Wind Turbines. *Eng. Technol. Ref.* **2014**, *1*, 922. [CrossRef]
5. Lombardi, D.; Bhattacharya, S.; Wood, D.M. Dynamic soil–structure interaction of monopile supported wind turbines in cohesive soil. *Soil Dyn. Earthq. Eng.* **2013**, *49*, 165–180. [CrossRef]
6. Damgaard, M.; Bayat, M.; Andersen, L.; Ibsen, L. Assessment of the dynamic behaviour of saturated soil subjected to cyclic loading from offshore monopile wind turbine foundations. *Comput. Geotech.* **2014**, *61*, 116–126. [CrossRef]
7. Bouzid, D.A.; Bhattacharya, S.; Otsmane, L. Assessment of natural frequency of installed offshore wind turbines using nonlinear finite element model considering soil-monopile interaction. *J. Rock Mech. Geotech. Eng.* **2018**, *10*, 333–346. [CrossRef]
8. Alkhoury, P.; Soubra, A.H.; Rey, V.; Aït-Ahmed, M. A full three-dimensional model for the estimation of the natural fre-quencies of an offshore wind turbine in sand. *Wind Energy* **2021**, *24*, 699–719. [CrossRef]
9. Kjørlaug, R.A.; Kaynia, A.M. Vertical earthquake response of megawatt-sized wind turbine with soil-structure interaction effects. *Earthq. Eng. Struct. Dyn.* **2015**, *44*, 2341–2358. [CrossRef]
10. Corciulo, S.; Zanoli, O.; Pisanò, F. Transient response of offshore wind turbines on monopiles in sand: Role of cyclic hy-dro–mechanical soil behaviour. *Comput. Geotech.* **2017**, *83*, 221–238. [CrossRef]
11. Zuo, H.; Bi, K.; Hao, H. Dynamic analyses of operating offshore wind turbines including soil-structure interaction. *Eng. Struct.* **2018**, *157*, 42–62. [CrossRef]
12. Galvín, P.; Romero, A.; Solís, M.; Domínguez, J. Dynamic characterisation of wind turbine towers account for a monopile foundation and different soil conditions. *Struct. Infrastruct. Eng.* **2017**, *13*, 942–954. [CrossRef]
13. Kausel, E. Early history of soil–structure interaction. *Soil Dyn. Earthq. Eng.* **2010**, *30*, 822–832. [CrossRef]
14. Reese, L.C. Non-dimensional solutions for laterally loaded piles with soil modulus assumed proportional to depth. In Proceedings of the 8th Texas Conference on Soil Mechanics and Foundation Engineering, Austin, TX, USA, 14–15 September 1956.
15. McClelland, B.; Focht, J. Soil modulus for laterally loaded piles. *J. Soil Mech. Found. Div.* **1956**, *82*, 1081-1–1081-22. [CrossRef]
16. Darvishi-Alamouti, S.; Bahaari, M.R.; Moradi, M. Natural frequency of offshore wind turbines on rigid and flexible monopiles in cohesionless soils with linear stiffness distribution. *Appl. Ocean Res.* **2017**, *68*, 91–102. [CrossRef]
17. Andersen, L.V.; Vahdatirad, M.; Sichani, M.T.; Sørensen, J.D. Natural frequencies of wind turbines on monopile founda-tions in clayey soils—A probabilistic approach. *Comput. Geotech.* **2012**, *43*, 1–11. [CrossRef]
18. Adhikari, S.; Bhattacharya, S. Vibrations of wind-turbines considering soil-structure interaction. *Wind Struct. Int. J.* **2011**, *14*, 85–112. [CrossRef]

19. Adhikari, S.; Bhattacharya, S. Dynamic Analysis of Wind Turbine Towers on Flexible Foundations. *Shock Vib.* **2012**, *19*, 37–56. [CrossRef]
20. Bhattacharya, S.; Adhikari, S. Experimental validation of soil–structure interaction of offshore wind turbines. *Soil Dyn. Earthq. Eng.* **2011**, *31*, 805–816. [CrossRef]
21. Nogami, T.; Paulson, S.K. Transfer matrix approach for nonlinear pile group response analysis. *Int. J. Numer. Anal. Methods Geomech.* **1985**, *9*, 299–316. [CrossRef]
22. Zhang, L.; Gong, X.-N.; Yang, Z.-X.; Yu, J.-L. Elastoplastic solutions for single piles under combined vertical and lateral loads. *J. Cent. South Univ. Technol.* **2011**, *18*, 216–222. [CrossRef]
23. Zhang, L.; Zhao, M.; Zou, X. Behavior of Laterally Loaded Piles in Multilayered Soils. *Int. J. Géoméch.* **2015**, *15*, 06014017. [CrossRef]
24. Wang, M.; Wang, Z.; Zhao, H. Analysis of Wind-Turbine Steel Tower by Transfer Matrix. In Proceedings of the 2009 International Conference on Energy and Environment Technology, Guilin, China, 16–18 October 2009.
25. Huang, M.; Zhong, R. Coupled horizontal-rocking vibration pf partially embedded pile groups and its effect on resonance of offshore wind turbine structures. *Chinsese J. Geotech. Eng.* **2014**, *36*, 286–294.
26. Anoyatis, G.; Mylonakis, G.; Lemnitzer, A. Soil reaction to lateral harmonic pile motion. *Soil Dyn. Earthq. Eng.* **2016**, *87*, 164–179. [CrossRef]
27. Anoyatis, G.; Lemnitzer, A. Dynamic pile impedances for laterally–loaded piles using improved Tajimi and Winkler for-mulations. *Soil Dyn. Earthq. Eng.* **2017**, *92*, 279–297. [CrossRef]
28. Shuqing, W.; Bingchen, L. *Wave Mechanics for Ocean Engineering*; China Ocean University Press: Qingdao, China, 2013. (In Chinese)
29. Xin-xin, L.; Da-peng, S.; Xue-jiao, X.; Hao, W.; Yu-cheng, L. Experimental study of irregular wave force loads on high rise pile platform. *J. Waterw. Harb.* **2013**, *34*, 277–284. (In Chinese)
30. Dobry, R.; Gazetas, G. Dynamic Response of Arbitrarily Shaped Foundations. *J. Geotech. Eng.* **1986**, *112*, 109–135. [CrossRef]
31. Mylonakis, G.; Gazetas, G. Lateral Vibration and Internal Forces of Grouped Piles in Layered Soil. *J. Geotech. Geoenvironmental Eng.* **1999**, *125*, 16–25. [CrossRef]
32. Luan, L.; Zheng, C.; Kouretzis, G.; Ding, X. Dynamic analysis of pile groups subjected to horizontal loads considering coupled pile-to-pile interaction. *Comput. Geotech.* **2020**, *117*, 103276. [CrossRef]
33. Van Binh, L.; Ishihara, T.; Van Phuc, P.; Fujino, Y. A peak factor for non-Gaussian response analysis of wind turbine tower. *J. Wind Eng. Ind. Aerodyn.* **2008**, *96*, 2217–2227. [CrossRef]
34. Veritas, N. *Environmental Conditions and Environmental Loads*; Det Norske Veritas Oslo: Høvik, Norway, 2000.
35. Jara, F.A.V. *Model Testing of Foundations for Offshore Wind Turbines*; Oxford University: Oxford, UK, 2006.
36. Kaynia, A.M. *Dynamic Stiffness and Seismic Response of Pile Groups*; Massachusetts Institute of Technology: Cambridge, MA, USA, 1982.

Journal of
Marine Science
and Engineering

MDPI

Article

Topology Optimization-Driven Design for Offshore Composite Wind Turbine Blades

Jian Song *, Junying Chen, Yufei Wu and Lixiao Li *

College of Civil and Transportation Engineering, Shenzhen University, Shenzhen 518060, China
* Correspondence: jiansong@szu.edu.cn (J.S.); llxiao2021@gmail.com (L.L.)

Abstract: With the increase in wind turbine power, the size of the blades is significantly increasing to over 100 m. It is becoming more and more important to optimize the design for the internal layout of large-scale offshore composite wind turbine blades to meet the structural safety requirements while improving the blade power generation efficiency and achieving light weight. In this work, the full-scale internal layout of an NREL 5 MW offshore composite wind turbine blade is elaborately designed via the topology optimization method. The aerodynamic wind loads of the blades were first simulated based on the computational fluid dynamics. Afterwards, the variable density topology optimization method was adopted to perform the internal structure design of the blade. Then, the first and second generation multi-web internal layouts of the blade were reversely designed and evaluated in accordance with the stress level, maximum displacement of blade tip and fatigue life. In contrast with the reference blade, the overall weight of the optimized blade was reduced by 9.88% with the requirements of stress and fatigue life, indicating a better power efficiency. Finally, the vibration modal and full life cycle of the designed blade were analyzed. The design conception and new architecture could be useful for the improvement of advanced wind turbines.

Keywords: offshore wind turbine blade; composites; computational fluid dynamics; topology optimization; fatigue life

Citation: Song, J.; Chen, J.; Wu, Y.; Li, L. Topology Optimization-Driven Design for Offshore Composite Wind Turbine Blades. *J. Mar. Sci. Eng.* **2022**, *10*, 1487. https://doi.org/10.3390/jmse10101487

Academic Editors: Eugen Rusu, Kostas Belibassakis and George Lavidas

Received: 28 March 2022
Accepted: 1 August 2022
Published: 13 October 2022

Publisher's Note: MDPI stays neutral with regard to jurisdictional claims in published maps and institutional affiliations.

Copyright: © 2022 by the authors. Licensee MDPI, Basel, Switzerland. This article is an open access article distributed under the terms and conditions of the Creative Commons Attribution (CC BY) license (https://creativecommons.org/licenses/by/4.0/).

1. Introduction

Wind energy is one of the most mature forms of renewable energy and is an effective strategy to alleviate energy shortages, reduce environmental pollution and improve climate conditions [1–3]. The utilization of onshore wind energy has encountered bottlenecks due to the restriction of land wind resources, noise and environmental pollution. Hence, in the last two decades, offshore wind farms have been rapidly growing. In 2021, new installations of more than 6 GW were generated all over the world [4]. For the offshore wind turbine, the blade is one of the most important components to convert wind kinetic energy into electrical energy. However, approximately 20% of the failures in wind turbine components occurs in the blades [5,6]. Nowadays, the blade length in offshore wind turbines has dramatically grown over 100 m, resulting in major concerns about the blades' resistance to damage over a life period of 20–25 years [7]. Compared to the onshore wind turbine blades, offshore wind turbine blade damage may happen in different parts due to static, vibration and fatigue loadings [8,9]. Moreover, it is great of importance to lighten the weight of the blades to reduce transportation costs.

Extensive studies for the design of the optimal offshore wind turbine blades have been carried out in recent years [10–13]. Each blade includes skin and web structures, where the skin structure determines the aerodynamic characteristics and the web structure provides the stiffness and strength requirements of the blade. Naung et al. [14] proposed a highly efficient nonlinear frequency-domain solution approach for elaborating the aerodynamic and aeromechanical performances of an oscillating wind turbine blade aerofoil. In comparison to the time-domain method, the frequency-domain method was not only accurate but also computationally very efficient as the computation time was declined by 90%. Furthermore, Naung et al. [15] also investigated the influence of the wake of neighboring turbines on the aerodynamics of a wind turbine within windfarms based on the aforesaid frequency-domain method. The corresponding conclusions had been also obtained, namely that the frequency-domain solution method provided accurate results while reducing the computational cost by one to two orders of magnitude in comparison with the conventional time-domain method. In addition, Nakhchi and Naung et al. [16] also developed direct numerical simulations (DNS) to reveal the aerodynamic performance, transition to turbulence, and to capture the laminar separation bubble occurring on a wind turbine blade. Mamouri et al. [17,18] designed different shapes of wind turbine airfoil by incorporating entropy generation analysis. Chen et al. [19] performed the lay-up thickness size optimization for a 2 MW composite wind turbine blade with the objective of mass saving based on the particle swarm optimization method. Yet, only the blade stiffness and blade tip displacement were considered, and the extension of weight reduction was limited in accordance with the change of lay-up thickness. Ghiasi et al. [20] carried out the optimization design for the lay-up selection of a composite wind turbine blade with the objective of maximum stiffness. The results found that gradient-based methods were faster than others, but the optimal solution may be a local optimal value. In addition to the composite skin optimization, Zhang et al. [21] compared the maximum deformation, frequency and stress between the single-web and twin-web structures inside an 8 MW wind turbine blade, and the results showed that it was better to choose the twin-web structure form for large-size wind turbine blades. Liao et al. [22] employed an improved particle swarm algorithm (PSA) with the FAST program to optimize the thickness and location of the layers in the spar caps of wind turbine blades. The comparison of the results between the optimal and reference blades indicated that the optimization method was a feasible strategy to obtain the global optimization solution. The aforementioned studies focused on the local structural optimization design of wind turbine blades, whereas investigations on the structural optimization design of blades with the objective of weight reduction are considerably insufficient.

In the structural optimization design, topology optimization design is an effective way to obtain a reasonable internal layout of the blade, which can provide a new configuration and solution for engineering structural design [23]. Nowadays, the main topology optimization methods include the variable thickness method [24], homogenization method [25] and variable density method [26]. Generally, the variable thickness method and homogenization method are basically used for comparatively simple structures. For the variable density method, it has been integrated in finite element simulation software, i.e., ANSYS Workbench. Joncas et al. [27] adopted the topology optimization method to design the end part configuration of a thermoplastic composite wind turbine blade under waving moment and pendulum moment loading. Burton et al. [28] also adopted the topology optimization method to design the inner-surface structures of a wind turbine blade, and the optimization configuration was a non-prismatic structure. Yu et al. [29] designed a novel honeycomb-filled main beam cavity of a wind turbine blade based on the variable density topology optimization method [26], and a reasonable layout of honeycomb cells was obtained and the weight of the optimal blade was reduced by 8.41%. Zhang et al. [30] used the variable density topology optimization method to optimize the thickness and location of the main beam and twin-web of a wind turbine blade, and the optimal configuration showed that the webs play a key supporting role for maintaining the aerodynamic shape of the blade and the overall weight of blade obviously decreased though the dimension and location of the inner webs. However, the challenges remain about how many webs should be placed in the blades and the layout of the related webs.

To address this, Aage et al. [31] adopted the variable density topology optimization method to successfully design the internal structure of an aeroplane wing based on the full-scale internal structure. From this viewpoint, in this work, we report the design of the internal layout of a 5 MW offshore wind turbine blade with the objective of maximum stiffness using the variable density topology optimization method. The aerodynamic loads of blades were obtained though computational fluid dynamics (CFD) simulation. Afterwards, the internal structure of the blade was optimized using the variable density topology optimization, and two multi-web internal layouts were obtained through the reverse design inspired by the topology optimization results. Finally, by validating the performance indexes with respect to stress level, maximum displacement and fatigue life, the multi-web structural layout of the second generation optimal wind turbine blade was an optimal feasible structure, which answered the key scientific issues of how many webs should be placed inside the blade and where to array the related webs. We hope the design method and findings could provide novel and efficient routes to high-performance offshore wind turbine blades.

2. Analytical Preliminaries

2.1. Topology Optimization Method

The variable density topology optimization problem [32] can be expressed as follows:

$$\begin{gathered}\min_{\rho_e}\phi(\rho) = U^T KU = \sum_{e=1}^{N} (\rho_e)^P u_e^T k_0 u_e \\ s.t.\begin{cases} KU = F \\ \sum\limits_{e=1}^{N} \rho_e V_e \leq V^*, 0 \leq \rho_{\min} \leq \rho_e \leq 1 \end{cases}\end{gathered} \quad (1)$$

where ϕ means a sum of each element's compliance. U, K and F mean the global displacement, stiffness matrix and force vectors, respectively. k_0 and u_e mean the element stiffness matrix and displacement vector. ρ means the design variable vector, viz element density vector. N means the element amount of design domain. V_e and V^* mean the unit element volume and total volume for the design domain. To make sure of the numerically stable iteration, $\rho_{\min} = 0.001$ is chosen as the lower limitation of design variable. Additionally, a convolution-type filtering operation is used to filter the holes, viz.

$$\rho_e = \frac{\sum_{i \in N_e} w(r_i, r_e) v_i x_i}{\sum_{i \in N_e} w(r_i, r_e) v_i} \quad (2)$$

where $N_e = \{i | \|r_i - r_e\| \leq R\}$ means a neighborhood set within the radius, R. r_i and r_e mean the spatial central coordinate of elements i and e, respectively. $w(r_i, r_e) = R - \|r_i - r_e\|$ means the weighting function. v_i is the volume of element i.

2.2. CFD Model of Reference Composite Wind Turbine Blade

2.2.1. Control Equation, Geometric Model of Composite Wind Turbine Blade and Flow Field for CFD Simulation

In this work, the Navier–Stokes equation (RANS) based on Reynolds stress averaging was used to solve the flow field of the wind turbine blade, see Equation (1).

$$\frac{\partial}{\partial t}(\rho \vec{u}) + \nabla \cdot (\rho \vec{u}\vec{u}) = -\nabla p + \nabla\left(\mu\left(\nabla\vec{u} + \nabla\vec{u}^T\right) - \frac{2}{3}\mu(\nabla \cdot \vec{u})\right) + \vec{F} \quad (3)$$

where $\vec{F}$ is the external force applied to the fluid.

The wind turbine in this work is the NREL offshore 5 MW baseline wind turbine developed by National Renewable Energy Laboratory (NREL) [33]. The length of composite blade in the NREL 5 MW machine is about 61.63 m. The blade is artificially divided into 17 airfoils from the root to the tip, namely: Cylinder, Du40, Du35, Du30, Du25, Du21 and NACA64, respectively. Note: The detailed geometric parameters corresponding to the aforesaid airfoils can be found in Table S1, Supporting Information (SI).

The geometric model of the composite wind turbine blade can be established as follows: ① Translate the aerodynamic center of the airfoil to the coordinate origin; ② Rotate and transform the airfoil coordinate in accordance with the twist and chord; ③ Calculate the 3D coordinate of nodes using the following equation:

$$\begin{cases} x' = c \cdot \frac{|x - x_{\text{Aero}}|}{x - x_{\text{Aero}}} \sqrt{(x - x_{\text{Aero}})^2 + y^2} \cos\left(\arctan\frac{y}{x - x_{\text{Aero}}} + \beta\right) \\ y' = c \cdot \frac{|x - x_{\text{Aero}}|}{x - x_{\text{Aero}}} \sqrt{(x - x_{\text{Aero}})^2 + y^2} \sin\left(\arctan\frac{y}{x - x_{\text{Aero}}} + \beta\right) \\ z = r \end{cases} \quad (4)$$

where x and y mean the normalized coordinates; x' and y' mean the 3D coordinates; x_{Aero} means the aerodynamic center of the airfoil to the coordinate origin; c and β mean the chord and twist. Based on the aforesaid method, the 3D geometric model of composite wind turbine blade can be established using the Siemens NX 10.0 software, see Figure 1a,b.

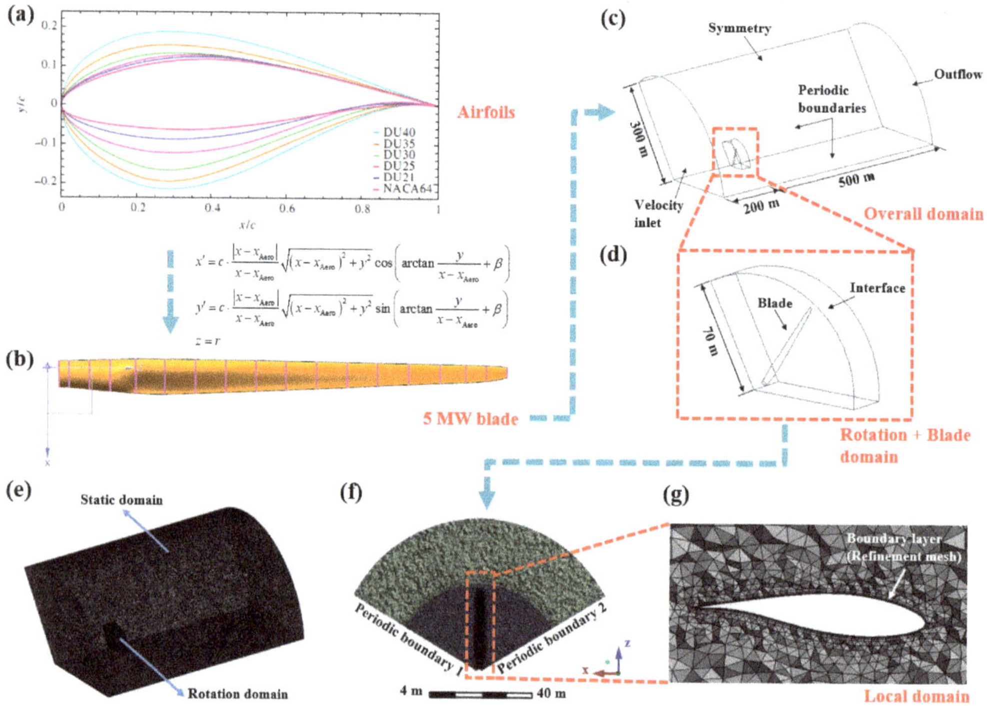

Figure 1. Establishment of 5 MW wind turbine blade. (**a**) Normalized section data of airfoils. (**b**) Wind turbine blade. (**c**) Geometric overall domain. (**d**) Geometric rotation and blade domains. (**e**,**f**) Stereogram and side view of fluent mesh models. (**g**) Local fluent mesh view of blade.

The flow domain for the CFD simulation of a composite wind turbine blade was defined as: ① The effects of tower and nacelle were not considered in this work; ② One-third model was chosen for the CFD simulation to reduce the computation time. Figure 1c exhibits the dimensions and boundary conditions of the CFD model, where the model consists of a rotation domain and a stationary domain. The data can be passed though the interface. The radiuses of the rotation and stationary domains were 70 m and 300 m, respectively. The inlet was 200 m from the hub center and the outlet was 500 m from the hub center.

The fluent mesh model for the CFD simulation of the composite wind turbine blade was obtained as follows: The aforesaid domains were meshed using unstructured tetrahedral element type based on the ANSYS Meshing tool. Note: although some scholars [34,35] meshed blades using the structured mesh, the unstructured mesh for blade and fluid is available by calculating Y+ values at different speeds and comparing the results of the output torque and power curves, which can obtain reasonable simulation results, see Figure 2. The mesh sizes for the rotation and stationary domains were 3 m and 1 m, respectively. Furthermore, the meshes in the symmetry surfaces, named Side_wail1 and Side_neil1, Side_neil1 and Side_nei2, were completely controlled by the periodic mesh matching using the periodic boundary constraint command to ensure that the periodic surface nodes correspond to each other, see Figure 1e,f. In order to better simulate the flow near the wall surface of the blade, the meshes around the blade surfaces were refined and fifteen expansion layers with a growth rate of 1.9 were set. The height of the first layer was 1×10^{-5} m, see Figure 1g.

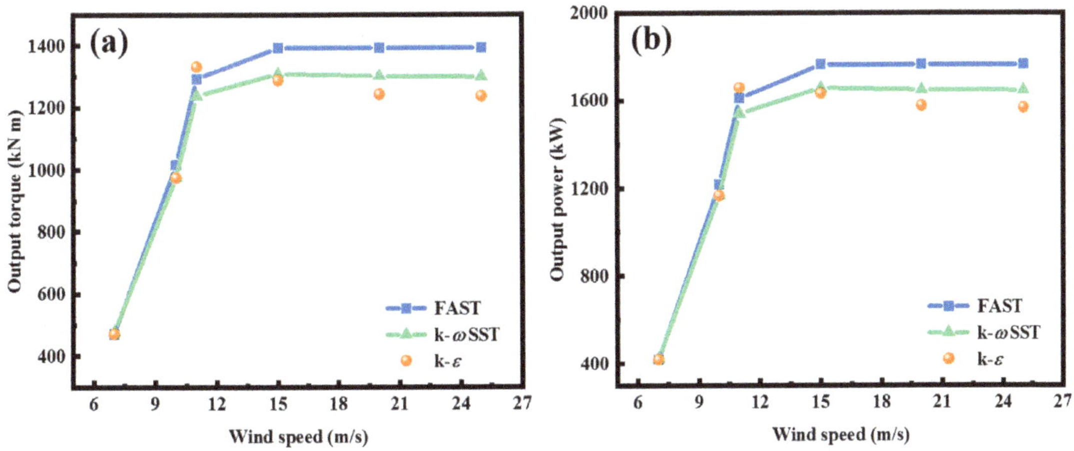

Figure 2. Output responses of composite wind turbine blade based on the k-ω SST and k- models and FAST. (**a**) Output torque response. (**b**) Output power response. Note: the simulation model is one third of wind turbine.

2.2.2. Mesh Quality and Independence Verifications

The mesh quality of the constructed CFD model was checked by using the y^+ value [36], which can be calculated as:

$$y^+ = u_* y / \nu \tag{5}$$

where u_* is the friction speed; y is the nearest wall; ν is the kinematic viscosity of fluid. If the y^+ is less than 1, the mesh quality is reasonable. The y^+ values were calculated under the wind speeds of 7 m/s, 11.4 m/s, 15 m/s, 20 m/s and 25 m/s, respectively, see Figure S1, SI. All the values were less than 1, indicating that the constructed CFD meshing models under different wind speeds were reasonable. Apart from the validation of mesh quality, the mesh independence was also carried out to find out an appropriate mesh size in this work. The case under the wind speed of 11.4 m/s was chosen, where the wind turbine speed is 12.1 RPM and the pitch angle is 0°. Four mesh sizes of 0.3 m, 0.2 m, 0.1 m and 0.07 m were adopted to mesh the blade surface, and the interface surface and outer surface meshes were set as 3 m and 6 m, respectively. After meshing, the amount of elements corresponding to the aforementioned mesh sizes were 3.63 million, 4.17 million, 578 million and 808 million, respectively. Table S2 (SI) shows the relationship between mesh number and calculated wind turbine torque. In view of the computation time and accuracy, the mesh size of 0.1 m was taken for the blade surface in the subsequent study of this work.

2.2.3. Calculation Method for the CFD Simulation

Considering that the subsequent topology optimization of wind turbine blade is a static solution question, the moving reference frame (MRF) method was adopted in this work. Specifically, the rotation speed of the wind turbine is attached to the rotation domain, and the flow fluxes originated from the stationary domain can be translated by the interface. Compared to the non-constant slip grid method, the MRF method is a constant calculation method, which is simpler, faster with less computation time to produce acceptable results in a short time. The method has been used in CFD simulations of static wind turbine blade [37,38].

2.3. *Finite Element Model of Reference Composite Wind Turbine Blade*

2.3.1. Mechanical Properties of Composite Materials of Wind Turbine Blades

Glass fiber reinforced resin unidirectional composites were used as the skin material of blade in this work. The mechanical properties and corresponding allowable values of composites can be found in Table 1.

Table 1. Mechanical properties and allowable values.

Mechanical Properties of Composites				Allowable Values of Composites			
Items	Values	Items	Values	Items	Values	Items	Values
E_x	4.5×10^4 MPa	μ_{xz}	0.3	TS_x	1.1×10^3 MPa	CS_z	−120 MPa
E_y	1.0×10^4 MPa	G_{xy}	5.0×10^3 MPa	TS_y	35 MPa	GS_{xy}	80 MPa
E_z	1.0×10^4 MPa	G_{yz}	3.8×10^3 MPa	TS_z	35 MPa	GS_{yz}	46 MPa
μ_{xy}	0.3	G_{xz}	5.0×10^3 MPa	CS_x	−680 MPa	GS_{xz}	80 MPa
μ_{yz}	0.4	ρ	2000 kg·m^3	CS_y	−120 MPa	-	-

Note: "*TS*", "*CS*" and "*GS*" indicate the tension, compression and shear strength.

2.3.2. Finite Element Model of Wind Turbine Blade

As the wind turbine blade has a complex geometric configuration, the ANSYS Meshing was used to generate tetrahedral elements for the internal room of the blade in this work. Furthermore, to weaken the influence of mesh dependence on the blade, the mesh size was chosen as 0.07 m. Finally, the total number of elements and nodes were 4.85 million and 6.69 million. The corresponding model can be found in Figure S2, SI.

2.4. *Design and Validation Cases of Wind Turbine Blade*

Three load cases are considered in this work, viz.:

(1) DLC-1 (Worst working case)

According to the IEC 61400-1 standard [39], the reference wind turbine is in the 1A IWC wind class, which means it can only withstand wind gusts up to 21% of its rated speed. Therefore, the first load case is when the wind turbine is operating under rated case in severe gusts, in which the gusts suddenly occur so that the blades do not have enough time to change pitch. As it is of great importance for the safety of turbine, this case is also used as the design condition for topology optimization in this work. The related parameters are given in Table 2.

Table 2. Operation parameters in different load cases.

Load Case	Rotation Speed (RPM)	Pitch Angle (°)	Wind Speed (m/s)
DLC-1 (Design)	12.1	0	13.8
DLC-2	12.1	23.3	25
DLC-3	0	89.3	37.5

Note: In DLC-1 case, wind speed = Rated wind speed + maximum allowable gust (11.4 + 21% × 11.4); in DLC-2 case, wind speed = cut-out wind speed; in DLC-3 case, wind speed = speed of typhoon.

(2) DLC-2 (Cut-out wind speed working case)

This case is the cut-out wind speed condition. This is the highest wind speed that the wind turbine can reach before opening the propeller to the downwind position and shutting it down. In this case, blade tip deflection is relatively small. The tip deflection acceleration is even less than that in DLC-1, and therefore the inertial load can be ignored in this case. In this work, this case was used as a verification design case, and the related parameters are listed in Table 2.

(3) DLC-3 (Shutdown working case)

This case is the parking brake condition, at which the blade is in the downwind position and the incoming wind is typhoon speed (37.5 m/s). Here, the downwind position is the pitch angle of the blade that is not affected by any torque. In this work, this case was also adopted as a verification design case, and the relevant parameters are shown in Table 2.

3. Results

3.1. CFD Simulation of Wind Turbine Blade

In this work, two turbulence models, viz. k-ω SST and k-ε, were adopted to elaborate the rationality of CFD simulation. Furthermore, the rationality analyses in terms of output torque and power were compared with the results obtained from the FAST software. Figure 2a illustrates the output torque curves based on the k-ω SST and k-ε models and FAST as the increase in wind speed, and the corresponding data are also listed in Table 3. It can be seen from Figure 2a and Table 3 that the predicted torque values of turbine blade based on the k-ω SST and k-ε turbulence models under the different wind speeds were close to the values based on the FAST. The k-ω SST model had a higher prediction accuracy in comparison with the k-ε model. Compared to the values obtained from the FAST, the errors of output torque based on the k-ω SST model in the wind speed range of 7–25 m/s were less than 7%. The predicted values for the k-ε turbulence model at high wind speeds were relatively low, and the related errors in the wind speed range of 15–25 m/s were within 7–12%.

Table 3. Output torques based on the k-ω SST and k-ε models and FAST.

Wind Speed (m/s)	Pitch Angle (°)	Wind Turbine Torque				
		FAST	k-ω SST	Error	k-ε	Error
7	0	472.70	479.05	1.34%	473.61	0.19%
10	0	1017.03	976.36	4.00%	976.36	4.00%
11	0	1293.77	1238.20	4.30%	1333.51	3.07%
15	10.45	1393.37	1309.79	6.00%	1289.05	7.49%
20	17.47	1393.37	1302.96	6.49%	1244.32	10.70%
25	23.57	1393.37	1300.84	6.64%	1237.86	11.16%

Apart from the output torque performance, the output power performance of the turbine blade based on the k-ω SST and k-ε models and FAST was also evaluated. The output power can be calculated as the following formula:

$$P = M\omega \tag{6}$$

where M is the torque of blade; ω is rotation speed. Figure 2b and Table 4 show the output power values of the turbine blade based on the k-ω SST and k-ε models and FAST, respectively. It can be noted that both the models simulated the output power performance relatively well, and the maximum errors were basically within 10%. However, the error based on the k-ε model at the wind speed of 25 m/s was 11.16%, probably resulting in a relatively dangerous result. Consequently, the k-ω SST turbulence model was adopted in this work.

Table 4. Output power based on the k-ω SST and k-ε models and FAST.

Wind Speed (m/s)	Pitch Angle (°)	Wind Turbine Power				
		FAST	k-ω SST	Error	k-ε	Error
7	0	419.20	424.86	1.35%	420.03	0.2%
10	0	1217.67	1168.75	4.02%	1168.75	4.02%
11	0	1611.07	1541.70	4.31%	1660.39	3.06%
15	10.45	1765.53	1659.65	6.00%	1633.37	7.49%
20	17.47	1765.57	1651.00	6.49%	1576.69	10.70%
25	23.57	1765.57	1648.30	6.64%	1568.50	11.16%

3.2. Structural Responses of Wind Pressure on the Surface of Wind Turbine Blade

According to the design case shown in Table 2 where the pitch angle of the wind turbine blade was adjusted to 0° and the meshing strategy discussed in Section 2.2.2, the structural responses of wind pressure on the surface of blade were obtained based on the k-ω SST turbulence model (Section 3.1). Figure 3 illustrates the pressure surface and suction surface contours of turbine blade under the aerodynamic external load. From Figure 2, the pressure was mainly concentrated in the blade root and trailing edge, while the suction force in the middle of the blade was mainly focused on the blade edge. The maximum pressure and suction force were 2723 and 5729 Pa, respectively.

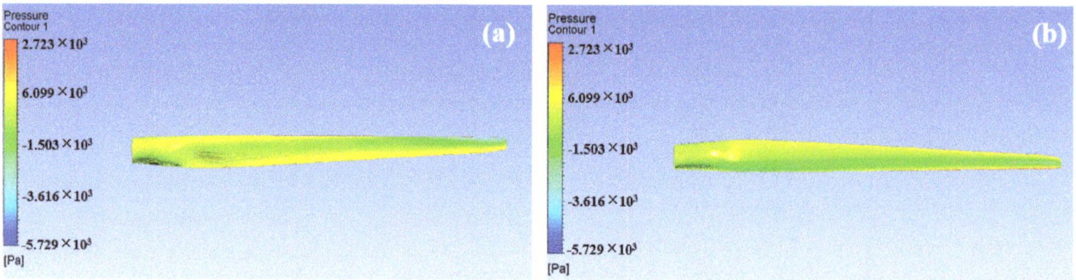

Figure 3. (**a**) Pressure surface contour. (**b**) Suction surface contour.

3.3. Topology Optimization Results of Internal Configuration of Wind Turbine Blade

Figure 4 exhibits the topology optimization results of the wind turbine blade. The outer shell of the blade was completely retained (Figure 4a), and in the internal configuration appeared obvious strip-shaped "gap" and "hole" regions along the axial direction of the blade (Figure 4b), which indicated several webs should be set in these regions. Furthermore, from the left and right enlarged views, some vertical webs should be retained in order to improve the bending resistance. Figure 4c shows the specific internal views cut from 12 various locations from the root to the tip of blade. More obviously, some web-like structures were retained though the topology optimization design.

Based on the aforesaid analysis, the preliminary design of the webs inside the blade in accordance with the topology optimization (Figure 4b,c) was subsequently carried out, see Figure 5. Taking into account seven cross-sections at different locations along the axis direction, the internal web structure of the blade from the root to tip generally changes from the single-web mode to twin-web mode and then to the single-web mode (Figure 5a,b), respectively. However, there are some discontinuous regions among the transition regions of webs, see Figure 5a. Consequently, we proposed to connect the discontinuous region "①" by using a twin-web structure and also arranging the twin-web structure in the region "②", but other regions were filled by the single-web structure. Four webs were reversely designed and the specific sizes of the web cross-sections were determined by the seven cross-sections shown in Figure 5b, in which the corresponding locations along the blade were at 6.70, 17.13, 28.18, 50.28, 56.69, 70.30 and 74.71%, respectively. Ultimately, according to the dimensions of the different cross-sections shown in Figure 5b, the first generation turbine blade inspired by topology optimization was constructed by Boolean operation between the designed web structures and blade boundary, see Figure 5c. The corresponding finite element model of the blade is exhibited in Figure 5c.

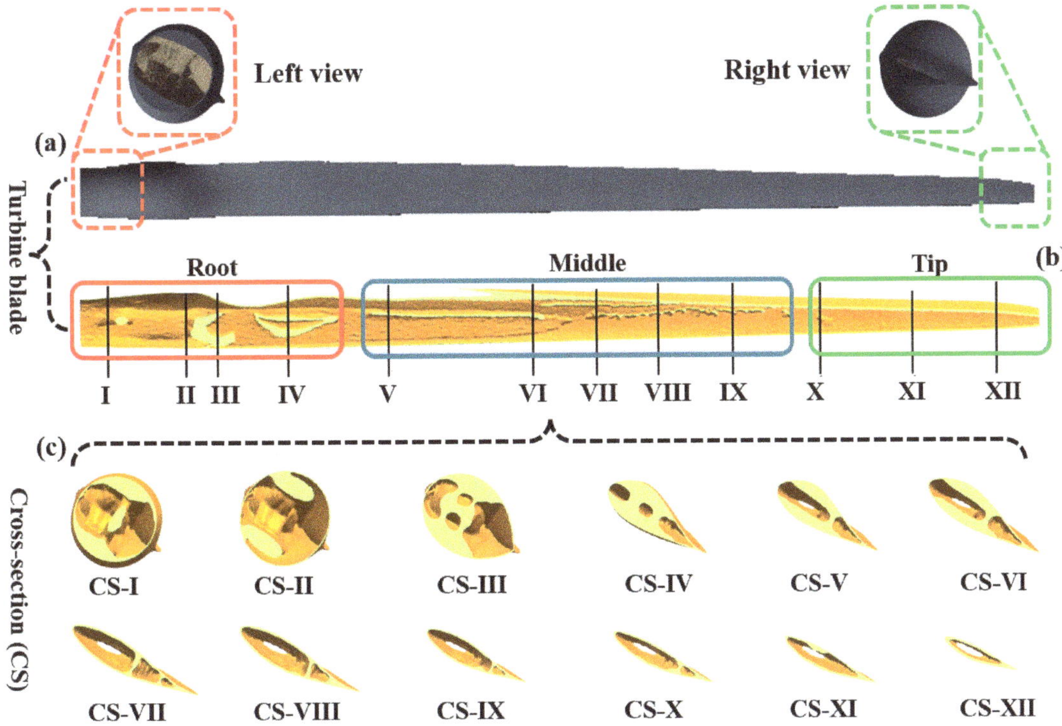

Figure 4. Topology optimization results of the wind turbine blade. (**a**) External shell structure. Inset: the enlarged left and right views. (**b**) Retained materials in the internal blade. (**c**) Cross-section views at different axis locations.

3.4. Validation of Performance Indexes for the First Generation Wind Turbine Blade

The designed first generation wind turbine blade was firstly validified in the DLC-1 case (Table 2) to ensure the feasibility of the reverse design structure. If it was a feasible solution, the design was further validified in the DLC-2 and DLC-3 cases. Otherwise, the aforesaid design strategy was repeated until all the work cases were satisfied.

Figure 6 and Table 5 show the simulation results of the first generation wind turbine blade in the DLC-1 case. From Figure 6a and Table 5, the stress levels in the X direction were the largest, resulting from the waving force (i.e., wind speed incoming flow direction). The maximum tensile and compressive stresses were 45.56 and 49.72 MPa, which are less than the corresponding allowable stress of GFRP. The maximum tensile and compressive stresses in the Y direction were 27.49 and 16.76 MPa, and the values in the Z direction were 17.66 and 12.72 MPa, respectively. The aforementioned stress values were within the allowable stress ranges. Moreover, the maximum displacement of the blade tip was 4.38 m, which was also less than 9 m. Thus, the overall stress and displacement indexes met the design requirements. Figure 6b illustrates the stress contours of the web of the first generation blade. The stresses in the X, Y and Z directions were less than the overall stresses, indicating that the reverse designed web structure satisfied the stress requirements. Figure 6c exhibits the local stress contours of the web in the X, Y and Z directions. It can be seen that the high stress levels were concentrated at the web notch location, which was due to the discontinuity of the designed web. Likewise, all the local stresses were within the requirements of stress. Based on the above-mentioned results, the preliminary design of the first generation turbine blade driven by the topology optimization was reasonable, but the blade is prone to fatigue damage in the long-term service. Thus, the fatigue life should be also considered as an important evaluation index of safety. Figure 6d and Table 5 demonstrate the fatigue life results in the X, Y and Z directions. The minimum fatigue lives in the X and Y directions were 1.27×10^9 and 6.45×10^8, both of which meet the fatigue life requirements. However, from Figure 6d, the fatigue life levels at the discontinuous location of the web were considerably lower. The minimum fatigue life in the Z direction was 8.97×10^7, which does not meet the requirement. Therefore, the first generation blade structure at the local position should be further modified. In this work, we directly connected the middle discontinuous region using a web structure and the second generation wind turbine blade was generated, see Figure 6e.

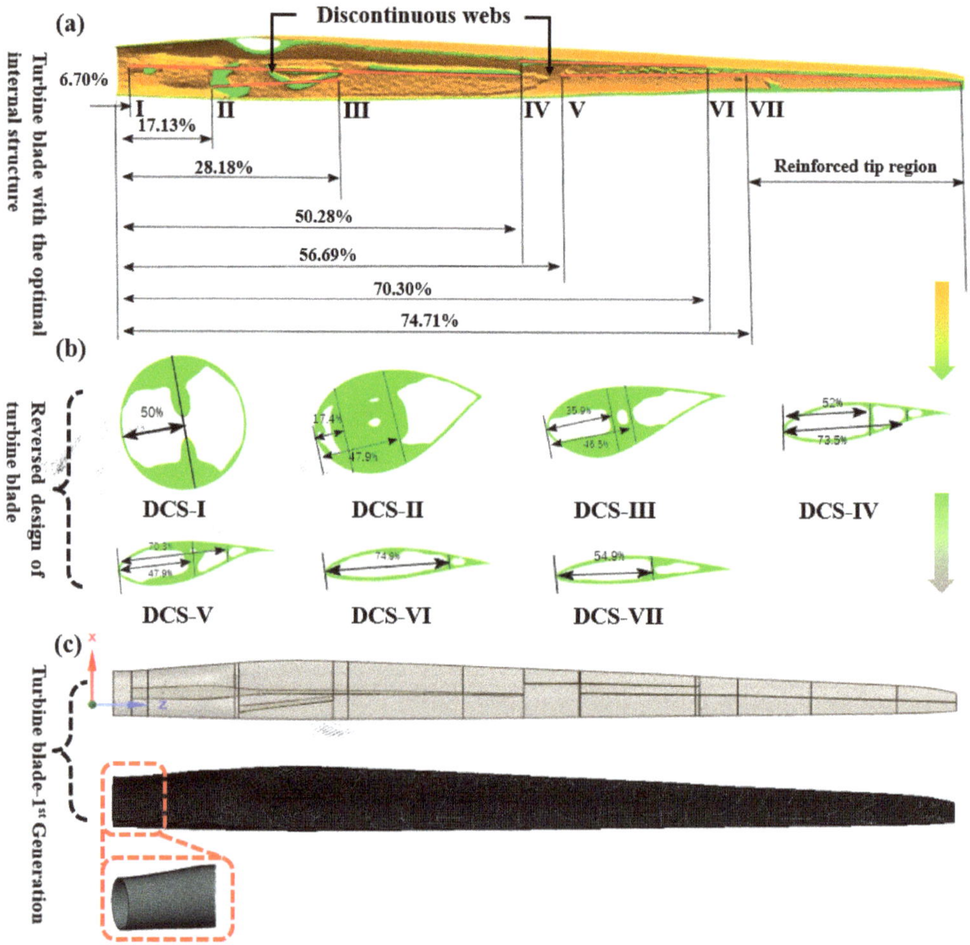

Figure 5. Reverse design of the wind turbine blade. (**a**) Turbine blade with the optimal internal structure. (**b**) Reverse design of turbine blade inspired by the topology optimization. (**c**) First generation turbine blade.

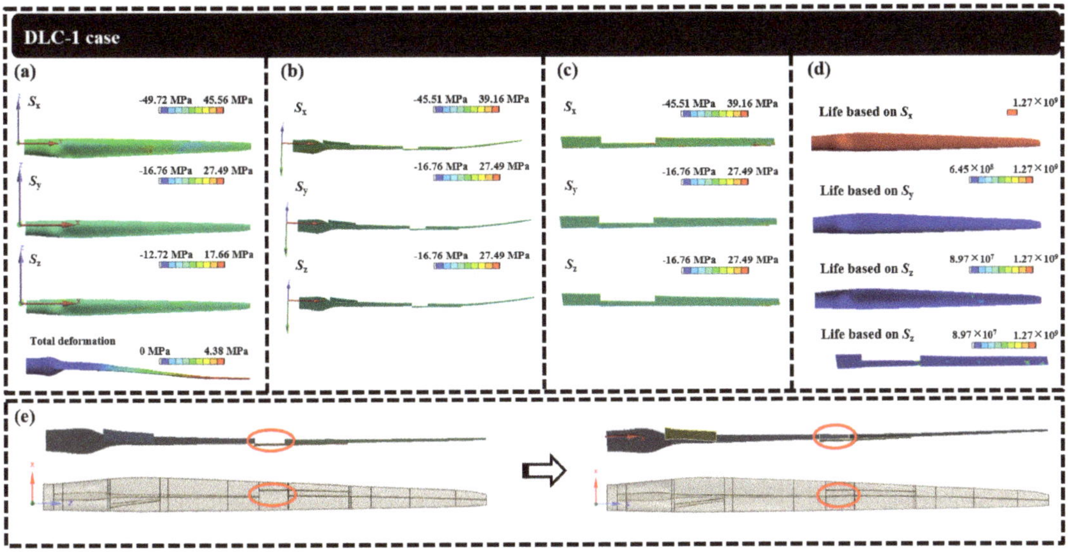

Figure 6. Validation of performance indexes for the first generation wind turbine blade in the DLC-1 case. (**a**) Stress and deformation contours of overall turbine blade. (**b**) Stress contours of internal structure. (**c**) Stress contours of localized regions of internal structure. (**d**) Fatigue life contours of whole turbine blade (**e**) The second generation design of the wind turbine blade.

Table 5. Simulation results of the first generation wind turbine blade in the DLC-1 case.

Direction	Overall Stress of Blade		Localized Stress of Web		Min. Fatigue Life	Disp. of Tip
	Max. TS	Max. CS	Max. TS	Max. CS		
X	45.56 MPa	49.72 MPa	39.16 MPa	45.51 MPa	1.27×10^9	
Y	27.49 MPa	16.76 MPa	27.49 MPa	16.76 MPa	6.45×10^8	4.38 m
Z	17.66 MPa	12.72 MPa	27.49 MPa	16.76 MPa	8.97×10^7	

Figure 7 shows the simulation results of the second generation wind turbine blade in the DLC-1, 2 and 3 cases. Table 6 compared results of performance indexes between the first and second generation wind turbine blades in the DLC-1 case. Compared with the performance indexes of the first generation blade in the DLC-1 case, the overall maximum tensile stress of the blade in the X direction was slightly changed, but the maximum compressive stress in the Y direction was reduced by 18.79%. It is worth noting that the maximum localized tensile and compressive stresses of the web in the Z direction were remarkably decreased by 97.05 and 95.17%, respectively. Furthermore, the maximum displacement of the blade tip was 4.03 m, which was also reduced by 7.99%. All the stress and displacement indexes met the design requirements. Importantly, after connecting the discontinuous region by the additional web, the minimal fatigue life of the web was 2.25×10^8, significantly improved by 150.84%, which achieved the requirement of fatigue life.

In addition, the overall stress contours of the second blade in the X, Y and Z directions were similar to those of the first blade (Figure 7a,b). The stress levels in the X direction were higher than those in the Y and Z directions. From Figure 7c, the stress levels at the local region in three directions were relatively uniform, indicating that the stress concentration had been accommodated by adding additional web structure in this region. Furthermore, the maximum tensile stresses were still less than the allowable ones. In summary, the design strategy for the internal structure of the second generation wind turbine blade is a feasible solution. In the following discussion, the performance indexes of the blade are further evaluated in the DLC-2 and DLC-3 cases.

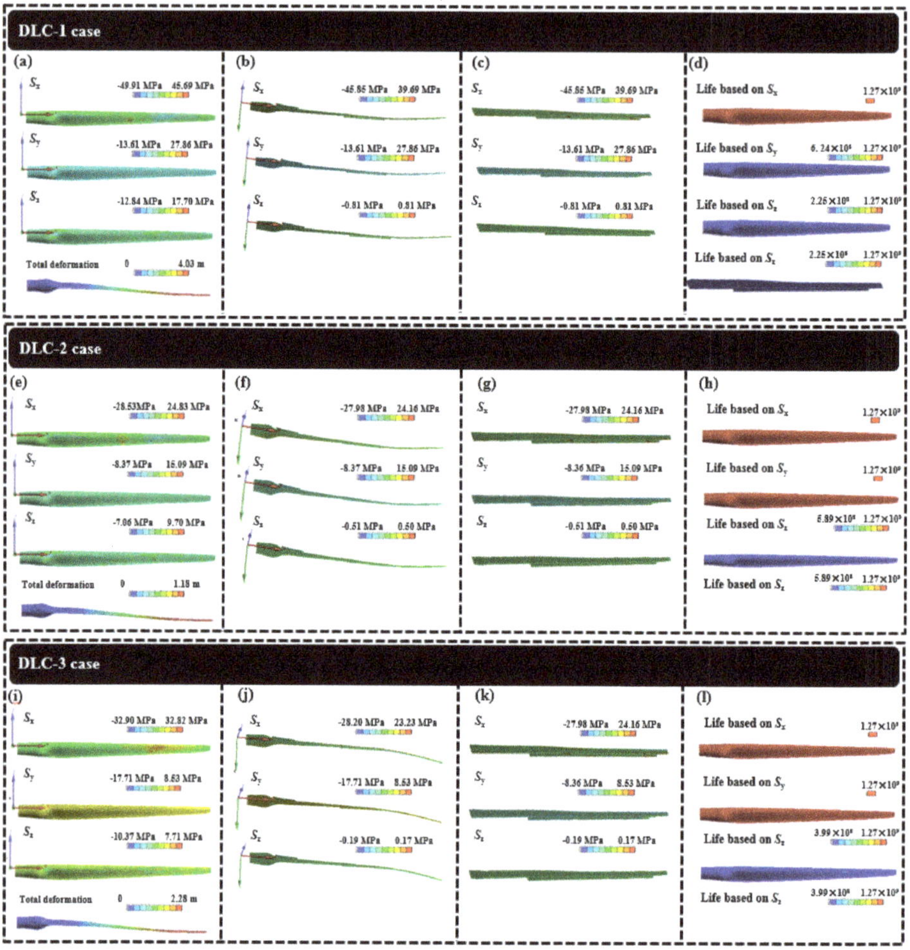

Figure 7. Validation of performance indexes for the second generation wind turbine blade in the DLC-1, DLC-2 and DLC-3 cases. (**a**,**e**,**i**) Stress and deformation contours of whole turbine blade. (**b**,**f**,**j**) Stress contours of internal structure. (**c**,**g**,**k**) Stress contours of localized regions of internal structure. (**d**,**h**,**l**) Fatigue life contours of whole turbine blade.

Figure 7 and Table 7 exhibit the validation results of the second blade in the X, Y and Z directions in the DLC-2 case. The overall stress distribution contours of the blade in the X, Y and Z directions were totally consistent with those of the first blade, and the maximum tensile and compressive stresses meet the stress requirements. The maximum displacement was only 1.18 m, which also meets the design allowable value. Compared to the performance indexes of the first blade, the stress and deformation responses of the second blade were relatively lower. It should be noted that the wind turbine obtained an additional torque in this case, and the outer part of the blade had a negative angle of attack which to some extent counteracted the internal lift. Although the turbine operated at a higher wind speed, the aerodynamic load acting at the blade tip was weaker than that in the DLC-1 case. Additionally, the minimal fatigue life also achieved the design requirement. Consequently, all the performance indexes of the second turbine blade met the design requirements, confirming that it was also a feasible solution in the DLC-2 case. In addition, Figure 7 and Table 7 also show the validation results of the second blade in the X, Y and Z directions in the DLC-3 case. In this case, the wind turbine was suffering from typhoon conditions, and the blade was in the down pitch stop state. The maximum displacement of blade tip was 2.28 m and the minimum fatigue life was 3.90×10^8, both of which met the design requirements. Furthermore, it can be seen from the stress contours that the middle stress levels in the X direction were slightly higher, giving an indication of stress concentration in the region. However, those in the Y and Z directions were relatively uniform. All the stress indexes were within the requirements. Therefore, the second turbine blade was also a feasible design in the DLC-3 case.

Table 6. Compared results of performance indexes between the first and second generation wind turbine blades in the DLC-1 case.

Item	Overall Stress of Blade		Localized Stress of Web		Min. Fatigue Life	Disp. of Tip
	Max. TS	Max. CS	Max. TS	Max. CS		
First X	45.56 MPa	49.72 MPa	39.16 MPa	45.51 MPa	1.27×10^9	4.38 m (first); 4.03 m (second); Δ = −7.99%
Second X	45.77 MPa	49.91 MPa	36.69 MPa	45.85 MPa	1.27×10^9	
Δ X	+0.46%	+0.38%	−6.31%	+0.75%	0%	
First Y	27.49 MPa	16.76 MPa	27.49 MPa	16.76 MPa	6.45×10^8	
Second Y	27.86 MPa	13.61 MPa	27.86 MPa	13.61 MPa	6.24×10^8	
Δ Y	+1.35%	−18.79%	+1.35%	−18.79%	−3.26%	
First Z	17.66 MPa	12.72 MPa	27.49 MPa	16.76 MPa	8.97×10^7	
Second Z	17.70 MPa	12.84 MPa	0.81 MPa	0.81 MPa	2.25×10^8	
Δ Z	+0.23%	+0.94%	−97.05%	−95.17%	+150.84%	

Table 7. Simulation results of the second generation wind turbine blade in the DLC-1, 2 and 3 cases. Unit in stress: MPa, in displacement of tip: m.

Case	Direction	Overall Stress of Blade		Localized Stress of Web		Min. Fatigue Life	Disp. of Tip
		Max. TS	Max. CS	Max. TS	Max. CS		
DLC-1	X	45.77	49.91	36.69	45.85	1.27e9	4.03
	Y	27.86	13.61	27.86	13.61	6.24e8	
	Z	17.70	12.84	0.81	0.81	2.25e8	
DLC-2	X	24.83	28.53	24.16	27.98	1.27e9	1.18
	Y	15.09	8.37	15.09	8.36	1.27e9	
	Z	9.70	7.06	0.50	0.51	5.98e8	
DLC-2	X	32.82	32.90	24.16	27.98	1.27e9	2.28
	Y	8.53	17.71	15.09	8.36	1.27e9	
	Z	7.71	10.37	0.50	0.51	3.90e8	

Note: "Disp" means displacement. "TS" and "CS" mean tensile stress, compressive stress.

In summary, a novel turbine blade with the optimal web structure guided by the topology optimization was accomplished.

4. Discussion

4.1. Comparison of Performance Indexes between the Novel and Reference Turbine Blades

Table 8 lists the compared results of the performance indexes between the novel and reference turbine blades. Overall, the stress levels of the novel blade were lower than those of the reference blade. Note: a positive value in Table 8 represents the increase in performance index. The displacement values of the novel blade in various load cases were larger than those of the reference blade, indicating that a more flexible blade was obtained in this work. Importantly, the weight of the novel blade was reduced by 9.88% relative to the reference blade, which is a significant benefit in decreasing the cost of turbine blades. Therefore, the novel wind turbine blade driven by topology optimization in this work has predictably better power efficiency than the reference blade without the loss of load-bearing capacity.

Table 8. Compared results of performance indexes between the novel and reference turbine blades. Unit in stress: MPa, in displacement of tip: m, in overall weight: kg.

Case	Dir.	Item	Overall Blade Stress Max. TS	Overall Blade Stress Max. CS	Localized Web Stress Max. TS	Localized Web Stress Max. CS	Disp. of Tip	Overall Weight
DLC-1	X	Ref. blade	32.70	30.60	14.30	16.00	3.42 (Ref.) 4.03 (New) Δ: 17.84%	21,247 (Ref.); 19,148 (New) Δ: −9.88%
		Nov. blade	45.769	49.91	36.69	45.85		
		Δ	39.97%	63.10%	156.57%	186.56%		
	Y	Ref. blade	13.10	19.50	13.10	19.50		
		Nov. blade	27.86	13.61	27.86	13.61		
		Δ	112.67%	−30.21%	112.67%	−30.21%		
	Z	Ref. blade	19.30	13.60	0.15	0.18		
		Nov. blade	17.70	12.84	0.81	0.81		
		Δ	−8.29%	−5.59%	440.00%	350.00%		
DLC-2	X	Ref. blade	18.90	15.70	8.85	9.70	1.08 (Ref.) 1.18 (New) Δ: 9.26%	
		Nov. blade	24.83	28.53	24.16	27.98		
		Δ	31.38%	81.72%	172.99%	188.45%		
	Y	Ref. blade	7.15	10.40	7.15	10.40		
		Nov. blade	15.09	8.37	15.09	8.36		
		Δ	111.05%	−19.52%	111.05%	−19.62%		
	Z	Ref. blade	10.80	6.14	0.07	0.08		
		Nov. blade	9.70	7.06	0.50	0.51		
		Δ	−10.19%	14.98%	614.29%	537.50%		
DLC-3	X	Ref. blade	23.80	25.50	6.08	5.61	1.94 (Ref.) 2.28 (New) Δ: 17.53%	
		Nov. blade	32.82	32.90	24.16	27.98		
		Δ	37.90%	29.02%	297.37%	398.75%		
	Y	Ref. blade	7.67	5.29	7.67	5.29		
		Nov. blade	8.53	17.71	15.09	8.36		
		Δ	11.21%	234.78%	96.74%	58.03%		
	Z	Ref. blade	9.96	7.42	0.09	0.09		
		Nov. blade	7.71	10.37	0.50	0.51		
		Δ	−22.59%	39.76%	455.56%	466.67%		

Note: "Dir.", "Ref." and "Nov." mean "Direction", "Reference" and "Novel", respectively.

4.2. Modal Analysis of the Novel Wind Turbine Blade

As the decrease of blade weight, the vibration problem of the novel wind turbine blade should be discussed to further identify the dynamic properties. In this work, the modal analysis was also carried out based on the novel blade, see Figure 8 and Table 9. It can be seen from Figure 8, the first six orders of the novel turbine blade include: first order waving vibration, second order pendulum vibration, third order waving vibration, fourth order waving vibration, fifth order waving vibration and sixth waving pendulum, which are similar to those of the reference blade. The vibration types of the blade are mainly dominated by waving and pendulum vibrations. Moreover, from Table 9, it can be seen that the first six order frequencies were well in agreement with those of the reference blade, indicating that the designed internal layout is reasonable.

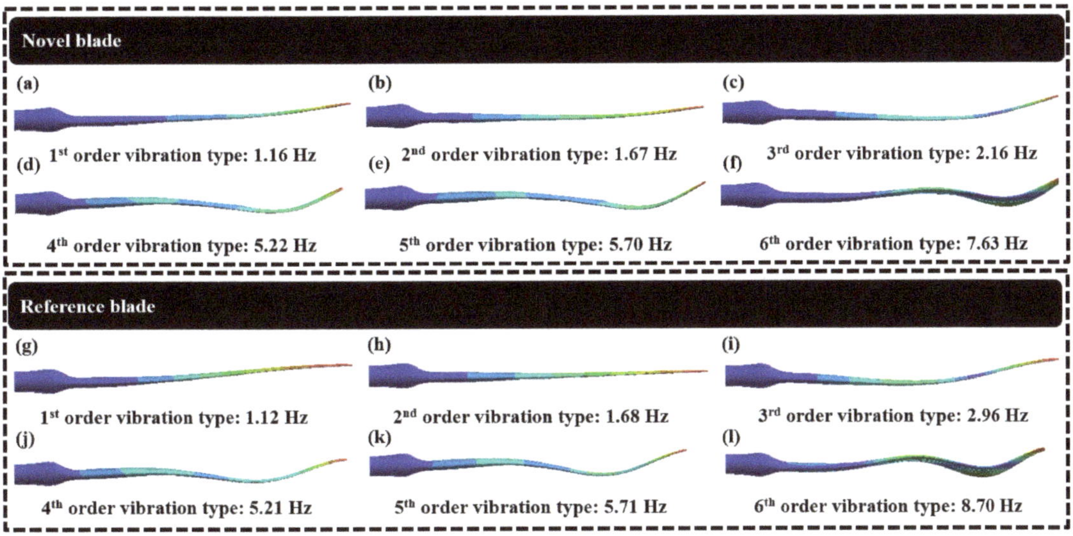

Figure 8. Vibration types of the novel and reference turbine blades.

Table 9. Comparison of frequency between the novel and reference turbine blades.

Order	Frequency (Hz)		Order	Frequency (Hz)	
	Ref. Blade	Nov. Blade		Ref. Blade	Nov. Blade
1	1.12	1.16	4	5.21	5.22
2	1.68	1.67	5	5.71	5.70
3	2.96	2.16	6	8.70	7.63

4.3. Full Life Cycle Assessment of the Novel Wind Turbine Blade

The service life of a wind turbine is generally more than 20 years. During long-term service, wind turbine blades are always subjected to complex aerodynamic loads induced by wind, with the result that it is very susceptible to fatigue damage and failure. Hence, to ensure the long-term service safety, a full life-cycle assessment of the novel wind turbine blade should be discussed in this work. According to the wind speed data of a full year in Guangdong Province, the wind speed range from 5–25 m/s was considered in this work, see Table 10. The statistical duration of wind speed in an hour can be obtained through the Weibull distribution, see Table S3, SI. Based on the finite element analysis given in Section 3, the stress responses of the novel blade were obtained, see Table 10. Considering the S-N curve of GFRP used as the shell composite material (Table S4, SI), the fatigue lives of turbine blades corresponding to the wind speed were calculated though the Goodman curve, see Table 10. Afterwards, the fatigue damage with respect to each wind speed was obtained via the ratio of stress range in the Weibull distribution and the related fatigue life (Table 10). Finally, the full life of the novel blade over 20 years was evaluated based on the linear P-M accumulative damage theory, viz:

$$Y = \frac{N}{N' \times \omega \times 60} = \frac{1/(\sum \gamma_i / N_i)}{N' \times \omega \times 60} = \frac{1/(\frac{0.197}{1.70\times 10^8} + \frac{0.15}{1.60\times 10^8} + \frac{0.101}{1.05\times 10^8} + \frac{0.061}{1.08\times 10^8})}{7250 \times 12.1 \times 60} = 21.9 \text{ Year} \tag{7}$$

where Y is the full life; ω is the rated speed of turbine blade, taken as 12.1 RPM (Table 2); N' is the sum of duration of wind speed in hours; γ_i is the stress range Weibull distribution; N_i is the fatigue life corresponding to wind speed. Consequently, the full life over the 20 years is 21.9 years, which meets the design requirement of 20 years.

Table 10. Wind speed distribution and fatigue damage in a year.

Wind Speed (m/s)	Duration of Wind Speed (h)	Max. Stress (MPa)	Min. Stress (MPa)	Stress Range Weibull Distribution	Fatigue Limitation (MPa)	Fatigue Life	Fatigue Damage
5	1526	12.84	3.85	0.210	40.4	$>2 \times 10^8$	0
7	1613	21.18	6.35	0.222	40.4	$>2 \times 10^8$	0
9	1425	42.43	12.73	0.197	40.4	1.70×10^8	1.16×10^{-9}
11	1090	42.64	12.79	0.150	40.4	1.60×10^8	9.38×10^{-10}
13	734	47.86	14.36	0.101	40.4	1.05×10^8	9.62×10^{-10}
15	439	42.54	12.76	0.061	40.4	1.08×10^8	5.65×10^{-10}
17	235	36.24	10.87	0.032	40.4	$>2 \times 10^8$	0
19	113	33.89	10.17	0.016	40.4	$>2 \times 10^8$	0
21	49	23.87	7.16	0.007	40.4	$>2 \times 10^8$	0
23	19	28.33	8.50	0.003	40.4	$>2 \times 10^8$	0
25	7	33.3	9.99	0.001	40.4	$>2 \times 10^8$	0

5. Conclusions

This work develops an innovative multi-web internal layout for the offshore wind turbine blade in accordance with the variable density topology optimization method, which theoretically answers the proposed scientific issues about how many webs need to be used inside the blade and where the related webs should be laid out. The following conclusions can be summarized as follows:

1. The surface pressure was obtained based on the CFD simulation, and the two turbulence models, viz. k-ω SST and k-ε, were adopted. By comparing the output torques and power, the k-ω SST model was chosen to calculate the surface pressure distribution. Moreover, the simulation results obtained from the CFD were also validated in comparison with those calculated from the FAST.
2. The topology optimization model was established based on the full-scale internal structure of offshore wind turbine blade, considering stress, displacement and fatigue life constraints.
3. After the full-scale topology optimization, two multi-web layouts were theoretically obtained driven by the optimal topological configuration for the first time. By validation, the second generation optimal blade completely met all the requirements and the weight was reduced by 9.88% relative to the reference blade, which was a significant benefit in decreasing the cost of turbine blades.
4. Vibration modal and full life cycle of the novel blade were also evaluated. The first six vibration types of the novel blade were consistent with those of the reference blade, further indicating that the designed internal layout was reasonable. Moreover, the full life cycle of the novel blade is 21.9 years, theoretically verifying that the novel blade is able to service more than 20 years in the given sea domain.

Supplementary Materials: The following supporting information can be downloaded at: https://www.mdpi.com/article/10.3390/jmse10101487/s1, Figure S1: y^+ values for evaluating the mesh quality under the different wind speeds; Figure S2: Illustration of the relationship between the amount of element and torque; Table S1: Distributed blade aerodynamic properties in NREL 5 MW wind turbine blade; Table S2: Mesh independence analysis; Table S3: Statistic duration of wind speed in hour; Table S4: S-N data of GFRP.

Author Contributions: J.S.: writing—original draft, reviewing, project administration and supervision; J.C.: writing—original draft, reviewing and validation; Y.W.: review, editing and validation; L.L.: conceptualization and supervision. All authors have read and agreed to the published version of the manuscript.

Funding: The support for this research has been provided by the National Natural Science Foundation of China (Grant No. 51905350), the Shenzhen Science and Technology Program (Grant No. KQTD20200820113004005), Shenzhen Key Laboratory of Structure Safety and Health Monitoring of Marine Infrastructures (Grant No. ZDSYS20201020162400001) and the National Natural Science Foundation of Guangdong Province (Grant No. 2022A1515011499) are gratefully acknowledged.

Institutional Review Board Statement: Not applicable.

Informed Consent Statement: Not applicable.

Data Availability Statement: Not applicable.

Conflicts of Interest: The authors declare that they have no known competing financial interests or personal relationships that could have appeared to influence the work reported in this paper.

References

1. Oh, K.Y.; Nam, W.; Ryu, M.S.; Kim, J.Y.; Epureanu, B.I. A review of foundations of offshore wind energy convertors: Current status and future perspectives. *Renew. Sustain. Energy Rev.* **2018**, *88*, 16–36. [CrossRef]
2. Liu, P.; Meng, F.R.; Barlow, C.Y. Wind turbine blade end-of-life options: An economic comparison. *J. Clean. Prod.* **2022**, *180*, 106202. [CrossRef]
3. Mishnaevsky, L.; Johansen, N.F.; Fraisse, A.; Faster, S.; Jensen, T.; Bendixen, B. Technologies of Wind Turbine Blade Repair: Practical Comparison. *Energies* **2022**, *15*, 1767. [CrossRef]
4. Dyrholm, M. *GWEC | Global Wind Report*; Global Wind Energy Council: Brussels, Belgium, 2021.
5. Chou, J.S.; Chiu, C.K.; Huang, I.K.; Chi, K.N. Failure analysis of wind turbine blade under critical wind loads. *Eng. Fail. Anal.* **2013**, *27*, 99–118. [CrossRef]
6. Liu, Y.; Hajj, M.; Bao, Y. Review of robot-based damage assessment for offshore wind turbines. *Renew. Sustain. Energy Rev.* **2022**, *158*, 112187. [CrossRef]
7. Rafiee, R.; Hashemi, M.R. Failure analysis of a composite wind turbine blade at the adhesive joint of the trailing edge. *Eng. Fail. Anal.* **2021**, *121*, 105148. [CrossRef]
8. Roh, C.; Ha, Y.J.; Ahn, H.J.; Kim, K.H. A Comparative Analysis of the Characteristics of Platform Motion of a Floating Offshore Wind Turbine Based on Pitch Controllers. *Energies* **2022**, *15*, 716. [CrossRef]
9. Guo, Y.; Wang, H.; Lian, J. Review of integrated installation technologies for offshore wind turbines: Current progress and future development trends. *Energy Convers. Manag.* **2022**, *255*, 115319. [CrossRef]
10. Asim, T.; Islam, S.Z.; Hemmati, A.; Khalid, M.S. A Review of Recent Advancements in Offshore Wind Turbine Technology. *Energies* **2022**, *15*, 579. [CrossRef]
11. Ibrahimbegovic, A. Flexible Blades Wind-Turbines: Giant Installations and System-of-Systems Approach to Optimizing Wind-Energy Farms. In *CIGOS 2021, Emerging Technologies and Applications for Green Infrastructur*; Springer: Berlin/Heidelberg, Germany, 2022; pp. 3–27.
12. Zhu, J.; Cai, X.; Ma, D.; Zhang, J.; Ni, X. Improved structural design of wind turbine blade based on topology and size optimization. *Int. J. Low-Carbon Technol.* **2022**, *17*, 69–79. [CrossRef]
13. Chen, J.; Kim, M.H. Review of Recent Offshore Wind Turbine Research and Optimization Methodologies in Their Design. *J. Mar. Sci. Eng.* **2021**, *10*, 28. [CrossRef]
14. Naung, S.W.; Erfanian Nakhchi Toosi, M.; Rahmati, M. An Experimental and Numerical Study on the Aerodynamic Performance of Vibrating Wind Turbine Blade with Frequency-Domain Method. *J. Appl. Comput. Mech.* **2021**, *7*, 1737–1750.
15. Naung, S.W.; Nakhchi, M.E.; Rahmati, M. High-fidelity CFD simulations of two wind turbines in arrays using nonlinear frequency domain solution method. *Renew. Energy* **2021**, *174*, 984–1005. [CrossRef]
16. Nakhchi, M.E.; Naung, S.W.; Rahmati, M. High-resolution direct numerical simulations of flow structure and aerodynamic performance of wind turbine airfoil at wide range of Reynolds numbers. *Energy* **2021**, *225*, 120261. [CrossRef]
17. Mamouri, A.R.; Khoshnevis, A.B.; Lakzian, E. Entropy generation analysis of S825, S822, and SD7062 offshore wind turbine airfoil geometries. *Ocean Eng.* **2019**, *173*, 700–715. [CrossRef]
18. Mamouri, A.R.; Khoshnevis, A.B.; Lakzian, E. Experimental study of the effective parameters on the offshore wind turbine's airfoil in pitching case. *Ocean Eng.* **2020**, *198*, 106955. [CrossRef]
19. Chen, J.; Wang, Q.; Shen, W.Z.; Pang, X.; Li, S.; Guo, X. Design, Structural optimization study of composite wind turbine blade. *Mater. Des.* **2013**, *46*, 247–255. [CrossRef]
20. Ghiasi, H.; Fayazbakhsh, K.; Pasini, D.; Lessard, L. Optimum stacking sequence design of composite materials Part II: Variable stiffness design. *Compos. Struct.* **2010**, *93*, 1–13. [CrossRef]
21. Zhang, Z.Y. Structural Design of Large Horizontal Axis Wind Turbine Blade Based on Fluid-Structural Interation. Ph.D. Thesis, Lanzhou University of Technology, Lanzhou, China, 2014.
22. Liao, C.C.; Zhao, X.L.; Xu, J.Z. Blade layers optimization of wind turbines using FAST and improved PSO Algorithm. *Renew. Energy* **2012**, *42*, 227–233. [CrossRef]
23. Song, J.; Zhang, Y.K.; Guo, X.; Hao, H.; Wen, W.D.; Cui, H.T. Topology and shape optimization of twin-web turbine disk. *Struct. Multidiscip. Optim.* **2022**, *65*, 1–20. [CrossRef]
24. Rossow, M.P.; Taylor, J.E. A finite element method for the optimal design of variable thickness sheets. *AIAA J.* **1973**, *11*, 1566–1569. [CrossRef]
25. Bendsøe, M.P.; Kikuchi, N. Generating optimal topologies in structural design using a homogenization method. *Comput. Methods Appl. Mech. Eng.* **1988**, *71*, 197–224. [CrossRef]

26. Mlejnek, H.P.; Schirmacher, R. An engineer's approach to optimal material distribution and shape finding. *Comput. Methods Appl. Mech. Eng.* **1993**, *106*, 1–26. [CrossRef]
27. Joncas, S. Thermoplastic Composite Wind Turbine Blades: An Integrated Design Approach. Ph.D. Thesis, Delft University of Technology, Delft, The Netherlands, 2010.
28. Buckney, N.; Pirrera, A.; Green, S.D.; Weaver, P.M. Structural efficiency of a wind turbine blade. *Thin-Walled Struct.* **2013**, *67*, 144–154. [CrossRef]
29. Yu, X. Optimal Design of Vertical Axis Wind Turbine Blade Structure with Honeycomb Core Layer Inside the Main Beam. Ph.D. Thesis, Tianjin University of Technology, Tianjin, China, 2021.
30. Zhang, Y.Y.; Zheng, Y.J.; Chen, Y.Y.; Sun, M.M.; Tang, J.M. Structural topology optimization design of wind turbine blade. *Shanxi Archit.* **2017**, *43*, 203–204.
31. Aage, N.; Andreassen, E.; Lazarov, B.S.; Sigmund, O. Giga-voxel computational morphogenesis for structural design. *Nature* **2017**, *550*, 84–86. [CrossRef]
32. Zuo, W.; Saitou, K. Multi-material topology optimization using ordered SIMP interpolation. *Struct. Multidiscip. Optim.* **2017**, *55*, 477–491. [CrossRef]
33. Jonkman, J.; Butterfield, S.; Musial, W.; Scott, G. *Definition of a 5-MW Reference Wind Turbine for Offshore System Development*; National Renewable Energy Lab: Golden, CO, USA, 2009.
34. Mamouri, A.R.; Lakzian, E.; Khoshnevis, A.B. Entropy analysis of pitching airfoil for offshore wind turbines in the dynamic stall condition. *Ocean Eng.* **2019**, *187*, 106229. [CrossRef]
35. Mahrooghi, A.; Lakzian, E. Optimization of Wells turbine performance using a hybrid artificial neural fuzzy inference system (ANFIS)—Genetic algorithm (GA). *Ocean Eng.* **2021**, *226*, 108861. [CrossRef]
36. Davidson, A.A.; Salim, S.M. CFD Modelling of Rotating Annular Flow Using Wall y+, International MultiConference of Engineers and Computer Scientists. *Trans. Eng. Technol.* **2018**, *1*, 318–330.
37. Manatbayev, R.; Baizhuma, Z.; Bolegenova, S. Numerical simulations on static Vertical Axis Wind Turbine blade icing. *Renew. Energy* **2021**, *170*, 997–1007. [CrossRef]
38. Son, C.; Kim, T. Development of an icing simulation code for rotating wind turbines. *J. Wind Eng. Ind. Aerodyn.* **2020**, *203*, 104239. [CrossRef]
39. Madsen, P.H.; Risø, D.J.R.N.L. *Introduction to the IEC 61400-1 Standar*; Technical Standard; Risø National Laboratory, Technical University of Denmark: Kongens Lyngby, Denmark, 2008.

processes

MDPI

Article

Study on the Aerodynamic Performance of Floating Offshore Wind Turbine Considering the Tower Shadow Effect

Danmei Hu *, Liwei Deng and Li Zeng

College of Energy and Mechanical Engineering, Shanghai University of Electric Power, Shanghai 200090, China; dengliwei_suep@163.com (L.D.); zengli1832507667@sina.com (L.Z.)
* Correspondence: hudanmei@shiep.edu.cn

Abstract: The aerodynamic performance of the floating offshore wind turbine (FOWT) is obviously affected by the motion of the platform, and becomes much more complicated considering the effect of tower shadow. In view of this, this paper aims at investigating the aerodynamic performance of the floating offshore wind turbine with and without a tower under the three most influential motions (surge, pitch and yaw) by computational fluid dynamic (CFD). The results show that the power of the wind turbine is reduced by 1.58% to 2.47% due to the tower shadow effect under the three motions, and the pressure difference distribution is most obviously interfered by the tower shadow effect under yaw motion and concentrates at the root and tip of the blade. In addition, the degree of interference of the tower shadow effect on the wake flow field is different under the three motions, resulting in a more complex wake structure. These conclusions can provide a theoretical basis and technical reference for the optimal design of floating offshore wind turbines.

Keywords: floating offshore wind turbine; tower shadow effect; computational fluid dynamic; aerodynamic performance; surge; pitch; yaw

Citation: Hu, D.; Deng, L.; Zeng, L. Study on the Aerodynamic Performance of Floating Offshore Wind Turbine Considering the Tower Shadow Effect. *Processes* **2021**, *9*, 1047. https://doi.org/10.3390/pr9061047

Academic Editor: Eugen Rusu

Received: 3 May 2021
Accepted: 11 June 2021
Published: 15 June 2021

Publisher's Note: MDPI stays neutral with regard to jurisdictional claims in published maps and institutional affiliations.

Copyright: © 2021 by the authors. Licensee MDPI, Basel, Switzerland. This article is an open access article distributed under the terms and conditions of the Creative Commons Attribution (CC BY) license (https://creativecommons.org/licenses/by/4.0/).

1. Introduction

With the consumption of traditional fossil fuels and serious environmental pollution, more and more countries in the world begin to attach importance to the development of renewable energy. As a kind of green, pollution-free and abundant renewable energy, wind energy has great potential for development [1]. Originally, wind turbines were mainly built on land, but due to the large noise of onshore wind turbines and relatively few wind and land resources, the wind power industry gradually developed to the sea [2]. From a worldwide point of view, the sea wind is rich in resources, and the wind is stable, which has become the irresistible general trend [3]. With the large-scale capacity of the offshore wind turbine, the size of the wind turbine and the height of the tower are increasing, resulting in a more obvious tower shadow effect [4]. Therefore, it is particularly important to study the unsteady aerodynamics under the tower shadow effect of FOWT.

Offshore wind power generation has obvious advantages over onshore wind power generation; however, FOWTs produce six degrees-of-freedom (6-DOF) [5] motions under the combined action of wind, waves and currents, resulting in more complex aerodynamics compared with onshore and offshore fixed wind turbines. In recent years, many scholars have studied the aerodynamic performance and structural characteristics of FOWTs from the aspects of experiments, models, research methods, etc. The blade element momentum (BEM) theory with two widely used engineering dynamic inflow models was used by Vaal [6] to investigate the effect of a periodic surge on the wind turbine. Ma [7] investigated the effects of the control system of the wind turbine and the motion of the floating platform on the blade aerodynamic performance during the representative typhoon time history. However, the BEM methods used to research aerodynamics lack the characteristics of capturing the physical details of the flow field, which the reasonability and accuracy are suspectable. Farrugia [8] used the results from the free-wake vortex simulations to analyze the wake characteristics of the wind turbine under floating conditions. Complex wake phenomena under the influence of extreme wave conditions were observed. Wen et al. [9,10] also used the FVM method for numerical simulation and found that the power coefficient overshoot is caused by the time lag between the output power and the wind farm power. The time lag and the resultant power coefficient overshoot increase as the platform surge frequency increases. In pitch motion, with the increase of the reduced frequency, the mean power output decreases at a low tip speed ratio and increases at a high tip speed ratio. Salehyar [11] investigated the dynamic response of a spar-buoy-based floating wind turbine to non-periodic disturbances through a coupled aero-hydro-elastic numerical model, observing the ability of the wind turbine to recover to the balanced position after being disturbed. Lin [12] investigated the aerodynamic characteristics in system pitch and surge motions and the asymmetric and complicated wake was observed. Tran et al. [13–16] performed the CFD simulation based on the dynamic mesh technique and an advanced overset moving grid method, respectively, to accurately consider the aerodynamic loads of a three-dimensional wind turbine. Results summarized the comparisons of different aerodynamic analyses under periodic surge, pitch and yaw motions to show the potential differences between the applied numerical methods. Chen [17] proposed a model of a spar-buoy and a semi-submersible floating wind turbine to compare with the experimental results of the two. Nguyen [18] studied the fully coupled motion of FOWT and applied the six degrees-of-freedom solid motion solver to multi-motion coupling to study the coupling of the surge, pitch and yaw motions. Huang [19] discussed the dynamic response of the local relative wind speed and local angle of attack of the blade section and the wind–wave force acting on the floating platform to reveal the interaction mechanism between the aerodynamic load and the motion of the platform with different degrees of freedom. Fang [20,21] applied a 1:50 model FOWT to explore the aerodynamics and characteristics of its wake under surge and pitch motions. Chen [22,23] investigated the aerodynamic characteristics of the wind turbine under surge–pitch coupling and pitch–yaw coupling by the combination of dynamic mesh and sliding mesh. The results show that the fluctuation of the overall aerodynamic performance of the wind turbine dramatically with the increase of amplitude and frequency. Sivalingam [24] examined the predictions of numerical codes by comparing them with experimental data of a scaled floating wind turbine.

In addition to the research on changes in wind turbine performance caused by platform motions, a series of studies have also focused on the tower shadow effect of the wind turbine. Kim [25] and Quallen [26] found that the diameter of the tower has a greater effect on the wind turbine than the gap between the tower and the blade, and the root of the blade is more affected by the tower. Ke [27] proposed an effective method for calculating the aerodynamic load and aeroelastic response of a large wind turbine tower blade coupling structure under the yaw condition, taking full account of wind shear, tower shadow, aerodynamic interaction and rotation effect. Noyes [28] used unsteady aerodynamic experiments to analyze the influence of tower shadow effect on aerodynamic loads and blade-bending moments of downwind two-blade wind turbines. Zhang [29] found that the maximum displacement and Mises stress increase with the increase of the average wind speed under the tower shadow effect. Li [30] investigated the aeroelastic coupling effect under periodic unsteady inflows, indicating that the tower shadow effect causes dramatic changes in the tilting moment, thrust force and output power when the blade rotates in front of the tower. Wu [31] investigated the unsteady flow characteristics in the tip region of the blade, observing the static pressure distribution of different blades near the leading edge of the tip is very different due to the influence of turbulence intensity and tower shadow effect.

According to the above literature review, the influence of the six degrees-of-freedom motions of the platform on the FOWTs has been studied to a certain extent, and the investigations on the tower shadow effect are mostly focused on the land-based fixed wind turbine. Nevertheless, the effect of platform motions combined with the tower shadow effect is rarely mentioned. This paper aims to investigate the unsteady aerodynamic characteristics of FOWTs under surge, pitch and yaw motions based on the tower shadow effect. The unsteady dynamic numerical simulation of the aerodynamic characteristics of the full-size wind turbine model in the rotating process was carried out using a UDF (user-defined function) and embedded sliding mesh, and the unsteady Reynolds-Averaged Navier–Stokes equations (RANS) and SST k-ω turbulence model are adopted. Considering the effect of the tower shadow effect, the power, thrust and the pressure distribution of blade sections under surge, pitch and yaw motions are compared and analyzed, and the near wake and far wake flow fields of the wind turbine are analyzed.

2. Model and Numerical Methods

2.1. Governing Equations and Turbulent Model

The three laws of mass conservation equation, momentum conservation equation and energy conservation equation need to be followed in fluid mechanics. For incompressible fluids, the continuity equation and the momentum equation (Navier–Stokes equation) can be used to describe the law of conservation of mass and momentum of the fluid. Regodeseves [32] and Burmester [33] validated the accuracy of the model in simulating the aerodynamic characteristics of FOWTs by comparing the simulation results of the model with the experimental data.

The definition of the continuity equation, Reynolds equation and scalar Φ time-averaged transport equation expressed in the tensor form are expressed as follows:

$$\frac{\partial \rho}{\partial t} + \frac{\partial}{\partial x_i}(\rho u_i) = 0, \tag{1}$$

$$\frac{\partial}{\partial t}(\rho u_i) + \frac{\partial}{x_j}(\rho u_i u_j) = -\frac{\partial p}{\partial x_i} + \frac{\partial}{x_j}(\mu \frac{\partial u_i}{\partial x_j} - \rho \overline{u'_I u'_j}) + S_i, \tag{2}$$

$$\frac{\partial(\rho\phi)}{\partial t} + \frac{\partial(\rho u_j \phi)}{\partial x_j} = \frac{\partial}{\partial x_j}\left(\Gamma \frac{\partial \phi}{\partial x_j} - \rho \overline{u'_j \phi'}\right) + S, \tag{3}$$

$$\tau_{ij} = -\rho \overline{u'_I u'_j}, \tag{4}$$

where ρ is the air density, t is the time, u_i and u_j represent the Reynolds mean velocity components of fluid, p is the pressure, μ denotes the coefficient of dynamic viscosity, S is the generalized source term (i, j = 1, 2, 3) and Γ represents the diffusivity. τ_{ij} corresponds to six different Reynolds stress terms, which is defined as Reynolds stress.

The SST *k-w* model combines the advantages of the *k-w* model in the near-wall region and the far-field calculation of the *k-ε* model, further modifies the turbulent viscosity and adds an orthogonal diffusion term, which can well predict the separation of the fluid under the negative pressure gradient. For the aerodynamic analysis of the wind turbine in this paper, the SST *k-w* model has obvious advantages, which is used in the later simulation.

The transport equation expression of SST *k-w* is as follows:

$$\frac{\partial(\rho k)}{\partial t} + \frac{\partial(\rho u_i k)}{\partial x_i} = \frac{\partial}{\partial x_j}\left(\Gamma_k \frac{\partial k}{\partial x_j}\right) + \overline{G}_k - Y_k + S_k, \tag{5}$$

$$\frac{\partial(\rho w)}{\partial t} + \frac{\partial(\rho u_i w)}{\partial x_i} = \frac{\partial}{\partial x_j}\left(\Gamma_w \frac{\partial w}{\partial x_j}\right) + G_w - Y_w + D_w + S_w, \tag{6}$$

where $\overline{G}_k$ is the turbulent energy caused by the average velocity gradient, G_w is the turbulent dissipation rate, Γ_k and Γ_w represent the effective diffusivity of k and w caused by turbulence, respectively, Y_k and Y_w are the dissipation of k and ω, D_w is the orthogonal divergence, both S_k and S_w represent user-defined source items.

2.2. Floating Foundation

The floating foundation wind turbine is affected by wind, waves and currents in the marine environment, which produces 6-DOF motions. As shown in Figure 1, the motion of the platform includes three rotational components (pitch about X, roll about Y and yaw about Z) and three translational components (sway in X, surge in Y and heave in Z). This paper chooses a Spar floating wind turbine foundation for its larger draft, small vertical excitation force and heave motion [34].

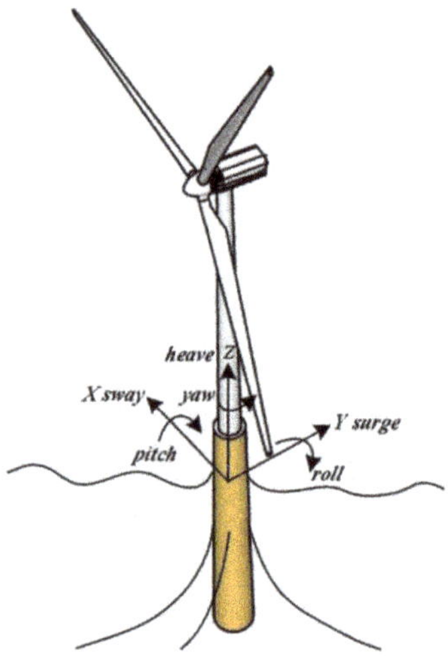

Figure 1. 6-DOF motions of a wind turbine floating foundation.

2.3. Wind Turbine Model

The NREL 5 MW wind turbine was selected as the research object, which is a three-bladed, upwind wind turbine and the power control adopts variable speed and variable pitch method. The diameter of the wind turbine is 126 m, and the hub height is 90 m. From root to tip, DU series airfoils of different thicknesses are used, and NACA series airfoils are used in the tip part. This study considers the existence of the tower, the bottom diameter of the tower is 6 m, and the top diameter is 3.87 m. The numerical simulation of the wind turbine is carried out under the rated operating conditions; the incoming wind speed is 11.4 m/s with the rotational speed of 12.1 r/min.

2.4. Flow Field and Boundary Conditions

As shown in Figure 2, the flow field calculation domain is divided into the external static domain and internal rotation domain. The rotational domain is used to define the rotation of the wind turbine relative to the external domain under the action of the inflow wind. The origin of the coordinate system is located in the center of the hub. The rotational domain is a cylinder with a diameter of 140 m and a height of 8 m. Considering the vastness of the sea area, the calculation domain should be divided large enough to reduce the influence of the boundary on the calculation accuracy [35]. The external flow field is a combination of a hemisphere with a diameter of 6.3 R (R is 63 m radius of the wind turbine rotor) and a cuboid with a width of 90 m. The inlet distance is 3.2 R from the rotation domain, and the outlet distance is 8 R from the rotation domain.

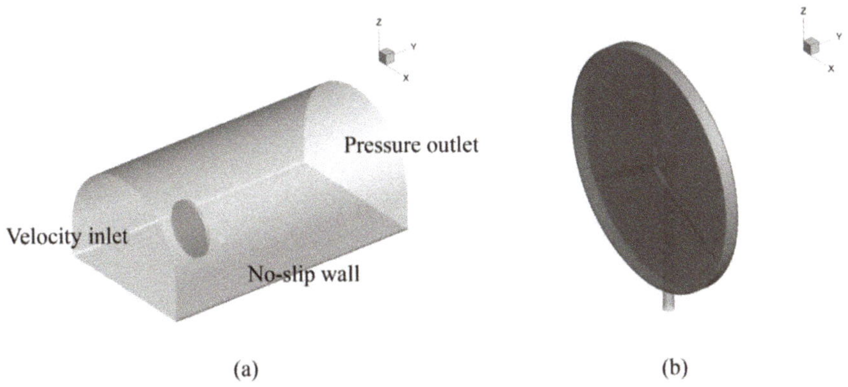

Figure 2. Computational domain of wind turbine flow field: (**a**) external domain; (**b**) rotational domain.

Due to the nested motion in the blade rotation domain, it is necessary to use the embedded sliding mesh to define its rotation. The area where the inner and outer domains are in contact with each other is set as the interface. There is a relative slip between the interfaces and the flow field information is transmitted. The inlet and outlet of the external flow field are set as velocity inlet and pressure outlet, respectively. Additionally, the surface of the blade and the surrounding boundary of the external flow field are set as non-slip walls.

2.5. Computational Mesh

The internal and external flow field of the wind turbine is divided into unstructured grids by using MESH software. Figure 3 presents the grid generation in the overall computational domain. As the blade structure of the wind turbine model is complex and the tip position of the blade is too sharp, the grid size of the rotation region close to the blade needs to be set smaller, and the surface grid of the blade is further refined, which meets the requirements of the SST *k-w* turbulence model. As shown in Figure 3a, there are a total of 8.27 million grids in the whole flow field, of which the internal flow field grid is 6.21 million and the external field grid is 2.06 million.

2.6. Computational Methods

The flow field of the wind turbine is simulated in ANSYS Fluent software, and the unsteady Reynolds-Averaged Navier–Stokes equations (RANS) is solved by using the SST *k-w* turbulence model considering the transition effects on the blade surface. The sliding grid method is used to simulate the rotation of the wind turbine, which is set as a transient calculation. The pressure-based solver is used, the SIMPLEC algorithm is selected for pressure-velocity coupling, in which the second-order upwind scheme is used for pressure term, convection term, turbulent kinetic energy equation and turbulent dissipation rate. Based on the steady calculation results, the unsteady calculation for 20 s is carried out, the additional calculation for 40 s is carried out in the case of platform motion, and the last platform motion period after convergence is taken as the analysis result. The residual of the continuity equation is reduced by at least four orders of magnitude in the process of calculation.

Figure 1 has defined the form of platform motions, including surge, pitch and yaw, which are the main factors affecting the aerodynamic performance of FOWTs. According to the situation of the literature [36], the suitable amplitude and frequency parameters of the floating foundation are selected, and the main motion parameters are shown in Table 1. In order to realize the above platform motions, the additional speed change is compiled by the DEFINE_PFORILE macro of UDF. It is assumed that the motion of the FOWT floating foundation is harmonic; its expression can be written as:

$$\beta_i(t) = A_i \sin(\varepsilon_i t), \tag{7}$$

where $\beta_i(t)$, A_i and ε_i (i = surge, pitch, yaw) denote the angle/displacement, amplitude and frequency of the platform motion.

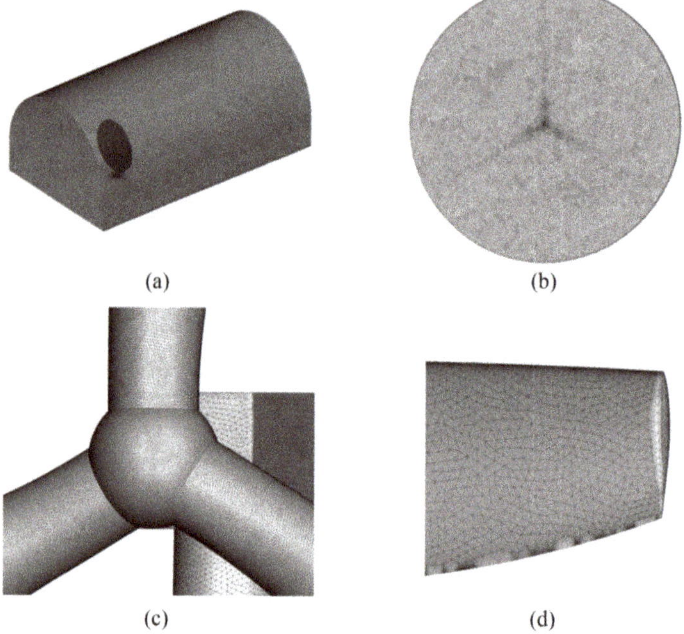

Figure 3. Computational mesh; (**a**) External domain; (**b**) Internal domain; (**c**) Hub surface mesh; (**d**) Blade surface mesh.

Table 1. Floating foundation motion conditions.

Conditions	Motion	Displacement	Amplitude	Frequency
1	surge	−5–5 m	/	0.05 Hz
2	pitch	/	−5–5°	0.05 Hz
3	yaw	/	−15–15°	0.05 Hz

The motion of the floating foundation is described by the matrix between coordinate systems [37]. The incoming wind speed is set to $V = [V_x, V_y, V_z]^T$, where V_x, V_y, V_z represents the partial velocity in the direction of X, Y and Z, respectively. the relative inflow velocity of the platform during surge motion is converted into V_S, which can be written as

$$V_S = V(t) + \begin{bmatrix} 0 \\ \beta'_{surge}(t) \\ 0 \end{bmatrix} = \begin{bmatrix} 0 \\ V_y + \varepsilon_{surge} A_{surge} \cos(\varepsilon_{surge} t) \\ 0 \end{bmatrix}, \tag{8}$$

where $\beta'_{surge}(t)$ denotes the velocity of surge motion at each moment.

The pitch motion corresponds to the rotation β_{pitch} of the floating platform around the x-axis, and the incoming flow velocity is converted to the relative velocity V_P, which can be presented as

$$V_P = \begin{bmatrix} 1 & 0 & 0 \\ 0 & \cos\beta_{pitch} & \sin\beta_{pitch} \\ 0 & -\sin\beta_{pitch} & \cos\beta_{pitch} \end{bmatrix} V = \begin{bmatrix} 0 \\ V_y\cos(A_{pitch}\sin(\varepsilon_{pitch}t)) \\ -V_y\sin(A_{pitch}\sin(\varepsilon_{pitch}t)) \end{bmatrix} \quad (9)$$

The yaw motion corresponds to the rotation β_{yaw} of the floating platform around the z-axis, and the incoming flow velocity is converted to the relative velocity V_Y, which can be presented as

$$V_Y = \begin{bmatrix} \cos\beta_{yaw} & \sin\beta_{yaw} & 0 \\ -\sin\beta_{yaw} & \cos\beta_{yaw} & 0 \\ 0 & 0 & 1 \end{bmatrix} V = \begin{bmatrix} V_y\sin(A_{yaw}\sin(\varepsilon_{yaw}t)) \\ V_y\cos(A_{yaw}\sin(\varepsilon_{yaw}t)) \\ 0 \end{bmatrix} \quad (10)$$

3. Numerical Model Verification

3.1. Model Grid Independence Verification

Table 2 exhibits the numerical value of the wind turbine torque under different meshing precision, which are refined along the circumference of wind turbine blades. The number of simulated grids is as follows: 2.86 million, 4.77 million, 6.71 million, 8.50 million and 10.08 million.

Table 2. Torque of different gird sizes.

Cells Number (Million)	2.86	4.77	6.71	8.50	10.08	Reference Value
Torque (N m)	3.54×10^6	3.82×10^6	3.90×10^6	3.98×10^6	4.03×10^6	4.01×10^6
Relative error (%)	11.72	4.74	2.74	0.75	0.50	

It can be found that with the refinement of the mesh, the error value of torque decreases constantly, and the decrease becomes smaller and smaller. The mesh of 8.50 million is selected for in-depth research due to the consideration of calculation accuracy and efficiency.

3.2. Model Power Verification

Due to the lack of feasible experimental data, the experimental results of Jonkman [38] were selected for verifying the accuracy of the simulation. The cut-in wind speed of the NREL 5 MW wind turbine is 3 m/s. When the wind speed reaches more than 3 m/s, the wind turbine starts to operate. The rated wind speed is 11.4 m/s, and the rated speed is 12.1 rpm. When the wind speed exceeds the rated wind speed, the constant power control is realized by changing the pitch angle and reducing the lift-drag ratio.

Figure 4 illustrates the comparison of stable power between numerical simulation and experimental data under different incoming wind speeds. All of the results exhibited a high consistency before the incoming wind speed reached 11.4 m/s, indicating that the current simulation method has good accuracy and reasonable results.

3.3. Validation Considering the Tower under the Unsteady Condition

It should be noted that the change of additional wind speed caused by the motion of the platform is synchronously added to the inlet velocity. The numerical model is verified with and without tower under unsteady calculation, respectively. A sufficient degree of steady condition calculation is carried out for the numerical model under rated conditions first, and the results of the steady calculation are taken as the initial flow field under unsteady wind speed conditions.

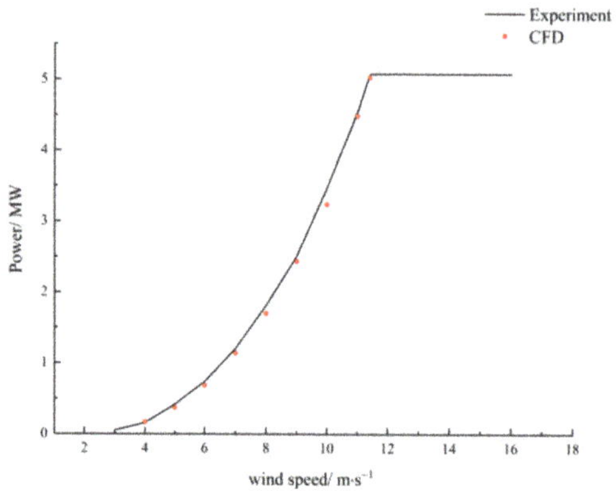

Figure 4. Power comparison of different wind speeds.

The uniform wind speed is 11.4 m/s, which is set at the inlet of the flow field. The speed of the wind turbine is 12.1 rpm. The time-step corresponding to 10° azimuth increment with 40 pseudo-time sub-iterations is 0.137741 s. The rotation time of the wind turbine for 4 cycles is calculated under unsteady conditions. Figure 5 shows the power fluctuation with and without the tower and it indicated that the power of the wind turbine tends to be stable after the wind turbine rotates for three cycles. Furthermore, a periodic rotation variation of the wind turbine in stable operation was observed locally. Compared with the non-tower, the existence of the tower makes the power decrease dramatically with an azimuth of 120° apart. This phenomenon is consistent with Wen [39]'s research, which verifies the influence of the tower shadow effect on the wind turbine.

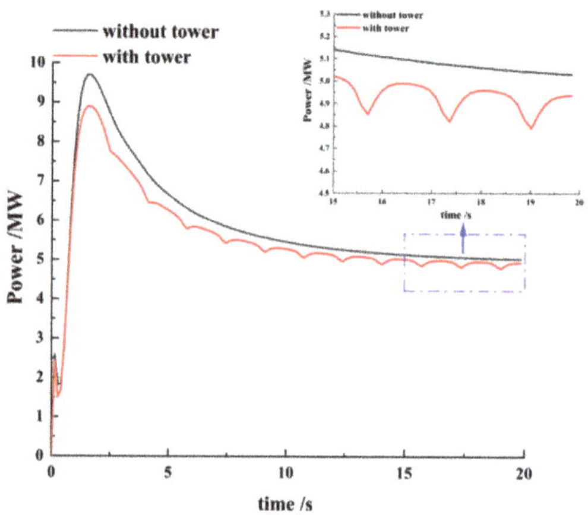

Figure 5. Power variation under unsteady rated condition.

4. Results and Discussion

Due to the obstruction of the tower to the airflow, the interference effect of the tower on the blades is different when the blades rotate to different azimuth angles. This paper takes the blades rotating to the azimuth directly above the tower as the starting point. The schematic diagram of the azimuth angle of the wind turbine blades is shown in Figure 6. θ is the angle between the two blades, γ is the influence area of the tower shadow effect and the wind turbine rotates counterclockwise.

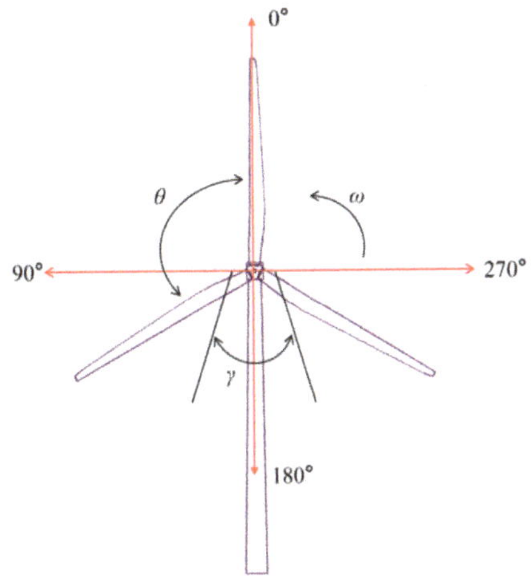

Figure 6. Schematic diagram of wind turbine blade-rotation phase angle.

4.1. Unsteady Aerodynamic Analysis under Surge Motion

4.1.1. Total Performance Analysis

Figure 7 illustrates the total power and thrust comparison with and without the tower under surge motion. It can be observed that the fluctuation period of power and thrust is consistent with the harmonic surge. The equilibrium position divides the power and thrust curve into rising and falling parts, mainly because the power and thrust are proportional to the relative wind speed. For f = 0.05 Hz, A_s = 5 m, as shown in Figure 7b, the maximum power of the wind turbine without the tower is 6.98 MW, while the wind turbine with the tower is 6.85 MW. Both of these were larger than the rated power, while the existence of the tower reduced the maximum wind turbine power by 1.86%. The minimum power was 3.37 MW and 3.24 MW, respectively, and the power of the wind turbine with the tower was reduced by 3.86%. The average power generation was 5.12 MW and 5.01 MW, respectively; the latter was reduced by 2.15%. Further, the power fluctuated at a frequency thrice that of the rotation frequency under the combination of surge and tower shadow, while the surge motion played a leading role in the influence of power fluctuation. As shown in Figure 7c,d, the trend of the axial thrust fluctuation was similar to the power. The peak value of thrust without the tower was 888.3 KN; the wind turbine with the tower was 881.4 KN. Additionally, the valley value of thrust was 772.7 KN and 764.7 KN, respectively. It can be seen that the peak and valley values of thrust were basically the same with or without the tower. In addition, the average thrust was 772.7 KN and 764.7 KN, separately; the latter was reduced by 1.04%. It can be concluded that the influence of the tower shadow effect on thrust was less than that on power under the same condition of surge motion.

To further explore the variation mechanism of the aerodynamic load of the wind turbine under platform motion, the airfoil-induced velocity distribution under surge motion was analyzed, as shown in Figure 8. The surge motion of the platform produces an additional induced speed V_{ind} to the wind turbine, which is obtained by superposing the surge speed with the free flow wind speed. When the platform moves forward, the V_{ind} is opposite to the incoming wind speed, and the relative wind speed increases. When the platform moves backward, the V_{ind} is in the same direction as the incoming wind speed, and the relative wind speed decreases. V_{ind} can be decomposed into chord velocity V_c and radial velocity V_r on the rotating plane, the magnitude and direction of the relative velocity V_{rel} acting on the rotating plane of the airfoil changes and the angle of attack changes accordingly. This theoretically expounds the essence of the fluctuation of the aerodynamic performance of the wind turbine under surge motion.

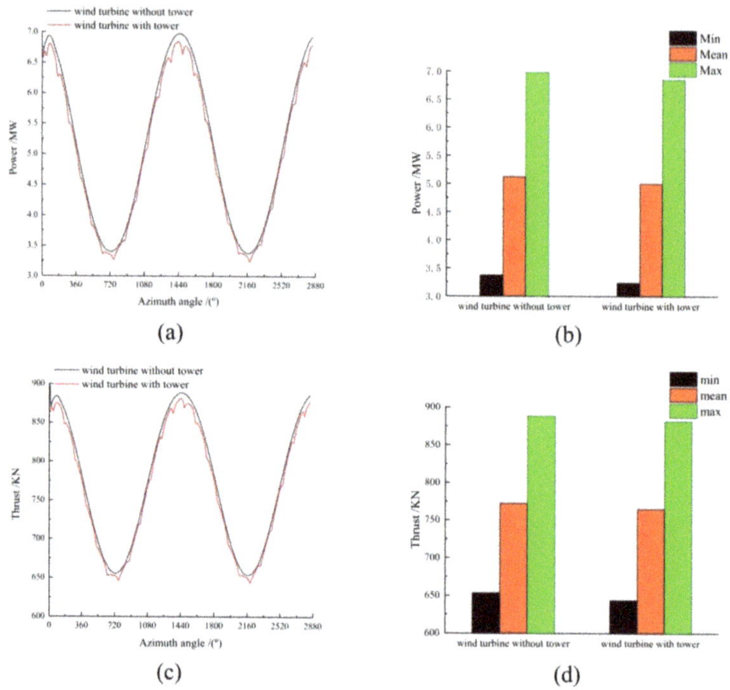

Figure 7. Total aerodynamic comparison of surge motion; (**a**) power versus azimuth angle; (**b**) extreme and average values of power; (**c**) thrust versus azimuth angle; (**d**) extreme and average values of thrust.

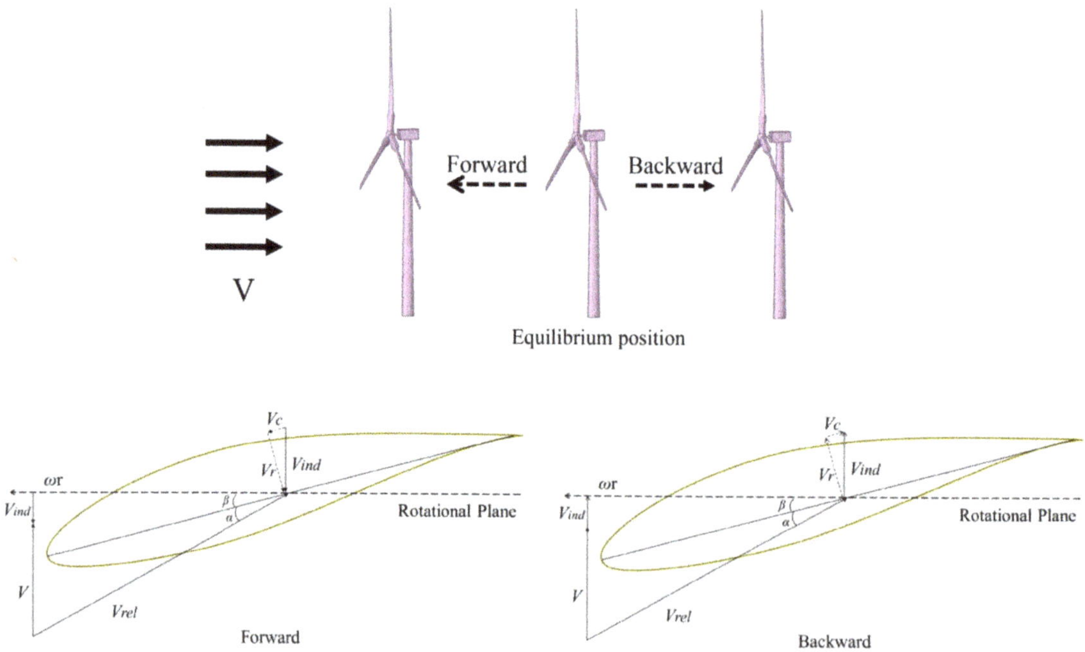

Figure 8. Schematic diagram of airfoil-induced velocity under surge motion.

4.1.2. Distribution of Pressure on the Blade Surface

Figure 9 shows the surge amplitude of the platform motion with reference to time. The second period of stable surge motion of the wind turbine is selected as the research object. Two typical positions were selected to analyze the pressure distribution on the blade surface.

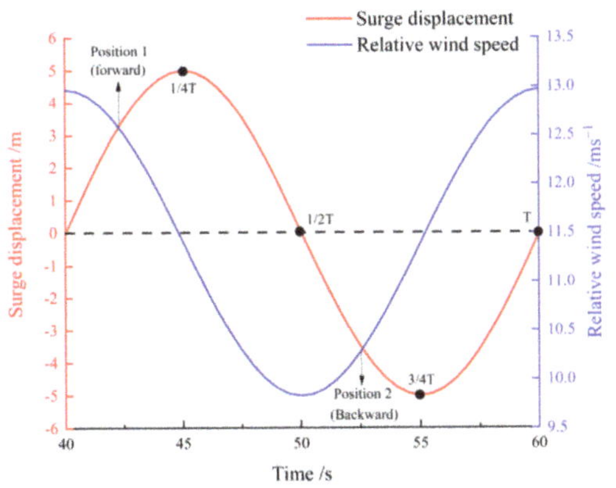

Figure 9. The two typical positions during surge motion.

The most fundamental influence of the tower shadow effect on the aerodynamic performance of the wind turbine is the interference of the tower on the blade; the blade surface pressure is the basic parameter to characterize the aerodynamic performance of the blade. This section selects two typical positions in which the blade rotates to the front of the tower under surge motion, and analyzes the pressure distribution under the root (r/R = 0.32), middle (r/R = 0.63) and tip (r/R = 0.94) sections of the blade. The abscissa is dimensionless as x/c (the abscissas of different points on the section/chord length of the section).

Figure 10 shows the distribution of pressure in each section of the blade with and without the tower at two typical positions in a surge cycle. It can be found that the pressure difference distribution at position 1 at the corresponding section is greater than that at position 2. It can be explained that the V_{rel} at each blade section reaches larger values due to the forward surge velocity of the platform, while the V_{rel} decreases when the platform surges backward. From the numerical point of view, the maximum pressure difference of the wind turbine without the tower at 0.32 R, 0.63 R, 0.94 R section is 1952 Pa, 4240 Pa and 7910 Pa, respectively. It can be seen that the closer to the tip of the blade, the greater the pressure difference on the blade surface. In addition, the tower shadow effect mainly affects the negative pressure value of the suction leading edge and the absolute value of the maximum negative pressure difference in each section is reduced by 10.56%, 7.61% and 5.36%, respectively. It can be inferred that the closer to the tip of the blade, the less obvious the interference of the tower shadow effect on the negative pressure of the suction surface.

4.1.3. Near Flow Field of Each Section of Blade

Figure 11 shows the interaction between the different blade sections and the tower when the blade rotates to the shadow area of the tower under surge motion. Position 1 was selected for further analysis. For the convenience of analysis, the variation of pressure fields near the blade surface and the distribution of absolute velocity streamline are discussed, respectively.

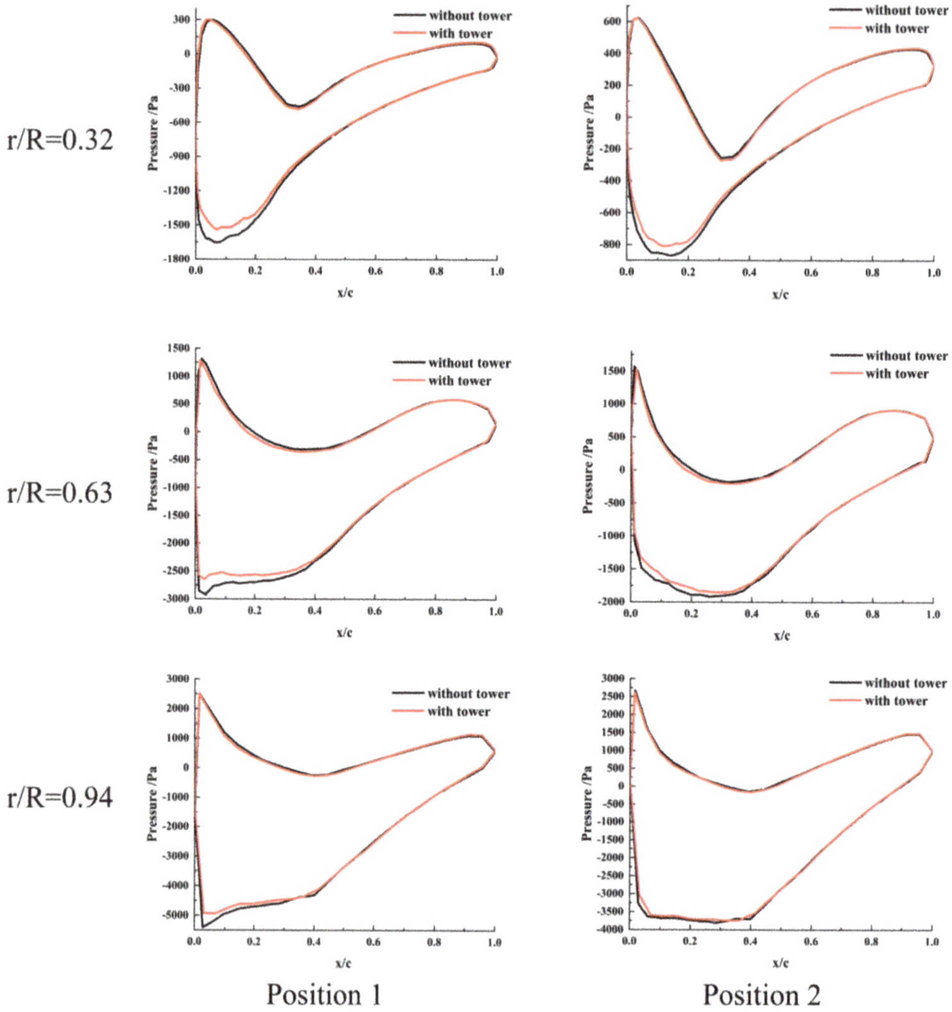

Figure 10. Distribution of pressure in each section of blade at position 1 and position 2.

By comparing the pressure field of the wind turbine with and without the tower at the same cross-section, it is seen that there is an obvious negative pressure field behind the suction surface of the blades. On the root and middle section of the blade, it can be observed that the suction leading edge negative pressure field of the wind turbine without the tower is larger and the negative pressure value is lower, while the pressure surface is basically the same, and the interference of the tower on the pressure field in the tip section is relatively small. These phenomena are in good agreement with the blade surface pressure distribution results described in the previous section. In addition, the whole negative pressure field at the root and middle section of the blade is obviously compressed due to the interference of the tower, which leads to a decrease in the pressure difference between the pressure surface and suction surface of the blade, thus reducing the overall work capacity of the blade. This further explains that the average power of the wind turbine decreases due to the influence of the tower shadow effect in the above results.

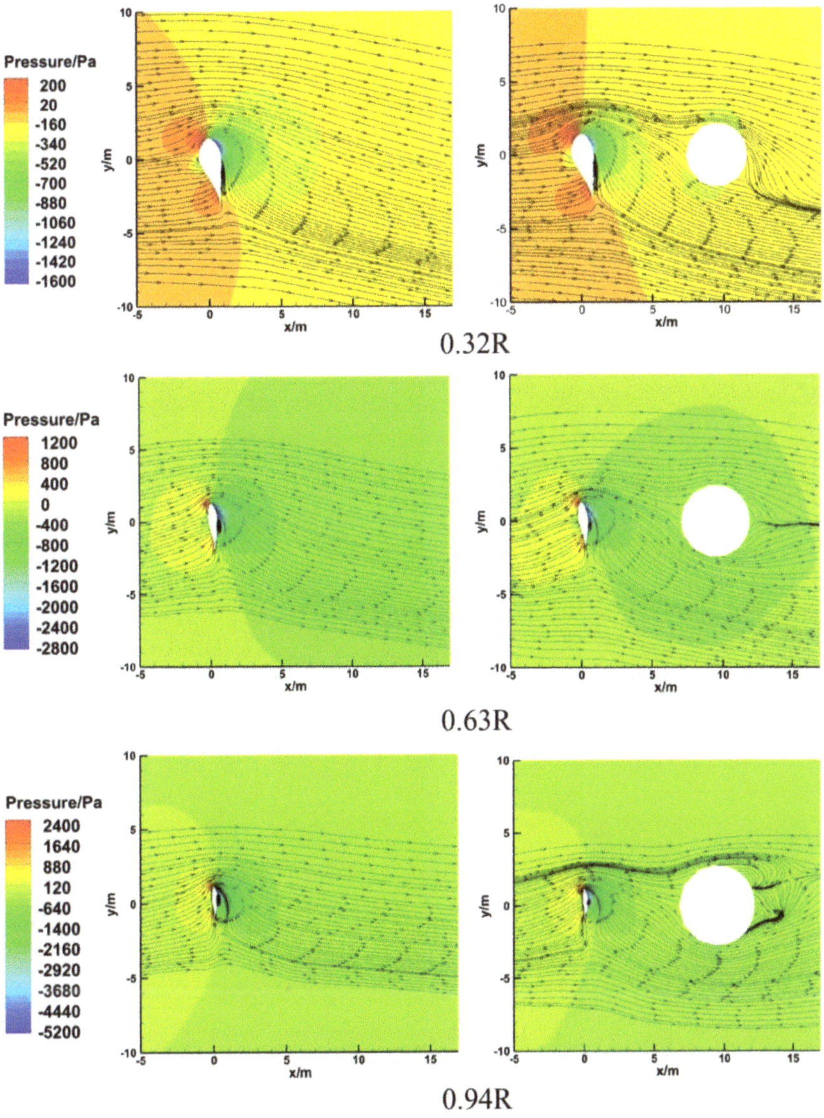

Figure 11. Streamline and pressure contours on different blade sections under surge motion.

Meanwhile, the streamline change of the wind turbine with the tower is not obvious compared with the wind turbine without the tower under surge motion, and the main change is concentrated in the wake of the tower. In the 0.32 R section, the streamline behind the tower shifts greatly, and the tower shadow effect is the most obvious in the interference of the flow field. In the 0.63 R section, the streamline on both sides of the tower shifts, and the flow field in this section is affected by both the tower shadow effect and the enhanced blade rotation effect. In the 0.94 R section, there is an obvious stall separation in the flow field behind the tower, and the separation point shifts to the direction of blade rotation. The area near the root of the blade is the main area affected by the tower shadow effect under surge motion, and the effect of the tower shadow effect weakens on the near wake flow field with the increase of blade height.

4.1.4. Wake Field behind the Wind Turbine

The main load source of the wind turbine is axial flow; the wake flow characteristics are particularly important for the analysis of wind turbine aerodynamics. By comparing the wake field with or without the tower, this section further illustrates the influence of platform motion on the aerodynamics of the wind turbine.

Figure 12 shows the velocity distribution of the wake field under platform surge motion. The symmetry of the wake field is disturbed and the flow field is slightly compressed. When the wind turbine moves forward, the average velocity of the wake field is greater than that of the backward motion. Considering the effect of the tower shadow, the influence range of the high-speed wake behind the hub expands, but it has little influence on the tip vortex, and the rear of the tower is accompanied by a large vortex shedding range and obvious vortex motion.

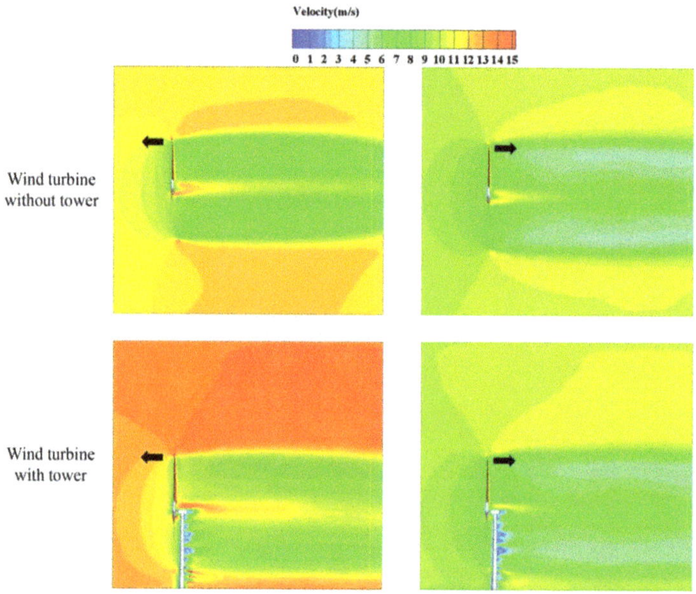

Figure 12. Velocity contours of wake field under surge motion.

4.2. Unsteady Aerodynamic Analysis under Pitch Motion

4.2.1. Total Performance Analysis

Figure 13 shows the overall power and thrust variation of the wind turbine with and without the tower. The pitch amplitude corresponds to the fluctuation of power and thrust values. For f = 0.05 Hz, A_p = 5°, as shown in Figure 13a, the power of the wind turbine fluctuates violently due to the existence of the tower, and it decreases sharply three times in one rotation cycle of the wind turbine. As shown in Figure 13b, the maximum power of the wind turbine without the tower was 5.16 MW, while the wind turbine with the tower was 5.11 MW. The minimum power of the former was 5.01 MW, while that of the latter s 4.70 MW. It can be seen that pitch motion slightly increases the peak and valley values of power, but the extreme value of the wind turbine decreases due to the existence of the tower. The average power generation was 5.06 MW and 4.98 MW, respectively; the latter was reduced by 1.58%. It can be found that the pitch motion of the platform increases the power of the wind turbine, but under the combination of pitch and tower shadow effect, the power fluctuation is larger and the increase is smaller. As shown in Figure 13c,d, the fluctuation trend of axial thrust is similar to that of power. The peak thrust and valley thrust of the wind turbine without the tower are 780.7 KN and 771.6 KN, and those with the tower are 777.9 KN and 751.1 KN, respectively. The average thrust was 774.5 KN and

769.0 KN, separately; the latter was reduced by 0.71%. That is, in terms of numerical values, the tower shadow effect slightly reduces the average power, while the average thrust is almost unchanged under pitch motion.

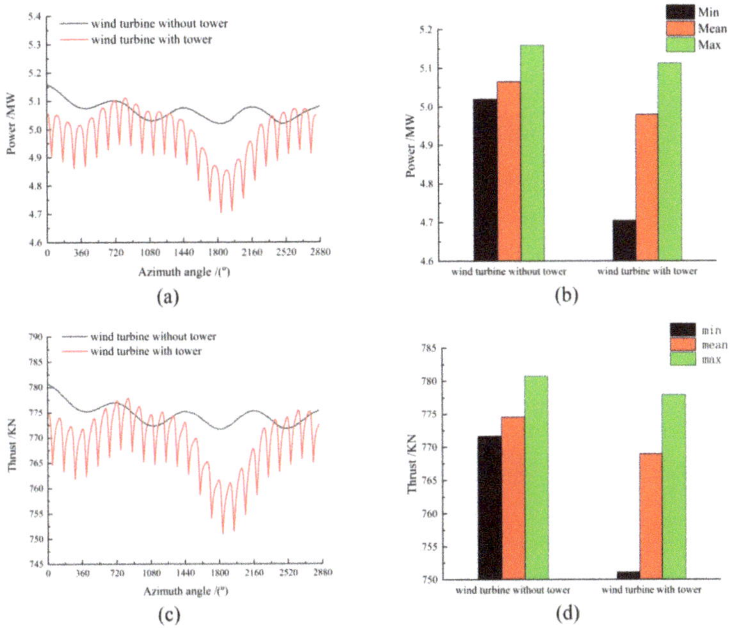

Figure 13. Total aerodynamic comparison of pitch motion; (**a**) Power versus azimuth angle; (**b**) Extreme and average values of power; (**c**) Thrust versus azimuth angle; (**d**) Extreme and average values of thrust.

Figure 14 shows the induced velocity distribution of airfoil under pitch motion. The pitch motion of the platform changes the angle between the incoming wind speed and the rotating plane of the wind turbine and produces an additional induced velocity V_{ind}. V_{ind} can be decomposed into chord velocity V_c and radial velocity V_r on the rotating plane, in which the direction of chord velocity component V_c depends on the positive or negative angle of pitch motion, which changes the relative velocity and direction of the rotating plane of the airfoil, and the angle of attack changes correspondingly. It leads to the fluctuation of the overall aerodynamic performance of the wind turbine under pitch motion.

4.2.2. Distribution of Pressure on the Blade Surface

Figure 15 shows the pitch amplitude and angular velocity of the platform motion with reference to simulation time. The second period of stable pitch motion of the wind turbine was selected as the research object, and the two typical positions of the wind turbine moving forward and backward to the front of the tower were selected for further analysis.

Figure 16 shows the pressure distribution in each section of the blade with and without the tower at two typical positions in a pitch cycle. In terms of numerical values, the maximum pressure difference of the wind turbine without the tower at 0.32 R, 0.63 R and 0.94 R sections is 1711 Pa, 3630 Pa and 7030 Pa, respectively. It can be found that the pressure difference of each section of the blade under pitch motion is smaller than that of surge motion. Comparing the pressure difference distributions of different sections at two typical positions, it can be seen that the pressure difference distributions on the pressure surface and the suction surface are basically the same. The change of the pressure difference is not obvious under pitch motion. Besides, affected by the tower shadow effect, the absolute value of the maximum negative pressure difference at the leading edge of suction in each section of the blade decreases by 9.02%, 7.31% and 4.93%, respectively. Compared with surge motion, it can be seen that under pitch motion, the interference of the tower shadow effect on each section of the blade is also mainly concentrated in the root and tip of the blade, but the degree of interference is less than that of surge motion.

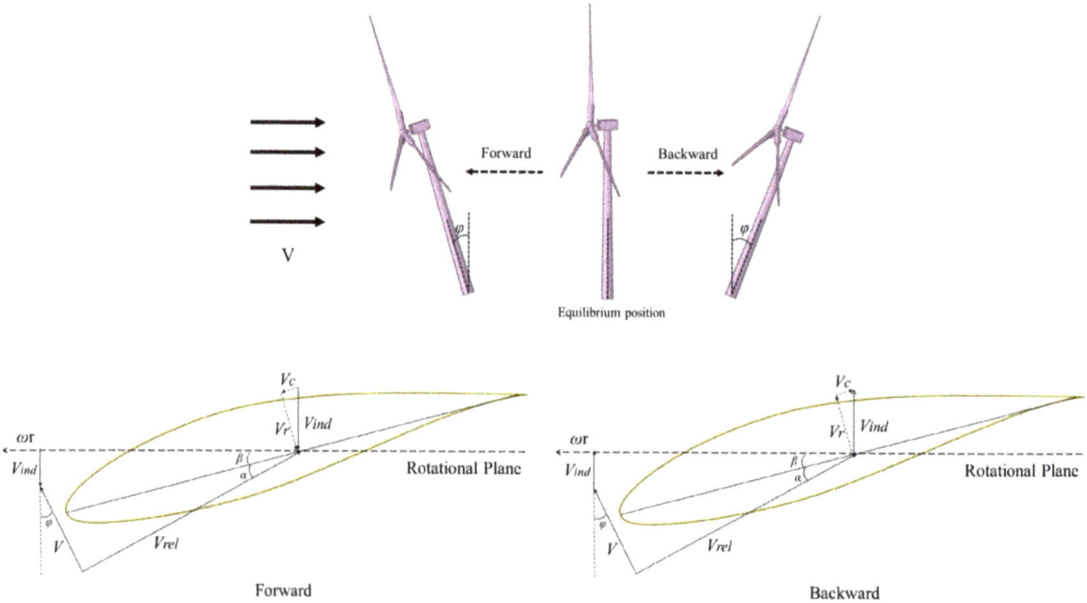

Figure 14. Schematic diagram of airfoil-induced velocity under pitch motion.

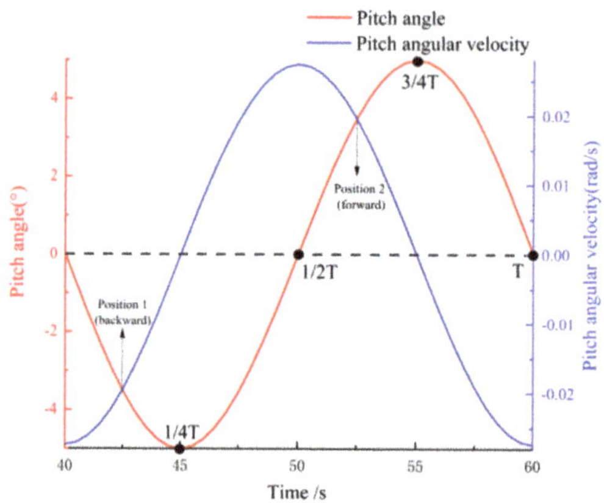

Figure 15. The two typical positions during pitch motion.

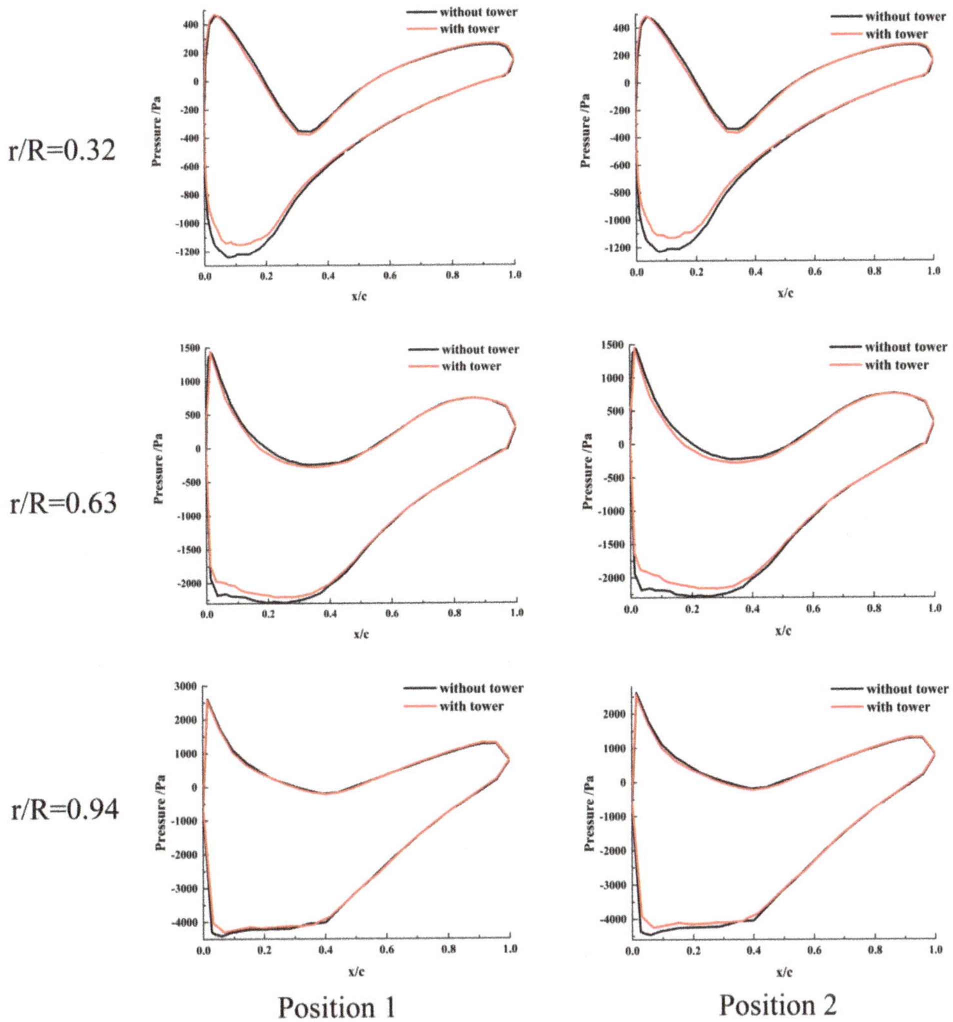

Figure 16. Distribution of pressure in each section of blade at position 1 and position 2.

4.2.3. Near Flow Field of Each Section of Blade

Figure 17 shows the interaction between the different blade sections and the tower when the blade rotates to the shadow area of the tower under pitch motion. The pitch motion of the platform to the rear and the rotation of the wind turbine to position 1 directly in front of the tower are selected for further analysis.

Comparing the pressure field on different blade sections with or without the tower under pitch motion, it can be seen that the pressure difference distribution of the blade surface is still concentrated on the leading edge of the blade. Under the influence of the tower shadow effect, the range of the negative pressure field near the suction surface of 0.32 R section and 0.63 R section is reduced, and the absolute value of negative pressure is smaller, but the negative pressure field of the suction surface of 0.94 R section has no obvious change. This further shows that the influence of the tower shadow effect on the pressure field is mainly concentrated in the root and the middle of the blade.

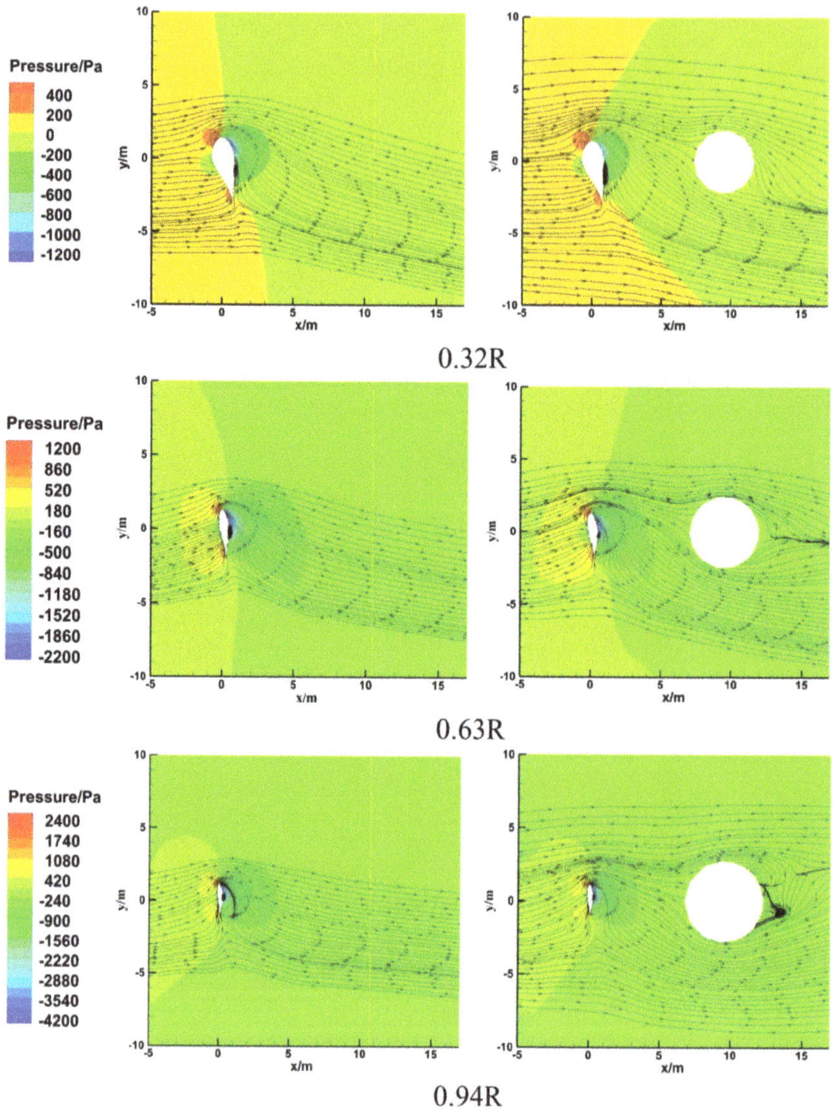

Figure 17. Streamline and pressure contours on different blade sections under pitch motion.

Observing the streamline distribution shown in Figure 17, the main difference is still concentrated in the near wake flow field of the tower. It can be seen that the distribution of the streamline at 0.32 R section and 0.63 R section is similar under surge motion, but there is an obvious stall separation vortex under the pitch motion at 0.94 R section, because compared with the surge motion moving only in the horizontal direction, the pitch motion applies a relative partial velocity in the vertical direction to the wind turbine when the azimuth of the wind turbine changes. The angle between the direction of the incoming flow velocity and the rotating plane of the wind turbine is constantly changing, which makes the flow field more complex.

4.2.4. Wake Field behind the Wind Turbine

Figure 18 shows the velocity distribution of the wake field under platform pitch motion. The two moments when the wind turbine is in the balanced position and pitches forward and backward, respectively, are selected for further analysis. It can be seen that when the wind turbine without the tower is in the equilibrium position, the mutual interference between the blade and the wake flow field is small, and the wake is more stable compared with surge motion. Nevertheless, the symmetry of the flow field behind the same wind turbine with the tower is affected, and there is an obvious low-speed disturbance zone behind the tower, which increases the complexity of the wake.

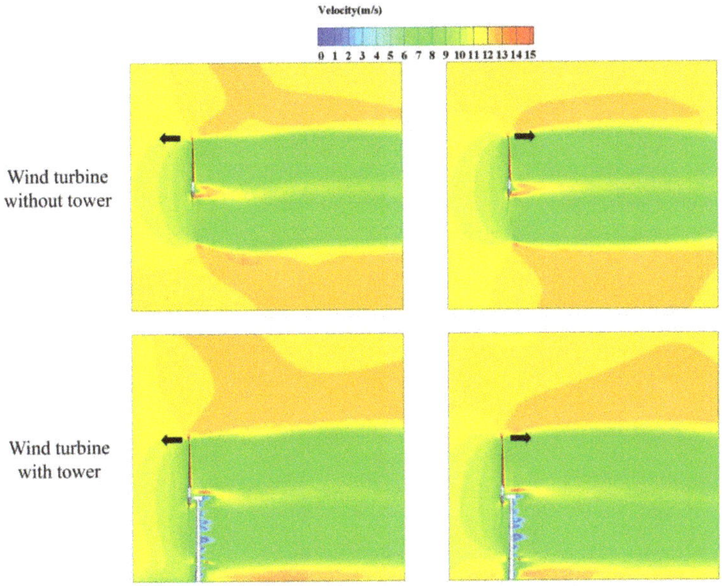

Figure 18. Velocity contours of wake field under pitch motion.

4.3. *Unsteady Aerodynamic Analysis under Yaw Motion*

4.3.1. Total Performance Analysis

Figure 19 shows the total power and thrust comparison with and without the tower under yaw motion. Similarly, the yaw amplitude corresponds to the fluctuation of power and thrust values. For f = 0.05 Hz, $A_y = 15°$, as shown in Figure 19a, it can be seen that the power of the wind turbine with the tower is always lower than that of the wind turbine without the tower under the same azimuth, and the power rises after three sharp drops in a wind turbine rotation cycle, which is the result of the combined action of the tower shadow effect and yaw motion. As shown in Figure 19b, under the effect of tower shadow, the maximum, average and minimum power of wind turbines are reduced by 1.94%, 2.47% and 4.16%, respectively. It can be concluded that the tower shadow effect aggravates the change of power extremum under yaw motion. As shown in Figure 19c,d, the fluctuation of the axial thrust value is consistent with that of the power value similarly; that is, they reach their respective extremes at the same azimuth. Under the effect of tower shadow, the maximum, average and minimum thrust of the wind turbine is reduced by 0.76%, 1.1% and 1.86%, respectively. It can be deduced that under yaw motion, the amplitude of the power drop is obviously affected by the tower shadow effect, while the thrust decreases slightly.

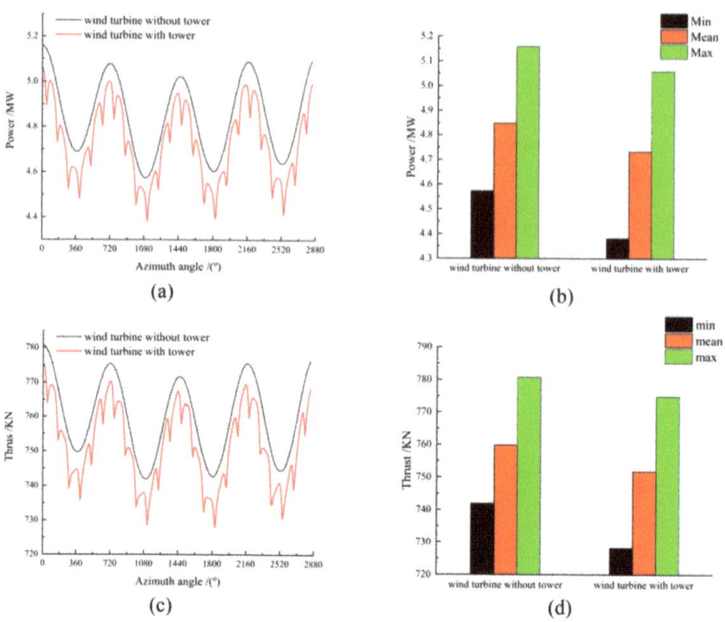

Figure 19. Total aerodynamic comparison of yaw motion; (**a**) Power versus azimuth angle; (**b**) Extreme and average values of power; (**c**) Thrust versus azimuth angle; (**d**) Extreme and average values of thrust.

Figure 20 shows the induced velocity distribution of airfoil under yaw motion. The yaw motion of the platform produces an additional induced velocity perpendicular to the rotation plane, and the direction of the V_{ind} can be determined according to the right-hand rule. When the platform rotates to the left, the angle between the direction of the inflow velocity and the direction of V_{ind} is an obtuse angle, and its relative inflow velocity increases; on the contrary, when the platform rotates to the right, the angle between the direction of the inflow velocity and the direction of V_{ind} is an acute angle, and its relative inflow velocity decreases. V_{ind} can also be decomposed into chord velocity V_c and radial velocity V_r on the rotating plane. *Vr* changes the rotation effect of the blade, and V_c causes periodic fluctuations in the angle of attack and the sectional load of the blade.

4.3.2. Distribution of Pressure on the Blade Surface

Figure 21 shows the yaw amplitude and angular velocity of the platform motion with reference to simulation time. The second period of stable yaw motion of the wind turbine and two typical positions in front of the tower were selected for further analysis.

Figure 22 shows the pressure distribution in each section of the blade with and without the tower at two typical positions in a yaw cycle. The azimuth moment of 180° and 900° rotation of the wind turbine is selected to study when the wind turbine is at a certain yaw angle to both sides and the wind turbine is directly in front of the tower. In terms of numerical values, the maximum pressure difference of the wind turbine without the tower at 0.32 R, 0.63 R, 0.94 R sections is 1620 Pa, 3455 Pa and 6890 Pa, respectively. It can be found that the pressure difference of each section of the blade under yaw motion is smaller than that of pitch motion. Comparing the pressure distributions of the two kinds of wind turbines at two typical positions, it can be found that the distributions are consistent. The change of pressure difference at the different positions of yaw motion is also not obvious; however, affected by the tower shadow effect, the absolute value of the maximum negative pressure difference at the leading edge of suction in each section of the blade decreases by 11.01%, 8.23% and 5.80%, respectively. It can be inferred that under yaw motion, the interference degree of the tower shadow effect on the negative pressure on the suction surface of the blade is greater than that of surge motion.

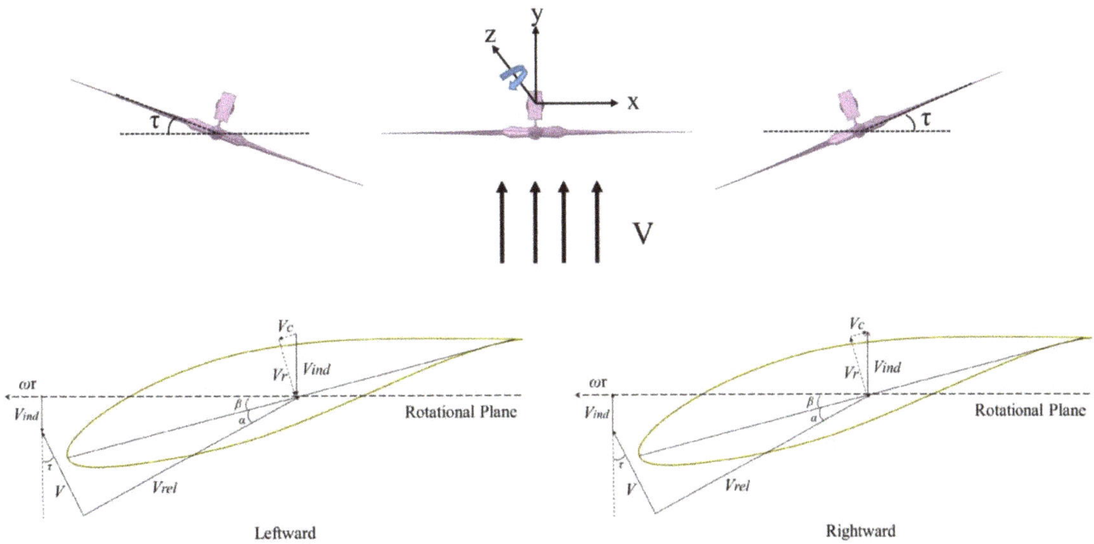

Figure 20. Schematic diagram of airfoil-induced velocity under yaw motion.

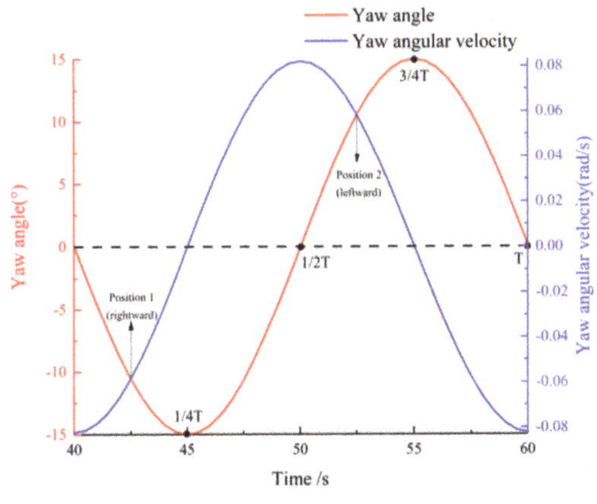

Figure 21. The two typical positions during yaw motion.

4.3.3. Near Flow Field of Each Section of Blade

Figure 23 shows the interaction between the different blade sections and the tower when the blade rotates to the shadow area of the tower under yaw motion. Position 1 was selected for further analysis.

Under yaw motion, the distribution trend of the pressure field in each section of the blade with or without the tower is similar to that of surge and pitch motions; the maximum negative pressure is still concentrated near the leading edge of the suction of the blade, and the tower compresses the negative pressure field and reduces the absolute value of the maximum negative pressure. The main influence range of the tower shadow effect is in the root and middle of the blade. From a numerical point of view, the blade surface pressure difference of the same section under yaw motion is lower than that of surge and pitch motions, which further shows that the work capacity of the wind turbine under yaw motion is the worst among the three motions.

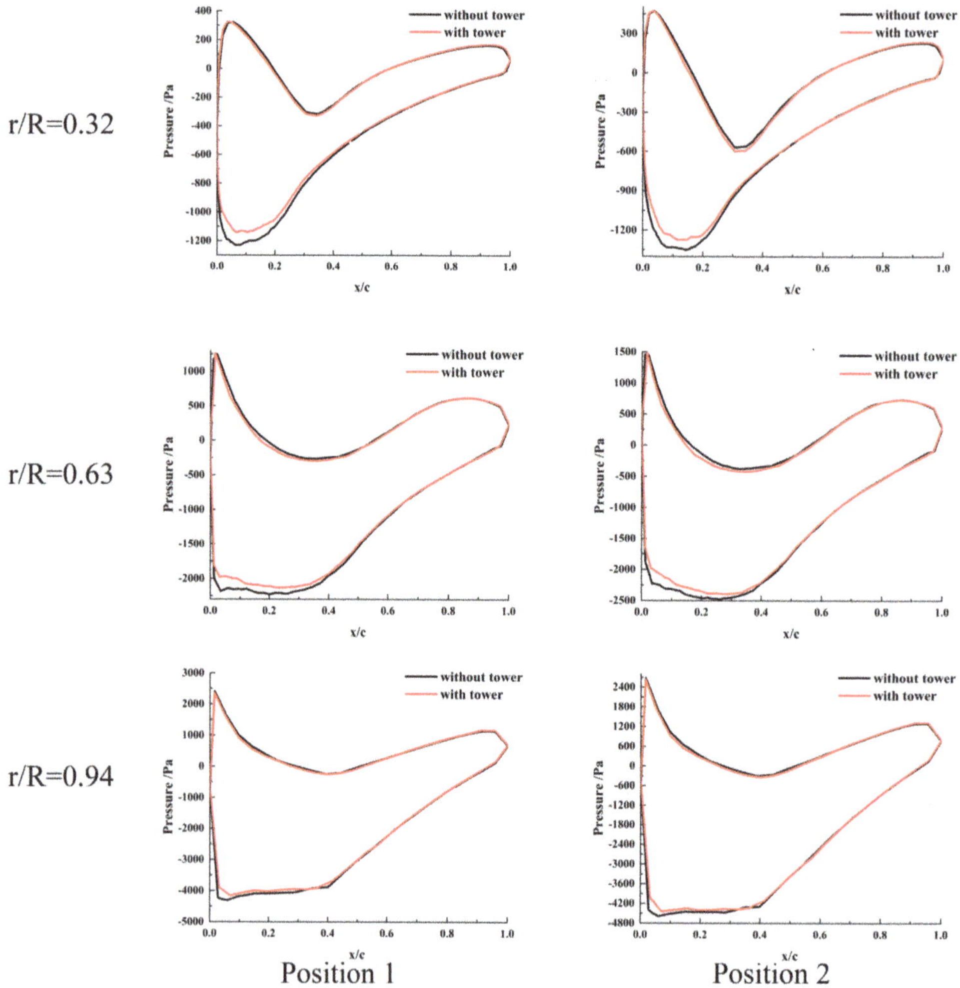

Figure 22. Distribution of pressure in each section of blade at position 1 and position 2.

By comparing the streamline of each section of the blade with or without the tower, the influence area of the tower shadow effect is still mainly concentrated in the wake flow field of the tower, and the streamline shifts behind the tower toward the trailing edge of the blade at 0.32 R section, which is mainly the interference of the tower. The streamline at 0.63 R section shifts in the opposite direction on both sides of the tower, which is the result of the joint action of the tower shadow effect and the enhanced rotation effect. At 0.94 R section, in addition to the weakening of the tower shadow effect and the enhancement of the rotation effect, the yaw motion changes the upwind area of the incoming flow and the rotating plane of the wind turbine, resulting in additional induced tangential velocity. The streamline behind the tower shifts to a greater extent along the rotation direction of the wind turbine.

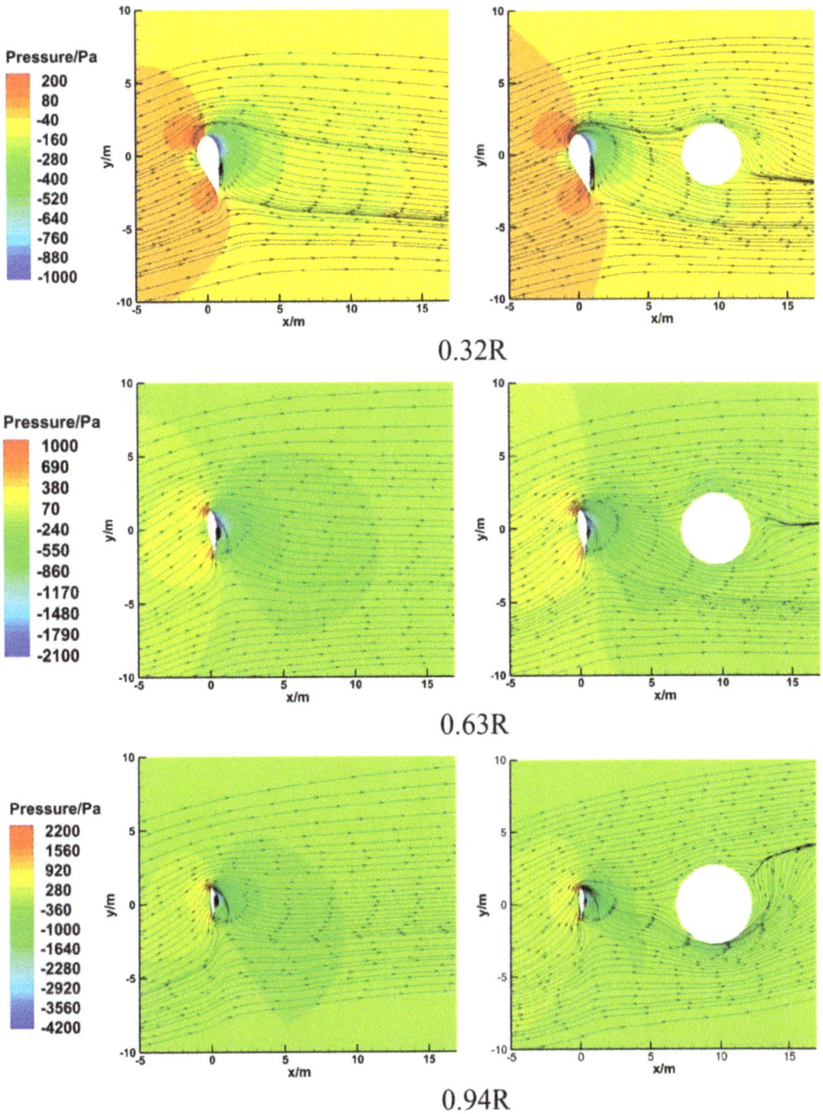

Figure 23. Streamline and pressure contours on different blade sections under yaw motion.

4.3.4. Wake Field behind the Wind Turbine

Figure 24 shows the velocity distribution of the wake field under platform yaw motion. The two moments when the wind turbine is in the equilibrium position and begins to yaw to the left and right side was selected for further analysis. The wake area of the wind turbine without the tower shows a nearly symmetrical distribution under the balanced position of yaw motion. Meanwhile, the high-speed wake area near the hub is smaller than that under pitch motion; it can be concluded that the yaw motion could reduce the induced velocity behind the hub. Furthermore, under yaw motion, the tower shadow effect similarly leads to the asymmetric distribution of the wake field.

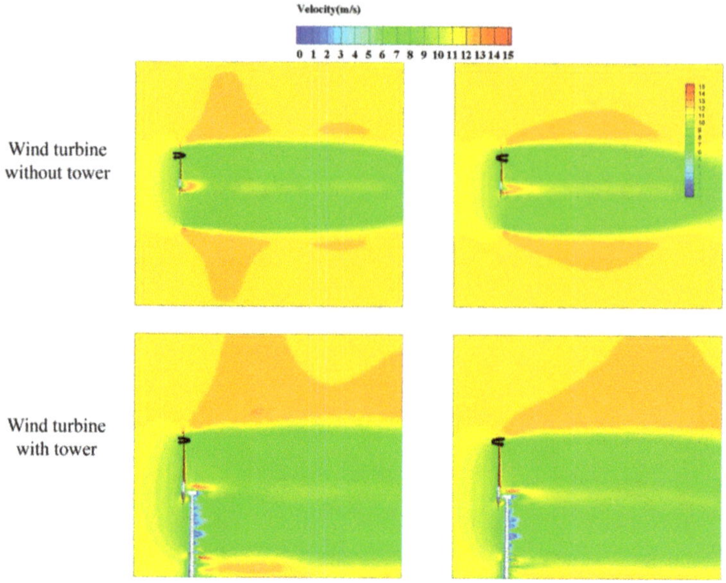

Figure 24. Velocity contours of wake field under yaw motion.

5. Conclusions

This paper investigated the aerodynamic performance of NREL 5 MW FOWT with a rigid blades turbine considering the tower shadow effect under surge, pitch and yaw motions, respectively. The motion form of the platform is equivalent to the change of relative velocity of the wind turbine. The dynamic inlet wind speed is compiled for unsteady numerical simulation using the UDF function. The accuracy of the numerical simulation was verified by proper computational grids. Three independent platform motions and two models with or without the tower were considered for calculation. The results illustrate the wind turbine's aerodynamics, including power, thrust, pressure distribution and flow field. The conclusions can be drawn as follows:

(1) The fluctuation frequency of power and thrust is always consistent with the motion frequency of the platform. The power is more obviously affected by the tower shadow effect than the thrust, in which the decrease of power is the largest under yaw motion, second-largest under surge motion, and smallest under pitch motion, with a decreased range of 1.58–2.47%.

(2) The influence of the tower shadow effect on the pressure difference of the wind turbine is mainly concentrated at the suction leading edge of the blade under different platform motions, and the interference ability of the tower from the root to the tip of the blade weakens along the blade-spreading direction. The pressure difference under yaw motion is most obviously interfered with by the tower, and the average maximum negative pressure is reduced by 8.35%, which is not conducive to the output power of FOWTs.

(3) For the pressure field, the tower shadow effect obviously compresses the range of the negative pressure field of the root and middle sections of the blade, while the negative pressure field of the tip section is less affected. For the near wake flow field, the wake of the tower at the root section is the most seriously interfered by the tower; the influence of the tower shadow effect decreases with the increase of the blade height and the additional induced velocity produced by yaw and pitch motions makes the near wake flow field of the tower more complicated.

(4) The wake field changes most violently under surge motion; the wake flow field is relatively stable under pitch and yaw motions. Besides, the tower shadow effect leads to

the increase of the velocity gradient near the hub, and the influence range of the high-speed wake of the tower wind turbine hub is the farthest under surge motion.

In view of these findings, the platform motion and the tower shadow effect have a great impact on the steady operation of FOWTs, and the combination of the two causes greater interference to the flow field. However, this paper only considers the aerodynamic characteristics of FOWTs under single-degree-of-freedom (surge, pitch and yaw) motion, and future research on factors such as heave, roll, sway and multi-degree-of-freedom coupling motions are worthy of more in-depth discussion.

Author Contributions: Conceptualization, L.D. and L.Z.; Formal analysis, L.D.; Investigation, L.D. and L.Z.; Methodology, L.D.; Project administration, D.H.; Software, L.D.; Supervision, D.H.; Validation, L.D.; Writing—original draft, L.D.; Writing—review and editing, L.D. All authors have read and agreed to the published version of the manuscript.

Funding: This research was funded by the Science and Technology Commission of Shanghai Municipality and the National Natural Science Foundation of China, grant number 18DZ1202302 and 50706025.

Institutional Review Board Statement: Not applicable.

Informed Consent Statement: Not applicable.

Data Availability Statement: The data are not publicly available.

Acknowledgments: This research work is sponsored by the Science and Technology Commission of Shanghai Municipality (18DZ1202302). Additionally, the support of the National Natural Science Foundation of China under project No. 50706025 is gratefully acknowledged.

Conflicts of Interest: The authors declare no conflict of interest.

Abbreviations

FOWT	Floating Offshore Wind Turbine
CFD	Computational Fluid Dynamic
6-DOF	Six Degrees-of-Freedom
BEM	Blade Element Momentum
FVM	Free Vortex Method
UDF	User-Defined Function
RANS	Reynolds-Averaged Navier–Stokes equations
V	Incoming Wind Speed

Nomenclature

V_S	the relative inflow velocity under surge motion
V_P	the relative inflow velocity under pitch motion
V_Y	the relative inflow velocity under yaw motion
θ	the angle between two blades
γ	the influence area of the tower shadow effect
ω	the rotational speed of the wind turbine
V_{ind}	the induced velocity of platform motion
V_{rel}	the relative velocity
$\mathrm{A_s}$	surge amplitude
$\mathrm{A_p}$	pitch amplitude
$\mathrm{A_y}$	yaw amplitude

References

1. Cheng, P.; Huang, Y.; Wan, D.C. A numerical model for fully coupled aero-hydrodynamic analysis of floating offshore wind turbine. *Ocean Eng.* **2019**, *173*, 183–196. [CrossRef]
2. Shen, X.; Hu, P.; Chen, J.E.; Zhu, X.C.; Du, Z.H. The unsteady aerodynamics of floating wind turbine under platform pitch motion. *Proc. Inst. Mech. Eng. Part A J. Power Energy* **2018**, *232*, 1019–1036. [CrossRef]

3. Wu, X.N.; Hu, Y.; Li, Y.; Yang, J.; Duan, L.; Wang, T.G.; Adcock, T.; Jiang, Z.Y.; Gao, Z.; Lin, Z.L.; et al. Foundations of offshore wind turbines: A review. *Renew. Sustain. Energ. Rev.* **2019**, *104*, 379–393. [CrossRef]
4. Cai, X.; Gu, R.R.; Pan, P.; Zhu, J. Unsteady aerodynamics simulation of a full-scale horizontal axis wind turbine using CFD methodology. *Energy Convers. Manag.* **2016**, *112*, 146–156. [CrossRef]
5. Sebastian, T.; Lackner, M. Analysis of the Induction and Wake Evolution of an Offshore Floating Wind Turbine. *Energies* **2012**, *5*, 968–1000. [CrossRef]
6. De Vaal, J.B.; Hansen, M.O.L.; Moan, T. Effect of wind turbine surge motion on rotor thrust and induced velocity. *Wind Energy* **2014**, *17*, 105–121. [CrossRef]
7. Ma, Z.; Li, W.; Ren, N.A.X.; Ou, J.P. The typhoon effect on the aerodynamic performance of a floating offshore wind turbine. *J. Ocean Eng. Sci.* **2017**, *2*, 279–287. [CrossRef]
8. Farrugia, R.; Sant, T.; Micallef, D. A study on the aerodynamics of a floating wind turbine rotor. *Renew. Energy* **2016**, *86*, 770–784. [CrossRef]
9. Wen, B.R.; Dong, X.J.; Tian, X.L.; Peng, Z.K.; Zhang, W.M.; Wei, K.X. The power performance of an offshore floating wind turbine in platform pitching motion. *Energy* **2018**, *154*, 508–521. [CrossRef]
10. Wen, B.R.; Tian, X.L.; Dong, X.J.; Peng, Z.K.; Zhang, W.M. On the power coefficient overshoot of an offshore floating wind turbine in surge oscillations. *Wind Energy* **2018**, *21*, 1076–1091. [CrossRef]
11. Salehyar, S.; Li, Y.; Zhu, Q. Fully-coupled time-domain simulations of the response of a floating wind turbine to non-periodic disturbances. *Renew. Energy* **2017**, *111*, 214–226. [CrossRef]
12. Lin, L.; Wang, K.; Vassalos, D. Detecting wake performance of floating offshore wind turbine. *Ocean Eng.* **2018**, *156*, 263–276. [CrossRef]
13. Tran, T.; Kim, D.; Song, J. Computational Fluid Dynamic Analysis of a Floating Offshore Wind Turbine Experiencing Platform Pitching Motion. *Energies* **2014**, *7*, 5011–5026. [CrossRef]
14. Tran, T.T.; Kim, D.H. The aerodynamic interference effects of a floating offshore wind turbine experiencing platform pitching and yawing motions. *J. Mech. Sci. Technol.* **2015**, *29*, 549–561. [CrossRef]
15. Tran, T.T.; Kim, D.H. The platform pitching motion of floating offshore wind turbine: A preliminary unsteady aerodynamic analysis. *J. Wind Eng. Ind. Aerodyn.* **2015**, *142*, 65–81. [CrossRef]
16. Tran, T.T.; Kim, D.H. A CFD study into the influence of unsteady aerodynamic interference on wind turbine surge motion. *Renew. Energy* **2016**, *90*, 204–228. [CrossRef]
17. Chen, J.H.; Hu, Z.Q.; Wan, D.C.; Xiao, Q. Comparisons of the dynamical characteristics of a semi-submersible floating offshore wind turbine based on two different blade concepts. *Ocean Eng.* **2018**, *153*, 305–318. [CrossRef]
18. Wu, C.-H.K.; Vinh-Tan, N. Aerodynamic simulations of offshore floating wind turbine in platform-induced pitching motion. *Wind Energy* **2017**, *20*, 835–858. [CrossRef]
19. Huang, Y.; Wan, D.C. Investigation of Interference Effects Between Wind Turbine and Spar-Type Floating Platform Under Combined Wind-Wave Excitation. *Sustainability* **2020**, *12*, 246. [CrossRef]
20. Fang, Y.; Duan, L.; Han, Z.L.; Zhao, Y.S.; Yang, H. Numerical analysis of aerodynamic performance of a floating offshore wind turbine under pitch motion. *Energy* **2020**, *192*, 116621. [CrossRef]
21. Fang, Y.; Li, G.; Duan, L.; Han, Z.L.; Zhao, Y.S. Effect of surge motion on rotor aerodynamics and wake characteristics of a floating horizontal-axis wind turbine. *Energy* **2021**, *218*, 119519. [CrossRef]
22. Chen, Z.W.; Wang, X.D.; Guo, Y.Z.; Kang, S. Numerical analysis of unsteady aerodynamic performance of floating offshore wind turbine under platform surge and pitch motions. *Renew. Energy* **2021**, *163*, 1849–1870. [CrossRef]
23. Chen, Z.W.; Wang, X.D.; Kang, S. Effect of the Coupled Pitch-Yaw Motion on the Unsteady Aerodynamic Performance and Structural Response of a Floating Offshore Wind Turbine. *Processes* **2021**, *9*, 290. [CrossRef]
24. Sivalingam, K.; Martin, S.; Wala, A.A.S. Numerical Validation of Floating Offshore Wind Turbine Scaled Rotors for Surge Motion. *Energies* **2018**, *11*, 2578. [CrossRef]
25. Kim, H.; Lee, S.; Lee, S. Influence of blade-tower interaction in upwind-type horizontal axis wind turbines on aerodynamics. *J. Mech. Sci. Technol.* **2011**, *25*, 1351–1360. [CrossRef]
26. Quallen, S.; Tao, X. An Investigation of the Blade Tower Interaction of a Floating Offshore Wind Turbine. In Proceedings of the 25th International Offshore and Polar Engineering Conference, Big Island, HI, USA, 21 June 2015.
27. Ke, S.T.; Wang, T.G.; Ge, Y.J.; Tamura, Y. Aerodynamic loads and aeroelastic responses of large wind turbine tower-blade coupled structure in yaw condition. *Struct. Eng. Mech.* **2015**, *56*, 1021–1040. [CrossRef]
28. Noyes, C.; Qin, C.; Loth, E.; Schreck, S. Measurements and predictions of wind turbine tower shadow and fairing effects. *J. Wind Eng. Ind. Aerodyn.* **2018**, *179*, 297–307. [CrossRef]
29. Zhang, J.; Zhen, G.; Xing, Z.; Liang, G.; Hu, D. Dynamic analysis of offshore wind turbine blades under the action of wind shear and fluctuating wind. *J. Vibroeng.* **2018**, *20*. [CrossRef]
30. Li, Z.W.; Wen, B.R.; Dong, X.J.; Peng, Z.K.; Qu, Y.G.; Zhang, W.M. Aerodynamic and aeroelastic characteristics of flexible wind turbine blades under periodic unsteady inflows. *J. Wind Eng. Ind. Aerodyn.* **2020**, *197*, 104057. [CrossRef]
31. Wu, W.M.; Liu, X.F.; Dai, Y.J.; Li, Q.Y. An in-depth quantitative analysis of wind turbine blade tip wake flow based on the lattice Boltzmann method. *Environ. Sci. Pollut. Res.* **2020**, *13*. [CrossRef]

32. Regodeseves, P.G.; Morros, C.S. Unsteady numerical investigation of the full geometry of a horizontal axis wind turbine: Flow through the rotor and wake. *Energy* **2020**, *202*, 117674. [CrossRef]
33. Burmester, S.; Vaz, G.; el Moctar, O. Towards credible CFD simulations for floating offshore wind turbines. *Ocean Eng.* **2020**, *209*, 107237. [CrossRef]
34. Chen, D.; Gao, P.; Huang, S.S.; Fan, K.; Zhuang, N.; Liao, Y.D. Dynamic response and mooring optimization of spar-type substructure under combined action of wind, wave, and current. *J. Renew. Sustain. Energy* **2017**, *9*, 063307. [CrossRef]
35. Wang, X.D.; Ye, Z.L.; Kang, S.; Hu, H. Investigations on the Unsteady Aerodynamic Characteristics of a Horizontal-Axis Wind Turbine during Dynamic Yaw Processes. *Energies* **2019**, *12*, 3124. [CrossRef]
36. Jonkman, J.M. *Dynamics Modeling and Loads Analysis of an Offshore Floating Wind Turbine*; National Renewable Energy Lab. (NREL): Golden, CO, USA, 2007.
37. Hu, D.M.; Ji, S.Q.; Wu, Z.X. Study on Aerodynamic Performance of Offshore Floating Wind Turbine with Fusion Winglets. *J. Sol. Energy Eng. Trans. ASME* **2020**, *142*, 051007. [CrossRef]
38. Jonkman, J.M.; Butterfield, S.; Musial, W.; Scott, G. *Definition of a 5MW Reference Wind Turbine for Offshore System Development*; Office of Scientific & Technical Information Technical Reports; National Renewable Energy Lab. (NREL): Golden, CO, USA, 2009.
39. Wen, B.R.; Wei, S.; Wei, K.X.; Yang, W.X.; Peng, Z.K.; Chu, F.L. Power fluctuation and power loss of wind turbines due to wind shear and tower shadow. *Front. Mech. Eng.* **2017**, *12*, 321–332. [CrossRef]

energies

Article

Reliability Updating of Offshore Wind Substructures by Use of Digital Twin Information

Dawid Augustyn [1,2,*], Martin D. Ulriksen [3] and John D. Sørensen [1]

1 Department of the Built Environment, Aalborg University, 9220 Aalborg, Denmark; jdas@build.aau.dk
2 Ramboll Energy, 6700 Esbjerg, Denmark
3 Department of Energy Technology, Aalborg University, 6700 Esbjerg, Denmark; mdu@et.aau.dk
* Correspondence: dja@build.aau.dk

Abstract: This paper presents a probabilistic framework for updating the structural reliability of offshore wind turbine substructures based on digital twin information. In particular, the information obtained from digital twins is used to quantify and update the uncertainties associated with the structural dynamics and load modeling parameters in fatigue damage accumulation. The updated uncertainties are included in a probabilistic model for fatigue damage accumulation used to update the structural reliability. The updated reliability can be used as input to optimize decision models for operation and maintenance of existing structures and design of new structures. The framework is exemplified based on two numerical case studies with a representative offshore wind turbine and information acquired from previously established digital twins. In this context, the effect of updating soil stiffness and wave loading, which constitute two highly uncertain and sensitive parameters, is investigated. It is found that updating the soil stiffness significantly affects the reliability of the joints close to the mudline, while updating the wave loading significantly affects the reliability of the joints localized in the splash zone. The increased uncertainty related to virtual sensing, which is employed to update wave loading, reduces structural reliability.

Keywords: offshore wind substructures; reliability updating; probabilistic fatigue assessment; digital twins; uncertainty quantification

Citation: Augustyn, D.; Ulriksen, M.D.; Sørensen, J.D. Reliability Updating of Offshore Wind Substructures by Use of Digital Twin Information. *Energies* **2021**, *14*, 5859. https://doi.org/10.3390/en14185859

Academic Editor: Eugen Rusu

Received: 5 August 2021
Accepted: 9 September 2021
Published: 16 September 2021

Publisher's Note: MDPI stays neutral with regard to jurisdictional claims in published maps and institutional affiliations.

Copyright: © 2021 by the authors. Licensee MDPI, Basel, Switzerland. This article is an open access article distributed under the terms and conditions of the Creative Commons Attribution (CC BY) license (https://creativecommons.org/licenses/by/4.0/).

1. Introduction

The offshore wind industry has experienced significant growth over the last decade [1]. As a result, the number of offshore wind turbines operating in Europe has reached 5402 in 2020 [2], with much more planned to be installed worldwide in the close future [3]. The typical lifetime of an offshore wind turbine ranges between 20 and 25 years, which means that over the coming years a large number of these structures reach their intended lifetime, and operators will have to take actions regarding their assets. Potential actions, denoted as decision models, can be to decommission, re-power, perform inspections, or extend lifetime. An optimal decision depends on what specific business model the operator pursues, but, regardless of the business aspect, an accurate and precise estimation of the structural reliability is key in making such a decision [4].

A digital twin-defined as a digital replica of a physical asset [5,6]-can help us to assess the structural integrity of existing structures more accurately and precisely compared to predictions from generic design practices because consistent and updated information of the structure is available. This has been successfully demonstrated in the oil and gas industry [7,8], in aerospace engineering [9], and in the offshore wind industry as well [10]. In fact, a number of wind standardization committees, including Det Norske Veritas (DNV) [11,12], International Electrotechnical Commission (IEC) [13], and Federal Maritime and Hydrographic Agency (BSH) [14], are working on design recommendations on how to use measurement data and inspection information to optimize decision models for existing wind turbines. Currently, a key missing aspect is how to use the improved structural models contained in digital twins to subsequently improve the decision models.

Although fully physics-based digital twins have not yet been applied to improve decision models for wind turbines, some publications already indicate how measurement data can be used to achieve such an improvement. Nielsen and Sørensen [15] applied dynamic Bayesian networks to calibrate a Markov deterioration model based on past inspection data of wind turbine blades. Ziegler and Muskulus [16] investigated the feasibility of lifetime extension for offshore wind monopile substructures, with particular focus on identifying important parameters to monitor during the operational phase of the turbines. Leser et al. [17] presented a general framework for fatigue damage estimation based on in situ measurements. Mai et al. [18] focused on prediction of the remaining useful lifetime of wind turbine support structure joints using met-ocean in situ data. Augustyn et al. [19] extended a conceptual framework for updating decision models based on information from a digital twin, initially proposed by Tygesen et al. [7], to be applied to offshore wind substructures. In the framework, a digital twin is established with an updated structural and load model, and subsequently the digital twin is used to quantify uncertainty and update the structural reliability.

In the present paper, we outline the framework by Augustyn et al. [19] beyond its conceptual level and propose a probabilistic method for updating the structural reliability of offshore wind turbine substructures based on new information obtained from digital twins. Depending on the information type available, various methods for updating reliability can be used [20]. If information on the structural integrity becomes available, for example, by an inspection of joints to identify potential cracks, risk-based inspection methods can be applied [21–24]. Even though the inspection planning methodology is matured and well-proven in industrial applications [25], its feasibility for the majority of offshore wind applications is questionable due to the profound inspection costs [26]. A more economically feasible alternative, in the form of condition-based monitoring, is typically investigated for offshore wind applications [27,28]. In this context, condition monitoring data can be applied to identify structural damage, and then the resulting integrity information can be employed for updating reliability [29]. Application studies have been presented for mechanical components in turbine [30] and wind turbine blades [4]. However, in these studies, the condition monitoring data merely provide structural integrity information at a global level-that is, if damage is present or not. In the present study, we aim at enhancing the spatial resolution of the integrity assessment and hereby provide information at a local (joint) level. Consequently, this paper proposes a framework where condition monitoring data are used to update structural models; these updated models are subsequently used to update structural reliability, including uncertainty stemming from the updating procedure.

The contribution of this paper consists of: (1) proposing a method on how the uncertainties related to the structural dynamics and load modeling in fatigue damage accumulation can be quantified and updated based on updated distribution functions of model parameters, which can be acquired with the aid of a digital twin. Subsequently, (2) we present a framework where the updated uncertainty is used to update the structural reliability based on a well-established probabilistic model [31,32]. Generally, the framework can be used for optimization of operation and maintenance of existing turbines and design of new structures. The framework is exemplified based on two numerical case studies, in which digital twins established in previous studies by the authors [33,34] are included.

The remainder of this paper is organized as follows. In Section 2, we outline the concept of structural reliability estimation and convey the motivation for the proposed structural reliability updating framework, which is presented in Section 3. The two following sections address the numerical case studies used to exemplify the framework for existing and new substructures; Section 4 describes the setup of the case studies and Section 5 presents the appertaining results. Finally, this paper closes with concluding remarks in Section 6.

2. Background and Problem Statement

A wind turbine consists of structural components, for which reliability analysis is performed using structural reliability theory [35], and electrical/mechanical components, for which classical reliability models can be used, with the main descriptor being the failure rate or the mean time between failure (MTBF). Regardless of the component type being addressed in the reliability analysis, a probabilistic model describing the component's integrity is required. The reliability of electrical/mechanical components is typically modeled by a Weibull model for the time to failure and the components are assumed to be statistically independent. Using, for example, failure tree analysis (FTA) and failure mode and effect analysis (FMEA), system reliability models can be established and the reliability update can be performed when new information becomes available [36–38]. In the present paper, jacket-type steel wind turbine substructures are considered, so structural reliability techniques are required to model loads, resistances, and model uncertainties and to account for the correlation between the components. The fatigue damage is often design driving for the structural components of offshore wind substructures, such as joints. In this instance, fatigue damage accumulation can be expressed in terms of probability of failure or, equivalently, by the reliability index [39].

Let $g(t)$ be the fatigue limit state at year $t \in \mathbb{N}$ for an offshore wind substructure; then [32,40],

$$g(t) = \Delta - \sum_{i=1}^{l} \sum_{j=1}^{z} \frac{N_{i,j} p_i t}{K \Delta s_{i,j}^{-m}} (X_d X_l X_s)^m, \tag{1}$$

where Δ is the fatigue resistance and the double summation expresses the accumulated fatigue damage. In particular, Δ is a stochastic variable representing the limit value of the accumulated fatigue damage estimated using, for example, SN curves, including the uncertainty related to application of Miner's rule for linear fatigue damage accumulation. In the expression for the fatigue damage, p_i is the yearly probability of occurrence for sea state i (including wind and wave parameters), $N_{i,j}$ is the number of cycles for the ith sea state and jth stress range $\Delta s_{i,j}$, and K and m are the parameters related to the SN curve, with m being the Wöhler exponent [41]. The uncertainties related to the SN curve approach are included by modeling K as a stochastic variable. X_d, X_l, and X_s are stochastic variables that model the uncertainties associated with the structural dynamics, load modeling and stress concentration.

If $g(t) \leq 0$, the limit state is exceeded and the structure fails, while $g(t) > 0$ implies that the structure is safe. The probability of fatigue failure in the time interval $t \in [0, T]$, $P_f(t) = P(g(t) \leq 0)$ can be estimated by first-order and second-order reliability methods [39] or, as is the case in this paper, by Monte Carlo methods [42]. The corresponding reliability index, β, can be computed as $\beta(t) = -\Phi^{-1}\left(P_f(t)\right)$, where Φ is the standard normal distribution function. The annual reliability index, $\Delta\beta$, can be calculated analogically assuming a reference period of one year.

We note that (1) X_d and X_l may be correlated, and, in this instance, they should be modeled by a joint probability density function with correlation coefficient ρ and (2) a linear formulation of the limit state equation can be readily generalized for a bi-linear formulation of the SN curve. The parameters in model (1) are elaborated in Section 2.1.

2.1. Uncertain Parameters and Their Modeling

The uncertainty modeling related to structural reliability due to fatigue damage is summarized in Figure 1. In the framework proposed in Section 3, we focus on updating stochastic variables related to structural dynamics and loading uncertainty, as schematically indicated by the dark blue boxes in Figure 1. The remaining part of the uncertainty (the light blue boxes in Figure 1) can be quantified based on experiments and data. This is not considered in the proposed framework, but a brief discussion is provided in the present subsection for the sake of completeness.

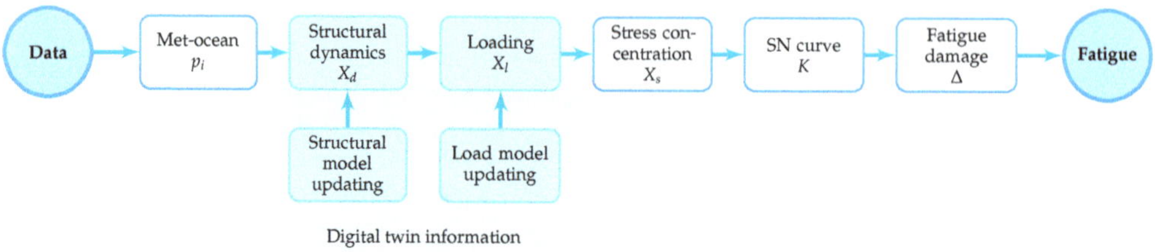

Figure 1. Stochastic variables modeling uncertainty in fatigue damage accumulation. The stochastic variables from the probabilistic model (1) are represented by separate boxes. The light blue boxes indicate stochastic variables estimated based on generic, design-based recommendations. The dark blue boxes indicate stochastic variables that can be quantified and updated based on new information from a digital twin.

2.1.1. Met-Ocean Model

The joint probability distributions of the wind-wave climate is discretized by a finite number of short-term sea state simulations including random wind and wave seeds to model a stochastic process [40]. Met-ocean uncertainty is included in (1) by the yearly probability of each sea state, denoted p_i. The met-ocean uncertainty can be quantified if long-term climate parameters are monitored [18,43].

2.1.2. Structural Dynamics

Estimating dynamic system properties is associated with uncertainties [44]. The uncertainties stem from environmental and operational variability, non-stationary sea states (fluctuating mean sea water level), time-variant structural conditions (corrosion, scour), output noise, and the formulation of the structural model, including modeling of highly uncertain parameters such as soil stiffness, joint stiffness and damping. We note that the output noise relates to the noise in the acceleration and/or strain signals, which is propagated through system identification procedures and results in uncertainty of the updated structural model parameters [44]. The structural dynamics uncertainty is included in (1) through the stochastic variable X_d.

2.1.3. Loading

Depending on the location of the wind turbine, the loading may include the following exogenous sources and their inherent uncertainties:

- Hydrodynamic loading-uncertainty related to calculating wave loads that stems from different wave theories (linear vs. non-linear), Morison's equation, stretching and mass and drag coefficients.
- Aerodynamic loading-uncertainty related to calculating wind loads that stems from wind turbulence, wake model, and shear coefficient.
- Ice loading-uncertainty related to calculating ice loads, for example, ice thickness, ice crushing strength and ice failure regime.
- Earthquake loading-uncertainty related to calculating earthquake loads, for example, earthquake acceleration profile, structural response, soil-structure integration, and force transfer.

If loading uncertainty is quantified based on information from digital twins, the main part of the uncertainty is related to obtaining the structural response due to external loading. This response is typically estimated based on virtual sensing methods, which are associated with uncertainties [34,45]. The loading uncertainty is included in (1) through the stochastic variable X_l.

2.1.4. Stress Concentration

Stress ranges in specific locations can be estimated based on simplified parametric equations, for example, Efthymiou [46] or detailed finite element (FE) models. The stress concentration uncertainty is included in (1) through the stochastic variable X_s. The stress concentration uncertainty can be quantified if a detailed FE model is used to establish hot spot stresses [47] or if hot spot stresses are measured directly.

2.1.5. SN Curve

The uncertainty in parameter estimation from the SN curve approach [48] is included in (1) through the stochastic variable K and the deterministic parameter m. If a bi-linear SN curve is used, then stochastic variables are used to model the two branches of the SN curve. The SN curve uncertainty can be quantified if fatigue testing is performed [48].

2.1.6. Fatigue Damage

Uncertainties related to the accumulated fatigue damage model (Miner's rule [49]) and the crack propagation method (Paris–Erdogan [50] or fracture mechanics) is included in (1) by modeling the resistance, Δ, as a stochastic variable.

2.2. Current State-of-Practice for Reliability Updating

Design standards define a specific level of reliability that offshore wind substructures must fulfill, for example, a target annual reliability index of $\Delta\beta = 3.3$ in IEC 61400-1 [31,40]. Reliability levels indicated in standards assume a generic level of uncertainty representative for all types of substructures and locations. Because the uncertainty is assumed to cover a wide range of structures and locations, the resulting design is, in many cases, conservative. The level of conservatism can be quantified when new information specific to a particular structure becomes available. One way of obtaining such information is by means of digital twins, which can be used to quantify the uncertainty and subsequently update the structural reliability.

3. Structural Reliability Updating Framework

In this paper, we propose a probabilistic framework in which digital twin information is used to update the uncertainties associated with the fatigue damage accumulation, which are then used to update the structural reliability. In particular, we use the updated parameters from the established digital twins to quantify the model uncertainties of the structural dynamics, X_d, and load modeling, X_l. The updated uncertainties are quantified based on a forward propagation method, which allows quantifying separate uncertainty sources stemming from specific model parameters. Having updated the relevant uncertainty contributions from the updated model parameters, the reliability is updated based on the linear probabilistic limit state Equation (1). Finally, the updated reliability serves as a decision basis for a decision model update. A schematic illustration of the framework is seen in Figure 2, and steps one to six are described in Sections 3.1–3.6.

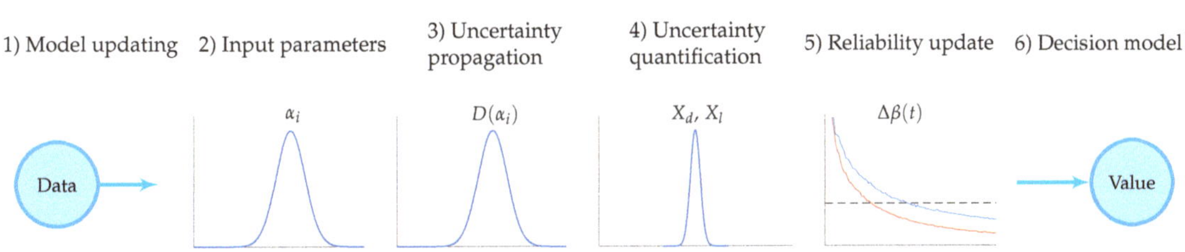

Figure 2. Structural reliability updating framework based on information from a digital twin. Updated parameters from the digital twin are used to quantify uncertainty in fatigue damage accumulation. Subsequently, the structural reliability is updated.

3.1. Model Updating

It is assumed that an updated structural model (step one) is available, which can be obtained based on well-established model formulation and updating procedures [51].

3.2. Input Parameters

The distribution functions of the updated model parameters (available from step one) are used in step two as input for the uncertainty quantification procedure. The stochastic variables reflect both the aleatory and epistemic uncertainties, which constitute the updated X_d and X_l uncertainties.

3.3. Uncertainty Propagation

The effect of the updated model parameters on the fatigue damage accumulation is established by a Monte Carlo uncertainty propagation method [52], as indicated in step three in Figure 2. Based on the uncertainty in the input parameters (i.e., the distribution functions of the updated numerical model parameters), we obtain the distribution of fatigue damage, hence quantifying the uncertainties in fatigue damage due to the updated model parameters. The uncertainty quantification procedure is described next. The aim is to express the uncertainty as a stochastic variable multiplied to the fatigue stress ranges.

The uncertainty in fatigue damage accumulation due to an uncertain parameter, $\alpha_j \in \boldsymbol{\alpha}$, can be quantified by simulating n realizations from this parameter's distribution function and calculating the corresponding fatigue damage. When calculating fatigue damage, the remaining parameters are assumed to be deterministic. Moreover, the fatigue damage is calculated assuming one sea state parameter. In this way, the introduced uncertainty is solely governed by the variability of α_j, hence quantifying this parameter's contribution to the fatigue damage accumulation uncertainty. For example, a distribution function of updated soil stiffness implies structural dynamics uncertainty, while a distribution function of an updated inertia coefficient in Morison's equation implies loading uncertainty.

Among a number of uncertainty quantification methods [53], a Bayesian framework [54] is recommended by a number of standard committees, for example, IEC and Joint Committee on Structural Safety (JCSS), due to its sound theoretical basis and wide range of applicability. However, a main challenge in the Bayesian framework is the requirement of a prior distribution on the parameters to be quantified. In the context of offshore wind uncertainties, information on prior distributions is not available in the background documents for the above mentioned standards and committees. Consequently, in the proposed framework, we implemented a simplified method where we start with the uncertainty modeling consistent with the design standard of wind turbines [40], and subsequently we quantify the uncertain parameters already included in (1) using the maximum likelihood method.

Assuming the fatigue damage, modeled as a stochastic variable depending on the uncertain parameter α_j, is normally distributed, $D(\alpha_j) \sim \mathcal{N}\left(\mu_{D_j}, \sigma^2_{D_j}\right)$, the fatigue damage distribution (mean value μ_{D_j} and standard deviation σ_{D_j}) can be found through the maximum likelihood method, where the likelihood is defined as

$$L\left(\mu_{D_j}, \sigma_{D_j}\right) = \prod_{i=1}^{n} \frac{1}{\sqrt{2\pi}\sigma_{D_j}} \exp\left(-\frac{1}{2}\left(\frac{D_i - \mu_{D_j}}{\sigma_{D_j}}\right)^2\right), \tag{2}$$

with D_i being the fatigue damage associated with the ith realization of α_j computed based on the updated structural model contained in the digital twin.

The log-likelihood function becomes

$$\ln L\left(\mu_{D_j}, \sigma_{D_j}\right) = -n \ln\left(\sqrt{2\pi}\sigma_{D_j}\right) - \sum_{i=1}^{n} \frac{1}{2}\left(\frac{D_i - \mu_{D_j}}{\sigma_{D_j}}\right)^2, \tag{3}$$

and the optimal parameters are found to be

$$\underset{\mu_{D_j}, \sigma_{D_j}}{\operatorname{argmax}} \ln L\left(\mu_{D_j}, \sigma_{D_j}\right). \tag{4}$$

3.4. Uncertainty Quantification

The procedure outlined in the previous subsection quantifies uncertainty in fatigue damage accumulation. However, the probabilistic model (1) requires uncertainty in stress ranges rather than in the fatigue damage. Therefore, it is now described how uncertainty in fatigue damage can be transformed into uncertainty in stress ranges, as indicated in step four in Figure 2.

The fatigue damage accumulation, D, is proportional to the stress ranges, Δs, according to $D \propto \Delta s^m$ (assuming a linear SN curve), from which it follows $\Delta s \propto D^{1/m}$. The stress range distribution parameters can be computed from Monte Carlo simulations. Alternatively, assuming the damage distribution function is normal, the stress range distribution's mean, $\mu_{\Delta s}$, and coefficient of variation (CoV), $c_{\Delta s}$, can be approximated as

$$\mu_{\Delta s} = \mu_i^{1/m} \tag{5}$$

and

$$c_{\Delta s} = \frac{c_i}{m}, \tag{6}$$

where μ_i and c_i are the mean and CoV of the fatigue damage distribution due to the uncertainty associated with α_j.

3.5. Reliability Update

The quantified and updated uncertainties can be consistently included in the probabilistic framework to update the reliability level. The probabilistic model (1) is used to derive an annual reliability level, $\Delta\beta(t)$, given the updated uncertainties. This procedure is indicated in the fifth step in Figure 2, where two reliability curves (with and without using information from a digital twin) are schematically presented. The outcome of the reliability update (increase or decrease) depends on the outcome of uncertainty quantification (increased or decreased).

3.6. Decision Models

Given new information from digital twins becomes available (either during operation or already in the design stage), the decision models can be updated as indicated by the last step in Figure 2. The digital twin information can be included based on Bayesian decision theory [24,55]. For existing structures, an operation and maintenance decision plan can be optimized based on an updated reliability level, for example, an updated inspection plan or lifetime reassessment. More specifically, a reliability-based inspection planning technique can be implemented [56] and some of the inspections can be removed (if any were planned during the lifetime of the structure in question) or new inspections can be included if the structural integrity is compromised. For new structures, the expected outcome of a future digital twin can be used to optimize structures already at the design stage (before the digital twin information becomes available) by the use of Bayesian pre-posterior theory [54].

4. Case Study Setup

To demonstrate an application of the proposed framework, we consider an example where information from a digital twin of an offshore wind jacket substructure is used to update the structural reliability of the substructure. The numerical models of the substructure and the turbine are described in Sections 4.1.1 and 4.1.2, followed by a description of the analyzed load case scenarios in Section 4.1.3. Based on the simulation results (in the form of stress range distributions), the structural reliability of selected joints is calculated in Section 4.2.1 by assuming a generic level of uncertainty. The results are nominal and are, in Section 5, compared with the results obtained by using digital twin information.

4.1. Modeling

We simulate a numerical model of a 7 MW jacket-supported turbine using the procedure outlined by Nielsen et al. [57] and applied by, e.g., Augustyn et al. [34]. The simulation procedure consists of the following steps: (1) the substructure model and corresponding wave loading are reduced to a Craig–Bampton superelement [58] with 30 internal modes accounting for internal substructure dynamics. A convergence study has been performed to ascertain that the reduced model (including 30 modes) adequately captures the relevant modal parameters of the non-reduced system. Subsequently, (2) the wind loading is computed through aero-elastic analyses, in which the substructure superelement is included. Finally, (3) the force-controlled recovery run outlined by Nielsen et al. [59] is performed, where the response of the substructure is recovered and relevant measurements are extracted. The applied model is formulated using state-of-the-art modeling approaches included in a typical design procedure for jacket substructures, and the model has been validated to accurately and precisely represent the structural dynamics of a combined substructure and wind turbine system [60,61].

4.1.1. Substructure

The jacket substructure and its appertaining wave loading were modeled using ROSAP (Ramboll Offshore Structural Analysis Programs), version 53 [62]. The jacket substructure considered in this study, which is depicted in Figure 3, has a total height of approximately 75 m. The substructure comprises three legs, each with a diameter ranging between 1.2 and 1.7 m, and four brace bays, each with a diameter ranging between 0.8 and 1.1 m. The substructure model includes, i.a., soil-pile interaction, local joint flexibility, scour, marine growth and appurtenance masses. The water depth is 55 m and the soil conditions are characterized as clay. The substructure includes 50 m grouted piles. The soil-structure interaction is modeled by the use of soil curves linearized according to the API method [63]. The structural damping was modeled according to a Rayleigh model [64] with 0.5% and 1% modal damping in the first and second bending modes, respectively.

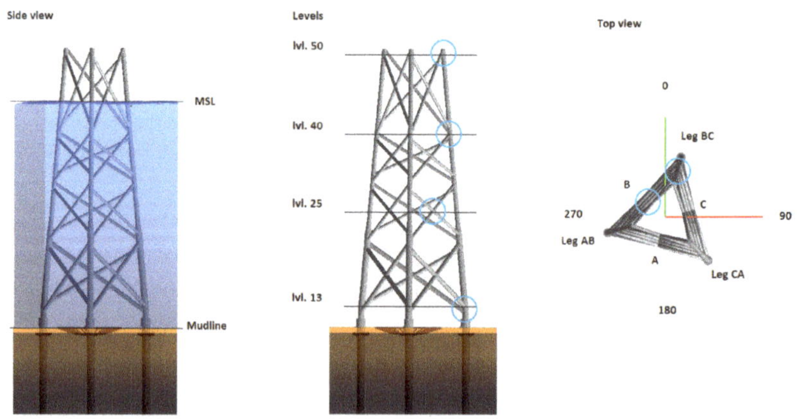

Figure 3. Substructure model used in the case studies. (**A**) Side view, (**B**) side view with indication of levels (blue circles indicate joints analyzed in the case studies) and (**C**) top view with indication of directions, side and leg names. NB: a wind turbine model is not shown in the figure.

The locations of the selected joints considered in the case studies are indicated in Figure 3B. The joint levels range between 13 (mudline) to 50 (top of the jacket). Results for sides B and C of the jacket, see Figure 3C, are provided. The joints are named in the following way: 50CL, where 50 indicates the level, C is the jacket side, and L indicates the lower element in the joint.

4.1.2. Wind Turbine

The substructure carries a representative 7 MW turbine, which is modeled in LACflex aero-elastic code [65]. The turbine includes a 90 m tubular tower with a diameter ranging between 4 and 6 m. Along the tower, three concentrated masses are assumed to emulate the effect of secondary-structures. The aero-elastic code employs a modal-based representation of the turbine (including the tower, rotor and blades). An aerodynamic damping contribution is included through the fluid-structure interaction when calculating aero-elastic forces. The wind turbine model was originally developed for industrial purposes, where it was applied in commercial projects. A rather similar model (albeit a 5 MW turbine instead of 7 MW), which adheres to the same modeling principles, has been applied in other studies on structural dynamics of wind turbines [34,66].

4.1.3. Load Cases

In this study, we consider the fatigue failure mode in the normal operating condition (design load case (DLC) 1.2 [40]). For a typical offshore wind jacket substructure, this DLC accounts for most of the fatigue damage [67].

The met-ocean parameters applied in this study are derived based on measurements from a representative North Sea site [68] and are summarized in Table 1. The wind speed ranges between 4 and 31 ms^{-1}, resulting in $n_b = 15$ wind speed bins. For each wind speed bin, representative wave parameters, i.e., the significant wave height and peak period, are assigned. The significant wave height ranges from 0.1 to 7.9 m while the peak period ranges from 3.0 to 9.6 s. The met-ocean parameters along with their yearly probability of occurrence are derived from a site-specific joint probability distribution function, which is a common design practice [40]. A total of $n_d = 12$ wind directions are analyzed (wind and waves are assumed fully aligned). For each wind speed, a total of $n_{TI} = 5$ turbulence intensity quantiles, namely, $q \in [q_{10}, q_{30}, q_{50}, q_{70}, q_{90}]$, are considered. The quantiles for each wind speed are calculated based on the Weibull distribution according to the IEC standard [40] for turbulence class B. The turbulence intensities for the given site ranges from 0.09 to 0.31. The fatigue damage is scaled with the corresponding turbulence intensity quantile probability, hence representing the target Weibull distribution. Every load case (wind speed, wave height, peak period and turbulence intensity) is simulated with $n_s = 6$ seeds. The total number of load cases analyzed is $n_t = n_b n_d n_{TI} n_s = 5400$.

Table 1. Load case definitions according to IEC [40] and representative site-specific parameters.

Turbine State	DLC	Wind Speed, U (ms^{-1})	Turbulence, TI (-)	Wave Height, H_s (m)	Wave Period, T_p (s)	Direction (deg)
Operational	1.2	4–31	0.31–0.09	0.1–7.9	3.0–9.6	0–330

4.2. Nominal Results

The structural reliability of selected joints of the jacket substructure is evaluated based on model (1) and the variables are summarized in Table 2. The stress ranges, $\Delta\sigma$, and number of cycles, N, were obtained from simulations. The SN curves for tubular joints in air and in seawater with cathodic protection are used according to [48]. The SN curve for the air environment are applied to the joint at level 50. For the remaining joints, the SN curve for seawater with cathodic protection is applied. For tubular joints exposed to seawater with cathodic protection, negative inverse slopes of $m_1 = 3$ and $m_2 = 5$ and intercepts of $\log K_{c_1} = 12.18$ and $\log K_{c_2} = 16.13$ are assumed to calculate the characteristic SN curve. For tubular joints in air environment, the following values can be used: $\log K_{a1} = 12.48$ and $\log K_{a2} = 16.13$, while assuming the same m values as for seawater environment. The mean SN curve for the probabilistic analysis was calculated from the characteristic SN curve's intercepts assuming a standard deviation of 0.20 [48].

Table 2. Variables used in the probabilistic model to estimate fatigue damage accumulation in the nominal case [32].

Variable	Distribution	Mean	CoV	Std. Dev.	Ref.
Δ	N	1.00	0.30	N/A	[69]
$\log K_{c1}$	N	12.58	N/A	0.20	[48]
$\log K_{c2}$	N	16.53	N/A	0.20	[48]
$\log K_{a1}$	N	12.88	N/A	0.20	[48]
$\log K_{a2}$	N	16.53	N/A	0.20	[48]
m_1	D	3	N/A	N/A	[48]
m_2	D	5	N/A	N/A	[48]
X_d	LN	1.00	0.10	N/A	[31]
X_l	LN	1.00	0.10	N/A	[70,71]
X_S	LN	1.00	0.05	N/A	[70]

Distribution: N-normal, LN-logNormal, D-deterministic.

4.2.1. Annual Reliability

The annual reliability index as a function of time, $\Delta\beta(t)$, is calculated based on the state-of-the-art probabilistic methods described in Section 2. The limit state Equation (1) was applied using the standard-based variables provided in Table 2. The reliability indices are presented in Figure 4 and Table 3 and are denoted as the nominal results. The results represent the situation where no additional knowledge from a digital twin is available. The results are provided for 10 selected joints, which are typically critical for a jacket design.

The structure is designed to have a fatigue lifetime of 25 years. The fatigue lifetime ends when the annual reliability index reaches the target value $\Delta\beta = 3.3$, which serves as the basis for reliability-based calibration of safety factors in recognized design codes [31,40]. For the considered case study, the design driving joints are 13BU and 40CU with a lifetime of 25 and 27 years. Joint 13BU is located close to the mudline, while joint 40CU is located slightly below the splash zone. Joints 40CL, 40BL, 25BU and 25BL have a lifetime between 50 and 100 years, while the remaining joints have a lifetime above 100 years.

Table 3. Fatigue lifetime derived based on probabilistic model (1) and stochastic variables presented in Table 2.

Joint	Fatigue Lifetime (Years)
50CL	>100
50BL	>100
40CL	54
40BL	77
40CU	>100
40BU	25
13CU	27
13BU	>100
25BL	98
25BU	86

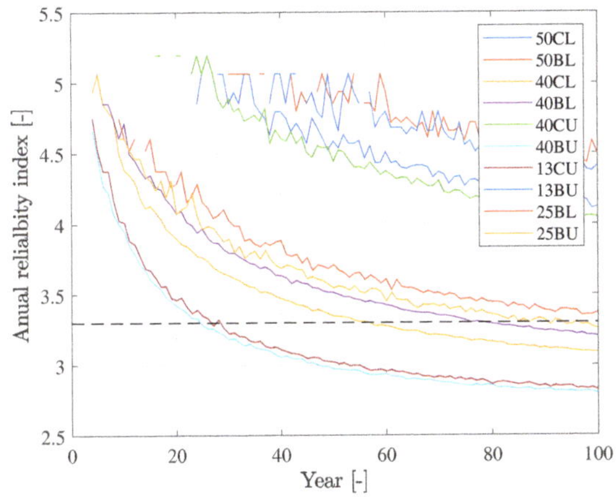

Figure 4. Structural reliability as function of time for the nominal model.

5. Case Study Results

In this section, we exemplify how new information from digital twins can be included in the proposed framework to quantify uncertainty and subsequently update structural reliability for the particular case study. We use information from previously established digital twins [33,34]. The effect of structural dynamics uncertainty, X_d, is investigated based on a model updating study presented in [33], where the soil stiffness, k_s, was calibrated based on in situ measurements. The effect of loading uncertainty, X_l, is investigated based on a virtual sensing study [34], where modal expansion was used to estimate unmeasured field quantities. The results are presented and discussed based on two design driving joints, namely, 13CU and 40BU.

5.1. Updating Structural Dynamics Uncertainty

In this subsection, we present the updated structural reliability based on an updated structural dynamics uncertainty. First, we present a sensitivity study on updating soil stiffness, followed by a case study based on in situ soil stiffness calibration [33].

5.1.1. Soil Stiffness Sensitivity

The effect of updating thesoil stiffness mean value, μ_{k_s}, for joint 13CU is presented in Figure 5 and in Table 4. It is assumed that new information from a digital twin is obtained; in this particular case, the mean value of uncertainty related to structural dynamics, μ_{X_d}, is updated. The results are derived by using the limit state Equation (1) with the standard-based variables provided in Table 2 and updated values for μ_{X_d}.

As seen in Figure 5, the soil stiffness has a significant impact on the fatigue lifetime. Updating the soil stiffness by a factor of 0.5 (resulting in reducing the mudline pile stiffness by half) results in a reduction in lifetime by a factor of 0.3. In contrast, increasing the soil stiffness by a factor of 2.0 results in a lifetime increase by more than fourfold (>100 years). The effect of updating soil stiffness on joint 40BU is negligible, as indicated in Figure 6.

Note that in Figures 5 and 6 (and the other figures describing structural reliability as a function of time), the reliability generally decreases with time, albeit non-monotonically in some cases. For example, consider the green curve in Figure 5, where a local increase in reliability around year 20 is observed. This is due to a limited number of Monte Carlo simulations, but we note that this limitation does not qualitatively affect the conclusions drawn from the analyses.

Table 4. Fatigue lifetime derived for different distributions of X_d.

k_s	13CU			40BU		
	μ_{X_d}	CoV X_d	Lifetime	μ_{X_d}	CoV X_d	Lifetime
0.50	1.20	0.10	7	0.98	0.10	26
0.75	1.10	0.10	15	0.99	0.10	25
1.00	1.00	0.10	25	1.00	0.10	25
1.25	0.90	0.10	85	1.01	0.10	25
1.50	0.80	0.10	>100	1.02	0.10	24
2.00	0.70	0.10	>100	1.04	0.10	22

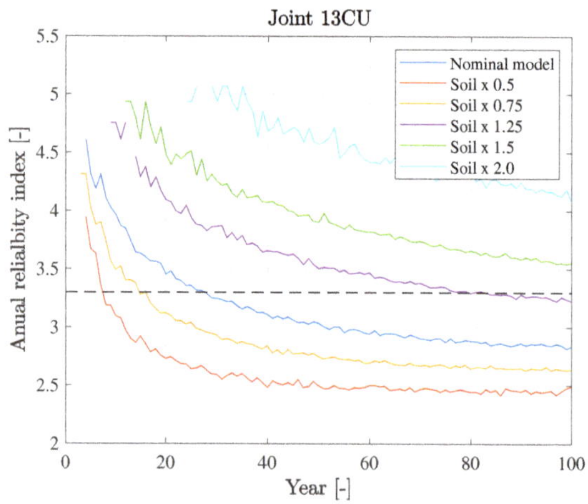

Figure 5. Impact of updating soil stiffness on structural reliability-joint 13CU.

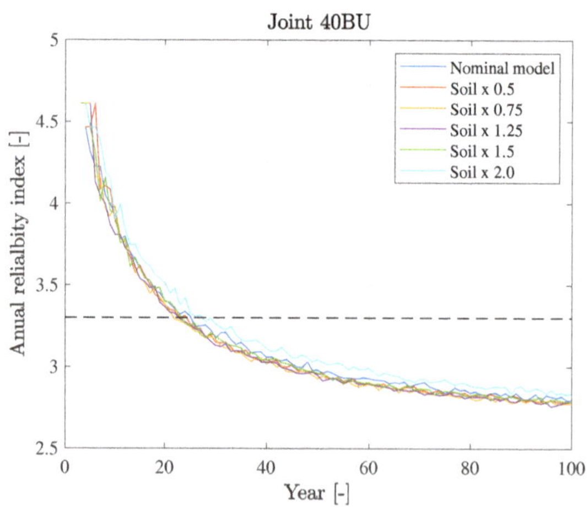

Figure 6. Impact of updating soil stiffness on structural reliability-joint 40BU.

5.1.2. Reliability Update-Soil Stiffness

Based on the results presented in [33], we assume the soil stiffness distribution function after the update can be approximated by a normal distribution with mean value of 4.7 and CoV = 0.12, i.e., $k_s \sim \mathcal{N}(4.7, (4.7 \times 0.12)^2)$. The soil stiffness uncertainty is propagated through the numerical model, and the uncertainty on stress ranges was estimated according to the method presented in Section 3.3. It was assumed, for illustrative purposes, that the soil stiffness is the only uncertain parameter affecting the X_d uncertainty, i.e., $\alpha = k_s$. The X_d uncertainty is quantified and its updated value was applied together with the nominal uncertainty values for the remaining stochastic variables in (1). The updated X_d distribution (mean value and CoV) as a result of the soil updating is presented in Table 5.

The soil stiffness update results in a reduction in the mean value of X_d for all joints except three joints in the splash zone (joints 40CL, 40BL and 40CU). The CoV of X_d is reduced for all joints because the CoV of X_d is reduced from the initial value of 0.10 for all joints. The structural reliability after the soil update is presented in Figure 7 alongside the lifetime compared to the nominal model presented in Table 5. After the soil update, we can observe an increase in fatigue life in four joints close to the mudline (13CU and 13BU) and in the lowest X-joint (25BL and 25BU). Compared to the nominal model, we can conclude that for both critical joints (40BU and 13CU), the fatigue lifetime is increased after the update. Note that the fatigue lifetime in joint 40BL is reduced despite a reduced CoV. That is due to the fact that for this joint, two opposite effects of the soil update are merged; namely, the positive effect of the reduced CoV (0.006 vs. 0.10) and the negative effect of the increased mean value (1.07 vs. 1.00).

The general conclusion holds that if both the mean value and CoV are reduced, then the fatigue lifetime is increased, while if both of the values are increased, then the opposite result holds. If either mean or CoV is reduced while the other is increased, the fatigue lifetime can either increase or decrease depending on the extent of the increase/decrease in mean value and CoV.

Table 5. Effect of updating soil stiffness on fatigue lifetime.

Joint	μ_{X_d}	CoV X_d	Lifetime (Years)	Compared to Table 3
50CL	0.98	0.004	>100	N/A
50BL	0.98	0.005	>100	N/A
40CL	1.04	0.004	62	+8
40BL	1.07	0.006	69	−8
40CU	1.05	0.003	>100	N/A
40BU	0.98	0.005	44	+19
13CU	0.65	0.058	>100	+
13BU	0.56	0.057	>100	N/A
25BL	0.90	0.013	>100	+
25BU	0.92	0.012	>100	+

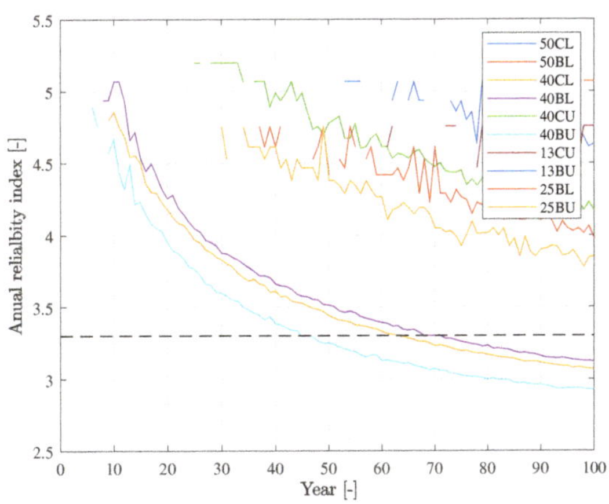

Figure 7. Structural reliability after the soil stiffness update. (k_s update based on the study in [33]).

5.2. Loading Uncertainty Update

In this subsection, we investigate the effect of updating loading uncertainty on the structural reliability. First, we present a sensitivity study on wave loading calibration, followed by updating the reliability based on load calibration using two virtual sensing configurations. The virtual sensing study is presented based on uncertainty quantified in [34]. In this subsection, the X_l uncertainty is updated based on an updated C_m parameter. It is assumed, similarly as in Section 5.1, that only one uncertain parameter affects the uncertainty modeling, i.e., $\boldsymbol{\alpha} = C_m$.

5.2.1. Wave Loading Sensitivity

The effect of updating the wave loading coefficient, C_m, on the structural reliability of joint 13CU is presented in Figure 8 and in Table 6. The mean value of the wave loading coefficient is modified by a factor of 0.8–1.2, which results in modifications of the loading uncertainty. It is assumed that new information from the digital twin is obtained; in this particular case, the mean value of uncertainty related to loading uncertainty, μ_{X_l}, is updated. The results are derived by using the limit state Equation (1) with the standard-based variables provided in Table 2 and updated values for μ_{X_l}.

The wave loading modification has a medium impact on the fatigue lifetime. Updating the wave loading by a factor of 0.8 (reducing the inertia-induced wave loading by 20%) results in an increased lifetime by a factor of 1.6. Increasing the wave loading by a factor of 1.2 results in reducing the lifetime by a factor of 0.7. The effect of updating wave loading on joint 40BU is more pronounced, as indicated in Figure 9. For this joint, reducing the wave loading by 20% results in a lifetime increase by more than fourfold (>100 years), while a wave loading increase by 20% results in a lifetime reduction by a factor of 0.3.

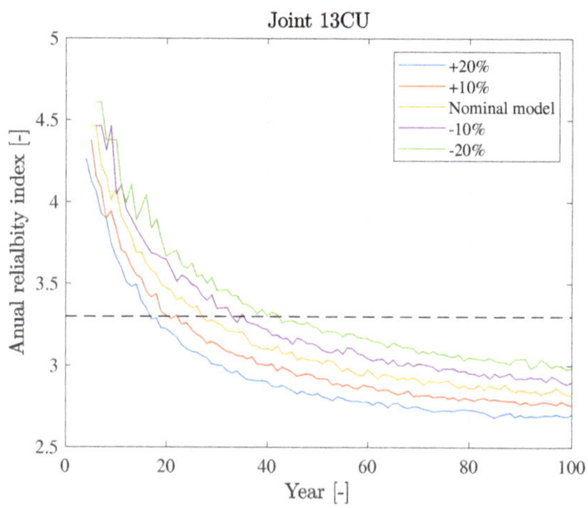

Figure 8. Impact of updating wave loading on structural reliability-joint 13CU.

Table 6. Fatigue lifetime derived for different distributions of X_l.

C_m	13CU			40BU		
	μ_{X_l}	CoV X_l	Lifetime	μ_{X_l}	CoV X_l	Lifetime
1.2	1.06	0.10	17	1.20	0.10	7
1.1	1.03	0.10	21	1.10	0.10	12
1.0	1.00	0.10	25	1.00	0.10	25
0.9	0.97	0.10	33	0.90	0.10	55
0.8	0.94	0.10	42	0.80	0.10	>100

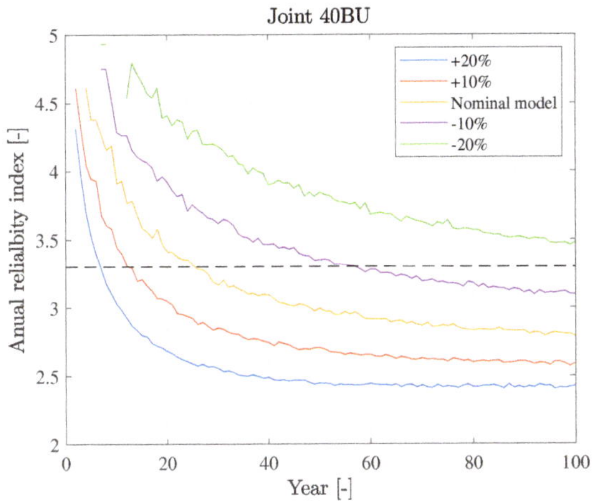

Figure 9. Fatigue lifetime derived for different distributions of X_l-joint 40BU.

5.2.2. Reliability Update-Virtual Sensing Uncertainty

The virtual sensing uncertainty quantified for two virtual sensing configurations are considered based on results presented in [34]. The following virtual sensing uncertainty configurations are used: (1) basic setup: CoV = 0.10 and (2) extended setup: CoV = 0.05, while the mean value for both setups is assumed to be 1.00. The basic setup includes only acceleration sensors above the water level, while the extended one, in addition, includes sub-sea acceleration sensors and a wave radar sensor. It is assumed that the virtual sensing uncertainty are combined with the nominal X_l uncertainty. Furthermore, it is assumed, for illustrative purposes, that the mean value of X_l equals 0.9. The X_l distribution parameters used in this study are summarized in Table 7 for joints 40CU and 13BU.

The results for joint 40BU are presented in Figure 10. As each model update configuration results in the same mean value update so the only difference in the stochastic model is the CoV, the higher the CoV, the shorter lifetime we should derive. This is confirmed in the results as the direct sensing method (measuring directly), with CoV = 0.00 resulting in a lifetime of 60 years, followed by the extended virtual sensing method (lifetime of 50 years and CoV = 0.05), while the most uncertain method (basic virtual sensing with CoV = 0.10) results in a fatigue lifetime of 40 years. In this case, each configuration derives a fatigue lifetime larger than the nominal one, i.e., 25 years. However, this is not the case for joint 13CU, where the fatigue lifetime using the basic virtual sensing configuration is 22 years, as depicted in Figure 11. Even though the mean value of the update results in reduced fatigue damage ($\mu_{X_l} = 0.97$ for this case), the negative effect of increased uncertainty (CoV $X_l = 0.14$) results in a fatigue lifetime reduction of 3 years.

Table 7. Fatigue lifetime updated based on various uncertain wave loading calibration methods. X_l distribution is updated (mean and CoV).

Configuration	13CU			40BU		
	μ_{X_l}	CoV X_l	Lifetime	μ_{X_l}	CoV X_l	Lifetime
Nominal	1.00	0.10	25	1.00	0.10	25
Basic	0.97	0.14	22	0.90	0.14	40
Extended	0.97	0.11	30	0.90	0.11	50
Direct sensing	0.97	0.10	35	0.90	0.10	60

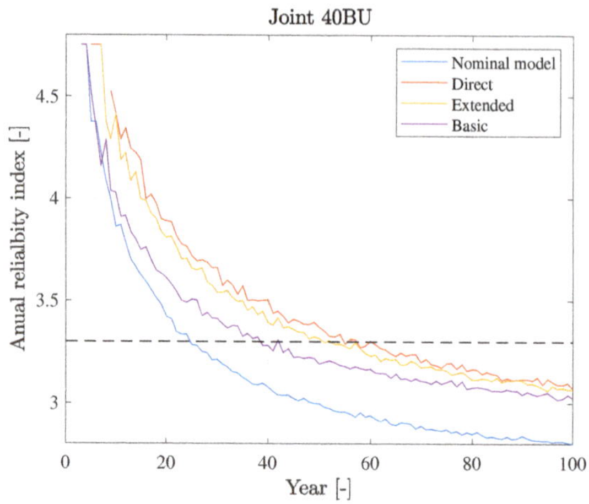

Figure 10. Impact of updating wave loading based on uncertain virtual sensing methods-40BU.

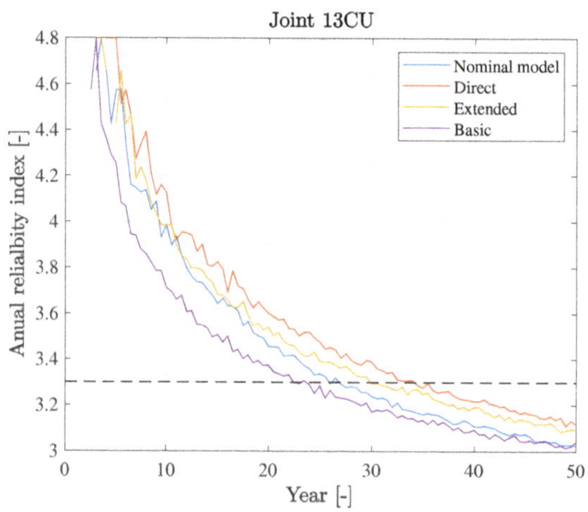

Figure 11. Impact of updating wave loading based on uncertain virtual. sensing methods-13CU.

5.3. Uncertainty Correlation

In the previous subsections, the X_d and X_l uncertainties were investigated separately, hence neglecting a potential correlation. In this subsection, we consider updating both X_d and X_l with varying correlation coefficients. The correlation can stem from interaction between the structural dynamics and loading parameters. For example, the loading parameters can be calibrated based on responses from a previously updated structural model.

We assume the structural and loading uncertainties are quantified based on new information from a digital twin, resulting in updated mean values of structural and load uncertainties: $\mu_{X_d} = 0.80$ and $\mu_{X_l} = 0.97$ and using the reference uncertainty level CoV = 0.10. The updated uncertainty value corresponds to increasing the soil stiffness by 50%, $k_s = 1.5$, and reducing the wave loading coefficient by 10%, $C_m = 0.9$. The results are presented for joint 13CU.

Three scenarios of correlation between X_d and X_l are investigated: (1) $\rho = 0$ (no correlation), which can be the case if the load calibration was performed without using information from the updated structural model, (2) $\rho = 1$ (full correlation), when, for example, load calibration using mode shapes from an updated structural model and (3) an intermediate case with $\rho = 0.5$, where both analytical and measured mode shapes were used for load calibration.

The structural reliability calculated for various scenarios is presented in Figure 12. The nominal setup yields a fatigue lifetime of 25 years, while the updated uncertainty results in a fatigue lifetime ranging between 23 and 48 years, where the difference stems solely from varying correlations. The largest fatigue lifetime is obtained when assuming no correlation, while the lowest lifetime is derived for full correlation. Note that despite reducing the mean values of X_d and X_l, the fatigue lifetime is reduced compared to the nominal result for the full correlation case. The results are in line with expectations, because positive correlation increases the combined $X_d X_l$ uncertainty.

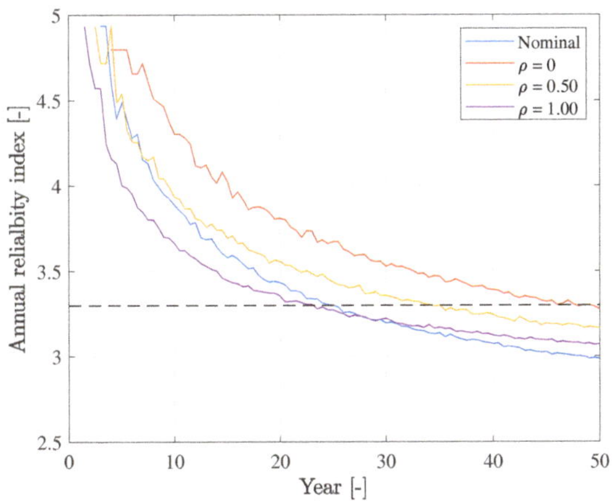

Figure 12. Impact of X_d and X_l correlation on structural reliability.

5.4. Application for New Structures

Assuming a number of digital twins for similar structures have been established in the past, we can, by applying the proposed framework, obtain a distribution function of $X_d X_l$, which indicates what is the expected outcome of updating the structural and load model. This knowledge can be used at the design stage, resulting in an optimized design given the expected model update is realized. However, the updated information may be at a preliminary stage of validation and therefore subject to some degree of uncertainty, i.e., the expected model update outcome only represents our (best) knowledge. Hence, we must confirm our expectation by performing model updates during the structural lifetime and consider all potential outcomes of the experiment (model update) in the design stage. This is accounted for by preparing a decision rule, which for any outcome introduces an action that guarantees that the wind turbine has a sufficient reliability level until the intended lifetime is reached. The proposed application is based on Bayesian pre-posterior decision theory [54] and has, in the offshore wind industry, been applied in, for example, optimization of operation and maintenance of wind turbines [55].

In the following, an illustrative example is presented for this application to new structures. Assume that, based on previous digital twins, we obtain a prior distribution function of quantified uncertainties, $X_f = X_d X_l$. This prior distribution can be regarded as the future (yet to be realized) distribution of the updated uncertainties and can be used already at the design stage.

For the sake of illustration, we assume that the future outcome of model updates can be modeled as $X_f \sim \mathcal{N}(0.9, (0.9 \times 0.05)^2)$, as depicted in Figure 13. The prior distribution is used together with model (1) to design the optimized structure. This is obtained by assuming that the generic structural dynamics and loading uncertainty are substituted with the expected uncertainty quantified based on the future experiment, $X_d X_l = X_f$. The decision models are derived based on (1), where, depending on the outcome of the model update, different values of X_f are assumed. The X_f values are summarized in Table 8. As a result, we derive an optimized structure, which has sufficient reliability until the intended lifetime is reached. This is indicated in Figure 14 by the blue curve. In the design, we have used the prior distribution of the updated uncertainty and assumed that updating the model is performed during the operation of the structure to confirm our expectation (obtain the posterior distribution). The point in time when updating the model must be performed can be derived by applying model (1) with the nominal uncertainty from Table 2, as shown in Figure 13 with the orange curve. Finally, we derive a point when the structure reaches the target reliability level and some action is needed to confirm its structural reliability. This is indicated by the orange curve in Figure 14.

When the model update time is reached, updating of the model is performed. As a result of model update, we can obtain one of the three outcomes for X_f, which will have an impact on the decision models, as depicted in Figure 15. In particular, we have the following potential outcomes:

- Most likely: the mean value of the derived model update is close to the mean value of the prior distribution assumed in the design stage, $\mu_{X_d X_l} = \mu_{X_f}$. In such a case, the structure is fit for operation for the intended lifetime and no further action is required. This scenario is indicated by the green line in Figure 15.
- Unlikely positive: the mean value is less than the value assumed in the design stage, $\mu_{X_d X_l} < \mu_{X_f}$. This results in a longer lifetime than expected and no further action is required. This scenario is indicated by the yellow line in Figure 15.
- Unlikely negative: the mean value is greater than the value assumed in the design stage, $\mu_{X_d X_l} > \mu_{X_f}$. This results in a shorter lifetime than expected and action is required to ensure a sufficient reliability during the intended lifetime of the structure. This scenario is indicated by the dashed red line in Figure 15.

Given the expected or positive outcome of updating the model is realized, no further action is required. However, if the outcome of updating the model is unexpectedly negative, the following mitigation actions can be considered to ensure the required level of reliability during the intended lifetime: (1) strengthening or (2) curtailing of the wind turbine (thereby reducing fatigue damage) and operating until the end of the intended lifetime. If it is economically infeasible to continue the operation of a particular turbine given the model updating outcome, one can consider premature decommissioning. The reliability level after the mitigation action is performed as indicated by the solid red line in Figure 15.

Table 8. Pre-posterior stochastic model.

Case	μ_{X_f}	CoV X_f	Comment	Information
Pre-posterior design	0.9	0.05	Prior knowledge on X_f	Generic design
Determine model update time	1.0	0.14	Using no extra information from digital twin	Generic design
Model updating (expected outcome)	0.9	0.05	The same as prior knowledge, lifetime as expected, no action	Digital twin
Model updating (positive outcome)	0.85	0.05	Positive outcome, longer lifetime than expected, potential for lifetime extension	Digital twin
Model updating (negative outcome)	0.95	0.05	Negative outcome, shorter lifetime than expected	Digital twin
Model updating (negative outcome + mitigation)	0.9	0.05	Mitigation (extra cost) required, after mitigation expected (or longer) lifetime achieved	Digital twin + mitigation

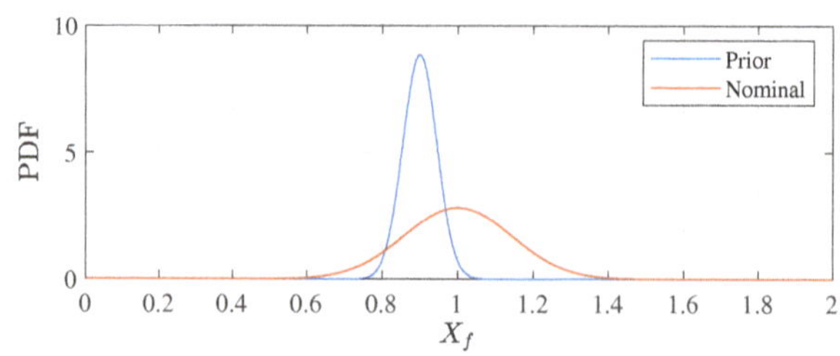

Figure 13. Stochastic model for pre-posterior design.

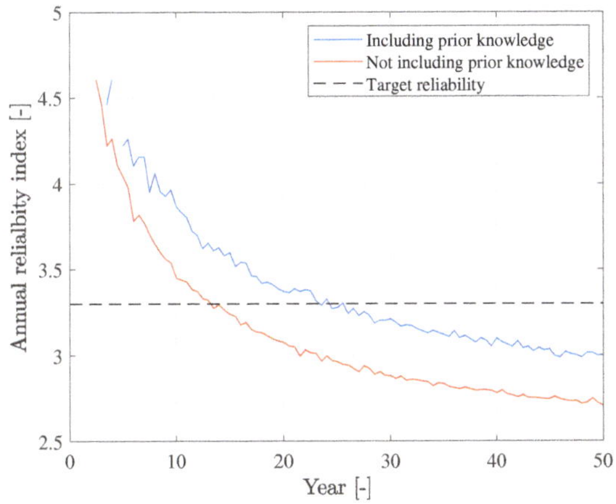

Figure 14. Benefit of including pre-posterior design (including prior knowledge on X_f).

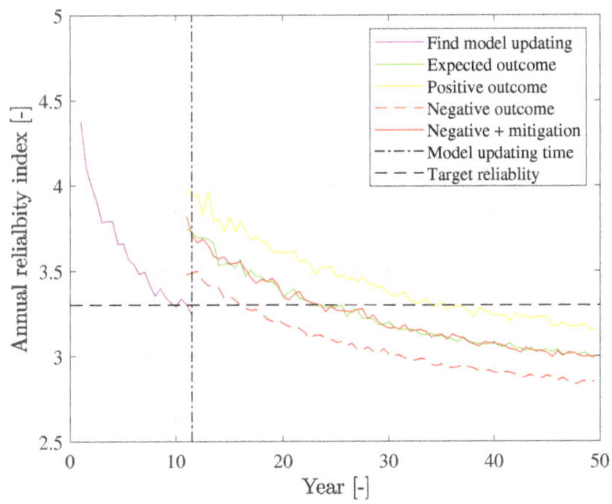

Figure 15. Pre-posterior design at inspection time.

6. Conclusions

In this paper, we propose a probabilistic framework for updating structural reliability of offshore wind substructures based on new information from digital twins. The digital twin information is consistently included in the framework by updating the uncertainty related to structural dynamics and load modeling and propagating this uncertainty to the fatigue damage accumulation. The resulting uncertainty is then converted into uncertainty of the stress ranges, which is included in a probabilistic model on structural reliability. The proposed framework is applicable to offshore wind substructures whose lifetimes are governed by fatigue damage accumulation.

The framework is applied to two case studies, where the potential for improved decision models for existing and new structures is demonstrated. In the former case, updating soil stiffness and wave loading is considered to investigate the potential for lifetime extension of fatigue critical joints. In the latter case, the framework is applied to optimize new structures by using Bayesian pre-posterior theory for future wave load calibration.

Author Contributions: Conceptualization, D.A., M.D.U. and J.D.S.; methodology, D.A., M.D.U. and J.D.S.; formal analysis, D.A., M.D.U. and J.D.S.; writing—original draft preparation, D.A., M.D.U. and J.D.S.; writing—review and editing, D.A., M.D.U. and J.D.S. All authors have read and agreed to the published version of the manuscript.

Funding: Innovationsfonden: 8053-00146B; Rambøll Fonden: A0.

Institutional Review Board Statement: Not applicable.

Informed Consent Statement: Not applicable.

Data Availability Statement: The study does not report any data.

Acknowledgments: The work presented herein is financially supported by Ramboll Foundation, Ramboll Energy and Innovation Fund Denmark. The financial support is highly appreciated. The first author would like to thank colleagues from Ramboll for their continuous support.

Conflicts of Interest: The authors declare no conflict of interest.

References

1. U.S. Department of Energy. *2018 Offshore Wind Technologies Market Report;* Technical Report; U.S. Department of Energy: Washington, DC, USA, 2018.
2. Wind Europe. *Offshore Wind in Europe. Key Statistics 2020*; Technical Report; Wind Europe: Brussels, Belgium, 2020.
3. Global Wind Energy Council. *Global Offshore Wind Report 2020*; Technical Report; Global Wind Energy Council: Brussels, Belgium, 2020.
4. Nielsen, J.S.; Tcherniak, D.; Ulriksen, M.D. A case study on risk-based maintenance of wind turbine blades with structural health monitoring. *Struct. Infrastruct. Eng.* **2021**, *17*, 302–318. [CrossRef]
5. Grieves, M. *Digital Twin: Manufacturing Excellence through Virtual Factory Replication*; White Paper; Michael W. Grieves, LLC: Winooski, VT, USA, 2015.
6. Grieves, M.; Vickers, J. Digital Twin: Mitigating Unpredictable, Undesirable Emergent Behavior in Complex Systems. In *Transdisciplinary Perspectives on Complex Systems: New Findings and Approaches*; Springer International Publishing: Cham, Switzerland, 2017; pp. 85–113. [CrossRef]
7. Tygesen, U.; Jepsen, M.; Vestermark, J.; Dollerup, N.; Pedersen, A. The True Digital Twin Concept for Fatigue Re-Assessment of Marine Structures. In Proceedings of the ASME 2018 37th International Conference on Ocean, Offshore and Arctic Engineering OMAE2018, Madrid, Spain, 17–22 June 2018. [CrossRef]
8. Pedersen, E.B.; Jørgensen, D.; Riber, H.J.; Ballani, J.; Vallaghé, S.; Paccaud, B. True Fatigue Life Calculation Using Digital Twin Concept and Operational Modal Analysis. In Proceedings of the Twenty-Ninth (2019) International Offshore and Polar Engineering Conference, Honolulu, HI, USA, 16–21 June 2019; Volume 1.
9. Kritzinger, W.; Karner, M.; Traar, G.; Henjes, J.; Sihn, W. Digital Twin in manufacturing: A categorical literature review and classification. *IFAC-PapersOnLine* **2018**, *51*, 1016–1022.
10. Sivalingam, K.; Sepulveda, M.; Spring, M.; Davies, P. A Review and Methodology Development for Remaining Useful Life Prediction of Offshore Fixed and Floating Wind turbine Power Converter with Digital Twin Technology Perspective. In Proceedings of the 2018 2nd International Conference on Green Energy and Applications (ICGEA), Singapore, 24–26 March 2018; pp. 197–204. [CrossRef]
11. DNVGL-ST-0262. *Lifetime Extension of Wind Turbines;* Technical Report; Det Norske Veritas Germanischer Lloyd: Høvik, Greater Oslo, Norway, 2016.
12. DNVGL-ST-0263. *Certification of Lifetime Extension of Wind Turbines;* Technical Report; Det Norske Veritas Germanischer Lloyd: Høvik, Greater Oslo, Norway, 2016.
13. IEC 61400-28 ED1. *Wind Energy Generation Systems-Part 28: Through Life Management and Life Extension of Wind Power Assets (Committee Draft);* Technical Report; International Electrotechnical Commission: Geneva, Switzerland, 2021.
14. BSH-7005. *Standard Design. Minimum Requirements Concerning the Constructive Design of Offshore Structures within the Exclusive Economic Zone (EEZ);* Technical Report; Bundesamt für Seeschifffahrt und Hydrographie: Hamburg, Germany, 2015.
15. Nielsen, J.S.; Sørensen, J.D. Bayesian Estimation of Remaining Useful Life for Wind Turbine Blades. *Energies* **2017**, *10*, 664. [CrossRef]
16. Ziegler, L.; Muskulus, M. Fatigue reassessment for lifetime extension of offshore wind monopile substructures. *J. Phys. Conf. Ser.* **2016**, *753*, 092010. [CrossRef]
17. Leser, P.E.; Warner, J.E.; Leser, W.P.; Bomarito, G.F.; Newman, J.A.; Hochhalter, J.D. A digital twin feasibility study (Part II): Non-deterministic predictions of fatigue life using in-situ diagnostics and prognostics. *Eng. Fract. Mech.* **2020**, *229*, 106903. [CrossRef]
18. Mai, Q.A.; Weijtjens, W.; Devriendt, C.; Morato, P.G.; Rigo, P.; Sørensen, J.D. Prediction of remaining fatigue life of welded joints in wind turbine support structures considering strain measurement and a joint distribution of oceanographic data. *Mar. Struct.* **2019**, *66*, 307–322. [CrossRef]

19. Augustyn, D.; Tygesen, U.T.; Ulriksen, M.D.; Sørensen, J.D. Data-Driven Design and Operation of Offshore Wind Structures. In Proceedings of the Twenty-Ninth (2019) International Offshore and Polar Engineering Conference, Honolulu, HI, USA, 16–21 June 2019; Volume 1, pp. 491–498.
20. Straub, D. Reliability updating with equality information. *Probabilistic Eng. Mech.* **2011**, *26*, 254–258. [CrossRef]
21. Zhang, R.; Mahadevan, S. Model uncertainty and Bayesian updating in reliability-based inspection. *Struct. Saf.* **2000**, *22*, 145–160. [CrossRef]
22. Schneider, R.; Thöns, S.; Straub, D. Reliability analysis and updating of deteriorating systems with subset simulation. *Struct. Saf.* **2017**, *64*, 20–36. [CrossRef]
23. Straub, D.; Faber, M.H. Risk based inspection planning for structural systems. *Struct. Saf.* **2005**, *27*, 335–355. [CrossRef]
24. Sørensen, J.D. Framework for risk-based planning of operation and maintenance for offshore wind turbines. *Wind Energy* **2009**, *12*, 493–506. [CrossRef]
25. Singh, M.; Markeset, T. A methodology for risk-based inspection planning of oil and gas pipes based on fuzzy logic framework. *Eng. Fail. Anal.* **2009**, *16*, 2098–2113. [CrossRef]
26. Hevia-Koch, P.; Klinge Jacobsen, H. Comparing offshore and onshore wind development considering acceptance costs. *Energy Policy* **2019**, *125*, 9–19. [CrossRef]
27. Netland, Ø.; Sperstad, I.B.; Hofmann, M.; Skavhaug, A. Cost-benefit Evaluation of Remote Inspection of Offshore Wind Farms by Simulating the Operation and Maintenance Phase. *Energy Procedia* **2014**, *53*, 239–247. EERA DeepWind' 2014, 11th Deep Sea Offshore Wind R&D Conference. [CrossRef]
28. Ulriksen, M.D. Damage Localization for Structural Health Monitoring: An Exploration of Three New Vibration-Based Schemes. Ph.D. Dissertation, Aalborg University, Aalborg, Denmark, 2018.
29. Simoen, E.; De Roeck, G.; Lombaert, G. Dealing with uncertainty in model updating for damage assessment: A review. *Mech. Syst. Signal Process.* **2015**, *56–57*, 123–149. [CrossRef]
30. Nielsen, J.S.; Sørensen, J.D. Computational framework for risk-based planning of inspections, maintenance and condition monitoring using discrete Bayesian networks. *Struct. Infrastruct. Eng.* **2018**, *14*, 1082–1094. [CrossRef]
31. Sørensen, J.D.; Toft, H.S. *Safety Factors-IEC 61400-1 ed. 4-Background Document*; Technical Report, DTU Wind Energy-E-Report-0066(EN); Technical University of Denmark (DTU): Lyngby, Denmark, 2014.
32. Velarde, J.; Kramhøft, C.; Sørensen, J.D.; Zorzi, G. Fatigue reliability of large monopiles for offshore wind turbines. *Int. J. Fatigue* **2020**, *134*, 105487. [CrossRef]
33. Augustyn, D.; Smolka, U.; Tygesen, U.T.; Ulriksen, M.D.; Sørensen, J.D. Data-Driven Model Updating of an Offshore Wind Jacket Substructure. *Appl. Ocean. Res.* **2020**, *104*, 102366. [CrossRef]
34. Augustyn, D.; Pedersen, R.R.; Tygesen, U.T.; Ulriksen, M.D.; Sørensen, J.D. Feasibility of modal expansion for virtual sensing in offshore wind jacket substructures. *Mar. Struct.* **2021**, *79*, 103019. [CrossRef]
35. Madsen, H.; Krenk, S.; Lind, N. *Methods of Structural Safety*, 2nd ed.; Dover Publications: New York, NY, USA, 2006.
36. Tavner, P. *Offshore Wind Turbines: Reliability, Availability and Maintenance*; Institution of Engineering and Technology: Stevenage, UK, 2012.
37. Arabian-Hoseynabadi, H.; Oraee, H.; Tavner, P. Failure Modes and Effects Analysis (FMEA) for wind turbines. *Int. J. Electr. Power Energy Syst.* **2010**, *32*, 817–824. [CrossRef]
38. Nguyen, T.A.; Min, D.; Choi, E.; Lee, J.W. Dependability and Security Quantification of an Internet of Medical Things Infrastructure based on Cloud-Fog-Edge Continuum for Healthcare Monitoring using Hierarchical Models. *IEEE Internet Things J.* **2021**. [CrossRef]
39. Madsen, H.O.; Krenk, S.; Lind, N.C. *Methods of Structural Safety*; Courier Corporation: Washington, DC, USA, 2006.
40. IEC-61400-1:2019. *Wind Energy Generation Systems-Part 1: Design Requirements*; Technical Report; International Electrotechnical Commission: Geneva, Switzerland, 2019.
41. Szala, G.; Ligaj, B. Effect of the Exponent in the Description of Wöhler Fatigue Diagram on the Results of Calculations of Fatigue Life. *Key Eng. Mat.* **2014**, *598*, 231–236. [CrossRef]
42. Naess, A.; Leira, B.; Batsevych, O. System reliability analysis by enhanced Monte Carlo simulation. *Struct. Saf.* **2009**, *31*, 349–355. [CrossRef]
43. Hübler, C.; Gebhardt, C.G.; Rolfes, R. Methodologies for fatigue assessment of offshore wind turbines considering scattering environmental conditions and the uncertainty due to finite sampling. *Wind Energy* **2018**, *21*, 1092–1105. [CrossRef]
44. Reynders, E.; Pintelon, R.; Roeck, G. Uncertainty bounds on modal parameters obtained from stochastic subspace identification. *Mech. Syst. Signal Process.* **2008**, *22*, 948–969. [CrossRef]
45. Iliopoulos, A.; Weijtjens, W.; Van Hemelrijck, D.; Devriendt, C. Fatigue assessment of offshore wind turbines on monopile foundations using multi-band modal expansion. *Wind Energy* **2017**, *20*, 1463–1479. [CrossRef]
46. Efthymiou, M.; Durkin, S. Stress concentrations in T/Y and gap/overlap K-joints. In Proceedings of the 4th International Conference on Behaviour of Offshore Structures, Delft, The Netherlands, 1–5 July 1985; p. 12.
47. Lee, J.M.; Seo, J.K.; Kim, M.H.; Shin, S.B.; Han, M.S.; Park, J.S.; Mahendran, M. Comparison of hot spot stress evaluation methods for welded structures. *Int. J. Nav. Archit. Ocean Eng.* **2010**, *2*, 200–210. [CrossRef]
48. DNVGL-RP-C203. *Fatigue Design of Offshore Steel Structures*; Technical Report; Det Norske Veritas Germanischer Lloyd: Høvik, Greater Oslo, Norway, 2016.

49. Miner, M.A. Cumulative Damage in Fatigue. *J. Appl. Mech.* **2021**, *12*, A159–A164. [CrossRef]
50. Paris, P.; Erdogan, F. A Critical Analysis of Crack Propagation Laws. *J. Basic Eng.* **1963**, *85*, 528–533. [CrossRef]
51. Friswell, M.I.; Mottershead, J.E. *Finite Element Model Updating in Structural Dynamics*; Kluwer Ademic Publishers: Amsterdam, The Netherlands, 1995.
52. Anderson, G. Error propagation by the Monte Carlo method in geochemical calculations. *Geochim. Cosmochim. Acta* **1976**, *40*, 1533–1538. [CrossRef]
53. ISO 13587:2012. *Three Statistical Approaches for the Assessment and Interpretation of Measurement Uncertainty*; Technical Report; International Electrotechnical Commission: Geneva, Switzerland, 2012.
54. Raiffa, H.; Schlaifer, R. *Applied Statistical Decision Theory*; Wiley & Sons: Hoboken, NJ, USA, 2000. [CrossRef]
55. Nielsen, J.J.; Sørensen, J.D. On risk-based operation and maintenance of offshore wind turbine components. *Reliab. Eng. Syst. Saf.* **2011**, *96*, 218–229. [CrossRef]
56. Onoufriou, T. Reliability based inspection planning of offshore structures. *Mar. Struct.* **1999**, *12*, 521–539. [CrossRef]
57. Nielsen, M.B.; Jensen, J.F.; Harper, C.; Knudsen, L.S.; Pedersen, R.R. State-of-the-art framework for design of offshore wind jacket foundations. *Steel Constr.* **2019**, *12*, 209–214. [CrossRef]
58. Craig, R.R.; Bampton, M.C.C. Coupling of substructures for dynamic analyses. *AIAA J.* **1968**, *6*, 1313–1319. [CrossRef]
59. Nielsen, M.B.; Jensen, J.F.; Augustyn, D.; Pedersen, R.R. Efficient response recovery procedures for detailed design of jacket foundations. In *Insights and Innovations in Structural Engineering, Mechanics and Computation*, 1st ed.; Taylor & Francis: London, UK, 2016; Volume 1, pp. 2060–2065. [CrossRef]
60. van der Valk, P.; Voormeeren, S. An overview of modeling approaches for complex offshore wind turbine support structures. In Proceedings of the ISMA2012-USD2012, Leuven, Belgium, 17–19 September 2012; pp. 1–4.
61. Passon, P. Offshore Wind Turbine Foundation Design. Ph.D. Thesis, DTU Wind Energy: Roskilde, Denmark, 2015.
62. Ramboll. *ROSAP-Ramboll Offshore Structural Analysis Package, Version 53*; Technical Report; Ramboll: Esbjerg, Denmark, 2018.
63. API-RP-2A. *Recommended Practice for Planning, Designing and Constructing Fixed Offshore Platforms*; Technical Report; American Petroleum Institute: Washington, DC, USA, 2014.
64. Liu, M.; Gorman, D. Formulation of Rayleigh damping and its extensions. *Comput. Struct.* **1995**, *57*, 277–285. [CrossRef]
65. Ramboll. *LACflex-Aeroelastic Simulation Tool*; Technical Report; Ramboll: Aarhus, Denmark, 2018.
66. Augustyn, D.; Cosack, N.; Ulriksen, M. On the influence of environmental and operational variability on modal parameters of offshore wind support structures. *Mar. Struct.* **2021**. submitted.
67. Kelma, S.; Schaumann, P. Probabilistic Fatigue Analysis of Jacket Support Structures for Offshore Wind Turbines Exemplified on Tubular Joints. *Energy Procedia* **2015**, *80*, 151–158. [CrossRef]
68. Anonymous. *Met-Ocean Report*; Technical Report; Confidential Publisher: New York, NY, USA, 2016.
69. Rasmus Folsø, S.O.; Parmentier, G. Reliability-based calibration of fatigue design guidelines for ship structures. *Mar. Struct.* **2002**, *15*, 627–651. [CrossRef]
70. Sørensen, J. Reliability-based calibration of fatigue safety factors for offshore wind turbines. *Int. J. Offshore Pol. Eng.* **2012**, *22*, 234–241.
71. JCSS. *Probabilistic Model Code*; Technical Report; Joint Committee on Structural Safety: Copenhagen, Denmark, 2001.

energies

MDPI

Article

A Study on Directly Interconnected Offshore Wind Systems during Wind Gust Conditions

Cathal W. O'Donnell [1,*], Mahdi Ebrahimi Salari [2] and Daniel J. Toal [1]

[1] Centre for Robotics & Intelligent Systems, University of Limerick, V94 T9PX Limerick, Ireland; Daniel.Toal@ul.ie
[2] MaREI Centre, University College Cork, P43 C573 Cork, Ireland; MEbrahimiSalari@ucc.ie
[*] Correspondence: Cathal.W.ODonnell@ul.ie

Abstract: An investigation of the effects of wind gusts on the directly interconnected wind generators is reported, and techniques toward the mitigation of the wind gust negative influences have been proposed. Using a directly interconnected system approach, wind turbine generators are connected to a single synchronous bus or collection grid without the use of power converters on each turbine. This bus can then be transformed for transmission onshore using High Voltage Alternating Current, Low-Frequency Alternating Current or High Voltage Direct Current techniques with shared power conversion resources onshore connecting the farm to the grid. Analysis of the potential for instability in transient conditions on the wind farm, for example, caused by wind gusts is the subject of this paper. Gust magnitude and rise time/fall time are investigated. Using pitch control and the natural damping of the high inertial offshore system, satisfactory overall system performance and stability can be achieved during these periods of transience.

Keywords: direct interconnection; wind gust; offshore wind; power generation

Citation: O'Donnell, C.W.; Ebrahimi Salari, M.; Toal, D.J. A Study on Directly Interconnected Offshore Wind Systems during Wind Gust Conditions. *Energies* **2022**, *15*, 168. https://doi.org/10.3390/en15010168

Academic Editors: Eugen Rusu, Kostas Belibassakis and George Lavidas

Received: 16 November 2021
Accepted: 24 December 2021
Published: 27 December 2021

Publisher's Note: MDPI stays neutral with regard to jurisdictional claims in published maps and institutional affiliations.

Copyright: © 2021 by the authors. Licensee MDPI, Basel, Switzerland. This article is an open access article distributed under the terms and conditions of the Creative Commons Attribution (CC BY) license (https://creativecommons.org/licenses/by/4.0/).

1. Introduction

Offshore wind will play a significant role in both Ireland's and Europe's decarbonisation plans. Ireland's large offshore territory, coupled with high wind availability across each season [1,2], make it an ideal candidate for offshore wind development. In line with the National Energy and Climate plan, 5 GW of offshore wind is planned for deployment in Ireland by 2030 [3]. According to the Sustainable Energy Authority of Ireland (SEAI)'s wind energy road map, Ireland has potential to far exceed this 5 GW of offshore wind with a predicted installed capacity of 30 GW by 2050 [4]. This growth prediction coincides with a general decrease in onshore wind farm planning applications. Harper et al. have evaluated the regulatory effects of wind turbine planning and financing in the United Kingdom [5]. This study identified onshore wind as having a 44% success rate compared with 89% in the offshore wind sector.

Wind gust analysis has been extensively performed for traditional wind turbine systems. Turbulence and wake effects and extreme load predictions for horizontal axis wind turbines have been studied by Brand et al. [6–8]. The effect of wind gusts on vertical axis wind turbines using Computational Fluid Dynamics (CFD) has been examined by Onol et al. [9]. The distribution of extreme gusts has been previously investigated for traditionally interconnected wind turbines by Cheng et al. [10]. Gust detection and prediction methods using Doppler LiDAR are an area of current development for wind farms [11]. The use of LiDAR for wake management has also been explored showing a wind farm power increase of 7.552% with a reduction in downwind turbulence [12].

This rapid expansion within the sector leaves an opportunity for the development of new interconnection technology such as the Direct Interconnection Technique (DIT) which is considered in this paper. This technique is a method of integrating renewable generation first proposed by Pican et al., 2011 [13]. This technique of integration minimises the utilisation of Back to Back Power Converters (B2BC) in offshore turbines by connecting each turbine to a common offshore synchronous bus which can then be transmitted back to shore by High Voltage Alternating Current (HVAC), High Voltage Direct Current (HVDC) or Low-Frequency Alternating Current (LFAC) [14,15]. The power conversion equipment can be relocated to a single offshore site allowing for better access and optimisation or where transmission constraints permit, relocated entirely onshore.

Power electronic conversion systems exhibit a high failure rate among wind turbine subassemblies [16,17]. This resultant downtime, coupled with the difficulty and cost of servicing offshore turbines [18], demonstrates the potential that DIT has for improving reliability and reducing costs associated with offshore wind. While detection and protection methods can aid in reducing power electronic converter failures [19–21], offshore maintenance of these power converters has been noted as a critical element in the levelised cost of energy [22], given the requirement for transport of parts and technicians to these offshore locations. According to a case study conducted by Su et al. failure of electrical subsystems accounted for the third highest rate of failure, accounting for 14% and 26% of total failures for the two farms studied. This accounted for 301 hours of downtime in project 1 and 693 h in project two [23].

DIT begins by spinning a pilot generator connected to the offshore bus establishing the bus reference voltage and frequency. Each subsequent generator is then spun up and connected to the bus with the pilot generator governing system frequency and voltage, and load sharing controllers optimising behaviour on subsequent generators. This high inertia system electrical bus is then transmitted onshore through the use of HVAC, LFAC or HVDC as required and grid interconnection is performed by a large scale B2BC. This method has also been extended to Airborne Wind Energy (AWE) systems by Salari et al. 2018 [24]. The difference between traditional interconnection and direct interconnection can be observed in Figures 1 and 2.

In the case of the traditional interconnection, each generator is effectively separated from the local wind farm bus by the B2BC in the wind tower. This facilitates separation of the generator, and transients caused by wind gusts for example, from the local farm bus, and the individual B2BC provides a means of dealing with transient conditions on the generator side [25]. With DIT, as multiple generators are directly interconnected to the same bus, any gust generated transient condition experienced initially by a leading-turbine-to-wind will affect the interconnected system of generators.

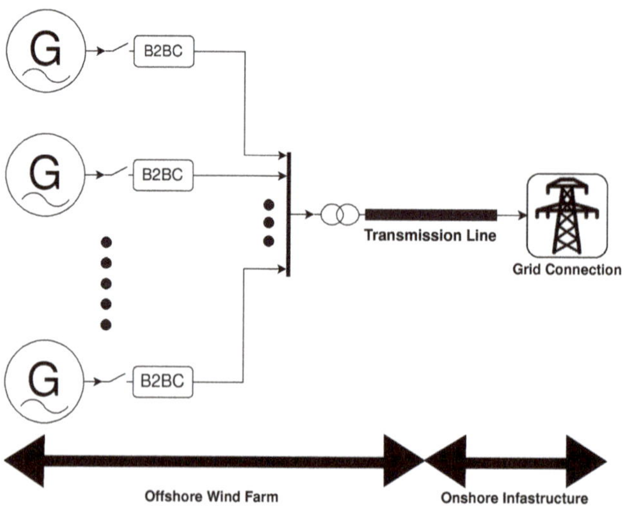

Figure 1. Traditional Interconnection.

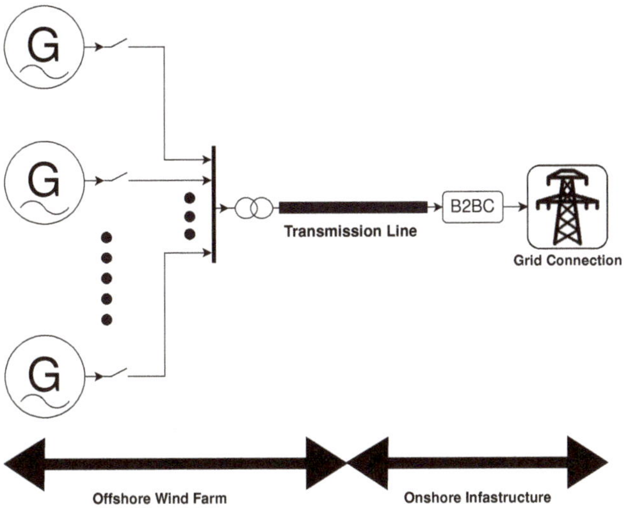

Figure 2. Direct Interconnection.

This paper studies the effects of wind gusts by simulating a directly interconnected wind farm and introducing a wind gust to a leading turbine. Gusts of varying types, magnitudes and transient times are applied to a leading turbine and overall system responses are investigated as described in detail in the simulation methodology section. Gust tolerance levels are measured and discussed with comparison to real-world coastal wind data.

2. Simulation Methodology

Gusts modelled in this study are created through the use of IEC standardised descriptions of gusts available in IEC 61400-3-1:2019 [26]. This offshore wind turbine standard in section 6.4.3.1 directly refers back to IEC 61400-1:2019 [27], an onshore wind turbine standard that describes the mean wind speed based wind gust profiles. In this standard, five extreme wind conditions are proposed: Extreme Operating Gusts (EOG), Extreme Direction Change (EDC), Extreme Coherent Gusts (ECG), Extreme Coherent Gusts with direction change (ECD) and Extreme Wind Shear (EWS). This study focuses on EOG and ECG. ECGs are considered as the increase in wind speed is sustained once the maximum gust speed is reached. This gust profile is useful for identifying any saturation limits of the system whereas the Mexican hat shape of EOGs present the greatest rate of change during the gust and therefore challenge the system due to the maximum rate of change of pitch angle. This worst-case analysis approach does not consider direction change of the wind, as the greatest amount of energy and therefore the most challenging input for the directly interconnected bus, occurs when the wind is directly incident on the turbine blades [28]. All gusts are applied with the hub facing directly into the wind with no yaw control considered. Future work may include an analysis of the other conditions. EOG and ECG profiles are generated using Equations (1) and (2).

$$u(z,t) = \begin{cases} \bar{U}(z) - 0.37U_{gust}sin(\frac{3\pi t}{T})(1-cos(\frac{2\pi t}{T})) & 0 \leq t \leq T \\ \bar{U}(z) & otherwise \end{cases} \quad (1)$$

As defined in IEC 61400-1, U_{gust} is the hub height magnitude defined by extreme wind speed recurrences for a particular site along with other physical factors such as rotor diameter. This U_{gust} factor is varied along with the period T to peak gust speed and settling time. $\bar{U}(z)$ is the average wind speed upon which the wind gust is superimposed.

$$u(z,t) = \begin{cases} \bar{U}(z) & t < 0 \\ \bar{U}(z) + 0.5U_{cg}(1 - cos(\frac{\pi t}{T})) & 0 \leq t \leq T \\ \bar{U}(z) + U_{cg} & t > T \end{cases} \quad (2)$$

In Equation (2), $\bar{U}(z)$ is the average wind speed, U_{cg} is the magnitude of the coherent gust and T represents the rise time of the gust. Once the gust is complete the wind speed remains at this new value of $\bar{U}(z) + U_{cg}$. A sample of each gust type with the same rise time of five seconds is shown in Figure 3.

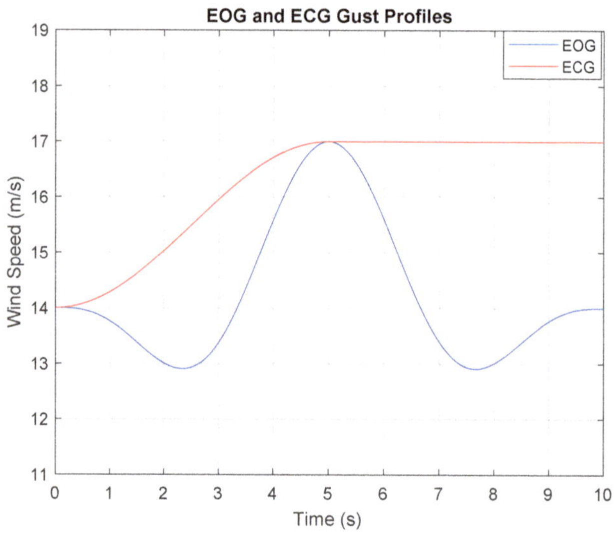

Figure 3. EOG and ECG profiles with rise times of 5 s.

The characteristic time for ECGs/EDCs stated in IEC 61400-1:2019 6.3.3.6 is 10 s [27]. This represents a rise time of 10 s between the base wind speed and the maximum wind speed value. However, in the case of EOGs in section 6.3.3.3, the characteristic time is 10.5 s. This time represents the length of the entire gust. Therefore the corresponding rise time would be 5.25 s. For this study, we round down this rise time to 5 s and consider rise times of 3 s, 5 s and 10 s to investigate these gusts on a like for like basis. These values are representative of the gust profiles as described in the standard, but also push beyond the values to investigate system limits.

Each turbine utilises a Permanent Magnet Synchronous Generator (PMSG) with a rated power of 800 kVA and is based on a real-world turbine characterisation [29]. The rated speed was selected at 14 m/s. The simulation begins with all wind turbines interconnected as per the direct interconnection algorithm described in [13]. Turbine 5 is selected as the pilot generator responsible for the maintenance of the frequency and voltage of the interconnected bus. This turbine in the real world would be selected as a turbine towards the centre of the wind park, thus minimising the risk of this turbine being the first to experience a wind disturbance. Each turbine has its own local dump load for spin up and disconnection from the main bus. A simulation diagram is shown in Figure 4. Disconnection could be required for dispatching down, in line with Transmission System Operator (TSO) instructions [30] or during times when the wind speed is outside of cut in and cut out speeds.

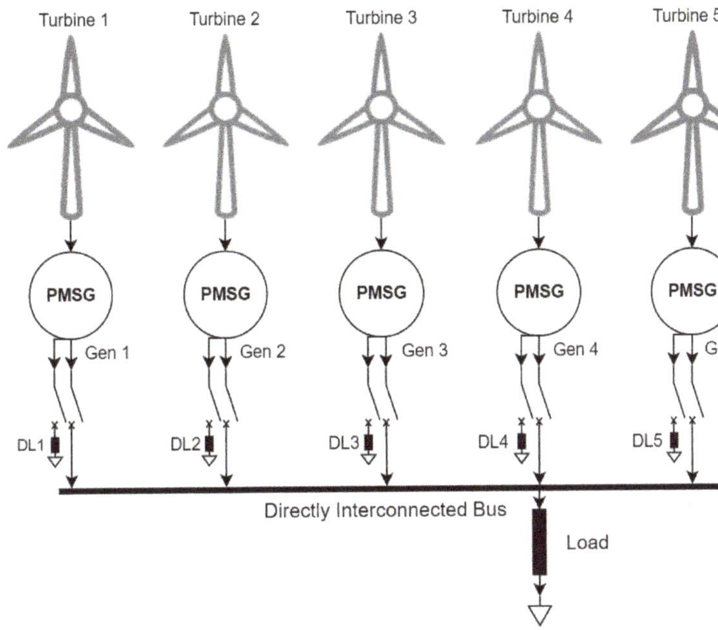

Figure 4. Simulation Diagram Showing the five turbines with their respective dump loads, the directly interconnected bus and the main load for the farm.

2.1. Simulation Model

2.1.1. Permanent Magnet Synchronous Generator model

As defined in [31] the PMSG model behaviour is described in Rotor Reference Frame (RRF) as follows:

$$\frac{d}{dt}i_d = \frac{1}{L_d}v_d - \frac{R}{L_d}i_d + \frac{L_q}{L_d}p\omega_m i_q \tag{3}$$

$$\frac{d}{dt}i_q = \frac{1}{L_q}v_q - \frac{R}{L_q}i_q - \frac{L_d}{L_q}p\omega_m i_d - \frac{\lambda p\omega_m}{L_q} \tag{4}$$

$$T_e = \frac{3}{2}p[\lambda i_q + (L_d - L_q)i_d i_q] \tag{5}$$

where R is the resistance of the stator windings, p is the number of pole pairs, T_e is electrical torque, L_d and L_q are the dq axis inductances, ω_m is angular velocity of the rotor, λ is amplitude of induced flux, v_d and v_q are dq voltages and i_d and i_q are dq currents. Equations (3) and (4) represent the ouput currents and volatges in dq frame and Equation (5) calculates electromagnetic torque.

L_q and L_d represent the relation between the phase inductance and the rotor position due to the saliency of the rotor. For a round rotor, there is no variation in the phase inductance therefore $L_d = L_q = \frac{L_{ab}}{2}$.

2.1.2. Wind Turbine Model

The Wind turbine is modelled using the Matlab Simulink wind turbine model with a nominal mechanical output power of 800 kW and a base wind speed of 14 m/s. The output of this block is applied to the generator shaft in per unit of generator ratings. We assume a direct drive system where mechanical efficiency (η_m) is 1. This wind turbine characteristic can be seen in Figure 5.

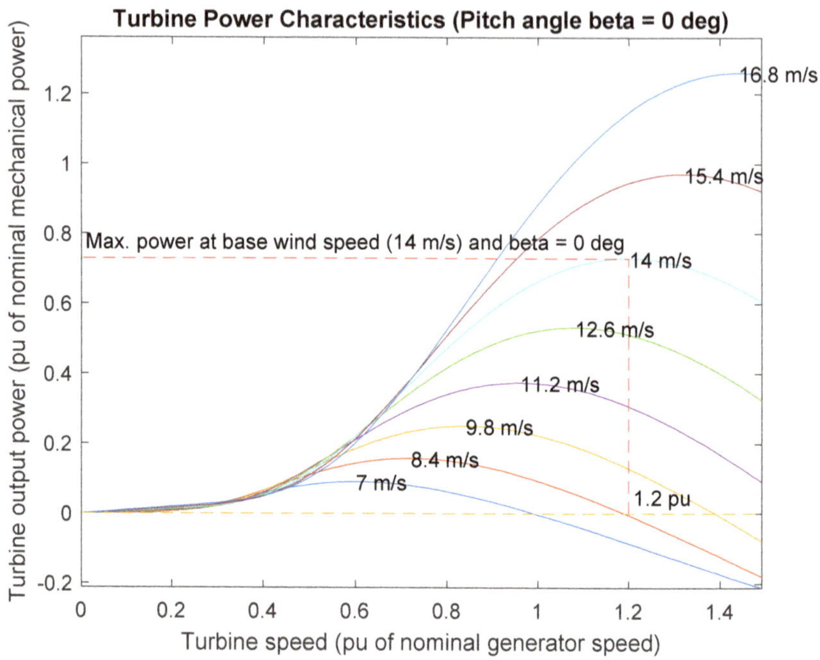

Figure 5. Wind Turbine Model Power Characteristic Curve.

A PID blade pitch angle controller is used with a rate of change limitation of eight degrees per second. This is to account for the fact that pitch angle cannot be varied instantaneously. This rate of change limitation value can be found in the NREL 5 MW reference wind turbine report [32]. A full control and simulation diagram can be found for both the pilot generator in Figure 6 and for non-pilot generators in Figure 7. For the pilot generator, the control system utilises a frequency setpoint and feedback loop to maintain the farm bus frequency. This pilot generator is set to a chosen power level and excluded from the farm power control loop. Non-pilot generators use a power reference and feedback loop to vary their active power contribution to the bus. The setpoint for these turbines is determined by a farm power level supervisor which takes the current farm power level and set point and distributes individual power levels to the turbines. For further information on the direct interconnection algorithm, see [13,24,29].

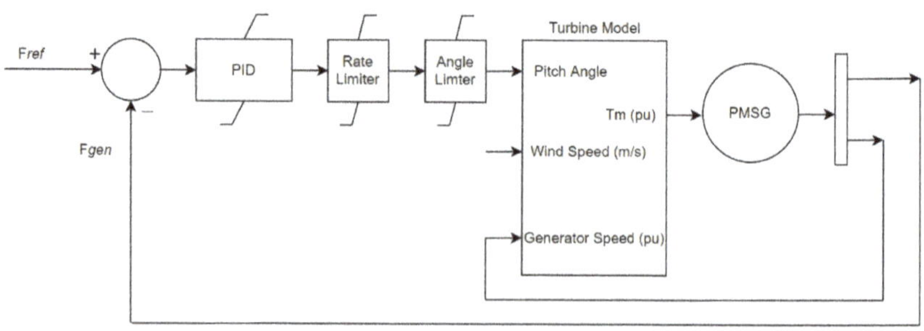

Figure 6. Control and Simulation Diagram for the Pilot Generator.

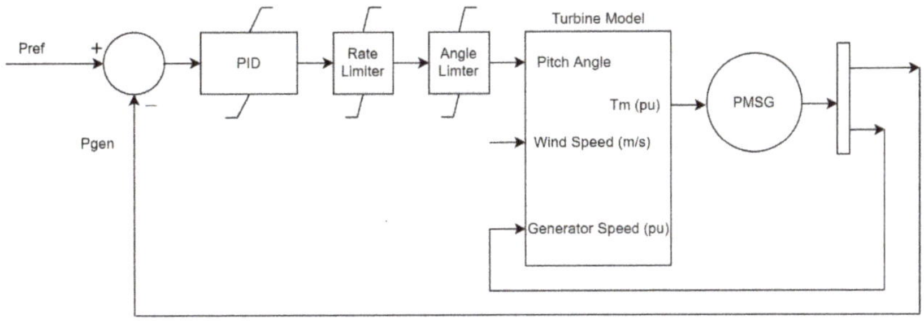

Figure 7. Control and Simulation Diagram for the Non-Pilot Generators.

2.2. Simulation Parameters

The simulation parameters utilised in this study are provided by the company JSPM, a subsidiary of the Areva group, and represent a real-world PMSG turbine [33]. These values are applied to each PMSG model in the simulation. This characterisation is employed to provide a consistent simulation analysis approach for DIT. These parameters displayed in Table 1 as used by Pican and Ebrahimi Salari [24,29], provide a basis for comparison of DIT in varying configurations and conditions. The farm size selection of 5 turbines is presented as the base number of turbines which would realistically be deployed in the field. Larger farms could be made up of a single directly interconnected bus or multiple strings of directly interconnected buses, each consisting of varying numbers of turbines due to transmission, resource availability or geographical constraints [34–36]. A larger number of interconnected generators, similar to the traditional AC power gird, will facilitate better sharing of disturbances and simplify the frequency and power response of the system.

Table 1. Simulation Parameters [33].

Parameter	Value
PMSG number of pole pairs	45
PMSG nominal frequency (Hz)	18.6
PMSG stator resistance (mΩ)	47
PMSG flux linkage (Wb)	6.86
Main load resistance (Ω)	0.1
Rotational Speed (RPM)	24.8

2.3. Gust Factor and Variation Limits

Gust Factor is a representation of the peak average τ second wind speed as a fraction of the T seconds moving average wind speed [37]. This is shown in Equation (6), where $U_{max,\tau}$ is the maximum τ second moving average wind speed in a T-second averaging period and $\overline{U_T}$ is the T-second average wind speed. Typical values for τ are 1–10 s with common values for T are 10 min to 1 h [37]. The gust factors of all test gusts applied are calculated and analysed.

$$G_{T,\tau} = \frac{U_{max,\tau}}{\overline{U_T}} \tag{6}$$

The tolerance threshold for both measured parameters of the simulation is selected as 5% (±2.5%). This threshold is selected to closely follow current grid connection codes on the farm side of the power converter [38]. This paper shows the differing gird requirements that the power conversion system of wind turbines are required to comply with. By limiting variation of frequency and active power to 5%, the power conversion system will be able to ensure grid interconnection compliance [30].

3. Results

3.1. Baseline System Response

The baseline system response is generated applying the reference wind gust to the leading turbine while all pitch controllers on turbines 1–5 are disabled. This shows the reaction of the interconnected system of wind turbines in the absence of controllers assisting in dealing with gust disturbances. As shown in Figure 8 without any pitch control the system behaves similar to a single synchronous machine causing the collective bus frequency to increase while active power is inserted at the lead turbine. This relatively small disturbance causes both the bus active power and bus frequency responses to vary outside the 5% (±2.5%) tolerance threshold.

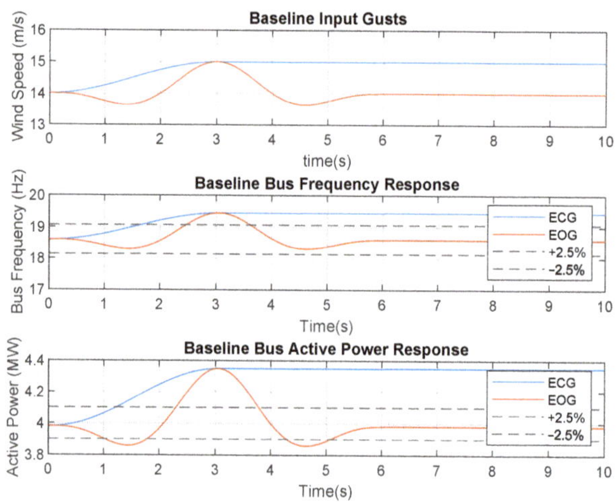

Figure 8. Baseline system test for both ECG and EOG. $\Delta T = 3$ s V_{gust} max = 15 m/s.

3.2. Extreme Coherent Gust Responses

The following section displays simulations results for Extreme Coherent Gusts (ECG) as described in the IEC Standard [27]. The test gust is applied to generator one with the system at a steady state at time zero. All generators are synchronised and interconnected to the main bus before time zero and have reached a steady state. Extreme coherent gust simulations are preformed at ΔT values of 3 s, 5 s and 10 s respectively. Example test gusts for $\Delta T = 3$ s are displayed in Figure 9.

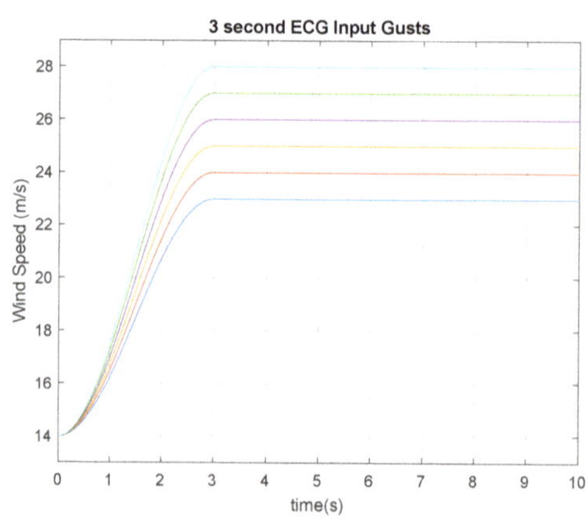

Figure 9. 3 s ECG test gusts applied.

For each rise time V_{gust} max is varied from 15 m/s to 30 m/s in 1 m/s increments. The corresponding gust factors of these gusts can be calculated by Formula (6), where $U_{max,\tau}$ is the maximum τ second moving average wind speed in a T-second averaging period and $\overline{U_T}$ is the T-second average wind speed [37]. The gust factors for the input ECG test gusts applied are displayed in Table 2 assuming a ten minute moving average base wind speed $\overline{U_T}$ = 14 m/s and $\tau = 1$ s.

Table 2. ECG Input Gust Factors.

V_{gust} max	Gust Factor
23 m/s	1.64
24 m/s	1.71
25 m/s	1.79
26 m/s	1.86
27 m/s	1.93
28 m/s	2.00
29 m/s	2.07
30 m/s	2.14

Figure 10 displays the frequency responses of each gust measured at the offshore bus. As can be clearly seen the 28 m/s gust response exceeds the limit of 5% (±2.5%) variation. This is due to the rate of change limitation of pitch angle variation of turbines. With an 8 degree per second maximum rate the pitch control is not capable of maintaining the 5% maximum variation. However, it can be seen that the system can damp the variation and return to steady state in all of the input gust cases. The 27 m/s gust also approaches the negative 2.5% limit but does not exceed it and therefore can be taken to be the maximum boundary limit with regard to our frequency response criteria.

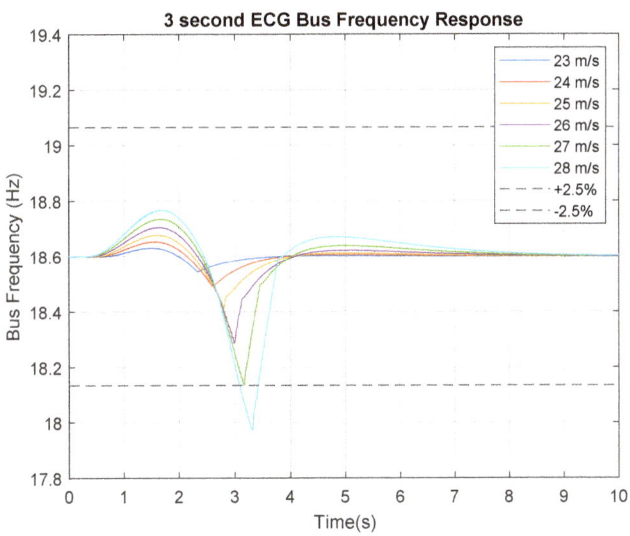

Figure 10. System frequency response to 3 s ECG gusts.

As can be seen in Figures 10 and 11, the active power and the frequency response are directly linked. As the 28 m/s ECG is rejected due to the frequency response criteria it can already be discounted. The 26 m/s, 27m/s and 28 m/s responses all fall outside the negative boundary leaving the 25 m/s as the maximum boundary within the limit with regard to the active power criteria. It can therefore be said that for the system modelled any ECG with ΔT of 3 s and magnitude up to and including 25 m/s can be tolerated.

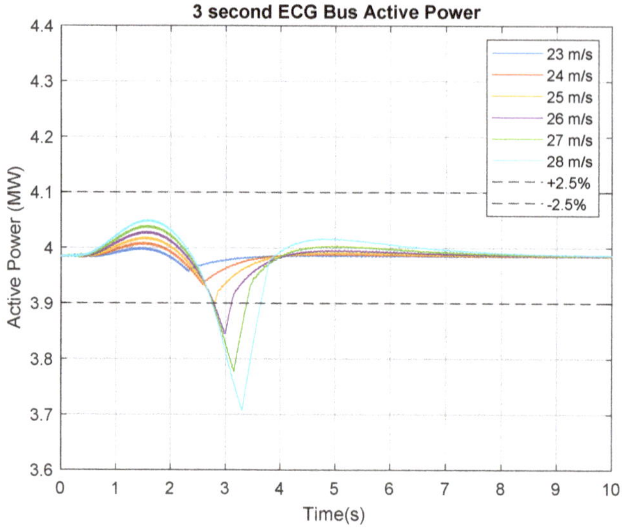

Figure 11. Bus active power response to 3 s ECG gusts.

The remaining simulations for ECGs with $\Delta = 5$ s and $\Delta = 10$ s all show performance within the 5% tolerance level. The 8 degrees/s of pitch angle control is capable of damping response without becoming saturated. The results of all the simulations are tabulated in Table 3. The light blue segments denote the respective criteria are satisfied while dark blue denotes that one or both of the $\pm$2.5% threshold levels have been exceeded. Considering the 5 s and 10 s rise time simulations, it can be observed that ECGs up to 30 m/s can be tolerated by the system. The maximum gust factors for these events of 1.64 through 2.14 are well beyond the gust factors measures at the coastal wind site in Frøya [37]. The 3 s ECG is within limits up to and including a V_{Gust} max of 25 m/s. With a gust factor of 1.79 from Table 2, this 25 m/s gust is well above the measured gust factors at this site with a mode value of 1.20.

Table 3. This table displays the results of all ECG simulations completed. Light Blue demonstrates the respective responses remain within the $\pm$2.5% boundary limitations with dark blue showing the criteria has not been met. The minimum and maximum values of both frequency and active power reached during each gust are displayed.

ΔT	3s				5s				10s			
Peak Gust Velocity	Frequency (Hz)		Active Power (MW)		Frequency (Hz)		Active Power (MW)		Frequency (Hz)		Active Power (MW)	
	Min	Max	Min	Max	Min	Max	Min	Max	Min	Max	Min	Max
<23m/s	18.555	18.632	3.957	4.000	18.598	18.602	3.980	3.988	18.599	18.601	3.982	3.987
24 m/s	18.493	18.654	3.934	4.009	18.598	18.602	3.980	3.988	18.599	18.601	3.982	3.987
25 m/s	18.411	18.678	3.898	4.019	18.597	18.602	3.980	3.989	18.599	18.601	3.982	3.987
26 m/s	18.288	18.705	3.845	4.029	18.598	18.602	3.980	3.989	18.599	18.601	3.982	3.987
27 m/s	18.136	18.735	3.777	4.039	18.598	18.602	3.980	3.989	18.599	18.601	3.982	3.987
28 m/s	17.973	18.768	3.707	4.050	18.598	18.603	3.980	3.989	18.599	18.601	3.982	3.987
29 m/s	17.808	18.801	3.635	4.061	18.598	18.603	3.980	3.989	18.599	18.601	3.982	3.987
30 m/s	17.593	18.879	3.543	4.100	18.596	18.603	3.979	3.989	18.599	18.601	3.982	3.987

3.3. Extreme Operating Gust Responses

This section outlines the Extreme Operating Gust responses for ΔT values of 3 s, 5 s and 10 s. The same initial conditions of synchronisation and steady state are utilised with the gust being applied to turbine 1 at time t = 0. The V_{Gust} max values are incremented by 1 m/s from 15 m/s to 30 m/s. Example 3 s EOG input gusts can be seen in Figure 12.

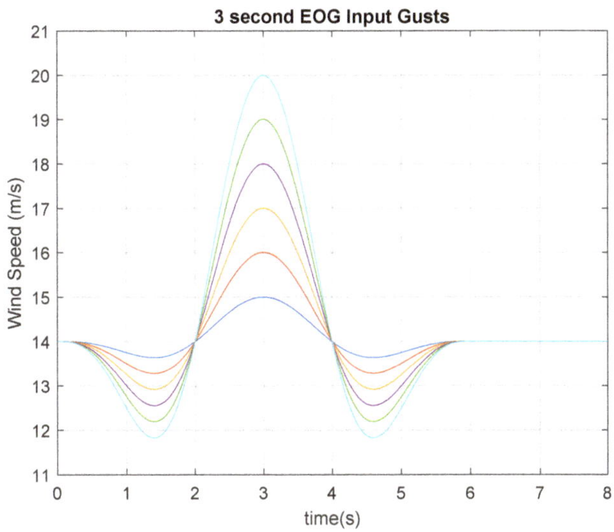

Figure 12. 3 s EOG test gusts applied.

The corresponding gust factors for the 3 s, 5 s and 10 s rise time EOGs are calculated by integrating (1) with limits of ±0.5 s of the peak gust time giving the sliding window of $\tau = 1$ s and are displayed in Table 4. The 10 min moving average wind speed $U_T = 14$ m/s.

Table 4. EOG Input Gust Factors.

V_{gust} max	Gust Factor 3 s	Gust Factor 5 s	Gust Factor 10 s
15 m/s	1.035	1.037	1.039
16 m/s	1.126	1.137	1.141
17 m/s	1.189	1.205	1.212
18 m/s	1.252	1.273	1.283
19 m/s	1.315	1.341	1.353
20 m/s	1.378	1.410	1.424
21 m/s	1.441	1.478	1.494
22 m/s	1.504	1.546	1.565

Considering Figure 13, it can clearly be seen that the wind farm struggles to maintain electrical frequency through EOGs, when compared with ECGs of the same magnitude displayed in the previous section. This is to be expected as now the leading turbine first experiences a dip in wind speed prior to the sharp rise to V_{gust} max. It can be observed that EOGs with magnitudes greater than 18 m/s lead to a violation of the 5% pk-pk limitation on bus frequency. The initial negative dip in wind speed preceding the rise causes a greater dV_{Gust}/dt which saturates the 8 degree per second rate of change limitation on the pitch controller.

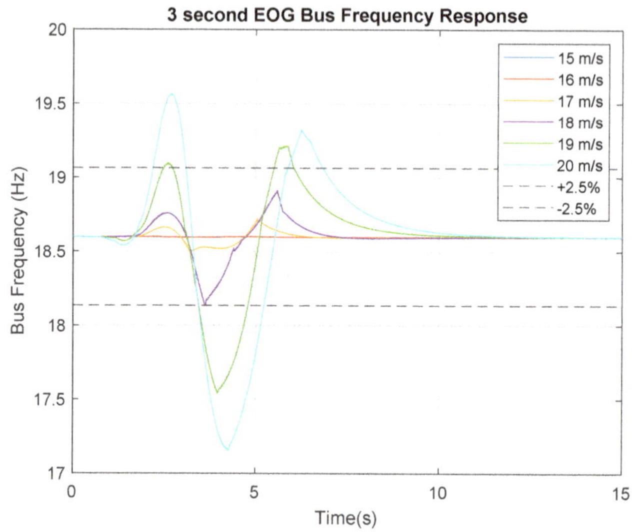

Figure 13. System frequency response to 3 s EOG gusts.

Figure 14 displays the active power variation on the main bus through the event. It can be seen that the 18 m/s gust displayed in purple, while within tolerance levels for frequency variation, fails to remain within 5% limitation on active power. However, as the active power only exceeds this limitation by 100 kW, it is possible that it could be considered tolerable in some electrical power conversion systems, particularly those which incorporate storage. This simulation assumes that all wind turbines remain connected to the bus throughout the transience however in the higher cases of V_{gust} max, it is likely that the turbine would be forced to disconnect from the main bus. This case however is outside the scope of this study and may be explored in future work.

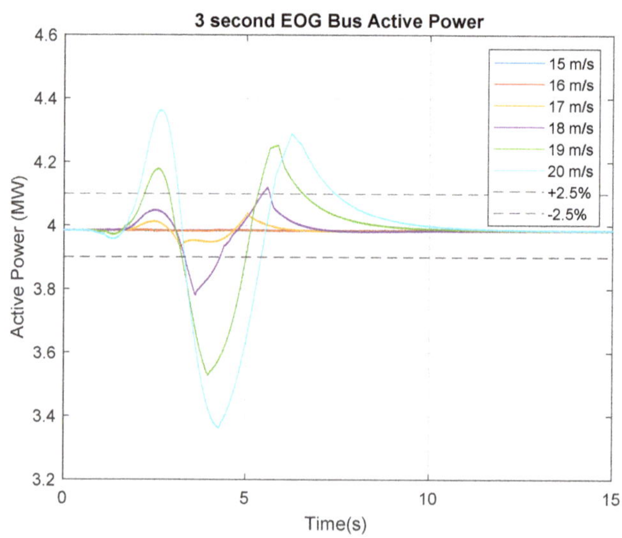

Figure 14. Bus active power response to 3 s EOG gusts.

The system frequency responses as shown in Figure 15 display a similar trend to that of the three second EOG tests. We can see however that the 19 m/s is within tolerable limits with the 20 m/s forming the boundary condition with regard to system frequency.

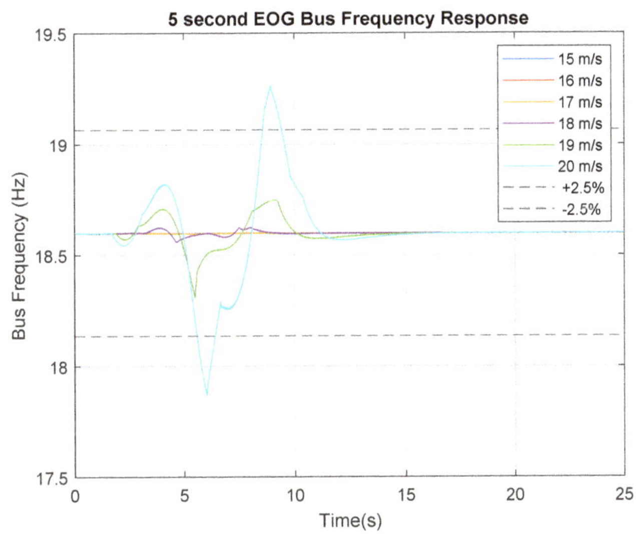

Figure 15. EOG Bus Frequency Responses $\Delta T = 5$ s.

Figure 16 displays the bus active power variation for the 5 s EOG tests. It can be observed that the 19 m/s EOG trace shown in green falls outside the negative 2.5% variation limit for a short period of time. For the purposes of this study, this will be declared outside the tolerance range.

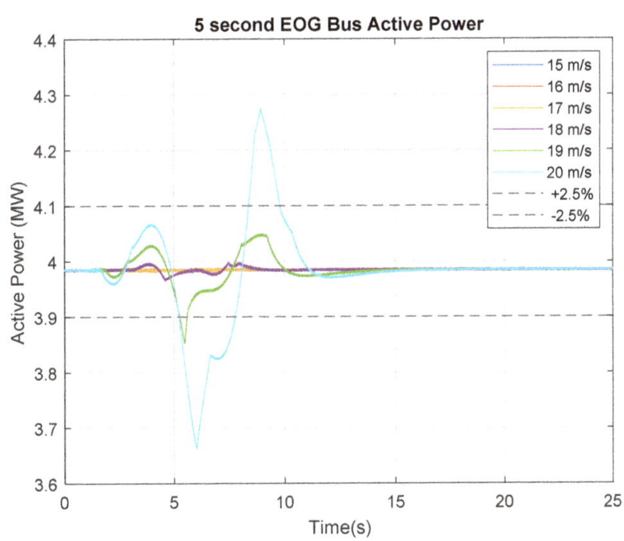

Figure 16. EOG Bus Active Power Responses $\Delta T = 5$ s.

All test runs are displayed in Table 5. The light blue denotes the output remains within the respective boundary condition with dark blue showing that one or both of the ±2.5% boundaries have been exceeded.

Table 5. This table displays the results of all EOG simulations completed. Light Blue demonstrates the respective responses remain within the ±2.5% boundary limitations with dark blue showing the criteria has not been met.

ΔT	3s				5s				10s			
Peak Gust Velocity	Frequency (Hz)		Active Power (MW)		Frequency (Hz)		Active Power (MW)		Frequency (Hz)		Active Power (MW)	
	Min	Max	Min	Max	Min	Max	Min	Max	Min	Max	Min	Max
15m/s	18.597	18.602	3.981	3.987	18.598	18.602	3.821	3.987	18.599	18.601	3.982	3.987
16 m/s	18.596	18.603	3.979	3.982	18.598	18.602	3.982	3.988	18.599	18.601	3.982	3.987
17 m/s	18.485	18.725	3.930	4.042	18.597	18.603	3.980	3.989	18.599	18.601	3.982	3.987
18 m/s	18.137	18.913	3.780	4.122	18.561	18.625	3.965	3.998	18.598	18.602	3.982	3.988
19 m/s	17.542	19.215	3.528	4.254	18.310	18.749	3.854	4.050	18.575	18.649	3.972	4.009
20 m/s	17.159	19.566	3.362	4.365	17.870	19.268	3.663	4.276	18.506	18.716	3.939	4.037
21 m/s	17.135	20.086	3.347	4.569	17.300	19.672	3.420	4.443	18.355	18.791	3.869	4.066
>22m/s	16.493	20.640	3.117	4.784	16.879	19.683	3.241	4.447	18.086	18.916	3.749	4.113

Analysing Table 5, it can be observed that the EOG gusts present a much greater challenge to the DIT bus parameters than ECGs. The 10 s rise time EOGs are the most effectively controlled which is to be expected as they have the lowest rate of change of V_{gust} max. The gust factors for these gusts are higher than the gust factors for shorter rise time gusts of the same magnitude. This is due to the wind speed cresting the maximum point for a greater time on either side of the maximum, therefore increasing the 1 s sliding average value. For a rise time of 10 s, the maximum gust factor which was successfully controlled by blade pitch angle control is 1.424. This gust factor is significantly below the 10 s for ECGs of 2.14 and above. Comparing this to the findings of Bardal et al., it can be observed that gust factors of 1.4 and above at the 100 m hub height are very rare [37]. As the average hub heights of modern offshore turbines are greater than 100 m, the 100 m data is the most relevant to this study.

If we consider the 3 and 5 s EOG data the corresponding boundary gust factors of 1.189 and 1.273 are within the range of values experienced offshore [37], however, the majority of gusts in the study fall below these values. This study also includes gust factors of gusts which may have occurred during times when the average wind speed may have been above the typical cut out speed of the turbine of 25 m/s and therefore the farm would not have been operating [39]. Gusts of this nature that do occur during the operation of a DIT wind farm would require further mitigation techniques outside of pitch angle control to maintain the 2.5% variation parameter studied.

4. Discussions & Conclusions

Extreme Operating and Coherent wind gust responses for directly interconnected systems have been investigated and discussed. It has been shown that through the use of pitch control on individual turbines the majority of wind gusts can be tolerated and the boundaries of this tolerance have been identified. The respective gust factors for these gust events have been calculated and compared to real coastal wind data [37]. These boundaries as presented in Tables 3 and 5 form the basis for further study on DITs interconnection to the grid. Power converter design and location can be investigated to further improve the gust tolerance of DIT systems. Additional analysis of large wind data sets will provide estimates of the frequency of gusts with gust factors greater than the tolerance levels described, facilitating comparison of DIT and traditionally interconnected wind systems in terms of capacity factor, capital expenditure (CapEx) and operational expenditure (OpEx).

Extreme operating gusts pose a greater challenge when compared to the extreme coherent gust conditions due to the higher rate of change in wind speed occurring throughout the gust. This study has not used B2BC which ordinarily provide a means on an individual turbine by turbine basis, of dealing with variations on the wind side while maintaining power on the grid side within specified limits of frequency and voltage. In the proposed DIT topology it is still intended to use B2BCs for a number of turbines as shown in Figure 2. Employing the farm level B2BC control and the pitch control as analysed in this paper will facilitate a greater tolerance range of gusts for DIT systems and will be the subject of future work. It is possible that with wind prediction methods such as LiDAR and more sophisticated machine learning-based control systems, that the boundaries could be further improved thereby reducing the load on the pitch control system and the power conversion systems down steam of the interconnected bus.

In conclusion, the Direct Interconnection Technique has been shown to be capable of tolerating wind gust conditions. The boundary of tolerance has been established and methods for further improvement have been proposed.

Author Contributions: C.W.O. carried out the reported research work, writing the paper and revisions. M.E.S. and D.J.T. supervised the research work, revisions and editing. The content of this paper is discussed by the authors and they all contributed to the final article. All authors have read and agreed to the published version of the manuscript.

Funding: This publication has emanated from research supported by the Science Foundation Ireland under the MaREI Centre research programme (Grant No. 12/RC/2302, and 14/SP/2740) and LERO Science Foundation Ireland grant 13/RC/2094. It is also co-funded under the European Regional Development Fund through the Southern and Eastern Regional Operational Programme to MaREI (www.marei.ie (accessed on 20 December 2021)) and Lero (www.lero.ie (accessed on 20 December 2021)) centres.

Institutional Review Board Statement: Not applicable.

Informed Consent Statement: Not applicable.

Data Availability Statement: Data Available on request.

Conflicts of Interest: The authors declare that the publication of this article has no conflict of interest.

References

1. Gallagher, S.; Tiron, R.; Whelan, E.; Gleeson, E.; Dias, F.; McGrath, R. The nearshore wind and wave energy potential of Ireland: A high resolution assessment of availability and accessibility. *Renew. Energy* **2016**, *88*, 494–516. [CrossRef]
2. Jung, C.; Schindler, D.; Buchholz, A.; Laible, J. Global Gust Climate Evaluation and Its Influence on Wind Turbines. *Energies* **2017**, *10*, 1474. [CrossRef]
3. Department of the Environment, Climate and Communications. *Ireland's National Energy and Climate Plan 2021–2030*; Government of Ireland: Dublin, Ireland, 2021. Avaliable online: https://www.gov.ie/en/publication/0015c-irelands-national-energy-climate-plan-2021-2030/ (accessed on 20 December 2021).
4. SEAI. Wind Energy Roadmap 2011–2050. 2011. Available online: https://www.seai.ie/publications/Wind_Energy_Roadmap_2011-2050.pdf (accessed on 20 December 2021).
5. Harper, M.; Anderson, B.; James, P.A.B.; Bahaj, A.S. Onshore wind and the likelihood of planning acceptance: Learning from a Great Britain context. *Energy Policy* **2019**, *128*, 954–966. [CrossRef]
6. Brand, A.J.; Peinke, J.; Mann, J. Turbulence and wind turbines. *J. Phys. Conf. Ser.* **2011**, *318*, 072005. [CrossRef]
7. Moriarty, P.J.; Holley, W.E.; Butterfield, S. Effect of Turbulence Variation on Extreme Loads Prediction for Wind Turbines. *J. Sol. Energy Eng.* **2002**, *124*, 387–395. [CrossRef]
8. Ebrahimi, A.; Sekandari, M. Transient response of the flexible blade of horizontal-axis wind turbines in wind gusts and rapid yaw changes. *Energy* **2018**, *145*, 261–275. [CrossRef]
9. Onol, A.O.; Yesilyurt, S. Effects of wind gusts on a vertical axis wind turbine with high solidity. *J. Wind. Eng. Ind. Aerodyn.* **2017**, *162*, 1–11. [CrossRef]
10. Cheng, P.W.; Bierbooms, W.A.A.M. Distribution of extreme gust loads of wind turbines. *J. Wind. Eng. Ind. Aerodyn.* **2001**, *89*, 309–324. [CrossRef]
11. Zhou, K.; Cherukuru, N.; Sun, X.; Calhoun, R. Wind Gust Detection and Impact Prediction for Wind Turbines. *Remote Sens.* **2018**, *10*, 514. [CrossRef]
12. Dhiman, H.S.; Deb, D.; Muresan, V.; Balas, V.E. Wake Management in Wind Farms: An Adaptive Control Approach. *Energies* **2019**, *12*, 1247. [CrossRef]
13. Pican, E.; Omerdic, E.; Toal, D.; Leahy, M. Direct interconnection of offshore electricity generators. *Fuel Energy Abstr.* **2011**, *36*, 1543–1553. [CrossRef]
14. Meng, Y.; Yan, S.; Wu, K.; Ning, L.; Li, X.; Wang, X.; Wang, X. Comparative economic analysis of low frequency AC transmission system for the integration of large offshore wind farms. *Renew. Energy* **2021**, *179*, 1955–1968. [CrossRef]
15. Rahman, S.; Khan, I.; Alkhammash, H.I.; Nadeem, M.F. A Comparison Review on Transmission Mode for Onshore Integration of Offshore Wind Farms: HVDC or HVAC. *Electronics* **2021**, *10*, 1489. [CrossRef]

16. Spinato, F.; Tavner, P.J.; Bussel, G.J.W.v.; Koutoulakos, E. Reliability of wind turbine subassemblies. *IET Renew. Power Gener.* **2009**, *3*, 387–401. [CrossRef]
17. Zhao, Y.; Li, D.; Dong, A.; Kang, D.; Lv, Q.; Shang, L. Fault Prediction and Diagnosis of Wind Turbine Generators Using SCADA Data. *Energies* **2017**, *10*, 1210. [CrossRef]
18. Ren, Z.; Verma, A.S.; Li, Y.; Teuwen, J.J.E.; Jiang, Z. Offshore wind turbine operations and maintenance: A state-of-the-art review. *Renew. Sustain. Energy Rev.* **2021**, *144*, 110886. [CrossRef]
19. Hossain, M.L.; Abu-Siada, A.; Muyeen, S.M. Methods for Advanced Wind Turbine Condition Monitoring and Early Diagnosis: A Literature Review. *Energies* **2018**, *11*, 1309. [CrossRef]
20. Lu, B.; Sharma, S.K. A Literature Review of IGBT Fault Diagnostic and Protection Methods for Power Inverters. *IEEE Trans. Ind. Appl.* **2009**, *45*, 1770–1777. [CrossRef]
21. Qiao, W.; Lu, D. A Survey on Wind Turbine Condition Monitoring and Fault Diagnosis—Part I: Components and Subsystems. *IEEE Trans. Ind. Electron.* **2015**, *62*, 6536–6545. [CrossRef]
22. Johnston, B.; Foley, A.; Doran, J.; Littler, T. Levelised cost of energy, A challenge for offshore wind. *Renew. Energy* **2020**, *160*, 876–885. [CrossRef]
23. Su, C.; Yang, Y.; Wang, X.; Hu, Z. Failures analysis of wind turbines: Case study of a Chinese wind farm. In Proceedings of the 2016 Prognostics and System Health Management Conference (PHM-Chengdu), Chengdu, China, 19–21 October 2016; pp. 1–6. [CrossRef]
24. Ebrahimi Salari, M.; Coleman, J.; Toal, D. Power Control of Direct Interconnection Technique for Airborne Wind Energy Systems. *Energies* **2018**, *11*, 3134. [CrossRef]
25. Li, S.; Haskew, T.A.; Swatloski, R.P.; Gathings, W. Optimal and Direct-Current Vector Control of Direct-Driven PMSG Wind Turbines. *IEEE Trans. Power Electron.* **2012**, *27*, 2325–2337. [CrossRef]
26. IEC. *EN IEC 61400-3-1:2019 Part 3-1: Design Requirements for Fixed Offshore Wind Turbines*; International Electrotechnical Commission: Geneva, Switzerland, 2019.
27. IEC. *EN IEC 61400-1:2019 Wind Energy Generation Systems—Part 1: Design Requirements*; International Electrotechnical Commission: Geneva, Switzerland, 2019.
28. Yang, J.; Fang, L.; Song, D.; Su, M.; Yang, X.; Huang, L.; Joo, Y.H. Review of control strategy of large horizontal-axis wind turbines yaw system. *Wind Energy* **2021**, *24*, 97–115. [CrossRef]
29. Pican, E.; Omerdic, E.; Toal, D.; Leahy, M. Analysis of parallel connected synchronous generators in a novel offshore wind farm model. *Energy* **2011**, *36*, 6387–6397. [CrossRef]
30. Eirgrid. *General Conditions of Connection and Transmission Use of System*; Eirgrid Group: Dublin, Ireland, 2013.
31. Grenier, D.; Dessaint, L.A.; Akhrif, O.; Bonnassieux, Y.; Le Pioufle, B. Experimental nonlinear torque control of a permanent-magnet synchronous motor using saliency. *IEEE Trans. Ind. Electron.* **1997**, *44*, 680–687. [CrossRef]
32. Jonkman, J.; Butterfield, S.; Musial, W.; Scott, G. *Definition of a 5-MW Reference Wind Turbine for Offshore System Development*; Technical Report NREL/TP-500-38060, 947422; National Renewable Energy Laboratory: Golden, CO, USA, 2009; [CrossRef]
33. Leroy, F. *JSPM Filial of AREVA NP*; Areva, S.A.: Courbevoie, France, 2011.
34. Bahirat, H.J.; Mork, B.A.; Høidalen, H.K. Comparison of wind farm topologies for offshore applications. In Proceedings of the 2012 IEEE Power and Energy Society General Meeting, San Diego, CA, USA, 22–26 July 2012; pp. 1–8. [CrossRef]
35. Sedighi, M.; Moradzadeh, M.; Kukrer, O.; Fahrioglu, M. Simultaneous optimization of electrical interconnection configuration and cable sizing in offshore wind farms. *J. Mod. Power Syst. Clean Energy* **2018**, *6*, 749–762. [CrossRef]
36. Wu, Y.; Zhang, S.; Wang, R.; Wang, Y.; Feng, X. A design methodology for wind farm layout considering cable routing and economic benefit based on genetic algorithm and GeoSteiner. *Renew. Energy* **2020**, *146*, 687–698. [CrossRef]
37. Bardal, L.M.; Sætran, L.R. Wind Gust Factors in a Coastal Wind Climate. *Energy Procedia* **2016**, *94*, 417–424. [CrossRef]
38. Singh, B.; Singh, S.N. Wind Power Interconnection into the Power System: A Review of Grid Code Requirements. *Electr. J.* **2009**, *22*, 54–63. [CrossRef]
39. Dupont, E.; Koppelaar, R.; Jeanmart, H. Global available wind energy with physical and energy return on investment constraints. *Appl. Energy* **2017**, *209*, 322–338. [CrossRef]

Journal of
Marine Science and Engineering

MDPI

Article

Environmental Compatibility of the Parc Tramuntana Offshore Wind Project in Relation to Marine Ecosystems

Koldo Diez-Caballero [1], Silvia Troiteiro [2,*], Javier García-Alba [3,†], Juan Ramón Vidal [1], Marta González [2], Sergi Ametller [2] and Raquel Juan [2]

[1] Tecnoambiente S.L., 08290 Cerdanyola del Vallès, Spain; koldo.diezcaballero@tecnoambiente.com (K.D.-C.); joanramon.vidal@tecnoambiente.com (J.R.V.)
[2] SENER Ingeniería y Sistemas S.A., 08290 Cerdanyola del Vallès, Spain; marta.gonzalez@sener.es (M.G.); sergi.ametller@sener.es (S.A.); raquel.juan@sener.es (R.J.)
[3] IHCantabria—Instituto de Hidráulica Ambiental de la Universidad de Cantabria, 39011 Santander, Spain; javier.garciaalba@unican.es
* Correspondence: silvia.troiteiro@sener.es
† Contribution of the hydrodynamical studies on the Golf de Roses.

Abstract: Parc Tramuntana is the first offshore wind project being promoted in the Catalonian waters, and due to this newness, it has generated a strong social debate surrounding expected environmental and socioeconomic impacts traditionally associated to marine wind farms, as there are no relevant references in this area. The objective of this report is to provide a specific analysis of some of the main potential impacts, based on detailed information and quantitative data, in order to place these impacts in a realistic context and determine their actual magnitude. This analysis is fed by diverse and detailed studies carried out over the last two years to assess the environmental impact of the project, in accordance with current regulations. According to environmental impact assessment, which is based on a standardized methodology, the impact of the project is objectively qualified as MODERATE on vectors such as turbidity and sedimentation, underwater noise, hydrodynamic circulation or the alteration of electromagnetic fields, and NOT SIGNIFICANT on aspects such as the proliferation of invasive exotic species. As this is an ongoing assessment process, this report presents initial conclusions that do not yet address all possible impacts. Nevertheless, the authors stress the importance of framing the debate on offshore wind in Catalonia in the context of the urgency of the climate emergency and its inevitable impacts on the natural environment.

Keywords: offshore wind impacts; turbidity; sediment dispersion; electromagnetic fields; acoustic impact; marine hydrodynamics; invasive species

Citation: Diez-Caballero, K.; Troiteiro, S.; García-Alba, J.; Vidal, J.R.; González, M.; Ametller, S.; Juan, R. Environmental Compatibility of the Parc Tramuntana Offshore Wind Project in Relation to Marine Ecosystems. *J. Mar. Sci. Eng.* **2022**, *10*, 898. https://doi.org/10.3390/jmse10070898

Academic Editors: Barbara Zanuttigh and Eugen Rusu

Received: 18 May 2022
Accepted: 21 June 2022
Published: 29 June 2022
Corrected: 26 December 2022

Publisher's Note: MDPI stays neutral with regard to jurisdictional claims in published maps and institutional affiliations.

Copyright: © 2022 by the authors. Licensee MDPI, Basel, Switzerland. This article is an open access article distributed under the terms and conditions of the Creative Commons Attribution (CC BY) license (https://creativecommons.org/licenses/by/4.0/).

1. Introduction

Currently Catalonia, as well as the surrounding regions and countries, faces a situation of climate emergency, a situation that has been recognized both by the Generalitat de Catalunya [1], the Spanish Government [2], and the European Parliament [3].

It is hence a situation that involves the adoption of urgent measures to reduce the carbon emissions for which our energy and production model is responsible. It is therefore not only advisable, but essential, among other actions, to propose important changes in Catalonia's energy generation model, which involve the progressive replacement of non-renewable sources, which currently provide more than the 80% of the electricity generated, with renewable sources, including offshore wind. This strategy has been developed in Catalonia through the *Pacte Nacional per a la Transició Energètica de Catalunya* (National Agreement for the Energy Transition of Catalonia) and has been embodied in the Preliminary Draft of the Energy Transition Law, which contemplates large, medium, and small-scale renewable electricity generation, mainly by local sources (Strategy n° 12) [4].

According to the energy prospective studies for Catalonia with a 2050 horizon (PROENCAT 2050) [5], in order to meet the energy and environmental objectives Catalonia has been committed to comply, by 2030 it will be necessary to incorporate up to 12,000 MW of renewable production into the system (of which 1000 MW should correspond to offshore wind), a figure that should grow to more than 61,000 MW by 2050 (with a total contribution of 3500 MW from offshore wind). This new renewable capacity should make it possible to supply a demand characterized by a greater electrification of the economy, facing the expected closure of nuclear power plants and drastically reducing the current dependence on non-renewable sources, of foreign origin and with a high impact on greenhouse gas emissions.

Once the need to develop offshore wind energy in Catalonia has been assumed, its spatial development should be approached on the basis of criteria that incorporate and allow the complexity of the marine environment to be assessed, weighing the advantages and disadvantages of each area and the benefits and impacts of offshore wind farms on the whole of the territory and Catalan society.

Marine spatial planning plays an important role in this task. In recent years, the Spanish Ministry for Ecological Transition and the Demographic Challenge (*MITERD*) has been working, in coordination with many other Spanish state administrations and each of the autonomous communities of the Atlantic and Mediterranean arc, to develop Marine Spatial Management Plans (POEM) [6], which are currently under review following the presentation of their draft for public information and consultation, in order to incorporate the relevant modifications resulting from the allegations received, as a step prior to their approval by Royal Decree.

It should be noted that these plans are not arbitrary or based solely on economic or territorial interests, but that their approach responds to the assessment of a set of criteria that ensure that the combined pressure of activities in the marine environment is maintained at levels compatible with the achievement of good environmental status (GES), and that they do not compromise the capacity of marine ecosystems to respond to changes induced by human activity. An ecosystem approach has therefore been followed for its definition, considering both the interactions between land and sea, as well as the expected changes resulting from climate change.

Besides that, the challenges that the development of offshore wind in Catalonia and Spain face are not few, largely due to the newness of this type of project in Spain. One of these challenges is the adaptation of new projects to a constantly evolving regulatory framework to incorporate this new technology into the Spanish energy model, but also with a certain social rejection, characteristic of any change in strategy and implementation of new technologies for the first time in the country. The influence of this regulatory framework, especially regarding the environmental impact assessment process of this kind of project was already analyzed by Salvador et al. (2018) [7].

From the point of view of the possible social reticence to the development of offshore wind power in Catalonia by some economic stakeholders and some members of the scientific community, the social debate on this new technology is often justified on the basis of the lack of specific references of similar projects in operation in the Mediterranean that allow to accurately foresee the possible impacts that this type of projects can generate in the area of implementation, generating uncertainty.

While onshore and fixed founded offshore wind farms are quite widely installed around the world and provide relevant references on their impacts on the environment along their entire life-cycle (see e.g., Verma et al., 2022 [8] or Kouloumpis and Azapagic, 2022 [9]), and the adverse environmental impacts of different wind generation technologies have been already assessed (mainly in onshore context) [10], there are still many effects that cannot be directly extrapolated to floating offshore wind farms, producing a significant knowledge gap regarding the impacts of this technology in the environment.

This lack of applicable references means that some voices tend to assimilate as their own, without a rigorous analysis, other offshore wind experiences in the North Sea or the Baltic, where both the technological characteristics of wind farms (mostly with fixed foundations) and the environment (depth, dynamics, ecosystems, etc.) are very different from those of the projects proposed in Catalonia.

It should be noted in this regard that, although there are currently no floating offshore wind projects in operation, the French Government has already taken the first step to be the pioneers in the Mediterranean, approving the installation of a pilot floating offshore wind farm, called "*Eoliennes Flottantes du Golfe du Lion*" (EFGL), and two other pilot farms ("*Eolmed*" and "*Provence Grand Large*") are in progress.

The EFGL wind farm consists of three 10 MW turbines, to be installed 16 km off the coast of Leucate, and within the perimeter of a protected area of the Natura 2000 Network, the marine natural park of the Gulf of Lion, the due Environmental Impact Study having been carried out and its impact considered compatible with the conservation objectives of this protected area. The other two wind farms are also planned with three turbines each, with power ratings between 8 and 10 MW, and at a distance from the coast of between 14 and 18 km.

In addition, on 14 March 2022, the French government confirmed its intention to build an additional 500 MW of floating offshore wind power in the Mediterranean by 2030.

2. Motivation

In this context, given the urgency of developing new renewable generation projects, due to the limited time available to meet emission reduction targets, and the periods required for the processing and technical development of projects, several developers specialized in offshore wind have begun to work, both in Catalonia and in other regions of Spain, to respond in the imminent future to the demand for these types of facilities.

One of these developments, pioneered in Catalonia in terms of technical maturity and coordination with the territory, Parc Tramuntana, proposes the installation of a floating offshore wind farm off the coast of the Empordà.

This project is being developed with the premise of integrating the participation of the territory from the earliest stages of conception, applying all the way and in all aspects of the design the best existing practices at the state-of-the-art level to ensure compliance with the so-called Precautionary Principle, which supports the adoption of protective measures against the possibility of a technological risk to the environment, without yet having a definitive scientific proof of such risk.

Likewise, the project, which has not yet been submitted for environmental processing, will follow all the way the environmental and administrative procedures required under current legislation, which includes the preparation and processing of the due Environmental Impact Study, including the process of public information and consultation with all the competent administrations.

The process leading to the selection of the most suitable site and the design proposal of the offshore wind facility has been and will continue to be a living and evolving process. It has started from an initial approach, based on the energy demand, the availability of connection to the electricity grid, the conditioning factors imposed by the legislation in force (environmental, sectorial, town planning, etc.), maritime and airspace planning and the characteristics of the site, and has evolved (in size and technical solutions) through the incorporation of additional requirements and conditioning factors derived from the conversations held with different stakeholders of the territory and society.

This report analyzes some of the aspects and potential impacts of the project that have generated some debate in the scientific community. This is the case of the recent publication of the article by Lloret et al. (2022) in the journal Science of the Total Environment [11], intending to serve as a reference for the evaluation of environmental impacts related to floating wind technology in the western Mediterranean. This purpose is also oriented towards introducing accuracy in the data analysis and promoting the debate with some sectors who systematically express opposition to this type of renewable generation project without a detailed analysis of its impacts on the environment.

3. General Description of the Parc Tramuntana Project

The Tramuntana Floating Offshore Wind Farm Project consists of the installation of an offshore renewable energy generation farm on the continental shelf of the province of Girona, at a distance of approximately 24 km from the coast of the Bay of Roses, in a range of depths between 120 and 180 m and on thick silty and detritic seabed, with a total absence of rocky outcrops. The turbine closest to the coast is located 14 km from Cap de Creus.

The wind farm is located within the area designated in the draft of the *POEM* for the Levantine-Balearic demarcation as a priority area for offshore wind (LEBA-2, Figure 1), being the only area identified in Catalonia as suitable for this type of activity. This site also partially coincides with a permanent closed area for trawling and is located outside protected natural areas.

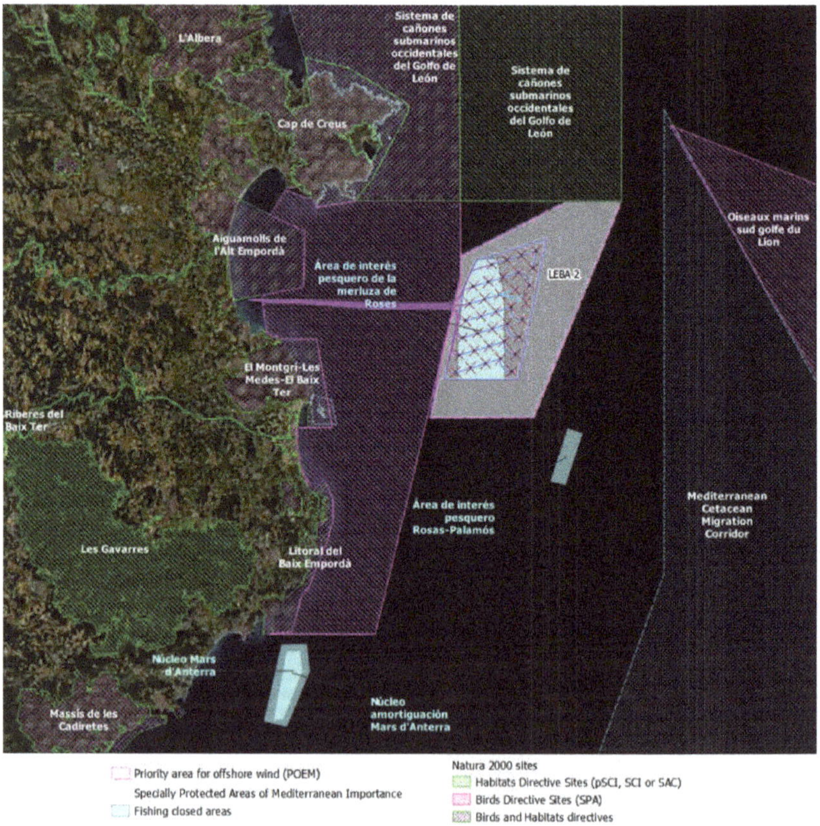

Figure 1. Location of the Tramuntana Floating Offshore Wind Farm Project in relation to environmental protected sites and other uses of maritime space.

The proposed solution consists of the following elements (Figure 2):

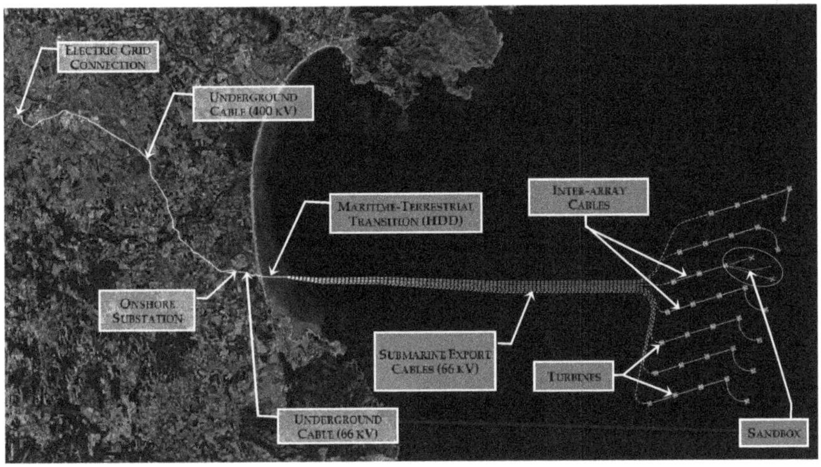

Figure 2. General layout of the Tramuntana Floating Offshore Wind Farm Project.

- Wind farm: it will have a total of 36 renewable energy generation positions, arranged in 7 rows perpendicular to the coast, in ENE direction (~N73° E). Initially, it will consist of 33 turbines of 15 MW and an R + D + i platform equipped with 3 positions for testing experimental devices (sandbox), with a total combined power of 500 MW. When fully developed, this sandbox will be replaced by 2 additional turbines of 15 MW each, with a maximum power of 525 MW;
- Submarine electrical cabling system for the interconnection of devices (inter-array cables) and subsequent export of the energy generated to shore (export cables). The export cables, which will be buried, run 24 km to the coast, in the municipality of Sant Pere Pescador;
- Maritime-terrestrial transition to the cable landing area on the coast, by means of HDD technique, and connection between the submarine cables and the terrestrial cables by transition boxes at the landing point;
- Terrestrial underground cabling (66 kV) for the connection between the transition boxes and the electrical substation, located approximately 1 km from the coast;
- Transformer substation, to raise the export voltage to 400 kV and provide the export electric circuit with the appropriate characteristics for connection to the onshore power grid.
- Terrestrial underground high-voltage conduction system (400 kV), to discharge the energy generated to the Spanish grid through the Santa Llogaia substation, near Figueres.

The estimated net energy production is approximately 1800 GWh/year, equivalent to 45% of the current consumption of the province of Girona, both household and industrial. The project includes a research platform, which will be developed in collaboration with the *Institut de Recerca d'Energia de Catalunya* (IREC) and will serve as a sandbox for different lines of R + D + i in the field of offshore renewable energy harnessing, as well as in other compatible uses of marine space such as aquaculture, fishing recovery, etc.

The wind farm takes up an area of approximately 95 km^2 where floating turbines and their anchoring systems are located. This area represents less than 0.12% of the surface of the Exclusive Economic Zone (EEZ) corresponding to Catalonia.

The 15 MW turbines have a total height of 261 m and a rotor diameter of 236 m, and are installed with a separation between turbines of between 1.2 and 2.5 km. They are attached to the seabed by 3 or 4 catenaries, each about 650 m long, attached to an anchor that remains buried about 14 m below the seabed.

The turbines are connected to each other by dynamic inter-array cables, which partially rest on the seabed, and from the end of each row of turbines comes an export cable, buried between 1.5 and 2 m below the seabed, which carries the generated energy to land. In the land–sea transition zone, a solution using small diameter perforations (Horizontal Directional Drilling or HDD) has been adopted to facilitate the landing of the cable without altering the seabed or the biological communities settled on it.

From the grounding point to the connection with the electrical grid, all the high voltage conduits are projected buried, to minimize the environmental and landscape impact.

Keeping this development proposal as an objective, since it is considered necessary to provide the Catalan system with this minimum capacity of marine renewable generation, the possibility of a phased development of the project is also proposed, so that a single array can be initially executed together with the sandbox facility. This would allow to advance in the installation of the general energy export infrastructures, and to have for a certain period of time a testing and demonstration facility to validate these types of wind farms in the Mediterranean and to verify their environmental and socioeconomic compatibility.

4. Methods of Analysis

In the analysis of impacts carried out in the Environmental Impact Study of the project, still under development, an identification and quantification methodology was used, which is based on the methodology proposed by *Conesa* (1995) [12], in accordance with the requirements included in Annex VI of Law 21/2013, modified by Law 9/2018, regarding the identification, quantification, and assessment of the foreseeable significant effects of the projected activities on the environment.

It is important to differentiate, regarding impacts, those occurring during the construction phase, which has a more limited duration in time, from those occurring during the operational phase, which may be prolonged throughout the service life of the facility. In addition, due to the nature of this project, which extends over both the marine and terrestrial areas, it is necessary to differentiate between the two areas when assessing the environmental factors affected.

In this report, a preview of the assessment of some of the main impacts on the marine environment, related to the conservation of biodiversity and the good environmental status of the area, is made in order to dispel doubts regarding some of the potential impacts that represent a major concern for the authors of the article of Lloret et al. (2022) [11]. It is important to mention that, for each of the potential impacts preliminarily identified by means of a Leopold matrix, each of the following attributes were characterized (Table 1).

Table 1. Impact classification table.

Attribute		Degree	Description	Value
Nature (Sign)	Beneficial (+) or detrimental (-) nature of the impact.	Beneficial impact	Improvement in the current situation	+
		Detrimental impact	Loss in current situation	-
Extension (EX)	Theoretical area of influence of the impact in relation to the project's surroundings (% of area with respect to the surroundings in which the effect is manifested).	Isolated		1
		Partial		2
		Extensive		4
		Total		8
		Critical		12
Persistence (PE)	Time of permanence of the effect from its appearance, after which the initial conditions prior to the action are recovered.	Sporadic	<1 year	1
		Temporary	1–10 years	2
		Permanent	>10 years	4
Synergy (YES)	Reinforcement of two or more single effects caused by simultaneous actions.	No synergies	-	1
		Synergist	Moderate synergism	2
		Very synergist	Highly synergistic	4
Effect (EF)	Cause–effect relationship, form of manifestation of the effect on a factor as a consequence of an action.	Indirect	It has an immediate impact on the relationship of one environmental factor to another	1
		Direct	With immediate effect on an environmental component	4
Recoverability (MC)	Possibility of total or partial reconstruction and return to the initial conditions prior to the action, by means of human intervention (introduction of corrective measures).	Short-term/immediately recoverable	<1 year	1
		Recoverable in the medium term	1–10 years	2
		Mitigable	Alteration that can be reduced by corrective measures	4
		Irrecoverable	Alteration impossible to repair	8
Intensity (IN)	Degree of impact of the action on the factor, in the specific area in which it acts.	Short		1
		Medium		2
		High		4
		Very high		8
		Total		12
Moment (MO)	Manifestation period or time elapsing between the occurrence of the action and the beginning of the effect.	Long term	>10 years	1
		Medium term	1–10 years	2
		Short term	<1 year	3
		Immediate	0	4

Table 1. *Cont.*

Attribute		Degree	Description	Value
Reversibility (VR)	Possibility of returning to the initial conditions prior to the action, by natural means, once the action ceases to act on the environment.	Short term	<1 year	1
		Medium term	1–10 years	2
		Long term	10–15 years	3
		Irreversible	>15 years	4
Accumulation (AC)	Progressive increase in the manifestation of the effect, when the action that generates it persists continuously or repeatedly.	Simple	No side effects or cumulative effects	1
		Cumulative	It increases in severity over time	4
Periodicity (PR)	Regularity of manifestation of the effect.	Irregular	Appears in and irregular manner	1
		Periodic	Appears periodically	2
		Continuous	Appears in a constant manner over time	4

The combination of the valuation of the attributes described above for each analyzed impact was carried out in accordance with the methodology proposed by Conesa (1995) [12], according to the following formula:

$$I = 3IN + 2EX + MO + PE + RV + SI + AC + EF + PR + MC \quad (1)$$

results in a standardized and objective characterization of its significance, which may be COMPATIBLE, MODERATE, SEVERE, or CRITICAL impact. In order to properly understand how these impact categories are defined, their definitions are included below:

- Compatible Environmental Impact (C): impact whose recovery is immediate after the cessation of the activity, not requiring protective or corrective actions. $I \leq 25$;
- Moderate Environmental Impact (M): impact whose recovery does not require intensive protective or corrective practices, and in which the achievement of the initial environmental conditions requires a certain amount of time. $25 < I \leq 50$;
- Severe Environmental Impact (S): impact in which the recovery of environmental conditions requires the application of protective or corrective measures, and in which, even with these measures, recovery requires a long period of time. $50 < I \leq 75$;
- Critical Environmental Impact (Cr): impact whose magnitude is greater than the acceptable threshold. This results in a permanent loss of the quality of environmental conditions, with no possible recovery, even with the adoption of protective and corrective measures. $I > 75$.

It is worth noting that the analysis focused on a set of negative impacts, leaving in the background the more positive aspects of the project, mainly related to the contribution to the reduction of GHG emissions, which will indirectly help mitigate the serious deterioration of marine ecosystems associated with climate change.

5. Results and Discussion of the Analysis of the Main Impacts of Offshore Wind, Applied to Parc Tramuntana

5.1. Analysis of the Impact of the Project on Water Turbidity and Sedimentation

The increase in water turbidity is one of the main impacts detected, and it happens both during the execution phase, where it is of greater importance, and during the operation phase. The effects that sediment resuspension can cause indirectly in the transport and deposition of suspended sediment were also analyzed.

The installation of anchors, chains, and cables on sedimentary substrate, and the consequent sediment resuspension, are likely to cause direct effects on water quality (turbidity, alteration of the trophic and chemical states), and in turn indirect effects on the biota that may be affected by the turbidity plume, particularly hard-bottom filtering and suspension-feeding species. It should be noted that no such seabed has been detected in the project area, the seabed affected being composed entirely of fine sands and muds.

In general, regarding the alteration in water turbidity, the main effects on marine fauna are a decrease in visibility (affecting the behavior of certain species in their ability, among other aspects, to capture prey or detect predators) or, if prolonged over time, affecting the feeding and breathing capacity of suspension-feeding and filter-feeding animals, including fish, whose gills may suffer physical damage due to the presence of abrasive particles in the water.

Reduced light penetration also affects the depth of the photic zone (the layer where sufficient light reaches for photosynthesis) and thus the primary production capacity of algae and phanerogams.

The subsequent deposition of suspended sediment on the seabed would have its greatest impact on sessile species, particularly on those species most sensitive to sedimentation.

Considering these effects, the area with the greatest sensitivity and potential impact due to turbidity is considered to be the *Cymodocea nodosa* meadow in the shallower area, at depths below 20 m, close to the exit of the route through the HDD. In the remaining area affected by the evacuation route and the area where the wind turbines will be installed, no significant presence of structuring sessile species was detected.

The increase in turbidity may also have an associated impact on water quality, when possible pollutants present in the sediment are mobilized. This potential impact is dismissed, given that the analysis of sediment samples taken in the area affected by the project shows no evidence of significant contamination.

In the Tramuntana project, during construction, the foreseeable increase in turbidity is mainly associated with the activities of drilling execution (HDD) in the maritime-terrestrial zone and export submarine cable burial by jetting for its protection. Both actions are temporary (with a permanence of the impact in the order of hours) and their effects on turbidity are reversible. In this phase, anchoring activities of anchor and mooring lines are short-time events, and have a lower capacity to generate turbidity, as well as the laying of inter-array cables, a part of whom can rest directly on the seabed, without the need for burial.

During operation of the wind farm, since most of the submarine cables are buried and immobile, the only contribution to increased turbidity on the seabed is that associated with the wind turbine anchoring systems, due to the resuspension of sediments from the seabed that may be produced by the movements of the mooring lines under the effect of waves and wind.

5.1.1. Impact of HDD on Turbidity and Sedimentation (Construction Phase)

The HDD is a small diameter borehole that is drilled between the entrance pit, located on land behind the beach dunes, and the exit point in the sea 1770 m away from the previous one, at a depth of about 16 m. This is the technical solution proposed to allow the landing of the submarine cables without affecting the seabed surface and the biological communities present on it, some *Cymodocea nodosa* meadows of high ecological value.

These boreholes are drilled using drill heads operated from land and assisted by a jack-up structure (platform with legs) near the exit point (Figure 3).

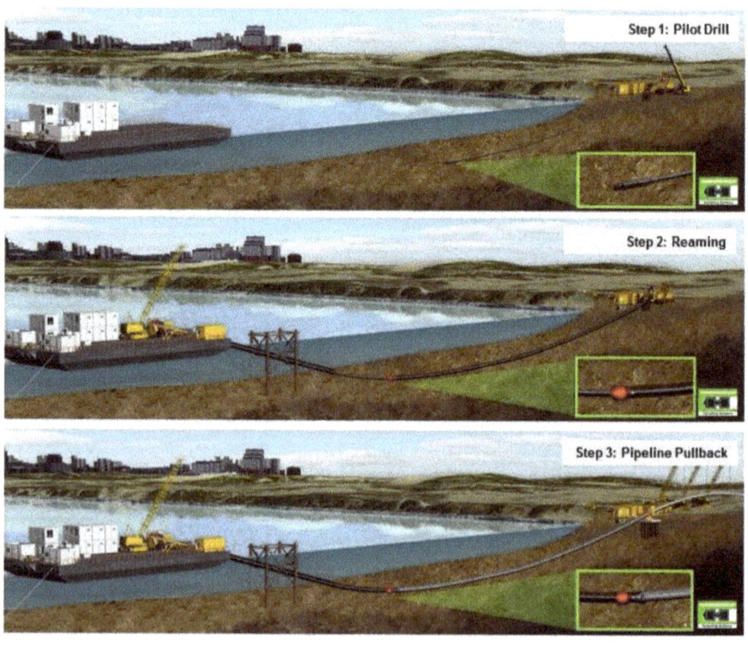

Figure 3. HDD drilling scheme for land–sea transition. Source: Herrenknecht.

The drilling system requires the use of bentonite fluid (composed of water, bentonite, which is a natural clay, and a small proportion of polymeric additives, wetting and dispersing agents), to facilitate drilling and to extract the excavated material, which is constantly recirculated and recycled to an onshore separation plant. Only at the exit of the drilling head at the end of drilling path is the loss of a limited volume of this fluid foreseeable. This fluid would temporarily increase turbidity around the exit point.

To reduce the dispersion and sedimentation of these bentonite clays around the exit point, containment measures are planned to be implemented by temporarily casing the exit point, to allow collection and recycling of the clays.

Based on the dispersion studies of the most unfavorable bentonite fluid emission scenarios, carried out by means of the MOHID numerical model, the predicted impact, characterized by the exceeding of the reference turbidity level (defined in accordance with Additional Provision IX of Law 22/1988 of 28 July 1988 on Coasts as 1.5 times the normal average of suspended solids measured in pre-operational state, i.e., 6.75 mg/L), is temporary (lasting less than 2 h), of low intensity (with maximum values not exceeding 40 mg/L), and limited to the immediate surroundings (distances of approximately 30 m).

5.1.2. Impact of Jetting Operations on Turbidity (Construction Phase)

The jetting technique consists of opening a trench in the seabed by applying pressurized water jets that causes the soil beneath and around the cable to fluidize, allowing the cable to sink through the suspended sediments to the bottom of the trench, to the required burial depth (Figure 4).

Figure 4. Example of subsea cable burial by jetting. Source: CT Offshore.

This technique minimizes the impact footprint and sediment resuspension, being the cable installation and burial method that generates less spatial impact (reduced trench width) and less temporal impact (reduced execution time) of those analyzed in the project.

This burial technique would be applied from the drilling exit point (HDD) to the wind farm site, about 24 km away.

To assess the effect of this operation, the sediment dispersion generated per m of advance was simulated again by the MOHID numerical model, considering a conservative hypothetical situation, in which it is assumed that all the sediment in a 2 m × 1 m trench is resuspended. The results of suspended solids concentration in the water column under this assumption show that the time duration of significant concentrations (above 6.75 mg/L) near the bottom is, in all cases, low: 3 h in the shallowest areas and those closest to the phanerogam meadows (<20 m) and up to 5 h in the deepest ones (120 m).

The impacts are therefore temporary (no more than 3–5 h above the reference threshold of 6.75 mg/L), of low intensity (maximum values below 40 mg/L), and limited again to the immediate surroundings (distances of less than 30 m), and mostly in areas of soft, non-vegetated substrate, so that the effects of the temporary increase in turbidity in the benthic zone on the physical environment and biota are considered compatible, temporary, and of low magnitude. By way of comparison, it should be noted that these levels of suspended solids and their persistence are equivalent to those usually produced during sea storms around the meadows.

Sedimentation in the shallow zone is limited to the area near the trench, and at a distance of more than 30 m does not exceed 2 cm. In the 30 m closest to the trench, 3 cm of thickness can occasionally be reached, between 30 and 60 m distance, 2 cm are not exceeded, and at more than 60 m, 1 cm is not exceeded.

It should be emphasized that the numerical modeling carried out is very conservative, since it considers the suspension of all the material present in the trench. In real observations made in other projects for laying and burying submarine power cables [13,14], the volume of resuspended material has been less than 30% of the trench volume, and the mobilized material accumulates mostly in a short radius around the trench (at a distance of less than 10 m).

5.1.3. Turbidity Impact of the Movement of Turbine Mooring Lines (Operation Phase)

The anchoring systems for floating turbines are composed of three to four mooring lines, consisting of chains connecting the floating platforms with anchors buried more than 10 m below the seabed (Figure 5). Part of these catenaries rest on the seabed to counteract the movement of the turbines under the effect of wind and waves with its weight.

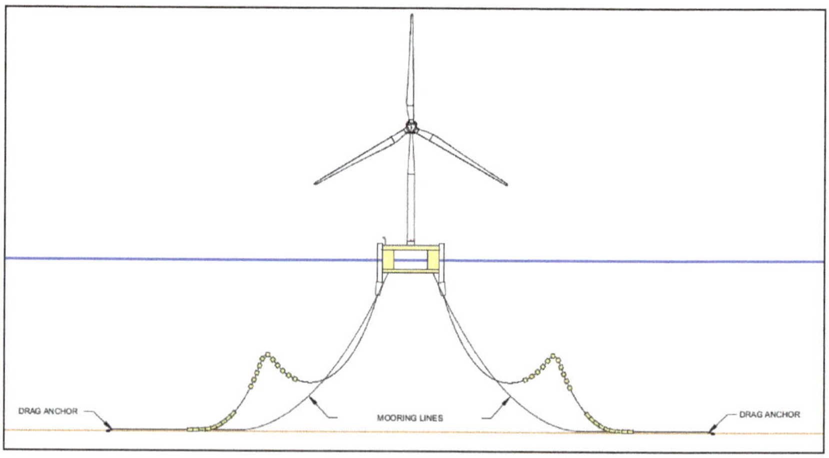

Figure 5. Schematic of the subsea mooring and cabling system of a floating turbine. Source: SENER.

A part of this resting section, approximately 330 m long, will be subject to certain movements that may cause the sediment to resuspend along its contact with the bottom, which may have indirect effects on the biota present in the area, in this case mostly benthic macrofauna that colonize the soft bottoms, as well as demersal species (e.g., hake).

It is estimated that the area affected by this effect is of the order of 1650 m^2 for each catenary (along the resting section and with a maximum arc of movement of 4°), or up to 6600 m^2 for each turbine (considering 4 mooring lines). Considering the area affected by the 35 planned turbines (0.23 km^2) compared to the total area where the wind farm would be installed (about 95 km^2), the area affected by this impact is about 0.24%. By contrast, the area currently subject to resuspension and turbidity on the seabed due to trawling activities in the Gulf of Roses (assuming a fleet of 21 vessels with an average operation of 180 days per year, with 5 h of fishing periods) is estimated at 950 km^2 (95,000 Ha). In addition, this activity generates turbidity values much higher than those associated with the mobility of the mooring lines of a wind farm [11].

Based on the studies carried out, analyzing the possible movements in the catenaries, low speed movements are expected and the effects are not likely to reach a distance greater than 5 m at both sides of the chain, but will be limited to the furrow of maximum amplitude of the footprint considered for the catenary. The level of turbidity generated would therefore be close to that observed due to natural variability, and lower than that usually produced during episodes of strong currents or anthropogenic activity (e.g., if compared to the effect on turbidity produced by trawling, which turbidity plumes can exceed 100 m of thickness and reach suspended solids concentration of 200 mg L^{-1} close to the sea bottom [15]).

The temporal effects of turbidity, although significant, will therefore be of low magnitude and low periodicity occurrence, considering the slowed movement of the chains. These effects on the physical environment are considered compatible, considering the adaptive capacity of the species that colonize the affected sedimentary bottoms.

5.1.4. Overall Assessment of the Impact on Turbidity and Sedimentation

During the construction phase, the impact on the increase in turbidity and sedimentation is considered significant in the shallow area, considering that it happens in waters of protected natural areas (RN2000) and the presence of the marine phanerogam *Cymodocea nodosa* meadow in the landing area, despite the fact that the effect is temporary (hours) and the rate and spatial extent of sedimentation is limited.

By assessing the different descriptors of each impact (turbidity and sedimentation), according to the described methodology, a Moderate impact rating is obtained for both during the construction phase, as shown in Table 2:

Table 2. Overall assessment of the impact on turbidity and sedimentation during the construction phase.

Attribute	Turbidity		Sedimentation	
	Characterization	**Rating**	**Characterization**	**Rating**
Sign	Negative	(−)	Negative	(−)
Extension	Partial	2	Partial	2
Persistence	Temporary	2	Permanent	4
Synergy	Synergistic	2	Synergistic	2
Effect	Direct	4	Direct	4
Recoverability	Immediate	1	Immediate	1
Intensity	Medium	2	Low	1
Moment	Immediate	4	Immediate	4
Reversibility	Short period	1	Medium period	2
Accumulation	Simple	1	Cumulative	4
Periodicity	Continuous	4	Continuous	4
TOTAL RATING	MODERATE	−29	MODERATE	−32

In terms of the impact on this vector during the operation phase, the expected impacts on turbidity are very limited, restricted to less than 1% of the project area, which in relation to the current impact of trawling in the area occupied by the park where this activity would cease (about 600 ha, equivalent to 6% of the park's surface), represents, both in terms of the extent and magnitude of the impact, a relative improvement in terms of turbidity conditions and recoverability of the seabed from sediment resuspension. From the modeling carried out, it is concluded that the levels of suspended solids in the water column derived from the movement of the chains are scarcely significant (less than 2 mg/L at distances of less than 25 m) at all times, being considered of low intensity and persistence (levels of less than 1 mg/L in less than 2 h).

The assessment of these impacts in the operation phase is summarized in Table 3:

Table 3. Overall assessment of the impact on turbidity and sedimentation during the operation phase.

Attribute	Turbidity		Sedimentation	
	Characterization	**Rating**	**Characterization**	**Rating**
Sign	Negative	(−)	Negative	(−)
Extension	Partial	2	Partial	2
Persistence	Permanent	4	Permanent	4
Synergy	Synergistic	2	Synergistic	2
Effect	Direct	4	Direct	4
Recoverability	Immediate	1	Immediate	1
Intensity	Low	1	Low	1
Moment	Immediate	4	Immediate	4
Reversibility	Short period	1	Medium period	2
Accumulation	Simple	1	Cumulative	4
Periodicity	Continuous	4	Continuous	4
TOTAL RATING	MODERATE	−28	MODERATE	−32

5.2. *Analysis of the Project's Impact on Underwater Noise*

The project has the potential to alter the acoustic environment of the area, both during the construction and operation phases.

In the study area, background underwater noise (reference situation), which is related to natural sources (e.g., wind) and artificial sources (e.g., maritime traffic, professional fishing), was measured by means of PAM (Passive Acoustic Monitoring) equipment. The recorded noise values reach peaks (SPL) of 156–159 dB re 1 μPa and cumulative averages (SEL) of 134–139 dB re 1 μPa at the wind farm proposed location. The greatest contribution is from maritime traffic, highlighting the important fishing activity linked to the port of Roses.

Underwater noise has the potential to alter the acoustic environment of the area and affect several species, especially marine mammals, with special attention to cetaceans present in the study area (mainly bottlenose dolphins), and chelonians such as loggerhead turtles (with less presence in the area).

The acoustic impact will depend on the distance to the source of the receptor. Considering the distribution of cetaceans in the study area, observed during the monitoring campaigns carried out during the last year, the main type of detected cetaceans are dolphins (bottlenose and striped dolphins), associated with a medium frequency hearing range (150–160 kHz), and to a lesser extent species with a low frequency hearing range (7–35 kHz), such as the fin whale.

In relation to the potential effect on turtles (loggerhead turtle), a reference level RMS of 166–175 dB re 1 μP [16] is considered to be the range for behavioral change effects, based on experimental measurements with species in captivity.

According to *MAGRAMA* [17], the reference levels for the definition of exclusion zones are 160 and 180 dB rms, corresponding to thresholds for which behavioral changes and physiological damage are detected in cetaceans, respectively.

5.2.1. Noise Impact during the Construction Phase

During the construction phase, the main source of noise is associated with the vessels that will install the wind turbines and their anchoring systems and lay the cable. The underwater noise emission is proportional to the speed of the vessels, which, due to the nature of the work to be carried out, will generally be low (between 3 and 5 knots).

Other sources of noise identified in this phase are the use of acoustic devices for positioning during cable laying or anchor installation (echo sounders, sonars, acoustic positioning systems), the use of pumping equipment and the pressurized water jetting system (jetting), displacement along the bottom, sliding (dragging), and finally machinery on board for lifting and lowering equipment.

One of the main differences between the projected wind farm (with floating foundations) and most wind farms in the world (which have fixed foundations) is that in this case no pile driving activities are carried out for the execution of foundations. Pile hammering is the main source of underwater noise impact identified by the scientific community in association with offshore wind farms, having been analyzed in a large number of studies that show its impact on certain marine species, particularly cetaceans. Consequently, this important impact does not happen in a project such as Parc Tramuntana, as no impulsive, high-energy noise will be generated during construction.

Thus, the noise that may be generated during the construction phase does not differ much from noise associated with other types of maritime works, associated with low frequencies (e.g., navigation), and is of a temporary nature.

According to the consulted literature, the noise generated by cable-laying ships in shallow waters is of the order of 164–188 dB re 1 μPa, at 1 m from the source, acting at frequencies between 0.7 and 50 kHz: surface ship 180 dB (CEDA, 2011) [18], 164–170 dB (Nexans Skagerrak). In relation to underwater activities, a reference of burial works at 1 m from the source is available, with emissions up to 188.5 dB (at 11 kHz) [19] or 174 dB [20]. It should also be noted that modern newly built vessels reduce acoustic emissions and vibrations from engine operation very significantly.

The effects of the construction phase on this variable are considered significant, although the existing risk of direct effects is limited to the presence of wildlife in the vicinity of the noise source, noting that the negative effects would be temporary and that noise generation would be progressive, with a chasing effect that avoids further damage to the animals. The assessment of this impact in the installation phase is therefore moderate (Table 4):

Table 4. Assessment of the impact on submarine noise during the construction phase.

Attribute	Characterization	Rating
Sign	Negative	(−)
Extension	Extensive	4
Persistence	Temporary	2
Synergy	Synergistic	2
Effect	Direct	4
Recoverability	Immediate	1
Intensity	Medium	2
Moment	Immediate	4
Reversibility	Short period	1
Accumulation	Simple	1
Periodicity	Continuous	4
TOTAL RATING	MODERATE	−28

As the main corrective measure, the use of modern workboats with low noise emission certifications (e.g., Silent-E) is proposed [21].

5.2.2. Noise Impact during the Operation Phase

During the operation phase of the wind farm, the main source of underwater noise generation will be the wind turbines themselves, which will produce continuous noise due to the movement of the blades, transmitted mainly through vibrations along the floating platform and the bottom-anchoring chains, and to a much lesser extent directly through the water surface, since the change in medium (air/water) produces a significant attenuation of atmospheric sound.

There will also be transient noise related to the movement of the anchor chains on the bottom due to the movement of the floating turbines.

In addition to the noise from the wind turbines, there will be noise from the vessels responsible for maintenance/repair tasks, similar in nature to that expected during the construction phase. The impact of these vessels will have a limited frequency (20 days/year), similar to those currently caused by marine and fishing traffic.

In order to characterize the impact of underwater noise produced by the turbines, the existing literature on the subject was analyzed, coming from the scarce experiences in other floating offshore wind farms, specifically the studies and measurements of underwater noise in the Hywind Tampen wind farm [22], the first floating technology wind farm installed off the coast of Norway, in the North Sea.

In this wind farm, the noise recorded during operation was characterized as continuous and low frequency (25, 50, and 125 Hz). The results did not exceed in any case an SPL of 160 dB re 1μPa (reference level adopted as a threshold to determine the limit of the cetacean disturbance zone, according to *MAGRAMA* recommendations [17]), and the isophone range of 120 dB re 1 μPa, assimilable to the background noise recorded at the Hywind Tampen wind farm site), occurs, according to the developed acoustic modeling on the basis of measurements, at circa 2 km from the wind turbines, with a maximum noise footprint of approximately 40 km^2.

Similar experiences in the environmental monitoring phase of a fixed foundation (jackets) wind farm operating at a depth of 25 m off New York Bay, the Block Island Wind Farm [23], show maximum noise values generated during operation of up to 120 dB re 1 μPa at a distance of 50 m from the turbines, at a site with a background noise of 110 dB re 1 μPa.

These levels are considered an approximation to the underwater acoustic footprint that may be produced by Parc Tramuntana, whose increase in the acoustic scenario around the wind turbines will be cumulative depending on the number of wind turbines and the environmental conditions of the surroundings. It is estimated that this acoustic increase will be attenuated to the background noise levels existing at a distance of approximately 2 km or less (since in the case of Tramuntana the background noise is more intense than that existing in the other reference wind farms analyzed).

The expected RMS levels from this distance (around 120 dB re 1 uPa) are significantly lower than the background noise and the thresholds that would cause annoyance to the most sensitive species identified (dolphins and sea turtles), which are not highly abundant in the wind farm site area.

After this analysis, the effects of the operation phase on this variable are considered significant, of partial extension (affecting the area near the wind farm's footprint), and of medium magnitude due to the acoustic levels generated. The risk of direct effects is limited to the presence of fauna in the vicinity of the noise source, which is irregular and periodically distributed. The assessment of this impact during the operation phase is shown in Table 5:

Table 5. Assessment of the impact on submarine noise during the operation phase.

Attribute	Characterization	Rating
Sign	Negative	(−)
Extension	Extensive	4
Persistence	Permanent	4
Synergy	Synergistic	2
Effect	Direct	4
Recoverability	Immediate	1
Intensity	Medium	2
Moment	Immediate	4
Reversibility	Short period	1
Accumulation	Cumulative	4
Periodicity	Continuous	4
TOTAL RATING	MODERATE	−38

The applicable measures are the monitoring of acoustic levels and the recording of cetacean activity in the area, through visual censuses and the installation of hydrophones, to verify the levels and activity of potentially affected fauna.

5.3. Analysis of the Project's Impact on Electromagnetic Fields (EMF)

This potential impact is only likely to occur during the operation phase, as it is associated with the transmission of electricity generated by the wind farm through the submarine cables.

The interconnection cables (inter-array) installed in the offshore wind farm and the export cables will operate in alternating current at 66 kV–50 Hz. An analysis of the EMF generated by these cables was carried out using numerical modeling based on the *Biot–Savart* analytical calculation for the different cable segments (both the dynamic inter-array cables, which connect the turbines to each other, and the buried export cables, which connect the wind farm to shore). This model adopted a conservative calculation, considering the most unfavorable possible scenario (cable operation at full load), and without considering the shielding effects due to the cable protection armor.

This simulation of EMF levels (see Figure 6) obtains magnetic field levels (B) for the maximum inter-array cables at the surface of the conductors of 90 μT and, for the evacuation cable, a maximum level of 5 μT on the seabed.

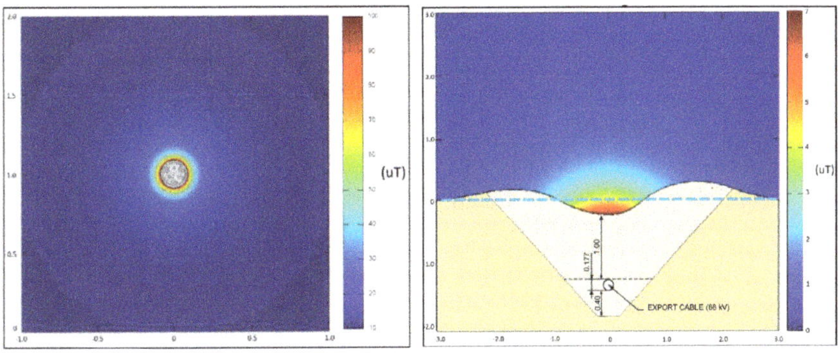

(a) (b)

Figure 6. EMC modeling for the inter-array (**a**) and export (**b**) cables.

The electric field (E) induced by the 66 kV conductors will be zero on the outside of the cables, since it is blocked by the metallic screen of the cable itself.

The potential effects of EMF on aquatic fauna include the possible disorientation of migratory species that use the terrestrial magnetic field for orientation during navigation, behavioral alteration with attraction or repulsion effects (barrier effect), as well as potential physiological damage at the cellular level.

As a reference, it should be noted that normal values of the earth's geomagnetic field can range from 20 to 75 μT, depending on the geographical location.

In order to assess which of these effects could be relevant to the project, a large amount of literature was consulted to analyze the effects of EMF on different species (Table 6):

Table 6. Documented effects of EMF on marine species.

Species	Threshold Value	Effect	Source
Eel (*Anguilla anguilla*)	5 μT (direct current)	Course deviation during migration	Westerberg and Begout-Anras (2000) [24]
Turtle (*Caretta caretta*)	4.9 μT_	Disorientation of juveniles during migrations periods	Fuxjager et al. (2011) [25]
Mussel (*Mytilus galloprovincialis*)	300–1000 μT_	Alteration of cellular processes for adults in the laboratory	Ottaviani et al. (2002) [26] Malagoli et al. (2004) [27]
Shrimp (*Crangon crangon*)	10–100 μT_ 800 μT_	Behavioral disturbance in adults Attraction effect on adults	Hutchinson et al. (2018) [28]
Fish and macroinvertebrates	1–3.2 mT_	No significant long-term behavioral effects detected	Woodruff et al. (2012) [29] Bochert and Zettler (2004) [30]
Elasmobranchs	91 μV/m	Alteration of behavior by attraction or repulsion	CMACS (2003) [31] Gill and Taylor (2001) [32]
	60,000 μV/m	Narcosis	Smith (1974) [33]
Elasmobranchs and rajoids	1/8 to 8 Hz	Electrosensitivity within the frequency range	Kalmijin (2000) [34] Walker et al. (2003) [35]

In relation to species sensitive to electric fields, elasmobranchs would be the most potentially affected group, whose ability to detect these fields is very sensitive, detecting levels of less than 0.5 V/m (5 nV/cm).

Among the species considered in the bibliography, the most important because of their potential presence and migratory habits are the loggerhead turtle (*Caretta caretta*), the bottlenose dolphin (*Tursiops truncatus*), devil fish (*Mobula mobular*), and the main protected species of elasmobranchs with a potential presence in the study area due to their conservation status: common thresher (*Alopias vulpinus*), basking shark (*Cetorhinus maximus*), tope shark (*Galeorhinus galeus*), bluntnose sixgill shark (*Hexanchus griseus*), shortfin mako shark (*Isurus oxyrinchus*), smooth-hound (*Mustelus* spp.), blue shark (*Prionace glauca*), and spiny dogfish (*Squalus acanthias*).

The maximum magnetic fields generated by the project are within the range of values detectable by marine fauna.

The field generated by the inter-array cables in the wind farm area (up to 90 μT) will be detectable by potentially present species of interest. Considering that the cables analyzed do not alter the electric field, significant effects on elasmobranchs are dismissed, although their behavior will be analyzed during the operation phase. This effect is localized in the closeness of the cables, as the magnetic field levels attenuate rapidly a few meters away.

The potential effects on fauna are considered locally limited; it should be noted that the cable's footprint in the sea bottom is approximately 10,000 m^2, less than 1% of the wind farm's area, and barely perceptible (<5 μT) at short distances (<1 m). Due to the extent of these fields and the magnitude of EMF and derived effects, they are not considered to have a significant barrier effect on migratory species.

In the case of the fields associated with the export cables (<5 μT), these are of limited magnitude, close to the levels of natural electromagnetic disturbances that regularly happen (*Nyqvist* et al., 2020) [36], and are therefore expected to be barely perceptible by potentially affected fauna, as the cable is buried deep enough for the field to attenuate to levels that are practically imperceptible at the seabed surface.

It should be noted that in EMF monitoring surveys of buried electrical interconnections, levels of the order of nT are usually detected. Considering its continuous route to the coast, the species of interest potentially affected would be the loggerhead turtle, in the event of passing through to nest on nearby beaches, and the bottlenose dolphin, due to its local presence. However, due to the magnitude of the impact, as described above, the effect on these species is considered insignificant. These values can also be extrapolated to the subway cable line through the HDD.

On the other hand, it should be noted that it is foreseeable that in the operating phase the EMF generated will be lower than that modeled, considering the maximum regular operating loads contemplated and the additional shielding provided by the cable protection armor; according to the literature consulted, these levels would not be capable of causing physiological damage or significant changes in the behavior of the fauna present in the project area.

Anyway, the increase in EMF during the wind farm's operation phase was considered a significant effect, mainly in the wind farm's footprint area (where the inter-array cables are located). This impact is considered moderate, according to the following attribute classification (Table 7).

Table 7. Assessment of the impact of EMF on marine fauna.

Attribute	Characterization	Rating
Sign	Negative	(−)
Extension	Partial	2
Persistence	Permanent	4
Synergy	Simple	1
Effect	Direct	4
Recoverability	Immediate	1
Intensity	Medium	2
Moment	Immediate	4
Reversibility	Short period	1
Accumulation	Simple	1
Periodicity	Continuous	4
TOTAL RATING	MODERATE	−30

Among the main mitigation measures, in addition to the design of the cables to reduce EMF through shielding and burial in the evacuation route, it is proposed to monitor EMF levels in the inter-array cables and in the evacuation route, as well as the monitoring of cetaceans, turtles, pelagic communities (including elasmobranchs), and benthic macrofauna.

5.4. *Analysis of the Project's Impact on Marine Circulation and Nutrient Distribution*

Once the offshore wind farm is in operation, the operation of the wind turbines could generate changes in the atmospheric and oceanic dynamics due to the capture and modification of wind energy. On the other hand, the floating platforms, with a draft of −15 m, could generate a certain wave reduction effect or local alteration of the currents.

The main expected effect could be the reduction in wind energy in the area of the wind farm, associated to the wake effect, which consists of the reduction of wind speed and kinetic energy downwind of the wind farm.

It is considered that the capacity of the underwater structures to alter the hydrodynamic regime of the Gulf of Roses is very low, considering the depth at which the wind farm is located (>100 m) and that the installed structures (floating platform, chains, anchors, and inter-array cables) are not large enough to produce a significant reduction in current flows, stratification, or vertical transport phenomena in the water column.

Although these effects may happen, they will be very localized in the immediate surroundings of the structures and of little entity at the level of the Gulf of Roses (where they will probably be imperceptible), which does not confer the ability to become cumulative or synergistic downstream. The effect of this potential reduction in the wind and hydrodynamic regime on the development of species and habitats is assumed to be low, due to the expected low incidence of the reduction in physical effects on the column and seabed of the installed wind turbines, as well as the capacity of biological adaptation of the species potentially affected to small changes in the circulation.

This impact has first been assessed on the basis of a critical analysis of the existing literature, and a rigorous study by IHCantabria about the impact of the project on wind fields and marine currents has also been undertaken, by means of numerical modeling. This study, whose methodology has been presented to the Spanish National Research Council (CSIC) experts for their approval, and no objections have been received from them, will provide quantitative information on the foreseeable variation in current speeds in the area of the Gulf of Roses.

As a result of the bibliographic analysis, some scientific publications that analyze and document wind field alteration effects were identified, mainly based on wind farms located in the North Sea [37]. In some cases, it is suggested that these alterations could lead to increases in precipitation, although the predicted changes are so small in magnitude that it is difficult to distinguish them from natural variability. According to Platis et al. (2018) [38], the wake effect is greater under stable atmospheric conditions than under turbulent conditions. When conditions are turbulent, such as those associated with the Tramuntana winds that will feed the wind farm, the wake effect is restricted to the local area inside the wind farm and the immediate surroundings.

It should also be noted that in general, regarding wind disturbance, the numerical models give overestimated values with respect to the field measurements. This divergence between modeled and measured values indicates deficiencies in the knowledge of the effects. Measurements made in the "Horn Rev" and "Nysted" fields in the Baltic Sea show reductions of 8–9% immediately after the field, and 2% at distances between 5 and 20 km [39]. As for the regional effect, it is estimated to be small, the energy loss in the first kilometer of the atmosphere being 0.007%. The conclusion is that floating wind farms may have a minor to moderate impact on atmospheric and oceanic dynamics (depending on the location and size of the wind farm), although there is insufficient specific knowledge of the cascading effects of large-scale atmospheric and oceanic processes to reduce the current uncertainty.

In relation to the possible effect of the alteration of currents on the distribution of nutrients, Van Berkel et al. (2020) [40] point to the potential upwelling effect associated with the wake, causing the upwelling of deep-water mass to the surface. However, this effect depends on the relative direction between the wind and the coast, being a frequent phenomenon on the Atlantic coast but not so in the western Mediterranean.

It should be noted that the bibliographic references in which appreciable variations in the current regime have been observed as an indirect consequence of the presence of a wind farm correspond to cases in shallow areas, with fixed foundation wind farms, and with a high concentration of turbines, being scenarios not very comparable with the project scenario.

The results of the hydrodynamical study by means of numerical modeling developed by IHCantabria show that the effect of the wind farm over the surface currents and those averaged over the water column in the coastal area is very small, with an averaged difference in the currents magnitude below 1 cm/s and maximum values of 1 cm/s during 98% of the time.

In the proximities of the wind farm in the open sea, the differences obtained, in both surface and averaged water column, remain below 1 cm/s at the north, east, and west from the farm, while at the south this difference is up to 2 cm/s, which represents an averaged variation inferior to 3.5% of the average speed registered in the study area by the buoy of *Puertos del Estado*.

In any case, to confirm the potential impact, the mitigation measures and Environmental Monitoring Plan propose monitoring the wind and current regime of the involved area, as well as monitoring the thermohaline structure of the main indicators of the water masses (temperature, salinity, turbidity), nutrients in the water column, and monitoring the evolution of the main pelagic and benthic communities in the affected area, whose evolution will be analyzed with a Before-After Control-Impact (BACI) study approach.

The impact is considered significant, permanent (during the life of the wind farm), and synergistic. Due to the uncertainty of the effect and the fact that mitigation measures cannot be developed, a medium magnitude is assigned to it. The resulting classification is shown at Table 8:

Table 8. Assessment of the impact on marine hydrodynamics.

Attribute	Characterization	Rating
Sign	Negative	(−)
Extension	Extensive	4
Persistence	Permanent	4
Synergy	Synergic	2
Effect	Direct	4
Recoverability	Immediate	1
Intensity	Medium	2
Moment	Immediate	4
Reversibility	Short period	1
Accumulation	Cumulative	4
Periodicity	Continuous	4
TOTAL RATING	MODERATE	−38

5.5. Analysis of the Impact of the Project on the Proliferation of Invasive Alien Species

Another aspect that has been identified by scientists as a possible threat to marine biodiversity, in association with offshore wind farms, is their potential to promote the proliferation of invasive alien species.

According to the available knowledge by the scientific community on the proliferation of allochthonous species in the Mediterranean, either introduced intentionally or by chance by different ways, it is currently considered that this phenomenon could be one of the main causes of biodiversity loss in the Mediterranean. The number of invasive alien species inventoried in this sea in 2012 reached almost a thousand (more than 5% of the species present in this sea), with more than 300 identified in the western basin [41].

There is a wide variety of species with the potential to become invasive, although organisms with the ability to attach to hard surfaces (biofouling), including mollusks, are generally the most notorious. The known introduction ways are numerous, one of the most relevant being the opening of the Suez Canal in 1869, which involved communication between the Red Sea and the Mediterranean Sea, favoring the migration of numerous species from the Red Sea to the Eastern Mediterranean, some of which have come to spread throughout the Mediterranean basin [42].

Other known ways of anthropogenic introductions are the transport of sessile species attached to the hull of merchant ships or other structures moving from one place to another, the discharge of ballast water, which carries planktonic and nektonic organisms (some of which are larval stages of invasive species), the introduction of species of interest for mariculture and fish farming (which often carry parasitic organisms), the intentional or accidental release of specimens by aquariums, or even the recent proliferation of microplastics in the seas [43].

Regarding this threat, it is also worth noting the effect that climate change is likely to have on this phenomenon, since it is expected to affect, by modifying water temperature, the structure of marine communities, providing more opportunities for exotic species to disperse and compete with and displace native species [42].

Based on current knowledge, and in relation to the structures and actions foreseen in the project, the following possible routes of entry of allochthonous species into the project area are contemplated:

- Through adherence to the structures of the foundation floaters, during their towing from the manufacturing or assembly port:
  The structures are expected to be manufactured either in western Mediterranean ports, with a low probability of introduction of new species, or from the Atlantic ports of the Iberian Peninsula. The rest of the elements that will remain on the seabed (submarine cables, anchors, and chains) are not expected to be exposed to marine colonization outside the park, since they will be transported aboard ships.

In both cases, the main means of transport will be semi-submersible vessels between the manufacturing port and the wind turbine assembly port (e.g., in Tarragona), so possible adhesions will only occur during transport between the port of Tarragona and the wind farm site. The probability of introduction of allochthonous species is therefore very low, and comparatively lower than that which may be due to the transit of merchant ships coming from the Strait of Gibraltar or the Suez Canal to ports in the western Mediterranean.

- By attachment to working vessels during construction (cable-layers, tugboats, etc.): These vessels are highly specialized and generally work in all the world's oceans, so there is a possibility that they may carry allochthonous species attached to their hulls, with potential for invasion and displacement of native species.
  However, this is not a significant threat, in relation to the number of vessels with which it is associated, compared to the intense maritime traffic of merchant ships, pleasure craft, fishing boats, or cruise ships that sail the waters of the western Mediterranean from other seas.
- By releasing ballast water from the foundation platforms:
  The floating turbine foundation platforms have ballast systems that involve filling or emptying tanks integrated into their structure to contribute to the stability of the whole. The partial filling of these tanks will be carried out at the assembly port (Tarragona), ending at the location of the wind farm.
  It is therefore not foreseeable, as has been argued above in the case of the adhesion entry route, that these ballast waters will constitute a significant entry route for invasive non-native species.

Finally, the potential effect of the floating structures themselves, as well as the anchoring system and the inter-array cables, which will be the only elements exposed to marine colonization, should be evaluated as possible elements of attraction and settlement of invasive alien marine species. This effect, far from introducing new species, could only contribute to a greater growth of the species present in the area. Its effect would be equivalent to that provided by any structure capable of providing a substrate for the settlement of sessile organisms, especially if it is located within the photic zone, where photosynthesis occurs and with it the greatest growth of biomass. That is, it would be equivalent to that associated with boat hulls, buoys, fishing gear, artificial reefs, or shallow rocky bottoms.

It should also be noted, in the case of platforms, that these structures are protected against corrosion and biofouling by means of cathodic protection systems and/or impressed current, which significantly minimizes biological adhesion. Therefore, only the surface of the chains and inter-array cables would be exposed to colonization at a relevant level.

It is possible that the presence of the floating structures may generate slight changes in the behavior of pelagic species, which may be attracted or repelled in their immediate environment depending on their feeding, predatory, shelter, or reproduction habits. However, these effects are not considered to be relevant in terms of compromising the conservation of these species, especially when compared to other pressures on these species in the project area, such as trawling or maritime traffic.

For all of the above reasons, the effect of the project on the proliferation of invasive alien species was considered not significant for the purposes of impact assessment.

6. Conclusions

The social controversy related to the introduction of offshore wind power in Catalonia has scaled to a scientific debate about its environmental compatibility, and specifically in relation to the Parc Tramuntana project. Recent publications have contributed to sustaining this debate, without being adequately based on rigorous analysis of both the characteristics of the project and its specific location within the Gulf of Roses, so it has been considered appropriate to provide an exhaustive analysis of the main impacts identified by the scientific community in relation to offshore wind farms.

The aim of this report was therefore to dispel some of the uncertainty that motivates this opposition, based on detailed information and quantitative data, in order to place these impacts in a realistic context and determine their true magnitude.

This is possible because a series of diverse and detailed studies have been developed over the last two years to assess the environmental impact of the project, which are relatively advanced and ready to provide some well-founded conclusions. In the meantime, the study continues to be prepared and, once the relevant studies are completed, it will be submitted as part of the documentation required by current legislation for the environmental processing of the project. Although this procedure contemplates a 30-day public information phase, it is desired to maintain a less limited channel of dialogue and prior communication, for which reason the promoter commits to advance as far as possible the conclusions that emerge from the environmental studies and to make them available to any interested person through the project's web page.

It should be noted that both the project and the corresponding environmental impact study are being developed in accordance with current environmental processing regulations and based on the Scope Document submitted by the General Directorate for Environmental Quality and Assessment of the Ministry of the Environment. Likewise, the design of the wind farm's infrastructure is in accordance with the conditions established by urban and sectorial regulations.

Based on the studies carried out, which are based on a standardized methodology, the impact of the project is objectively classified as MODERATE on vectors such as turbidity and sedimentation, underwater noise, hydrodynamic circulation, or the alteration of electromagnetic fields, and NOT SIGNIFICANT on aspects such as the proliferation of invasive exotic species (Table 9).

Table 9. Summary of analyzed environmental impact ratings and characterization.

Impact Assessment	Phase	Rating (-)	Characterization
Turbidity	Construction Operation	29 28	MODERATE MODERATE
Sedimentation	Construction Operation	32 32	MODERATE MODERATE
Submarine noise	Construction Operation	28 38	MODERATE MODERATE
EMF on marine fauna	Construction Operation	N/A 30	N/A MODERATE
Marine hydrodynamics	Construction Operation	N/A 38	N/A MODERATE
Proliferation of invasive exotic species	Construction Operation	N/A N/A	N/A N/A

Impacts on turbidity and sedimentation are expected along the evacuation cables track during construction stage and only in the close vicinity of turbine moorings during the operational stage.

Submarine noise is expected also during both stages, mainly associated to construction equipment and ships during construction and to both maintenance ships and turbine structure vibrations during operation, although emission levels are not expected to be higher than current background noise in the area or to cause severe damage to sensitive species present in the area.

EMF are only expected in operational stage, being significant mainly in the close vicinity of the floating inter-array cables. Generated fields are expected to be detectable by sensitive species, but not strong enough to cause severe alterations in their behavior.

The impact on marine hydrodynamics, based on numerical modeling, is expected to be low, limited to changes in currents magnitude of 1–2 cm/s in the vicinity of the wind farm during operational stage.

Finally, the effect of the project on the potential proliferation of invasive species was considered not significant, as there are small chances for the introduction of alien species associated to project structures, being irrelevant when compared with other identified introduction vectors in the area.

In general, the studies carried out allow to anticipate a low impact of the floating offshore wind farm project on marine biodiversity in the area of implementation, which makes it compatible with the conservation of such biodiversity.

In addition, most of the significant impacts identified allow the application of measures to mitigate their effects, as well as their follow-up during the environmental monitoring derived from the environmental processing, which will make it possible to know the evolution of the wind farm's effects on the environment, as well as the adoption of additional measures if necessary.

These initial conclusions address only certain environmental aspects of the project that have been identified as being of greatest concern to some members of the scientific community. As the environmental impact study progresses, it is expected that conclusions regarding other potential environmental or socioeconomic impacts will also be published.

It is clear that any project, of whatever nature, introduces certain changes in the environment and therefore has an associated impact. However, it is equally true that it is imperative to slow down climate change by all available means, among which the transformation of the energy generation model to a renewable model is a priority.

It should be remembered that the effects of climate change predicted by the IPCC for the coming decades (many of which are already occurring today, including in the Mediterranean) include an increase in the temperature of the planet and its oceans, the extinction and anomalous migration of species, colonization by invasive species, the modification of marine currents and acidification of the seas, droughts, and catastrophic climatic phenomena.

Therefore, the suitability or otherwise of offshore wind power in Catalonia should not be evaluated solely from the perspective of its effect on biodiversity, especially considering that climate change itself will certainly introduce much more drastic changes in Mediterranean ecosystems than those that can be attributed to any individual project, resulting in more severe, lasting, and irreversible impacts on marine biodiversity.

Thus, from an equanimous and global vision, the assessment of a project such as Parc Tramuntana should be carried out with scientific rigor and without falling into the tunnel vision that only focuses on the local negative impacts, without placing them in context. Because, according to environmentalists [44], even if it is accepted that, like any other renewable energy project or any other type of project, it may have a certain level of impact on local biodiversity, the relationship between this impact and the protection of biodiversity throughout the Mediterranean is largely in favor of the latter.

Finally, in response to the arguments calling for a halt to marine renewables under a precautionary principle that only greater scientific knowledge can overcome, it is necessary to indicate that unfortunately the time available for the study is limited, since the climate emergency is already a reality and requires an urgent response. It is therefore a priority to join efforts to ensure the optimal response to the climate challenge, as is being done in neighboring countries.

Author Contributions: Conceptualization, K.D.-C. and S.T.; methodology, K.D.-C.; software, J.G.-A.; formal analysis, J.G.-A.; investigation, K.D.-C. and S.T.; supervision, S.A.; writing—original draft preparation, S.T. and K.D.-C.; writing—review and editing, S.A., R.J., M.G. and J.R.V.; project administration, M.G. All authors have read and agreed to the published version of the manuscript.

Funding: This research received no external funding.

Institutional Review Board Statement: Not applicable.

Informed Consent Statement: Not applicable.

Data Availability Statement: Not applicable.

Conflicts of Interest: Mr. Koldo Diez-Caballero and Mr. Juan Ramón Vidal are employees of Tecnoambiente S.L., an environmental consultant that has developed the hydrographic, geophysical and environmental surveys of the project area, and is developing the Environmental Impact Assessment studies of the offshore wind farm project. Ms. Silvia Troiteiro, Ms. Marta González, Mr. Sergi Ametller and Ms. Raquel Juan are employees of SENER Ingeniería y Sistemas, S.A., a company of the SENER Group. The offshore wind farm mentioned in the article is being developed by BlueFloat Energy International, S.L.U. together with SENER Renewable Investments, S.L. (a company of the SENER Group) as minority partner. Ms. Silvia Troiteiro, Ms. Marta González, Mr. Sergi Ametller and Ms. Raquel Juan have participated in the elaboration of the study referred to in the article and have had access to all information produced by such study. Mr. Javier García-Alba is an employee of the IHCantabria, a non-profit research organization that has developed the specialized hydrodynamic studies to determine the impact of the wind farm on marine dynamics.

References

1. Acord de Govern de la Generalitat de Catalunya, de 14 de Maig de 2019, de Aprovació de la Declaració de Emergència Climàtica a Catalunya. Available online: https://govern.cat/govern/docs/2019/05/14/15/12/9ff53be9-20cc-4ce1-9c7c-cdd5dc762187.pdf (accessed on 26 April 2022).
2. Acuerdo del Consejo de Ministros el 21 de Enero de 2020 Por el que se Aprueba la Declaración del Gobierno Ante la Emergencia Climática y Ambiental. Available online: https://www.miteco.gob.es/es/prensa/declaracionemergenciaclimatica_tcm30-506551.pdf (accessed on 26 April 2022).
3. European Parliament Resolution of 28 November 2019 on the Climate and Environment Emergency (2019/2930(RSP)). Available online: https://www.europarl.europa.eu/doceo/document/TA-9-2019-0078_EN.pdf (accessed on 26 April 2022).
4. Institut Català de l'Energia. Llei de Transició Energètica de Catalunya i Transformació de l'Institut Català d'Energia at l'Agència d'Energia de Catalunya. Available online: http://icaen.gencat.cat/ca/participacio/llei-de-transicio-energetica-de-catalunya-i-transformacio-de-linstitut-catala-denergia-en-lagencia-denergia-de-catalunya/ (accessed on 26 April 2022).
5. Institut Català de l'Energia. Prospectiva Energètica de Catalunya 2050 (PROENCAT 2050). Documento Resumen Presentado el 4 de Febrero de 2022. Available online: http://icaen.gencat.cat/es/l_icaen/prospectiva_planificacio/ (accessed on 26 April 2022).
6. Ministerio para la Transición Ecológica y el Reto Demográfico. Proyecto de Real Decreto/2021 por el que se Aprueban los Planes de Ordenación del Espacio Marítimo de las Cinco Demarcaciones Marinas Españolas. 2021. Available online: https://www.miteco.gob.es/es/costas/temas/proteccion-medio-marino/ordenacion-del-espacio-maritimo/default.aspx (accessed on 26 April 2022).
7. Salvador, S.; Gimeno, L.; Sanz Larruga, F.J. The influence of regulatory framework on environmental impact assessment in the development of offshore wind farms in Spain: Issues, challenges and solutions. *Ocean. Coast. Manag.* **2018**, *161*, 165–176. [CrossRef]
8. Verma, S.; Akshoy, R.P.; Nawshad, H. Selected Environmental Impact Indicators Assessment of Wind Energy in India Using a Life Cycle Assessment. *Energies* **2022**, *15*, 3944. [CrossRef]
9. Kouloumpis, V.; Azapagic, A. A model for estimating life cycle environmental impacts of offshore wind electricity considering specific characteristics of wind farms. *Sustain. Prod. Consum.* **2022**, *29*, 495–506. [CrossRef]
10. Chowdhury, N.E.; Shakib, M.A.; Xu, F.; Sayedus, S.; Islam, M.R.; Bhuiyan, A.A. Adverse environmental impacts of wind farm installations and alternative research pathways to their mitigation. *Clean. Eng. Technol.* **2022**, *7*, 100415. [CrossRef]
11. Lloret, J.; Turiel, A.; Solé, J.; Berdalet, E.; Sabatés, A.; Olivares, A.; Gili, J.M.; Vila-Subirós, J.; Sardá, R. Unravelling the ecological impacts of large-scale offshore wind farms in the Mediterranean Sea. *Sci. Total Environ.* **2022**, *824*, 153803. [CrossRef]
12. Conesa, V. *Guía Metodológica para la Evaluación del Impacto Ambiental*, 4th ed.; Mundi-Prensa: Madrid, Spain, 2010.
13. Seacon. *Sediment Spillage during Array Cable Installation at Nysted Offshore Wind Farm*; Report 0402-1-1-L001 rev.1; GS SEACON APS: Ringsted, Denmark, 2005.
14. Elliott, J.; Smith, K.; Gallien, D.R.; Khan, A.A. Observing Cable Laying and Particle Settlement During the Construction of the Block Island Wind Farm. In *Final Report to the U.S. Department of the Interior*; Bureau of Ocean Energy Management, Office of Renewable Energy Programs: Washington, DC, USA, 2017; p. 225.
15. Martín, J.; Puig, P.; Palanques, A.; Ribó, M. Trawling-induced daily sediment resuspension in the flank of a Mediterranean submarine canyon. *Deep. Sea Res. Part II Top. Stud. Oceanogr.* **2014**, *104*, 174–183. [CrossRef]
16. McCauley, R.D.; Duncan, A.J.; Penrose, J.D.; McCabe, K.A. Marine Seismic Surveys: A Study of Environmental Implications. *APPEA J.* **2000**, *40*, 692–708. Available online: http://citeseerx.ist.psu.edu/viewdoc/download?doi=10.1.1.624.3763&rep=rep1&type=pdf (accessed on 26 April 2022). [CrossRef]

17. Ministerio de Agricultura, Alimentación y Medio Ambiente. Documento Técnico Sobre Impactos y Mitigación de la Contaminación Acústica Marina. Madrid. 2012, p. 146. Available online: https://www.miteco.gob.es/es/costas/temas/proteccion-medio-marino/doc-tecnico-impactos-mitigacion-contaminacion-acustica-marina_tcm30-157028.pdf (accessed on 26 April 2022).
18. CEDA. Ceda Position Paper: Underwater Sound in Relation to Dredging. *Terra Aqua* **2011**, *125*, 23–28.
19. Taormina, B.; Bald, J.; Want, A.; Thouzeau, G.; Lejart, M.; Desroy, N.; Carlier, A. A review of potential impacts of submarine power cables on the marine environment: Knowledge gaps, recommendations and future directions. *Renew. Sustain. Energy Rev.* **2018**, *96*, 380–391. [CrossRef]
20. Nedwell, J.R.; Workman, R.; Parvin, S.J. *The Assessment of Likely Levels of Piling Noise at Greater Gabbard and Its Comparison with Background Noise, Including Piling Noise Measurements Made at Kentish Flats*; Report No. 633R0115; Subacoustech Environmental Limited: Southampton, UK, 2005.
21. DNVGL. Rules for Classification, Ships. Part 6 Additional Class Notations, Chapter 7 Environmental Protection and Pollution Control. Edition January 2017. Available online: https://rules.dnv.com/docs/pdf/DNV/ru-ship/2017-01/DNVGL-RU-SHIP-Pt6Ch7.pdf (accessed on 26 April 2022).
22. Martin, B.; MacDonnell, J.; Vallarta, J.; Lumsden, E.; Burns, R. HYWIND Acoustic Measurement Report: Ambient Levels and HYWIND Signature. Technical Report for Statoil by JASCO Applied Sciences. 2011. Available online: https://www.equinor.com/content/dam/statoil/documents/impact-assessment/hywind-tampen/Equinor-Hywind-Acoustic-Measurement-Report-JASCO-00229-December-2011.pdf (accessed on 26 April 2022).
23. HDR. Field Observations during Wind Turbine Operations at the Block Island Wind Farm, Rhode Island. Final Report to the U.S. Department of the Interior, Bureau of Ocean Energy Management, Office of Renewable Energy Programs. OCS Study BOEM 2019-028. 2019; p. 281. Available online: https://espis.boem.gov/final%20reports/BOEM_2019-028.pdf (accessed on 26 April 2022).
24. Westerberg, H.; Bégout, M.L. Orientation of silver eel (*Anguilla anguilla*) in a disturbed geomagnetic field. Advances in Fish Telemetry. In Proceedings of the 3rd Conference on Fish Telemetry, Norwich, UK, 1 January 2000; pp. 149–158.
25. Fuxjager, M.; Eastwood, B.; Lohmann, K. Orientation of Hatchling Loggerhead Sea Turtles to Regional Magnetic Fields along a Transoceanic Migratory Pathway. *J. Exp. Biol.* **2011**, *214*, 2504–2508. Available online: https://www.researchgate.net/publication/51489400_Orientation_of_hatchling_loggerhead_sea_turtles_to_regional_magnetic_fields_along_a_transoceanic_migratory_pathway (accessed on 26 April 2022). [CrossRef]
26. Ottaviani, E.; Malagoli, D.; Ferrari, A.; Tagliazucchi, D.; Conte, A.; Gobba, F. 50 Hz Magnetic fields of varying flux intensity affect cell shape changes in invertebrate immunocytes: The Role of potassium ion channels. *Bioelectromagnetics* **2002**, *23*, 292–297. [CrossRef] [PubMed]
27. Malagoli, D.; Gobba, F.; Ottaviani, E. 50 Hz magnetic fields activate mussel immunocyte p38 MAP kinase and induce HSP70 and 90. *Comp. Biochem. Physiol. C Toxicol. Pharmacol.* **2004**, *137*, 75–79. [CrossRef] [PubMed]
28. Hutchison, Z.; Sigray, P.; He, H.; Gill, A.; King, J.; Gibson, C. *Electromagnetic Field (EMF) Impacts on Elasmobranch (Shark, Rays, and Skates) and American Lobster Movement and Migration from Direct Current Cables (No. BOEM 2018-003)*; Bureau of Ocean Energy Management (BOEM): Narragansett, RI, USA, 2018. Available online: https://espis.boem.gov/final%20reports/5659.pdf (accessed on 26 April 2022).
29. Woodruff, D.L.; Ward, J.A.; Schultz, I.R.; Cullinan, V.I.; Marshall, K.E. Effects of Electromagnetic Fields on Fish and Invertebrates. In *Task 2.1.3: Effects on Aquatic Organisms Fiscal Year 2011 Progress Report-Environmental Effects of Marine and Hydrokinetic Energy (No. PNNL-20813 Final)*; Pacific Northwest National Laboratory (PNNL): Richland, WA, USA, 2012. Available online: https://www.pnnl.gov/main/publications/external/technical_reports/PNNL-20813Final.pdf (accessed on 26 April 2022).
30. Bochert, R.; Zettler, M.L. Long-Term Exposure of Several Marine Benthic Animals to Static Magnetic Fields. *Bioelectromagnetics* **2004**, *25*, 498–502. Available online: https://www.researchgate.net/publication/8338761_Long-term_exposure_of_several_marine_benthic_animals_to_static_magnetic_fields (accessed on 26 April 2022). [CrossRef] [PubMed]
31. CMACS. A Baseline Assessment of Electromagnetic Fields Generated by Offshore Windfarm Cables. COWRIE Report EMF-01-2002-66. 2003. Available online: https://tethys.pnnl.gov/sites/default/files/publications/COWRIE_EMF_Offshore_Cables.pdf (accessed on 26 April 2022).
32. Gill, A.B.; Taylor, H. The Potential Effects of Electromagnetic Fields Generated by Cabling between Offshore Wind Turbines upon Elasmobranch Fishes. Countryside Council for Wales 2001. CCW Contract Science Report No 488. Available online: https://tethys.pnnl.gov/sites/default/files/publications/Gill%20and%20Taylor%202001.PDF (accessed on 26 April 2022).
33. Smith, E. Electro-physiology of the electrical shark-repellant. *Trans. Korean Inst. Electr. Eng.* **1974**, *65*, 1–20.
34. Kalmijn, A.J. Detection and Processing of Electromagnetic and Near-field Acoustic Signals in Elasmobranch Fishes. *Philos. Trans. R. Soc. Lond. Board Biol. Sci.* **2000**, *355*, 1135–1141. [CrossRef]
35. Walker, M.M.; Diebel, C.E.; Kirschvink, J.L. Detection and use of the Earth's Magnetic Field by Aquatic Vertebrates. In *Sensory Processing in Aquatic Environments*; Collins, S.P., Marshall, N.J., Eds.; Springer: New York, NY, USA, 2003; pp. 53–74.
36. Nyqvist, D.; Durif, C.; Johnsen, M.G.; de Jong, K.; Forland, T.N.; Sivle, L.D. Electric and Magnetic Senses in Marine Animals, and Potential Behavioral Effects of Electromagnetic Surveys. *Mar. Environ. Res.* **2020**, *155*, 104888. [CrossRef]
37. Farr, H.; Ruttenberg, B.; Walter, R.; Wang, Y.-H.; White, C. Potential environmental effects of deepwater floating offshore wind energy facilities. *Ocean. Coast. Manag.* **2021**, *207*, 16. [CrossRef]
38. Platis, A.; Siedersleben, S.; Bange, J.; Lampert, A.; Bärfuss, K.; Hankers, R.; Canadillas, B.; Foreman, R.; Schulz-Stellenfleth, J.; Djath, B.; et al. First in situ evidence of wakes in the far field behind offshore wind farms. *Sci. Rep.* **2018**, *8*, 2163. [CrossRef]

39. Christiansen, M.B.; Hasager, C.B. Wake effects of large offshore wind farms identified from satellite SAR. *Remote Sens. Environ.* **2005**, *98*, 251–268. [CrossRef]
40. Van Berkel, J.; Burchard, H.; Christensen, A.; Mortensen, L.O.; Svenstrup Petersen, O.; Thomsen, F. The Effects of Offshore Wind Farms on Hydrodynamics and Implications for Fishes. *Oceanography* **2020**, *33*, 108–117. [CrossRef]
41. Zenetos, A.; Gofas, S.; Morri, C.; Rosso, A.; Violanti, D.; García Raso, E.; Cinar, M.E.; Almogi-Labin, A.; Ates, A.S.; Azzurro, E.; et al. Alien species in the Mediterranean Sea by 2012. A contribution to the application of European Union's Marine Strategy Framework Directive (MSFD). Part 2. Introduction trends and pathways. *Mediterr. Mar.* **2012**, *13*, 328–352. [CrossRef]
42. Otero, M.; Cebrian, E.; Francour, P.; Galil, B.; Savini, D. Monitoreo de Especies Marinas Invasoras at áreas Marinas Protegidas (AMP) del Mediterráneo: Estrategia y Guía Práctica Para Gestores. UICN 2013. p. 136. Available online: https://portals.iucn.org/library/efiles/documents/2013-008-Es.pdf (accessed on 26 April 2022).
43. Subías-Baratau, A.; Sànchez-Vidal, A.; Di Martino, E.; Figuerola, B. Marine biofouling organisms on beached, buoyant and benthic plastic debris in the Catalan Sea. *Mar. Pollut. Bull.* **2022**, *175*, 113405. [CrossRef] [PubMed]
44. Pastor, X. L'eòlica Marina, Una Aliada de la Biodiversitat. Published at el Diari de Girona, on 6 March 2022. Available online: https://www.diaridegirona.cat/opinio/2022/03/06/l-eolica-marina-aliada-biodiversitat-63484707.html (accessed on 26 April 2022).

Journal of
Marine Science and Engineering

MDPI

Article

A Numerical Simulation of a Variable-Shape Buoy Wave Energy Converter

Shangyan Zou [1,†] and Ossama Abdelkhalik [2,*,†,‡]

1 School of Civil and Constructional Engineering, Oregon State University, Corvallis, OH 97331, USA; shangyan.zou@oregonstate.edu
2 Department of Aerospace Engineering, Iowa State University, Ames, IA 50011, USA
* Correspondence: ossama@iastate.edu
† These authors contributed equally to this work.
‡ 2241 Howe Hall, 537 Bissell Road, Ames, IA 50011, USA.

Abstract: Wave energy converters (WECs) usually require reactive power for increased levels of energy conversion, resulting in the need for more complex power take-off (PTO) units, compared to WECs that do not require reactive power. A WEC without reactive power produces much less energy, though. The concept of Variable Shape Buoy Wave Energy Converters (VSB WECs) is proposed to allow continuous shape-change aiming at eliminating the need for reactive power, while converting power at a high level. The proposed concept involves complex and nonlinear interactions between the device and the waves. This paper presents a Computational Fluid Dynamics (CFD) tool that is set up to simulate VSB WECs, using the ANSYS 2-way fluid–structure interaction (FSI) tool. The dynamic behavior of a VSB WEC is simulated in this CFD-based Numerical Wave Tank (CNWT), in open sea conditions. The simulation results show that the tested device undergoes a significant deformation in response to the incoming waves, before it reaches a steady-state behavior. This is in agreement with a low-fidelity dynamic model developed in earlier work. The resulting motion is significantly different from the motion of a rigid body WEC. The difference in the motion can be leveraged for better energy capture without the need for reactive power.

Keywords: ocean wave energy; flexible wave energy converters; Computational Fluid Dynamics; fluid–structure interaction; highly nonlinear dynamics

Citation: Zou, S.; Abdelkhalik, O. A Numerical Simulation of a Variable-Shape Buoy Wave Energy Converter. *J. Mar. Sci. Eng.* **2021**, *9*, 625. https://doi.org/10.3390/jmse9060625

Academic Editor: Eva Loukogeorgaki

Received: 10 May 2021
Accepted: 31 May 2021
Published: 4 June 2021

Publisher's Note: MDPI stays neutral with regard to jurisdictional claims in published maps and institutional affiliations.

Copyright: © 2021 by the authors. Licensee MDPI, Basel, Switzerland. This article is an open access article distributed under the terms and conditions of the Creative Commons Attribution (CC BY) license (https://creativecommons.org/licenses/by/4.0/).

1. Introduction

Wave energy conversion has been attracting more research attention due to the rising demand for renewable energy and its advantages including high power density [1], large energy potential, and consistency. Yet, wave energy conversion is still in the pre-commercial phase. In the last decade, the dynamics and control of WECs were significantly investigated. Starting from linear models, which are widely applied in developing optimal controls [2–5], to nonlinear hydrodynamics (especially nonlinear FK force) [6,7], to high-fidelity CNWT simulations of wave energy converters [8,9]. In recent years, significant research effort has been made to improve the economic index of wave energy conversion. Yet, some challenges still exist; one of which is the need for reactive power to optimize energy production. Some conventional optimal control methods [4] aim to achieve resonance between the device and ocean waves to maximize energy production; yet a complex PTO unit would be required to produce reverse power flow, which represents a high expenditure and the generated power quality is poor (large fluctuation) [10,11]. On the other hand, using damping control does not require reactive power, but the performance of the WEC is degraded in terms of power production.

The concept of VSB WECs was recently proposed to address this challenge. The VSB WEC has a'soft' buoy, allowing relatively rapid shape-changing in response to pressure variations. This is unlike the conventional rigid WECs and unlike the concept of variable geometry WECs [12–15] in which a rigid WEC changes shape occasionally and slowly in response to changing wave conditions. The VSB WEC involves nonlinear interaction between the device and the waves (continuous deformation in response to the waves). This interaction can be leveraged to produce more power without the need for reactive power. The motion of the device can be excited due to the deformation as pointed out in reference [16], which employs a physics-based low-fidelity model.

Since the proposed device involves complex and nonlinear hydrodynamics, a tool for high-fidelity simulation is developed in this paper to investigate the behavior of VSB WEC. Due to the nature of this problem, a FSI tool that can handle variable-shape (morphing) structure is needed in the CNWT. This is unlike most existing high-fidelity WEC simulation tools [8], which can handle only rigid body FSI. The FSI capability employed in this paper implements 2-way FSI; the data is transferred between the solid domain (finite element analysis (FEA) model) and the fluid domain (CFD model). As far as the simulation software is concerned, ANSYS® and OpenFOAM are both widely applied in developing CNWTs and solving FSI [17]. The ANSYS 2-way FSI (system coupling simulation) is used in this paper. As detailed in this paper, this type of problem requires large mesh motions; this is accommodated in this paper using the diffusion-based mesh smoothing method.

The main contribution of this paper is to develop a high-fidelity simulation tool that enables the investigation of the recently proposed VSB WECs, which are characterized by their nonlinear hydrodynamics. The tool presented in this paper will enable the validation of other low fidelity models for VSB WECs that are usually used for WEC control design and analysis. The paper is organized as follows: numerical modeling is introduced in Section 2 and mesh generation is introduced in Section 3. Sections 4 and 5 present the simulation results and discussion, respectively. Finally, the conclusion is drawn in Section 6 with the plan for future work.

2. Numerical Modeling

The details of the numerical simulation approach applied in this study are introduced in this section. A schematic view of the proposed device and the computational domain is presented in Figure 1. As shown in the figure, the dimensions of the computational domain are 80 × 60 × 60 m. The height of the Free Surface Level (FSL) is 40 m measured from the global coordinate system, which is also shown in the figure. Since the proposed device is allowed to change the shape in response to the waves, a significant interaction between the device and waves is expected. By applying ANSYS 2-way FSI, the data transfers between solid and fluid are created on the interface between solid and surrounding fluids (denoted as FSI 1) and the interface between solid and internal gas (denoted as FSI 2). The device initially has a shape of a hollow sphere with a radius of 2 m and a thickness of 0.3 m.

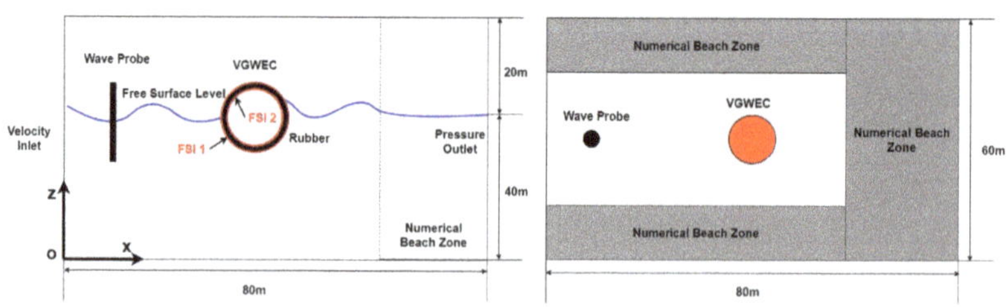

Figure 1. Computational domain layout.

2.1. Governing Equations

Three stratified phases including the environmental air, environmental water and chamber gas are applied in this simulation. Although the environmental air and water are assumed to be incompressible, the chamber gas is assumed to be compressible since the chamber has a large volumetric change due to the interaction between VSB WEC and waves. The conservation equations for unsteady compressible fluid simulation can be written as:

$$\begin{aligned}
&\frac{\partial \rho}{\partial t} + \nabla \cdot (\rho \vec{v}) = 0 \\
&\frac{\partial}{\partial t}(\rho \vec{v}) + \nabla \cdot (\rho \vec{v}\vec{v}) = -\nabla p + \nabla \cdot (\bar{\bar{\tau}}) + \rho \vec{g} + \vec{F} \\
&\frac{\partial}{\partial t}(\rho E) + \nabla \cdot (\vec{v}(\rho E + p)) = \nabla \cdot (k_e \nabla T)
\end{aligned} \tag{1}$$

where ρ is the fluid density, $\vec{v}$ is the flow velocity vector in the continuity equation. In the momentum equation, $\bar{\bar{\tau}}$ is the stress tensor, and $\rho \vec{g}$ represents the gravitational body force. Since in this study, the wave absorption zone is included near the pressure outlet, the source term $\vec{F}$ is added in the momentum conservation due to the applied numerical beach treatment. Further, in the energy equation, E is the energy, and k_e is the effective thermal conductivity. A transient Reynolds Averaged Navier Stokes (RANS) solver is applied with the governing equations discretized over the fluid domain by using finite volume method. The fluid simulation (CFD) is further coupled with transient structural analysis (FEA).

2.2. Physics Modeling

The turbulence model applied in this study is the realizable $k - \epsilon$ model, which is considered robust and computational efficient [17]. Compared to the widely applied standard $k - \epsilon$, the applied model provides more robustness, which is more suitable in this study since the interaction between the device and the waves is highly nonlinear. The transport equations for turbulent kinetic energy (k) and dissipation rate (ϵ) will be solved. Details about the mathematical modeling of the k-epsilon turbulent model can be found in [18,19]. Furthermore, since the chamber gas has a negligible motion with respect to the device, the internal fluid domain is considered to be laminar.

The Volume of Fluid (VOF) method is applied to simulate the multi-phase flow. Environmental air is assumed to be the primary phase, and the environmental water and chamber gas are both assumed to be the secondary phase. To track the interface (e.g., free surface) between different phases, the volume fraction equation is applied:

$$\frac{\partial(\alpha_q \rho_q)}{\partial t} + \nabla \cdot (\alpha_q \rho_q \vec{v}_q) = 0 \tag{2}$$

where ρ_q and α_q are the density and volume fraction of the qth fluid (phase q). The volume fraction α_q varies between 0 and 1 in a cell. For each cell, it is possible that:

$$\begin{aligned}
&\alpha_q = 0 \text{ if the cell is empty of phase } q \\
&0 < \alpha_q < 1 \text{ if the cell contains interface} \\
&\text{between phase } q \text{ and other phases} \\
&\alpha_q = 1 \text{ if the cell is full of phase } q
\end{aligned} \tag{3}$$

Furthermore, the summation of the volume fraction of all phases should be unity:

$$\sum_{q=1}^{n} \alpha_q = 1 \tag{4}$$

To further sharpen the free surface between environmental air and water, the interfacial anti-diffusion is enabled by adding a negative diffusion source term in the volume fraction equation [20]:

$$\frac{\partial \alpha_q}{\partial t} + \nabla \cdot (\vec{v}_q \alpha_q) = -\nabla \cdot (\vec{v}_c \alpha_q (1 - \alpha_q)) \tag{5}$$
$$\vec{v}_c = \gamma |\vec{v}_q| \frac{\nabla \alpha_q}{|\nabla \alpha_q|}$$

where $\vec{v}_c$ and $\vec{v}_q$ represent the compression velocity normal to the interface and cell velocity, respectively, and γ is the compression factor.

Moreover, the large mesh motion due to the highly nonlinear interaction is accounted by diffusion-based smoothing dynamic mesh method. The pressure–velocity coupling problem is solved by the coupled algorithm to achieve aggressive convergence in transient simulation of the proposed challenging model. The density, momentum, turbulence, and energy are spatially discretized based on the second-order upwind scheme. The applied transient formulation is first-order implicit.

2.3. Domain and Boundary Conditions

The boundary conditions applied in this study are summarized in Figure 2. The waves are generated from the velocity inlet, and the downstream boundary is specified as a pressure outlet. Symmetry boundary conditions are applied for the top, bottom, and sides boundary conditions. The purpose of setting the symmetry boundary conditions is to construct an open sea condition for the simulation where the deep water and infinite air assumptions are applied. As addressed in [21], velocity inlet and slip wall are also considered as appropriate selections for the top, bottom, and sides boundary conditions if the boundaries are located far away from the device (which is the case in this study). The interface between the surrounding fluids and the solid and between the chamber gas and the solid are both defined as adiabatic walls. A mesh will be morphed around the solid to account for the motion and deformation of the device.

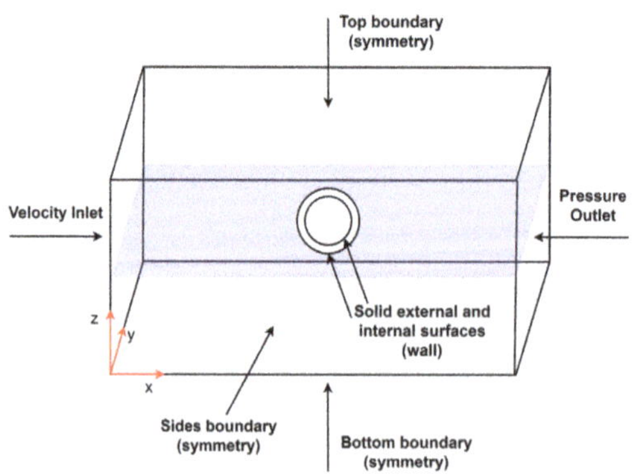

Figure 2. Boundary conditions applied in the simulation.

Since open sea conditions are applied in this study, the generated waves are assumed to be deep water waves with regard to a water depth 60 m. To prevent the drifting of the device, the flow current velocity is assumed to be zero. The Person–Monskowitz (PM) wave spectrum is applied as the irregular wave spectrum:

$$S_{PM}(\omega) = \frac{5}{16} \frac{H_s^2 \omega_p^4}{\omega^5} e^{-\frac{5\omega_p^4}{4\omega^4}} \tag{6}$$

where H_s is the significant wave height and $\omega_p = \frac{2\pi}{T_p}$ is the peak frequency. The irregular wave that has a significant height of $H_s = 1$ m and a peak period of $T_p = 6$ s is applied in this paper given the fact that typical peak period of ocean waves varies from 6 s to 12 s. The frequency range of the wave is from 0.2 rad·s^{-1} to 4 rad·s^{-1} and the number of frequencies is 20. Moreover, deep water waves are normally multidirectional waves, thus the directional spreading function also needs to be specified. The widely applied cos-2s directional spreading function [22–24] is applied in this study:

$$G(\theta) = G(s)\cos^{2s}(\frac{\pi}{2\theta_s}(\theta - \theta_p)) \tag{7}$$

where s is the frequency independent cosine exponent, which is a positive integer. θ_p is the mean wave heading angle, and θ_s is angle spreading from θ_p. In this paper, s is selected as 30 for deep water waves as suggested in [25]. The mean wave heading angle is 0° and the spreading angle is 90°, which is found to be a good assumption for directional spreading when $s > 5$ [25]. Additionally, the number of angular components is chosen as 10. The resulting wave spectrum is the multiplication of the frequency spreading and directional spreading:

$$S(\omega, \theta) = S_{PM}(\omega)G(\theta) \tag{8}$$

The final wave field will be generated based on the wave spectrum by taking superposition of different wave components:

$$\eta(r,t) = \sum_{i=1}^{N_f}\sum_{j=1}^{N_a} a_{ij}cos(k_{x,ij}x + k_{y,ij}y - \omega_i t + \phi_{ij}) \tag{9}$$

where $a = \sqrt{2\int_{\theta-\frac{\delta\theta}{2}}^{\theta+\frac{\delta\theta}{2}}\int_{\omega-\frac{\delta\omega}{2}}^{\omega+\frac{\delta\omega}{2}} S(\omega,\theta)d\omega d\theta}$ is the amplitude of each wave component. Further, $k_{x,ij} = k_i\cos(\theta_j)$ and $k_{y,ij} = k_i\sin(\theta_j)$, where k_i is the wave number of the ith wave component (which has a frequency of ω_i) and θ_j is the jth wave heading. Random phase shift ϕ_{ij} is used in generating the wave field, which is uniformly distributed in $[0, 2\pi]$.

A multiple-directional numerical beach (as shown in Figure 1) is implemented in the vicinity of the downstream and the sides to suppress the wave reflection. A damping sink term is added in the momentum equation for this damping zone:

$$S = -[C_1\rho V + \frac{1}{2}C_2\rho \mid V \mid V]f(z)f(x) \tag{10}$$

where $C_1 = 116.24631$/s and $C_2 = 3192.6571$/m are the linear and quadratic damping resistance, respectively. V is the vertical velocity, z is the distance measured from FSL, and x is the distance along the flow direction. The length of the multi-directional beach in the downstream (beach direction $[1,0,0]$) is specified as twice the wave length ($D = 2\lambda$) [26,27] and the length of the damping zone in the sides (beach direction $[0,1,0]$ and $[0,-1,0]$) is defined as one wave length ($D = \lambda$).

3. Mesh Generation

Mesh quality is critical for free surface flows and the proposed challenging model, which involves highly nonlinear interaction between VSB WEC and the waves. Although a structured mesh provides better efficiency and accuracy in the simulation, it has the difficulty in tracking curved or complex geometries. On the other hand, the unstructured mesh is suitable to track complex geometries and change the mesh size locally, while a significant number of cells are required to ensure accuracy [28]. Therefore a hybrid mesh generation is performed in the fluid domain using ANSYS FLUENT meshing application. The structured mesh is applied in the background with a larger element size and the unstructured mesh is applied near the hull of the device and the internal gas domain with smaller element size. Though hybrid mesh has the advantages of computationally efficient and tracking features accurately, the resulting mesh is non-conformal due to different mesh topologies applied in different zones. Thus, the meshes are matched in the faces that are meshed with different topologies in the calculation.

The mesh of the overall computational domain is presented in Figure 3. As shown in the figure, the unstructured mesh is applied in a refined region, which has a box shape and internal gas domain with a smaller mesh size. The mesh is also locally refined in the regions that require extra accuracy. It is clearly visible in Figure 3 that the mesh in the expected free surface is refined. In addition, as shown in Figure 4, the mesh is also refined in the interface between surrounding fluids and solid and the interface between the internal fluid and the solid to capture the geometry and motion of the device. Moreover, inflation layers are placed in the boundary layer along the surface (with respect to the surrounding fluids) of VSB WEC to capture the rapid velocity change. As shown in Figure 5, the unstructured mesh is also applied in the solid domain to capture the geometry. Since the applied geometry is a simple hollow sphere, no local mesh refinement is used in the solid domain.

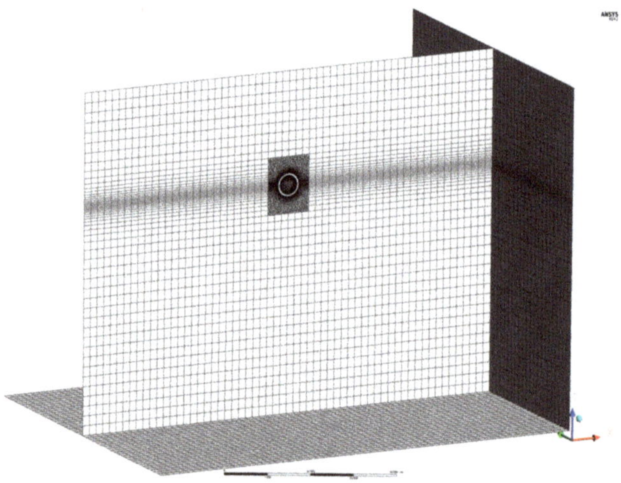

Figure 3. Mesh generation of the overall computational domain. Faces selected for presentation are bottom, outlet, and a cross-sectional face located at $y = 30$ m.

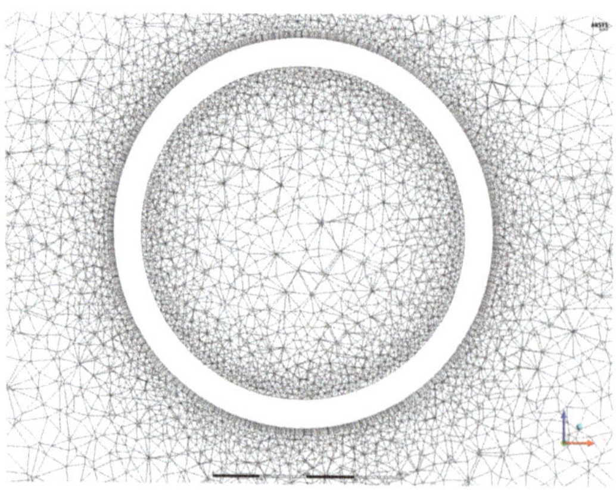

Figure 4. Mesh generation around the body surface with the mesh refinement at the interfaces and the boundary layer.

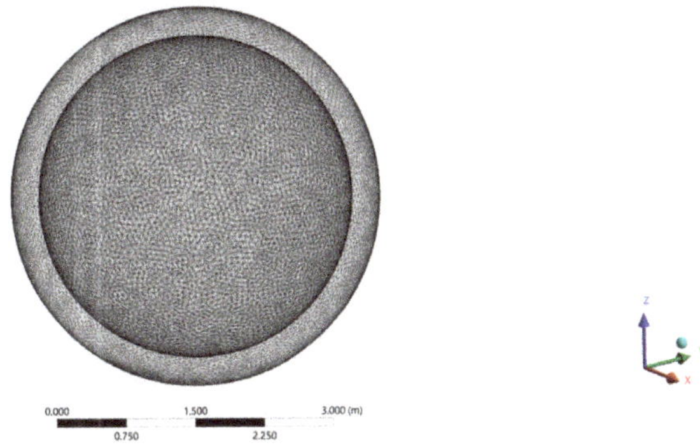

Figure 5. Mesh generation for the solid domain.

To select an appropriate grid resolution, a mesh sensitivity analysis is performed with three different levels of mesh resolutions (coarse, medium, and fine). The device is assumed to be rigid body (fixed-shape) in this analysis, and the details of each grid resolution are summarized in Table 1. The structured mesh is applied in the background, which occupies around 99.8% of the fluid domain, while only 18.4% of the cells of the fluid domain are formed in the background when the medium mesh is applied. On the other hand, nearly 81.6% of the cells (unstructured mesh) are concentrated in less than 0.2% of the fluid volume. The presented mesh arrangement allows a significantly reduced number of cells in the background to accelerate the calculation and keeps the accuracy of the simulation near the wave and structure interaction. The three-dimensional motions of the center of gravity (CoG) of the device (assumed rigid body) versus different mesh resolutions are plotted in Figures 6–8. As shown in the figures, the medium mesh is considered to be an appropriate mesh resolution since it provides close accuracy as the fine mesh while significantly saves the computational cost. Notably, the motion of the device in the y-direction is predicted inaccurately by applying the coarse mesh. The corresponding mesh size applied in different domains of the medium mesh is summarized in the following. The grid resolution applied on the body surfaces is 0.06 m (most refined) and applied in the background is 1.7 m. Moreover, the mesh is refined in the Refined Region (unstructured) with a mesh size of 0.42 m and the free surface is locally refined with a grid resolution of 0.17 m. The resulting minimum cell volume and maximum cell volume of the fluid domain are 2.12×10^{-6} m^3 and 3.25 m^3, respectively. The similar mesh size (medium mesh) will be applied for the following simulations of VSB WEC (without assuming fixed-shape).

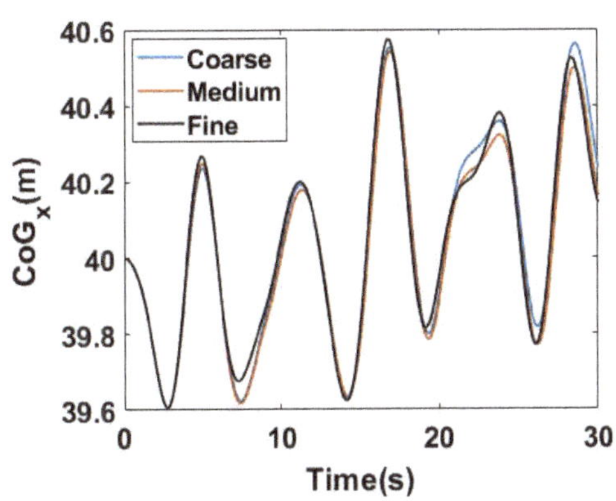

Figure 6. x-directional motion of the center of gravity of the device with different mesh resolutions.

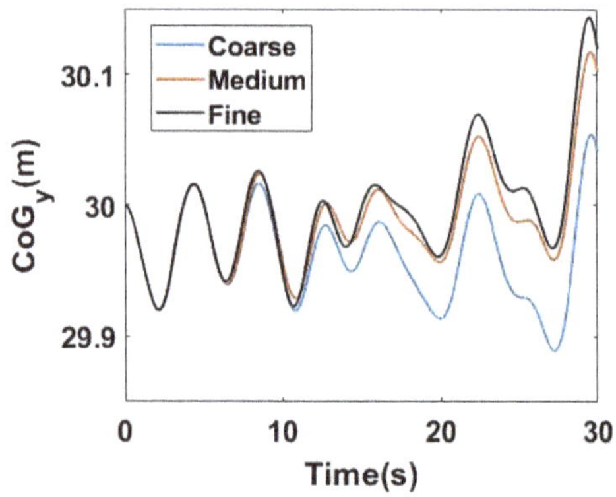

Figure 7. y-directional motion of the center of gravity of the device with different mesh resolutions.

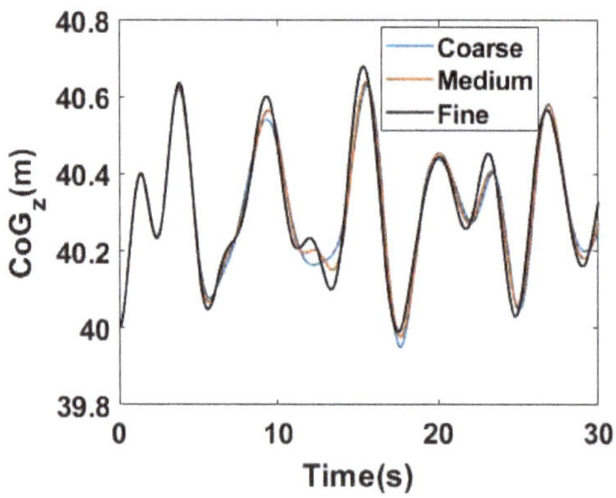

Figure 8. z-directional motion of the center of gravity of the device with different mesh resolutions.

Table 1. The breakdown of different types of mesh applied in mesh sensitivity analysis.

Mesh Types	Number of Cells		Total
	Background	Refined Region	
Coarse	112,536	281,661	394,197
Medium	144,300	640,014	784,314
Fine	331,996	1,628,346	1,960,342

4. Numerical Results

The system coupling simulations are carried out on Iowa State University's High-Performance Computer (HPC). The number of processors used in the parallel simulation is 36 and the corresponding elapsed time for the simulation of VSB WEC with medium mesh is approximately 96.6 hrs including post-processing. To guarantee the flow current number is close to 1, the time step is selected as 0.01 s and a total of 3000 steps will be solved (simulation end time is 30 s). Although, a more robust analysis (may be conducted in the future) can be conducted by significantly extending the simulation timeseries (e.g., 20 min simulation for varied wave conditions). Giving that the computational cost of this simulation is extremely high, only a short time period (30 s) will be simulated and analyzed. Additionally, the applied simulation end time covers part of the steady-state responses, which provides meaningful results for both transient and steady-state analysis.

4.1. Free Decay Test

To validate the proposed simulator, a free heave decay test is first performed. The device (assumed rigid body in the decay test) is initially placed 1.3 m away from the equilibrium point with no initial velocity. As shown in Figure 9, the natural decay period of the device is found to be around 2.6 s. Moreover, the energy given to the device can be successfully dissipated by the implemented multi-directional numerical beach, and the device stays at the equilibrium without being excited by reflected waves (absorbed).

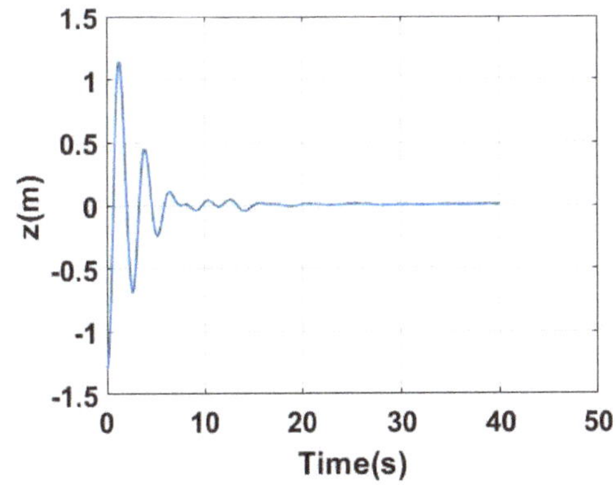

Figure 9. Heave decay test.

4.2. Irregular Wave Generation

As described in Section 2.3, multi-directional irregular waves will be applied in this study to simulate open sea conditions. A wave probe is placed between the inlet and the device (at $y = 30$ m) to monitor the free surface elevation (as shown in Figure 1). As suggested in [29], the distance between the inlet and the wave probe is $1L = 11$ m. The surface elevation measured at the free surface ($\alpha_{air} = 0.5$) by the wave probe is shown in Figure 10.

The wave pattern on the free surface is presented in Figures 11–14 at different time instants when the most significant interaction between waves and the device happens and at the end of the simulation. It is visible in the figure that the wave pattern (which is multidirectional) is significantly affected around the device due to the strong interaction. Although, a wavy water surface is assumed initially, the wave pattern near the outlet and the sides is flattened during the simulation (as shown in Figure 14). To better illustrate the influence of the wave forces on the motion and deformation of VSB WEC, Figures 15 and 16 show the dynamic pressure field around the device at two different time instants. A large dynamic pressure acts on the bottom of the device at $t = 0.5$ s with a peak pressure around 2.89×10^3 Pa, which not only causes a motion, but also a deformation of the hull (compress the device). At $t = 1.5$ s, there is also a significant deformation of the bottom half of the device (mainly at the sides) and peak dynamic pressure is around 0.696×10^3 Pa acts on the wet surface in upstream. By presenting these figures, we can claim the challenging interaction between the waves and the device is successfully captured by the proposed CFD model.

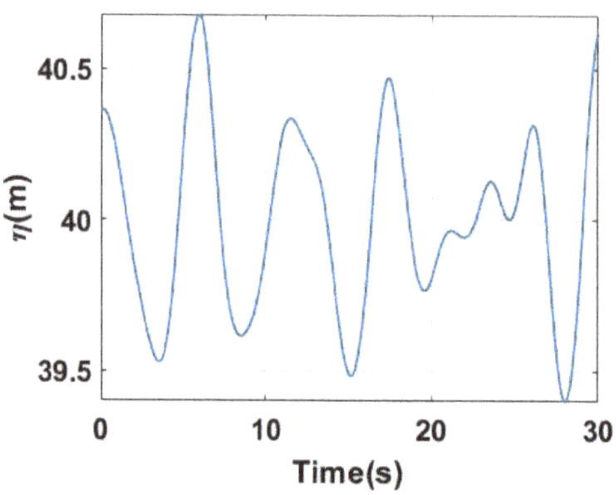

Figure 10. Free surface elevation measured at the wave probe.

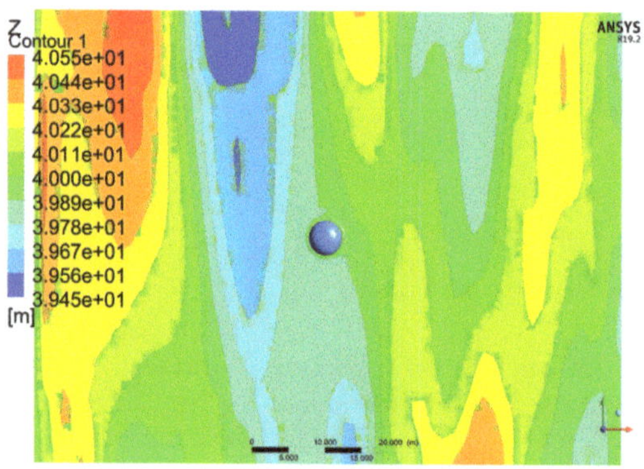

Figure 11. Wave pattern on the free surface at $t = 0.5$ s.

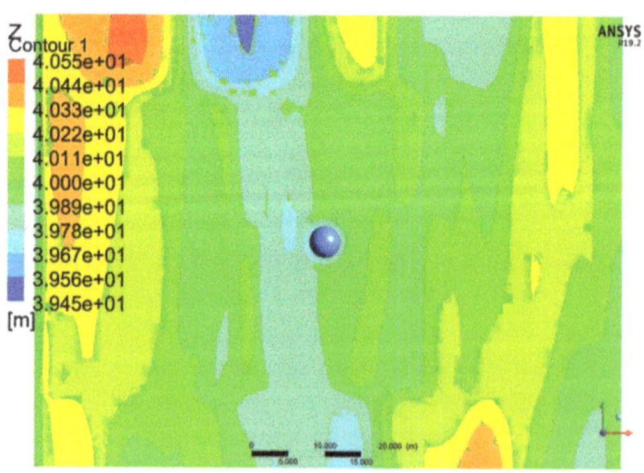

Figure 12. Wave pattern on the free surface at $t = 1$ s.

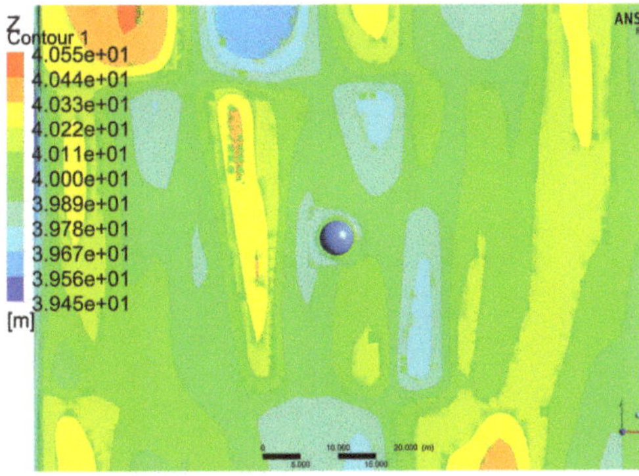

Figure 13. Wave pattern on the free surface at $t = 1.5$ s.

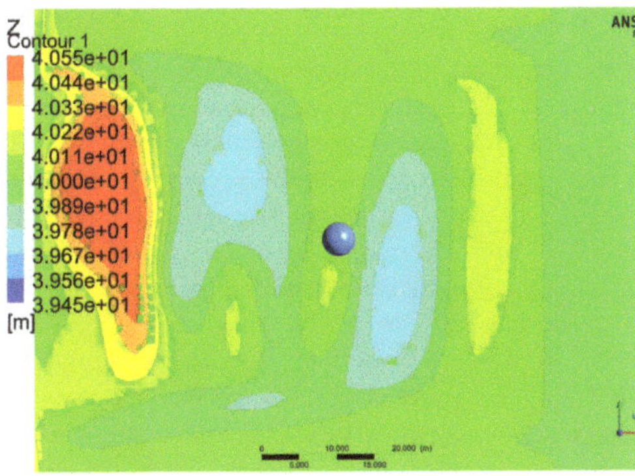

Figure 14. Wave pattern on the free surface at $t = 30$ s.

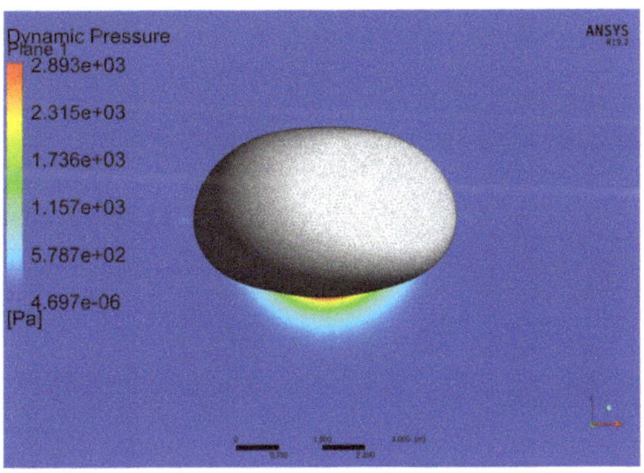

Figure 15. Dynamic pressure field measured around the device at $t = 0.5$ s.

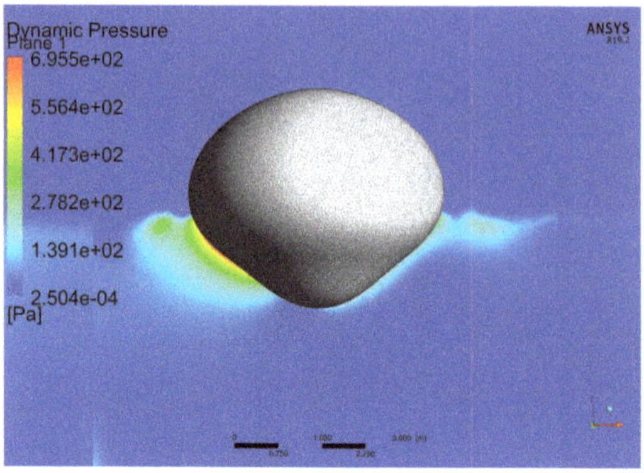

Figure 16. Dynamic pressure field measured around the device at $t = 1.5$ s.

4.3. VSB WEC Deformation

Since the proposed device allows significant deformation in response to the waves, it is interesting to study the geometry variation in free motion. As shown in Figure 17, although the geometry of the device varies significantly from 0 s to 1.5 s, the shape of the device at 1.5 s indicates a trend of recovering to the original shape (at 0 s). Thus, a periodic change of the shape is expected, which is also reasonable since the gas chamber is a restoring mechanism. This periodic phenomenon can be better explained in the following figures. Figures 18 and 19 show the vertical motion of two nodes where the U=pper Node denotes the node defined originally at $(40, 30, 42)$ m and the Lower Node denotes the node defined at $(40, 30, 38)$. Figure 18 shows the individual motion of two nodes with respect to their original location. For a conventional fixed-shape buoy WEC (FSB WEC), the motions of these two nodes are expected to be identical since rigid body motion is assumed. Although, in this study, their motion patterns are different. The relative vertical distance between these two nodes is plotted in Figure 19. The original distance between these two nodes is 4 m, which will be kept as a constant for a FSB WEC. While the vertical distance of these two nodes for the VSB WEC changes periodically and finally reaches a steady-state deformation (approximately when $t > 15$ s). The natural period of steady-state deformation is around 1.96 s, which is almost one third of the wave peak period. The relation between the natural period of the shape deformation and the wave period can be more rigorously studied in the future by simulating the dynamic behavior of VSB WEC with varied wave conditions, which may benefit future optimization of the proposed device.

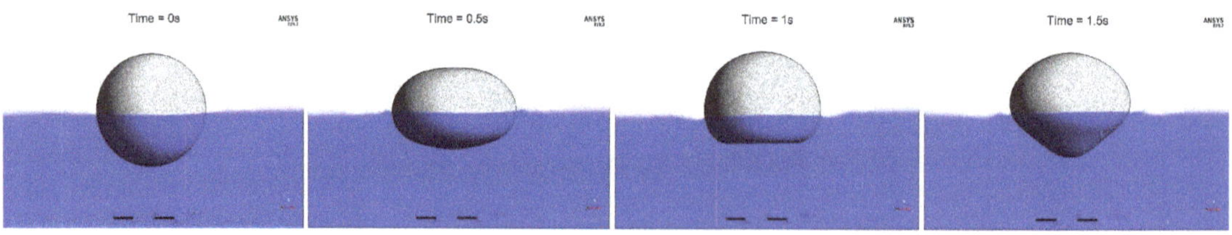

Figure 17. Snapshots of the deformation of VSB WEC at different time instants.

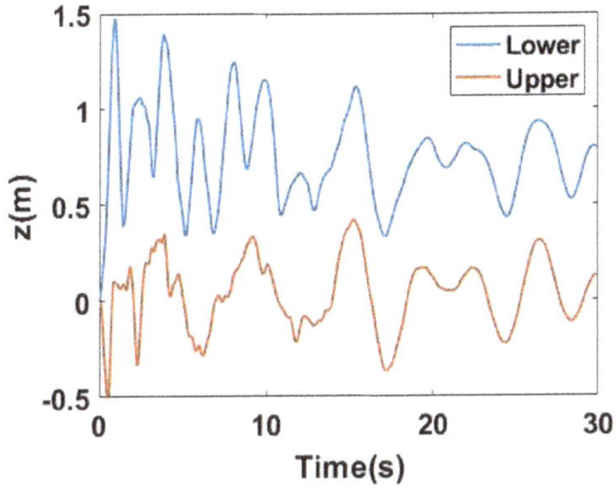

Figure 18. Motions of the upper node and the lower node with respect to their original location.

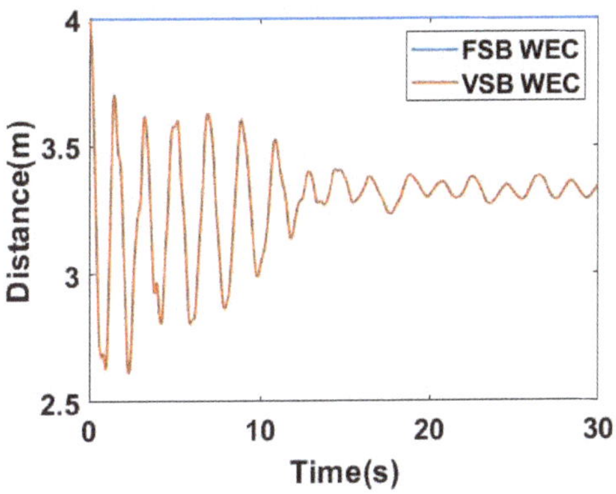

Figure 19. Relative vertical distance between the upper node and the lower node during operation.

4.4. VSB WEC Motion

The motion of VSB WEC is presented in detail in this section and it will be compared with FSB WEC motion under the same wave condition to highlight the difference in motions introduced by changing the shape. The motion difference can be later controlled and utilized to produce more wave power. Additionally, the mass and dimensions of FSB WEC are the same as VSB WEC (undeformed shape). The motions of the center of gravity of VSB WEC and FSB WEC in x, y, and z directions are shown from Figures 20–22, respectively. The motion of VSB WEC significantly differs from FSB WEC in three dimensions both in terms of phase and magnitude. The difference in vertical motion between the two devices is presented in Figure 23. The maximum difference is around 0.26 m, and the maximum z-directional motion of FSB WEC is around 0.64 m, which indicates a maximum 40.6% change in the heave motion caused by the introduction of geometry variation. This large difference represents a room for improving the performance of wave energy conversion by introducing appropriate control.

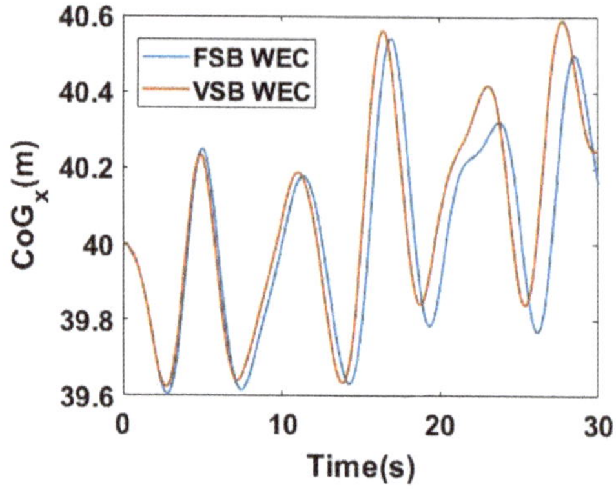

Figure 20. x-directional motion of center of gravity compared between FSB WEC and VSB WEC.

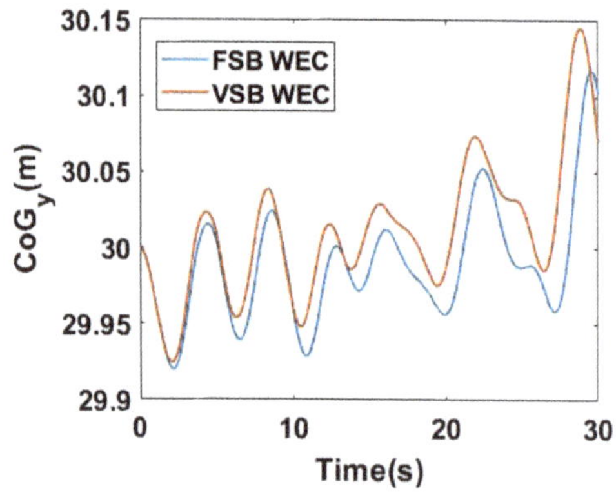

Figure 21. y-directional motion of center of gravity compared between FSB WEC and VSB WEC.

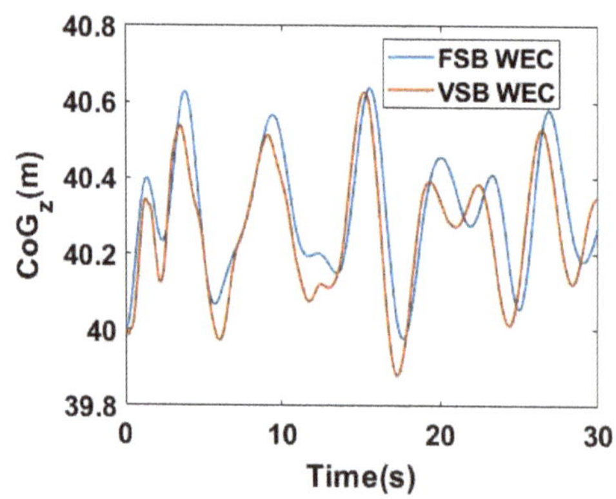

Figure 22. z-directional motion of center of gravity compared between FSB WEC and VSB WEC.

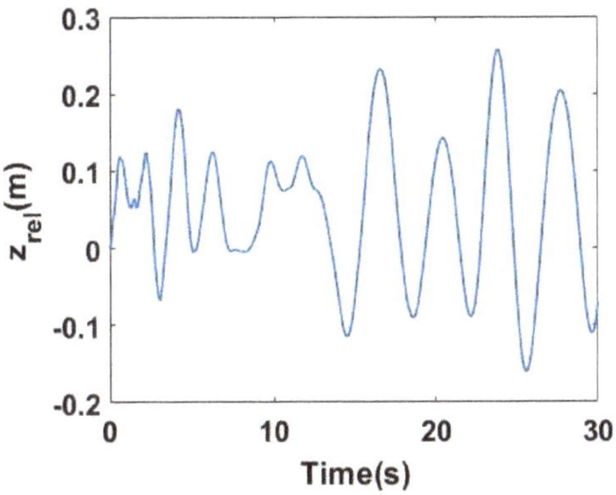

Figure 23. Difference in motions of FSB WEC and VSB WEC in heave direction.

5. Discussion

In this paper, a 3D numerical simulation architecture is introduced to simulate the fluid–structure interaction of the proposed VSB WEC. The simulation results show the proposed device has a significant shape-changing due to the highly nonlinear interaction between waves and the device. The resulting motion of VSB WEC also significantly differs from the motion of a conventional FSB WEC with the same mass and dimensions. It is noted that the design of VSB WEC can be further optimized in terms of energy extraction since this paper focuses on presenting a framework of the high-fidelity simulation of this device. The applied design is simple in terms of geometry and mechanism. For instance, a possible design presented in [16] is a VSB WEC that has a rigid body part (cylindrical shape) and a shape-changing part (truncated conical shape). A control valve is included to control the pressure oscillation in the gas chamber. As found in [16], a restoring mechanism (gas chamber in this study) is required to keep a steady-state deformation of the device in order to harvest more energy. The presented framework is applicable to simulate the performance of other designs of VSB WEC.

Hyperelastic material is applied in this study since the proposed device is required to be 'soft'. The material applied in the simulation obeys Neo-Hookean hyperelasticity with an initial shear modulus of 0.1 Mpa and an initial bulk modulus of 2000 Mpa, and the density of the material is 1000 kg$\cdot$m^{-3}. The corresponding strain–stress curve of the applied material is shown in Figure 24. Unlike vulcanized rubber [30] (initial shear modulus is 0.41 Mpa and initial bulk modulus is 414.5 Mpa), which only allows small deformation, the material used in this study is 'softer', which allows more deformation. In addition, unlike neoprene rubber (initial shear modulus is 0.027 Mpa and initial bulk modulus is 13.86 Mpa), which is too soft such that the deformation is too large to keep a stable simulation. More advanced dynamic mesh techniques need to be applied to address this challenging mesh motion. The study of the effect of different materials is interesting, though beyond the scope of this paper, therefore a material that has the hyperelastic properties between vulcanized rubber and neoprene rubber is applied such that the simulation is stable with a considerable deformation of the device.

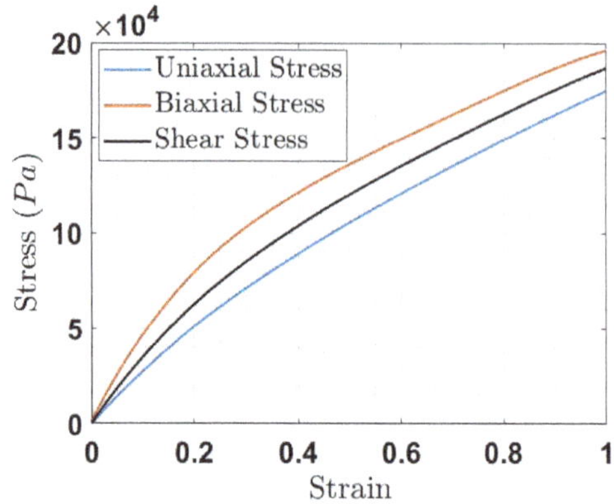

Figure 24. Material properties of the applied hyperelastic material.

6. Conclusions

A high-fidelity CFD-based numerical wave tank simulation for a VSB WEC is presented in this paper. This numerical tool is demonstrated to be able to simulate a VSB spherical WEC. This highly nonlinear interaction between the device and the waves is simulated using a 2-way FSI technique. Open sea conditions are applied in this study by assuming infinite air, deep water, and multi-directional waves. The numerical results in this paper capture the shape deformation in response to the varying surface pressure from the simulated ocean waves. It was shown that the VSB spherical WEC exhibits a transient response period before it reaches a steady-state motion and deformation. It was shown that this resulting motion of the VSB WEC significantly differs from that of a FSB WEC of the same shape as that of the non-deformed VSB WEC. This difference in response characteristics (difference in motion trajectories) is key for investigating the advantage of the VSB WEC power production over the FSB WEC.

Author Contributions: S.Z. implementation of numerical experiment, investigation, data analyses, visualization, original draft; O.A. conceptualization of the study, investigation, funding acquisition, refining of draft. All authors have read and agreed to the published version of the manuscript.

Funding: This research is funded by National Science Foundation (NSF), USA, under Grant Number 2023436.

Institutional Review Board Statement: Not applicable.

Informed Consent Statement: Not applicable.

Data Availability Statement: Not applicable.

Acknowledgments: This material is based upon work supported by the National Science Foundation (NSF), USA, under Grant Number 2023436.

Conflicts of Interest: The authors declare that they have no conflict of interest.

References

1. Clément, A.; McCullen, P.; Falcão, A.; Fiorentino, A.; Gardner, F.; Hammarlund, K.; Lemonis, G.; Lewis, T.; Nielsen, K.; Petroncini, S.; et al. Wave energy in Europe: Current status and perspectives. *Renew. Sustain. Energy Rev.* **2002**, *6*, 405–431. [CrossRef]
2. Li, G.; Weiss, G.; Mueller, M.; Townley, S.; Belmont, M.R. Wave energy converter control by wave prediction and dynamic programming. *Renew. Energy* **2012**, *48*, 392–403. [CrossRef]
3. Bacelli, G.; Ringwood, J.V. Numerical optimal control of wave energy converters. *IEEE Trans. Sustain. Energy* **2014**, *6*, 294–302. [CrossRef]
4. Falnes, J. *Ocean Waves and Oscillating Systems: Linear Interactions Including Wave-Energy Extraction*; Cambridge University Press: Cambridge, UK, 2002.

5. Zou, S.; Abdelkhalik, O.; Robinett, R.; Bacelli, G.; Wilson, D. Optimal control of wave energy converters. *Renew. Energy* **2017**, *103*, 217–225. [CrossRef]
6. Giorgi, G.; Ringwood, J.V. Computationally efficient nonlinear Froude–Krylov force calculations for heaving axisymmetric wave energy point absorbers. *J. Ocean Eng. Mar. Energy* **2017**, *3*, 21–33. [CrossRef]
7. Penalba, M.; Mérigaud, A.; Gilloteaux, J.C.; Ringwood, J.V. Influence of nonlinear Froude–Krylov forces on the performance of two wave energy points absorbers. *J. Ocean Eng. Mar. Energy* **2017**, *3*, 209–220. [CrossRef]
8. Yu, Y.H.; Li, Y. Reynolds-Averaged Navier–Stokes simulation of the heave performance of a two-body floating-point absorber wave energy system. *Comput. Fluids* **2013**, *73*, 104–114. [CrossRef]
9. Anbarsooz, M.; Passandideh-Fard, M.; Moghiman, M. Numerical simulation of a submerged cylindrical wave energy converter. *Renew. Energy* **2014**, *64*, 132–143. [CrossRef]
10. Hong, Y.; Waters, R.; Boström, C.; Eriksson, M.; Engström, J.; Leijon, M. Review on electrical control strategies for wave energy converting systems. *Renew. Sustain. Energy Rev.* **2014**, *31*, 329–342. [CrossRef]
11. Penalba, M.; Ringwood, J.V. A review of wave-to-wire models for wave energy converters. *Energies* **2016**, *9*, 506. [CrossRef]
12. Tom, N.; Lawson, M.; Yu, Y.H.; Wright, A. Preliminary analysis of an oscillating surge wave energy converter with controlled geometry. In Proceedings of the 11th European Wave and Tidal Energy Conference, Nantes, France, 6–11 September 2015.
13. Tom, N.; Lawson, M.; Yu, Y.H.; Wright, A. Spectral modeling of an oscillating surge wave energy converter with control surfaces. *Appl. Ocean Res.* **2016**, *56*, 143–156. [CrossRef]
14. Tom, N.; Lawson, M.; Yu, Y.; Wright, A. Development of a nearshore oscillating surge wave energy converter with variable geometry. *Renew. Energy* **2016**, *96*, 410–424. [CrossRef]
15. Kelly, M.; Tom, N.; Yu, Y.H.; Thresher, R.; Abbas, N. Development of the Second-Generation Oscillating Surge Wave Energy Converter With Variable Geometry. In Proceedings of the 27th International Ocean and Polar Engineering Conference, San Francisco, CA, USA, 25–30 June 2017;
16. Zou, S.; Abdelkhalik, O. Modelling Of A Variable-Geometry Wave Energy Converter. *IEEE J. Ocean. Eng.* **2020**. [CrossRef]
17. Windt, C.; Davidson, J.; Ringwood, J.V. High-fidelity numerical modelling of ocean wave energy systems: A review of computational fluid dynamics-based numerical wave tanks. *Renew. Sustain. Energy Rev.* **2018**, *93*, 610–630. [CrossRef]
18. Fluent, A. *ANSYS Fluent Theory Guide, Release 2019 R2*; ANSYS Inc.: Canonsburg, PA, USA, 2019.
19. Alberello, A.; Pakodzi, C.; Nelli, F.; Bitner-Gregersen, E.M.; Toffoli, A. Three dimensional velocity field underneath a breaking rogue wave. In *Volume 3A: Structures, Safety and Reliability, Proceedings of the ASME 2017 36th International Conference on Ocean, Offshore, and Arctic Engineering, Trondheim, Norway, 25–27 June 2017*; V03AT02A009; American Society of Mechanical Engineers: New York, NY, USA, 2017.
20. Gupta, V.K.; Khan, M.; Punekar, H. Development and Application of Interfacial Anti-Diffusion and Poor Mesh Numerics Treatments for Free Surface Flows. In Proceedings of the 2015 IEEE 22nd International Conference on High Performance Computing Workshops, Bengaluru, India, 16–19 December 2015; pp. 12–18.
21. Tezdogan, T.; Demirel, Y.K.; Kellett, P.; Khorasanchi, M.; Incecik, A.; Turan, O. Full-scale unsteady RANS CFD simulations of ship behaviour and performance in head seas due to slow steaming. *Ocean Eng.* **2015**, *97*, 186–206. [CrossRef]
22. Lee, C.; Jung, J.S.; Haller, M.C. Asymmetry in directional spreading function of random waves due to refraction. *J. Waterw. Port Coastal Ocean Eng.* **2009**, *136*, 1–9. [CrossRef]
23. Guo, Q.; Xu, Z. Simulation of deep-water waves based on JONSWAP spectrum and realization by MATLAB. In Proceedings of the 2011 19th International Conference on Geoinformatics, Shanghai, China, 24–26 June 2011; pp. 1–4.
24. Simanesew, A.; Krogstad, H.E.; Trulsen, K.; Borge, J.C.N. Development of frequency-dependent ocean wave directional distributions. *Appl. Ocean Res.* **2016**, *59*, 304–312. [CrossRef]
25. Gilloteaux, J.; Babarit, A.; Clément, A. Influence of spectrum spreading on the SEAREV wave energy converter. In Proceedings of the International Ocean Energy Conference, Vancouver, BC, Canada, 29 September–4 October 2007.
26. Kamath, A.M. Calculation of Wave Forces on Structures Using Reef3d. Master's Thesis, Institutt for Bygg, Anlegg og Transport, Trondheim, Norway, 2012.
27. Lloyd, C.; O'Doherty, T.; Mason-Jones, A. Development of a wave-current numerical model using Stokes 2nd Order Theory. *Int. Mar. Energy J.* **2019**, *2*, 1–14. [CrossRef]
28. Beneš, P.; Kollárik, R. Preliminary computational fluid dynamics (CFD) simulation of EIIB push barge in shallow water. *Sci. Proc. Fac. Mech. Eng.* **2011**, *19*, 67–73. [CrossRef]
29. Luan, Z.; He, G.; Zhang, Z.; Jing, P.; Jin, R.; Geng, B.; Liu, C. Study on the Optimal Wave Energy Absorption Power of a Float in Waves. *J. Mar. Sci. Eng.* **2019**, *7*, 269. [CrossRef]
30. Treloar, L. Stress-strain data for vulcanised rubber under various types of deformation. *Trans. Faraday Soc.* **1944**, *40*, 59–70. [CrossRef]

Journal of
Marine Science and Engineering

MDPI

Article

Investigations of Hydraulic Power Take-Off Unit Parameters Effects on the Performance of the WAB-WECs in the Different Irregular Sea States

Mohd Afifi Jusoh, Zulkifli Mohd Yusop, Aliashim Albani, Muhamad Zalani Daud and Mohd Zamri Ibrahim *

Renewable Energy & Power Research Interest Group (REPRIG), Eastern Corridor Renewable Energy Special Interest Group, Faculty of Ocean Engineering Technology and Informatics, Universiti Malaysia Terengganu, Kuala Nerus 21030, Terengganu, Malaysia; mohd.afifi.jusoh@gmail.com (M.A.J.); zul_12521@yahoo.com (Z.M.Y.); a.albani@umt.edu.my (A.A.); zalani@umt.edu.my (M.Z.D.)
* Correspondence: zam@umt.edu.my; Tel.: +60-96683328

Abstract: Hydraulic power take-off (HPTO) is considered to be one of the most effective power take-off schemes for wave energy conversion systems (WECs). The HPTO unit can be constructed using standard hydraulic components that are readily available from the hydraulic industry market. However, the construction and operation of the HPTO unit are more complex rather than other types of power take-off, as many components parameters need to be considered during the optimization. Generator damping, hydraulic motor displacement, hydraulic cylinder and accumulator size are among the important parameters that influence the HPTO performance in generating usable electricity. Therefore, the influence of these parameters on the amount of generated electrical power from the HPTO unit was investigated in the present study. A simulation study was conducted using MATLAB/Simulink software, in which a complete model of WECs was developed using the Simscape fluids toolbox. During the simulation, each parameters study of the HPTO unit were separately manipulated to investigate its effects on the WECs performance in five different sea states. Finally, the simulated result of the effect of HPTO parameters on the amount of generated electrical power from the HPTO unit in different sea states is given and discussed.

Keywords: ocean wave energy; wave-activated-body; hydraulic power take-off

Citation: Jusoh, M.A.; Yusop, Z.M.; Albani, A.; Daud, M.Z.; Ibrahim, M.Z. Investigations of Hydraulic Power Take-Off Unit Parameters Effects on the Performance of the WAB-WECs in the Different Irregular Sea States. *J. Mar. Sci. Eng.* **2021**, *9*, 897. https://doi.org/10.3390/jmse9080897

Academic Editor: Eugen Rusu, Kostas Belibassakis and George Lavidas

Received: 13 July 2021
Accepted: 13 August 2021
Published: 20 August 2021

Publisher's Note: MDPI stays neutral with regard to jurisdictional claims in published maps and institutional affiliations.

Copyright: © 2021 by the authors. Licensee MDPI, Basel, Switzerland. This article is an open access article distributed under the terms and conditions of the Creative Commons Attribution (CC BY) license (https://creativecommons.org/licenses/by/4.0/).

1. Introduction

Ocean waves are considered as one large untapped and predictable renewable energy resource on earth. Ocean waves contain tremendous of usable energy and have the potential to contribute to a significant share of global renewable energy sources. Ocean wave energy has many advantages such as high energy density, high source availability, source predictability and low environmental impact compared to other renewable energy (RE) sources [1]. The energy density of ocean waves is the highest among all renewable energy sources, which is around 50–100 kW/m [2]. Approximately, 8000–80,000 TWh/year ocean wave energy is available globally [2]. Due to its advantages, electrical energy production from the ocean waves has received a great deal of attention over the past several decades. Numerous wave energy converter systems (WECs) with different harnessing methods have been invented to convert the kinetic energy contained in the ocean waves into usable electricity, as reported in [3–6]. The existing WECs can be classified into wave-activated-body (WAB), oscillating water column (OWC) and overtopping device based on their working principles.

The WAB wave energy converters (WAB-WECs) are also known as oscillating bodies wave energy converters and point absorber wave energy converters. WAB-WECs can be defined as a single body or multiple bodies devices being oscillated by the wave excitation force [7]. WAB-WECs covers a kinds of WEC and recent development of WAB-WECs around the world has been reported in [4,6,8,9]. In general, these WAB-WECs consist of three main subsystems, namely wave energy converter (WEC), power take-off (PTO) unit, and control system (CS) unit. WEC is a front-end device that absorbs the kinetic energy from the ocean waves. The absorbed energy is then converted to electricity through the PTO unit, whereas the control system unit is used to optimize the electrical energy produced from the WECs during its operation.

PTO is one of the most essential subsystems of WAB-WECs. In recent decades, a wide variety of PTOs have been designed, developed and experimentally tested for numerous types of WEC device, as reported in [10]. The various kinds of PTO concepts can be classified based on their main working principles, such as mechanical–hydraulic, direct mechanical, direct electrical drive, air and hydro turbine. The hydraulic power take-off (HPTO) unit is one of the most reliable and effective PTOs for the WECs [11,12]. The HPTO unit has excellent characteristics, such as high efficiency, wide controllability, well-suited to the low-frequency and large power density of ocean waves, etc. The HPTO unit can also be assembled using standard hydraulic components that are readily available from the hydraulic equipment suppliers. In [11], a review of the most popular HPTO concepts used in WAB-WECs is reported. From the study, the HPTO concepts can be classified into two main groups, i.e., a variable-pressure and a constant-pressure concept. The constant-pressure concept has received more attention. Based on the report in [13], the efficiency of the constant-pressure concept is much higher than that of the variable-pressure concept, which can reach up to 90%. According to [14], several crucial parameters may influence the efficiency of the HPTO unit, such as the mounting position and piston size of the hydraulic actuator, volume capacity and pre-charge pressure of the accumulator, displacement of hydraulic motor and damping coefficient of the electric generator.

Recently, several kinds of research into the HPTO in WAB-WECs have been published for various objectives, for example in [15–21]. From the literature, most of the studies have concentrated on the performance of the HPTO unit without investigating the influence of the important parameters of HPTO. Only a few studies have discussed this issue, e.g., [14,16]. In [14], the influence of the HPTO unit parameters on the power capture ability of two-raft-type of WEC was studied. However, the HPTO concept used in [14] is very different from the HPTO concept considered in the present study. The effects of the HPTO unit on the percentage of power reduction were not discussed in this study. Meanwhile, in [16], the sensitivity of the generator damping coefficient on the average generated electrical power was investigated. From the simulation results, the authors concluded that the generator damping coefficient is relatively sensitive to the changes in wave height and period. However, the sensitivity of the other HPTO parameters was not discussed in [16]. Since there is a lack of published articles on this issue, the present study proposes an investigation of the HPTO unit parameters on the performance of the WAB-WECs. The main objective of the study is to investigate the effect of these important parameters on the electrical power generated from the HPTO unit. The findings can be a useful reference to other researchers for improvement of WECs in future.

The remainder of this paper is organized as follows. Section 2 presents the design of the considered WECS and Section 3 describes the simulation study for investigation of the HPTO parameters. Section 4 presents the results and discussion. Finally, the conclusion and future work are discussed in Section 5.

2. Design of WEC with HPTO Unit

The WEC concept based on [20] is considered in the present study. The simplified concept is shown in Figure 1A. The WECs consist of a single floating body (floater) attached to the fixed body via a hinged arm, which is also known as the WEC device. This WEC device design is almost similar to the concepts used in [13,22–24]. The WEC device is unique due to its ability to convert both wave kinetic energy and wave potential by utilizing the pitch motion of the floater and hinged arm, as presented in Figure 1A. In this system, the WEC is connected to the fixed-body directed to the dominant wave direction to optimally absorb the kinetic energy from the ocean waves. The WEC device is then connected to the HPTO and the CS unit is placed in the HPTO house to convert the mechanical energy from the WEC device into usable electricity.

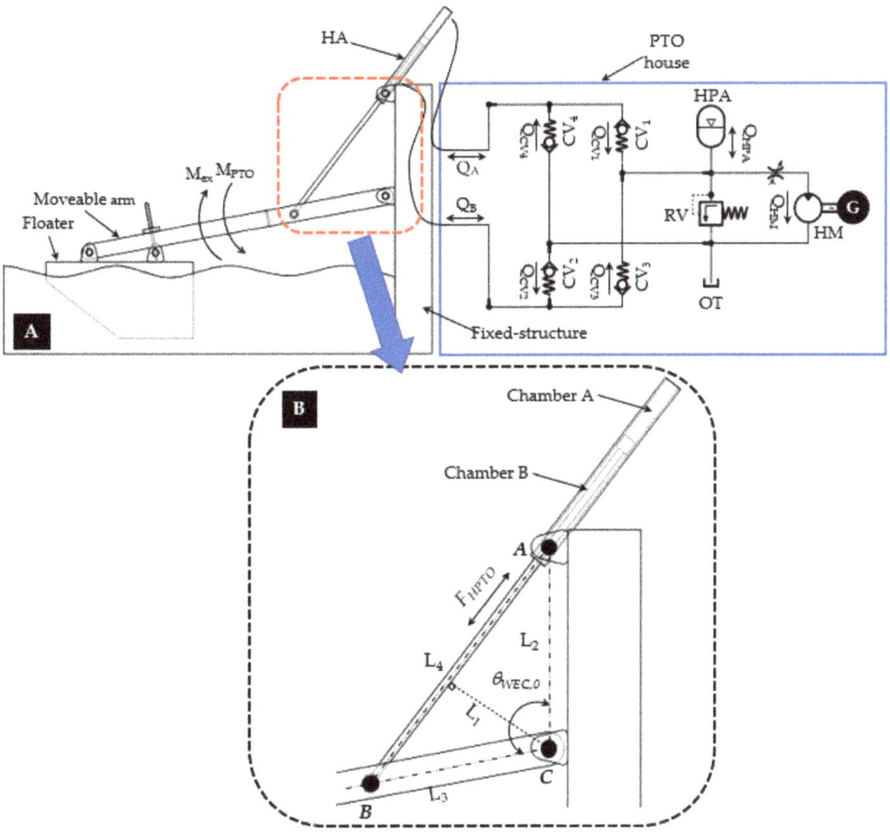

Figure 1. An illustration of WEC with the HPTO unit concept. (**A**) A complete layout design, and (**B**) Enlarge image of the interconnection between HA, floater's arm and fixed structure.

Figure 1A presents the simplified diagram of the HPTO unit which includes a hydraulic actuator (HA), hydraulic hose (HH), check valve rectifier module (CV_1, CV_2, CV_3 & CV_4), oil tank (OT), high-pressure and low-pressure accumulators (HPA & LPA), pressure relief valve (RV_1 & RV_2), hydraulic motor (HM) and electric generator (G). In this concept, double-acting with a single rod type of HA is considered. HA is used as a linear pump to absorb the mechanical energy from the reciprocating motion of the WEC. The piston rod of the HA is attached to the floater's arm using a rod end clevis while the barrel of the HA is attached to the fix-structure. Then, the HA is connected to the check valve rectifier module, an arrangement of four check valves in a bridge circuit configuration, as shown in Figure 1A. The check valve rectifier used in this HPTO unit is similar to the Graetz bridge concept, which is used for conversion of an alternating-current (AC) input into a direct-current (DC) output. For the HPTO, the check valve rectifier module is used to control the fluid flow direction (Q_A & Q_B) from the HA to the HM. Thus, the large chamber (chamber A) of the HA barrel is connected to the inlet and outlet of CV_1 and CV_4, while the small chamber (chamber B) of the HA barrel is terminated to the inlet and outlet of CV_2 and CV_3, respectively. The check valve rectifier module is then connected to the HPA and LPA. The HPA is included in the HPTO unit to constrain the pressure of the HM in the desired ranges. Finally, the generation module which consists of fixed-displacement of HM coupled to G is placed between the HPA and OT. In addition, the pressure relief valves RV_1 & RV_2 are placed to prevent the HPTO unit from over-pressurized fluid flows.

During the operation of WECs, the passing of ocean waves causes a WEC device to pitch upward and downward simultaneously. Then, the reciprocating motion of the WEC device causes back-and-forth motions of the piston and directly generates the high-pressure fluid flow from the HA chambers. During the upward movement of the WEC device, the high-pressure fluid flows from chamber A to chamber B through CV_1, HPA, HM, LPA, and CV_2. On the other hand, during the downward movement, the high-pressure fluid flows from chamber B to chamber A through the CV_3, HPA, HM, LPA, and CV_4 of the WEC device. The high-pressure fluid flowing through the HM causes the HM and G to rotate simultaneously in one direction and thus produces usable electricity. Overall, the speed and torque of the HM depend highly on the characteristics of ocean wave motions such as speed, frequency, wavelength, and amplitude [25]. They also depend on the important parameters of the HPTO unit itself.

2.1. Formulation of Hydrodynamic Pitch Motion of WEC

The hydrodynamic pitch motion of WEC in real waves can be formulated in the time domain to account for the non-linear effects such as the hydrodynamics of a floater, HPTO force, etc. So, the equation for the pitch motion of the WEC device can be expressed by:

$$M_D = M_{ex} - M_{rad} - M_{res} - M_{HPTO} \tag{1}$$

where M_D is the D'Alembert moment of inertia, and M_{ex}, and M_{rad} are the moments due to the diffracted and radiated ocean waves. M_{res} and M_{HPTO} are the moments due to hydrostatic restoring and HPTO unit interactions, respectively. The hydrodynamic pitch motion of the WEC equation above can be extended as follows:

$$(J_{WEC} + J_{add,\infty})\alpha_{WEC}(t) + \int_0^t k_{rad}(t-\tau)\ \omega_{WEC}(t) + k_{res}\ \theta_{WEC}(t) + M_{HPTO}(t) = \int_{-\infty}^{\infty} h_{ex}(t-\tau)\eta_W(\tau)d\tau \tag{2}$$

where J_{WEC} is the moment of inertia of WEC (includes floater and hinged arm), $J_{add,\infty}$ is the added mass at the infinite frequency. τ is the time delay. θ_{WEC}, ω_{WEC} and α_{WEC} are the angular position, angular velocity and angular acceleration of WEC during the pitch motion, respectively. $\theta_{WEC} = 0$ corresponds to the WEC device at rest. k_{rad} and k_{res} are the radiation impulse response function and the hydrostatic restoring coefficients. h_{ex} is the excitation force coefficient and η_W is the undisturbed wave elevation at the center point of the floater. The coefficients of k_{rad}, k_{res} and h_{ex} can be determined from the dynamic diffraction analysis using computational fluid dynamics (CFD) software, such as ANSYS/AQWA, as previously implemented in [20,26–30]. Alternatively, these coefficients also can be obtained using the boundary element method (BEM) toolbox in WAMIT software, as suggested in several studies [31–35].

Since the non-linear effect of the HPTO unit is considered in Equation (2), the moment due to the HPTO unit M_{HPTO} can be described using Equation (3). According to Figure 1B, F_{HPTO} is the feedback force of the PTO unit applied to the WEC device. L_1 is the perpendicular distance between HA and point C, which can be obtained using Equation (4), where L_2 and L_3 are the distance between points *A-C* and *B-C*, respectively. $L_{4,0}$ is the initial distance between points *A-B*. $\theta_{WEC,0}$ and θ_{WEC} are the initial and instantaneous angle of the WEC device. x_p is the linear displacement of the piston. Based on Equation (4), L_1 is always relatively changes according to the change of the arm angle, as illustrated in Figure 1B.

$$M_{HPTO} = F_{HPTO}L_1 \tag{3}$$

$$L_1 = \frac{L_2L_3\sin(\theta_{WEC,0} - \theta_{WEC})}{L_{4,0} + x_{HA}} \tag{4}$$

$$x_p = L_{4,0} - \sqrt{{L_2}^2 + {L_3}^2 - 2L_2L_3\cos(\theta_{WEC,0} - \theta_{WEC})} \tag{5}$$

2.2. Formulation of HPTO Unit

Technically, the force generated by the HPTO unit, F_{HPTO} is represented by the feedback force applied by the HA to the WEC device. The nonlinear F_{HPTO} is generated due to the dynamic pressure in chambers A and B (P_A & P_B) and the piston friction force (F_{fric}) of the HA. The F_{HPTO} can be expressed using Equation (6), where $A_{p,A}$ and $A_{p,B}$ are the sectional areas of the piston in chambers A and B, respectively. $A_{p,A}$ and $A_{p,B}$ can be obtained using Equations (7) and (8), where d_p and d_r are the piston and rod diameter of the HA. Meanwhile, the dynamics of P_A and P_B can be calculated according to the fluid continuity function, as described in Equations (9) and (10), where β_{eff} is the bulk modulus of the hydraulic fluid. Q_{CV1} to Q_{CV4} are the flow rates across the check valves CV_1 to CV_4. L_s, x_p and $\dot{x}_p$ are the stroke length, linear displacement and linear velocity of the piston, respectively.

$$F_{HPTO} = P_A A_{p,A} - P_B A_{p,B} + F_{fric} \tag{6}$$

$$A_{p,A} = \pi {d_p}^2/4 \tag{7}$$

$$A_{p,B} = \pi\left({d_p}^2 - {d_r}^2\right)/4 \tag{8}$$

$$\frac{d}{dt}P_A = \frac{\beta_{eff}}{A_{p,A}(L_s - x_p)}(\dot{x}_p A_{p,A} + Q_{CV4} - Q_{CV1}) \tag{9}$$

$$\frac{d}{dt}P_B = \frac{\beta_{eff}}{A_{p,B}(L_s - x_p)}(\dot{x}_p A_{p,B} + Q_{CV2} - Q_{CV3}) \tag{10}$$

The flow rates Q_{CV1} to Q_{CV4} can be generally calculated using Equation (11), where C_d and A_{CV} are the discharge coefficient and the working area of each check valve. P_{CVin} and P_{CVout} are the inlet and outlet pressure of each check valve. ρ_{oil} is the hydraulic oil density.

$$Q_{CV} = \begin{cases} C_d A_{CV}\sqrt{\frac{2}{\rho_{oil}}|P_{CVin} - P_{CVout}|} & ,if\ P_{CVin} > P_{CVout} \\ 0 & ,if\ P_{CVin} < P_{CVout} \end{cases} \tag{11}$$

On the other hand, the fluid volume (V_{HPA}) and flow rate (Q_{HPA}) which enters the accumulator can be calculated using Equations (12) and (13), where $V_{HPA,cap}$ is a capacity, $P_{HPA,0}$ is the initial pressure and $P_{HPA,in}$ is the inlet gauge pressure of HPA, and n is the specific heat ratio, respectively. The initial pressure in the accumulators depends on the pre-charge pressure of the nitrogen gas in the HPA bladder.

$$V_{HPA} = \begin{cases} V_{HPA,cap}\left[1 - \left(\frac{P_{HPA,0}}{P_{HPA,in}}\right)^{\frac{1}{n}}\right] & ,\ if\ P_{HPA,in} > P_{HPA,0} \\ 0 & ,\ if\ P_{HPA,in} \leq P_{HPA,0} \end{cases} \tag{12}$$

$$Q_{HPA} = \dot{V}_{HPA} = \begin{cases} \frac{1}{n}V_{HPA,cap}\left(1 - \frac{P_{HPA,0}}{P_{HPA,in}}\right)^{\frac{1-n}{n}} \frac{P_{HPA,0}\dot{P}_{HPA,in}}{{P_{HPA,in}}^2} & ,\ if\ P_{HPA,in} > P_{HPA,0} \\ 0 & ,\ if\ P_{HPA,in} \leq P_{HPA,0} \end{cases} \tag{13}$$

Meanwhile, the flow rate across the HM (Q_{HM}) and the actual torque of HM (τ_{HM}) can be obtained using Equations (14) and (15), where D_{HM}, ω_{HM} and ΔP_{HM} are the displacement, angular speed and the internal pressure difference of HM. $\eta_{HM,\ V}$ and $\eta_{HM,\ M}$ are the volumetric and mechanical efficiency of HM.

$$Q_{HM} = D_{HM}\omega_{HM}/\eta_{HM,V} \tag{14}$$

$$\tau_{HM} = \Delta P_{HM} D_{HM} \eta_{HM,M}/2\pi \tag{15}$$

Finally, the electric power generated by the electric generator (P_G) can be expressed using Equation (16), where ω_G, τ_G and η_G are angular speed, the torque and overall efficiency of the electrical generator. Since the electrical generator and HM are rotating simultaneously, the ω_G and τ_G are equal to ω_{HM} and τ_{HM}, respectively.

$$P_G = 2\pi\omega_G\tau_G\eta_G = 2\pi\omega_{HM}\tau_{HM}\eta_G \tag{16}$$

2.3. Main Parameters of HPTO Unit

As previously mentioned in [14], there are several important parameters affecting the performance of the HPTO unit such as piston diameter, the volume capacity of HPA, displacement of HM, etc. Table 1 provides the important parameters of the HPTO unit that are summarized from Equations (3)–(16). For some parameters, the higher and lower values can reduce the performance of the HPTO unit. To investigate this problem, a detailed study regarding the influence of these parameters on the overall performance of WECs is performed in the present study.

Table 1. Important parameters of the HPTO unit.

No.	Important Parameters of HPTO	Unit
1	Vertical mounting of HA, L_2	m
2	Piston diameter, d_p	m
3	Volume capacity of HPA, $V_{HPA,cap}$	L
4	Pre − charge gas pressure of HPA, $P_{HPA,0}$	bar
5	Displacement of HM, D_{HM}	cc/rev
6	Damping coefficient of the generator, d_G	Nm/(rad/s)

In the present study, the complete simulation investigation of HPTO unit parameters was implemented in the MATLAB/Simulink software (Version: 2018b). The detailed methodology of the present study is described in the following subsections.

3. Simulation Investigation of HPTO Unit Parameters

3.1. Simulation Set-up of WEC with HPTO Unit

The MATLAB/Simulink software was used in this study to model a complete WEC system shown in Figure 1. The WECs consists of two main parts, the WEC model and the HPTO model. The WEC model was developed using a mathematical function block based on Equations (1)–(5). The wave elevation data, η_W and F_{HPTO} are the inputs, while the x_p is the output of the WEC model. According to Equations (1) and (2), the hydrodynamics parameters of the WEC device such as k_{rad}, k_{res}, h_{ex}, added mass coefficient (k_{add}) and impulse response function (k_{imp}) are required to develop the considered WEC model. The hydrodynamic parameters of the WEC model from the previous study in [20] were used, since a similar WEC concept was considered in this study. In [20], these hydrodynamic parameters were obtained from the frequency domain analysis study that was carried out using ANSYS/AQWA software. The obtained hydrodynamic parameters are presented in Figure 2.

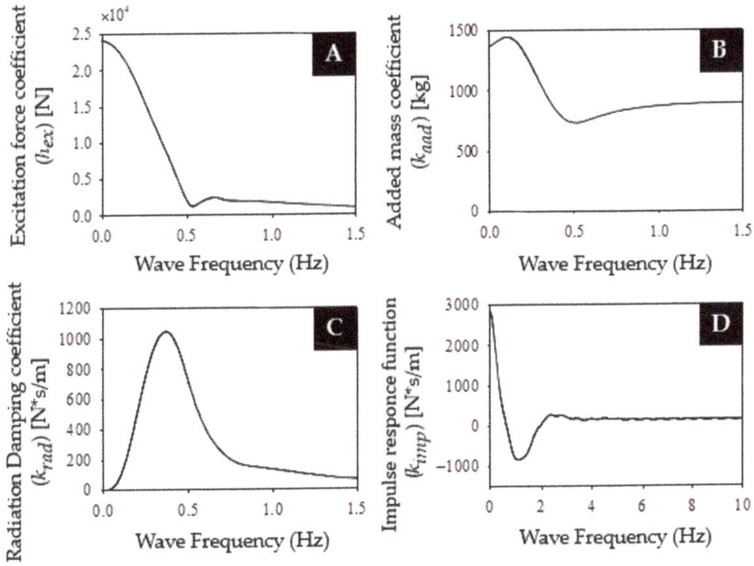

Figure 2. Hydrodynamic parameters of WEC device. (**A**) Excitation force coefficient, (**B**) Added mass coefficient, (**C**) Radiation damping coefficient, and (**D**) Impulse response function.

Meanwhile, the Simscape SimHydraulic toolbox in MATLAB/Simulink was used to develop the HPTO unit model. The snapshot of the developed model in MATLAB/Simulink is depicted in Figure 3. As illustrated in the figure, x_p is the input of the HPTO unit. Using x_p signal, the linear velocity of the piston, $\dot{x}_p$ is obtained using a first-order lag-based linear displacement to linear velocity converter. The double-acting hydraulic cylinder (DAC) component was used as HA. In the Simscape SimHydraulic toolbox, the DAC component is constructed based on the translational hydro-mechanical converter and translational hard stop blocks. The rod motion is limited with the mechanical translational hard stop block. The ideal force sensor block was connected to the HA rod to measure the F_{HPTO}. The pressure and flow rate sensor blocks were also connected to the HA to measure the dynamic pressure and flow rate of HA. To account for the friction loss along the pipe length and the fluid compressibility, the hydraulic pipeline blocks were used to connect some of the components, as illustrated in Figure 3. The thermodynamic transformation in the HPA was assumed to be isentropic, which is reasonable considering the cycle time in the device. Furthermore, for ease of control, a simple rotational damper with varying damping coefficients was used to represent the electric generator unit. In this way, the resistive torque imposed by the electric generator can be manipulated by varying the value of the damping coefficient (d_G). The generated electric power from the electric generator can be calculated using Equation (16). In this study, the initial parameters of the HPTO unit were manual tuned. However, these parameters are not optimal yet. The detailed specifications of each component in the developed HPTO model are provided in Table 2. Finally, the developed HPTO unit model in MATLAB/Simulink was then experimentally validated using an actual HPTO test rig in the dry lab environment.

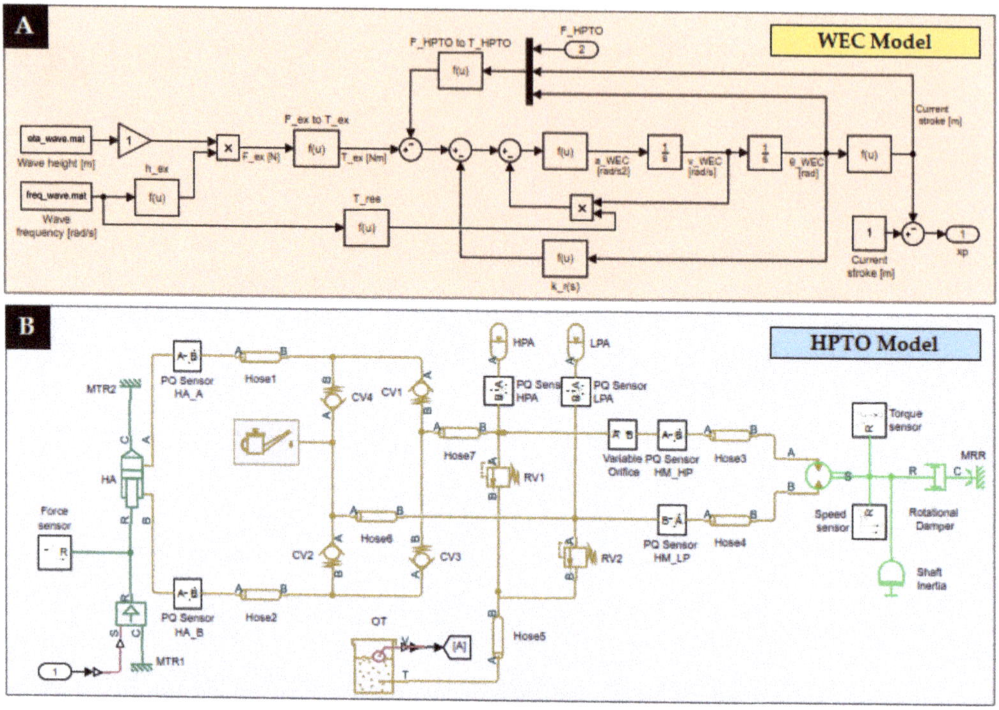

Figure 3. A complete model of WEC with HPTO unit in MATLAB/Simulink. (**A**) WEC device model, and (**B**) HPTO unit model.

Table 2. Detailed specifications of the developed HPTO unit.

Component Descriptions		Value	Unit
Hydraulic actuator	Piston diameter, d_p	0.035	m
	Piston rod diameter, d_r	0.022	m
	Stroke length, L_s	0.3	m
	Initial stroke length, $L_{s,0}$	0.15	m
	Vertical distance mounting, L_2	0.5	m
	Horizontal distance mounting, L_3	0.5	m
	Initial rod length (point A to B), L_4	0.766	m
High-pressure accumulator	Pre-charge pressure, $P_{HPA,0}$	46.9	bar
	Volume capacity, $V_{HPA,cap}$	2.8	L
	Adiabatic index, γ	1.4	-
Low-pressure accumulator	Pre-charge pressure, $P_{LPA,0}$	3.2	bar
	Volume capacity, $V_{LPA,cap}$	4.0	L
	Adiabatic index, γ	1.4	-
Hydraulic motor	Displacement, D_{HM}	8.4	cc/rev
	Nominal shaft angular velocity, $\omega_{HM,\ nom}$	200	rpm
	Volumetric efficiency at nominal condition, $\eta_{HM,\ V}$	0.92	-
	No-load torque, $\tau_{HM,\ no}$	0.05	Nm
Electric generator	Rated power, $P_{G,\ rated}$	100	W
	Rated speed, $\omega_{G,\ rated}$	200	rpm
	Rated torque, $\tau_{G,\ rated}$	6.0	Nm
	Damping coefficient, d_G	0.03	Nm/(rad/s)
	Moment of inertia	0.0036	kg/m^2
Fluid properties	Density, D_{oil}	50	kg/m^3
	Viscosity, Vis_{oil}	850	cSt

3.2. Simulation of WEC with HPTO Unit Using Five Irregular Sea States

The simulation analysis was started with the performance evaluation of the WEC with the HPTO model, using five different irregular wave input conditions. This simulation was intended to evaluate the effect of the significant wave height (H_W) and the peak wave period T_W on the electrical output power produced by the HPTO unit. Five different sea states were considered as summarized in Table 3. Sea state A is the nominal wave condition for the developed WEC with the HPTO model, in which the significant wave height and the peak wave period were set to 0.8 m and 2.5 s, respectively. Sea state B and C were considered for wave height case studies, while sea state D and E are for wave period case studies. For sea states B and C, the significant wave heights were set to 0.6 m and 1.0 m, which is ±0.2 m of the wave height in sea state A. Meanwhile, the peak wave periods for both states were maintained at the nominal value. For sea states D and E, the significant wave height was maintained at the nominal value, while the peak wave periods were set to 2.0 s and 3.0 s, which is ±0.5 s of the nominal wave period.

Table 3. The parameters of five different sea states.

Sea State	Significant		Remarks
	Wave Height, H_W (m)	Wave Period, T_W (s)	
A	0.8	2.5	Nominal wave condition
B	0.6	2.5	Wave height case
C	1.0		
D	0.8	2.0	Wave period case
E		3.0	

3.3. Investigation Studies of HPTO Unit Parameters

A further simulation proceeded with the investigation of the HPTO unit parameters study. As previously mentioned, the main purpose of the study is to investigate the influence of the important parameters on the performance of the HPTO unit. Thus, from this study, how the performance of the HPTO unit varies to its configuration parameters was discovered. To achieve the objective, several case studies were conducted on the developed simulation model of WECs. A summary of the case studies is provided in Table 4. The regulating condition of each case was selected based on the availability of the components from the hydraulic equipment market.

Table 4. Detailed of the case studies.

Case	Important Parameters of HPTO	Default Value	Regulating Condition		Unit
			Ranges	Step	
1	Vertical Mounting of HA, L_2	0.5	0.1–0.7	0.1	m
2	Piston diameter, d_p	0.035	0.025–0.060	0.005	m
3	Volume capacity of HPA, $V_{HPA,cap}$	2.8	0.5–10.5	2.0	L
4	Pre-charge gas pressure of HPA, $P_{HPA,0}$	46.9	20–80	10.0	bar
5	Displacement of HM, D_{HM}	8.4	6–18	2	cc/rev
6	Damping coefficient of the generator, d_G	0.287	0.1–1.3	0.2	Nm/(rad/s)

4. Results and Discussion

The results from the investigation simulations are presented and discussed in three sections. The first Section 4.1 presents the experimental validation of the developed WEC with the HPTO unit model. The second Section 4.2 provides the performance analysis of the developed WEC with the HPTO unit model in different sea states and the third Section 4.3 presents the finding from the investigations of the influence parameters on the HPTO unit performance.

4.1. Experimental Validation of the HPTO Model

Figure 4A shows the hardware in the loop (HIL) test rig of the HPTO unit. In general, the HIL test rig was developed based on the design of WECs in Figure 1 and several sensors were installed to monitor the variations of HPTO force, oil pressure, the oil level in the oil tank and as the hydraulic motor shaft speed, as illustrated in Figure 4B. The servo-electric actuator was also installed in the HIL test rig to replicate wave-induced relative pitch motion to drive the HPTO unit to capture the wave energy. A servo motor controller based on Labview/Arduino integration and the data acquisition system for data collection from the sensors were placed in the control system unit. The relative pitch motion generated by the electric actuator is according to the input wave state. To simulate the condition that is close to real-world application, the hydrodynamic parameters from the CFD analysis and the feedback HPTO force were considered to calculate the produced excitation force applied to the floater's arm. The sinusoidal wave input with the amplitude and period of 0.4 m and 2.5 s were considered for the validation of HPTO model. The captured image of maximum upward and downward motions of the HIL test rig during the experimental validation of the HPTO unit is depicted in Figure 4C,D.

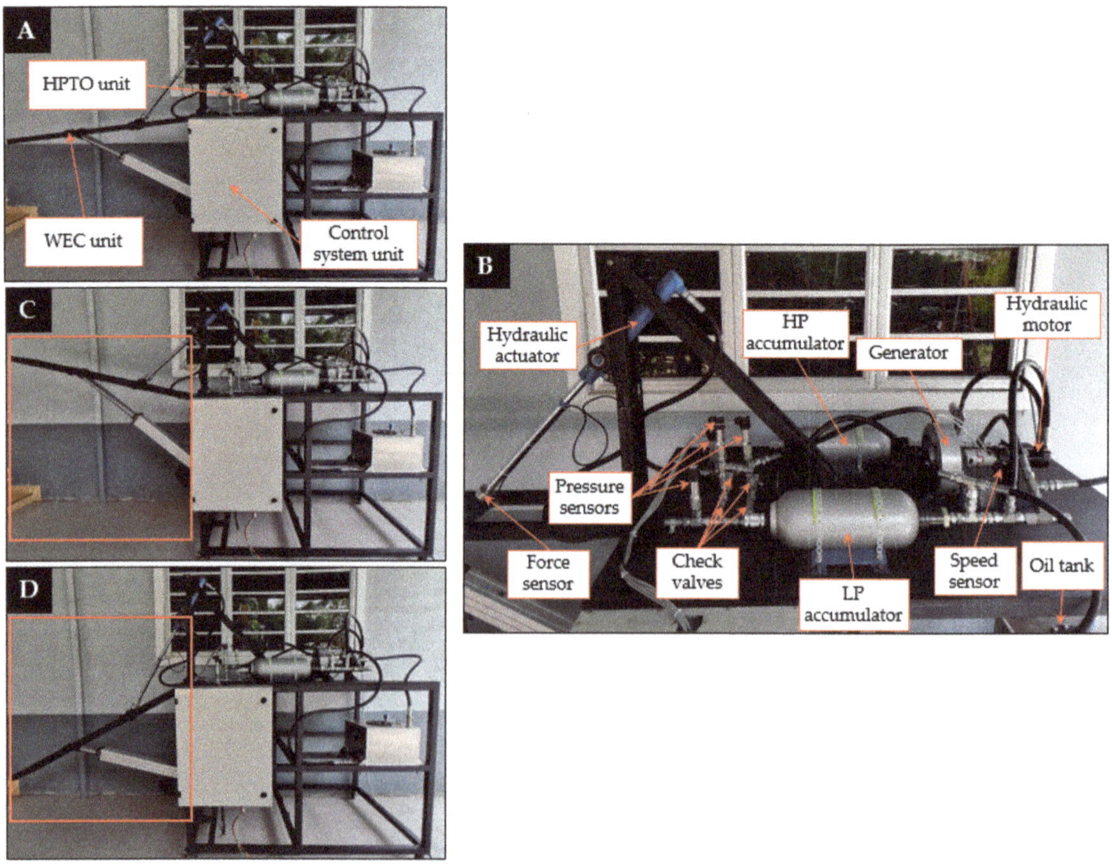

Figure 4. Experimental evaluation of WEC with hydraulic PTO unit. (**A**) A complete dry-lab test rig, (**B**) Enlarge image of HPTO unit setup, (**C**) Maximum upward, and (**D**) Maximum downward position of the WEC device.

Figure 5 shows the results of the behaviour of the HPTO unit in a regular wave condition with an amplitude and period of 0.4 m and 2.5 s. From the figure, it can be seen that the simulation results of the developed WEC with the HPTO unit model in

MATLAB/Simulink is in good agreement with the results obtained from the HIL test rig. However, slight differences and some fluctuations in hydraulic motor speed and HPTO force results were obtained during the experiment, as depicted in Figure 5A,B. These differences and fluctuations may be attributed to the errors in the manufacture and assembly of the test rig, the measuring errors of the transducers, the vibration of the hydraulic motor, and the leakage in the hydraulic motor, cylinders and joints. Such a good agreement presented in Figure 5A,B indicates that the developed WEC with HPTO unit model in MATLAB/Simulink presented in the present study would be effective and reliable as a tool for predicting the amount of power that can be generated from the ocean waves.

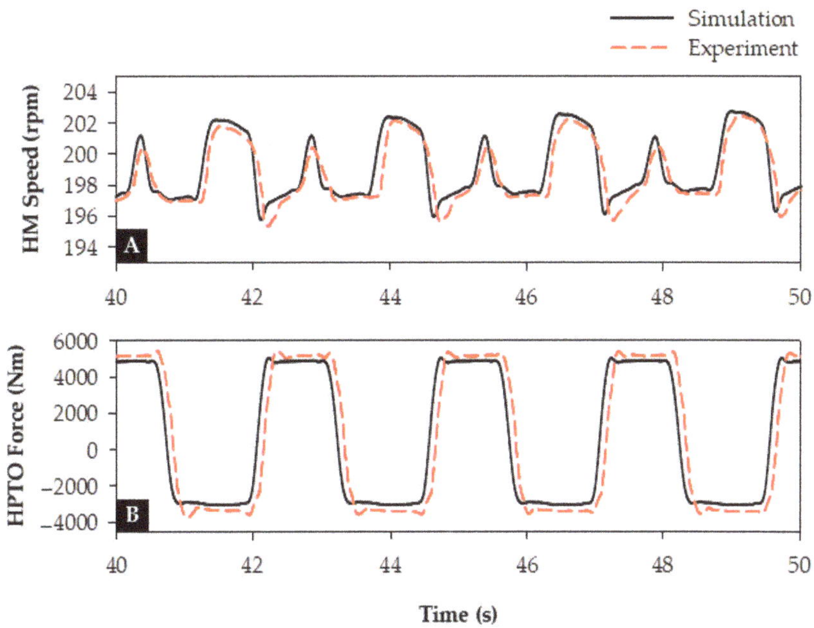

Figure 5. Behaviour of HPTO unit in a regular wave condition. (**A**) Speed of hydraulic motor and (**B**) HPTO force applied to the floater's arm.

4.2. Performance of WEC with HPTO Model in Five Irregular Sea States

The simulations of the developed WEC with the HPTO unit model using different sea states was first carried out in the present study. This simulation was intended to evaluate the performance of the developed model against the different wave heights and periods. The simulation was started with the nominal sea state (sea state A) and the results from the simulation are presented in Figures 6 and 7. Figure 6A shows the responses of WEC and the hydraulic cylinder piston against the irregular wave input in sea state A. The figure shows that the displacement of the WEC device was slightly lower than the wave elevation, particularly during the upward motion. This is due to influencing factors such as the hydrostatic restoring moment, the moment due to the HPTO unit and the initial moments of floater and arm [36]. Based on the results, the average displacement of the WEC device and hydraulic piston was 70% and 15% of the wave elevation. The figure also depicts that the displacements of the WEC device and piston were slightly delayed from the wave elevation. Figure 6B presents the profile of the HPTO force applied to the WEC device. On average, the HPTO forces applied to the WEC device during upward and downward motion equaled 3.64 kN and 1.99 kN, respectively. The unbalanced HPTO force applied to the WEC device is due to the unsymmetrical effective area of the piston. A larger effective area of piston produced a higher force rather than a smaller effective area piston.

Figure 6C shows the profile of HPA pressure. From the figure, the pressure of HPA reached up to 49 bar several times, which was a 4.5% increased from its pre-charge pressure setting.

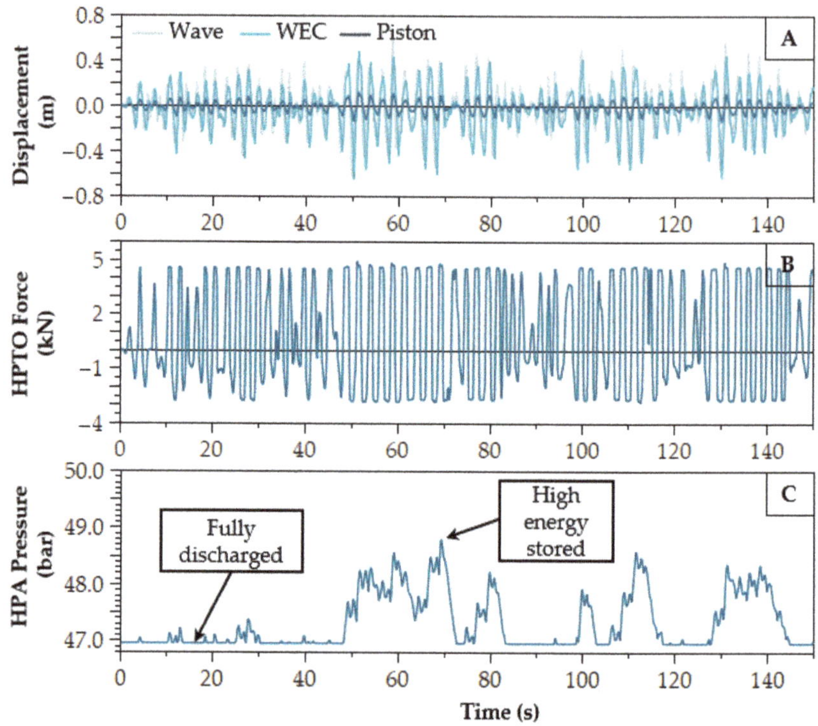

Figure 6. Performance of WEC with HPTO unit in sea state A, (**A**) Displacement of Wave, WEC and hydraulic cylinder piston, (**B**) HPTO force applied to WEC device, and (**C**) Pressure of high-pressure accumulator.

Meanwhile, Figure 7A,B show the pressure and speed profile of the HM. It can be seen from the figures that the pressure and the speed of the hydraulic motor reached up to 49 bar and 200 rpm. The smoothing effect of the HPA unit on the hydraulic motor pressure can be seen in Figure 7A. The HPA was able to reduce the fluctuation of the hydraulic motor pressure, particularly after 50 s of HPTO operation. Figure 7C illustrates the profile of the generated power from the HPTO unit. The average power generated from the generator was 70.9% of its rated capacity (100 W). The figure also demonstrates that some fluctuations exist in the generated power from the generator, particularly at the early stage of the operation. The comparison simulation results of the WEC with HPTO unit in each sea state are summarized in Table 5. From the table, the simulation result showed that the significant wave height and peak wave period were affected by the overall performance of the WEC with the HPO unit. First, the table reveals that the averaged angular displacement of the WEC device was increased and decreased, relatively, with increases and decreases of the significant wave height and peak wave period. From the table, the angular displacement of the WEC device in sea states B and D were reduced by 38% and 33% (upward) and 24% and 21% (downward) of its angular displacement in the nominal sea state. Meanwhile, the angular displacement in the sea states C and E were increased by 21% and 16% (upward) and 28% and 18% (downward), respectively. The increase and decrease of the angular displacements are due to the increase and decrease HPTO force applied to the WEC device, which can be obtained in Table 5. The increase and decrease of WEC displacement, relatively, also increase and decrease the generated output power from the HPTO unit. As can be seen, the average power generated from the HPTO unit in sea states B and D were decreased by 41% and 34% of power from sea state A, while 15.5% and 10.4% were increased in sea states C and E. Overall, the generated power from the HPTO unit in each sea state is below its rated capacity. Thus, several parameter optimization methods, as suggested in [20], can be further implemented to increase the generated power from the HPTO unit.

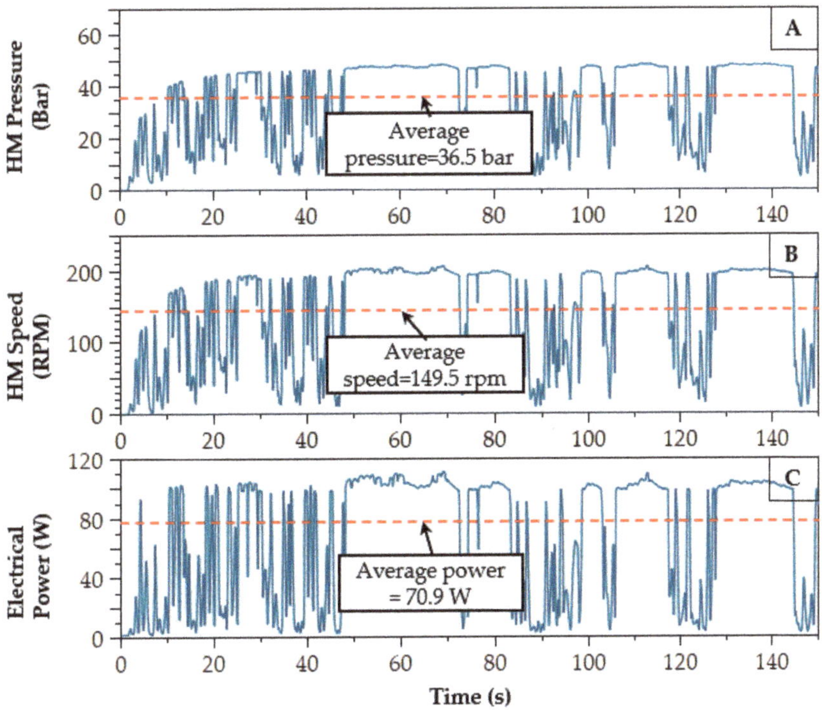

Figure 7. Performance of WEC with HPTO unit in sea state A (continue), (**A**) Pressure of hydraulic motor, (**B**) Speed of hydraulic motor, and (**C**) Electrical power generated from HPTO unit.

Table 5. Performance of WEC with HPTO unit in different sea states.

Performance Description (Unit)		**Sea State**				
		A	B	C	D	E
Averaged Angular Displacement of WEC (°)	upward	5.73	3.53	6.94	3.84	6.67
	downward	6.95	5.26	8.87	5.49	8.17
Averaged piston displacement (m)	upward	0.033	0.019	0.041	0.022	0.038
	downward	0.035	0.026	0.043	0.028	0.040
Averaged HPTO force (kN)	upward	3.64	3.49	4.45	3.54	4.09
	downward	1.99	1.63	2.42	1.73	2.24
Averaged operating pressure of HM (bar)	-	36.5	27.4	41.9	29.2	39.7
Averaged speed of generator (rpm)	-	149.5	109.1	173.6	117.2	164.4
Averaged generated electrical power (W)	-	70.9	41.7	81.8	47.1	78.2

4.3. Investigation Studies of HPTO Unit Parameters

In this subsection, the influence of each HPTO unit parameter on the generator power in five different sea states is discussed. From this subsection, how the power of the generator varies with the considered HPTO parameters and their corresponding to the sea states can be discovered in the following subsection.

4.3.1. Case 1: Position of Hydraulic Cylinder

The influence of HA position on the power of the generator was firstly investigated in this study. For this case, the position of HA was manipulated by adjusting the L_2 and L_3, as indicated in Figure 1. In this case, L_2 was incrementally varied by 0.1 m within the range of 0.1 to 0.7 m. To be fair, during adjustment of the HA position, the initial rod length of HA from point A to B (L_4) was maintained as default for every sequence. Figure 8 depicts that the averaged power generated from the generator varies with the vertical mounting distance of HA for different sea states. The figure clearly illustrates that the averaged power generated from the generator increases along with the increase of the horizontal mounting of HA for all sea states. Then, it decreases after reaching the optimal mounting position. From the figure, the optimal values for L_2 are different for each sea state. For the small wave height and period sea states (sea states B and D), the optimal value for L_2 is smaller than for the case of large wave height and period sea states. At the optimal mounting position, averaged generated power for the sea states A to E were around 72 W, 42 W, 84 W, 47 W and 79 W, respectively. At the lower L_2, averaged generated power for all sea states were significantly reduced compared to the bigger value of L_2, as depicted in Figure 8. For example, at L_2 equal to 0.1 m, the averaged generated power from the generator for the sea states A to E were reduced by 88%, 91%, 82%, 90% and 84% of their optimal values. While at L_2 equal to 0.7 m, the averaged power generated from the generator for sea states A to E were reduced by 43%, 53%, 32%, 51% and 38% of their optimal values, respectively. The percentage of averaged power reduction also indicates that the mounting position of HA relies on the wave height and wave period. The percentage of averaged power reduction shows that the mounting position of HA was more affected by low height and a small period for the sea states. Technically, the huge reduction of the averaged power generated at the lower L_2 is due to the larger HPTO force applied to the WEC device. So, from these investigation results, it can be suggested that the distance of L_2 should be equal or larger than L_3 to prevent huge losses due to the mounting position of HA.

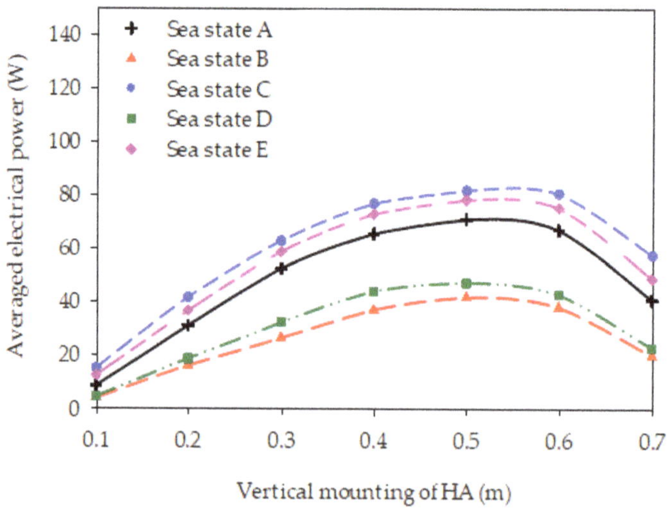

Figure 8. Averaged power of generator versus horizontal mounting distance corresponding to different sea states.

4.3.2. Case 2: Piston Size of the Hydraulic Actuator

The influence of the effective area of the piston in both hydraulic actuator chambers is also a concern in this study. As previously shown in Equation (6), the effective piston area relatively affects the amount of feedback HPTO force F_{HPTO} applied to the WEC device. From Equations (7) and (8), the effective area of the piston in both hydraulic chambers can be calculated via the diameter size of piston and rod (d_p and d_r). Since a double-acting with single rod hydraulic actuator was considered in this study, the effective piston area in chamber B is affected by the rod size. This means that the effective piston area in chamber A is larger than that in chamber B. To investigate, the wide range of d_p was used to examine the performance of the HPTO unit. In this case, the values of d_p was incrementally varied by 0.005 m within the range of 0.025 to 0.060 m. In this case, the minimum range was selected based on the smallest piston size that is currently available in the hydraulic equipment market. While the value of d_r remains as the initial parameter setting. Figure 9 presents the effect of the piston and rod diameter on the averaged power generated from the generator in different sea states. From the figure, it can be observed that the averaged generated power is influenced by the piston size of the hydraulic actuator. The figure depicts that the averaged generated power first increases with the increase of the piston diameter size and then starts to decrease after obtaining the optimal value of d_p. From the figure, it is clearly shown that the smaller significant wave height and peak period sea states were more affected by the piston size. The result shows that the average generated power for sea states B and D started to decrease after 0.045 m size of the piston, while for the bigger wave height and period sea states, such as sea state A, C and E, the average generated power started to decrease after 0.050 m and 0.055 m, respectively. This may be due to the lower wave forces during sea states B and D compared to wave forces for sea states A, C and E. Technically, more high-pressured fluid can be supplied to the hydraulic motor by a hydraulic actuator with a larger piston size. As a result, the average generated power at the optimal point for sea states A, C and E were found to be 94 W, 118 W and 110 W. Meanwhile, the average generated power at the optimal point for sea states B and D were found to be 64 W and 70 W, respectively.

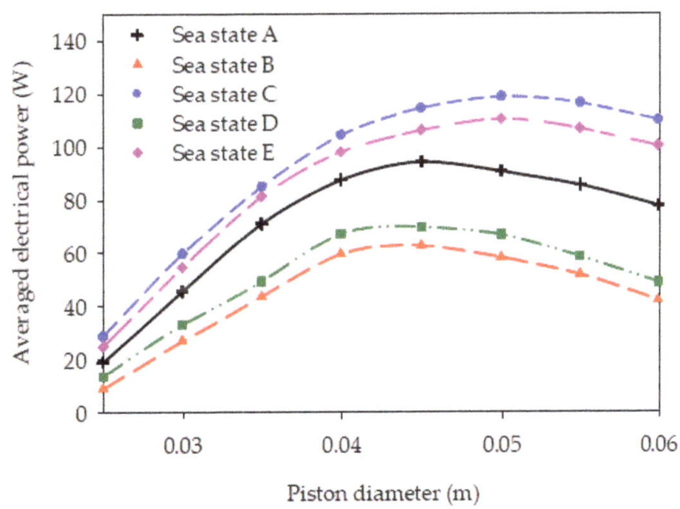

Figure 9. Averaged power of generator versus piston diameter corresponding to different sea states.

4.3.3. Case 3: Volume Capacity of HPA

The accumulator plays an important role in mitigating the power fluctuation during a dynamic process of wave power conversion. By using HPA and a robust control strategy, the HPTO unit enables conversion of the high fluctuating wave power into a smooth and continuous electrical power. Since the HPA is more important to the HPTO unit, a further investigation into the main parameters of HPA, such as volume capacity ($V_{HPA,cap}$) should be conducted. Hence, the effect of the volume capacity of HPA on the HPTO performance was explored in the present study. A wide range of $V_{HPA,cap}$ was used to evaluate its effect on the averaged generator power corresponding to the different sea states. In this case, the value of $V_{HPA,cap}$ was incrementally varied by 2 L within the range of 0.5 to 10.5 L, as previously mentioned in Table 4. Figure 10 presents the effect of $V_{HPA,cap}$ on the averaged generated power from the generator for the different sea states. The figure showed that the averaged power generated from the generator is influenced by the increase of $V_{HPA,cap}$, particularly for large significant wave height and peak period sea states, while there was a less significant effect for small significant wave height and peak period sea states. From the result, the average generated power, particularly for sea states C and E significantly reduced by the increase of $V_{HPA,cap}$. By changing the value of $V_{HPA,cap}$ from 0.5 L to 10.5 L, the average generated power for sea states C and E were reduced by 28% and 20%, respectively. This power reduction can be attributed to more energy accumulated in the HPA, rather than directly flowing to the hydraulic motor when the large capacity of the HPA is implemented.

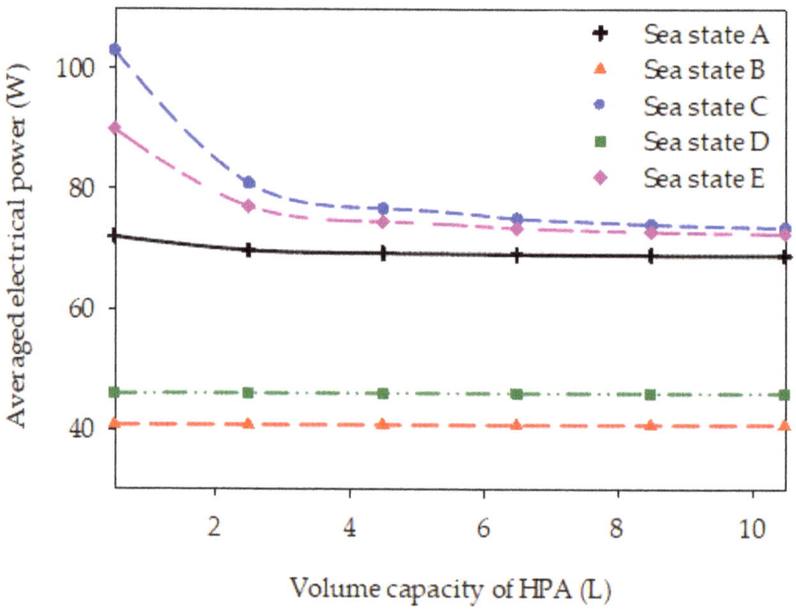

Figure 10. Averaged power of generator versus volume capacity of HPA corresponding to different sea states.

4.3.4. Case 4: Pre-Charge Pressure of HPA

The pre-charge pressure of HPA ($P_{HPA,0}$) is another important parameter of the HPTO. Technically, $P_{HPA,0}$ determines how much hydraulic fluid will remain accumulated in the HPA. The charging process of HPA begins when hydraulic fluid flows into the fluid chamber when the HPTO unit pressure is greater than the $P_{HPA,0}$. During charging, the gas is compressed to store energy. Once the HPTO unit pressure is below $P_{HPA,0}$ level, the high-pressure nitrogen gas in the ballast forces hydraulic fluid from the fluid chamber into the hydraulic motor. For such a dynamic process, the investigation of the effect of the $P_{HPA,0}$ on the power of the generator was considered. In this case, the value of $P_{HPA,0}$ was incrementally increase by 10 bar within the range of 20 to 80 bar. Figure 11 presents the variation of $P_{HPA,0}$ on the averaged generated power from the generator for different sea states. As can be seen in the figure, for all sea states the average power of the generator slightly increased, and then tended to be steady after the $P_{HPA,0}$ reached an optimal of $P_{HPA,0}$. The figure showed that a higher level of $P_{HPA,0}$ can be used for large significant wave height and peak period sea states. For example, the level of $P_{HPA,0}$ can be set up to 70 bar for sea states C and E, while, the highest $P_{HPA,0}$ for sea states, B and D was only up to 40 bar. This difference is due to the different operating pressure of HPTO in each sea state, in which the operating pressure of HPTO is higher in sea states C and E rather than sea state B and D. The averaged generated power at the optimal point for sea state A, C and E were reached up to 80 W, 124 W and 108 W, while for sea states B and D the averaged generated power at the optimal point only reached 43 W and 49 W, respectively.

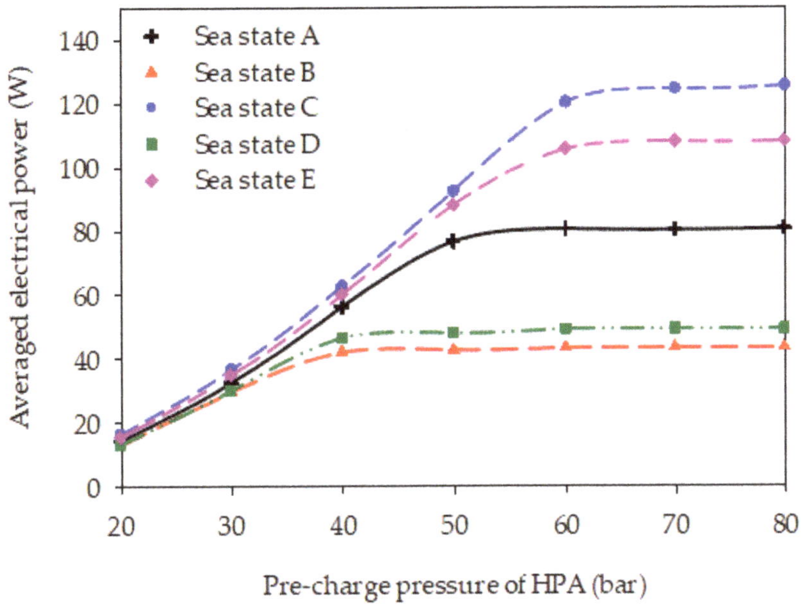

Figure 11. Averaged power of generator versus pre-charge pressure of HPA corresponding to different sea states.

4.3.5. Case 5: Displacement of Hydraulic Motor

The fluid displacement of the hydraulic motor (D_{HM}) is another important influencing parameter in the HPTO unit. Referring to Equation (14), D_{HM} directly affects the power and torque of the hydraulic motor. Therefore, it is necessary to explore the effect of the D_{HM} on the power of the hydraulic motor and generator. From the preliminary survey, the smallest size hydraulic motor currently available from the available hydraulic equipment for HPTO application 6 cc/rev. Thus, the variation range of D_{HM} was set within 6 to 20 cc/rev and the value of D_{HM} was incrementally varied by 2 cc/rev. Figure 12 illustrates the effect of D_{HM} on the averaged generated power from the generator corresponding to different sea states. From the figure, it is clearly shown that the averaged generated power increases with the increase in D_{HM} and then starts to decrease after reaching an optimal value of D_{HM} for all sea states. The figure shows that a higher value of D_{HM} can be implemented with a large significant wave height and peak period sea states. The result indicates that the value of D_{HM} can be considered up to 10 cc/rev for sea states C and E, but only up to 8 cc/rev for sea states B and D. This may be due to the high operating pressure of HPTO during the large significant wave height and peak period sea states. At the optimal value of D_{HM}, the average generated power for sea states A to E can reach up to 72 W, 44 W, 102 W, 50 W and 90 W, respectively. Apart from that, the result also shows that the overestimated value of D_{HM} was more affected in HPTO performance at the small significant wave height and peak period sea states. For example, the average generated power in sea states B and D was reduced by up to 59% and 60% once the value of D_{HM} was increased from the optimal (8 cc/rev) to 18 cc/rev; while for sea states C and E, the average generated power in sea states C and E were reduced by up to 46% and 47% once the value of D_{HM} was increased from the optimal (10 cc/rev) to 18 cc/rev. However, this was vice-versa during the underestimated value of D_{HM}, in which the HPTO performance in the large wave height and period sea states was more affected by the D_{HM}. The figure clearly shows that the averaged generated power for sea states C and E were significantly reduced by 53% and 50%. Therefore, the result reveals that the underestimation and overestimation of the D_{HM} can significantly reduce the power generated from the generator.

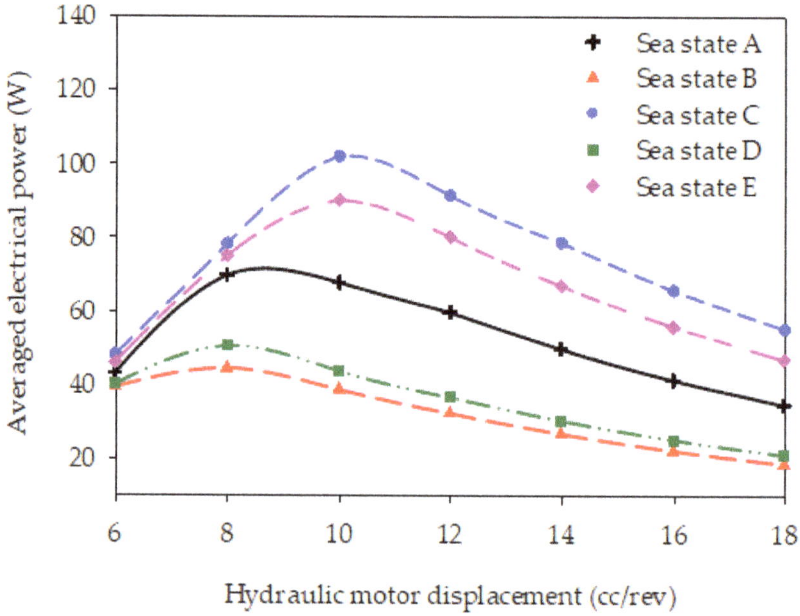

Figure 12. Averaged power of generator versus fluid displacement of HM corresponding to different sea states.

4.3.6. Case 6: Damping Coefficient of Electrical Generator

The damping coefficient (d_G) is different for each generator. As mentioned in [11], the d_G of each generator is subjected to its type, capacity, etc. Technically, generated output power is affected by the d_G. Since the generator is one of the major components of the HPTO unit, the effect of the d_G on the performance of the HPTO needs to be investigated. Thus, the effect of d_G on the generated power from the HPTO unit was investigated in this study. In this case, the d_G was varied from 0.1 to 1.2 Nm/(rad/s) with increments of 0.1 Nm/(rad/s) in each sequence. Figure 13 depicts the effect of d_G on the averaged generated power corresponding to five different sea states. From the figure, it can be seen that the optimal d_G that achieves the highest averaged power was sensitive to changes in significant wave height and peak wave period. During the nominal sea state, the optimal d_G was found to around 0.3 Nm/(rad/s), while, for the short and long peak wave period sea states (sea states B and C), the optimal d_G was found to be around 0.35 and 0.26 Nm/(rad/s). Meanwhile, for the small and large significant wave height sea states (sea states D and E), the optimal d_G was found at around 0.34 and 0.24 Nm/(rad/s), respectively. The result also indicates that the overestimation of d_G badly reduced the averaged generated power for all sea states, particularly sea states A, C and E. As can be seen in the figure, the average generated power for sea states A to E was reduced by 73%, 62%, 72%, 63% and 74% of its optimal value, respectively, due to overestimation of d_G by up to 1.3 Nm/(rad/s). Hence, from the results, it can be said that the generator damping coefficient needs to be optimally controlled to maximize power absorption from the ocean. Thus, the use of several damping control strategies, as suggested in [17], can be considered in the HPTO unit.

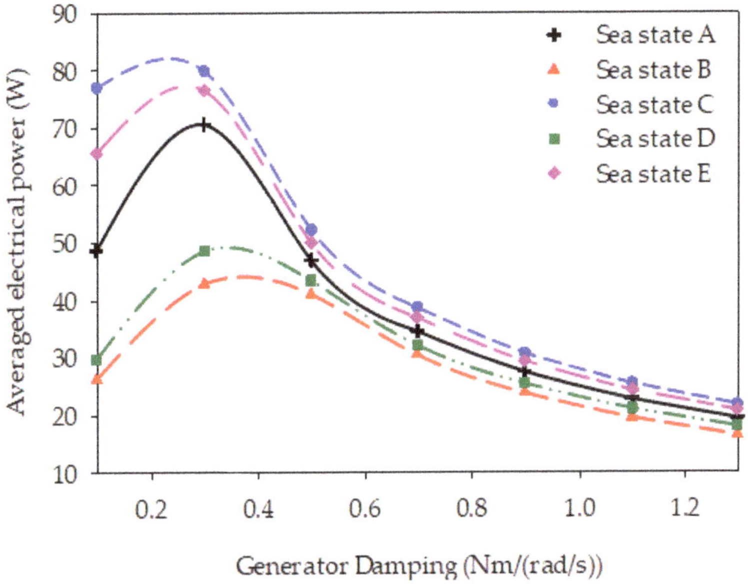

Figure 13. Averaged power of generator versus damping of generator corresponding to different sea states.

5. Conclusions

A comprehensive analysis of the effects of the important HPTO parameters on performance in generating usable electricity was conducted in the present study. Six critical parameters of the HPTO unit, including vertical mounting and the piston size of the HA, volume capacity and pre-charge pressure of HPA, displacement of HM and damping coefficient of the generator were considered. A simulation study was conducted using MATLAB/Simulink software, in which a complete model of WEC with the HPTO unit was developed using the Simscape fluids toolbox in MATLAB/Simulink. Five different irregular sea state inputs were used to evaluate the effect of each HPTO parameter against the different significant wave heights and peak periods. From the investigation, the following conclusion can be drawn:

(1) For case 1, the effect of the vertical mounting position (L_2) of the hydraulic actuator on the power generated by the generator was obtained. From the simulation result, it was found that the averaged power generated increased along with an increase of L_2 for all sea states and then decreased after reaching the optimal distance. At smaller L_2 (0.1 m), the averaged generated power for sea states A to E was reduced by 88%, 91%, 82%, 90% and 84% of their optimal values, respectively. The value of L_2 is sensitive to the significant wave height and peak wave period, in which the best of L_2 is larger during the large significant wave height and peak wave period sea state.

(2) For case 2, the averaged power generated was increased along with the increase of piston size (d_p) for all sea states. However, the average power was decreased for the over-sized of d_p state. The simulation result shows a d_p sensitive to the significant wave height and peak wave period. Thus, a large size of d_p should be used in the large significant wave height and peak wave period sea state.

(3) For case 3, the simulation results demonstrate that the volume capacity of HPA ($V_{HPA,cap}$) is less sensitive to changes in small significant wave height and peak wave period sea state, and inversely for a large significant wave height and peak wave period sea state, where the increase of $V_{HPA,cap}$ reduced the averaged power generated. This is due to more power accumulated in the HPA rather than directly flowing to

the hydraulic motor. Thus, an appropriate $V_{HPA,cap}$ should be selected based on the HPTO capacity to avoid this power reduction.

(4) For case 4, the simulation result shows that the pre-charge pressure of HPA ($P_{HPA,0}$) should be higher for the large significant wave height and peak wave period sea state rather than the small significant wave height and peak wave period sea state.

(5) For case 5, the investigation results reveal that the underestimated and overestimated hydraulic motor displacement (D_{HM}) was significantly sensitive to wave height and peak wave period. Thus, a variable displacement hydraulic motor with a robust control strategy should be considered.

(6) For case 6, the simulation results found that overestimated damping coefficient of the generator (d_G) hardly reduced the averaged generated power. Thus, d_G needs to be optimally controlled using appropriate damping control strategies to maximize power absorption from the ocean waves.

The present investigation studies may help researchers and engineers of WECs to improve the efficiency of their systems. The optimization of the critical parameters above is another attractive issue in terms of maximizing the generated power from the HPTO unit. Thus, it is suggested that further research regarding the HPTO parameter optimization using heuristic optimization algorithms should be conducted.

Author Contributions: M.A.J., conceptualization, methodology, software, data curation, analysis, writing—original draft; M.Z.I., writing—review and editing and supervision, project administration, funding acquisition; M.Z.D., conceptualization, methodology, writing—review and editing and supervision; Z.M.Y., software, data curation, analysis and writing—original draft; A.A. data curation, analysis and writing—review and editing. All authors have read and agreed to the published version of the manuscript.

Funding: This project was funded by the Ministry of Higher Education (MOHE) under the Fundamental Research Grant Scheme (FRGS/1/2019/TK07/UMT/01/1).

Institutional Review Board Statement: Not applicable.

Informed Consent Statement: Not applicable.

Data Availability Statement: Data available on request due to restrictions of privacy.

Acknowledgments: The authors would like to thank the Ministry of Higher Education (MOHE) and Universiti Malaysia Terengganu (UMT) for financial support for this research.

Conflicts of Interest: The authors declare no conflict of interest.

References

1. Melikoglu, M. Current status and future of ocean energy sources: A global review. *Ocean Eng.* **2018**, *148*, 563–573. [CrossRef]
2. Farrok, O.; Ahmed, K.; Tahlil, A.D.; Farah, M.M.; Kiran, M.R.; Islam, M.R. Electrical power generation from the oceanic wave for sustainable advancement in renewable energy technologies. *Sustainability* **2020**, *12*, 2178. [CrossRef]
3. Ahamed, R.; McKee, K.; Howard, I. Advancements of wave energy converters based on power take off (PTO) systems: A review. *Ocean Eng.* **2020**, *204*, 107248. [CrossRef]
4. Al Shami, E.; Zhang, R.; Wang, X. Point absorber wave energy harvesters: A review of recent developments. *Energies* **2019**, *12*, 47. [CrossRef]
5. Falcão, A.F. Wave energy utilization: A review of the technologies. *Renew. Sustain. Energy Rev.* **2010**, *14*, 899–918. [CrossRef]
6. Yusop, Z.M.; Ibrahim, M.Z.; Jusoh, M.A.; Albani, A.; Rahman, S.J.A. Wave-Activated Body Energy Converter Technologies: A Review. *J. Adv. Res. Fluid Mech. Therm. Sci.* **2020**, *76*, 76–104. [CrossRef]
7. Sheng, W.; Lewis, A. Power takeoff optimization for maximizing energy conversion of wave-activated bodies. *IEEE J. Ocean. Eng.* **2016**, *41*, 529–540. [CrossRef]
8. Drew, B.; Plummer, A.R.; Sahinkaya, M.N. A review of wave energy converter technology. *Proc. Inst. Mech. Eng. Part A J. Power Energy* **2009**, *223*, 887–902. [CrossRef]
9. Titah-Benbouzid, H.; Benbouzid, M. An up-to-date technologies review and evaluation of wave energy converters. *Int. Rev. Electr. Eng.* **2015**, *10*, 52–61. [CrossRef]
10. Kukner, A.; Erselcan, İ.Ö. A review of power take-off systems employed in wave energy. *J. Nav. Sci. Eng.* **2014**, *10*, 32–44.
11. Jusoh, M.A.; Ibrahim, M.Z.; Daud, M.Z.; Albani, A.; Yusop, Z.M. Hydraulic power take-off concepts for wave energy conversion system: A review. *Energies* **2019**, *12*, 4510. [CrossRef]

12. He, X.; Xiao, G.; Hu, B.; Tan, L.; Tang, H.; He, S.; He, Z. The applications of energy regeneration and conversion technologies based on hydraulic transmission systems: A review. *Energy Convers. Manag.* **2020**, *205*, 112413. [CrossRef]
13. Hansen, R.H.; Kramer, M.M.; Vidal, E.; Hansen, R.H.; Kramer, M.M.; Vidal, E. Discrete displacement hydraulic power take-off system for the wavestar wave energy converter. *Energies* **2013**, *6*, 4001–4044. [CrossRef]
14. Liu, C.; Yang, Q.; Bao, G. Influence of hydraulic power take-off unit parameters on power capture ability of a two-raft-type wave energy converter. *Ocean Eng.* **2018**, *150*, 69–80. [CrossRef]
15. Zou, S.; Abdelkhalik, O. Control of Wave Energy Converters with Discrete Displacement Hydraulic Power Take-Off Units. *J. Mar. Sci. Eng.* **2018**, *6*, 31. [CrossRef]
16. Ding, B.; Cazzolato, B.S.; Arjomandi, M.; Hardy, P.; Mills, B. Sea-state based maximum power point tracking damping control of a fully submerged oscillating buoy. *Ocean Eng.* **2016**, *126*, 299–312. [CrossRef]
17. Jianan, X.; Tao, X. MPPT Control of Hydraulic Power Take-Off for Wave Energy Converter on Artificial Breakwater. *J. Mar. Sci. Eng.* **2019**, *8*, 304. [CrossRef]
18. Liu, C.; Yang, Q.; Bao, G. Performance investigation of a two-raft-type wave energy converter with hydraulic power take-off unit. *Appl. Ocean Res.* **2017**, *62*, 139–155. [CrossRef]
19. Xuhui, Y.; Qijuan, C.; Zenghui, W.; Dazhou, G.; Donglin, Y.; Wen, J.; Weiyu, W. A novel nonlinear state space model for the hydraulic power take-off of a wave energy converter. *Energy* **2019**, *180*, 465–479. [CrossRef]
20. Jusoh, M.A.; Ibrahim, M.Z.; Daud, M.Z.; Yusop, Z.M.; Albani, A. An Estimation of Hydraulic Power Take-off Unit Parameters for Wave Energy Converter Device Using Non-Evolutionary NLPQL and Evolutionary GA Approaches. *Energies* **2020**, *14*, 79. [CrossRef]
21. Jusoh, M.A.; Ibrahim, M.Z.; Daud, M.Z.; Yusop, Z.M.; Albani, A.; Rahman, S.J.; Mohad, S. Parameters estimation of hydraulic power take-off system for wave energy conversion system using genetic algorithm. In *Proceedings of the IOP Conference Series: Earth and Environmental Science*; Institute of Physics Publishing: Bristol, UK, 2020; Volume 463, p. 12129.
22. Hansen, A.H.; Asmussen, M.F.; Bech, M.M. Model predictive control of a wave energy converter with discrete fluid power power take-off system. *Energies* **2018**, *11*, 635. [CrossRef]
23. Gaspar, J.F.; Calvário, M.; Kamarlouei, M.; Soares, C.G. Design tradeoffs of an oil-hydraulic power take-off for wave energy converters. *Renew. Energy* **2018**, *129*, 245–259. [CrossRef]
24. Gaspar, J.F.; Calvário, M.; Kamarlouei, M.; Guedes Soares, C. Power take-off concept for wave energy converters based on oil-hydraulic transformer units. *Renew. Energy* **2016**, *86*, 1232–1246. [CrossRef]
25. Penalba, M.; Sell, N.P.; Hillis, A.J.; Ringwood, J.V.; Penalba, M.; Sell, N.P.; Hillis, A.J.; Ringwood, J.V. Validating a wave-to-wire model for a wave energy converter—Part I: The hydraulic transmission system. *Energies* **2017**, *10*, 977. [CrossRef]
26. Shadman, M.; Estefen, S.F.; Rodriguez, C.A.; Nogueira, I.C.M. A geometrical optimization method applied to a heaving point absorber wave energy converter. *Renew. Energy* **2018**, *115*, 533–546. [CrossRef]
27. Coiro, D.P.; Troise, G.; Calise, G.; Bizzarrini, N. Wave energy conversion through a point pivoted absorber: Numerical and experimental tests on a scaled model. *Renew. Energy* **2016**, *87*, 317–325. [CrossRef]
28. Chen, Z.; Zhou, B.; Zhang, L.; Sun, L.; Zhang, X. Performance evaluation of a dual resonance wave-energy convertor in irregular waves. *Appl. Ocean Res.* **2018**, *77*, 78–88. [CrossRef]
29. Sun, P.; Li, Q.; He, H.; Chen, H.; Zhang, J.; Li, H.; Liu, D. Design and optimization investigation on hydraulic transmission and energy storage system for a floating-array-buoys wave energy converter. *Energy Convers. Manag.* **2021**, *235*, 113998. [CrossRef]
30. Liu, Z.; Qu, N.; Han, Z.; Zhang, J.; Zhang, S.; Li, M.; Shi, H. Study on energy conversion and storage system for a prototype buoys-array wave energy converter. *Energy Sustain. Dev.* **2016**, *34*, 100–110. [CrossRef]
31. Liang, C.; Ai, J.; Zuo, L. Design, fabrication, simulation and testing of an ocean wave energy converter with mechanical motion rectifier. *Ocean Eng.* **2017**, *136*, 190–200. [CrossRef]
32. Zurkinden, A.S.; Ferri, F.; Beatty, S.; Kofoed, J.P.; Kramer, M.M. Non-linear numerical modeling and experimental testing of a point absorber wave energy converter. *Ocean Eng.* **2014**, *78*, 11–21. [CrossRef]
33. Rahmati, M.T.; Aggidis, G.A. Numerical and experimental analysis of the power output of a point absorber wave energy converter in irregular waves. *Ocean Eng.* **2016**, *111*, 483–492. [CrossRef]
34. Jama, M.A.; Noura, H.; Wahyudie, A.; Assi, A. Enhancing the performance of heaving wave energy converters using model-free control approach. *Renew. Energy* **2015**, *83*, 931–941. [CrossRef]
35. Forehand, D.I.M.; Kiprakis, A.E.; Nambiar, A.J.; Wallace, A.R. A Fully Coupled Wave-to-Wire Model of an Array of Wave Energy Converters. *IEEE Trans. Sustain. Energy* **2016**, *7*, 118–128. [CrossRef]
36. Windt, C.; Davidson, J.; Ringwood, J.V. High-fidelity numerical modelling of ocean wave energy systems: A review of computational fluid dynamics-based numerical wave tanks. *Renew. Sustain. Energy Rev.* **2018**, *93*, 610–630. [CrossRef]

energies

MDPI

Article

Environmental Assessment of the Impacts and Benefits of a Salinity Gradient Energy Pilot Plant

Etzaguery Marin-Coria [1,*], Rodolfo Silva [1,*], Cecilia Enriquez [2], M. Luisa Martínez [3] and Edgar Mendoza [1]

[1] Instituto de Ingeniería, Universidad Nacional Autónoma de México, Ciudad Universitaria, Circuito Exterior S/N, Coyoacán, Mexico City 04510, Mexico; EMendozaB@iingen.unam.mx
[2] Campus Yucatan, Universidad Nacional Autónoma de México (UNAM), Mexico City 04510, Mexico; cenriqz@ciencias.unam.mx
[3] Instituto de Ecología, A.C. (INECOL), Xalapa, Veracruz 91073, Mexico; marisa.martinez@inecol.mx
* Correspondence: janeth.marin.coria@gmail.com (E.M.-C.); RSilvaC@iingen.unam.mx (R.S.)

Abstract: Although the technologies involved in converting saline gradient energy (SGE) are rapidly developing, few studies have focused on evaluating possible environmental impacts. In this work, the environmental impacts of a hypothetical 50 kW RED plant installed in La Carbonera Lagoon, Yucatan, Mexico, are addressed. The theoretical support was taken from a literature review and analysis of the components involved in the pressure retarded osmosis (PRO) and reverse electrodialysis (RED) technologies. The study was performed under a three-stage scheme (construction, operation, and dismantling) for which the stress-inducing factors that can drive changes in environmental elements (receptors) were determined. In turn, the possible modifications to the dynamics of the ecosystem (responses) were assessed. Since it is a small-scale energy plant, only local impacts are expected. This study shows that a well-designed SGE plant can have a low environmental impact and also be of benefit to local ecotourism and ecosystem conservation while contributing to a clean, renewable energy supply. Moreover, the same plant in another location in the same system could lead to huge modifications to the flows and resident times of the coastal lagoon water, causing great damage to the biotic and abiotic environment.

Keywords: salinity gradient energy; RED; PRO; coastal systems; stress factors; receptors; environmental impact

Citation: Marin-Coria, E.; Silva, R.; Enriquez, C.; Martínez, M.L.; Mendoza, E. Environmental Assessment of the Impacts and Benefits of a Salinity Gradient Energy Pilot Plant. *Energies* **2021**, *14*, 3252. https://doi.org/10.3390/en14113252

Academic Editor: Eugen Rusu

Received: 24 April 2021
Accepted: 28 May 2021
Published: 3 June 2021

Publisher's Note: MDPI stays neutral with regard to jurisdictional claims in published maps and institutional affiliations.

Copyright: © 2021 by the authors. Licensee MDPI, Basel, Switzerland. This article is an open access article distributed under the terms and conditions of the Creative Commons Attribution (CC BY) license (https://creativecommons.org/licenses/by/4.0/).

1. Introduction

As the supply of fossil fuels diminishes, the opportunity of switching to renewable sources of energy will put an end to some of the negative environmental impacts seen since the first industrial revolution [1,2]. The oceans are a major source of renewable energies, such as marine and tidal currents, wave energy, thermal, and salinity gradients, which can all be harnessed [3,4].

Chemical energy known as salinity gradients (SGE) or saline gradient potential (SGP) is available in coastal zones where two water flows of different saline content coincide, e.g., where a river meets the sea [5,6]. By controlling this mixture and capturing the energy before it is released, electricity can be produced without greenhouse gas emissions. It is possible to use only naturally occurring water flows, but it is also possible to employ hybrid systems, which use effluents of anthropic origin, such as residual waters from desalination plants [7–9]. Similarly, the effluent from wastewater treatment plants, of low salinity, could be used as input for an SGP system [10,11].

The methods for producing energy from a saline gradient are varied, but the most advanced methods are reverse electrodialysis (RED) and pressure retarded osmosis (PRO), both of which have already been tested outside the laboratory. Regarding RED, the companies WETSUS and REDstack have developed a 50 kW device in the Netherlands [12,13], while for PRO, the company Statkraft developed a 10 kW plant in Norway [14].

The International Energy Agency has reported that 15,102 TWh of electricity could be produced through salinity gradient in river mouths worldwide; that is 74% of global electricity consumption [15]. However, various physical and environmental limitations were not included in this estimation. Today, taking some of these restrictions into account, and counting only river mouths where this type of energy plant would be feasible, the estimation is 625 TWh, 3% of world consumption [5].

There have been many technological advances in PRO and RED around the world, but there is little information on the impacts the operation and maintenance of SGE plants could have on the functions of nearby ecosystems. The scientific literature surrounding the implementation of SGE at a given study site is scarce. Early works addressing environmental conditions to be monitored mention the amount of water to be extracted (defined as environmental flow, maximum extraction factor, extraction flow, design flow, the annual variation of flow, etc.), the physicochemical characteristics of the water, the physical and chemical characteristics of the input solutions (fresh, marine, treated) and other characteristics, such as salinity structure and temperature (temporal and annual variations) at the extraction and discharge sites [5,12,16–19]. Few works address the very important effects that the SGE implementation could cause on the sediment balance, care in the use of cleaning products (which when accidentally released pollute), and care and disposal of final effluents and membranes [16]. Other studies mention the importance of hydrodynamic studies and environmental forcings that may affect the thermohaline structure and therefore the amount of energy generated from the saline gradient [16]. Even so, there are very few case studies that mention potential environmental impacts. A study proposing a potential site for SGE at Lake Urmia, in Iran, (a Ramsar wetland with a Biosphere Reserve status with endemic species) only assessed in detail the economic implications of implementation [17]. One reason for this is that no operational devices exist.

Some papers mention that the impacts are similar to those of water treatment, desalination, or other renewable energy plants [1,18,20–22]. These works give an overview of potential impacts to habitat, local vegetation and associated fauna, water quality, sediment properties, and social issues related to fisheries and navigation rights and hydrodynamic modifications (changes in flows and their directions and mixing zones). All these impacts are caused by the location of the devices or their interactions with the environment. Specific work on the impact of saline gradient technology highlights potential impacts regarding water intake, final effluent disposal, and impacts associated with infrastructure [23]. However, the study in [22] summarises the overall potential impacts of SGE implementation using a three-phase scheme (construction, operation, and decommissioning).

This paper aims to present a scheme for an environmental impact assessment (EIA) that allows the identification of possible environmental impacts from the implementation of SGE in a coastal lagoon within an environmentally protected area. Through the description of stressors, receptors, and responses, an EIA is developed for the coastal system of La Carbonera, in the state of Yucatan, Mexico.

2. Materials and Methods

2.1. Study Area

The area considered in this work is La Carbonera lagoon, in the northeast of the Yucatan Peninsula, Mexico (21°13′41.80″–21°14′4.79″ N, 89°53′21.66″–89°54′0.45″ W) (Figure 1) [24]. This coastal lagoon lies in a region of karst characteristics and is approximately 16.5 km^2 in area. More detailed information concerning the geology of the region can be found in Appendix A. The channel connecting it to the sea, the Gulf of Mexico, is quite recent, a result of the passage of Hurricane Gilbert, in 1988. The fresh water of the system comes from submarine groundwater discharges, or springs, carrying continental water, which is the result of regional precipitation. The main inlet of fresh water, locally called a 'peten', is located southwest of the system. 'Peten' is a colloquial Mayan name that refers to islands of vegetation which are associated with freshwater springs that allow the development of perennial rainforests, frequently exposed to flooding [25]. It is a shallow lagoon, 0.30–2.0 m, and as such is very much influenced by the local atmospheric climate. The bathymetry of the lagoon and the coastal zone are presented in Appendix A. The mean annual rainfall is 1025 mm and there are three seasons: dry (March to June), rainy (July to October), and 'Nortes', characterised by a decrease in temperature, storm clouds, and heavy rains (November to February) [26–28]. The region is influenced by the transit of tropical storms during the summer months, which commonly intensify into hurricanes. Appendix A contains more detailed information on the climate of the study area.

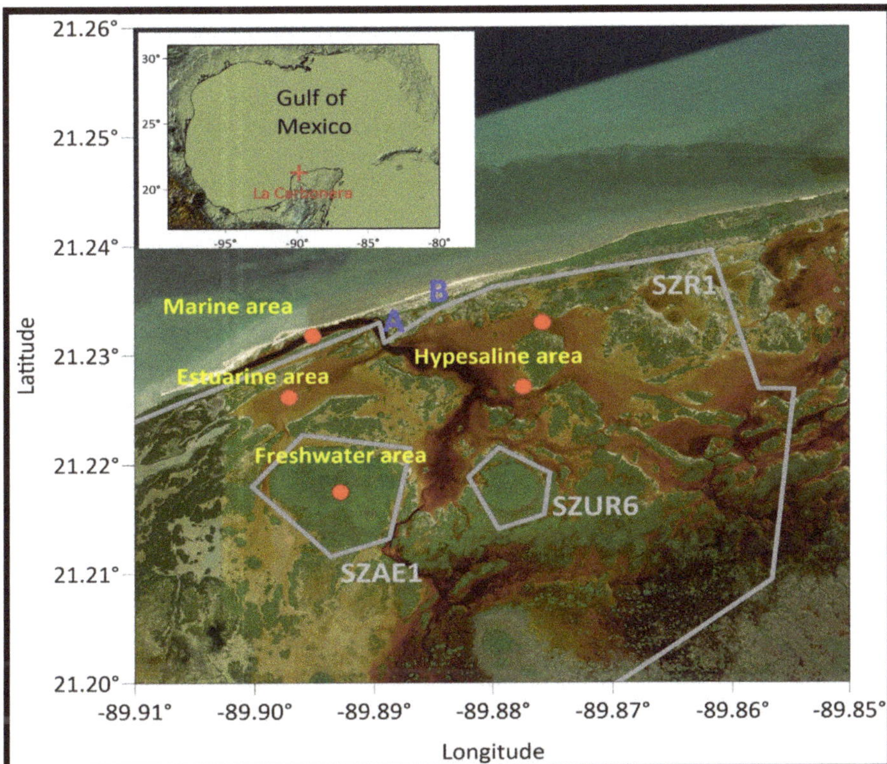

Figure 1. Location of La Carbonera lagoon, Yucatan, Mexico, showing areas with significant saline properties, the location of the CTD sensors (red dots), and sites proposed for the RED power plant (A and B). The polygons SZR1, SZAE1, and SZUR6 are areas with protection categories within the Cienegas and Manglares State Reserve of the North Coast of Yucatán (RECMY).

La Carbonera is part of a system of wetlands along the coast of Yucatan. It has a sand bar of 1.5 km in length, a sandy beach, and coastal dunes of medium height. Mangroves lie behind the dunes, around the lagoon, and sometimes further inland. There are springs and an area of swampland around the lagoon [29]. Geological data on the lagoon are given in Appendix A.4.

On the coast, the waves have low energy, except in the Nortes season or tropical storm or hurricane conditions, when the waves, currents, wind, and precipitation are extreme. The tidal regime is mixed, predominantly diurnal, with tides ranging from 0.40 m spring tides to 0.08 m neap tides. Appendix A has more information on the hydrodynamics, currents, and tides. The mean air temperature is between 24 °C and 26 °C, with variations of up to 10 °C throughout the day, minimums in December and January and maximums in July and August [30].

2.2. Legal Framework

This lagoon system is part of a Natural Protected Area (NPA) in the region known as the Cienegas and Manglares State Reserve of the North Coast of Yucatan (RECMY), which has a total surface area of 55,000 ha and covers several municipalities of the State of Yucatan. The aim of officially recognising the ecological importance of the region was to protect the coastal ecosystems there, which are mainly well-conserved mangroves. La Carbonera is in the west of the reserve, and within it, several polygons have been assigned categories according to their uses. The lagoon is within polygon SZR1, which has the category of Buffer Zone/Public Use. In this zone, current or future actions are given permits if they lead to sustainable development and, at the same time, create conditions to conserve the reserve's ecosystems in the long term. Recreation, leisure activities, group or individual tours are allowed in designated sites, approved for this purpose. Overnight stays and camping are permitted, as well as the development of low-impact infrastructure, in accordance with the Ecological Use Plan for the Coastal Territory of Yucatan. Such infrastructure must be subject to the corresponding authorisations and permits in terms of land and environmental use; including lodging infrastructure, walkways, trails, conditioning of water crossings, signage, and surveillance, which are aimed at sustainable use and the inspection and surveillance of such sites. The construction of new infrastructure, as well as actions that have an effect on hydrological flows, must comply with the relevant environmental impact requirements (Figure 1, polygon SZR1).

Within polygon SRZ1, there is a smaller polygon of category Core/Subzone Restricted Use Zone, SZUR6. This is a peten zone within the lagoon. It is better conserved, or little altered; an area that contains ecosystems, natural phenomena, and geohydrological processes of special interest, as well as species of flora and fauna that have special protection. In this polygon, only exceptional activities that do not modify the ecosystems and that are subject to strict control and supervision measures are allowed, subject to having a permit from the Reserve's authorities [31].

There is also another polygon with the category of Buffer Zone/Subzone of Special Use, SZAE1. This is where the main groundwater discharge to the lagoon is located. It is an area composed of vegetation mosaics with a certain degree of conservation. It contains natural resources which are essential for the social development of the inhabitants of the area. The exploitation of these resources must be carried out without damaging the ecosystem and without substantially modifying the landscape or causing irreversible environmental impacts, in appropriate ways, subject to limited, supervised load capacities. Ecotourism is therefore allowed if it is sustainable and compatible with the environment. For this, operators must have the corresponding permits and management plans, from the Reserve's authorities. The promotion of environmental management units for the intensive and extensive use of wild flora and fauna is also allowed, with the corresponding registers and authorised management plans. Similarly, artisanal and subsistence fishing activities are permitted, subject to surveillance and supervision, with fishing gear that has been authorised, for each case, in the specific sites of these subzones [31].

2.3. Geographical Boundaries and Current Uses

Within its geographic margins, La Carbonera lagoon has no human settlements or infrastructure around it, although it is located between two towns, to the west, Sisal, and to the east, Chuburna. The current anthropic activities in the lagoon are local fishing, small-scale tourism promoted in the surrounding localities, and scientific research by various local and national institutions.

In the scientific literature, some characteristics are mentioned which imply that monitoring is necessary before and after the implementation of SGE exploitation: a very marked saline structure (a horizontal gradient, in this case), low tidal amplitude, and substantial control of hydrodynamics with atmospheric forcings [16]. More specific biophysical details of La Carbonera are presented in the following sections.

2.4. Thermohaline Structure

The thermohaline structure of this lagoon, similar to many coastal systems, has important variations at different temporal scales throughout the year, giving variations in the potential energy obtained from SGE. According to the salinity and temperature characteristics, La Carbonera has four defined zones [32–35] (Figure 1) as follows:

- A permanent fresh water effluent in the southwest of the lagoon (<5 psu (practical salinity units) with an almost constant temperature (27 °C));
- An estuarine zone in the central west of the system (where the salinity concentration varies with tides between 5 and 35 psu);
- A marine zone in the mouth of the sea inlet (around 35 psu);
- A hypersaline zone in the east of the system (60–100 psu).

The temperatures in zones 2, 3, and 4 show seasonal (15–37 °C) and daily variations of up to 10 °C and 20 psu or more in salinity. This lagoon has a strong horizontal salinity gradient, with hypersaline characteristics in the east, marine in the centre, estuarine in the west, and a freshwater zone in the southwest. In Appendix A, the salinity patterns of the lagoon are shown.

2.5. Theoretical Potential for Generating SGE

Reyes-Mendoza et al. [35] have reported the theoretical potential of SGE available in La Carbonera lagoon, based on the thermohaline structure and environmental factors (evaporation, air temperature, etc.). There are three possible plant configurations to harness the saline gradient of the lagoon: freshwater/seawater (FW/SW), freshwater/hypersaline water (FW/HW), and seawater/hypersaline water (SW/HW). The energy available from these are the following:

- FW/SW: 0.244 ± 0.0889 kW, with a maximum of 0.527 kW in May, in the dry season;
- FW/HW: 1.111 ± 0.277 kW, which is the maximum average of the three plants' configurations;
- SW/HW: 0.413 ± 0.194 kW, with a maximum of 0.916 in December, in the Nortes season.

On the other hand, the great variations in the thermohaline structure determine the variability of the power potential throughout the year. For a capacity factor of 90%, the installed capacity for the FW/SW configuration should be 0.133 kWh, for the SW/HW configuration 0.25 kWh, and for the FW/HW configuration, 0.527 kWh [35]. These figures were obtained considering the mixing of 1 m^3/s of saline and fresh water.

2.6. Biological Characteristics

In this region, there are several species of economic importance (regional and local fisheries), ecosystemic and anthropogenic applications (medicinal, construction, tourism), or areas with some category of risk, according to the 059-Semarnat-2010 [36] or the International Union for Conservation of Nature (IUCN) (Table 1). Although many more species are reported, only those with anthropic importance and/or uses, or a protection category were assessed.

Table 1. Species reported in the north of Yucatan, detailing their ecological and commercial importance, anthropic uses, and protection category.

Scheme 37	Category of Protection	Importance and/or Uses
	VEGETATION [37,38] Mangrove	
Conocarpus erectus *Avicennia germinans* *Rhizophora mangle* *Laguncularia racemosa*	Threatened	Erosion control and soil conservation/construction. Key in the life cycle of various organisms (fish and crustaceans). Carbon reservoirs, natural filter for water quality.
	Coastal dune	
Cordia sebestena *Jacquinia macrocarpa*	Uncategorised	Ornamental, medicinal, substrate fixation
Bravaisia berlandieriana	Uncategorised	Construction, medicinal, substrate fixation
Metopium brownei	Uncategorised	Toxic, substrate fixation
Capparis incana *Gomphrena serrata* *Rivina humilis* *Capparis flexuosa*	Uncategorised	Medicinal, substrate fixation
Acanthocereus tetragonus *Sideroxylon americanum*	Uncategorised	Edible, substrate fixation
Pithecellobium keyense	Uncategorised	Substrate fixation
	Peten	
Ficus cotinifolia *Manilkara zapota* *Sabal yapa*	Uncategorised Uncategorised	Reclamation of degraded land Medicinal, construction, and handicrafts
	Lowland flooded forest	
Sporobolus pyramidatus *Solanum nigrum* *Haematoxylum campechianum* *Ipomoea carnea*	Uncategorised Uncategorised Uncategorised Uncategorised	Primary cover resistant to saline soils Edible Textile Medicinal, ornamental
	Seagrasses	
Thalassia testudinum	Minor concern	Key in the life cycle of various organisms (fish and crustaceans). Carbon reservoirs.
	FISHES [39]	
Fundulus persimilis *Fundulus grandissimus*	Subject to special protection	Endemic
Gambusia yucatana	Uncategorised	Endemic
Sphoeroides testudineus *Strongylura notata* *Harengula clupeola*	Uncategorised	Traded for bait, resident, abundant species (vital in the food chain)
Trachinotus falcatus *Lutjanus griseus* *Lutjanus synagris* *Floridichthys polyommus* *Archosargus probatocephalus* *Eucinostomus gula* *Eucinostomu argenteusMugil curema* *Mugil trichodon* *Hyporhamphus unifasciatus* *Chriodorus atherinoides*	Uncategorised	Marketed for local and regional consumption
Poecilia velifera	Subject to special protection	-
Aetobatus narinari	Almost threatened	Commercial importance in the region
	CRUSTACEANS [40]	
Callinectes sapidus	Uncategorised	Marketed for local and regional consumption
	BIRDS [41]	
Phoenicopterus ruber *Phalacrocorax brasilianus* *Cochlearius cochlearius* *Platalea ajaja* *Ardea alba*	Minor concern	For tourism (birdwatching)
Campylorhiynchus yucatanicus	Near threatened	Endemic
Egretta rufescens	Subject to special protection	For tourism (birdwatching)
	REPTILES [42]	
Crocodylus moreletii *Eretmochelys imbricate* *Chelonia mydas* *Caretta caretta*	Subject to special protection Critically endangered Endangered Endangered	For tourism (sighting) Tourism (controlled releases at turtle camps) Tourism (controlled releases at turtle camps) Tourism (controlled releases at turtle camps)
	ARTHROPODS [43]	
Limulus polyphemus	Near threatened	Medicinal

2.7. Environmental Impact Assessment

To ascertain the potential impacts of SGE implementation at La Carbonera, an Environmental Impact Assessment (IEA) was conducted [44–47]. The selection of the environmental impacts was based on the available scientific literature, such as [1,21–23].

The impact analysis used the [48] classification framework proposed for the impact assessment of other ocean renewable energies. The analysis was conducted for three phases: construction, operation, and decommissioning. This EIA considered stressors, receptors, and environmental responses and impacts. Stressors are those characteristics of the environment that may change due to the implementation of SGE in construction, operation, and decommissioning. Receptors are elements of the ecosystem with the potential for some form of response to the stressor, including the various biotic and abiotic components of the ecosystem with the potential to be affected. Effects, or responses, are how these receptors change, without indicating magnitude or significance. Finally, impacts address the severity, intensity, or duration of the effects and cover the direction of the effect, which can generate positive or negative outcomes.

The impacts of this technology are similar in the construction phase to those of seawater desalination plants, wastewater treatment plants, or other renewable energy plants. Since there is currently insufficient on-site experience on the impacts, the stressors considered were based on the analysis of the components of the PRO and RED technologies (Figure 2). This figure gives a general view of the main components of both technologies. In section A, the pipes and the filtering system are shown. In section B, depending on the technology used, and the expected production, the site will require several, repeated PRO or RED modules (membranes in both cases, in RED also electrolyte solutions), as well as hydraulic installations (turbines in PRO), sanitary installations and storage facilities for inputs and waste. Section C shows the electrical installations for power distribution/storage (should include cabling, towers, etc.). In both PRO and RED, a final by-product is a brackish water or seawater, depending on the scheme used (SW/FW, SW/HW, or FW/HW).

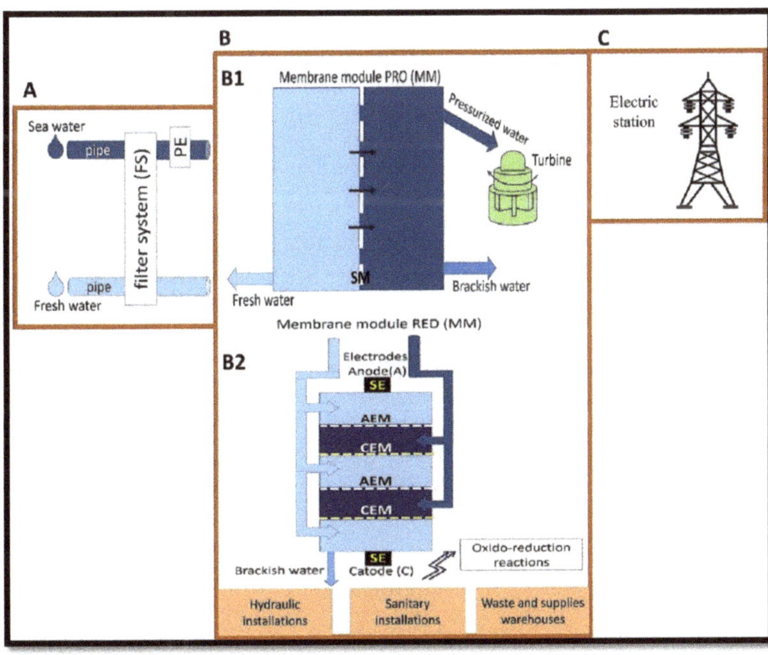

Figure 2. Components and processes involved in RED and PRO technologies that may produce environmental impacts in coastal systems. Abbreviations: PE (pressure exchanger), SE (electrolyte solution), AEM (anion exchange membrane), CEM (cation exchange membrane). The components by sections are shown (A, B and C).

For the determination of receptors, shown in Figure 2 (A–C), general impacts, similar to other engineering projects in coastal systems, are expected, including the following:

1. changes to land use (excavation, land newly used, etc.);
2. emission of pollutants (atmosphere, water, land, solid waste, etc.);
3. storage of waste (in situ, transport, waste sites, etc.);
4. overexploitation of resources (raw materials, energy consumption, water, flora, and fauna, etc.);
5. alterations in species composition and abundance (emigration, reduction in numbers, extinction, etc);
6. deterioration of the landscape (topography, vegetation, watercourses, surroundings, etc.);

Finally, from the revision of biotic and abiotic features of the system shown in the case study, an analysis of Stressors, Receptors, and general Responses to the implementation of SGE in potential systems was presented.

It is hoped that this work can serve as a guide for other cases elsewhere.

3. Results and Discussion

A general analysis of the environmental impacts of PRO and RED technologies in three phases (construction, operation, and decommissioning) is presented. These impacts can also be applied to other potential systems that use salinity gradients for energy production (river mouths, estuaries, coastal lagoons). The receptors highlight those characteristics of coastal environments that may be susceptible to change due to the SGE (Figure 3). The responses were analysed for each receptor, revealing the potential impacts of this technology (Figure 3).

In this overview of possible impacts on potential systems, it is important to mention that these may be minor or major impacts, depending on several factors, such as the size of the plant, the characteristics of each system (biotic and abiotic), and scheme used for energy harvesting (natural solutions or anthropic effluents). In the case of La Carbonera, the exploitation scheme proposed is in line with the regulations in force and adapted to the specific characteristics of the site.

The scheme for La Carbonera is the hypothetical implementation of net output of 50 kW RED power plant of the size and production of the RED prototype located in Afsluitdijk, in the Netherlands. This plant included three water storage tanks, two pretreatment systems (concentrated and dilute solution), and eight membrane stacks. The proposed infrastructure is therefore low impact.

For the size and energy production of this hypothetical plant, it is important to consider first that many of the environmental impacts in each phase will depend on the characteristics of the site where the plant is built. Therefore, some aspects taken into account to select a location, with consideration to possible responses, are reported in Figure 3 and include the following:

- Consider the proximity of the resource (freshwater, marine, hypersaline);
- Choose building sites on land with sparse vegetation;
- Consider smaller areas for excavation and the introduction of pipelines;
- Consider that permanent effects (such as changes to hydrodynamics) have greater effects on receptors than temporary effects (such as construction noise or increased turbidity) in resilient systems, such as coastal lagoons [41];
- Consider final effluent discharge area;
- Consider areas accessible to tourism;
- Consider species conservation.

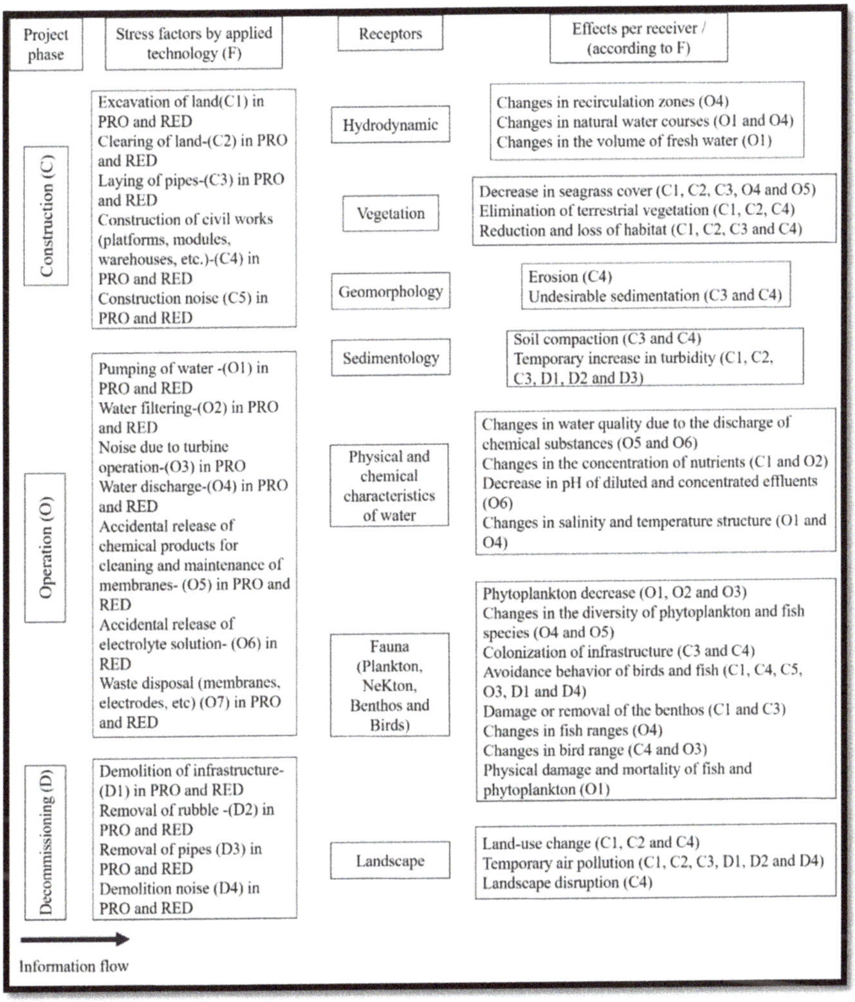

Figure 3. Stressors, receptors, and possible effects due to the implementation of SGE in La Carbonera lagoon in a three-stage scheme (construction, operation, and decommissioning) for the main components of the PRO and RED technologies. The letters indicate the number of stressors according to the phase of the project.

From these considerations, installing the RED plant behind a small jetty in the lagoon, A in Figure 1, would induce significantly greater environmental impacts than with option B of Figure 1. First, this result is because the excavation for the pipelines would be greater, as the hypersaline water must come further, and all the impacts associated with the various stressors (C1, C2, and C3, Figure 3) must be considered. Second, the pipeline would pass through several mangrove patches, and therefore, the vegetation here and its associated fauna would be severely affected. Thirdly, although the availability of seawater at this location (at the mouth of the lagoon) would be close to the RED plant, the interruption to the tidal flow entering the lagoon, due to the constant intake of this solution, would generate various negative impacts. One of the main impacts possible is that large areas of the hypersaline zone that depend on this sea–lake exchange could dry out. Exposed sediments would therefore be salinised, and large areas of mangroves would dry out. In addition, the interruption of this flow would change the hydrodynamic conditions that favour the distribution and abundance of the various fish species reported in Table 1.

Likewise, this would affect the sedimentation processes at the lagoon mouth. On the other hand, the discharge of effluent from this location into the sea would also imply greater excavation and the laying of pipes and damage to the mangrove and associated fauna (Figure 3).

Although the proposed infrastructure is small, the correct location of a plant and especially of the collecting and discharging zones may result in huge differences in terms of impacts: in this example, locating the plant at site A may generate several negative impacts that can be avoided with the location and design of plant at B (Figure 1). The design process, including location selection and analyses to be considered, are presented next.

3.1. Construction Phase

The proposed location for the RED plant is on the coastline between La Carbonera lagoon and the coastal dune, outside the main SRZ1 polygon, and under the jurisdiction of the Federal Maritime Terrestrial Zone (ZOFEMAT) (A, Figures 1 and 4). At this location, several elements need to be constructed (roads and other infrastructure related to basic services). In the literature, as an example, the design of a PRO plant with a production capacity of 50 kW net output covers approximately 7000 m^2 (the approximate size of a football field) [49], while the area of the Afsluitdijk RED plant covers approximately 2750 m^2 (measured from Google maps). The total area proposed at La Carbonera is 5250 m^2, the RED plant elements will be distributed in 3000 m^2 (Figures 4 and 5), which includes three tanks for the concentrated (hypersaline), diluted water (marine or fresh), and the resulting effluent (brackish marine). There will also be space for the membrane modules and for a test laboratory. Finally, spaces for storage, waste storage, sanitary facilities, and an electrical station will be added (Figure 5).

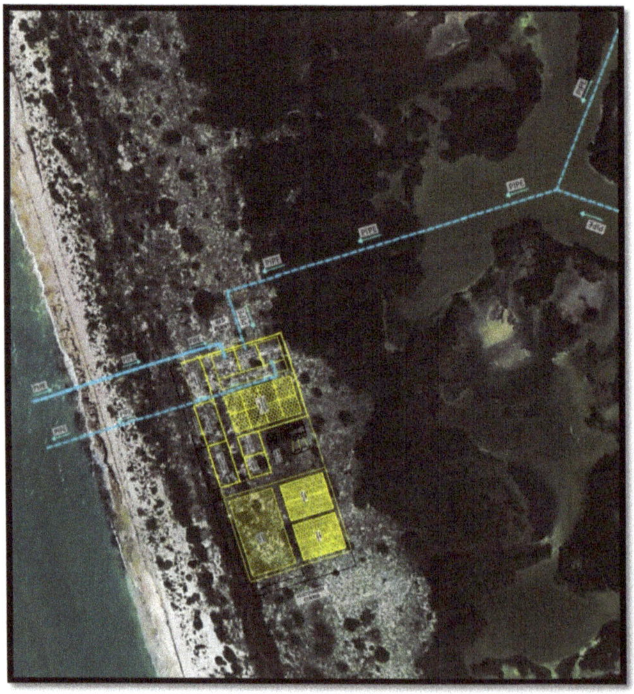

Figure 4. Potential location of the RED pilot power plant; the dotted lines indicate the pipelines.

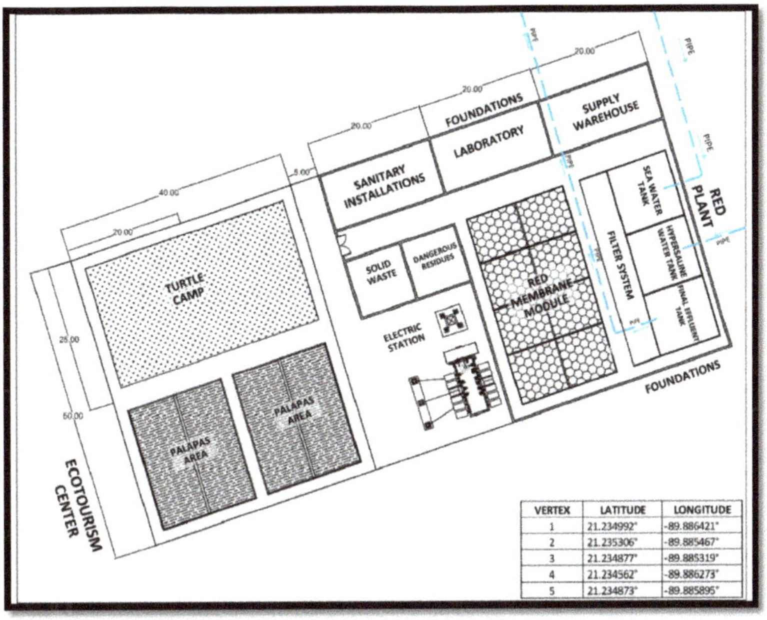

Figure 5. Distribution of space for the proposed RED plant and ecotourism centre at La Carbonera site, 5250 m^2.

La Carbonera, in spite of being a natural protected area, is becoming increasingly popular with tourists. This is encouraged since tourism is more profitable than fishing and agriculture. In line with regulations that allow minor infrastructure in the area, it is proposed that alongside the SGE plant a small ecotourism centre is built (with a total area of 2000 m^2), with cabins, viewpoints, rest areas, and a section for supplies, such as kayaks and boats. (Figures 4 and 5). This centre could include a turtle camp, as the area receives three protected species (*E. imbricata*, *C. mydas* and *C. caretta*, Table 1), arriving to lay their eggs. The incubation of the eggs, laid on the beach, would improve hatching success. The huts and nesting pens could be made using local palms (*Sabal yapa*, Table 1), widely used for this type of construction [50].

The ecotourism centre would benefit from the electricity generated at the RED plant. This energy could be used to charge mobile phones, cameras, or torches, in addition to the electricity needed for the turtle camp (mainly the nurseries), and in the future, could also supply energy for electric boats. The energy generated would be sufficient to make the plant self-sufficient in electricity. An ecotourism centre would provide well-paid jobs and also limit the pressures associated with the present mode of tourism; visitors coming into the area on day trips from other towns and villages which has brought various environmental problems to this area. The biological richness of the region is accessible in only a few places, and many features of the tourism model are misguided. Excessive growth in some of these areas has led to the infilling of swamp areas to build homes, damage to the coastal dunes due to road construction, and poor management of solid waste and residual water, in particular [42] (Figure 6). Using this site solely for ecotourism, scientific development, and species conservation could help to curb such chaotic, harmful growth around La Carbonera. In addition, the mangroves, swamps, and dune vegetation would be preserved because the ecotourism centre would not be built in these ecosystems. It would also promote environmental awareness among the local population.

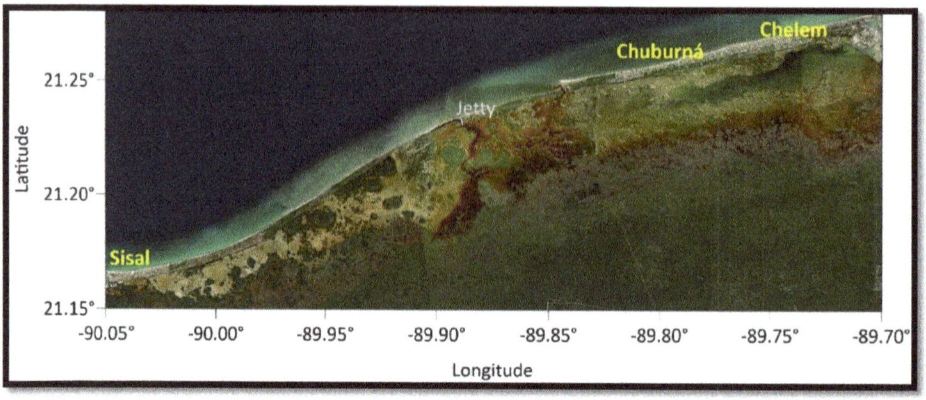

Figure 6. Aerial view of the villages around La Carbonera lagoon (the jetty is shown as a reference).

The stressors examined (Figure 3) include excavation, site preparation, placing pipelines, and the construction of platforms and modules. The impacts are similar to those of any other engineering project: removal of vegetation, habitat loss, erosion, unwanted sedimentation, soil compaction, temporary increase in turbidity, avoidance behaviour of birds and fish (due to construction noise), damage or removal of benthos, change of land use, temporary air pollution, and landscape disruption. Nevertheless, all of these will depend on the construction techniques and on the precautionary measures taken. The plant would be located on the seashore (Figure 2, section B), and as it would be affected by the tides, it is proposed that the RED plant be built on pillar-supported infrastructure (palafitte) that will allow water to flow beneath. These low-impact structures do not require extensive excavation [51] and would minimise some impacts on geomorphology and vegetation. They would also permit the free passage of wildlife, particularly useful in storm surge events. In terms of landscape disruption and land use change, as it is a small project, the impacts are expected to be minimal. Given that the site proposed for the plant will not modify the hydrodynamics of the lagoon, natural sediment transport changes are not expected. Additionally, any increase in turbidity, damage or elimination of benthos, erosion or soil compaction, and air pollution will only be temporary. In the case of vegetation and habitat loss, this location was chosen (Figure 4) precisely because there are no large patches of vegetation, except for some coastal dune species in the vicinity (Table 1).

During construction, in order to lay pipes, excavation work is required, which can lead to temporary increases in turbidity (due to sediment movement), loss of vegetation cover (terrestrial and aquatic) and associated habitat loss, removal of benthic organisms, and the release of nutrients and pollutants from the bottom of the lagoon (Figure 3).

The pumping system should be low cost and easy to maintain (to avoid algal blooms, oxygenation, and suspended particles). Three schemes are possible for the water intake in the lagoon: Marine Zone and Freshwater Zone (MW/FW); Marine Zone and Hypersaline Zone (MW/HW); or Freshwater Zone and Hypersaline Zone (FW/HW) (Figure 1). Water intakes should be close to the plant, both to minimise the impact of pumping losses on net power production and to avoid environmental impacts associated with some stressors (excavation) [52]. One option to avoid impacts from the excavation is to lay pipes on the surface and thus avoid the impacts associated with this stress factor. In the MW/FW and FW/HW schemes, the first problem is that the area where the fresh water emerges is within polygon SZAE1, and although ecotourism activities and the exploitation of flora and fauna can occur in this area with a permit, laying pipelines may not be compatible with the conservation laws of this polygon [31]. On the coasts of Yucatan, groundwater is used to supply fresh water to nearby settlements. Thus, the MW/FW and FW/HW schemes could be a possibility using wells or pumping to extract fresh or brackish water [53] in order to avoid negative effects on other processes and activities.

In the MW/HW scheme, which the plant is designed to use (Figure 4), hypersaline water would be taken from the hypersaline zone to the plant, and the pipeline would pass through parts of the SRZ1 polygon; however, low impact sustainable infrastructure would be used (Figure 4). Seawater would be brought to the plant directly from the sea, under the jurisdiction of the ZOFEMAT (Figure 4). Although another possibility for this region, where there is high radiation and evaporation, generating hypersaline conditions, would be to use seawater evaporation ponds. The volume of hypersaline water needed would therefore not be an impediment.

For the energy production expected, an inflow of 200 m^3/h of both solutions is required [52]. These inflows will pass through a hydraulic network, which includes the set of pipes, pumps, and other accessories (elbows, valves, etc.) that will allow the entry and exit of the solutions. For a flow of this volume, 10-inch diameter pipes are required; hence, the impacts associated with excavation will be less. The hypersaline water must be piped to the plant; options to avoid the mangrove patches in its path exist, and special attention must be paid to the protected mangrove species (*C. erectus*, *A. germinans*, *R. mangle* and *L. racemosa*) [36]. For the discharge pipe, the same factors apply. These species are important because they provide shelter and protection areas for birds and various species of fish, and they deliver ecosystem services such as filtering water discharges from the mainland to the sea [54] and protection from wind and waves that prevents coastal erosion in an area that is also affected by hurricanes [55].

In the construction phase of the proposed design, no large-scale pipes would be installed within the lagoon; therefore, the loss of seagrasses, the release of pollutants and nutrients from the bottom of the lagoon, and the removal of organisms from the benthos would be practically nil, and the increase in turbidity due to the movement of sediments would be only temporary (weeks). Regarding the seagrass ecosystem, in areas with high salinity, seagrasses are not found; therefore, in the water intake in the hypersaline zone, there will be no damage to seagrasses [56,57]. In the hypersaline zone, the depths are 50 cm or less (Figure 2), and therefore, the hydraulic network here should be able to pump hypersaline water from different points to prevent it from drying out.

On the other hand, regarding the avoidance behaviour of birds due to construction noise (Table 1), emblematic species such as the flamingo (*Phoenicopterus ruber*, Table 1) would probably modify their distribution in the area only temporarily (weeks). In addition, with RED technology, no turbines will be used in the operation phase; hence, noise pollution would only occur during construction.

3.2. Operation Phase

In this phase, the impacts associated with water pumping and water pretreatment (Figure 3) are the first that should be considered. Pretreatment of the water intake is crucial for the operation of a RED system [58]. Such a system must ensure low-cost performance and effective sediment filtration [52]. In the RED plant of Afsluitdijk, the filtration system is of the drum and gravity type, and even with small intakes of very good quality water cartridges, filters can be used [52]. The impacts associated with water pumping and pretreatment are related to the amount of fresh water and hypersaline water that will be taken from the system and pass through it. These impacts include changes to natural watercourses, as well as changes in the nutrients and salinity of the water. Possibly, and depending on the intensity, these alterations may affect native species and natural ecosystems. For instance, it is known that changes in salinity alter the reproductive and feeding behaviour of flamingos and horseshoe crabs (*Limulus polyphemus*) [59].

Another impact is the possible decrease in phytoplankton biomass retained in the filtration system. The latter is a concern at the Afsluitdijk pilot plant because large quantities of plankton, fish, and larvae must be filtered, which has ecological implications and may also have economic implications [60]. Since the biomass of microorganisms at the base of food chains is affected by this, it can lead to imbalances in the food chain and local fisheries [23,48].

Detailed hydrological studies are therefore needed to determine the amount of water that can be extracted from the lagoon and how much it can be altered without generating the above-mentioned impacts. In the case of the Afsluitdijk pilot plant, the intakes in the sea and in the lagoon area of 200 m^3/h, assuming a technical potential of 1 MJ/m^3 of seawater and freshwater, can produce up to 50 kW net power output [52]. In La Carbonera, taking the concentrated solution from the lagoon (hypersaline water) counteracts the effects on the biomass of microorganisms to a certain extent, since the biomass is reduced due to the hypersalinity of the area [59]. The diluted solution could be taken from wells on the coast, in which case there would be no phytoplankton, due to the lack of light. On the other hand, it could be taken directly from the sea, and this would have fewer microorganisms than that of the estuarine and marine zones of the lagoon [61].

Another stress factor is the disposal of the final by-product of the RED process—the brackish water. Even if the intended scheme is MW/HW, the water mix would have a similar or higher salinity than seawater. The change in salinity of the effluent must be calculated in the laboratory. The effluent must be discharged in an appropriate area, at the appropriate time, and the dispersion of this effluent by the hydrodynamic actions of the system should not alter the natural salinity patterns in that ecosystem [32,33].

Depending on the volume of the water, pump diffusers may be needed (alternating or slanted) in the hydraulic network, to distribute the flows in different directions within the lagoon, or into the sea [62]. The discharge of water used may not induce negative impacts at sites where the hydrodynamic performance and salinity concentrations are known prior to the design for the effluent flow. This is so at sites where there has been salinity deterioration as a result of previous anthropogenic activities.

The change in salinity is only one of the environmental conditions responsible for the variety and abundance of fish reported for this lagoon (Table 1) [39]. In addition, salinity indirectly affects the distribution of species through its role in water density and the resulting hydrodynamics [63].

The spatial/temporal variation of water masses and their salinity is important for the distribution of organisms, especially of fish, which only live under certain salinity ranges, according to their tolerance to this parameter [64–66]. This is important for the distribution of various marine species of commercial interest (Table 1). Discharging a saline effluent into the lagoon, of marine salinity or slightly higher, in the hypersaline zone, will lower the salinity in this zone and thus limit the amount of hypersaline water available for power generation. In consequence, the impact on different species may be significant.

It is important to mention that there are many characteristics that make this an environment that harbours great diversity in fishes, but high salinities have been associated with a lower richness and diversity of fish [67]. Thus, for species distributed in marine/estuarine environments, such as *S. testudineus*, *S. notata*, *H. clupeola*, *T. falcatus*, *L. griseus*, *L. synagris*, *F. polyommus*, *A. probatocephalus*, *E. gula*, *E. argenteus*, *M. curema*, *M. trichodon*, *H. unifasciatus*, *C. atherinoides* and *A. narinari*, (Table 1) [39], a potential decrease in salinity in the hypersaline zone would alter the extent of their distribution areas.

However, to avoid changes in salinity and resulting limitations in resources, it is better to discharge this effluent into the sea since, being saline and of a small volume, it will not have the same effects as if it were brine [68]. On the other hand, stressors such as the accidental release of cleaning and maintenance chemicals and electrolyte solutions must be regulated against any facility handling hazardous chemicals.

Both in the pretreatment of solutions and in the cleaning of membranes and facilities, products such as chlorine are used, which can be toxic to the environment [23,48]. Usually, chlorine is used to avoid degradation of the membranes caused by biological growth in them. There is evidence that even small amounts of chlorine (e.g., 0.1 ppm) can have ecological impacts which induce a significant reduction in the productivity of dragged phytoplankton and species diversity [23]. Similarly, electrolyte solutions should be handled with caution [69]. The electrolyte solutions are stored in the electrode compartment which also contains the electrodes and is sealed with membranes that generally have special properties to ensure the confinement of the electrolyte, recirculating it in a closed circuit [70]. There are reports of the toxicity of this type of element [23], but there is not much information on the toxic gases or compounds that are generated in the redox process within such a compartment (depending on the redox couple and the electrolyte used). Furthermore, in these compartments, some instability in pH control sometimes occurs since anion exchange membranes have a non-negligible proton transport number. This may allow an increase in the pH of the electrode solution, accompanied by a decrease in the pH of the effluent, and this may not be environmentally acceptable if it is too high [70]. Although the pH of the effluent would only change in the event of an accidental release, it is important to note that this parameter is an indicator of water quality. It affects the toxicity of certain compounds, such as ammonia, by controlling their ionisation, as well as the bioavailability of certain pollutants, such as heavy metals. For example, water with a pH range of 6.5 to 8.5 is suitable for many biological systems. Values of over 9.0 and lower than 5.8 limit the development and physiology of aquatic organisms [71].

Finally, it is worth noting the positive impacts that an SGE plant could have on La Carbonera. In the operational phase, the pipes will provide new spaces for colonization (bioincrustation) by sessile species [1]. These structures offer heterogeneity to the habitat and appropriate surfaces for algae and sessile organisms to colonise, especially on the muddy bottom. Fish and other invertebrates will be attracted by the hard surface, the shade, the changes in turbulence, the small spaces, and eventually, by the availability of food sources [48,72]. However, this type of infrastructure can also encourage the establishment of non-native species, invasive species, and blooms of harmful algae; therefore, the extent and composition of the colonisation are difficult to predict [73,74]. In the case of La Carbonera, pipes in the hypersaline zones would be easily colonised in the short term by barnacles, which live in the carbonated structures which stick to the surfaces and may cause deterioration. For this reason, their colonisation should be treated cautiously since they also damage RED membranes, encouraging the proliferation of microorganisms which impede the free passage of water or ions and thereby reduce the functionality of the system [75]. Although the production of 50 kW of electric energy from renewable sources is by no means a technological challenge, this work aims to provide electric power using the best technologies available, producing the smallest possible footprint, in harmony with the land use of this area (i.e., [76]).

With the present technological maturity of SGE techniques, these objectives do not yet yield low costs. Firstly, because the cost of the energy depends on the establishment of the market and industry; if the technological development is successful, the manufacturing of the parts will become cheaper. In the meantime, support from public funding is to be expected [77]. On the other hand, when environmental and social aspects are prioritised, a cost–benefit analysis based only on economic variables is not sufficient [78,79]. If the community face expensive-energy versus no-energy, their decision will be controversial.

For this specific case, the goal is not to urbanise the area or provide services that will promote urbanisation. The goal is to offer services that may assist environmental protection and conservation activities held in La Carbonera Lagoon and its surroundings where only low-density ecotourism is allowed. In this sense, a pilot plant using emergent technology is both suitable and affordable. Other technologies may not be feasible at this site. For example, although an established solar energy industry exists, the installation of solar panels may increase the air temperature around them, which is undesirable in environmentally sensitive areas. Additionally, a strategy for energy storage would be needed, thus increasing the final cost. There are similar disadvantages for wind energy, with additional construction difficulties [21]. Thus, an SGE plant, with the proposed additional facilities, seems to be a good strategy for promoting education, environmental conservation, and technology development.

3.3. Decommissioning Phase

In this phase, few impacts are expected. A RED plant built on palafittes should not require large-scale activity using heavy machinery. In the case of the pipes that have been colonised by algae and sessile organisms, their removal will eliminate this artificial habitat and also the organisms that installed themselves there. The soil is not expected to be severely impacted by the type of foundations used. On the other hand, the noise associated with demolition may temporarily affect the behaviour of birds and fish. Finally, the infrastructure of the proposed ecotourism centre should not be demolished since these facilities offer attractive spaces for tourism and assist in the conservation of species in the area.

4. Conclusions

A preliminary environmental assessment of the potential impacts of a 50 kW net output RED power plant using SGE on the coastal lagoon of La Carbonera was carried out. Many impacts depend on the size and location of the SGE plant and the water intake or on the technology to be implemented. In this case, a small-scale SGE power plant, using the RED technology, combined with an ecotourism centre, would provide renewable energy and protection of resources for the lagoon system and the nearby area around it, as well as providing well-paid jobs for local people that may eventually encourage an improvement in the care of coastal ecosystems.

Although several water schemes can be used in this ecosystem (FW/MW, FW/HW, and MW/HW), everything will depend on the availability of saline water. In both cases, there are alternatives; for instance, to use the dilute solution (FW) through wells on the coast, and the concentrated solution (HW) directly from the lagoon (according to hydrological studies) or through marine water evaporation ponds.

In terms of the potential impacts, the most concerning are the change in the volume of water in the lagoon and the disposal of the final effluent. While many negative potential responses could be expected, if the EIA covers information on the biotic and abiotic characteristics of the ecosystem at a given point in time, in addition to the site characteristics, many of these impacts could be minimised or avoided.

While the study in [22] suggested that most of the impacts due to the implementation of SGE occur in the construction phase, this study shows that although the impacts then may be very evident, even with mitigation measures, the ecosystems in La Carbonera could recover from these temporary effects. More damaging to the ecosystems would be the permanent and constant changes in characteristics which an SGE plant would induce, such as in the hydrodynamics (changes in water flows or volume) and salinity gradients. Therefore, it must be taken into account that since they are highly variable ecosystems that have close hydrological connectivity with the surrounding systems, any modification can have implications for neighbouring systems.

From the results of the Receptors and Responses analysis, the importance of monitoring these features during and after the construction phase must be underlined [79]. Many of the Receptors analysed are also present in other systems with potential for SGE. Even though methodologies for their characterisation already exist for most of them, it is still difficult to find quantitative criteria which demonstrate how positive or negative these Receptors are when SGE technology is applied.

Even so, this work offers the first attempt to evaluate the potential changes induced by SGE plants. This analysis for La Carbonera is important at the present time even though the technology discussed is as yet untested at a large scale, and there are no plants using it in Mexico. Small pilot plants, such as that suggested here, could offer insights for successful larger developments in the future.

Finally, although this system is healthy at present, an EIA would serve to minimise negative impacts. In this way, the benefits of SGE, zero greenhouse gas emissions, and the use of renewable sources could be successfully harnessed at La Carbonera.

Author Contributions: Conceptualisation, E.M.-C. and R.S.; methodology, E.M.-C. and M.L.M.; software, E.M.-C.; validation, R.S., C.E., M.L.M., and E.M.; formal analysis, E.M.-C. and R.S.; investigation, E.M.-C.; writing—original draft preparation, E.M.-C. and R.S.; writing—review and editing, E.M.-C., R.S., C.E., M.L.M., and E.M.; visualisation, E.M.-C. and R.S.; supervision, C.E., M.L.M., and E.M.; project administration, R.S.; All authors have read and agreed to the published version of the manuscript.

Funding: This research was funded by the CONACYT-SENER-Sustentabilidad Energética project: FSE-2014-06-249795 Centro Mexicano de Innovación en Energía del Océano (CEMIE-Océano).

Informed Consent Statement: Not applicable.

Data Availability Statement: Data sharing not applicable. No new data were created or analysed in this study. Data sharing is not applicable to this article.

Acknowledgments: We are grateful to Jill Taylor for the English revision, and to Roberto Rivera for support in managing Autocad. The authors thank the Centro Mexicano de Innovación en Energía del Océano (CEMIEOcéano) for providing support in the research in the CONACYT-SENER-Sustentabilidad Energética project: FSE-2014-06-249795 Centro Mexicano de Innovación en Energía del Océano (CEMIE-Océano). Finally, thanks to the Civil Engineering Postgraduate of the UNAM and CONACYT for the grant awarded for the development of this project.

Conflicts of Interest: The authors declare no conflict of interest.

Appendix A

Appendix A.1. Precipitation, Evaporation, and Atmospheric Temperature

The heaviest rains, associated with tropical systems in the peninsula, occur between June and October. In the winter, from November to February, it is also common for rain associated with the passing of cold fronts, or 'Nortes'. The annual precipitation varies between 444 mm and 1290 mm and is greater north to south, and east to west, in the state of Yucatan. The dry season, March to June, has very little rain, with high levels of solar radiation which generate extremely high temperatures [80].

As is typical in tropical areas, the evaporation exceeds the precipitation (approximately 1800 and 1290 mm/year, respectively). The atmospheric temperature has a defined annual oscillation, with maximums in summer. The annual average temperature is 26 °C, the maximum monthly average is around 36 °C, in May, and the minimum monthly average is 16 °C, in January [35,81].

Appendix A.2. Tides, Winds, and Waves

According to [82], the tides are diurnal, with higher high water = 0.590 m; mean high water = 0.461 m; mean water level = 0.326 m; mean low water = 0.238 m, and mean lower low water = 0.000 m.

As explained by [83], winds in the Yucatan Peninsula are mostly Trade winds, from the east and northeast, locally modified by a marked system of coastal breezes, with sustained averages of 5.5 m/s. The waves are of low energy, with a mean annual significant wave height of Hs = 0.78 m, mean peak period Tp = 4.6 s, and a predominant northeast direction (see Figure A1).

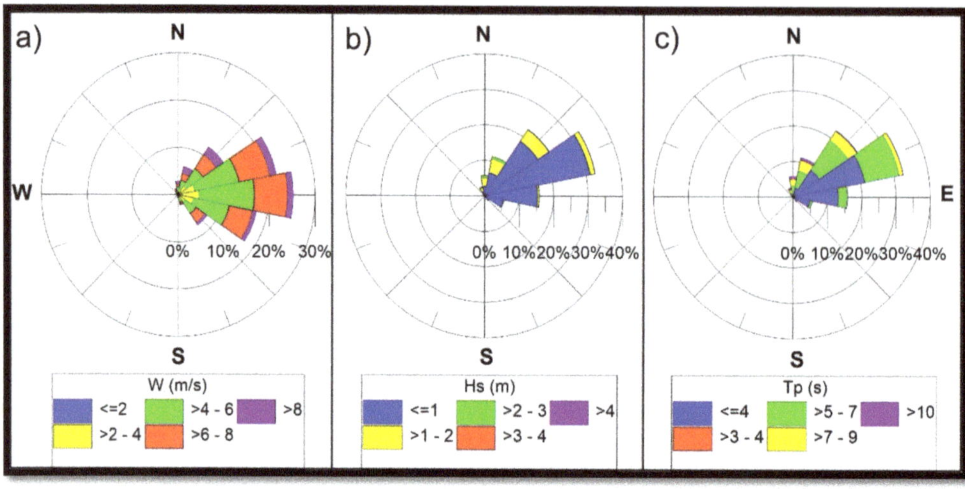

Figure A1. Roses showing the direction and magnitude of (**a**) wind velocity, (**b**) wave height, and (**c**) peak period. Reanalysis data from [83].

In addition, the region has atmospheric systems of short duration which modify these patterns: in summer there are tropical cyclones, which can become hurricanes, with sustained winds of up to 44 m/s, Hs > 10 m, and Tp > 12 s, and in winter, strong 'Nortes' can have winds of up to 15 m/s, Hs = 4.5 m, and Tp = 8.7 s [83].

Appendix A.3. Bathymetry

The inlet linking the lagoon with the sea has depths of 1.5–2 m, the shallow interior is only 0.5 m deep, with the exception of the channel that links the freshwater inflow area with the lagoon (southwest), where depths are 2 m close to the inflow, and 0.5 m towards the lagoon interior (Figure A2). The depths in the marine area show the shallow, extensive nature of the continental shelf off Yucatan.

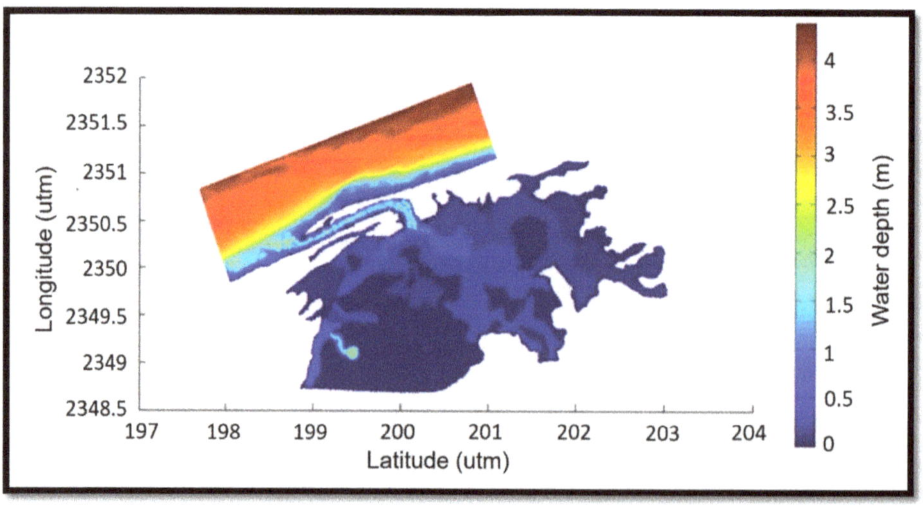

Figure A2. Bathymetry of the lagoon of La Carbonera and nearby marine area [33].

Appendix A.4. Hydrogeological Characteristics

La Carbonera is part of the karst system of the Peninsula of Yucatan (Batllori-Sampedro et al., 2006). In the lagoon, three types of sediment lie on top of the karst strata: sand, rocky sediment, and mud [24]. The orientation of the sandy beaches on the coast is predominantly north–northwest. The mean grain diameter is 0.2–0.5 mm [84,85].

The main hydrological feature of the coastal systems of the peninsula es that freshwater emerges from underground springs or seeps into lagoons and wetlands [86]. Once in the lagoon systems, it begins to mix and patterns of salinity and temperature develop between the fresh water and the water of the lagoon, producing a complex thermohaline circulation system [27]. Such is the case with the inflow in the southwest of the lagoon, where an estuarine gradient is produced.

Appendix A.5. Hydrodynamics

The main hydrodynamic drivers are the tides, the predominant winds of the region (Trade Winds, sea breezes, and Nortes). The study in [33] reported that inside the system, the residual currents are around 0.05 m/s, both when influenced by the local breezes as well as the winds (Figures A3 and A4). As can be seen, the tides entering through the inlet spread east, west, and southwards in the lagoon system (Figure A3). In the centre, south, and east of the lagoon, there are mixing zones, with higher values (close to 0.2 m/s) which are greater during 'Nortes' (Figure A4).

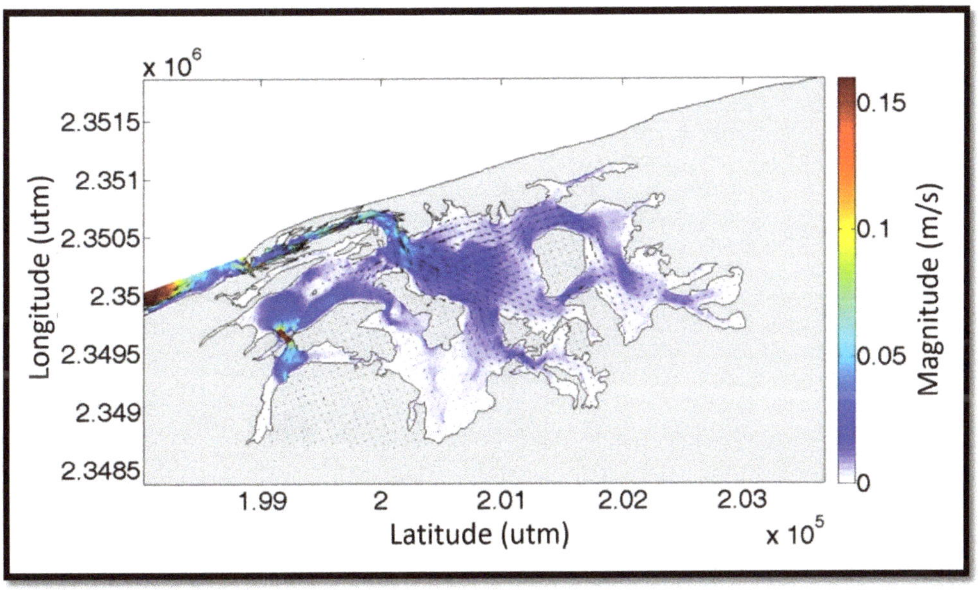

Figure A3. Example of the residual circulation pattern (m/s) during a spring tide with sea breeze forcing [33].

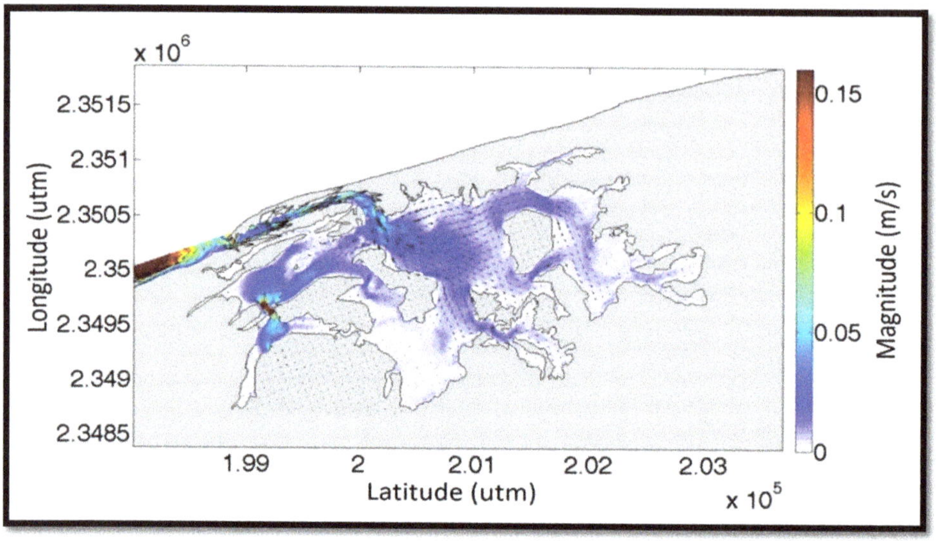

Figure A4. Example of the residual circulation pattern (m/s) during a spring tide with forcing from NE winds [33].

Appendix A.6. Salinity Patterns

The general patterns of salinity in the lagoon of La Carbonera are seen in Figure A5. During a tidal cycle, these patterns are very dynamic [33]. At high and low tides the estuarine, marine, and hiperhaline zones move, due to the influence of the tides and dominant winds, as well as the inflow from the underground spring, which can increase or decrease the salinity in the lagoon. At the highest tide level, and under the effects of strong winds, the seawater can move further into the hiperhaline zone. At low tide, brackish water can reach the mouth of the lagoon [33].

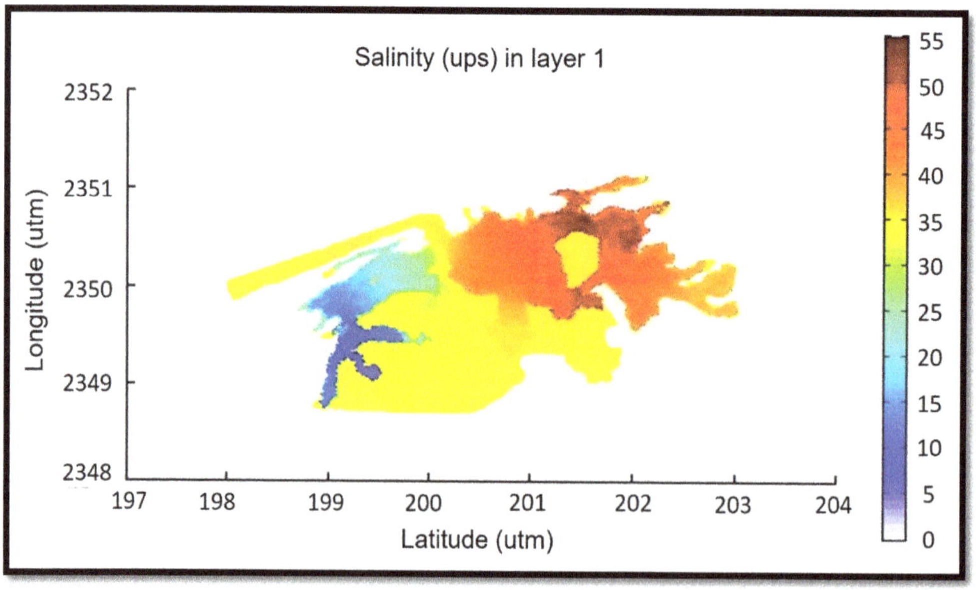

Figure A5. Example of salinity patterns in the La Carbonera lagoon. The hiperhaline zone can reach levels of over 80 ups [33].

References

1. Mendoza, E.; Lithgow, D.; Flores, P.; Felix, A.; Simas, T.; Silva, R. A framework to evaluate the environmental impact of OCEAN energy devices. *Renew. Sustain. Energy Rev.* **2019**, *112*, 440–449. [CrossRef]
2. Zhao, S.; Zou, L.; Tang, C.Y.; Mulcahy, D. Recent developments in forward osmosis: Opportunities and challenges. *J. Membr. Sci.* **2012**, *396*, 1–21. [CrossRef]
3. Copping, A.; Sather, N.; Hanna, L.; Whiting, J.; Zydlewski, G.; Staines, G.; Gill, A.; Hutchison, I.; O'Hagan, A.; Simas, T.; et al. *Annex IV State of the Science Report: Environmental Effects of Marine Renewable Energy Development Around the World*; TETHYS: Richland, WA, USA, 2016. Available online: https://tethys.pnnl.gov/publications/state-of-the-science-2016 (accessed on 25 March 2017).
4. Edenhofer, O.; Seyboth, K.; Creutzig, F.; Schlömer, S. On the Sustainability of Renewable Energy Sources. *Ann. Rev. Environ. Resour.* **2013**, *38*, 169–200. [CrossRef]
5. Alvarez-Silva, O.A.; Osorio, A.F.; Winter, C. Practical global salinity gradient energy potential. *Renew. Sustain. Energy Rev.* **2016**, *60*, 1387–1395. [CrossRef]
6. Jia, Z.; Wang, B.; Song, S.; Fan, Y. Blue energy: Current technologies for sustainable power generation from water salinity gradient. *Renew. Sustain. Energy Rev.* **2014**, *31*, 91–100. [CrossRef]
7. Ahmad, M.; Williams, P. Application of salinity gradient power for brines disposal and energy utilisation. *Desalin. Water Treat.* **2009**, *10*, 220–228. [CrossRef]
8. Tedesco, M.; Cipollina, A.; Tamburini, A.; van Baak, W.; Micale, G. Modelling the Reverse ElectroDialysis process with seawater and concentrated brines. *Desalin. Water Treat.* **2012**, *49*, 404–424. [CrossRef]
9. Zhu, Y.; Wang, W.; Cai, B.; Hao, J.; Xia, R. The salinity gradient power generating system integrated into the seawater desalination system. *IOP Conf. Ser. Earth Environ. Sci.* **2017**, *52*, 012067. [CrossRef]
10. Ye, M.; Pasta, M.; Xie, X.; Cui, Y.; Criddle, C.S. Performance of a mixing entropy battery alternately flushed with wastewater effluent and seawater for recovery of salinity-gradient energy. *Energy Environ. Sci.* **2014**, *7*, 2295–2300. [CrossRef]
11. Yip, N.Y.; Brogioli, D.; Hamelers, H.V.M.; Nijmeijer, K. Salinity Gradients for Sustainable Energy: Primer, Progress, and Prospects. *Environ. Sci. Technol.* **2016**, *50*, 12072–12094. [CrossRef] [PubMed]
12. Berrouche, Y.; Pillay, P. Determination of salinity gradient power potential in Québec, Canada. *J. Renew. Sustain. Energy* **2012**, *4*, 053113. [CrossRef]
13. Micale, G.; Cipollina, A.; Tamburini, A. 1—Salinity gradient energy. In *Sustainability Energy from Salinity Gradients*; Cipollina, A., Micale, G., Eds.; Woodhead Publishing: Sawston, UK, 2016; pp. 1–17. [CrossRef]
14. Sanvik, S.O.; Skihagen, S.E. Status of Technologies for Harnessing Salinity Power and the Current Osmotic Power Activities. Article to the 2008 Annual Report of the International Energy Agency Implementing Agreement on Ocean Energy Systems Annex I: Review, Exchange and Dissemination of Information on Ocean Energy Systems. 2008. Available online: http://www.iea-oceans.org/publications.asp?id=1 (accessed on 28 May 2017).
15. IEA. *Key World Energy Statitics*; International Energy Agency: Paris, France, 2013; p. 82.
16. Alvarez-Silva, O.; Osorio, A.F. Salinity gradient energy potential in Colombia considering site specific constraints. *Renew. Energy* **2015**, *74*, 737–748. [CrossRef]
17. Emdadi, A.; Gikas, P.; Farazaki, M.; Emami, Y. Salinity gradient energy potential at the hyper saline Urmia Lake—ZarrinehRud River system in Iran. *Renew. Energy* **2016**, *86*, 154–162. [CrossRef]
18. Karunarathne, H.D.S.S.; Walpalage, S. Applicability of Pressure Retarded Osmosis Power Generation Technology in Sri Lanka. *Energy Procedia* **2013**, *34*, 211–217. [CrossRef]
19. Ortega, S.; Stenzel, P.; Alvarez-Silva, O.; Osorio, A.F. Site-specific potential analysis for pressure retarded osmosis (PRO) power plants—The León River example. *Renew. Energy* **2014**, *68*, 466–474. [CrossRef]
20. HydroQuebec. *A Renewable Energy Option: Osmotic Power*; HydroQuebec: Montreal, QC, Canada, 2015; ISBN 978-2-550-72233-5.
21. Martínez, M.L.; Vázquez, G.; Pérez-Maqueo, O.; Silva, R.; Moreno-Casasasola, P.; Lopez-Portillo, J.; Mcgregor, I.; Hecke, L.G.; Hernández-Santana, J.; García-Franco, J.; et al. A systemic view of potential environmental impacts of ocean energy production. *Renew. Sustain. Energy Rev.* **2021**, Subbmitted.
22. Seyfried, C.; Palko, H.; Dubbs, L. Potential local environmental impacts of salinity gradient energy: A review. *Renew. Sustain. Energy Rev.* **2019**, *102*, 111–120. [CrossRef]
23. Papapetrou, M.; Kumpavat, K. 10—Environmental aspects and economics of salinity gradient power (SGP) processes. In *Sustainability Energy from Salinity Gradients*; Cipollina, A., Micale, G., Eds.; Woodhead Publishing: Sawston, UK, 2016; pp. 315–335. [CrossRef]
24. Bonilla-Gómez, J.L.; Badillo-Alemán, M.; Gallardo-Torres, A.; Chiappa-Carrara, X. Temporal variation, growth and natural mortality of two species of mojarras (Perciformes Gerreidae) from a tropical coastal lagoon: La Carbonera, Yucatan, Mexico. *Rev. Cienc. Mar. Costeras* **2013**, *5*, 57–67. [CrossRef]
25. Carmona-Escalante, A.; Guadarrama, P.; Ramos-Zapata, J.; Castillo-Argüero, S.; Montaño, N.M. Arbuscular mycorrhizal fungi associated with coastal vegetation in Chuburna, Yucatan, Mexico. *Trop. Subtrop. Agroecosyst.* **2013**, *16*, 431–443.

26. Carbajal, N. *Hidrodinámica y Transporte de Contaminantes y Sedimentos en el Sistema Lagunar Nichupté-Bojórquez, Quintana Roo*; Informe final SNIB-CONABIO Proyecto No. CQ063; Instituto Potosino de Investigación Científica y Tecnológica: Mexico City, Mexico, 2009.
27. Sánchez-Santillán, N.; Lanza-Espino, G.; Sánchez-Trejo, R. Analysis of the dynamic climatology of the Yucatan Peninsula, Mexico. In *Recursos Acuáticos Costeros del Sureste Vol. II. Fondo Mixto—CONACYT—Gobierno del Estado de Yucatán*; Consejo de Ciencia, Innovación y Tecnología del Estado de Yucatán: Mérida, Yucatán, Mexico, 2012; pp. 353–374. ISBN 978-607-9060-08-4.
28. Vidal, R. *Climatic Regions of Mexico*; Institute of Geography, National Autonomous University of Mexico: Mexico City, Mexico, 2005; 216p.
29. Palacios-Sánchez, S.E.; Vega-Cendejas, M.E. Cambios alimenticios en tres especies de Sphoeroides (Tetraodontiformes: Tetraodontidae) posterior al huracán Isidoro en Bocana de la Carbonera, Sureste del Golfo de México. *Rev. Biol. Trop.* **2010**, *58*, 1223–1235. [CrossRef]
30. Mariño, I.; Enriquez, C. *Estudios batimétricos, Hidrodinámicos y de Calidad de Agua en Lagunas Costeras de Yucatán*; Reporte Técnico, Fondo mixto—CONACYT—Gobierno del Estado de Yucatán, Clave: 66254; Centro de Investigación y Estudios Avanzados del Instituto Politécnico Nacional Unidad Mérida: Mérida, Yucatán, México, 2011; 121p.
31. DOEY. *Diario Oficial del Estado de Yucatán*; Decreto Número 285, 19 de Marzo de 2010; Gobierno del Estado: Mérida, Mexico, 2010.
32. Jerónimo, G.; Gómez-Valdés, J.; Badillo, M.; López-Aguilar, K.; Galindo, C.; Gallardo, A.; Loera, J.; Arrollo-Pedraza, L.; Chiappa-Carrara, X. Variación Estacional de Temperatura y Salinidad en la Laguna la Carbonera, Yucatán, 2009–2010. In *En Recursos Acuáticos Costeros del Sureste*; Vol. II. Fondo Mixto—CONACYT—Gobierno del Estado de Yucatán; Consejo Nacional de Ciencia y Tecnología: Mexico City, Mexico, 2012; pp. 71–88. ISBN 978-607-9060-08-4.
33. Marin, E. *Modelación de la Hidrodinámica de un Sistema Lagunar en Humedal Costero con Descargas de Agua Subterránea (DAS)*; Universidad Nacional Autónoma de México: Mexico City, Mexico, 2016.
34. Marin-Coria, E.; Enriquez, C. Variaciones de temperatura y salinidad en el sistema lagunar La Carbonera, Yucatán, México. *Rev. Digit. E-BIOS* **2016**, *4*, 35–45.
35. Reyes-Mendoza, O.; Alvarez-Silva, O.; Chiappa-Carrara, X.; Enriquez, C. Variability of the thermohaline structure of a coastal hypersaline lagoon and the implications for salinity gradient energy harvesting. *Sustain. Energy Technol. Assess.* **2020**, *38*, 100645. [CrossRef]
36. SEMARNAT. *Norma Oficial Mexicana NOM-059-Semarnat-2001. Protección Ambiental-Especies Nativas de México de Flora y Fauna Silvestres-Categorías de Riesgo y Especificaciones para su Inclusión, Exclusión o Cambio-Lista de Especies en Riesgo*; Diario Oficial de la Federación, 6 de Marzo del 2002, Segunda Sección; Secretaría de Medio Ambiente y Recursos Naturales: Ciudad de México, México, 2002.
37. Guadarrama, P.; Salinas-Peba, L.; Chiappa-Carrara, X.; Ramos-Zapata, J.A. Florística, composición y estructura de las comunidades vegetales de la porción occidental de la Reserva Estatal Ciénegas y Manglares de la Costa Norte de Yucatán. *Rev. Mex. Biodivers.* **2018**, *89*, 784–805. [CrossRef]
38. Torres, W.; Méndez, M.; Dorantes, A.; Durán, R. Estructura, composición y diversidad del matorral de duna costera en el litoral yucateco. *Bol. Soc. Bot. Méx.* **2010**, *86*, 37–51. [CrossRef]
39. Gallardo-Torres, A.; Badillo-Alemán, M.; Galindo-de Santiago, C.; Loera-Pérez, J.; Rioja-Nieto, R.; Chiappa-Carrara, X. Listado Taxonómico de los Peces de la Laguna Boca de la Carbonera, Yucatán: Un Primer paso para el Manejo y Evaluación de los Recursos Costeros del Norte de Yucatán. In *En Recursos Acuáticos Costeros del Sureste*; Vol. II. Fondo Mixto—CONACYT—Gobierno del estado de Yucatán; Consejo Nacional de Ciencia y Tecnología: Mexico City, Mexico, 2012; pp. 270–288. ISBN 978-607-9060-08-4.
40. Poot-López, G.R.; López-Rocha, J.A.; González-Salas, C.; Guillen-Hernández, S.; Villegas-Hernández, H. Sex related differences in density, selectivity and vulnerability of the Atlantic blue crab, Callinectes sapidus (Rathbun, 1896), in the southern Gulf of Mexico. *Reg. Stud. Mar. Sci.* **2019**, *32*, 100846. [CrossRef]
41. Chiappa-Carrara, X.; Enríquez, C.; Mariño, I.; Badillo, M.; Gallardo, A.; Yáñez, C.; Marin-Coria, E.; Arena, L.; Guadarrama, P.; López, K.; et al. Monitoreo ambiental de largo Plazo, herramienta para evaluar la Resiliencia de sistemas costeros. In *Caracterización Multidisciplinaria de la Zona Costera de Sisal*; Garza-Pérez, J.R., Ize Lema, I., Eds.; El Laboratorio Nacional de Resiliencia Costera: Yucatán, México, 2017.
42. Santoyo, A. *Esbozo Monográfico de Sisal, Yucatán*; Reporte Técnico; Laboratorio Nacional de Resiliencia Costera: Yucatán, México, 2017; 119p.
43. Sandoval-Gio, J.J.; Zamora-Bustillos, R.; Avilés-Ramírez, G.A.; Ortiz-León, H.J.; Rosas-Correa, C.O. First report of a spawning site of Limulus polyhemus at Ría Lagartos Biosphere Reserve, Yucatan, Mexico. *Rev. Bio Cienc.* **2016**, *5*, e354.
44. Coria, I.D. El estudio de impacto ambiental: Características y metodologías. *Invenio* **2008**, *11*, 125–135.
45. Dellavedova, M. Guía Metodológica para la Elaboración de una Evaluación de Impacto Ambiental. Universidad de la Plata. 2011. Available online: http://blogs.unlp.edu.ar/planeamientofau/files/2013/05/Ficha-N%C2%BA-17-Gu%C3%ADa-metodol%C3%B3gica-para-la-elaboraci%C3%B3n-de-una-EIA.pdf (accessed on 13 January 2021).
46. Fernández-Vitora, V. *Guía Metodológica para la Evaluación del Impacto Ambiental*; Editorial Mundi-Prensa: Madrid, Spain, 2010.
47. Hyman, E.L. Combining Facts and Values. In *Environmental Impact Assessment: Theories and Techniques*, 1st ed.; Routledge: New York, NY, USA, 1988; p. 322. [CrossRef]
48. Boehlertm, G.; Gill, A. Environmental and Ecological Effects of Ocean Renewable Energy Development: A Current Synthesis. *Oceanography* **2010**, *23*, 68–81. [CrossRef]

49. Andrade Soriano, E.N. *Desarrollo de una Metodología para Diseñar una Planta de Energía Osmótica*; Universidad Distrital Francisco José de Caldas: Bogotá, Colombia, 2015.
50. Caballero, J.; Pulido, M.; Martínez-Ballesté, A. *The Use of the Guano Palm (Sabal Yapa) in the Tourism Industry in Quintana Roo, Mexico*; Forest Products, Livelihoods and Conservation. Case Studies on Non-Timber Forest Products Management Systems; Alexiades, M., Shanley, P., Eds.; CIFOR-CGIAR: Bogor, Indonesia, 2004; Volume 3.
51. Bahamón, A.; Álvarez, A.M. *Palafito: De Arquitectura Vernácula a Contemporánea*; Parramón: Lleida, Spain, 2009; p. 144.
52. Cipollina, A.; Micale, G.; Tamburini, A.; Tedesco, M.; Gurreri, L.; Veerman, J.; Grasman, S. 5—Reverse electrodialysis: Applications. In *Sustainability Energy from Salinity Gradients*; Cipollina, A., Micale, G., Eds.; Woodhead Publishing: Sawston, UK, 2016; pp. 135–180. [CrossRef]
53. Graniel, E.; Vera, I.; González, L. Dinámica de la interfase salina y calidad del agua en la costa nororiental de Yucatán. *Ingeniería* **2004**, *8*, 15–25.
54. Osland, M.J.; Feher, L.C.; López-Portillo, J.; Day, R.H.; Suman, D.O.; Guzmán Menéndez, J.M.; Rivera-Monroy, V.H. Mangrove forests in a rapidly changing world: Global change impacts and conservation opportunities along the Gulf of Mexico coast. *Estuar. Coast. Shelf Sci.* **2018**, *214*, 120–140. [CrossRef]
55. Capurro, L. A large coastal ecosystem: The Yucatán Península. *Adv. Perspect.* **2002**, *22*, 69–75.
56. Palko, H. *Exploring Potential Sites for Salinity Gradient Renewable Energy on the North Carolina Coast and Evaluating the Potential Effects of Local Salinity Regime Variation on SAV Communities due to Reverse Electrodialysis Effluent*; University of North Carolina: Chapel Hill, NC, USA, 2017.
57. Salas, M.; Gómez-López, D.; Duque, G. Estructura de las praderas de Thalassia testudinum en un gradiente de profundidad en la Guajira, Caribe colombiano. *Bol. Investig. Mar. Costeras* **2010**, *39*, 381–395.
58. Vermaas, D.A.; Kunteng, D.; Saakes, M.; Nijmeijer, K. Fouling in reverse electrodialysis under natural conditions. *Water Res.* **2013**, *47*, 1289–1298. [CrossRef] [PubMed]
59. Wojtarowski, A.; Martínez, M.L.; Silva, R.; Vázquez, G.; Enriquez, C.; López-Portillo, J.; García-Franco, J.G.; MacGregor-Fors, I.; Lara-Domínguez, A.L.; Lithgow, D. Renewable energy production in a Mexican biosphere reserve: Assessing the potential using a multidisciplinary approach. *Sci. Total Environ.* **2021**, *776*, 145823. [CrossRef]
60. Didde, R. Pilot Plant Brings "Blue Energy" Closer. 2014. Available online: https://resource.wur.nl/en/show/Pilot-plant-brings-blue-energy-closer.htm (accessed on 23 February 2021).
61. Gómez, L. Aspectos ecológicos y biotecnológicos. *Rev. Cub. Quím.* **2013**, *19*, 3–20.
62. CONAGUA. *Manual de Agua Potable, Alcantarillado y Saneamiento. Guía para el Diseño de Emisores Submarinos*; CONAGUA: México City, México, 2007.
63. Enriquez, C.; Mariño-Tapia, I.; Jeronimo, G.; Capurro-Filograsso, L. Thermohaline processes in a tropical coastal zone. *Cont. Shelf Res.* **2013**, *69*, 101–109. [CrossRef]
64. Blaber, S. *Fish and Fisheries of Tropical Estuaries*; Chapman & Hall: London, UK, 1997.
65. Telesh, I.V.; Khlebovich, V.V. Principal processes within the estuarine salinity gradient: A review. *Mar. Pollut. Bull.* **2010**, *61*, 149–155. [CrossRef]
66. Vega-Cendejas, M.E.; Hernández de Santillana, M. Fish community structure and dynamics in a coastal hypersaline lagoon: Rio Lagartos, Yucatan, Mexico. *Estuar. Coast. Shelf Sci.* **2004**, *60*, 285–299. [CrossRef]
67. Arceo-Carranza, D.; Chávez-López, R. Cambio Climático, Estuarios e Hipersalinidad. In *Tópicos de Agenda para la Sostenibilidad de Costas y Mares Mexicanos*; Rivera-Arriaga, E., Sánchez-Gil, P., Gutiérrez, J., Eds.; Universidad Autónoma de Campeche: Campeche, Mexico, 2019; pp. 275–290.
68. Dawoud, M.A. Environmental Impacts of Seawater Desalination: Arabian Gulf Case Study. *Int. J. Environ. Sustain.* **2012**, *1*, 22–37. [CrossRef]
69. Aaberg, R.J. Osmotic power: A new and powerful renewable energy source? *Refocus* **2003**, *4*, 48–50. [CrossRef]
70. Scialdone, O.; Guarisco, C.; Grispo, S.; Angelo, A.D.; Galia, A. Investigation of electrode material—Redox couple systems for reverse electrodialysis processes. Part I: Iron redox couples. *J. Electroanal. Chem.* **2012**, *681*, 66–75. [CrossRef]
71. Pérez-Castillo, A.G.; Rodríguez, A. Indice fisicoquímico de la calidad de agua para el manejo de lagunas tropicales de inundación. *Rev. Biol. Trop.* **2008**, *56*, 1905–1918. [PubMed]
72. Garner, S.B.; Boswell, K.M.; Lewis, J.P.; Tarnecki, J.H.; Patterson, W.F. Effect of reef morphology and depth on fish community and trophic structure in the northcentral Gulf of Mexico. *Estuar. Coast. Shelf Sci.* **2019**, *230*, 106423. [CrossRef]
73. Chen, Q.; Yuan, H.; Chen, P. Short-term effects of artificial reef construction on the taxonomic diversity and eco-exergy of the macrobenthic faunal community in the Pearl River Estuary, China. *Ecol. Indic.* **2019**, *98*, 772–782. [CrossRef]
74. Yang, X.; Lin, C.; Song, X.; Xu, M.; Yang, H. Effects of artificial reefs on the meiofaunal community and benthic environment—A case study in Bohai Sea, China. *Mar. Pollut. Bull.* **2019**, *140*, 179–187. [CrossRef]
75. Vaselbehagh, M.; Karkhanechi, H.; Takagi, R.; Matsuyama, H. Biofouling phenomena on anion exchange membranes under the reverse electrodialysis process. *J. Membr. Sci.* **2017**, *530*, 232–239. [CrossRef]
76. Hernández-Fontes, J.V.; Martínez, M.L.; Wojtarowski, A.; González-Mendoza, J.L.; Landgrave, R.; Silva, R. Is ocean energy an alternative in developing regions? A case study in Michoacan, Mexico. *J. Clean. Prod.* **2020**, *266*, 121984. [CrossRef]
77. Ciarreta, A.; Espinosa, M.P.; Pizarro-Irizar, C. Is green energy expensive? Empirical evidence from the Spanish electricity market. *Energy Policy* **2014**, *69*, 205–215. [CrossRef]

78. Silva, R.; Chávez, V.; Bouma, T.J.; van Tussenbroek, B.I.; Arkema, K.K.; Martínez, M.L.; Oumeraci, H.; Heymans, J.J.; Osorio, A.F.; Mendoza, E.; et al. The Incorporation of Biophysical and Social Components in Coastal Management. *Estuaries Coasts* **2019**, *42*, 1695–1708. [CrossRef]
79. Silva, R.; Oumeraci, H.; Martínez, M.L.; Chávez, V.; Lithgow, D.; van Tussenbroek, B.I.; van Rijswick, H.F.M.W.; Bouma, T.J. Ten Commandments for Sustainable, Safe, and W/Healthy Sandy Coasts Facing Global Change. *Front. Mar. Sci.* **2021**, *8*. [CrossRef]
80. Herrera-Silveira, J.A. Lagunas Costeras de Yucatan (SE, México) Investigación, Diagnóstico y Manejo. *Ecotropicos* **2006**, *19*, 94–108.
81. INEGI. *Estudio Hidrológico de la Península de Yucatán*, 1st ed.; Instituto Nacional de Estadística, Geografía e Informática y Gobierno del Estado de Yucatán: Yucatán, México, 2012.
82. SEMAR. Available online: https://oceanografia.semar.gob.mx/Templates/grafnum_progreso.html (accessed on 20 May 2021).
83. Silva, R.; Ruiz, G.; Posada, G.; Pérez, D.; Rivillas, G.; Espinal, J.; Mendoza, E. *Atlas de Clima Marítimo de la Vertiente Atlántica Mexicana*; Instituto de Ingeniería, Universidad Nacional Autónoma de México: Mexico City, Mexico, 2008.
84. Alcerreca, J.C.; Silva, R.; Mendoza, E. Simple settling velocity formula for calcareous sand. *J. Hydraul. Res.* **2013**, *51*, 215–219. [CrossRef]
85. Cuevas Jiménez, A.; Euán Ávila, J.I.; Villatoro Lacouture, M.M.; Silva Casarín, R. Classification of Beach Erosion Vulnerability on the Yucatan Coast. *Coast. Manag.* **2016**, *44*, 333–349. [CrossRef]
86. Kachadourian-Marras, A.; Alconada-Magliano, M.M.; Carrillo-Rivera, J.J.; Mendoza, E.; Herrerías-Azcue, F.; Silva, R. Characterization of Surface Evidence of Groundwater Flow Systems in Continental Mexico. *Water* **2020**, *12*, 2495. [CrossRef]

Journal of
Marine Science and Engineering

MDPI

Article

Wind–Wave Coupling Effect on the Dynamic Response of a Combined Wind–Wave Energy Converter

Jinghui Li [1], Wei Shi [1,*], Lixian Zhang [1], Constantine Michailides [2] and Xin Li [3]

1 State Key Laboratory of Coastal and Offshore Engineering, Deepwater Engineering Research Center, Dalian University of Technology, Dalian 116024, China; jinghuili@mail.dlut.edu.cn (J.L.); 2248742388@mail.dlut.edu.cn (L.Z.)
2 Department of Civil Engineering and Geomatics, Cyprus University of Technology, Limassol 3036, Cyprus; c.michailides@cut.ac.cy
3 Institute of Earthquake Engineering, Faculty of Infrastructure Engineering, Dalian University of Technology, Dalian 116024, China; lixin@dlut.edu.cn
* Correspondence: weishi@dlut.edu.cn; Tel.: +86-411-8470-8709

Abstract: There is a huge energy demand from offshore renewable energy resources. To maximize the use of various renewable energy sources, a combined floating energy system consisting of different types of energy devices is an ideal option to reduce the levelized cost of energy (LCOE) by sharing the infrastructure of the platform and enhancing the power production capacity. This study proposed a combined concept of energy systems by combing a heave-type wave energy converter (WEC) with a semisubmersible floating wind turbine. In order to investigate the power performance and dynamic response of the combined concept, coupled aero-hydro-servo-elastic analysis was carried out using the open-source code F2A, which is based on the coupling of the FAST and AQWA tools by integrating all the possible environmental loadings (e.g., aerodynamic, hydrodynamic). Numerical results obtained by AQWA are used to verify the accuracy of the coupled model in F2A in predicting dynamic responses of the combined system. The main hydrodynamic characteristics of the combined system under typical operational conditions were examined, and the calculated responses (motions, mooring line tension and produced wave power) are discussed. Additionally, the effect of aerodynamic damping on the dynamic response of the combined system was examined and presented. Moreover, a second fully coupled analysis model was developed, and its response predictions were compared with the predictions of the model developed with F2A in order for the differences of the calculated responses resulted by the different modeling techniques to be discussed and explained. Finally, the survivability of the combined concept has been examined for different possible proposed survival modes.

Keywords: wind–wave energy structures; wind turbine; fully coupled analysis; hydrodynamic response; aerodynamic damping; wave energy converters

Citation: Li, J.; Shi, W.; Zhang, L.; Michailides, C.; Li, X. Wind–Wave Coupling Effect on the Dynamic Response of a Combined Wind–Wave Energy Converter. *J. Mar. Sci. Eng.* **2021**, *9*, 1101. https://doi.org/10.3390/jmse9101101

Academic Editor: Eugen Rusu

Received: 10 September 2021
Accepted: 1 October 2021
Published: 9 October 2021

Publisher's Note: MDPI stays neutral with regard to jurisdictional claims in published maps and institutional affiliations.

Copyright: © 2021 by the authors. Licensee MDPI, Basel, Switzerland. This article is an open access article distributed under the terms and conditions of the Creative Commons Attribution (CC BY) license (https://creativecommons.org/licenses/by/4.0/).

1. Introduction

In recent years, the field of offshore wind power generation has developed rapidly and has become the fastest-growing energy source for marine renewable energy. As of the end of 2020, the global wind power capacity reached 744 GW, with 50% increase over 2019 [1]. With the rapid development of offshore wind technology, offshore wind turbines are increasing in scale and size, and the scale of wind farms is also increasing. The national renewable energy laboratory (NREL) 5-MW reference wind turbine [2] and DTU 10-MW reference wind turbine [3] have been widely used in comparative studies. The largest offshore wind turbine MySE 16.0-242 16 MW with rotor diameter of 242 m was launched by Ming Yang Smart Energy in August of 2021. To further exploit the offshore wind from deep water, various floating offshore wind concepts were proposed including mainly spar, semisubmersible, tension leg platform (TLP), and barge. Compared to TLP and spar, the semisubmersible platform is more feasible in various water depths and has low installation costs of the mooring system. At present, the concepts of semisubmersible wind turbines mainly include WindFloat [4], Dutch Tri-floater [5], Windsea [6], Windflo [7], braceless [8], V-shaped [9], OC4-DeepCwind [10], and so on. Wave energy is also a large energy source, with a much higher power density than wind power, but this energy technology is not fully commercialized at present due to its high cost and reliability issue. WECs can generally be categorized as oscillating bodies, oscillating water columns, and overtopping devices [11].

By sharing space, supporting structures, cables, and other infrastructure, combining floating wind turbine system and WEC can not only reduce the cost of the device but also capture wind and wave energy and greatly improve the utilization of renewable energy of the ocean. At present, many studies on numerical simulations and model tests based on different floating offshore wind concepts (barge, TLP, spar, and semisubmersible) combined with WEC have been carried out. Aboutalebi et al. [12,13] proposed to install an oscillating water column WEC on the barge platform to reduce the fatigue movement of the platform. In Michailides et al. [14], the required wind–wave analysis for harsh environmental conditions has been examined. Ren et al. [15] carried out an experimental and numerical study of dynamic response of a new combined TLP wind turbine and wave energy converter. Bachynski et al. [16] proposed a TLP combined three-point absorber WECs system and studied the performance under operational and extreme sea cases. Russo et al. [17], Tomasicchio et al. [18], and Xu et al. [19] carried out experimental studies of the dynamic characteristics of Spar Buoy Wind Turbine and studied its dynamic behavior. Hu et al. [20] carried out the optimal design of WEC and dynamic characteristics analysis of a hybrid system combing a semisubmersible floating wind platform (WindFloat) and WECs. Sarmiento et al. [21] carried out a new floating semisubmersible structure combined with WECs (three oscillating water columns, OWC) in order to characterize the performance of the platform and OWCs. The MARINA Platform project funded by the European Union [22] proposed a combined system of semisubmersible platforms with point absorption and oscillating water column WEC. Peiffer et al. [23] and Aubault et al. [24] carried out coupled dynamic analysis through numerical models and experimental models. Muliawan et al. [25,26] proposed a comprehensive concept that combines a spar-type floating wind turbine with a torus WEC (named STC). Wan et al. [27–29] studied the dynamic response of an STC under typical operational conditions and extreme conditions based on numerical and experimental methods. Another concept that combines a floating semisubmersible wind turbine and a flap-type WEC (named SFC) was proposed. Michailides et al. [30–32] systematically studied the integrated operation of SFCs through numerical and experimental models. Ren et al. [33] carried out experimental and numerical studies on the hydrodynamic response of a new combined monopile wind turbine and a heaving-type WEC under typical operational conditions. In addition, Wang et al. [34] studied the hydrodynamic response of the combined system of the semisubmersible platform and heaving-type WEC.

With the development of the field of offshore wind turbines, the aero-hydro-servo-elastic coupling tool has become the key to solve the equation of motion and calculate the dynamic response of floating wind turbines. Different simulation tools have been developed so far. DeepLines [35,36], DARwind [37], Bladed, HAWC2, and FAST are the most well-known tools for fully coupled analysis of wind turbines. Jonkman et al. [38,39] developed a hydrodynamic module to consider the diffraction and radiation effects of floating platforms. Due to its open-source nature, it has also been used to participate in the development of a fully coupled framework. Kvittem et al. [40] combined AeroDyn with the non-linear finite element software SIMO/REFLEX. Shim [41] developed an interface that combines FAST with the fluid dynamic analysis tool CHARM3D. Recently, Yang et al. [42] developed a new aero-hydro-servo-elastic coupling framework based on FAST and AQWA (F2A) for dynamic analysis of FOWTs, combining the advantages of aero-elastic-servo of FAST and hydrodynamic analysis of AQWA [43].

In this study, with use of the recently proposed coupled analysis tool F2A, a fully coupled analysis is performed for a combined system consisting of a semisubmersible platform and a heaving-type WEC. A numerical model of the combined structure that was capable of simulating its motion and dynamic responses under different typical operational conditions was developed and used with F2A. A second model with use of a different analysis tool is developed for performing fully coupled analysis. Different types of analysis and comparison were performed to examine the fully coupled responses of the combined structure but also for emphasizing the effect of different modeling methods/techniques on the response quantities of the combined system. The time-domain dynamic response characteristics, mooring characteristics, and PTO-produced power of the combined system are examined and presented in this study. Under the excitation of only waves, there was small difference between the two numerical models, while under irregular wave and turbulent wind conditions, the difference between two models was larger. The impact of aerodynamic loads as well as aerodynamic damping is significant. When the wind velocity is small, aerodynamic damping has a significant effect on reducing the resonance of surge and pitch, and there is an obvious positive relationship. When the wind velocity increases, wind thrust has a greater impact than aerodynamic damping. The rationality behind the efficient use of the two examined numerical tools for studying combined concepts has been presented. Finally, an extreme condition is studied to ensure the survivability of the combined system. The results show that the two models had significant differences in dynamic motion and mooring force prediction under irregular wave and turbulent wind condition due to the wind–wave coupling effect at low frequency range.

2. Theoretical Background

Different from offshore oil and gas platform, a floating wind turbine experiences the aerodynamic load and hydrodynamic load simultaneously. These loads are both important in the system design at the same order. The basic theories used in the combined system analysis are described in this section.

2.1. Aerodynamic Loads and Aerodynamic Damping

Blade element momentum (BEM) theory is one of the efficient and most commonly used methods for calculating induced loads on wind turbine blades [44]. In this theory, the wind turbine blade is divided into many sections along the span direction, and these sections are called blade elements. The wind turbine blade is simplified as a finite element that is superimposed in the radial direction, so the three-dimensional aerodynamic characteristics of the wind turbine blade can be obtained by integrating the aerodynamic characteristics of the element in the radial direction [42].

The axial velocity V_2 and the circumferential velocity V_3 at the rotor plane are calculated as follows:

$$V_2 = (1 - a)V_1 \tag{1}$$

$$V_3 = \Omega r(1 + a') \tag{2}$$

where V_1 is the incoming wind velocity, a is the axial induction factor, and a' is the angular induction factor. The axial velocity and the circumferential velocity vector are combined to obtain the resultant velocity V_0. The lift force (L) is formed perpendicular to the resultant velocity, and the drag force (D) is formed in the same direction as the incoming velocity. They can be calculated by the following equations:

$$L = \frac{1}{2}\rho V_0^2 A C_L \tag{3}$$

$$D = \frac{1}{2}\rho V_0^2 A C_D \tag{4}$$

where A is the rotor sweeping area, ρ is the air density, C_L is the lift coefficient, and C_D is the drag coefficient.

The lift coefficient C_L and drag coefficient C_D of the aerofoil are projected in the normal and tangential directions, respectively, to obtain the normal force coefficient C_N and the tangential force coefficient C_T, as shown in the following equations:

$$C_N = C_L \cos\varphi + C_D \sin\varphi \tag{5}$$

$$C_T = C_L \sin\varphi - C_D \cos\varphi \tag{6}$$

Finally, the thrust force and torque on the blade section are calculated by the following equations:

$$dT = \frac{1}{2}\rho V_0^2 c C_N dr \tag{7}$$

$$dM = \frac{1}{2}\rho V_0^2 c C_T r dr \tag{8}$$

where c is the aerofoil chord length.

When studying aerodynamic loads, the influence of aerodynamic damping cannot be ignored. Analyzed at the micro level, the aerodynamic damping of the wind turbine essentially comes from the relationship between the aerodynamic load of the wind turbine blades and the inflow wind velocity. Without considering the platform motion and elastic structure deformation [45], the following equation can be used:

$$V_0 = V_1\sqrt{(1-a)^2 + \left[\frac{\Omega r}{V_0}(1+a')\right]^2} = V_1 f_1 \tag{9}$$

The factor f_1 indicates that the relative inflow wind velocity at the blade is simultaneously affected by the axial induction factor, angular induction factor, structural deformation of the wind turbine, and rotational velocity [46].

Considering the relative movement of the platform, the relative velocity is written as follows:

$$V_0 = V_1 f_1 - L\dot{x}_5\cos(x_5) - \dot{x}_1 \tag{10}$$

where x_5 is the pitch motion of the platform, $\dot{x}_5$ is the pitch velocity of the platform, and $\dot{x}_1$ is the surge velocity of the platform.

From Equation (7), the total thrust T on the blade can be obtained as follows:

$$T \infty V_0^2 \infty \left(V_1 f_1 - L\dot{x}_5\cos(\dot{x}_5) - \dot{x}_1\right)^2 \tag{11}$$

Ignoring the velocity components above the second order, the following equation is obtained:

$$\begin{aligned} T &\infty (V_1 f_1)^2 - 2V_1 f_1 L\dot{x}_5\cos(\dot{x}_5) - 2V_1 f_1\dot{x}_1 \\ &\infty (V_1 f_1)^2 - 2V_1 f_1 L\dot{x}_5 - 2V_1 f_1\dot{x}_1 \end{aligned} \tag{12}$$

The last two terms in the above equation represent aerodynamic damping and can be added to the left side of the equation of motion. The aerodynamic thrust is proportional to the square of the wind velocity, and the aerodynamic damping is proportional to the first power of the wind velocity. This shows that when the wind velocity is not particularly high, the effect of aerodynamic damping on the offshore floating wind turbine may be more obvious, and the relationship between them is positive. However, as the wind velocity increases, the influence of aerodynamic thrust may be much greater than that of aerodynamic damping. Of course, the aerodynamic damping force of offshore floating wind turbines is affected not only by wind velocity but also by other complex factors, such as aerodynamic induction factor. Thrust coefficient may even be affected by pitch angle and stall effect. Since F2A and AQWA were used for simulation in this study, the largest difference between them is whether considering aerodynamic damping.

2.2. Potential Flow Theory

The potential flow theory may be to calculate the hydrodynamic loads on marine structures. It is usually assumed that the fluid is non-rotating, non-viscous, and incompressible, and the fluid is assumed to be an ideal fluid [47,48]. Derived from the conservation of mass and the conservation of momentum, the governing equations of fluid motion, the Laplace equation are expressed as follows:

$$\nabla^2\phi = 0 \tag{13}$$

$$\frac{\partial u}{\partial t} + (u \cdot \nabla)u = -\nabla\left(gz + \frac{p}{\rho}\right) \tag{14}$$

where ϕ represents the three-dimensional velocity potential function and $\phi = \phi(x, y, z, t)$. The velocity potential can be decomposed into the following equation:

$$\phi = \phi^i + \phi^d + \phi^r \tag{15}$$

where ϕ^i is the velocity potential function of the incident wave, ϕ^d is the diffraction potential function, and ϕ^r is the radiation potential function.

For the fluctuation problem of linear periodic motion, the time factor can be separated by the variable separation method, and the velocity potential can be expressed as follows:

$$\phi = \mathrm{Re}\left[\varphi e^{-i\omega t}\right] \tag{16}$$

where ω is the angular frequency and Re represents the real part.

2.3. Viscous Loads

In potential flow theory, since the assumption is inviscidity, viscosity needs to be considered in practice. In this study, the Morrison model was added to consider the viscosity [48,49]. In AQWA, the viscosity was simulated by adding Morrison elements as follows:

$$f_d = 0.5C_d\rho D\mu|\mu| \tag{17}$$

where C_d represents the drag coefficient, and in this study, $C_d = 1.2$ was selected due to $d/L \geq 0.2$ and $H/d \leq 0.2$; H, d, and L represent the wave height, water depth, and wavelength, respectively; μ, D, and are the incoming flow velocity, structure diameter, and fluid density, respectively; and f_d represents the drag force on a unit height of the structure.

2.4. Equation of Motion

In this study, the wind and wave energy combined system consists of two bodies having in total 12 degrees of freedom (six degrees for each body), which requires comprehensive consideration of multiple degree-of-freedom systems [29]. The multi-body equation of motion is given as follows:

$$\begin{aligned}
&\begin{pmatrix} (M+m)_{11} & m_{12} \\ m_{21} & (M+m)_{22} \end{pmatrix}\begin{pmatrix} \ddot{x}_1(t) \\ \ddot{x}_2(t) \end{pmatrix} + \int_0^t \begin{pmatrix} \kappa(t-\tau)_{11} & \kappa(t-\tau)_{12} \\ \kappa(t-\tau)_{21} & \kappa(t-\tau)_{22} \end{pmatrix}\begin{bmatrix} \dot{x}_1(\tau) \\ \dot{x}_2(\tau) \end{bmatrix} d\tau \\
&+\begin{pmatrix} (R)_{11} & 0 \\ 0 & (R)_{22} \end{pmatrix}\begin{bmatrix} x_1(t) \\ x_2(t) \end{bmatrix} = \begin{bmatrix} f^{wind}(t) \\ 0 \end{bmatrix} + \begin{bmatrix} f_1^{wave}(t) \\ f_2^{wave}(t) \end{bmatrix} + \begin{bmatrix} f_1^{drag}(t) \\ f_2^{drag}(t) \end{bmatrix} \\
&+\begin{bmatrix} f_1^{interface}(t) \\ f_2^{interface}(t) \end{bmatrix} + \begin{bmatrix} F_1^{PTO}(t) \\ F_2^{PTO}(t) \end{bmatrix}
\end{aligned} \tag{18}$$

where m is the added mass matrix, x, $\dot{x}$, and $\ddot{x}$ are the displacement, velocity, and acceleration matrix in the time domain, respectively, $\kappa(\tau)$ is the retardation function, which is based on the added mass and potential damping matrix, and f is the summation of the external forces in time domain. The subscripts 1 and 11 refer to the variables of body 1 (braceless); subscripts 2 and 22 refer to the variables of body 2 (WEC); and subscripts 12 and 21 present the coupling terms between the braceless and WEC. The vertical (heave) quadratic damping of the braceless and WEC terms is modeled by the quadratic damping matrix on the left side of Equation (18). The term f^{wind} denotes wind load on the turbine rotor, while f^{wave} is the wave forces applied on the braceless platform and WEC. The interface forces $f_1^{interface}$ between the two bodies include horizontal contact forces and vertical friction forces. F^{PTO} is the PTO forces. Each term of the interface and PTO forces is applied on the two bodies with the same value but in different directions.

2.5. Brief Description of F2A

The baseline version of AQWA was incapable of predicting the aero-servo-elastic of floating offshore wind turbines, but it accepted time domain analysis of external forces implemented by dynamic link library (.dll). In order to enable AQWA to form a fully coupled analysis of floating offshore wind turbines, FAST was integrated in AQWA with some simulation function implemented [36]. Therefore, the coupling framework was "FAST2AQWA", denoted as F2A. The coupling of AQWA and FAST was accomplished by user_force.dll and source code subroutine of FAST. Related simulation capabilities of FAST were completely implemented in time domain analysis in the DLL that can be called by AQWA during the simulation.

3. Numerical Model of Combined System

The combined system was composed of a semisubmersible platform and a heaving-type WEC connected by a guide-roller system in the middle and an upper connecting system [50]. The wind turbine model used for this study is the NREL 5-MW baseline wind turbine. The main parameters of the 5-MW wind turbine are shown in Table 1 [2]. An illustration of the combined system is shown in Figure 1, and the main parameters of the combined system are shown in Table 2. In this study, the time-domain hydrodynamic simulation of the combined system was based on AQWA. Hydrodynamic panel models are shown in Figure 2.

Table 1. The main parameters of the 5-MW wind turbine.

Rotor-Nacelle-Assembly t	350
Center of Gravity (CoG) m	(−0.2,0.0,70)
Tower mass t	347.46
Total WT mass moment of inertia about X axis (I_{xx}) kg*m^2	3,770,000,000
Total WT mass moment of inertia about Y axis (I_{yy}) kg*m^2	3,660,000,000
Total WT mass moment of inertia about Z axis (I_{zz}) kg*m^2	112,000,000

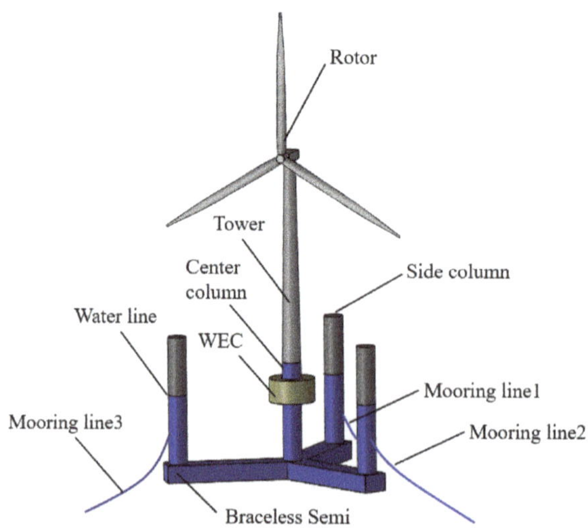

Figure 1. Sketch map of the combined system.

Table 2. The main parameters of the combined system.

	Parameters	Values
Semisubmersible platform	Semisubmersible mass t	9738
	Diameter of the central column m	6.5
	Diameter of the three side columns m	6.5
	Water displacement m^3	10,298
	Water depth m	200
	Operating draft m	30
	Center of semisubmersible m	(0,0,24.36)
WEC	Outer/Inner diameter m	16/8
	FigureHeight/Draft m	8/3.5
	Mass t	463.5
	Water displacement m^3	452.2
	Center of mass m	(0,0,1)

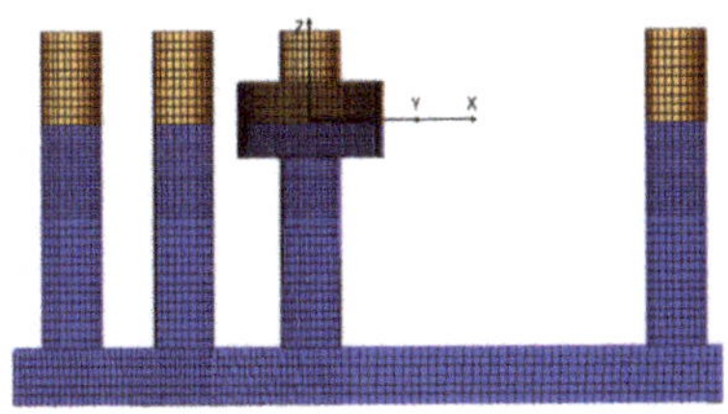

Figure 2. Panel models for the hydrodynamic analysis.

The PTO system (shown in Figure 3), which captures the wave energy through the relative heave motion of the semisubmersible platform and the WEC, was simplified as a linear spring and a linear damper. This model was accomplished by establishing a fender element in ANSYS/AQWA. Based on the discussion of damping coefficient of Fender and stiffness coefficient of linear spring [27–29,33,34,50], it was found that a large value of B_{pto} value may lead to air compressibility that cannot be ignored, while a small K_{pto} coefficient may ignore the influence of stiffness coefficient on the produced power [19]. Therefore, 1.5×10^6 Ns^2/m was selected for B_{pto} and 1 N/m for K_{pto} in this study. The force of PTO was calculated by the following equation:

$$F_{pto} = B_{pto} \cdot (v_2 - v_1) + K_{pto} \cdot (x_2 - x_1) \tag{19}$$

where B_{pto} and K_{pto} are the linear damping coefficient and linear spring stiffness coefficient, respectively; v_1 and x_1 are the velocity and displacement of the semisubmersible platform; and v_2 and x_2 are the velocity and displacement of the WEC, respectively.

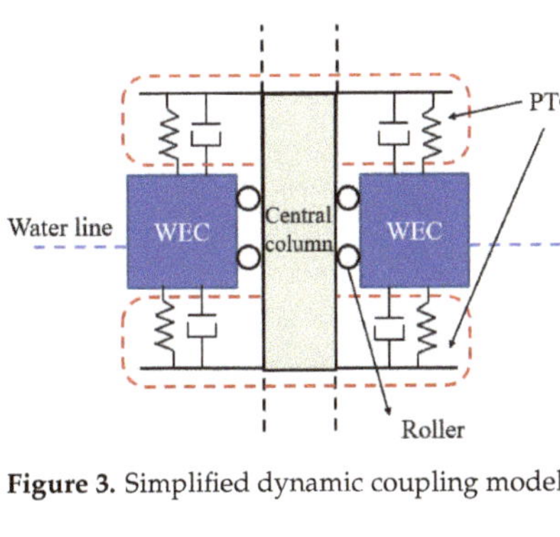

Figure 3. Simplified dynamic coupling model between the WEC and braceless.

Knowing the damping force of the PTO system and the relative velocity between the platform and WEC, the produced power of the WEC can be calculated through the following equation:

$$P_{PTO} = F_{PTO} \cdot (v_2 - v_1) \tag{20}$$

4. Results and Discussion

The simulations conducted in this study were primarily carried out under typical operational conditions and extreme conditions from a typical site at 61°21′ N latitude and 0°0′ E longitude near the Shetland Islands, northeast of Scotland, UK. The water depth at the site was 200 m [51]. The examined load cases are listed in Table 3. For regular wave case (LC1), the 1800 s–1900 s was used in the comparison to get rid of transient effect. For irregular wave cases (LC2–LC4), the total simulation time is 4600 s, and the first 1000 s has been excluded to avoid the transient effect. It should be noted that the time series results of the motion and force responses between 3500 to 3700 s are displayed to better present the difference between different codes.

Table 3. Load cases table.

Load Case	Wind Velocity U_{wind} (m/s)	Wave Height H/H_s (m)	Wave Period T/T_p (s)
LC1	/	2.0	9.0
LC2	8.4	2.0	14.8
LC3	11.4	2.4	10.9
LC4	23.8	5.5	13.5
LC5	50.0	13.8	19.2

4.1. Free Decay Test

A free decay test was performed in the numerical simulation to determine the natural frequencies of the surge, heave, and pitch of the platform. The natural frequencies are listed in Table 4. Figure 4 was the time domain curve of free decay test of platform. The results showed that the natural frequencies of the surge (Figure 4a) and heave (Figure 4b) of the platform were the same from two codes, and there is a difference of about 11.7% of the natural frequency of the platform pitch from two codes.

Table 4. Natural frequencies (rad/s) of platform.

	Surge	Heave	Pitch
F2A	0.078	0.256	0.239
AQWA	0.078	0.256	0.210

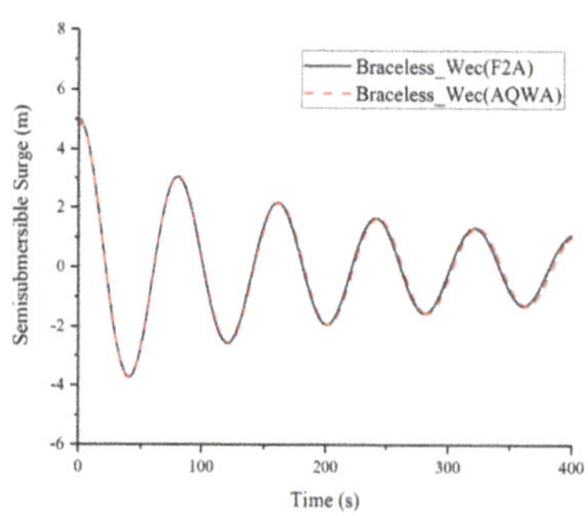

(a)

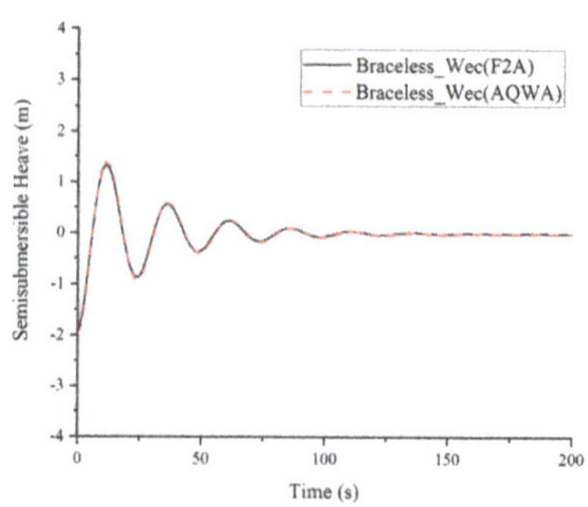

(b)

Figure 4. *Cont.*

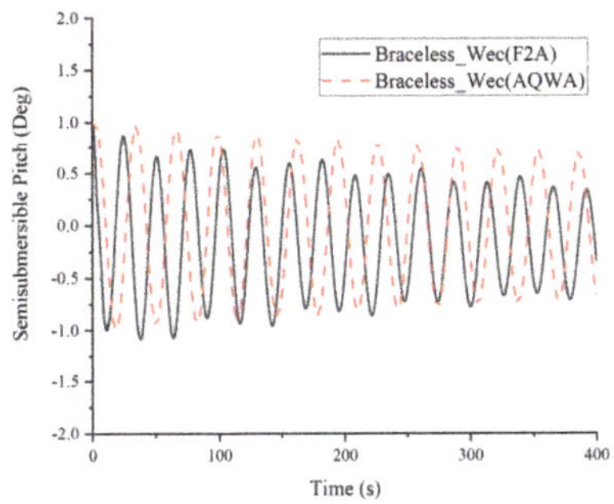

(c)

Figure 4. Time domain curve of free decay test of the semisubmersible platform: (**a**) Surge; (**b**) Heave; (**c**) Pitch.

4.2. Regular Wave Condition

In this study, the simulation of typical regular wave conditions (LC1) was performed, and the characteristics of the motion response, mooring response, and produced power of multiple bodies (F2A and AQWA simulation) were compared.

4.2.1. Motion Response

In this section, the motion response under a regular wave (H = 2.0 m, T = 9.0 s) was selected, and a motion response of three degrees of freedom (DOF) of the platform (surge, relative heave, pitch) is presented (Figure 5). The results were compared between the fully coupled framework F2A and the hydrodynamic software AQWA.

The surge motion of the platform is shown in Figure 5a. Compared with the multibody simulation results of F2A and AQWA, the amplitude was basically the same, but the simulation results of F2A at the wave peak were slightly larger. The relative heave motion responses simulated by F2A and AQWA were basically the same (Figure 5b). Slightly larger pitch response from F2A could be identified compared with AQWA results. Under only wave conditions, F2A and AQWA had good consistency in simulating multibody motion characteristics, especially in relative heave.

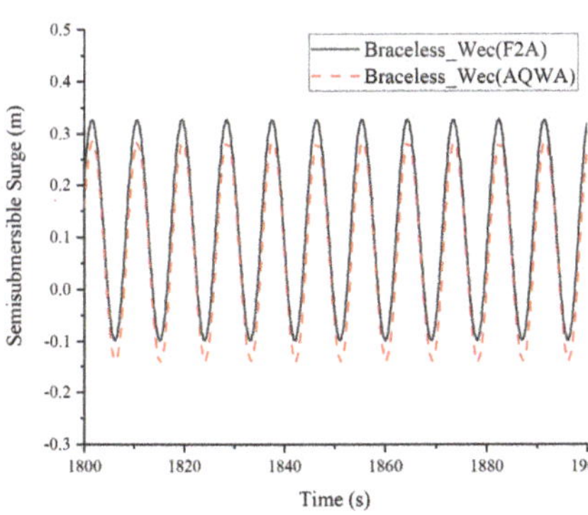

(a)

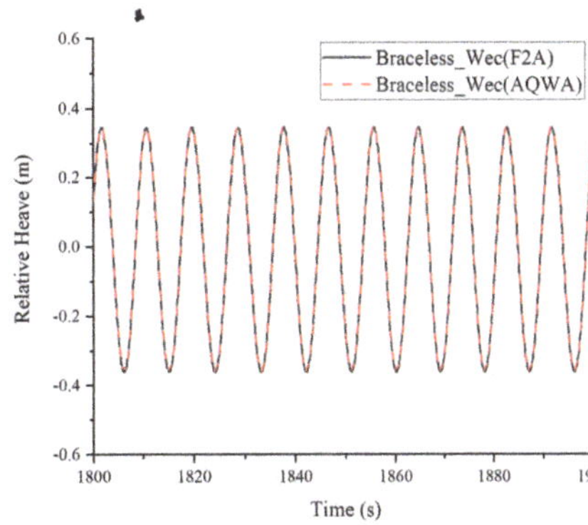

(b)

Figure 5. *Cont.*

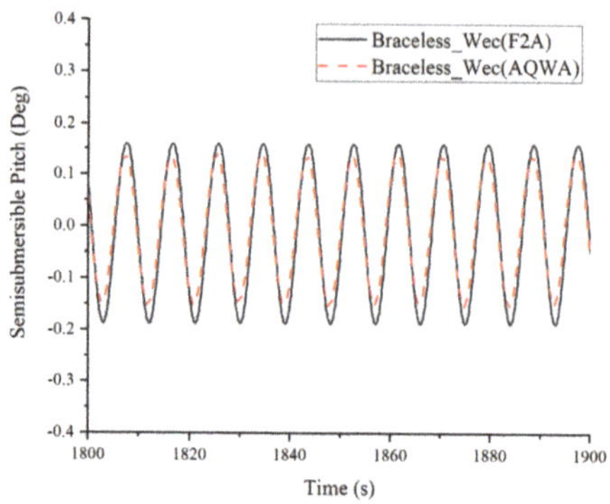

(c)

Figure 5. Time domain motion response of the semisubmersible platform: (**a**) Surge; (**b**) Relative Heave; (**c**) Pitch.

4.2.2. Mooring Line Force

Figure 6 shows the time-domain response of the mooring force of Mooring Line 1 (ML1) and Mooring Line 2 (ML2). The mooring line force of ML1 (downwind direction) was smaller than that of ML2 (upwind direction) when subjected to waves in the heading direction. Results from F2A have slightly larger mooring line force than those from AQWA simulation.

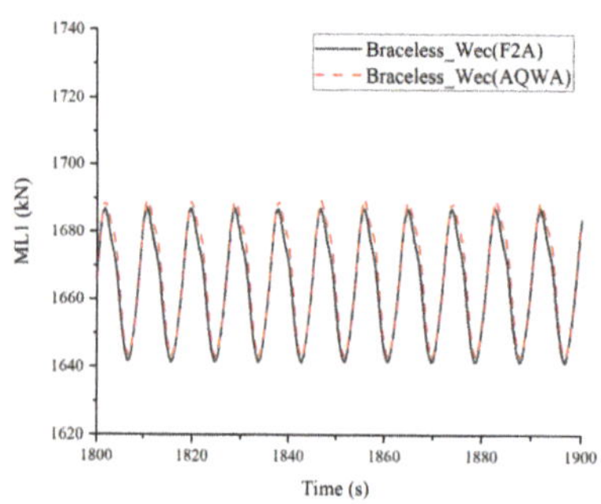

(**a**)

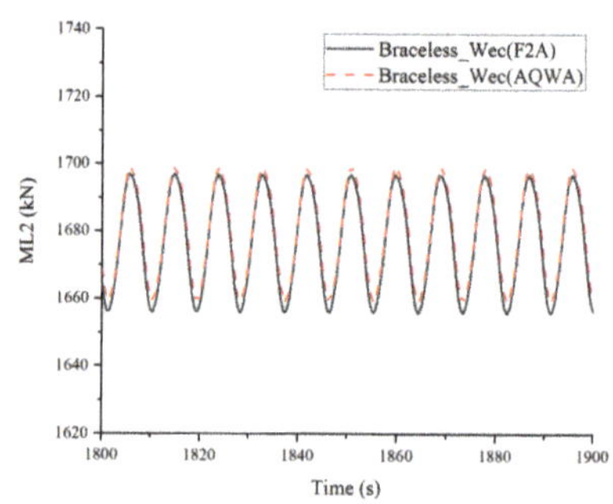

(**b**)

Figure 6. Mooring force of: (**a**) ML1; (**b**) ML2.

4.2.3. Produced Wave Power

The wave energy was captured by PTO system. The relative heave velocity from both codes are similar to each other (Figure 7a). From Equation (19), the damping force can be observed in Figure 7b. Based on the relative heave velocity and damping force, the produced wave power could be identified, which is similar from F2A results and AQWA results (Figure 7c).

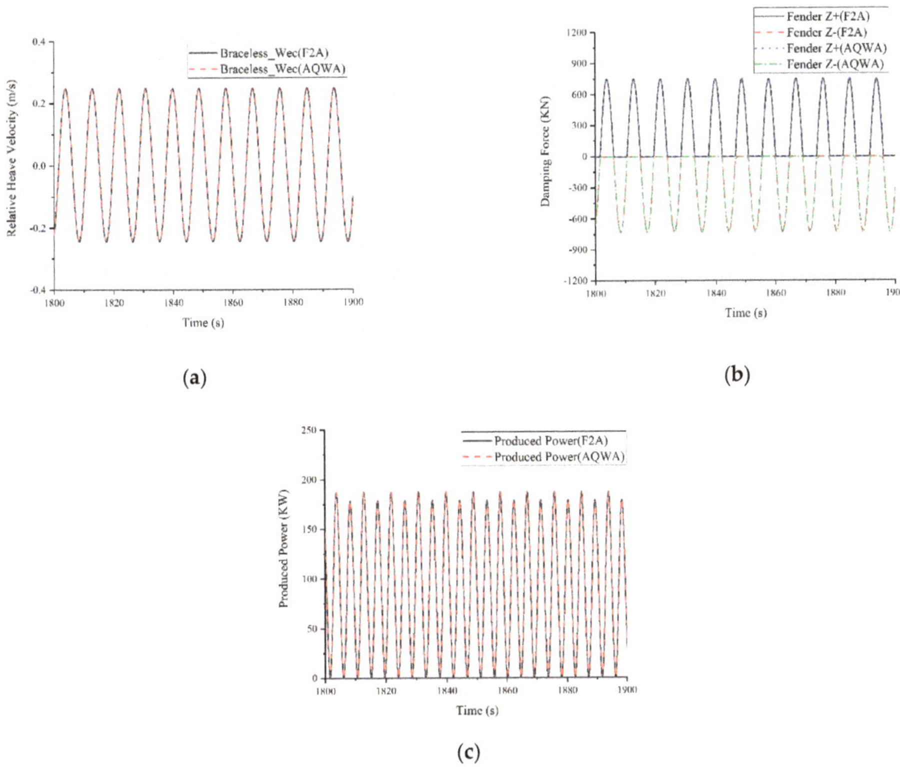

Figure 7. Time series of different responses of the combined structure under regular waves: (**a**) Relative Heave Velocity; (**b**) Damping Force; (**c**) Produced Power.

Under only wave conditions, F2A and AQWA had good consistency in simulations of the multibody dynamic response. This was because F2A integrates the AQWA hydrodynamic module.

4.3. Irregular Wave and Turbulent Wind Conditions

This section presents the motion response, mooring force, and produced power under irregular wave and turbulent wind conditions.

4.3.1. Motion Response

Figure 8 shows the motion response in surge, pitch, and heave directions under normal operation conditions in LC2 and LC3. The responses of surge and pitch in both load cases from F2A (Figure 8a,b) simulation are smaller than those predicted by AQWA (Figure 8e,f). This is due to the fact that the wind load in F2A and AQWA is differently implemented. In F2A, the wind field is pre-generated and both wind aerodynamic load and damping are considered due to the rotating turbine rotor. However, in AQWA, the wind load is implemented as an external load using dll function and no interaction and aerodynamic damping is included. The difference for the relative heave motions simulated by two tools was limited for both LC2 and LC3 (Figure 8c,d).

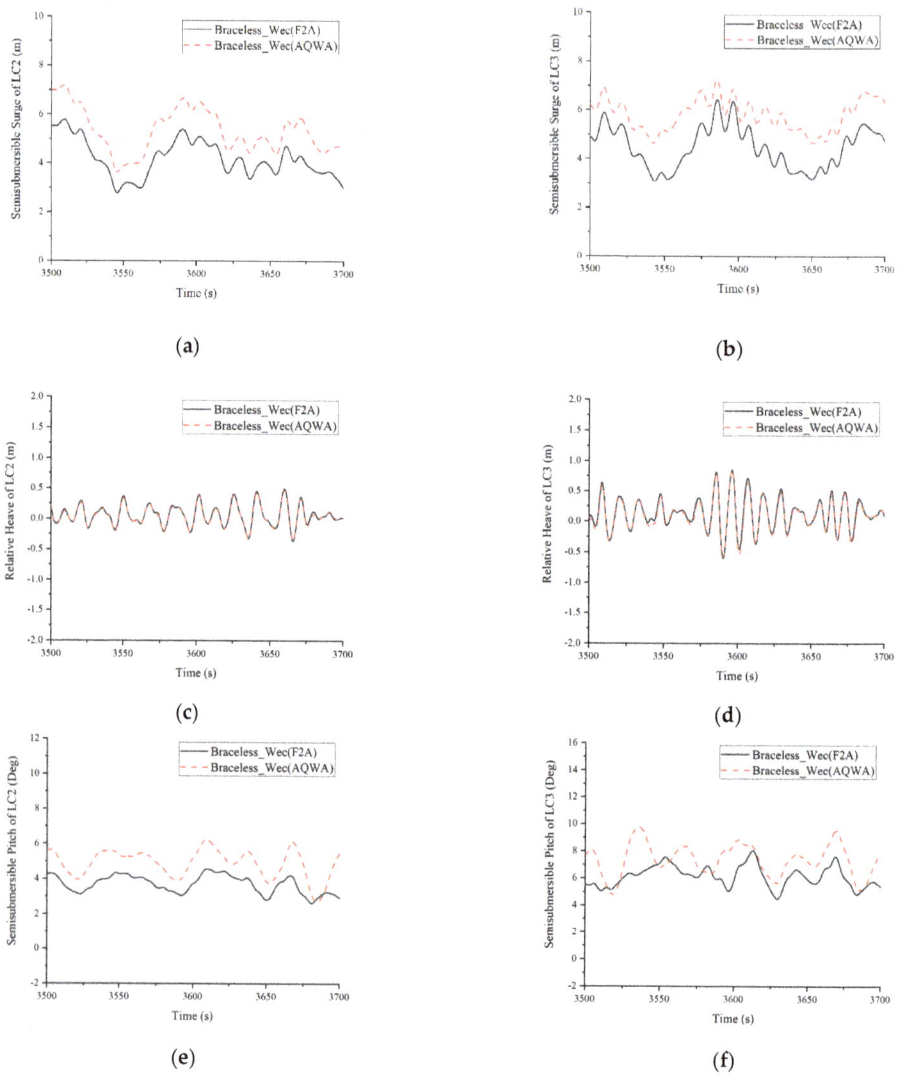

Figure 8. Time domain motion response of the semisubmersible platform: (**a**) Surge in LC2; (**b**) Surge in LC3; (**c**) Relative Heave in LC2; (**d**) Relative Heave in LC3; (**e**) Pitch in LC2; (**f**) Pitch in LC3.

Figure 9 shows motion response in LC4, in which the wind speed is larger than rated wind speed. Similar as the load case LC2 and LC3, the motion responses in surge and pitch direction from F2A are lower than those motions from AQWA due to the aerodynamic damping. The relative heave motions predicted from two codes are similar to each other.

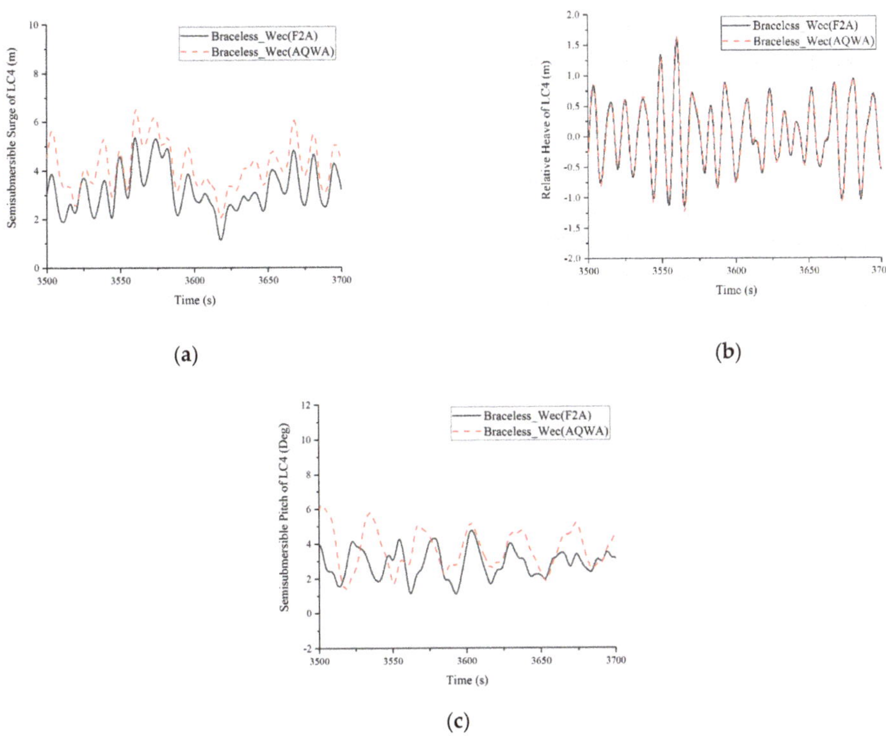

Figure 9. Motion responses of the semisubmersible platform in LC4: (**a**) Surge; (**b**) Relative Heave; (**c**) Pitch.

4.3.2. Mooring Line Force

The mooring line forces of ML1 and ML2 from LC2, LC3, and LC4 were compared for F2A results and AQWA in Figure 10. Basically, the mooring line force of ML2 in the upwind direction is larger than the mooring line force of ML1 in the downwind direction because wind and wave will drive the combined system to move in a downwind direction from its equilibrium position. Therefore, ML1 will get relaxed and ML2 will get tensioned. For the mooring line force of ML1, the reduction of the mooring force from its force at equilibrium position from F2A simulation is less than the reduction of the force from AQWA simulation due to the difference of the surge motions from two codes in all three load cases. Thus, the mooring line forces of ML1 from F2A are larger than those from AQWA (Figure 10a,c,e). However, for mooring line force of ML2, the increasing of the mooring force from its force at equilibrium position from F2A simulation is less than the increasing of force from AQWA simulation. Therefore, the mooring line forces of ML2 from F2A are less than those from AQWA (Figure 10b,d,f). Due to the pitch control above the rated wind speed (LC4), the wind thrust force is much smaller than that in LC2 (below rated wind speed) and LC3 (at rated wind speed). Therefore, the aerodynamic damping effect is much smaller. In this case (LC4), the contribution from wave load is much larger than wind load, and the discrepancy from two codes is minimal. Therefore, there is a slight difference in the mooring line force of ML2 for LC4.

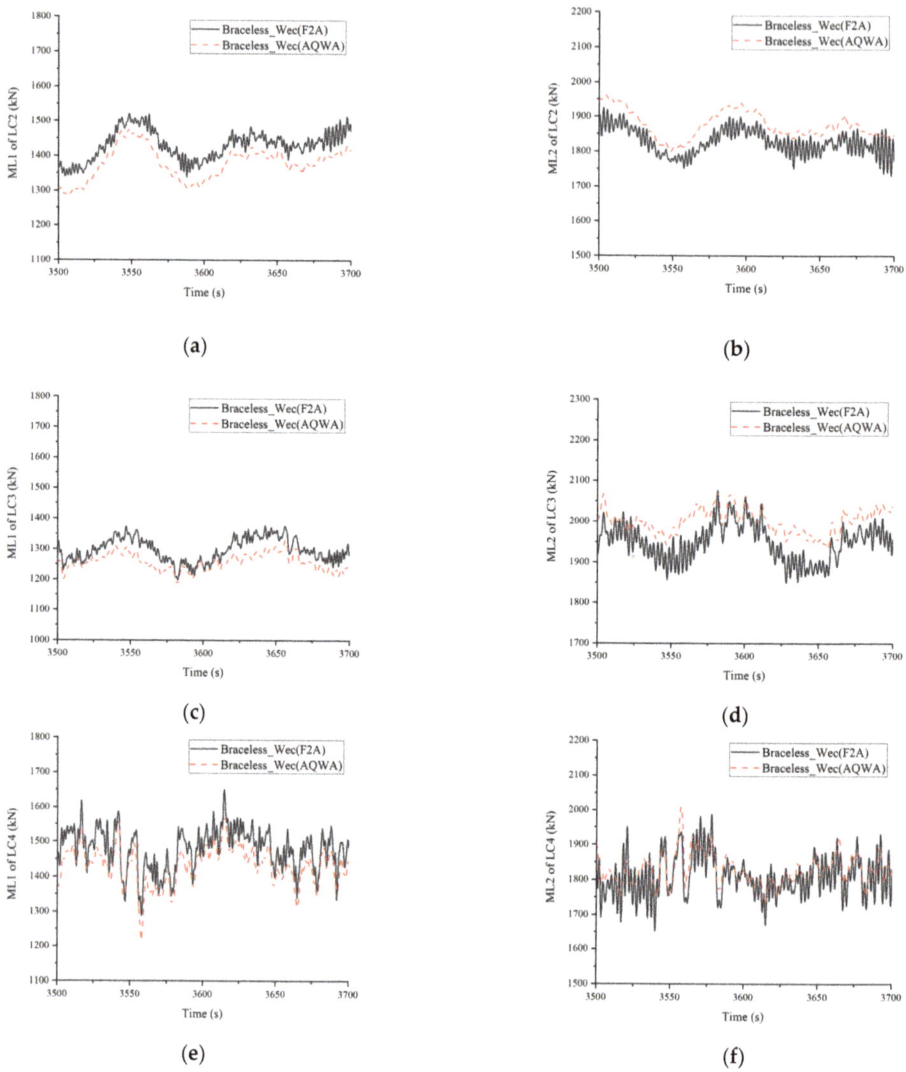

Figure 10. Mooring force of: (**a**) ML1 in LC2, (**b**) ML2 in LC2, (**c**) ML1 in LC3, (**d**) ML2 in LC3, (**e**) ML1 in LC4, (**f**) ML2 in LC4.

4.3.3. Produced Wave Power

The relative heave velocities between the semisubmersible platform and the WEC, the damping force and produced power from WEC in LC2, LC3, and LC4, are presented in Figure 11. Due to less effect of aerodynamic damping on heave motion, no significant difference is identified for the relative heave velocities, damping force, and produced power. When the wave height increases from LC2 to LC4, the relative heave velocity (Figure 11a,c,e) also increases due to the large relative heave motion in a severe sea state, which is beneficial to capture wave energy. Figure 11b,d,f present the vertical damping force. The F2A and AQWA simulation results were similar. As the wave height increases, the vertical damping force increases significantly. Figure 11g,h,i show the produced energy power from WEC. With the sea state moving from mild to severe (LC2 to LC4), more power can be produced from the wave. Figure 11i is the produced power under severe sea conditions (LC4). It can be seen the maximum power in LC4 is even as large as 3.5 MW (not shown in Figure 11f), which is comparable to the power produced from the wind turbine.

However, the mean produced power is less than 600 kW, and it shows obvious instability, while the produced power of WEC is much smaller. Therefore, the wind power production for NREL 5 MW WT will make the main contribution to the total power production of the combined system under the severe sea conditions.

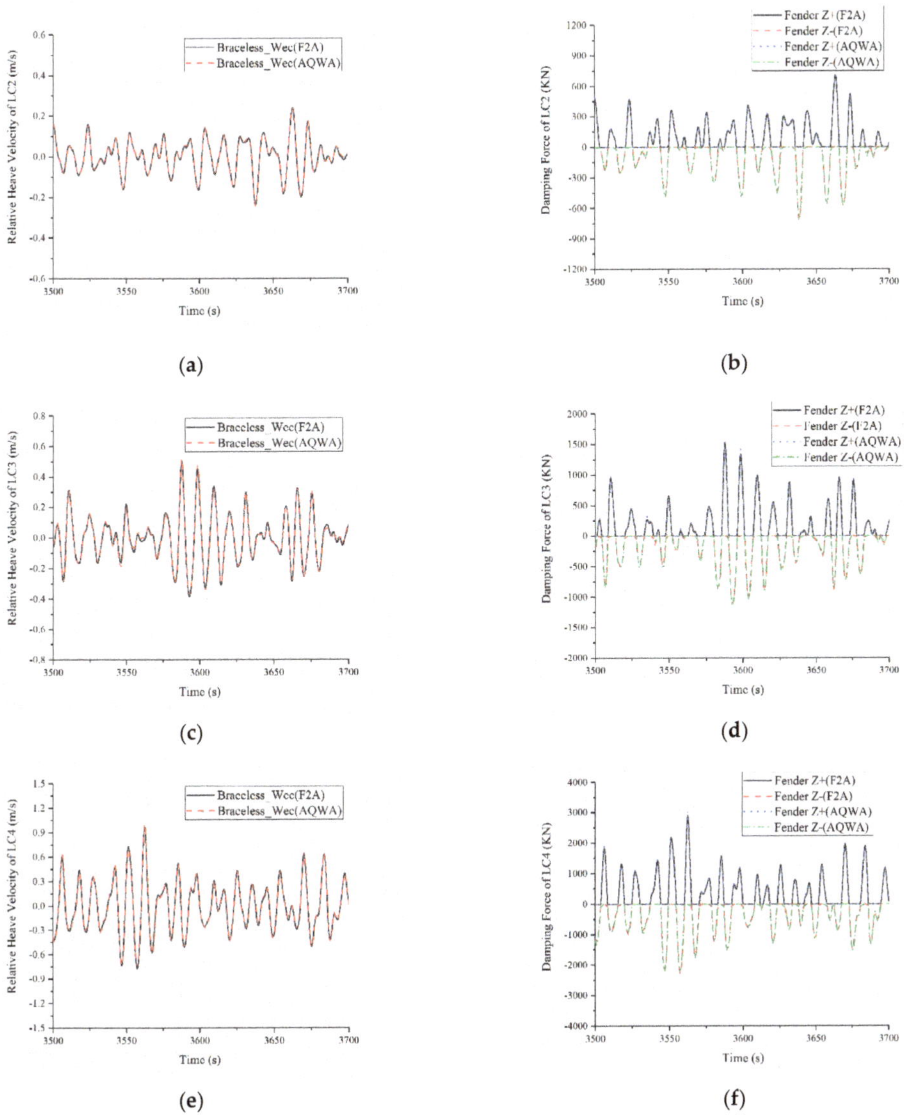

Figure 11. *Cont.*

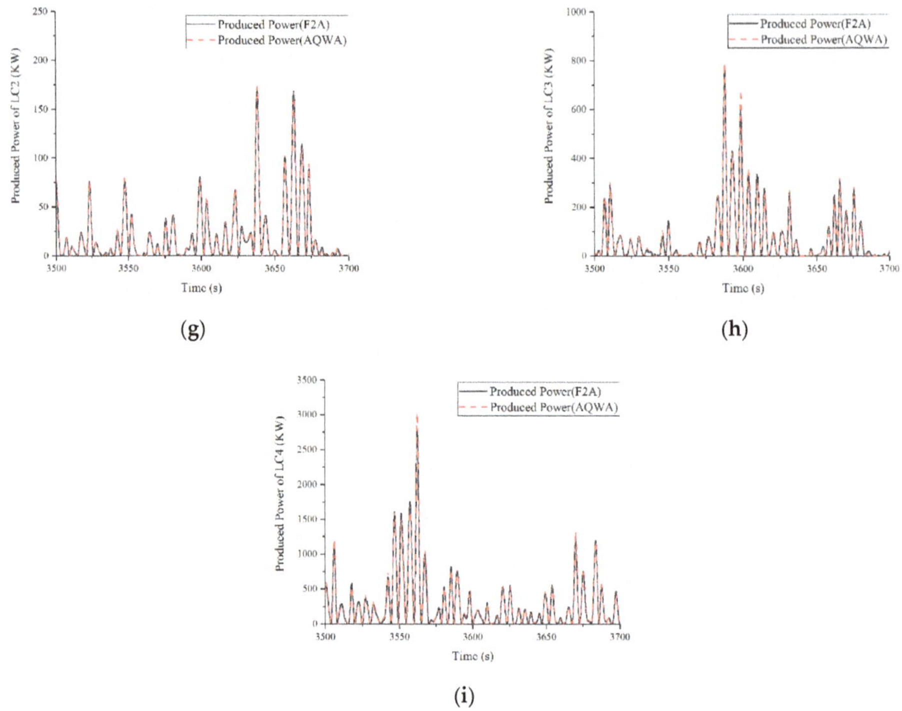

Figure 11. Time series of different responses of the combined structure: (**a**) Relative Heave Velocity of LC2, (**b**) Damping Force of LC2, (**c**) Relative Heave Velocity of LC3, (**d**) Damping Force of LC3, (**e**) Relative Heave Velocity of LC4, (**f**) Damping Force of LC4, (**g**) Produced Power of LC2, (**h**) Produced Power of LC3, (**i**) Produced Power of LC4.

4.3.4. Statistical Analysis

Figure 12 displays the statistics of responses in operational conditions from both F2A simulation and AQWA simulation. The maximum and minimum values are determined as the highest crest and lowest trough of the corresponding time series. Figure 12a–c show that the simulation results of F2A and AQWA were different in the surge and pitch motion, and the heave motion results were slightly different, which was consistent with the previous results. When the wind velocity was the rated wind velocity of the wind turbine (12.4 m/s), the surge and pitch motion amplitudes of the platform were the largest, and the heave motion was greatly affected by the hydrodynamic load. When the wave height increases (LC2–LC4), the heave amplitude increases accordingly. Figure 12d shows that under the three operational conditions, the statistical values of the mooring force of ML2 simulated by F2A and AQWA were slightly different. Similarly, when the wind velocity was the rated wind velocity of 12.4 m/s (LC3), the mooring force of ML2 was the largest. Figure 12e shows that as the sea state becomes worse (LC2–LC4), the damping force and the produced power increase sharply. By comparing, aerodynamic loads were confirmed to have significant effects on surge and pitch motion and mooring forces, while hydrodynamic loads had significant effects on the vertical responses (heave, damping force, and produced power).

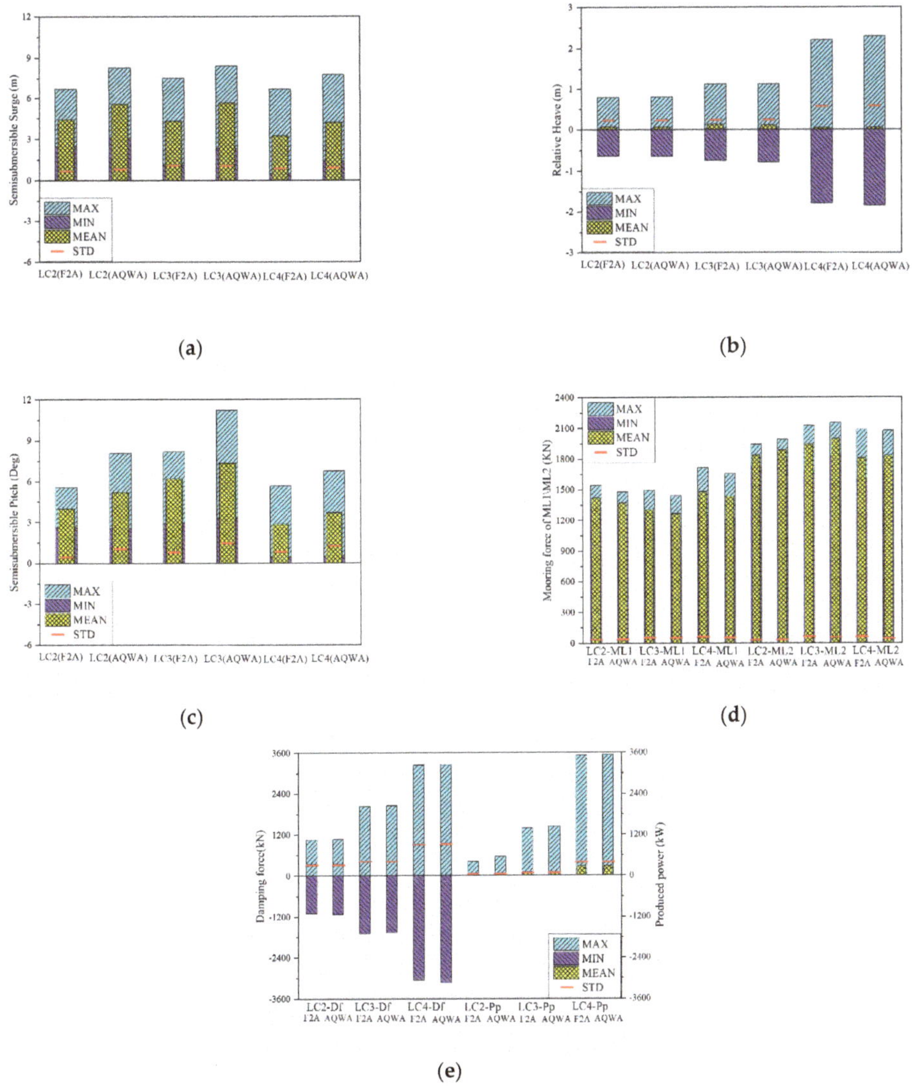

Figure 12. Comparison of the statistical results of in LC2, LC3, LC4: (**a**) Surge Motion; (**b**) Relative Heave Motion; (**c**) Pitch Motion; (**d**) Mooring Force of ML1 and ML2; (**e**) Damping Force (Df) and Produced Power (Pp).

4.4. Spectrum Analysis

In this section, the power spectral density (PSD) of platform motion response, mooring line force, damping force, and produced wave power were presented and compared for F2A simulation and AQWA simulation.

4.4.1. Motion Spectrum

Figure 13a,d,g display the PSD of the platform surge under the three load cases. The energy was mainly concentrated in the natural frequency of the surge and the wave frequency. For the mild sea state (LC2) in Figure 13a, there is a reduction for surge motion at the surge resonance peak in F2A simulation compared with AQWA simulation due to the aerodynamic damping effect. However, this reduction is not significant for worse sea state (LC3 and LC4) for surge motion in Figure 13d,g. For the relative heave motion, the responses are dominated by the frequency from 0.3 to 0.9 rad/s which is related to the wave peak frequency. Therefore, both F2A simulation and AQWA simulation are very similar to each other (Figure 13b,e,h). Figure 13c,f,i are the PSD of the platform pitch motion. Observing the PSD curves of surge and pitch, the reason for the difference in the low frequency area around pitch resonance peak was the influence of aerodynamic damping considered in F2A simulation. It was obvious that when the wind velocity was small, aerodynamic damping obviously reduced the resonance response in the low-frequency region but had little effect on the wave frequency range. This phenomenon conforms well to the known characteristics of the general damping effect. However, when the wind velocity gradually increased (LC2–LC4), the influence of air damping gradually decreased. This was because the aerodynamic thrust acting on the rotor increases rapidly with increasing wind velocity, offsetting the increase in the aerodynamic damping effect caused by surge. Among these effects, aerodynamic damping and wind velocity had a first-order relationship, and aerodynamic thrust and wind velocity had a quadratic relationship. Therefore, at low wind velocities, aerodynamic damping had a greater influence, and at high wind velocities, aerodynamic thrust had a more obvious influence. It can be found that aerodynamic damping had a greater impact on pitch than surge and a stronger reduction in pitch motion.

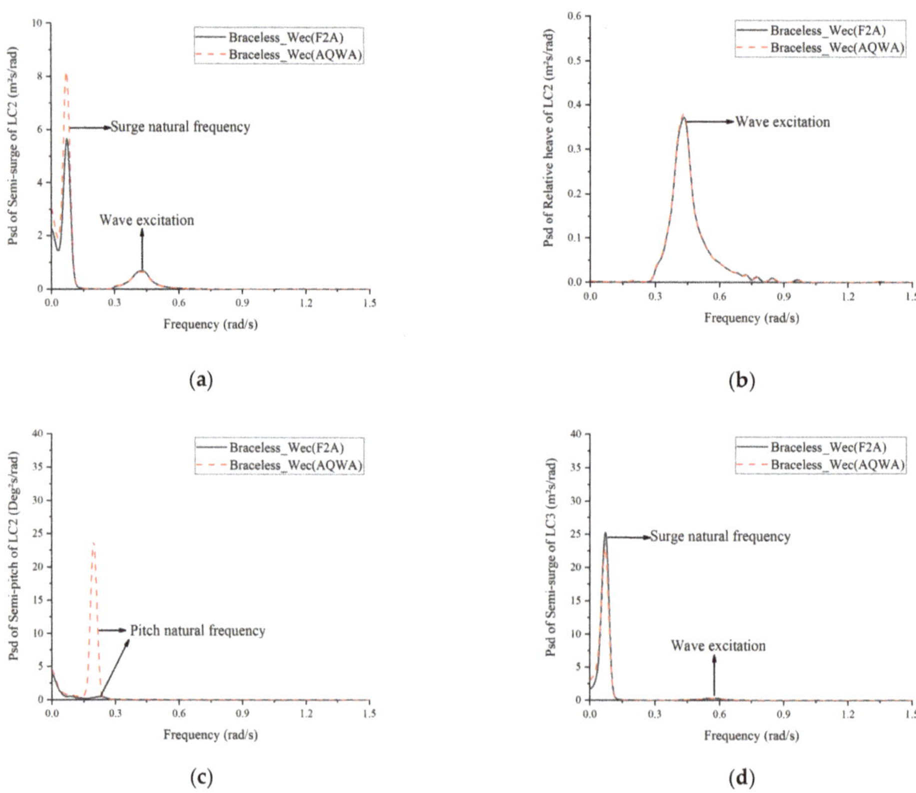

Figure 13. *Cont.*

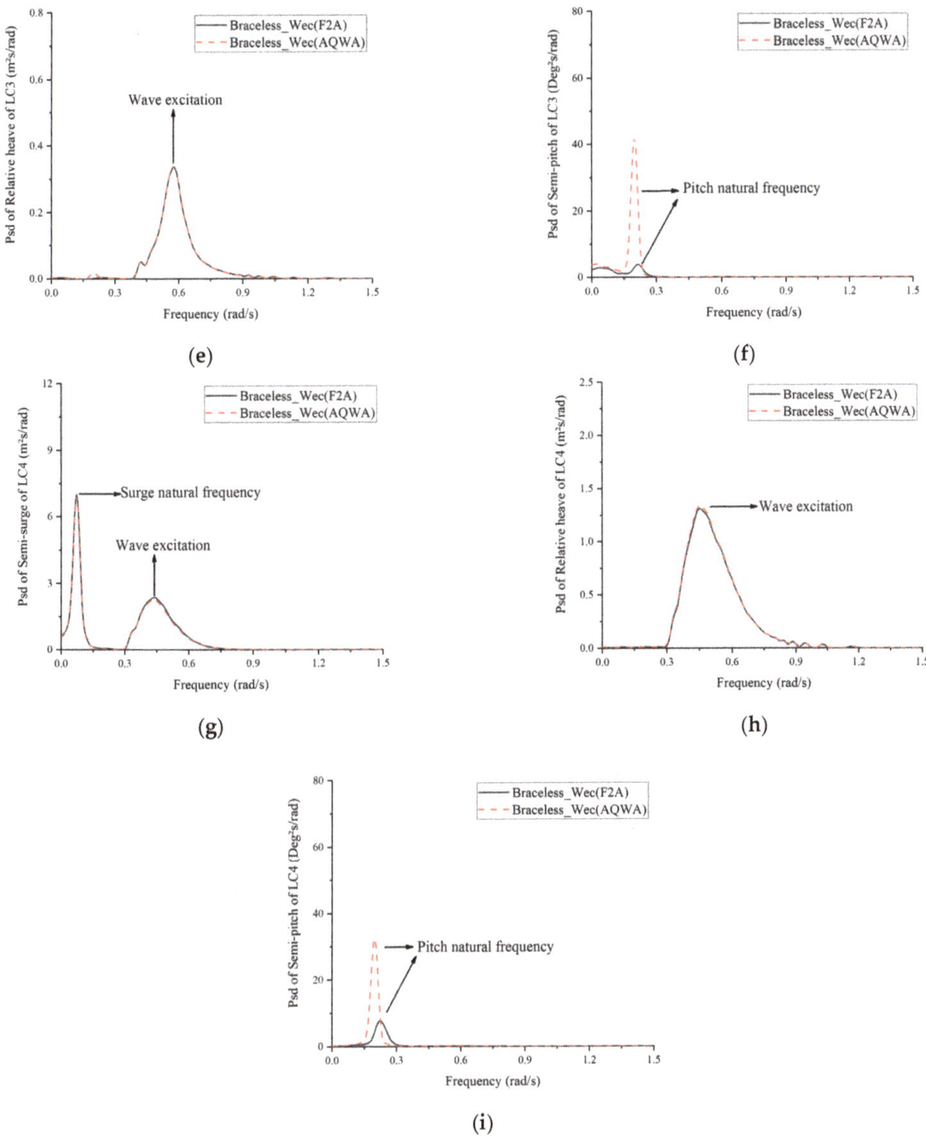

Figure 13. Comparisons of the PSD of the motion for LC2, LC3, and LC4: (**a**) Surge Motion of LC2; (**b**) Relative Heave Motion of LC2; (**c**) Pitch Motion of LC2; (**d**) Surge Motion of LC3; (**e**) Relative Heave Motion of LC3; (**f**) Pitch of LC3; (**g**) Surge Motion of LC4; (**h**) Relative Heave Motion of LC4; (**i**) Pitch Motion of LC4.

4.4.2. Mooring Tension Spectrum

The mooring line responses in the frequency domain in the head sea under different load cases from F2A simulation and AQWA simulation are shown in Figure 14. For all the load cases, the most significant contribution to the ML 1 and ML2 tension comes from the low-frequency region (surge mode response). The contributions from pitch mode response are also identified. For worse sea state (LC4), the contributions to the ML1 and ML2 tension from the wave frequency range from 0.3 rad/s to 0.9 rad/s are comparable to the contribution from surge mode response. Similar to the PSD of the motion response, in the F2A simulation, aerodynamic damping had a weakening effect on the response at surge and pitch natural frequencies. With the wind velocity increases (LC2–LC4), the aerodynamic damping effect gradually decreased.

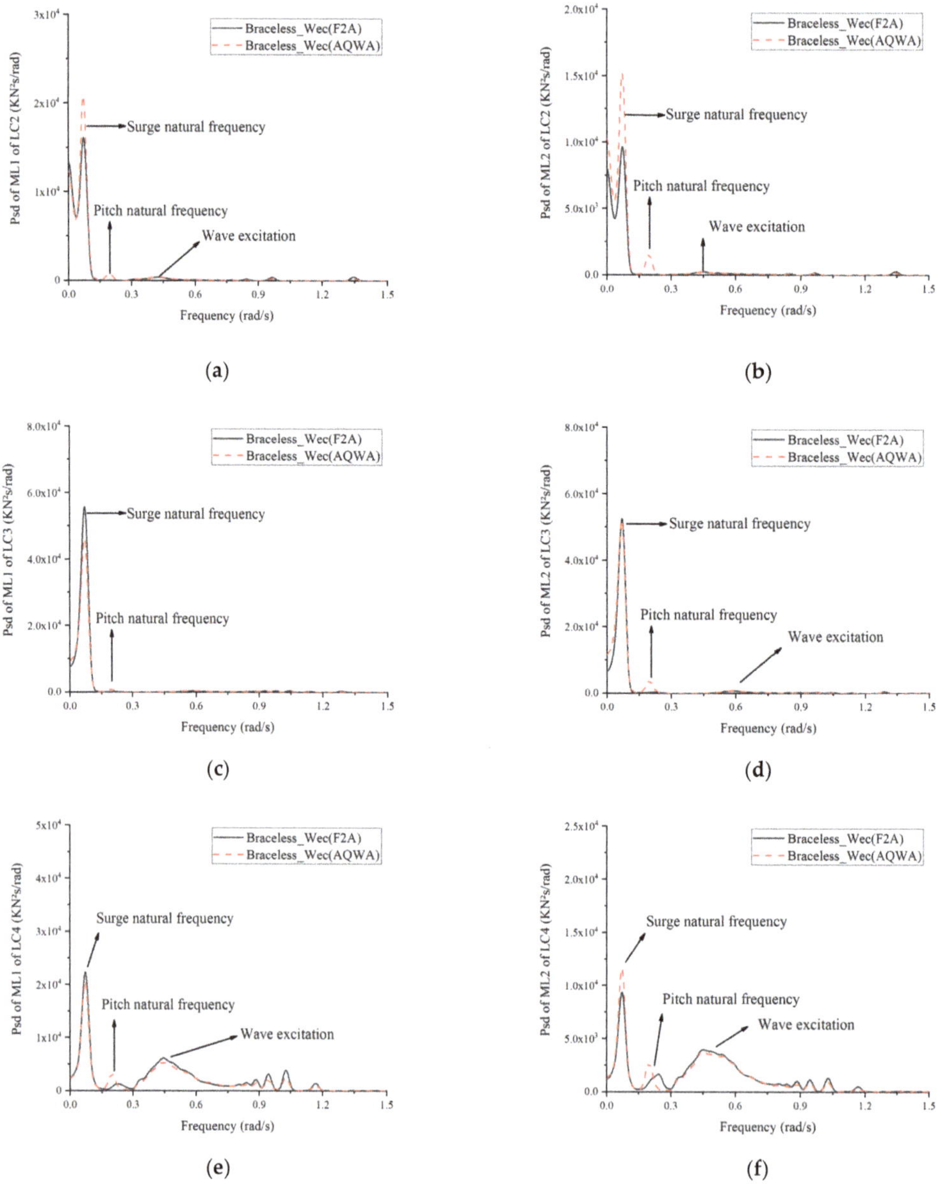

Figure 14. Comparisons of the PSD of the mooring line force of: (**a**) ML1 of LC2; (**b**) ML2 of LC2; (**c**) ML1 of LC3; (**d**) ML2 of LC3; (**e**) ML1 of LC4; (**f**) ML2 of LC4.

4.4.3. Damping Force and Produced Wave Power Spectrum

Figure 15 shows the damping force in the vertical direction and the produced wave energy power. The simulation results of the two simulation tools are slightly different. The energy was mainly concentrated in the wave frequency. Figure 15b,d,f are the PSDs of the produced power. The produced wave power frequencies are located in the double wave frequency and low frequency regions. There was no significant effect from aerodynamic damping.

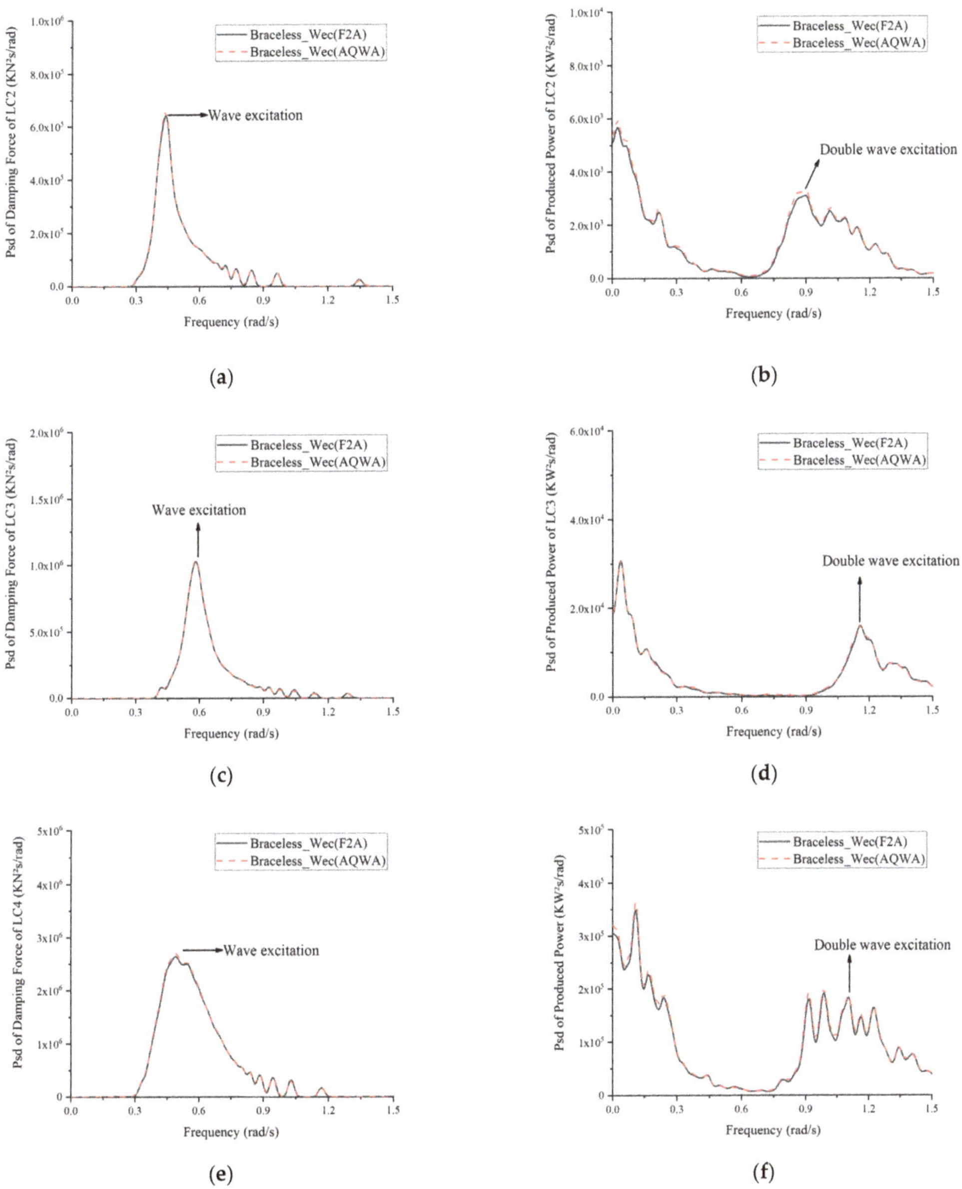

Figure 15. Comparisons of the PSD of: (**a**) Damping Force of LC2; (**b**) Produced Power of LC2; (**c**) Damping Force of LC3; (**d**) Produced Power of LC3; (**e**) Damping Force of LC4; (**f**) Produced Power of LC4.

4.5. Dynamic Responses in Extreme Conditions

Under extreme condition (LC5), the WEC and the semisubmersible platform were locked to each other, and there was no relative motion. The PTO system and the wind turbine were parked. As shown in Figure 16a, the WEC and the semisubmersible platform had the same heave motion and were in a locked state. Figure 16b,c show the PSDs of the platform surge and heave. The PSD of the platform surge was mainly dominated by the wave frequency and the natural frequency of the surge. The PSD of the platform heave was mainly dominated by the wave frequency. From the PSD of the pitch (Figure 16e), it can be seen that the energy was mainly concentrated on the wave frequency and the natural frequency of the pitch, but the resonance effect on the pitch simulated by AQWA was much smaller than the result of the F2A simulation. When using F2A simulation, the upper wind turbine was parked, and the pitch angle was set to 90°. Figure 16e,f show the mooring forces of ML1 and ML2. The energy was mainly dominated in the wave frequency range.

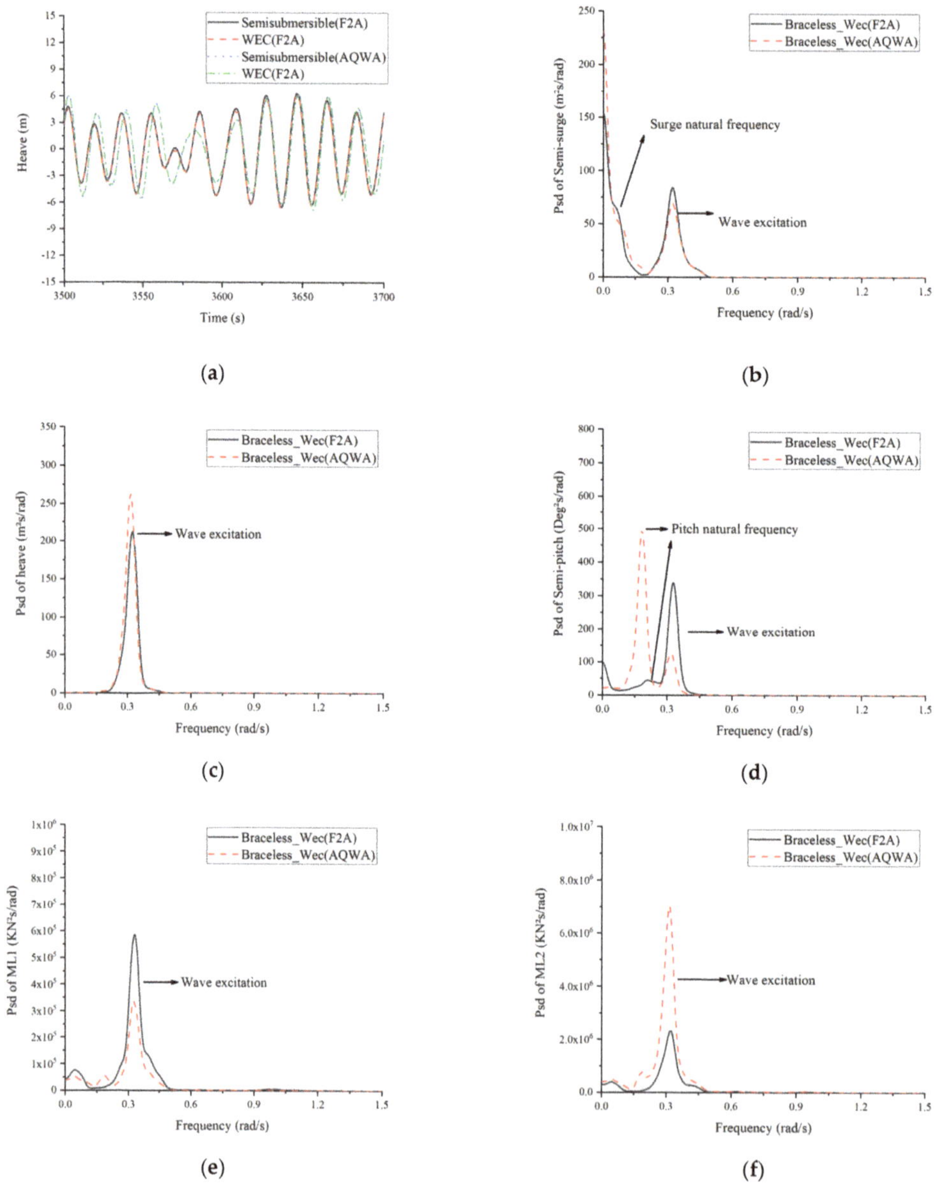

Figure 16. Motion and mooring force response of the platform under LC5: (**a**) Heave motion time series; (**b**) PSD of platform surge; (**c**) PSD of platform heave; (**d**) PSD of platform pitch; (**e**) PSD of ML1; (**f**) PSD of ML2.

5. Conclusions

In this study, through a recently proposed fully coupled analysis framework (F2A), a fully coupled analysis of a combined system consisting of a semisubmersible platform (braceless) and a heaving-type WEC was carried out. A numerical model of the combined structure that was capable of simulating its motion and dynamic responses under different typical operational and extreme conditions was developed and used. Different types of analysis and comparison were performed to examine the fully coupled responses of the combined structure. The primary results are summarized as follows:

(1) Under regular wave conditions, regardless of wind conditions, the simulation results of F2A and AQWA were basically similar in time domain motion, mooring force, and generated wave power. Since F2A integrated the AQWA hydrodynamic module, it shows that F2A has good consistency with AQWA.

(2) Under irregular wave and turbulent wind conditions, F2A and AQWA had significant differences in time domain motion and mooring force. It shows that F2A effectively reduces the dynamic response amplitude because of the aerodynamic damping effect in F2A; as the wave and wind velocity increased, the amplitude of time-domain motion and mooring force (ML2) are significantly affected by aerodynamic loads, and the dynamic response amplitudes are the largest at rated wind velocity. The trends of relative heave motion, damping force, and produced wave power were similar.

(3) Through the PSD analysis of the dynamic response, it was more obvious that aerodynamic damping used F2A effectively inhibited the resonance at the low frequency range, which was the largest difference compared to the prediction calculated with the simulation tool of AQWA. The relationship between aerodynamic damping and wind velocity was linear, and the relationship between aerodynamic thrust and wind velocity was quadratic. Therefore, when the wind'velocity was small, the influence of aerodynamic damping was significant, and especially the resonance of surge and pitch was significantly reduced. When the wind velocity increased, the aerodynamic thrust had a significant effect. The inhibition effect of aerodynamic damping was mainly concentrated in the low-frequency region and had no obvious effect on the resonance caused by the wave frequency. The resonance was mainly excited near the wave frequency; there was a slight difference between F2A and AQWA simulation results close to resonance.

(4) Under extreme conditions, the WEC was locked on a semisubmersible platform as a way of survival to withstand the impact of large wave force. From the PSD analysis of time domain motion and mooring force, the energy was mainly concentrated near the wave frequency, and the wave influence was obvious.

Author Contributions: Conceptualization, W.S., L.Z., C.M. and X.L.; methodology, W.S., C.M. and X.L.; investigation, J.L., L.Z., W.S. and C.M.; writing—first draft preparation, J.L., L.Z., W.S. and C.M.; review and editing, W.S., C.M. and X.L.; supervision, W.S. and C.M.; project administration, W.S.; funding acquisition, W.S and X.L. All authors have read and agreed to the published version of the manuscript.

Funding: This research was funded by the National Natural Science Foundation of China (Grant No. 52071058, 51939002). This work is also partially supported by LiaoNing Revitalization Talents Program (XLYC1807208) and special funds for promoting high quality development from department of natural resources of Guangdong province (GDNRC [2020]016).

Institutional Review Board Statement: Not applicable.

Informed Consent Statement: Not applicable.

Data Availability Statement: Not applicable.

Acknowledgments: The authors would like to thank the Yang Yang from Ningbo University for his valuable discussion of simulation in F2A software.

Conflicts of Interest: The authors declare no conflict of interest.

References

1. World Wind Energy Association. Available online: https://wwindea.org/worldwide-wind-capacity-reaches-744-gigawatts (accessed on 24 March 2021).
2. Jonkman, J.; Butterfield, S.; Musial, W.; Scott, G. *Definition of a 5-MW Reference Wind Turbine for Offshore System Development*; National Renewable Energy Laboratory (NREL): Golden, CO, USA, 2009.
3. Bak, C.; Zahle, F.; Bitsche, R.; Kim, T.; Yde, A.; Henriksen, L.C.; Hansen, M.H.; Blasques, J.P.A.A.; Gaunaa, M.; Natarajan, A. *The DTU 10-MW Reference Wind Turbine*; Danish Wind Power Research: Lyngby, Denmark, 2013.
4. Roddier, D.; Cermelli, C.; Aubault, A.; Weinstein, A. WindFloat: A floating foundation for offshore wind turbines. *J. Renew. Sustain. Energy* **2010**, *2*, 033104. [CrossRef]
5. Huijs, F.; de Bruijn, R.; Savenije, F. Concept design verification of a semi-submersible floating wind turbine using coupled simulations. *Energy Procedia* **2014**, *53*, 2–12. [CrossRef]
6. Lefranc, M.; Torud, A. Three wind turbines on one floating unit, feasibility, design and cost. In Proceedings of the Offshore Technology Conference, Houston, TX, USA, 2–5 May 2011.
7. Le Boulluec, M.; Ohana, J.; Martin, A.; Houmard, A. Tank testing of a new concept of floating offshore wind turbine. In Proceedings of the ASME 2013 32nd International Conference on Ocean, Offshore and Arctic Engineering, Nantes, France, 9–14 June 2013.
8. Luan, C.; Gao, Z.; Moan, T. Design and analysis of a braceless steel 5-MW semi-submersible wind turbine. In Proceedings of the ASME 2016 35st International Conference on Ocean, Offshore and Arctic Engineering, Busan, Korea, 19–24 June 2016.
9. Karimirad, M.; Michailides, C. V-shaped semisubmersible offshore wind turbine: An alternative concept for offshore wind technology. *Renew. Energy* **2015**, *83*, 126–143. [CrossRef]
10. Robertson, A.; Jonkman, J.; Masciola, M.; Song, H.; Goupee, A.; Coulling, A.; Luan, C. *Definition of the Semisubmersible Floating System for Phase II of OC4*; National Renewable Energy Laboratory (NREL): Golden, CO, USA, 2014.
11. Falcão, A.F.D.O. Wave energy utilization: A review of the technologies. *Renew. Sustain. Energy Rev.* **2010**, *14*, 899–918. [CrossRef]
12. Aboutalebi, P.M.; Zoughi, F.; Garrido, I.; Garrido, A.J. Performance analysis on the use of oscillating water column in barge-based floating offshore wind turbines. *Mathematics* **2021**, *9*, 475. [CrossRef]
13. Aboutalebi, P.M.; Zoughi, F.; Martija, I.; Garrido, I.; Garrido, A.J. Switching control strategy for oscillating water columns based on response amplitude operators for floating offshore wind turbines stabilization. *Appl. Sci.* **2021**, *11*, 5249. [CrossRef]
14. Michailides, C.; Gao, Z.; Moan, T. Response analysis of the combined wind/wave energy concept sfc in harsh environmental conditions. In Proceedings of the RENEW2014 1st International Conference on Renewable Energies Offshore, Lisbon, Portugal, 24–26 November 2014.
15. Ren, N.; Ma, Z.; Shan, B.; Ning, D.; Ou, J. Experimental and numerical study of dynamic responses of a new combined TLP type floating wind turbine and a wave energy converter under operational conditions. *Renew. Energy* **2020**, *151*, 966–974. [CrossRef]
16. Bachynski, E.E.; Moan, T. Point absorber design for a combined wind and wave energy converter on a tension-leg support structure. In Proceedings of the ASME 32nd International Conference on Ocean, Offshore and Arctic Engineering, Nantes, France, 9–14 June 2013.
17. Russo, S.; Contestabile, P.; Bardazzi, A.; Leone, E.; Iglesias, G.; Tomasicchio, G.R.; Vicinanza, D. Dynamic loads and response of a spar buoy wind turbine with pitch-controlled rotating blades: An experimental study. *Energies* **2021**, *14*, 3598. [CrossRef]
18. Tomasicchio, G.R.; Vicinanza, D.; Belloli, M.; Lugli, C.; Latham, J.P.; Iglesias Rodriguez, J.G.; Jensen, B.; Vire, A.; Monbaliu, J.; Taruffi, F.; et al. Physical model tests on spar buoy for offshore floating wind energy conversion. *Ital. J. Eng. Geol. Environ.* **2020**, *1*, 129–143.
19. Xu, X.; Day, S. Experimental investigation on dynamic responses of a spar-type offshore floating wind turbine and its mooring system behaviour. *Ocean Eng.* **2021**, *236*, 109488. [CrossRef]
20. Hu, J.; Zhou, B.; Vogel, C.; Liu, P.; Willden, R.; Sun, K.; Zang, J.; Geng, J.; Jin, P.; Cui, L.; et al. Optimal design and performance analysis of a hybrid system combing a floating wind platform and wave energy converters. *Appl. Energy* **2020**, *269*, 114998. [CrossRef]
21. Sarmiento, J.; Iturrioz, A.; Ayllón, V.; Guanche, R.; Losada, I.J. Experimental modelling of a multi-use floating platform for wave and wind energy harvesting. *Ocean Eng.* **2019**, *173*, 761–773. [CrossRef]
22. Marina Platform. Available online: https://www.msp-platform.eu/projects/marina-platform (accessed on 3 April 2020).
23. Peiffer, A.; Roddier, D.; Aubault, A. Design of a point absorber inside the WindFloat structure. In Proceedings of the ASME 2011 30th International Conference on Ocean, Offshore and Arctic Engineering, Rotterdam, The Netherlands, 19–24 June 2011.
24. Aubault, A.; Alves, M.; Sarmento, A.N.; Roddier, D.; Peiffer, A. Modeling of an oscillating water column on the floating foundation WindFloat. In Proceedings of the ASME 2011 30th International Conference on Ocean, Offshore and Arctic Engineering, Rotterdam, The Netherlands, 19–24 June 2011.
25. Muliawan, M.J.; Karimirad, M.; Moan, T. Dynamic response and power performance of a combined Spar-type floating wind turbine and coaxial floating wave energy converter. *Renew. Energy* **2013**, *50*, 47–57. [CrossRef]
26. Muliawan, M.J.; Karimirad, M.; Gao, Z.; Moan, T. Extreme responses of a combined spar-type floating wind turbine and floating wave energy converter (STC) system with survival modes. *Ocean Eng.* **2013**, *65*, 71–82. [CrossRef]
27. Wan, L.; Gao, Z.; Moan, T. Experimental and numerical study of hydrodynamic responses of a combined wind and wave energy converter concept in survival modes. *Coast. Eng.* **2015**, *104*, 151–169. [CrossRef]

28. Wan, L.; Gao, Z.; Moan, T.; Lugni, C. Experimental and numerical comparisons of a combined wind and wave energy converter concept under operational conditions. *Renew. Energy* **2016**, *93*, 87–100. [CrossRef]
29. Wan, L.; Gao, Z.; Moan, T.; Lugni, C. Comparative experimental study of the survivability of a combined wind and wave energy converter in two testing facilities. *Ocean Eng.* **2016**, *111*, 82–94. [CrossRef]
30. Michailides, C.; Gao, Z.; Moan, T. Experimental study of the functionality of a semisubmersible wind turbine combined with flap-type Wave Energy Converters. *Renew. Energy* **2016**, *93*, 675–690. [CrossRef]
31. Michailides, C.; Gao, Z.; Moan, T. Experimental and numerical study of the response of the offshore combined wind/wave energy concept SFC in extreme environmental conditions. *Mar. Struct.* **2016**, *50*, 35–54. [CrossRef]
32. Michailides, C.; Luan, C.; Gao, Z.; Moan, T. Effect of flap type wave energy converters on the response of a semi-submersible wind turbine in operational conditions. In Proceedings of the ASME 2014 33rd International Conference on Ocean, Offshore and Arctic Engineering, San Francisco, CA, USA, 8–13 June 2014.
33. Ren, N.; Ma, Z.; Fan, T.; Zhai, G.; Ou, J. Experimental and numerical study of hydrodynamic responses of a new combined monopile wind turbine and a heave-type wave energy converter under typical operational conditions. *Ocean Eng.* **2018**, *159*, 1–8. [CrossRef]
34. Wang, Y.; Zhang, L.; Michailides, C.; Wan, L.; Shi, W. Hydrodynamic response of a combined wind-wave marine energy structure. *J. Mar. Sci. Eng.* **2020**, *8*, 253. [CrossRef]
35. Le Cunff, C.; Heurtier, J.; Piriou, L.; Berhault, C.; Perdrizet, T.; Teixeira, D.; Ferrer, G.; Gilloteaux, J.C. Fully coupled floating wind turbine simulator based on nonlinear finite element method: Part I—Methodology. In Proceedings of the ASME 2013 32nd International Conference on Ocean, Offshore and Arctic Engineering, Nantes, France, 9–14 June 2013.
36. Perdrizet, T.; Gilloteaux, J.; Teixeira, D.; Ferrer, G.; Piriou, L.; Cadiou, D.; Heurtier, J.M.; Le Cunff, C. Fully coupled floating wind turbine simulator based on nonlinear finite element method: Part II—Validation Results. In Proceedings of the ASME 2013 32nd International Conference on Ocean, Offshore and Arctic Engineering, Nantes, France, 9–14 June 2013.
37. Chen, J.; Hu, Z.; Liu, G.; Wan, D. Coupled aero-hydro-servo-elastic methods for floating wind turbines. *Renew. Energy* **2019**, *130*, 139–153. [CrossRef]
38. Jonkman, J.M. Dynamics of offshore floating wind turbines-model development and verification. *Wind Energy* **2009**, *12*, 459–492. [CrossRef]
39. Jonkman, J.M. *Dynamics Modeling and Loads Analysis of an Offshore Floating Wind Turbine*; National Renewable Energy Laboratory (NREL): Golden, CO, USA, 2007.
40. Kvittem, M.I.; Bachynski, E.E.; Moan, T. Effects of hydrodynamic modelling in fully coupled simulations of a semi-submersible wind turbine. *Energy Procedia* **2012**, *24*, 351–362. [CrossRef]
41. Shim, S. Coupled Dynamic Analysis of Floating Offshore Wind Farms. Ph.D. Thesis, Texas A & M University, College Station, TX, USA, 2010.
42. Yang, Y.; Bashir, M.; Michailides, C.; Li, C.; Wang, J. Development and application of an aero-hydro-servo-elastic coupling framework for analysis of floating offshore wind turbines. *Renew. Energy* **2020**, *161*, 606–625. [CrossRef]
43. *AQWA Manual Release 19.0*; ANSYS Inc.: Canonsburg, PA, USA, 2018.
44. Moriarty, P.J.; Hansen, A.C. *AeroDyn Theory Manual*; National Renewable Energy Laboratory (NREL): Golden, CO, USA, 2005.
45. Chen, J.; Hu, Z. Experimental investigation of aerodynamic effect-induced dynamic characteristics of an OC4 semi-submersible floating wind turbine. *Proc. Inst. Mech. Eng. Part M J. Eng. Marit. Environ.* **2018**, *231*, 19–36. [CrossRef]
46. Karimirad, M.; Moan, T. Effect of aerodynamic and hydrodynamic damping on dynamic response of a spar type floating wind turbine. *Development* **2010**, *2*, 3.
47. Faltinsen, O. *Sea Loads on Ships and Offshore Structures*; Cambridge University Press: Cambridge, UK, 1993.
48. Li, Y.C.; Teng, B. *The Effect of Waves on Marine Buildings*, 3rd ed.; Ocean Publication: Beijing, China, 2015; pp. 270–273.
49. Morison, J.R.; Johnson, J.W.; Schaaf, S.A. The force exerted by surface waves on piles. *Pet. Trans. Aime* **1950**, *2*, 149–154. [CrossRef]
50. Wang, Y.; Shi, W.; Zhang, L.; Michailides, C.; Zhou, L. Hydrodynamic analysis of a floating hybrid renewable energy system. In Proceedings of the 30th International Society of Offshore and Polar Engineers, Shanghai, China, 14–19 June 2020.
51. Zhao, Y.; Yang, J.; He, Y.; Gu, M. Coupled dynamic response analysis of a multi-column tension-leg-type floating wind turbine. *China Ocean. Eng.* **2016**, *30*, 505–520. [CrossRef]

Journal of
*Marine Science
and Engineering*

MDPI

Article

3D Numerical Study of the Impact of Macro-Roughnesses on a Tidal Turbine, on Its Performance and Hydrodynamic Wake

Ilan Robin *, Anne-Claire Bennis and Jean-Claude Dauvin

Morphodynamique Continentale et Côtière (UMR CNRS 6143), Université de Caen Normandie, Rouen Normandie, 14000 Caen, France; anne-claire.bennis@unicaen.fr (A.-C.B.); jean-claude.dauvin@unicaen.fr (J.-C.D.)
* Correspondence: ilan.robin@unicaen.fr

Abstract: Biofouling is an important factor to consider when calculating the energetic efficiency of tidal farms. Despite the fact that biofouling effects have been widely investigated in the past for naval applications, very few studies concern tidal turbines. This paper proposes a numerical approach to assess the impact of biofouling on tidal turbines, which is efficient for testing many configurations. Two turbulence models are tested (RANS k-ω SST and LES Smagorinsky) for the motionless blade case to validate them. Then we chose to use the Smagorinsky model for the case of a complete tidal turbine rotor with realistically fouled blades. The pressure coefficient is strongly affected by the barnacle in the motionless blade case and the power coefficient is slightly degraded in the complete rotor case. Motionless blade cases do not represent the real biofouling behaviour for two reasons. First, sessile species settle in the down flow part of the chord where their impact is less important. Then, the surrounding turbulence provoked by the blades rotation in the rotor case reduces the impact of biofouling. In the wake, biofouling generates small vortexes that propagate into the larger ones, causing them to spread their energy.

Keywords: marine renewable energy; tidal energy; fluid–structure interaction; biofouling; turbulence; numerical modelling

Citation: Robin, I.; Bennis, A.-C.; Dauvin, J.-C. 3D Numerical Study of the Impact of Macro-Roughnesses on a Tidal Turbine, on Its Performance and Hydrodynamic Wake. *J. Mar. Sci. Eng.* **2021**, *9*, 1288. https://doi.org/10.3390/jmse9111288

Academic Editor: Eugen Rusu

Received: 14 October 2021
Accepted: 9 November 2021
Published: 18 November 2021

Publisher's Note: MDPI stays neutral with regard to jurisdictional claims in published maps and institutional affiliations.

Copyright: © 2021 by the authors. Licensee MDPI, Basel, Switzerland. This article is an open access article distributed under the terms and conditions of the Creative Commons Attribution (CC BY) license (https://creativecommons.org/licenses/by/4.0/).

1. Introduction

When a surface is submerged in natural water, it is submitted to colonisation by numerous species. The accumulations of biological organisms on this surface is called biofouling. The first step of this phenomenon is the colonisation by micro-organisms like bacteria or micro-algae into a thin layer a few millimetres thick (biofilm) [1]. This thin film serves as a base for larger sessile species such as barnacles and mussels [2], which often organise in habitats allowing the arrival of mobile organisms (shrimps, crabs, ...). The speed, the kind and the arrangement of the colonisation depend on many factors [3]: depth, salinity, kind of surface, region, temperature, etc. Larvae are also able to select the substrate suitable for their development depending on the water streaming and chemical properties. Moreover, specific conditions can lead to particularities among the species. For example, in the Alderney Race located between France and United Kingdom, where the tidal stream is extremely energetic, the species evolving in such conditions are mainly smaller, with more developed fixing organs and a smoother surface [4]. Some studies show that biofouling also occurs on moving solids like ship propellers or tidal turbines. The biofouling is the origin of the artificial reef effect created by off-shore artificial constructions such as the wind turbine farms described in [5]. However, despite this positive effect, biofouling is a real challenge for the marine renewable implantation in sea water. Ref. [6] shows its negative impact on boat hulls. The experimental study shows a higher resistance on heterogeneous rough parts of the hull, which is very dependent on the position of the roughnesses [7]. More generally, studies of the effects of small roughnesses on the flow along a flat plate are well-understood in both air [8] and water [9]. Small structures promote the boundary layer transition by making the boundary layer transition move upstream and also increase the rate of turbulence. With the development of renewable marine energies and some special cases in aviation, studies are aiming towards understanding the effects of various roughnesses on wing profiles [10]. Ref. [11] looks into the effect of ice accretion located on the leading edge and finds that the aerodynamic performance of the profile is significantly deteriorated by ice. Ref. [12] experimentally studied the effects of numerous cones distributed homogeneously over the surface of a blade in a wind tunnel. The study is carried out in stationary flow with solid cones ranging from 0.0035 c (chord) to 0.0285 c in height. Results show a strong increase in the drag. Ref. [13] investigates the aerodynamic impact of an isolated barnacle-shaped excrescence on a blade in a wind tunnel. The barnacle is 0.02 c high and is located at 60% of the chord. Three angles for the flow incidence are studied: 5°, 10° and 15° under two different operating modes: a stationary or oscillating blade. On the motionless blade, the barnacle has no impact on lift but considerably increases drag. The pressure coefficient field is very different between the 5° and 15° cases in the motionless case but seems very similar once the blade is set in motion.

Nevertheless, the marine environment at tidal sites is not easy to study in realistic conditions. So, the main parts of these studies are conducted in ideal conditions, easily replicable experimentally. In fact, marine currents are really strong and numerical approach is more suitable to predict biofouling in more realistic conditions. Most papers aim at developing models for small roughnesses. However with the development of CFD, more and more 2D studies are emerging: Ref. [14] investigates the effects of biofouling on a NACA0018 profile using numerous turbulence models (Reynolds Averaged Navier–Stokes (RANS): k-ϵ, k-ω SST (Shear Stress Transport) and Large Eddy Simulation (LES): Smagorinsky). RANS k-ω-SST and LES-Smagorinsky gave close results for the blade performances but RANS models are less efficient than LES models to compute the vortexes in the wake. The results show that if the biofouling species are placed on the first half of the blade chord, their effects are maximised. Conversely, when biofouling is in the second half, the effects are reduced.This phenomenon is explained by the isolated roughnesses that create an early detachment of the boundary layer. Nevertheless, biofouling generates three-dimensional effects in the flow [15,16]. A 2D study is therefore not sufficient to fully understand such effects.

To our knowledge, the only work relating to the impact of biofouling on the performance of tidal turbines applied to a complete turbine is [17], where biofouling is a surface finish applied with a wall function on the whole turbine. In this paper, we propose to investigate the impact of a realistic fouling on an entire turbine. After a short introduction, Section 2 explains the methodology and methods used to carry out the numerical tests. Section 3 is the validation of the model and its comparison with the experimental data of [13]. Section 4 shows our results that are discussed. Section 4 draws the conclusion.

2. Materials and Methods

The numerical simulations are divided into two parts. The first part is a 3D model of a single blade with only one barnacle. This aims to validate the numerical model using the experimental results. The second part relates to the modelling of a complete biofouled rotor. That allows us to study the impact of a realistic implantation of sessile species on the performance and the wake of the tidal turbine. Unlike the work of [17], biofouling is explicitly represented in the mesh by integrating the barnacles into the 3D structure and not by using a roughness model applied to the turbine.

2.1. Experimental Setups

Experiments were carried out by [13,18] and their data were used for comparison with the numerical results.

The results of the **motionless blade experiment** chosen for the validation of the numerical model comes from [13]. Their experimental method is described hereafter. The tests were carried out in the Handley–Page wind tunnel with dimensions of 1.61 m high ×

2.13 m wide × 2.74 m long. The air flow velocity was 45 m·s^{-1} with 2.5% turbulence. The blade is made from a NACA 63-619 foil of 0.55 m chord. It passes completely through the height of the wind tunnel. The barnacle is represented by a solid cone with a radius of 20 mm at its base and 10 mm at its top and a height of 11 mm, dimensions nearby to the *Balanus crenatus* found in the Alderney Race [4]. A total of 25 pressure orifices are positioned on and around the barnacle in order to follow the evolution of the pressure field. Three angles of attack are studied: 5°, 10° and 15°. Sensor HDI series gauge sensors measured the dynamic stall while pressure transducers were used to sample the pressure evolution along the blade. The experimental setup also allowed blade oscillation, but unsteady tests were not used here.

The **full rotor simulation** is built according to [18] and uses the IFREMER-LOMC[1] horizontal axis turbine. The water velocity was fixed to 0.8 m·s^{-1} with 3% of turbulence intensity. The rotor rotation was forced by a motor to fix the Tip Speed Ratio (TSR) to 4. TSR is defined as follows :

$$\lambda = \frac{|\Omega_x| R}{U_\infty}, \quad (1)$$

where Ω_x is the angular velocity, R is the radius of the tidal turbine and U_∞ is the inlet flow velocity. The rotor characteristics are presented in Table 1.

Table 1. General characteristics of the IFREMER-LOMC turbine

Turbine Profile	IFREMER-LOMC NACA 63418	Unit
Rotor Radius (R)	350	mm
Hub Radius	46	mm
Pitch	0	degrees
TSR	4	-

Torque sensors were used to measure the power and drag coefficients of the entire structure, including the hub. Laser Doppler Velocimeters (LDV) were used to monitor the wake and vortexes.

2.2. Governing Equations

The numerical model is used for validation (in air) and investigation (in water) cases. For the numerical simulations, the following hypothesis are considered:

(i) Air and water are considered as viscous fluids.
(ii) Both fluids are considered incompressible. This hypothesis can be questioned in the case of air, but the validation cases in air have a Mach number of 0.14. It is generally accepted that for a flow with a Mach number below 0.3, the fluid can be considered incompressible.
(iii) Gravity is neglected.
(iv) The study is carried out in the middle of the water column, so wave and bottom effects are neglected.

The 3D Navier–Stokes equations are suitable to solve the fluid motion under the incompressible assumption (with $i = 1, 2, 3$ representing the 3 directions in a Cartesian framework):

$$\frac{\partial u_i}{\partial x_i} = 0, \quad (2)$$

$$\frac{\partial u_i}{\partial t} + \frac{\partial u_i u_j}{\partial x_j} = -\frac{1}{\rho}\frac{\partial p}{\partial x_i} + \frac{1}{\rho} f_i + \nu \frac{\partial^2 u_i}{\partial x_j \partial x_j}, \quad (3)$$

where u_i is the fluid velocity in the i-direction, t is the time, ρ is the fluid density (kg·m^{-3}), p is the pressure, f_i represents the volumetric forces in the i-direction, and ν is the kinematic viscosity.

The RANS k-ω SST and LES Smagorinsky turbulence models give close results for the calculation of blade forces. However, the RANS approaches, including the k-ω SST model, known to be cheaper, do not give accurate results for the calculation of the wake. On the other hand, LES models like Smagorinsky offer better results for the generation and development of vortices in the wake but the computational cost is high. Thus, the both turbulence models are compared in this work.

2.2.1. Reynolds-Averaged Navier-Stokes Turbulence Model

In RANS models, the velocity u_i is decomposed into an averaged part ($\overline{u_i}$) and a fluctuating part (u_i') such as:

$$u_i = \overline{u_i} + u_i. \tag{4}$$

The low-frequency component is obtained by applying the Reynolds average ($\bar{\cdot}$) to the instantaneous velocity. This average is also applied to the pressure leading to the same decomposition.

In this framework, the Navier–Stokes equations are :

$$\frac{\partial \overline{u_i}}{\partial x_i} = 0, \tag{5}$$

$$\frac{\partial \overline{u_i}}{\partial t} + \frac{\partial \overline{u_i}\,\overline{u_j}}{\partial x_j} = -\frac{1}{\rho}\frac{\partial \overline{p}}{\partial x_i} + \frac{1}{\rho}\overline{f_i} + \nu \frac{\partial^2 \overline{u_i}}{\partial x_j \partial x_j} - \frac{\partial \overline{u_i' u_j'}}{\partial x_j}, \tag{6}$$

where $\overline{p}$ is the mean pressure.

The mean value $\overline{u_i}$ is considered as varying slightly in time compared to the variation of fluctuation whereas the mean value of u_i' is zero. In order to solve Equations (5) and (6), the knowledge of the Reynolds stress tensor $\overline{u_i' u_j'}$ is necessary. After approximating this term using a turbulence viscosity depending on both turbulence kinetic energy (k) and specific dissipation rate (ω), the evolution equations for k and ω need to be solved to determine these quantities [19]. They are initialised as :

$$k = \frac{3}{2}(I|U_\infty|)^2, \tag{7}$$

where I is the turbulence intensity and U_∞ is the reference velocity (undisturbed velocity),

$$\omega = \frac{k^{0.5}}{0.009^{0.25} L}, \tag{8}$$

where L is a reference length scale equal to the chord of the profile (c) for the motionless blade simulation and to the rotor radius (R) for the full rotor simulation.

2.2.2. Large Eddy Simulation Turbulence Model

For LES, flow characteristics are separated into two parts according to the turbulent scales by applying a mathematical filter ($\tilde{\cdot}$). $\tilde{u}_i$ is composed of the large eddies whose size is greater than the size of the filter. u^* carries the smaller eddies with a size inferior to the filter size. In the Smagorinsky turbulence model, the filter size is correlated with the mesh size and $\tilde{u}_i$ is solved explicitly by solving :

$$\frac{\partial \tilde{u}_i}{\partial t} + \frac{\partial \tilde{u}_i \tilde{u}_j}{\partial x_j} = \frac{1}{\rho}\left(-\frac{\partial \tilde{p}}{\partial x_i} + \tilde{f}_i\right) + (\nu + \nu_{sgs})\frac{\partial^2 \tilde{u}_i}{\partial x_j x_j} - \frac{\partial \tau_{ij}}{\partial x_i}, \tag{9}$$

where $\tilde{p}$ and $\tilde{f}_i$ are the filtered pressure and volumetric forces, respectively. ν_{sgs} is the turbulent eddy viscosity.In the original Smagorinsky model, ν_{sgs} is computed as :

$$\nu_{sgs} = (C_k \Delta)^2 \sqrt{2 S_{ij} S_{ij}}, \tag{10}$$

where Δ is the width of the filter and S_{ij} is the resolved-scale strain rate tensor. However, the Sub Grid Scale Kinetic Energy (SGS TKE) variation of the Smagorinsky model is used here. This choice is made because of the ability of the SGS TKE to evaluate the forces and model the flows in the near walls. The classical Smagorisky model does not allow modelling of the viscous sub-layer of the boundary layer without drastically increasing the number of cells [20]. Thus, ν_{sgs} is written:

$$\nu_{sgs} = C_e \Delta k_1^{0.5}, \tag{11}$$

where C_e is a constant of the SGS Kinetic Energy model constant and k_1 is the turbulent kinetic energy computed according to:

$$\frac{\partial k_1}{\partial t} + \tilde{u}_i \frac{k_1}{\partial x_i} = c_e \Delta_x \sqrt{k_1} \tilde{S_{ij}} \tilde{S_{ij}} - C_\epsilon \frac{k_1^{\frac{3}{2}}}{\Delta_x} - \frac{1}{\partial x_j}\left[\left(\nu + \frac{c_e}{\sqrt{k_1}} \Delta \sqrt{k_1}\right) \frac{\partial k_1}{\partial x_i}\right], \tag{12}$$

2.3. Boundary Conditions

The boundary conditions are given by the following equations:

$$u_i|_{\delta\Omega_1} = U_\infty x_1, \tag{13}$$

$$p|_{\delta\Omega_2} = p_\infty, \tag{14}$$

$$\frac{\partial u_i}{x_i} n_i|_{\delta\Omega_{3,4,5,6}} = 0, \tag{15}$$

$$u_i|_{walls} = 0, \tag{16}$$

where $\delta\Omega_1$ is the inlet, $\delta\Omega_2$ is the outlet and $\delta\Omega_{3,4,5,6}$ are the four other surfaces (bottom, up, front and back). *walls* represents solid structures, the blade surfaces here. n_i is the normal vector of the surface on which it is applied. p_∞ is the undisturbed pressure.

Equation (13) is a velocity inlet condition set to a constant value (U_∞) in the flow direction. Equation (14) describes the pressure outlet value (usually chosen to avoid to over-constrain the system). A slip velocity condition is considered on the surrounding surfaces to limit the side effects Equations (15) and (16) is a no-slip condition applied on the blades. This condition is only valid in the solid surface datum.

Meanwhile, the initial conditions were set to:

$$u_i|_\Omega = 0, \tag{17}$$

$$p|_\Omega = p_\infty, \tag{18}$$

where Ω is the computational domain.

2.4. Geometries, Meshes and Numerical Setups

For both cases, the 3D blade geometry is made using the open source software QBlade [21] that allows building of a blade using its various sections and twist. Two kinds of barnacles are studied to investigate the differences between them: a conical barnacle according to [13] experiment and a realistic barnacle generated using 3D digital imaging (Figure 1). The open CAD software Blender [22] is used to fix the barnacles to the structures.

SnappyHexMesh is a module of OpenFoam that generates unstructured meshes [19]. This module allows one to control the parameters of the mesh such as the number of refined layers near the walls, the size of the smallest computational cells, the skewness, the orthogonality, etc. Thus, all the meshes respect the following characteristics: skewness smaller than 4 and a non-orthogonality parameter lower than 60°. Near the walls, cells are always structured. The smallest cell length scale is 2.1875×10^{-4} c and 5.6×10^{-2} c for the biggest one. The meshes contain around 2 million cells for the motionless blade case and around 9 million for the full rotor simulation. The time step (ΔT) is computed by

OpenFoam using the $CFL < 0.5$ condition (Courant–Friedrichs–Lewy condition) on the computational domain to ensure the numerical stability of the code with:

$$CFL = u_{max} \cdot \left(\frac{\Delta T}{\Delta x_i} \right) < 0.5, \quad (19)$$

where u_{max} is the maximum velocity magnitude in the domain and Δx is the length of the local cell at the u_{max} position.

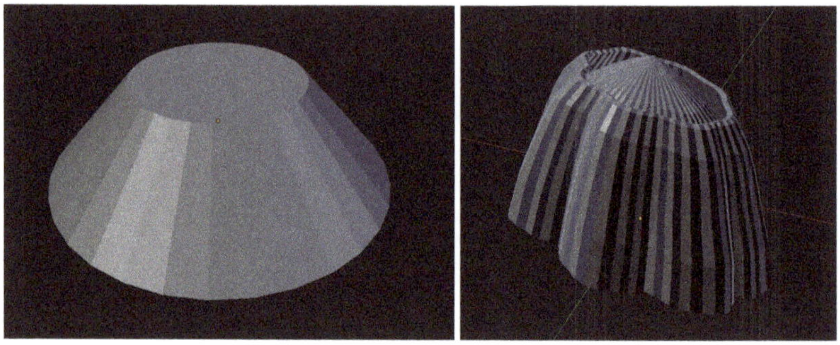

Figure 1. 3D structures of the conic (**left**) and realistic (**right**) barnacles.

2.4.1. Motionless Blade Simulation with a Single Barnacle

The blade structure made for this test is identical to that of the original experimental study [13]. This allows us to work on the validation of the full-scale model. The foil section is a 55 mm chord NACA 63-619. The barnacle is also placed at 60% of the chord at 40 mm from the centre of the blade in the y direction. The barnacle is thus located at 1/4 of the length of the blade. Half of the blade is used as a reference (clean blade), and the barnacle is placed in the middle of the second part of the blade (Figure 2). The experimental data show that the impact of the barnacle on the blade is limited to a few barnacle base diameters (0.3 c) around the barnacle. Thus, the barnacle should not impact the results of the clean part of the computational domain. Both barnacle geometries are tested and compared.

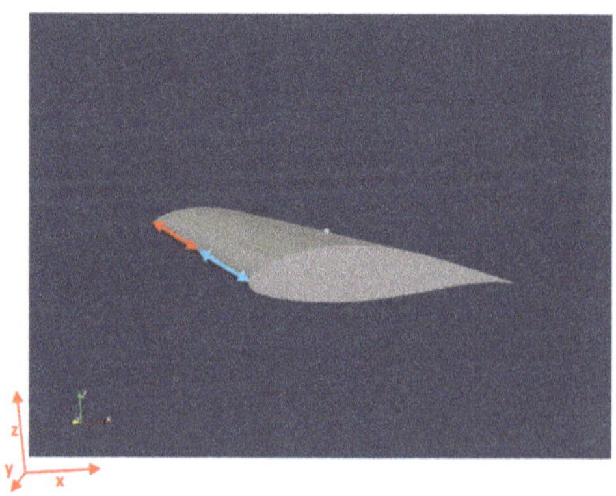

Figure 2. 3D geometry of the blade with one barnacle. The clean part of the blade is marked by the red arrow. The barnacle is in the middle of the section marked with the blue arrow.

Several sizes of computational domains were tested to remove the effects caused by boundary conditions for the smaller domain. Widths from 2 c to 8 c were tested and, after 3 c (1.65 m), numerical results were independent of the width. Thus the simulation channel is 1.60 m high × 8 m wide × 7.3 m long and limits the impact of the boundary conditions. The thinnest cells are located close to the blade walls to capture the boundary layer. The dimensionless wall distance, $y+$ is set to 1 on the clean section ($y+ = \frac{yu_\tau}{\nu}$, where y is the distance to the wall and u_τ is the friction velocity). Mesh is structured near the blade in six successive layers with an increase ratio of 1.3 between each layer. The wake expected position is refined using a refinement box to avoid the filtering by the mesh of the wake vortices. The refined mesh is shown in Figure 3. Irregularities on the 2D cut are due to 2D projections in 3D cells which are not distorted. The fluid used in motionless blade simulation is air (supposedly incompressible). The physical simulation parameters are given in Table 2.

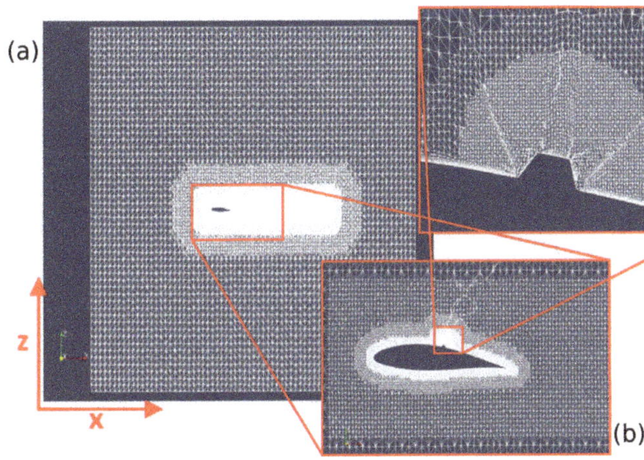

Figure 3. 3D geometry and mesh of the entire computational domain (**a**), around the blade (**b**), and around the conic barnacles (**c**). Distored cells are due to the cutting plane and do not represent the 3D cells.

Table 2. Summary of the physical parameters used in simulations of the motionless blade cases.

Parameter	Value	Unit
ρ	1.177	$kg{\cdot}m^{-3}$
ν	1.57×10^{-5}	$m^2{\cdot}s^{-1}$
U_∞	45	$m{\cdot}s^{-1}$
p_∞	1.013×10^6	Pa
k	1.898	$m^2{\cdot}s^{-2}$
ω	4.574	s^{-1}

Four angles of attack were tested and compared to experimental data (5°, 10°, 14°, 15°). The Reynolds number of the motionless blade cases (with the chord (c = 0.055 m) as reference length) is $Re_c = 1.5 \times 10^5$.

2.4.2. Full Rotor Simulation with a Realistic Barnacle Colonisation

In this section, a full rotor simulation is presented. The rotor hub is removed to limit the computation time. The turbine used in this work has been numerically studied previously for other subjects than biofouling (e.g., flow induced rotation) with clean blades [23]. Barnacles are fixed to the blades according to the realistic implantation on the blades of the AHH HS 1000 tidal turbine shown in [13] (Figure 4). We assume that the colonisation is identical on the three blades. The barnacles are settled on the downstream part of the blade, from 60% of the chord. Moreover, a large part of them are grouped in a patch. Indeed, the barnacles seem to favour the less energetic positions of the blades and their grouping contributes to protect them from the strongest currents. The chosen mesh for the clean case is the converged one used in [23]. It has been subjected to a mesh convergence study related to the forces applied to the rotor. For the fouled case, the general parameters of the mesh are kept, and the barnacles are taken into account as part of the solid structures. The computational domain is a cube with sides equal to 4 rotor diameters. The cells are twice as thin in the X-direction, which is the direction of the main velocity. Both meshes (clean and fouled) are composed of about 9 million points. Around the turbine, a 1.5 diameter refinement cylinder forms a moving part of the mesh. It is connected to the static zone by an Arbitrary Mesh Interface (AMI) which transfers fluid information from one zone to the other. The rotation of this cylinder generates the rotation of the rotor by sliding on the static zone. The mesh is shown in Figure 5.

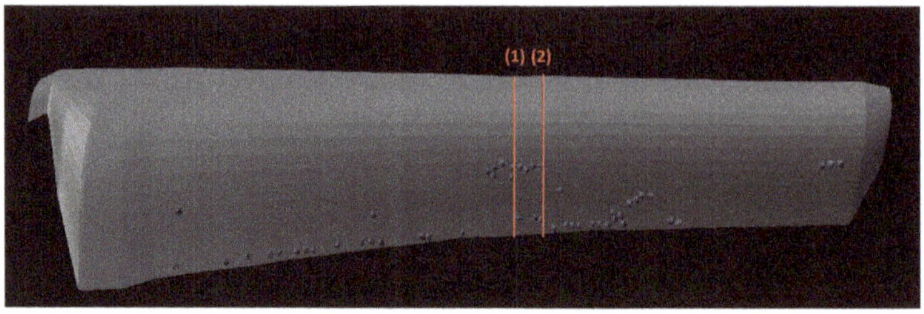

Figure 4. 3D geometry of one of the three blades of the rotor with barnacles. Red lines are cut positions for post-processing.

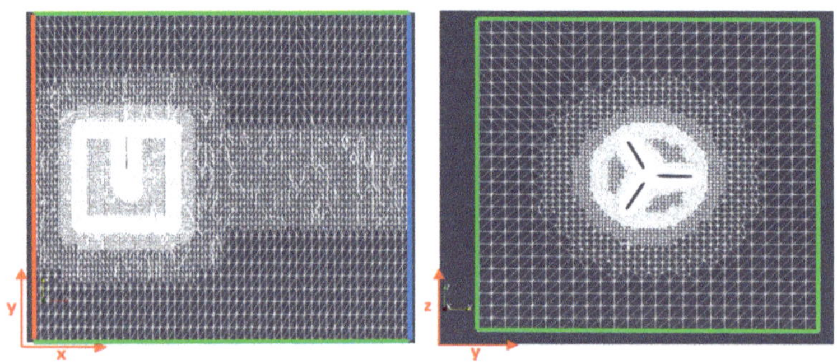

Figure 5. Views of the X–Y (**left**) and Y–Z (**right**) planes of the computational domain including the rotor geometry. Green lines are the no-slip boundary conditions, the red line is the inlet with the velocity condition and the blue line is the pressure outlet condition.

The full rotor is moving in the water. Physical and numerical parameters are given in Table 3.

Table 3. Summary of the physical parameters used in dynamic simulations of the rotor

Parameter	Value	Unit
ρ	1025	$kg{\cdot}m^{-3}$
ν	1.3×10^{-6}	$m^2{\cdot}s^{-1}$
U_∞	0.8	$m{\cdot}s^{-1}$
p_∞	0	Pa
Ω_R	9.143	$rad{\cdot}s^{-1}$
I_∞	0.03	-

Ω_R is the rotor's rotation speed and I_∞ is the turbulence intensity. The chord-based Reynolds number at the tip of the blades for the full rotor simulations is $Re_c = 1.7 \times 10^5$.

2.5. Test Case Summary

Four simulations were run for the single blade case with a single barnacle for four angles of incidence (5°, 10°, 14° and 15°). Two simulations were run with the full rotor rotating structure: one with clean blades and one with a realistic colonisation of barnacles. To help distinguish the various configurations described above, the Tables 4 and 5 provide a summary of the various cases and give some additional numerical parameters.

Table 4. Summary of the numerical cases.

Case Name	Fluid	Rotation	Barnacle	Angle of Attack
Blade 5°	Air	No	Only one	5°
Blade 10°	Air	No	Only one	10°
Blade 14°	Air	No	Only one	14°
Blade 15°	Air	No	Only one	15°
Rotor clean	Water	Yes	Realistic	0°
Rotor colonised	Water	Yes	Realistic	0°

Table 5. Numerical parameters summary.

Case Name	ΔT_{min} (s)	ΔT_{min}(s)	ΔX_{min} (m)	Total Running Time
Blade 5°	$\sim 10^{-7}$	$\sim 10^{-4}$	2.2×10^{-4}	1.26
Blade 10°	$\sim 10^{-7}$	$\sim 10^{-4}$	2.2×10^{-4}	1.21
Blade 14°	$\sim 10^{-7}$	$\sim 10^{-4}$	2.2×10^{-4}	1.16
Blade 15°	$\sim 10^{-7}$	$\sim 10^{-4}$	2.2×10^{-4}	1.15
Rotor clean	$\sim 10^{-9}$	$\sim 10^{-4}$	4.4×10^{-4}	2.41
Rotor colonised	$\sim 10^{-12}$	$\sim 10^{-4}$	2.6×10^{-4}	1.505

3. Results

This section is separated into two parts. Section 3.1 refers to the validation and comparison of the models with the experimental data with a single blade with only one barnacle on it. It is completed by a short analysis of the wake. Section 3.2 presents the comparison of two simulations with a rotor. The first simulation is the reference case with clean blades (the results are compared to experimental data) and the second is the case with a realistic colonisation (Figure 4).

3.1. Motionless Blade Simulation with One Barnacle

To compare the numerical model results to experimental data, the pressure field around the barnacles is taken at every fluid cell centre (along the blade surface) in the studied area at fixed time points chosen after the flow stabilisation. On Figure 6 is presented the opposite of the pressure coefficient C_p given by $-C_p = \frac{p - p_\infty}{-q}$, with $q = 0.5\rho U_\infty^2 = 1191.71$ Pa. X^* is the scaled position as $X^* = (x/c, y/c, z/c) = (x^*, y^*, z^*)$, where x, y and z are, respectively, the stream-wise, the span-wise and the vertical directions. Nevertheless, with the LES model, results are not averaged, which explains slight asymmetries on pressure fields (Figure 6). The blade curvature is suppressed by projecting all the cells in a plane parallel to the blade mean angle. The mesh is refined around the complex geometries and the shape of the barnacle appears in the field extraction process.

The effects of the numerical conic barnacle are very similar to the experimental ones (Figure 6) (experimental pressure fields are available in [13]) : in all cases, the barnacle is preceded by an over-pressure followed by a strong depression at the top of it. The flow change extends further downstream (3 radii) than upstream (2 radii). On the sides, the impact is felt up to 4 radii. Even with a numerical model, the perfect symmetry of the results is not guaranteed because the turbulence of the fluid creates slight variations in the flow that impact the distribution of the fluid pressure near the wall. The orders of magnitude of the C_p coefficient are the same as those measured experimentally. The main value in the field is 5.2% higher in the numerical results. The effect of the angle of attack on the pressure field is consistent with measurements: the higher the angle, the smaller the biofouling effect. The pressure field is almost unchanged for an angle of attack of 15°.

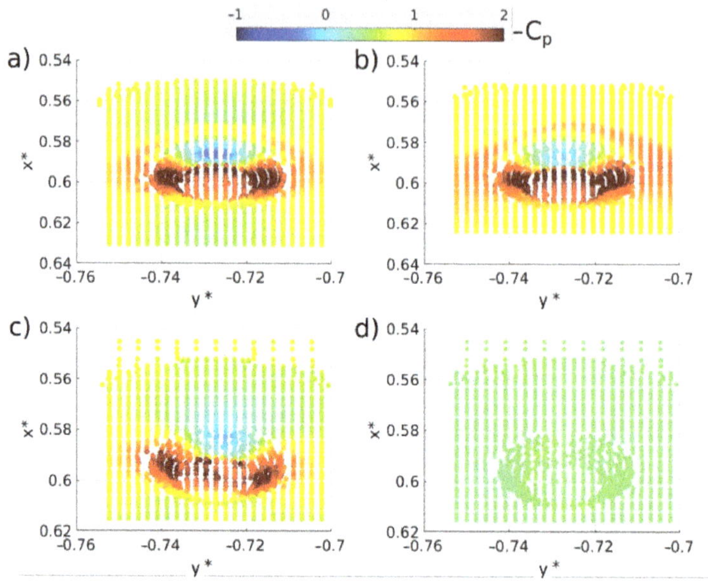

Figure 6. Opposite of the pressure coefficient ($-C_p$) around barnacle for an angle of attack of 5° (**a**), 10° (**b**), 14° (**c**) and 15° (**d**) on the blade surface. Results are from LES simulation.

To represent better the scales of the pressure variation and compare the models, the evolution of the opposite of C_p along the chord is plotted in Figure 7. The two models present different behaviours downstream of the barnacle. The k-ω SST model is better for the lowest angle of attack (5°), with the pressure increasing progressively along the chord as in the experimental data until it reaches its final value at the trailing edge. In contrast, the Smagorinsky model overestimates the pressure field which tends to decrease behind the barnacle. However, both turbulence models allow a good reproduction of the pressure drop in the fouling area.

With the 10° angle, the two turbulence models are closer in terms of mean value. However, the k-ω SST model better represents the overpressure in front of the barnacle. Downstream, both models underestimate the pressure along the blade.

The modelling is less accurate for 15°. Indeed, both turbulence models underestimate the impact of the barnacle on the flow. An additional computation is then performed to study the behaviour of the model in this critical range of values (Figure 8). At 14°, the simulated impact is more coherent with the measurements but some discrepancies are observed. We deduced that the experiment is highly sensitive to the angle of attack in the range between 14° and 15°. Small variations in the experiment or the 3D geometry can also interfere with results.

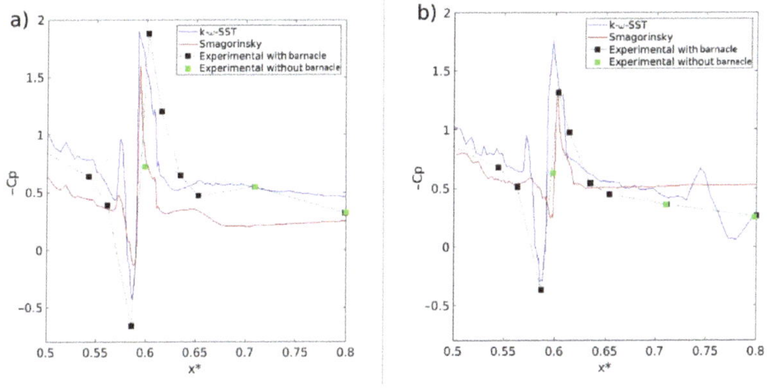

Figure 7. Opposite Pressure coefficient evolution ($-C_p$) according to the dimensionless x position (x^*) along the chord-wise position on the centre-line for an angle of attack of 5° (**a**) and 10° (**b**). Numerical results with k-ω SST (blue line) and Smagoginsky (red line) are presented for the fouled blade. Experimental values for clean and fouled blades are shown in green and black squares, respectively, from [13] data.

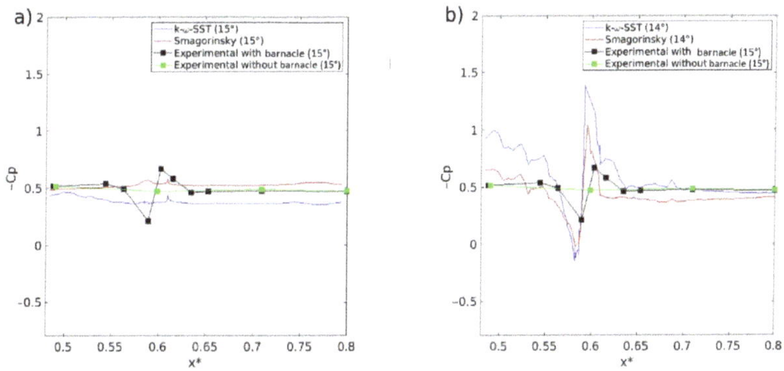

Figure 8. Opposite Pressure coefficient ($-C_p$) evolution according to the dimensionless x position (x^*) along the chord-wise position on the centre-line for an angle of attack of 15° (**a**) and 14° (**b**). Numerical results with k-ω SST (blue line) and Smagorinsky (red line) are presented for the fouled blade. Experimental values for clean and fouled blades are shown in green and black squares, respectively, (with an angle of attack of 15°) from [13] data.

Numerical simulation ensures a full C_p profile along the blade without having to invest in additional probes (Figure 9). For example, the small decrease in pressure before the overpressure ($0.56 < x^* < 0.57$) was not captured by the probes during the experimental session. This phenomenon only appears for low angles (up to 10°). Normal ($C_n = \frac{n}{q}$, where n is the forces normal to the blade per unit of span) and drag ($C_d = \frac{d}{q}$, where d is the pressure drag forces of the blade per unit of span) coefficients are computed (Figure 10). As shown in experimental data, the barnacle has no significant impact on C_n. The coefficient grows until it reaches the aerodynamic stall around 13° before decreasing with the angle. The drag coefficient is more impacted by the barnacle with an exponential increase for a mean angle greater than 10°. The barnacle causes an increase in this coefficient for low mean angles. However, when the angle continues to increase, the dynamic stall becomes more important and the effect of the barnacle fades. The numerical model reproduces this tendency. For the angle of attack of 15°, the pressure variations caused by the barnacle are almost zero. The simulation then shows results close to those expected for a clean blade. The experimental data still show an impact for this angle but the numerical results at 14° overestimate these variations. Thus, the model seems very sensitive to the angle of attack parameter.

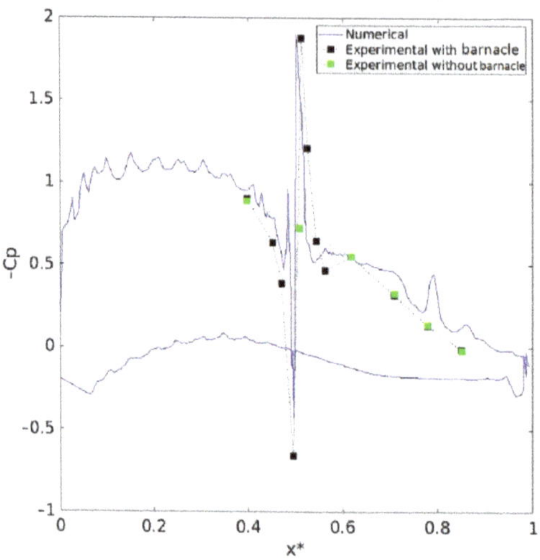

Figure 9. $-C_p$ evolution against the dimensionless x-position profile (including the lower face) at a fixed time point for numerical modelling with fouling (blue line) and experimental mean values for a clean blade (green squares) and a fouled blade (black squares) for an angle of attack of 5° from [13] data.

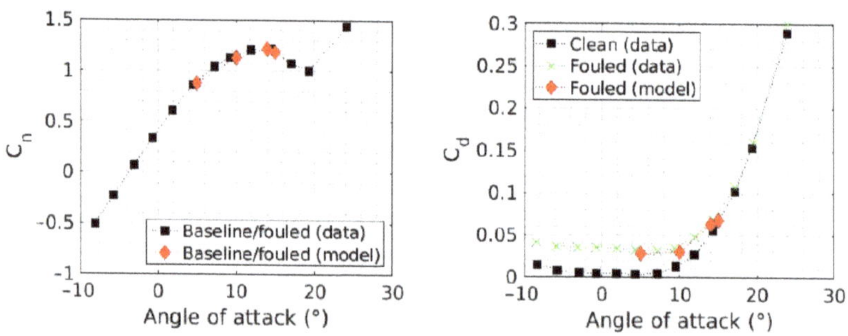

Figure 10. Normal (**Left**) and drag (**Right**) coefficients measurements against the mean angle for clean (black squares) and fouled (green crosses) blades from [13] data. Numerical results for fouled blades are represented by red lozenges.

The main difference between the two turbulence models is their ability to compute the wake. Figure 11 shows that the LES successfully separates the vortex releases from each other. The RANS model, which averages the physical quantities, only identifies the general shape of the wake. The intensity of the vortexes is also lower, indicating a higher numerical dissipation. Thus, the LES is chosen over the RANS for its better ability to represent the wake.

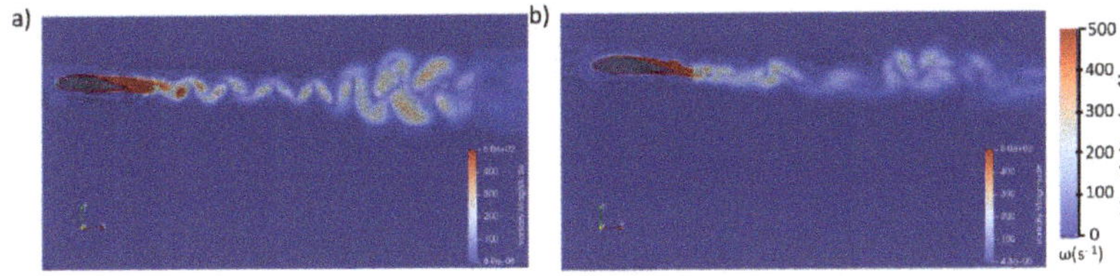

Figure 11. Magnitude of the vorticity around and behind the blade for the LES Smagorinsky (**a**) and the RANS k-ω SST turbulence model (**b**).

Figure 12 compares the time evolution of the wake of the clean part of the blade with the one in the plane of the barnacle with an angle of attack of 5°. In both cases, the first vortex is identical (T = 0.4 s) but, while the clean case starts to stabilise quickly with vortex releases alternating between the lower and upper surface, the barnacle case does not show vortexes of high vorticity intensity (>300 s^{-1}) during the first time steps. Once the wake is stabilised, the biofouling blade releases vortexes that propagate "upwards" in a regular manner. The clean blade, on the other hand, shows a turbulence structure similar to Von Karman vortex streets.

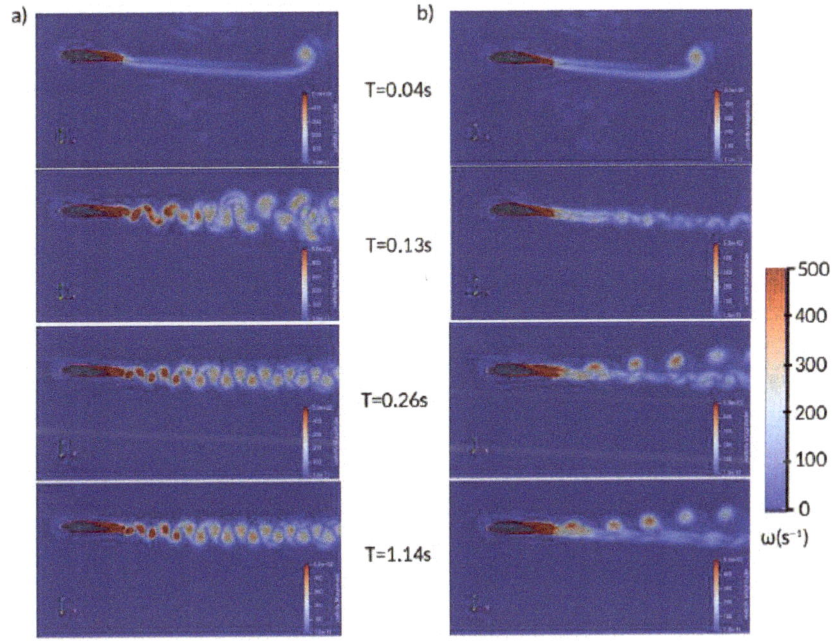

Figure 12. Magnitude of vorticity around and behind the blade at T = 0.04 s, T = 0.13 s, T = 0.26 s, and T = 1.14 s for cases without (**a**) and with (**b**) a barnacle.

Finally, the wake thickness is an interesting physical quantity to analyse: Figure 13 shows, as expected, that for the case without a barnacle as well as for the case with a barnacle, the wake thickness increases with the distance behind the blade. However, the behaviour of this increase is not the same in both cases. In the clean case, the increase is slower and follows a parabolic trend, while the case with the barnacle shows a faster and linear increase. Off the finer part of the mesh shown in Figure 3 which extends 4 chords downstream of the blade, the mesh is too coarse and diffuses the vortexes too quickly to follow the evolution of the wake thickness. It would be interesting to know if, further downstream, the wake thickness of the case without a barnacle eventually catches up with the one of the case with a barnacle.

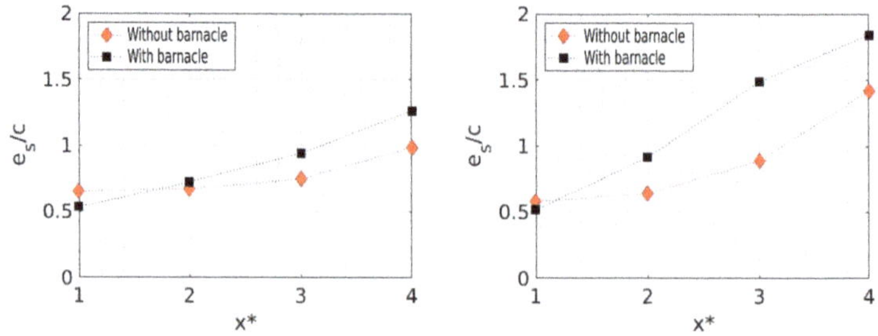

Figure 13. Dimensionless wake thickness as a function of dimensionless position in the wake without (red diamonds) and with (black squares) a barnacle for angles of attack of 5° (**left**) and 15° (**right**).

3.2. Full Rotor Simulation Simulation

3.2.1. Impact of Biofouling on Tidal Turbine Performances

The time evolution of the power and drag coefficients (C_{power} and C_d) for the complete rotor are shown in Figures 14 and 15. The realistic addition of the sessile species, according to [13], does not change the general behaviour of the turbine. However, a decrease in C_{power} by 1.6% is observed. It is explained by an early dynamic stall and the formation of re-circulation loops on the upper surface. However, under the chosen conditions, the barnacles do not seem to create any additional boundary layer stall, which transitions by itself relatively close to the leading edge This can be explained by the particular position of the barnacles: the individuals naturally fix themselves in a zone that is already less energetic, where it is easier to settle. This small drop of C_{power} may also be related to the small area colonised. If the blade was more fouled, with larger or more numerous species, the result might be more significant. In any case, the difference of the pressure coefficient is not sufficient to conclude to a performance loss.

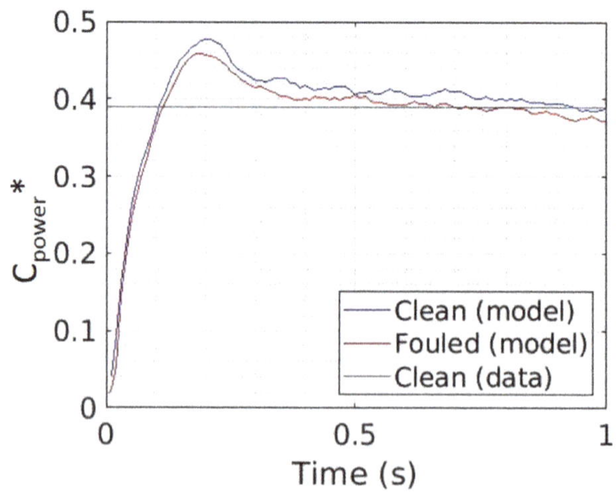

Figure 14. Time evolution of the corrected power coefficient (C^*_{power}). Measurements for a clean turbine are in black while numerical results for clean and fouled turbines are in blue and red, respectively.

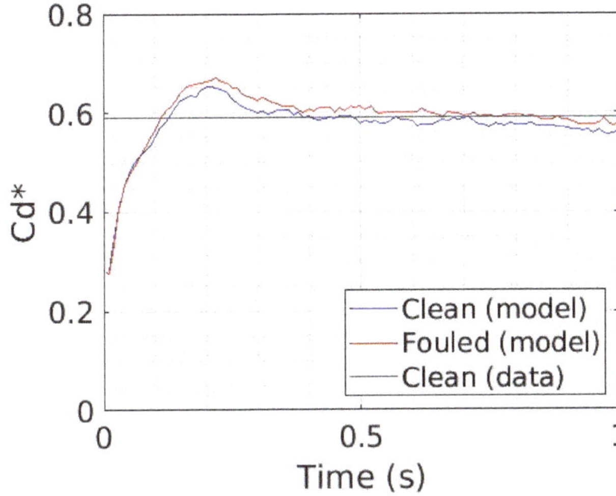

Figure 15. Same legend as for Figure 14 but for the corrected drag coefficient (C_d^*).

The drag coefficient increases by 7.5% (Figure 15). It is significantly less than for the motionless blade case simulation that reached a rising of 800% for low angles of attack. This result can be explained by two points. Barnacles do not take part in the dynamic stall, contrary to the motionless blade case where the barnacle is located at 60% of the chord. Then, the barnacles are not evenly distributed on the blade and remain relatively far from each other. 3D effects also play a role in the process: the vortexes generated by the more upstream barnacles are not directly sent into the wake as in 2D but continue to follow the blade on a different plane from the barnacle.

3.2.2. Impact of Biofouling on Tidal Turbine Wake

The chosen configuration does not allow us to see any significant impact of the colonisation on the wake of the tidal turbine. The fluid–structure interactions generated by the barnacles are small and are therefore quickly diffused and dissipated. The isovalues of the Q criteria show no significant differences in the wake or near wall. Nevertheless, a probe is placed downstream (1 diameter) of the turbine at the tip of the blade position (0.7, 0.35, 0) to study pressure, velocity and vorticity variations. The signals are relatively close for the clean and fouled cases. The amplitude of vorticity magnitude variation is lower in the biofouled case than in the clean one. The curve is also less smooth: showing that small vortexes regularly pass in the wake. A Discrete Fourier Transform (DFT) analysis of the vorticity signal is performed over 1.2 s with a time step of 0.005 s (240 samples) (Figure 16). The sample is one second long on the same time period for both cases. Both signals show a main harmonic around 2 Hz that corresponds to the tip vortex releases of the turbine. The intensity of the main harmonic (H1) is lower in the biofouled case than in the clean case. The first three harmonics are also slightly shifted (0.1 Hz) towards the high frequencies and less intense than in the clean case. In general, biofouling leads to an energetic decrease in the vortexes generated by the colonised surfaces.

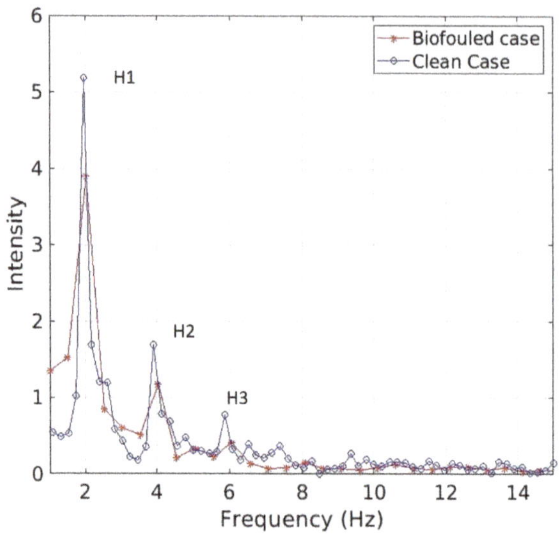

Figure 16. Discrete Fourier Transform of the vorticity numerical magnitude took one diameter behind the rotor at the tip of the blade position for a clean (blue line) and fouled (red line) tidal turbine. The probe appears on Figure 5.

A zoomed view of the blades allows us to understand how barnacles act on fluid to generate these vortexes (Figure 17). These figures confirm the fact that barnacles are behind the boundary layer transition. Nevertheless, in the case of a single barnacle, the vortexes generated by the fouling are directly sent downstream whereas in presence of a second barnacle in the same plan, the vortexes remain blocked between the two barnacles. The recirculation loop acts as a new surface over which the fluid flows. This shows that a single barnacle can have more effect on the wake than a couple in the same plane.

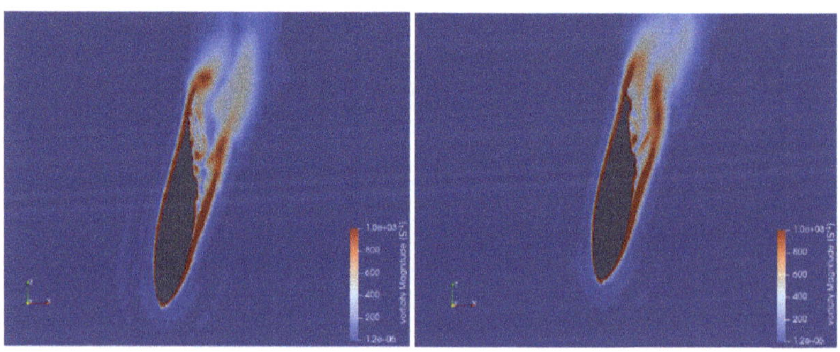

Figure 17. Zoomed view of two sections of one blade of the tidal turbine in the X–Z plan. The left refers to (1) and the right section refers to (2) on Figure 5.

4. Discussion and Conclusions

The three-dimensional study of explicit roughness raises difficult issues. The authors are aware that, despite the good results for the total forces, the pressure profiles shown in Figures 7–9, denote a non-physical behaviour (especially at the leading edge) even when averaged over time. This issue could perhaps be fixed using a finer mesh around the blade in the uniform straight flow. A mesh convergence study should therefore be carried out to overcome the issue.

For the full rotor, the authors were strongly constrained by the computational facilities. The mesh convergence was ensured by a preliminary study [23]. The numerical stability for the converged mesh requires very small time steps reaching (10^{-12} s). Thus, the simulations could only achieve three rotations in the clean case and two for the biofouled case. This produces drag and power coefficients which converge to a constant value, which allows us to conclude that the simulation is valid concerning the forces. However, the simulation time is too short to study a realistic behaviour of the wake, especially for the biofouled case. The DFT gives interesting results but should be performed on a longer time period.

The Reynolds number is also constant for the motionless blade and the full rotor simulations (around 1.6×10^5). A full scale turbine could have a higher *Re* in a realistic configuration. The increase in the *Re* would correspond to an increase in flow velocity relative to the blade profiles. However, experimental data from the motionless blade tends to show that the sooner the stall occurs, the less impact the biofouling has. The tests presented here may overestimate the actual losses due to biofouling.

This paper presents a numerical analysis of the impact of biofouling on turbines performances. Two turbulence models are compared to know which is the more suitable here. $k - \omega\ SST$ is better at predicting forces on the blade in the motionless blade cases (Figures 7 and 8) but the Smagorinsky model was used for the full rotor because of its capacity to compute the wake with accuracy (Figure 11).

Conclusions on the impact of biofouling on tidal turbines performances are close to the experimental results: for the motionless blade case, the barnacle does not impact the normal forces but highly increases the drag, especially for low mean angles. This phenomenon decreases when the angle continues to rise because of the natural stall of the profile that occurs upstream of the barnacle (Figures 6–8). No significant differences are noted between the conical and realistic barnacle structures. It is therefore recommended to work with the simplest model. A dynamic simulation of a full scale rotor is also performed with a realistic colonisation. Barnacles tend to be placed on the second part of the chord where the hydrodynamic stall creates a less energetic zone. This has the effect of greatly reducing their impact on the performance of the tidal turbine, which only loses less than 1% of its efficiency. The impact of the biofouling on the tidal turbine performance is clearly reduced in the full rotor case (Figures 14 and 15). First of all, the realistic position of barnacles (mostly in the second part of the chord) plays an important role on this result. Then, the fouled surface remains relatively low compared to the blade's surfaces. Lastly, the turbulence is quite different in the full rotor case; the blades generate big vortexes that propagate and possibly impact with the others. This increase in turbulence may be one of the reasons for the drop in the impact of the biofouling in realistic configurations. However, regardless of their position, biofouling generates drag (Figures 10 and 15). In the wake, the vortexes created by the biofouled structures are less energetic and diffuse more quickly (Figures 13 and 16). This phenomenon could even be advantageous for tidal farms where it is important that the downstream tidal turbines suffer little disturbance from the upstream tidal turbines in order to avoid rapid fatigue of the installations and a significant loss of production. However, it is important to remember that, although this is a realistic layout, there are as many configurations as there are geographical areas and therefore as many sessile species. Erect *Hydrozoans* that are not robust could have far greater effects than those highlighted here. More extensive colonisation could also change our results. Other implantation scenarios should now be explored. A characterisation and parametrisation of the biofouling will be considered to estimate the impact of the biofouling at different scale and development phases.

Author Contributions: Conceptualization, I.R. and A.-C.B.; methodology, I.R. and A.-C.B.; software, I.R.; investigation, I.R., A.-C.B.; writing—original draft preparation, I.R.; writing—review and editing, I.R., A.-C.B. and J.-C.D.; supervision, A.-C.B and J.-C.D.; funding acquisition, A.-C.B. and J.-C.D. All authors have read and agreed to the published version of the manuscript.

Funding: The research was supported by the "Région Normandie" through a PhD grant (no grant number) and by the "Université de Caen Normandie" (no grant number).

Institutional Review Board Statement: Not applicable.

Informed Consent Statement: Not applicable.

Data Availability Statement: The authors will provide OpenFoam files and result data on request.

Acknowledgments: I. Robin was supported by the "Région Normandie" through a PhD grant and by the "Université de Caen Normandie". A.-C. Bennis was funded by "Université de Caen Normandie". The authors are grateful to the CRIANN for the calculation facilities and technical support. Thank you Patrice Robin for the English reviewing. The authors are grateful to the CRIANN for the calculation facilities and technical support. We thank Patrice Robin for his English revision.

Conflicts of Interest: The authors declare no conflict of interest. The funders had no role in the design of the study; in the collection, analyses, or interpretation of data; in the writing of the manuscript; or in the decision to publish the results.

Abbreviations

The following abbreviations are used in this manuscript:

Parameters	Definitions	Units
D	Rotor diameter	m
R	Rotor radius	m
c	Blade chord	m
$\delta\Omega_1$	Computational domain inlet surface	-
$\delta\Omega_2$	Computational domain outlet surface	-
$\delta\Omega_{3,4,5,6}$	Computational domain side surfaces	-
ν	Kinematic viscosity	$m^2 \cdot s^{-1}$
ω	Rotor angular velocity	$rad \cdot s^{-1}$
p	Fluid pressure	Pa
u	Fluid velocity	$m \cdot s^{-1}$
ρ	Fluid density	$kg \cdot m^{-3}$
U_∞	Inlet velocity magnitude	$m \cdot s^{-1}$
p_∞	Undisturbed pressure	$m \cdot s^{-1}$
λ	Tip speed ratio	$m \cdot s^{-1}$
$\overline{(\cdot)}$	Mean values	-
$\tilde{(\cdot)}$	Filtered values	-
$y+$	Dimensionless wall distance	-
Ω_R	Rotor's rotation speed	$rad \cdot s^{-1}$
I_∞	Turbulence intensity	-
C_p	Dimensionless pressure coefficient	-
C_p^*	Corrected dimensionless power coefficient	-
C_d	Dimensionless drag coefficient	-

Notes

1 The French Research and Sea Exploitation Institute-Waves and Complex Environment Laboratory in Le Havre

References

1. Vinagre, P.A.; Simas, T.; Cruz, E.; Pinori, E.; Svenson, J. Marine Biofouling: A European Database for the Marine Renewable Energy Sector. *J. Mar. Sci. Eng.* **2020**, *8*, 495. [CrossRef]
2. Titah-Benbouzid, H.; Benbouzid, M. Marine Renewable Energy Converters Biofouling: A Critical Review on Impacts and Prevention. In Proceedings of the EWTEC, Nantes, France, 6–11 September 2015; pp. 1–8.
3. Kerr, A.; Beveridge, C.; Cowling, M.; Hodgkiess, T.; Parr, A.; Smith, M. Some physical factors affecting the accumulation of biofouling. *J. Mar. Biol. Assoc. UK* **1999**, *79*, 357 – 359. [CrossRef]
4. Foveau, A.; Dauvin, J.C. Surprisingly diversified macrofauna in mobile gravels and pebbles from high-energy hydrodynamic environment of the 'Raz Blanchard' (English Channel). *Reg. Stud. Mar. Sci.* **2017**, *16*, 188–197. [CrossRef]

5. Raoux, A.; Tecchio, S.; Pezy, J.; Lassalle, G.; Degraer, S.; Wilhelmsson, D.; Cachera, M.; Ernande, B.; Le Guen, C.; Haraldsson, M.; et al. Benthic and fish aggregation inside an offshore wind farm: Which effects on the trophic web functioning? *Ecol. Indic.* **2017**, *72*, 33–46. [CrossRef]
6. Lindholdt, A.; Dam-Johansen, K.; Olsen, S.; Yebra, D.; Kiil, S. Effects of biofouling development on drag forces of hull coatings for ocean-going ships: A review. *J. Coat. Technol. Res.* **2015**, *12*, 415–444. [CrossRef]
7. Song, S.; Ravenna, R.; Dai, S.; DeMarco Muscat-Fenech, C.; Tani, G.; Kemal Demirel, Y.; Atlar, M.; Day, S.; Incecik, A. Experimental investigation on the effect of heterogeneous hull roughness on ship resistance. *Ocean Eng.* **2021**, *223*, 108590. [CrossRef]
8. Heechan, J.; Seung Woo, L.; Seung Jin, S. Measurement of Transitional Surface Roughness Effects on Flat-Plate Boundary Layer Transition. *ASME J. Fluid Eng.* **2019**, *141*, 074501.
9. Balachandar, R.; Blakely, D. Surface roughness effects on turbulent boundary layers on a flat plate located in an open channel. *J. Hydraul. Res.* **2004**, *42*, 247–261. [CrossRef]
10. Chakroun, W.; Al-Mesri, I.; Al-Fahad, S. Effect of Surface Roughness on the Aerodynamic Characteristics of a Symmetrical Airfoil. *Wind Eng.* **2004**, *28*, 547–564. [CrossRef]
11. Gholamhosein, S.; Mirzaei, M.; Nazemi, M.M.; Fouladi, M.; Doostmahmoudi, A. Experimental study of ice accretion effects on aerodynamic performance of an NACA 23012 airfoil. *Chin. J. Aeronaut.* **2016**, *29*, 585–595.
12. Orme, J.; Masters, I.; Griffiths, R. Investigation of the effects of the biofouling on the efficiency of marine current turbines. In Proceedings of the MAREC2001,International Conference of Marine Renewable Energies, London, UK, 27–28 March 2001; pp. 91–99.
13. Walker, J.; Green, R.; Gillies, E.; Phillips, C. The effect of a barnacle-shaped excrescence on the hydrodynamic performance of a tidal turbine blade section. *Ocean Eng.* **2020**, *217*, 107849. [CrossRef]
14. Rivier, A.; Bennis, A.C.; Jean, G.; Dauvin, J.C. Numerical simulations of biofouling effects on the tidal turbine hydrodynamic. *Int. Mar. Energy J.* **2018**, *1*, 101–109. [CrossRef]
15. Picioreanu, C.; Vrouwenvelder, J.H.; van Loosdrecht, M. Three-dimensional modeling of biofouling and fluid dynamics in feed spacer channels of membrane devices. *J. Membr. Sci.* **2009**, *345*, 340–354. [CrossRef]
16. Demirel, Y.K.; Turan, O.; Incecik, A. Predicting the effect of biofouling on ship resistance using CFD. *Appl. Ocean Res.* **2017**, *62*, 100–118. [CrossRef]
17. Song, S.; Shi, W.; Demirel, Y.K.; Atlar, M. The effect of biofouling on the tidal turbine performance. In *Applied Energy Symposium*; MIT: Cambridge, MA, USA, 2019; pp. 1–10.
18. Mycek, P. Numerical and Experimental Study of the Behaviour of Marine Current Turbines. Ph.D. Thesis, University of Le Havre, Le Havre, France, 2013. (In French)
19. Greenshields, C. *OpenFoam User Guide Version 6*; OpenFOAM Foundation: London, UK, 2018.
20. Ramirez-Pastran, J.; Duque-Daza, C. On the prediction capabilities of two SGS models for large-eddy simulations of turbulent incompressible wall-bounded flows in OpenFOAM. *Cogent Eng.* **2019**, *6*, 1679067. [CrossRef]
21. Marten, D.; Peukert, J.; Pechlivanoglou, G.; Nayeri, C.; Paschereit, C. QBLADE: An Open Source Tool for Design and Simulation of Horizontal and Vertical Axis Wind Turbines. *Int. J. Emerg. Technol. Adv. Eng.* **2013**,*3*, 264–269.
22. Community, B.O. *Blender-a 3D Modelling and Rendering Package*; Blender Foundation, Stichting Blender Foundation: Amsterdam, The Netherlands, 2018.
23. Robin, I.; Bennis, A.C.; Dauvin, J.C. 3D Simulation with Flow-Induced Rotation for Non-Deformable Tidal Turbines. *J. Mar. Sci. Eng.* **2021**, *9*, 250. [CrossRef]

Journal of
Marine Science and Engineering

MDPI

Article

Numerical Performance Model for Tensioned Mooring Tidal Turbine Operating in Combined Wave-Current Sea States

Song Fu * and Cameron Johnstone

Energy Systems Research Unit, Department of Mechanical and Aerospace Engineering, University of Strathclyde, Glasgow G1 1XQ, UK; cameron.johnstone@strath.ac.uk
* Correspondence: song.fu@strath.ac.uk

Abstract: This study proposes the design of a tidal turbine station keeping system based on the adoption of a tensioned mooring system. Damping is introduced to investigate its effect on the reduction in the peak load experienced by tidal turbines during their operational lives in high-energy wave–current environments. A neutrally buoyant turbine is supported using a tensioned cable-based mooring system, where tension is introduced using a buoy fully submersed in water. The loads on the turbine rotor blades and buoy are calculated using a wave and current-coupled model. A modelling algorithm is proposed based on inverted pendulums, which respond to various sea state conditions, to study the behaviour of the system as well as the loads on blades. The results indicate that the tensioned mooring system reduces the peak thrust on the turbine and validates the applicability of the model.

Keywords: tidal turbine; modelling; mooring system; blade loading; wave–current interaction

Citation: Fu, S.; Johnstone, C. Numerical Performance Model for Tensioned Mooring Tidal Turbine Operating in Combined Wave-Current Sea States. *J. Mar. Sci. Eng.* **2021**, *9*, 1309. https://doi.org/10.3390/jmse9111309

Academic Editor: Eugen Rusu

Received: 14 September 2021
Accepted: 15 November 2021
Published: 22 November 2021

Publisher's Note: MDPI stays neutral with regard to jurisdictional claims in published maps and institutional affiliations.

Copyright: © 2021 by the authors. Licensee MDPI, Basel, Switzerland. This article is an open access article distributed under the terms and conditions of the Creative Commons Attribution (CC BY) license (https://creativecommons.org/licenses/by/4.0/).

1. Introduction

Tidal-stream energy can contribute significantly to global renewable energy generation, and the UK has an estimated 10–15% of the global harvestable tidal resources [1]. Tidal stream turbine (TST) technology has developed to a stage where first large-scale commercial facilities are being deployed. For example, the MeyGen project, which has exported 17 GWh to the grid as of June 2019 [2], is proving to be a reliable and economically viable renewable energy source. However, the durability of such projects is a complicated subject because the loads on TSTs vary widely owing to the unsteady marine environment. Therefore, it is challenging to achieve 10–25 year fatigue lives for turbine. There are six main types of tidal energy convertors (TEC), namely, horizontal axis turbine, vertical axis turbine, oscillating hydrofoil, enclosed tips (venturi), archimedes screw, and tidal kite [3].

The design of tidal turbine station keeping systems varies based on the turbine architecture being considered and the method of attachment to the seabed being employed. At present, gravity-based structures, drilled monopiles, and drilled pin pile tripods are three widely used support structures for tidal turbines. To make tidal current generation commercially competitive with the traditional types of energy, the industry must focus on reducing the cost of generation of the tidal-stream energy. Two main cost factors that must be targeted are the installation and maintenance of the equipment. Therefore, flexible mooring-based systems are being used for the station keeping of floating tidal turbines, such as CoRMaT [4] and Minesto 'Tidal Kite'.

Modelling methods to investigate the dynamics of a tensioned mooring-based turbine have been discussed in this paper. The analysis and control of the marine mooring and cable system are presented in [5], where the method is used to solve the dynamics of the ship and offshore platform mooring system. Mooring systems from the offshore oil and gas and ship industry have been developed and applied to design some wave energy converters [6,7]. Research shows that the single point mooring system is suggested to be applied in large dimension wave energy converters owing to the ability to minimise environmental loads [8]. However these approaches are applied generally to mooring lines that are not fully tensioned and connected to a floating structure on the surface of water.

A submerged floating tidal current hydrokinetic turbine system named GEMSTAR was presented by [9]. GEMSTAR is a project developed at the the University of Naples and the first prototype has been tested in the towing tank. It reports that problems may arise in the design of the mooring system and structural optimization, as a consequence of the high loads due to turbine thrust and required buoyancy. However, the methods to calculate the thrust, torque, the buoyancy and other dynamic characteristics of the tensioned mooring turbine have not been investigated so far. The objective of this research is to build up a numerical model to simulate these characteristics of the tensioned mooring turbine.

In this paper, the system is assumed to be an inelastic mooring, and it is modelled based on an inverted pendulum system. A coupled pendulum with an external drive is expected to experience complicated dynamics. The existence of irregular vibrations and both periodic and chaotic trajectories of a mathematical double pendulum system is proven in [10]. The stabilisation of the inverted pendulum, which is a highly nonlinear system, has been studied extensively for control education and research purposes. However, the moored turbine system is a quasi-dynamic system owing to the external forces. The external forces such as loading on the turbine rotor blades and buoy are calculated using a wave coupled blade element momentum theory (BEMT). The code was developed at the University of Strathclyde to analyse the loading occurring on a turbine rotor-drive train when operating in energetic wave–current flow conditions [11].

2. Methodology

The focus of this study is to present a methodology to assess the behaviour of a neutrally buoyant turbine supported from a tensioned cable-based mooring system, where tension is introduced using a buoy working as a damper and fully submersed in water. The schematic of the system in operation is depicted in Figure 1.

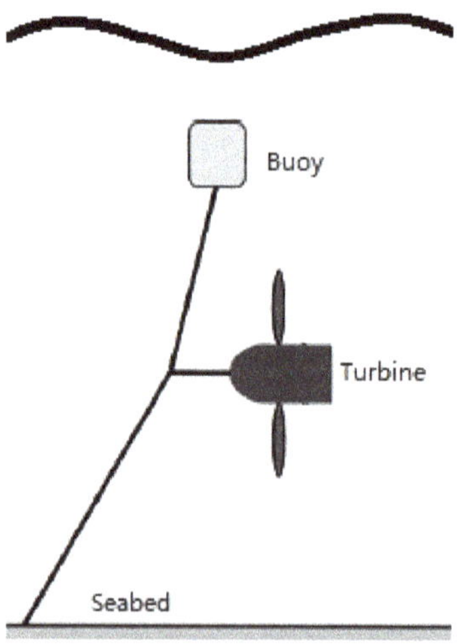

Figure 1. Schematic of tensioned mooring turbine in operation.

In order to solve the dynamics of the tensioned mooring turbine in a wave–current coupled environment efficiently. The tensioned mooring system is modelled as a special type of triple pendulum, called an inverted flail. It consists of three pendulums: the first one is attached to a fixed point considered to be an anchor and to its end mass; the other two pendulums are joined. An original flail system without the external drive and gravity field is depicted in Figure 2. This system was analysed in [12].

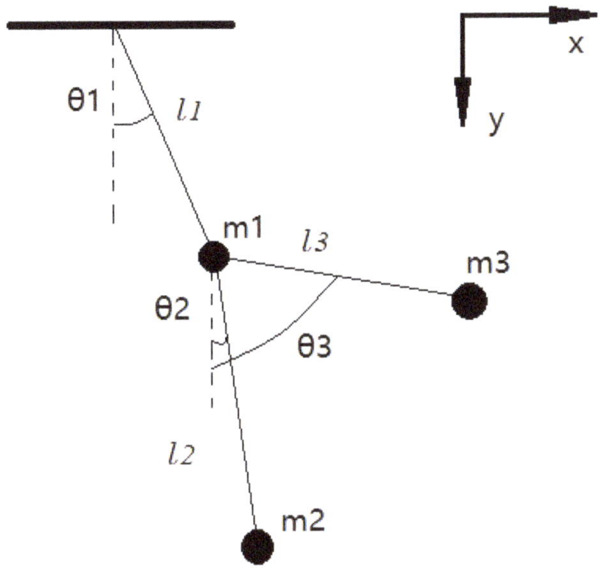

Figure 2. Geometry of flail pendulum.

Unlike the original flail system, the tensioned mooring supported turbine is driven by external forces as the loads on the turbine and buoy. A wave–current coupled BEMT was utilised to calculate the loads on the turbine rotor blades. The code was developed at the University of Strathclyde to analyse the loading on a turbine rotor-drive train when operating in energetic wave–current flow conditions [13]. In addition, owing to the turbine being able to move and respond to the moving flow field generated by the waves, the resulting motions due to flow field interactions must be taken into consideration. By coupling the tensioned mooring system with the forces obtained from BEMT as the external drive forces in wave–current environments, it is efficient to calculate the dynamics of the turbine and mooring lines together at each time step in a long simulation window.

2.1. Model of Mooring Supported Turbine

The mooring lines are assumed to continuously be in tension during operation. Therefore, this system can be modelled as an inverted flail pendulum in order to calculate its dynamics; Figure 3 provides the model for the three elements in flail pendulum. Equations of motion of the pendulum system can be derived using the following Lagrange's equation:

$$\frac{d}{dt}\left(\frac{\partial L}{\partial \dot{\theta}_i}\right) - \frac{\partial L}{\partial \theta_i} = Q_i, \quad (1)$$

where L = T − V is defined as the Lagrangian of the system, T is the kinetic energy and V the potential energy of system.

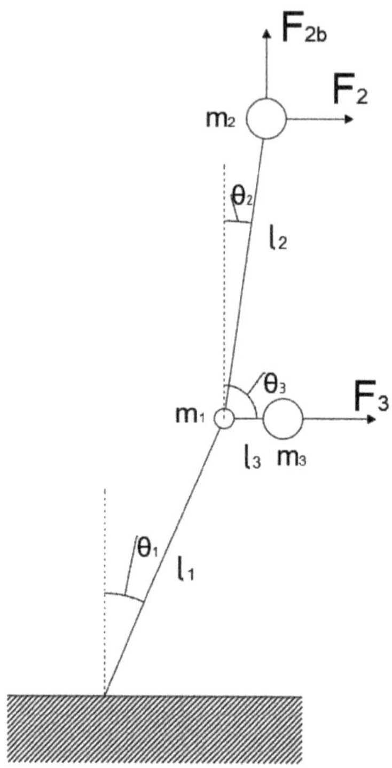

Figure 3. Multi-pendulum system with a finite number of rods and masses.

When an external force function Q_i is not considered, then the point of anchor is chosen, which is suspension of the first pendulum as the origin, and angles are measured from the vertical line; as shown in Figure 3, the Lagrangian of the system can be written as:

$$\begin{aligned} L = &\frac{1}{2}(m_1+m_2+m_3)l_1^2\dot{\theta}_1^2 + \frac{1}{2}m_2 l_2^2\dot{\theta}_2^2 + \frac{1}{2}m_3 l_3^2\dot{\theta}_3^2 \\ &+ m_2 l_1 l_2 \dot{\theta}_1\dot{\theta}_2 \cos(\theta_1-\theta_2) + m_3 l_1 l_3 \dot{\theta}_1\dot{\theta}_3 \cos(\theta_1-\theta_3) \\ &+ (m_1+m_2+m_3) g l_1 \cos\theta_1 + m_2 g l_2 \cos\theta_2 + m_3 g l_3 \cos\theta_3, \end{aligned} \tag{2}$$

where m_1 is the lumped mass of three mooring lines at the connection node, m_2 represents the mass of buoy, m_3 represents the mass of the turbine. l_1, l_2 and l_3 are the length of each segment. θ_1, θ_2 and θ_3 are generalised coordinates as shown in Figure 3.

It is assumed that the turbine and the buoy are neutrally buoyant, so the potential energy terms in Equation (2) can be eliminated. The new Lagrangian of the system becomes:

$$\begin{aligned} L = &\frac{1}{2}(m_1+m_2+m_3)l_1^2\dot{\theta}_1^2 + \frac{1}{2}m_2 l_2^2\dot{\theta}_2^2 \\ &+ m_2 l_1 l_2 \dot{\theta}_1\dot{\theta}_2 \cos(\theta_1-\theta_2) \\ &+ \frac{1}{2}m_3 l_3^2\dot{\theta}_3^2 + m_3 l_1 l_3 \dot{\theta}_1\dot{\theta}_3 \cos(\theta_1-\theta_3). \end{aligned} \tag{3}$$

Substituting Equation (3) into Equation (1) yields the Euler–Lagrange differential equations of the system:

$$\begin{aligned} &(m_1+m_2+m_3)l_1^2\ddot{\theta}_1 + m_2 l_1 l_2 \ddot{\theta}_2 \cos(\theta_1-\theta_2) \\ &- m_2 l_1 l_2 \dot{\theta}_2^2 \sin(\theta_1-\theta_2) + m_3 l_1 l_3 \ddot{\theta}_3 \cos(\theta_1-\theta_3) \\ &- m_3 l_1 l_3 \dot{\theta}_3^2 \sin(\theta_1-\theta_3) = Q_1 \end{aligned} \tag{4}$$

$$m_2 l_2^2 \ddot{\theta}_2 + m_2 l_1 l_2 \ddot{\theta}_1 \cos(\theta_1 - \theta_2) \\ - m_2 l_1 l_2 \dot{\theta}_1^2 \sin(\theta_1 - \theta_2) = Q_2 \tag{5}$$

$$m_3 l_3^2 \ddot{\theta}_3 + m_3 l_1 l_3 \ddot{\theta}_1 \cos(\theta_1 - \theta_3) \\ - m_3 l_1 l_3 \dot{\theta}_1^2 \sin(\theta_1 - \theta_3) = Q_3, \tag{6}$$

where Q_1, Q_2 and Q_3 are now generalised moments with respect to θ_1, θ_2 and θ_3. In this case, Q_3 equates to the momentum thrust develop by a turbine loading, where this is operating under combined wave and currents conditions. Q_2 relates to the buoyant forces occurring on the floater, which are considered in the form of buoyancy and wave excitation forces. Q_1 will be obtained from the relationship between Q_2 and Q_3. According to Anli and Ohlhoff [14,15], the generalised force can be obtained as:

$$Q_k = \sum_{i=1}^{n} F_i \frac{\partial r_i}{\partial q_k}, \tag{7}$$

where Q_k is the Generalised force associated with the kth Euler-Lagrange differential equation, F_i is the external force, r_i is the position of the point of application and q_k is the generalised coordinate.

Thus, substituting Equation (7) to the generalised coordinates with respect to θ_1, θ_2 and θ_3, the generalised moments for this system are given as:

$$Q_1 = F_3 l_1 \cos\theta_1 - F_{2b} l_1 \sin\theta_1 + F_2 l_1 \cos\theta_1 \tag{8}$$

$$Q_2 = F_2 l_2 \cos\theta_2 - F_{2b} l_2 \sin\theta_2 \tag{9}$$

$$Q_3 = F_3 l_3 \cos\theta_3. \tag{10}$$

When the generalized moments are obtained, the Euler–Lagrange differential equations of the system can be solved with given initial conditions. Substituting Equations (8)–(10) into Equations (4)–(6) then dividing by l_1, l_2 and l_3 yields:

$$(m_1 + m_2 + m_3) l_1 \ddot{\theta}_1 + m_2 l_2 \ddot{\theta}_2 \cos(\theta_1 - \theta_2) \\ - m_2 l_2 \dot{\theta}_2^2 \sin(\theta_1 - \theta_2) + m_3 l_3 \ddot{\theta}_3 \cos(\theta_1 - \theta_3) \\ - m_3 l_3 \dot{\theta}_3^2 \sin(\theta_1 - \theta_3) \\ = F_3 \cos\theta_1 - F_{2b} \sin\theta_1 + F_2 \cos\theta_1 \tag{11}$$

$$m_2 l_2 \ddot{\theta}_2 + m_2 l_1 \ddot{\theta}_1 \cos(\theta_1 - \theta_2) \\ - m_2 l_1 \dot{\theta}_1^2 \sin(\theta_1 - \theta_2) \\ = F_2 \cos\theta_2 - F_{2b} \sin\theta_2 \tag{12}$$

$$m_3 l_3 \ddot{\theta}_3 + m_3 l_1 \ddot{\theta}_1 \cos(\theta_1 - \theta_3) \\ - m_3 l_1 \dot{\theta}_1^2 \sin(\theta_1 - \theta_3) \\ = F_3 \cos\theta_3. \tag{13}$$

The external forces on the buoy and turbine are depicted in Figure 4. In this paper, the wave excitation forces on the turbine are assumed to be ignored.

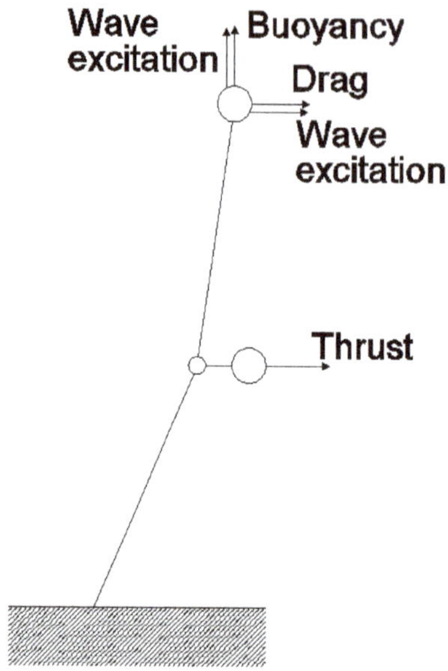

Figure 4. Forces on the system.

Wave excitation is considered to be a factor of the system. The buoy will be excited by the wave at a different magnitude based on its shape. The wave excitation in two directions can be defined as the exciting force and drift force, based Wu [16]. The thrust on the turbine can be obtained using an ESRU in-house BEMT code [11]. Modifications have been made in the original code owing to the relative velocity between the turbine and inflows:

$$\mathbf{U} = \mathbf{u} - \mathbf{U}_{\mathrm{T}}, \tag{14}$$

where $\mathbf{u}$ is the inflow velocity, which is calculated using the wave–current interaction model; $\mathbf{U}_{\mathrm{T}}$ is the inertia velocity of the turbine, which is the velocity generated by the motions of turbine in waves. It can be calculated in vertical and horizontal directions as:

$$\mathbf{U}_{\mathrm{T}} = \frac{\partial}{\partial \Delta t}(\mathbf{X}^{\mathrm{it}}_{\mathrm{turbine}} - \mathbf{X}^{\mathrm{it}-1}_{\mathrm{turbine}}), \tag{15}$$

where it is time step count number, $\mathbf{X}_{\mathrm{turbine}}$ is the turbine position and Δt is the time difference. The buoy is assumed to be spherical. The drag force is also calculated from the BEMT code. The relative velocity between the buoy and inflows must be considered:

$$\mathbf{V} = \mathbf{u} - \mathbf{U}_{\mathrm{B}}. \tag{16}$$

$\mathbf{u}$ can be calculated using the wave–current model based on the coordinates of the buoy. $\mathbf{U}_{\mathrm{B}}$ is the inertia velocity of the buoy:

$$\mathbf{U}_{\mathrm{B}} = \frac{\partial}{\partial \Delta t}(\mathbf{X}^{\mathrm{it}}_{\mathrm{buoy}} - \mathbf{X}^{\mathrm{it}-1}_{\mathrm{buoy}}). \tag{17}$$

2.2. Flow Diagram

The original in-house BEMT code for wave–current environments is based on a rigid supported turbine, where the position of the turbine does not change with time. However, the coordinates of the mooring supported turbine are variable with time and the relative velocity must be calculated using the relative motion between the turbine and wave–current inflow. Figure 5 depicts the main process of the simulation. The process nodes with the dark background are works based on this study, which are different from the original BEMT.

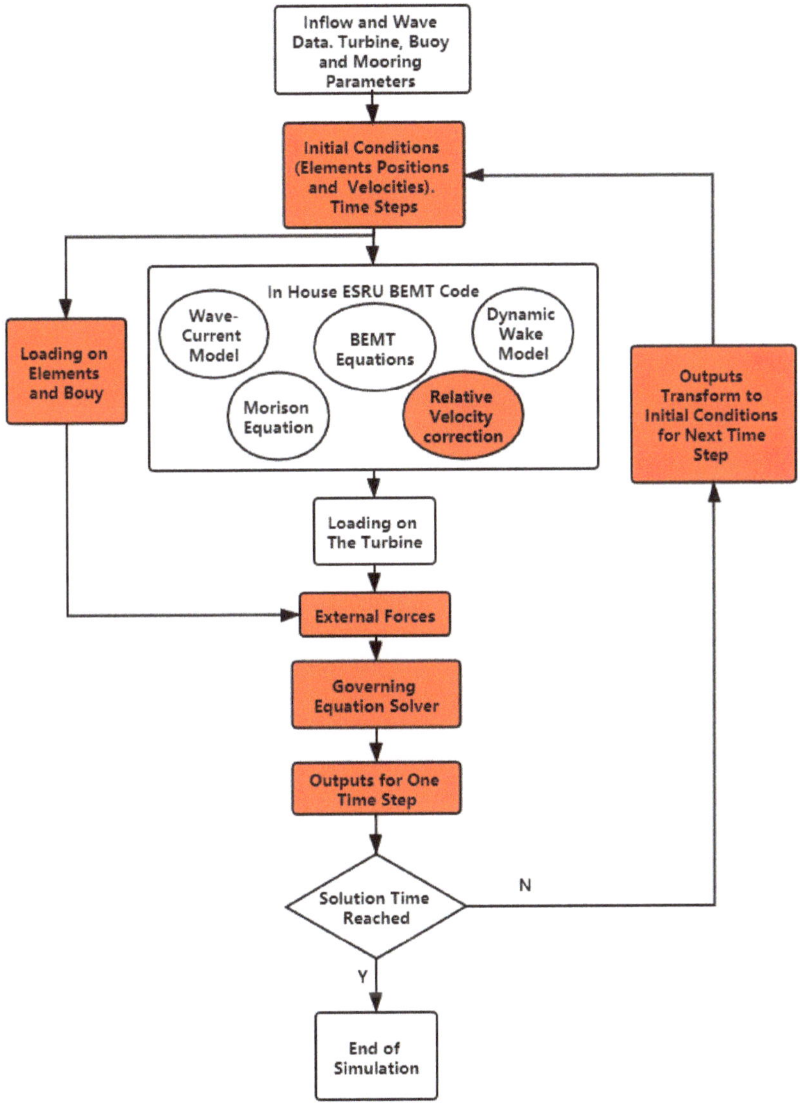

Figure 5. Solution process.

There are some variable input parameters that can be defined by users for different mooring turbine devices. The state of motion for the system in the first time step is defined by the initial conditions. The loads on each mooring element and buoy are obtained as the external forces, whereas the load on the turbine is calculated using the BEMT code while modifying the relative velocity. After the external forces are determined, the governing equation for the "flail" system with a finite number of segments can be solved by an ordinary differential equations solver to obtain the new state of motion of the system. The new values are used as the initial conditions for the next time step until the simulation culminates.

For the wave–current model module, calculations of the horizontal and vertical particle velocities are related to the horizontal and vertical coordinates of the turbine in different wave theories. The coordinates of the turbine and buoy hinge nodes are x_{turbine}, y_{turbine},

x_{buoy} and y_{buoy}. Because the vertical particle velocities are varied along blades, the blade element coordinates are:

$$x_{element} = x_{turbine} + h_{element}\cos\phi \quad (18)$$

$$y_{element} = y_{turbine} + h_{element}\sin\phi, \quad (19)$$

where $h_{element}$ is the element position on the blade and ϕ is the pitch angle of the turbine. Assume that the turbine is nearly horizontal during the operation and the pitch angle is 0 degree. The element coordinates become:

$$x_{element} = x_{turbine} \quad (20)$$

$$y_{element} = y_{turbine} + h_{element}. \quad (21)$$

Next, substitute these coordinates into wave models. Then, this module is modified to work for the mooring supported turbine.

The relative velocity modification of the inflow velocity must be considered in not only the BEMT equation but also the dynamic wake model and Morison equation.

Based on the methodology adopted, the dynamic inflow affects the BEMT model [17–19]. On a blade element bounded by radii R_1 and R_2 as shown in Figure 6, the momentum thrust equation depends on the time derivative of axial induction factor $\dot{a}$ as:

$$dF_A = 2\mathbf{u}a\dot{m} + \mathbf{u}m_A\dot{a}, \quad (22)$$

where $\dot{m}$ is the mass flow through the intersecting fluid annulus, a is the axial induction factor, and m_A is the apparent mass of the blade section.

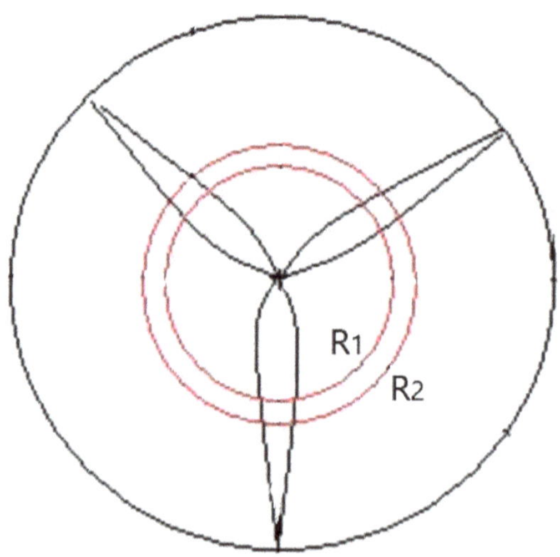

Figure 6. Blade element bounded by radii R_1 and R_2.

The mass flow through the annular element can be calculated as:

$$\dot{m} = \rho\mathbf{u}(1-a)dA, \quad (23)$$

where ρ is the density of water and $dA = \pi(R_2^2 - R_1^2)$.

For a turbine of radius R, Tuckerman [20] suggests that the apparent mass acting on the rotor can be approximated by an enclosing fluid ellipsoid, which can be expressed, through the use of potential flow theory, as:

$$m_{\mathrm{A}} = 8/3\rho R^3. \quad (24)$$

Substituting Equations (23) and (24) into Equation (22) and dividing each term by π, ρ, $\mathbf{u}$, and $\mathrm{d}A$ and multiplying by 2, then replacing the in flow velocity with the relative velocity U_x which is the horizontal component of relative velocity $\mathbf{U}$. The final form of the unsteady thrust coefficient for an annulus can be obtained as:

$$C_{\mathrm{FA}} = 4a(1-a) + \frac{16}{3\pi U_x}\frac{(R_2^3 - R_1^3)}{(R_2^2 - R_1^2)}\dot{a}. \quad (25)$$

Substituting thrust coefficient $C_{\mathrm{T}} = 4a(1-a)$ into Equation (25) gives:

$$C_{\mathrm{FA}} = C_{\mathrm{T}} + \frac{16}{3\pi U_x}\frac{(R_2^3 - R_1^3)}{(R_2^2 - R_1^2)}\dot{a}. \quad (26)$$

The last term on the right-hand side of Equation (26) can be used to calculate the additional force from the dynamic wake effects.

The inertial force caused by fluid acceleration, which is the added mass around a rotating blade section can be expressed as Morison equation, as presented by Buckland [21] and Chapman [22]. The inertial force per unit length, $\mathrm{d}l$ in the wave propagation direction on a submerged body can be calculated as:

$$\mathrm{d}F_{\mathrm{in}} = \rho C_{\mathrm{m}} A \frac{\partial \mathbf{u}}{\partial t}\mathrm{d}l, \quad (27)$$

where A is the cross horizontal sectional area parallel to the flow and C_{m} is the inertia coefficient, which is expressed as:

$$C_{\mathrm{M}} = 1 + C_{\mathrm{A}} = 1 + \frac{M_{\mathrm{A}}}{\rho A \mathrm{d}l}, \quad (28)$$

where M_{A} is the added mass for a blade element.

For blade elements, the added mass in axial and tangential directions can be approximated with that of a fixed pitched plate as per Theodorsen's theory [23]:

$$M_{\mathrm{A,axial}} = \rho\pi(\frac{c\sin\beta}{2})^2\mathrm{d}l \quad (29)$$

$$M_{\mathrm{A,tan}} = \rho\pi(\frac{c\cos\beta}{2})^2\mathrm{d}l, \quad (30)$$

which are the masses of the enclosing fluid cylinders with radii r of half the vertical and horizontal chord components c of the respective blade sections with section angle β [24].

Substituting Equations (28)–(30) into Equation (27) and plugging the components of relative velocity $\mathbf{U}$ into the equation gives the equations for the inertia forces in the axial and tangential directions for a blade element as:

$$\mathrm{d}F_{\mathrm{in,axial}} = \rho(1 + \frac{\pi((c\sin\beta)/2)^2}{A_\alpha})A_\alpha\frac{\partial U_x}{\partial t}\mathrm{d}r \quad (31)$$

$$\mathrm{d}F_{\mathrm{in,tan}} = \rho(1 + \frac{\pi((c\cos\beta)/2)^2}{A_\alpha})A_\alpha\frac{\partial U_y}{\partial t}\mathrm{d}r, \quad (32)$$

where A_α is the cross-sectional area of the airfoil at the blade section.

When the external forces based on the initial conditions for the first time step are calculated using the BEMT equations with the wave–current model, dynamic wake model, and Morison equations, the system Lagrange equation solver begins to solve the differential equations of motion for the mooring supported turbine. Next, the new values and angular velocities $\dot{\theta}_i$ for each segment are obtained to serve as the new initial conditions for the next time step. This loop continues until the time step reaches the end time of the simulation.

3. Initial Conditions and System Parameters

This study focuses primarily on external forces, which are buoyancy and wave–current-coupled forces. Sea states with regular and irregular waves were investigated. The wave data were collected by the UK Offshore Operators Association and were provided by British Oceanographic Data Center [25]. In this study, parameters of a 1 MW turbine and mooring system were applied to deep water. Parameters given below were fixed to control the number of variables. The material for the mooring line was chosen to be Dyneema, whose density is close to that of water. There were three sets of system parameters in each model, two of them were Dyneema mooring and one was steel mooring.

3.1. Initial Conditions

For the proposed model, it is necessary to define the initial condition for all segments, as listed in Table 1.

Table 1. Initial conditions for proposed model.

Initial Conditions		
$\theta_1 = \frac{\pi}{4}$	$\theta_2 = 0$	$\theta_3 = \frac{\pi}{2}$
$\dot{\theta}_1 = 0$	$\dot{\theta}_2 = 0$	$\dot{\theta}_3 = 0$

All angle values in Table 1 are in radian. The initial angles are set for every segment to represent an untensioned mooring line. This method can be applied to a general mooring system as well. The parameters for the turbine, buoy, and mooring line are shown in Table 2; the turbine diameter was set to be the same as that of the SIMEC Atlantis Energy tidal turbine AR2000 [26] and the weight was half of its weight. The definitions of l_1, l_2 and l_3 are shown in Figure 2.

Table 2. The parameters of the turbine, buoy and mooring line.

System Parameters			
m_{buoy}	5 t	rotor rotation speed	1.25 rad/s
m_{turbine}	80 t	airfoil profile	NRELs814 [27]
l_1	30 m	number of blade	3
l_2	15 m	mooring line segment length	0.5 m
l_3	3 m	mooring line material	Dyneema
buoy radius	3 m	number of blade element	20

3.2. Sea States

Table 3 lists the sea states investigated in the simulations to obtain the thrust and torque on the turbine. Steep and swell waves are investigated, comparing how wave excitation on the buoy affects the load on the turbine. The assumed hypothetical site for the generic turbine was chosen off the north east coast of the Orkney islands, Scotland. This site provided a flow speed of approximately 2.5 [m/s] and an average depth of 50 [m] [28].

Table 3. Sea states.

Sea States	1	2	3	Harsh Winter
H_s [m]	2.665	1.07	1.008	10.12
T_z [s]	6.135	11.07	4.653	10.06
water depth [m]	50	50	50	50
steepness (H/gT)	0.0719	0.0099	0.0237	0.1027
wave model	3rd-order	linear	2nd-order	3rd-order

A 3 min window of 0.1 [s] time step was simulated for each sea states.

4. Results

Firstly, the results with and without wave excitation on the buoy are compared in the same sea states. In sea state 1, a three-step approximate wave–current interaction model was applied in this simulation because its steepness was larger than 0.02 [29]. However, the hub height dropped to approximately 22.5 [m] during the operation compared with the original height, which was set at 30 [m] from the seabed. Therefore, the hub height for the rigid supported turbine was set equal to that of the mooring supported one. Figure 7 exhibits the result in sea state 1. These results indicate that the mooring supported turbine can more effectively reduce the peak thrust on the turbine when compared with the rigid supported turbine. Furthermore, the results from considering wave excitation on the buoy indicate a different waveform without wave excitation. The mean values of the three curves are close: 1.019 [MN] for the rigid supported turbine, 1.019 [MN] for that without wave excitation on the buoy, and 1.016 [MN] for that with wave excitation on the buoy. However, the standard deviation of the thrust on the mooring supported turbine considering wave excitation on the buoy is 28.23 [kN] and that without wave excitation is 16.90 [kN], which are 69.69% and 41.73% of that on the rigid structure value of 40.50 [kN].

The results for torque indicate a different trend compared with that of the thrust. Torques on the mooring-supported turbine with and without wave excitation on the buoy both have more fluctuations than that of the rigid supported turbine. However, the mean values are similar, that is, 364.91 [kN·m], 364.66 [kN·m], and 361.98 [kN·m] for rigid, without, and with, wave excitation on the buoy, respectively. The respective standard deviations are 18.19 [kN·m], 87.14 [kN·m], and 83.41 [kN·m]. This is because of modifying the relative velocity in the vertical direction, which is discussed later in this section.

Figure 8 exhibits the result for the turbines operating in sea state 2. It shows a favorable performance in the reduction of peak loading both in thrust and torque and the wave excitation on the buoy provides a positive effect in load reduction. The reason is that the buoy will oscillate more regularly in the wave with long wave periods. In this sea state, the mean values of thrust are 1.260 [MN], 1.265 [MN] and 1.265 [MN] for rigid supported turbine, mooring turbine without and with wave excitation on a buoy, respectively. The standard deviations are 47.35 [kN], 11.59 [kN] (24.5% of 47.35 [kN]) and 6.56 [kN] (13.9% of 47.35 [kN]) separately. The average torque values are 694.47 [kN·m], 699.92 [kN·m] and 699.06 [kN·m] and the standard deviations are 58.19 [kN·m], 24.98 [kN·m] and 20.86 [kN·m].

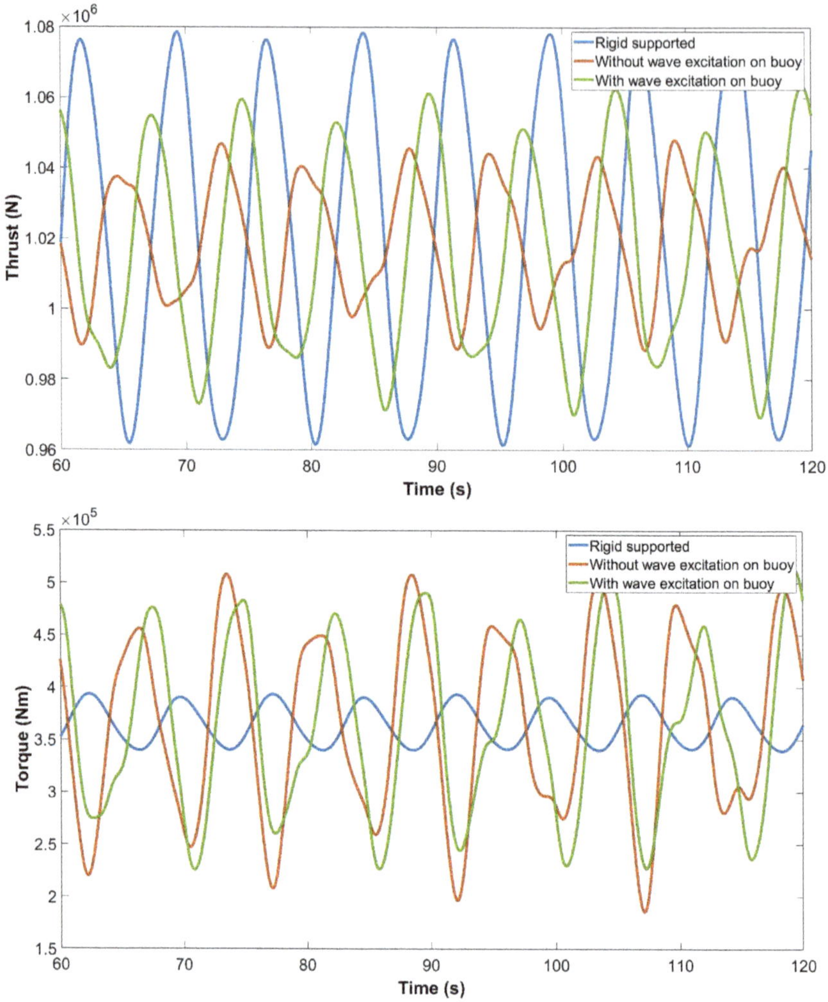

Figure 7. Thrust and torque in sea state 2.

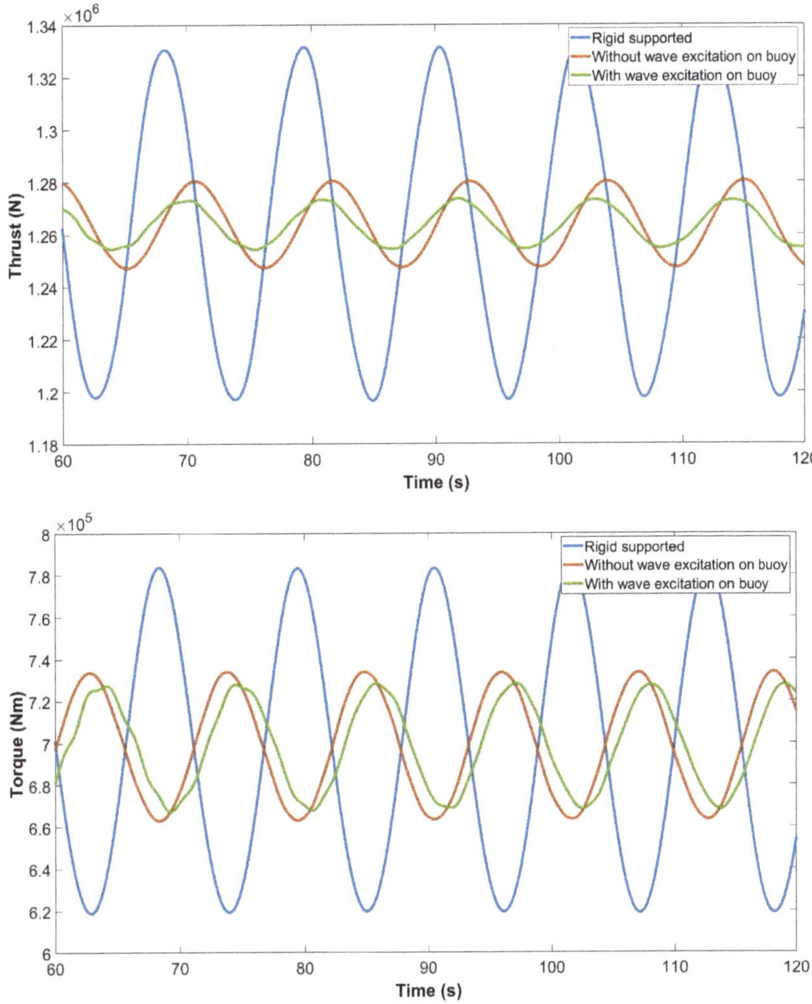

Figure 8. Thrust and torque in sea state 2.

In sea state 3, the performance of the system is different from all the sea states discussed so far as shown in Figure 9. The peak loading of thrust as a result of wave excitation on the buoy is larger when compared with the rigid supported turbine. The mean values of the thrust are 1.045 [MN], 1.044 [MN], and 1.040 [MN] for rigid supported turbine, mooring supported turbine without and with wave excitation on the buoy. The respective standard deviations are 5.82 [kN], 6.28 [kN] and 13.43 [kN]. The respective average torque values are 396.44 [kN·m], 394.84 [kN·m], and 390.25 [kN·m] and the standard deviations are 169.08 [kN·m], 20.73 [kN·m], and 35.48 [kN·m].

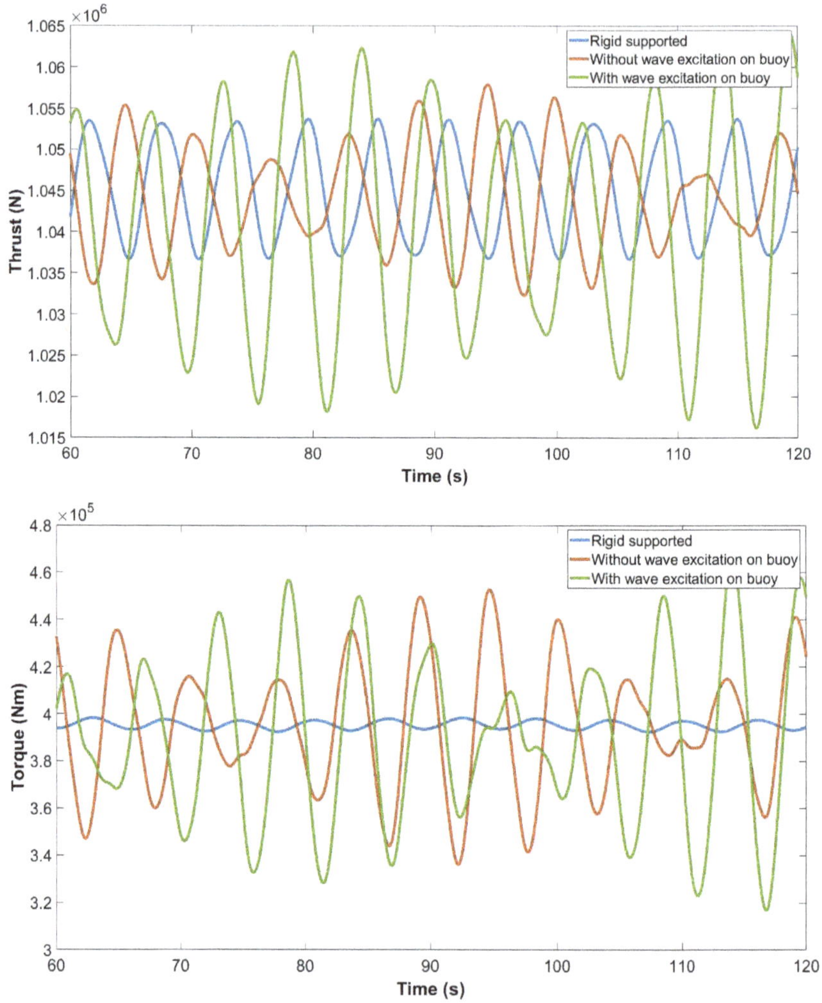

Figure 9. Thrust and torque in sea state 3.

The results for harsh winter conditions (the extreme storm sea state in the North Sea) are presented in Figure 10. The peak loading reduction in the thrust is similar to that in sea state 2. The mean values are 930.03 [kN], 897.42 [kN], and 894.36 [kN], respectively. The respective standard deviations are 390.80 [kN], 86.67 [kN] (22.2%), and 37.57 [kN] (9.6%). However, the torque has negative values, which indicate that the directions of the torque have reversed. In reality, this means that the turbine will be stalled due to displacement being significantly large enough for the velocities on blade elements to change (see Figure 11).

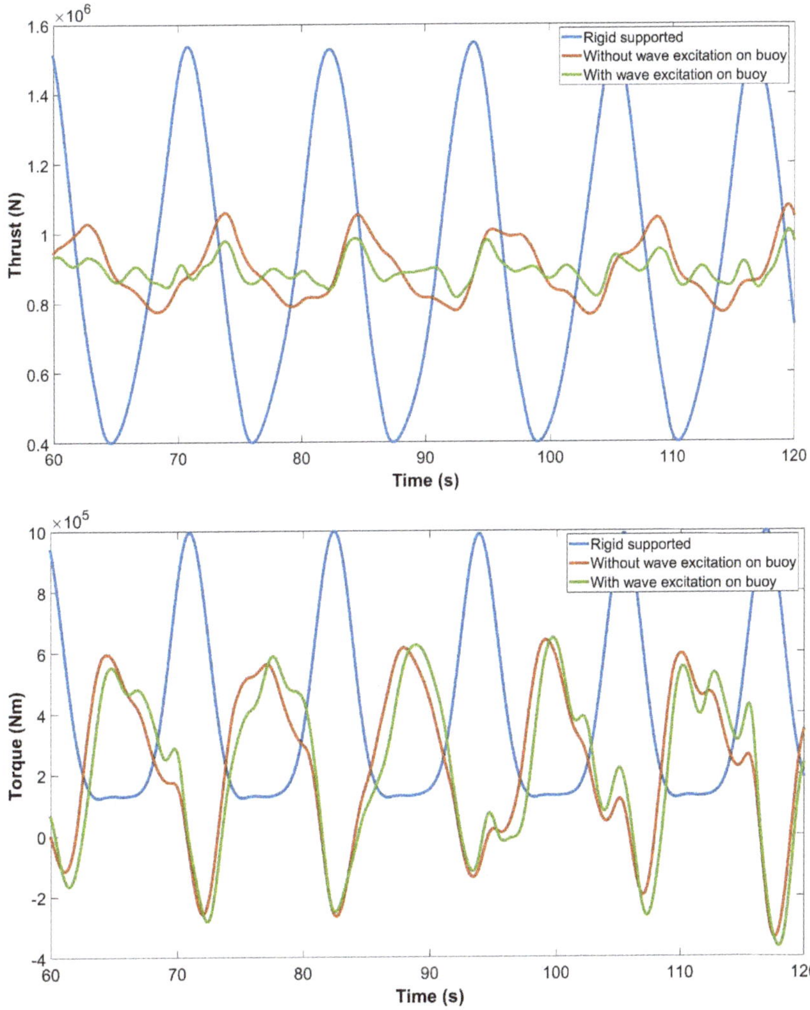

Figure 10. Thrust and torque in harsh winter.

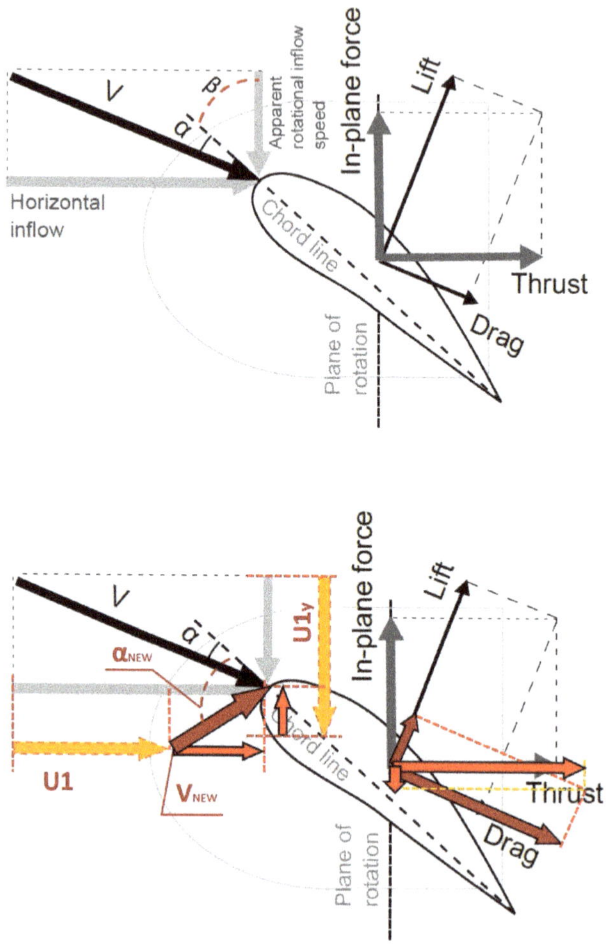

Figure 11. Inflow velocity vectors on blade sections in harsh winter.

5. Discussion

The thrust loading reduction in four listed sea states are favorable especially in the linear wave, but the reduction of torque on the rotor is not as satisfactory as the thrust; some findings are given in this section to discuss the noteworthy torque fluctuation.

5.1. Blade Section Velocities

In harsh winter sea states, the turbine moves heavily along the wave in the wave trough. The inertial velocity of the turbine, $\mathbf{U}_T$, is coupled with the wave–current velocity field and transferred to the local rotating blade coordinate system [24]. Two velocities, $U1$ and $U1y$, are obtained to modify the horizontal inflow and the apparent rotational inflow speed to calculate the relative velocity of the blade section. The new relative velocity results in a larger angle of attack, which generates a smaller lift force but a larger drag force on the blade section as shown in Figure 11. Therefore, the direction of the in-plane force is changed, which means the torque on the rotor has negative values and the blade is stalled. A blade profile that performs better than NRELs814 in larger angles of attack, or setting a pitch angle can be applied to avoid this phenomenon. However, this remains to be studied in the future.

5.2. Wavelength Factor

For sea state 3, the wavelength is 54.38 [m]. The water particle path reaches 27.19 [m] from the water surface, then it starts to decay dramatically. The turbine hub is at a height of approximately 21 [m] from the seabed and the turbine diameter is 20 [m], the upper part of the rotor is in the range of wave circular orbits, the lower part of the rotor is at the location where the wave decays rapidly as shown in Figure 12. This indicates that the wave affected approximately half of the rotor swept area during operation, but excited the buoy significantly. Furthermore, the load reduction for the mooring supported turbine does no better than that of the rigid supported turbine because the wave just reaches the hub of the turbine and affects half of the turbine; moreover, the superposition of the buoy and turbine oscillations in this sea state makes the performance unexpected. Further investigation of this kind of sea state will be part of future work.

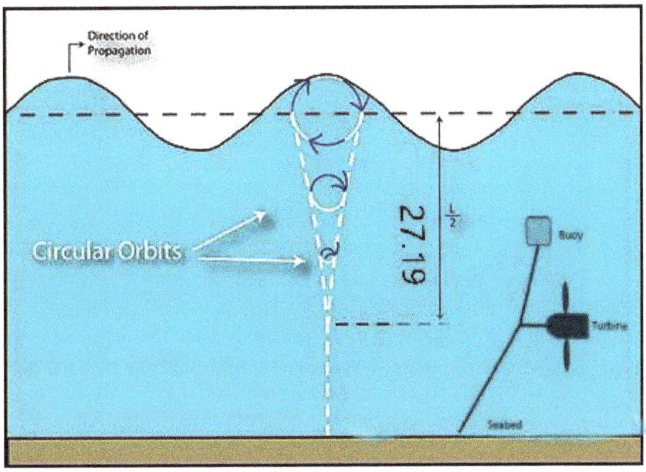

Figure 12. Water particle paths under waves in deep water and the turbine position.

5.3. Morison Effects

According to the results above, the variation of torque values on tension mooring supported turbines is larger than that of rigid supported ones in similar sea states. The reason for this is the added mass effects on blade sections and the inertia force added into the in-plane force to calculate the torque. The relative velocity term for a rigid supported turbine is only the wave particle velocity in the vertical direction transferred to the local rotating blade coordinate system. However, the inertia velocity of the turbine itself is coupled with the wave particle velocity in calculating the inertia force for the mooring supported turbine as in Equation (23); this results in an increase of inertia forces on blade sections in the mooring supported turbine. Then the total torque increases as the in-plane forces rise. Moreover, the method calculating the inertia force on the blade section for a rigid supported turbine may not be applicable for a turbine that can move in the water as a tension mooring supported turbine does.

Figure 13 presents the torque on the mooring supported turbine without Morison effects in sea state 1, and the torque reduces as expected. This indicates that the Morison effect, the added mass on the blade, plays a significant role in torque on a turbine that can move vertically. Moreover, the Morison effect module in BEMT could be improved in the future and the added mass effects on the mooring supported turbine blade should be further investigated.

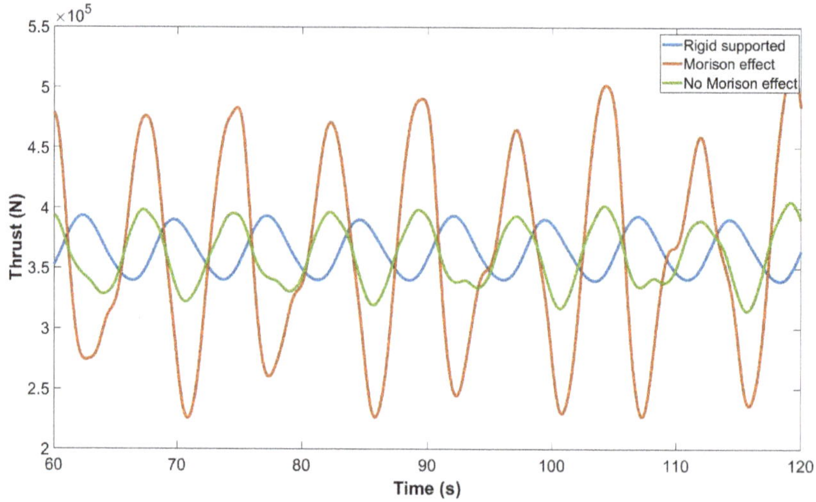

Figure 13. Morison effect in torque

6. Conclusions

The modified BEMT code for a tensioned mooring turbine itself is a very efficient tool and provides a fast simulation process. A 3 min simulation of a 0.1 [s] time step for 20 elements per blade and 0.5 [m] mooring line element lengths on a personal computer with 4 i7 cores takes approximately 1 h. The output contains various data, for example, inflow conditions on blade elements and forces on each element. The status of the turbine can be checked at every time step, and a simple animation can be generated for motions of the turbine. This is an efficient methodology for a highly dynamic system, such as a tensioned mooring supported turbine and other taut mooring systems, for checking its performance under various environmental conditions.

The dynamic marine climate causes significant load fluctuations on TSTs. This study found that the tensioned mooring supported system has a positive effect on thrust load reduction in most sea states. The system performance is satisfactory in swell waves, which have a long wavelength. Both thrust and torque fluctuations decrease significantly. The tensioned mooring-supported TSTs are useful for the swell environment or Stokes waves with long wavelengths. Therefore, installing this system can improve fatigue performance when compared with rigid supported TSTs.

In extreme conditions, such as harsh winter storms, the tensioned mooring supported TSTs significantly reduce the thrust loading; however, the turbine should be shut down and protected from potential damage to the blade due to negative torque. Furthermore, this stall phenomenon can be investigated by improving the blade design and adding a high angle of attack pitch angle to the blade so that the turbine can work under extreme conditions with a favourable load reduction.

The loading on the turbine is always calculated by a quasi-static method in most floating tidal turbine simulations, but the station keeping system uses a quasi-dynamic analysis. This study tried to combine the turbine loading calculation and dynamics of the mooring system by a modified BEMT code. However, this model can be enhanced with the addition of the torque calculation due to the added mass effect on blade sections caused by the quasi-static method, which results in the torque fluctuations. Further investigation should be done into the mass effects added to the mooring supported turbine blade.

In Stokes waves with short wavelengths, such as sea state 3 in the previous sections, the tensioned mooring supported TST can operate at the position where the wave particle paths can reach the whole turbine. For this, the diameter of the turbine must not to be too large and the buoy must supply a larger buoyancy when compared with the thrust loading on the turbine. The turbine can operate at an appropriate location in a water column. However, the load reduction in this condition is still not as satisfactory as under the other sea states; the performance is approaching that of the rigid supported turbine. Therefore, to obtain a satisfactory load reduction in thrust, the tensioned mooring supported system must be designed for application in environments where the water is not excessively deep.

The vertical velocity component of the wave motion had a large influence on the rotor out-of-plane bending moment of a turbine. Furthermore, the tension mooring supported turbine itself has a vertical velocity because it can move in a vertical direction, thus the out-of-plane bending moment may be greater than a rigid supported turbine. Furthermore, the torque on the rotor is also affected by the vertical velocity of the turbine itself due to the Morison effect. It is obvious that the vertical motion of the turbine has a significant impact on the design of the TST's drive train or internal components; therefore, investigations of the influence of the turbine's vertical velocity on the rotor out-of-plane bending moment should be further investigated in the future. On the other hand, future modifications to the original Morison effects may not be applicable to the blade elements on a turbine that can move in a vertical direction; therefore, modifications of Morison effects should be improved in the future. In extreme sea states, such as harsh winter storms, greater blade profiles and pitch angles should be investigated in order to avoid the negative torque. Adjustable rotor speed will be an another research target in the future and should be applied to the model in order to produce a reliable torque result.

Author Contributions: Conceptualization, S.F. and C.J.; methodology, S.F.; software, S.F.; validation, S.F. and C.J.; formal analysis, S.F.; investigation, S.F.; resources, S.F.; data curation, S.F.; writing—original draft preparation, S.F.; writing—review and editing, S.F. and C.J.; supervision, C.J.; project administration, C.J.; funding acquisition, C.J. All authors have read and agreed to the published version of the manuscript.

Funding: This research was funded by EPSRC, UKRI grant No. EP/K013319/1.

Institutional Review Board Statement: Not applicable.

Informed Consent Statement: Not applicable.

Data Availability Statement: Data sharing not applicable No new data were created or analyzed in this study. Data sharing is not applicable to this article.

Acknowledgments: The authors would like to thank EPSRC, UK RI for their support of this research via the award of Grant No. EP/K013319/1 Reducing the Costs of Marine Renewables via Advanced Structural Materials (ReC-ASM).

Conflicts of Interest: The authors declare no conflict of interest.

References

1. Johnstone, C.; Pratt, D.; Clarke, J.; Grant, A. A techno-economic analysis of tidal energy technology. *Renew. Energy* **2013**, *49*, 101–106. [CrossRef]
2. Simecatlantis. Projects. Available online: https://simecatlantis.com/projects/meygen/ (accessed on 13 June 2019).
3. Emec Tidal Devices. Available online: http://www.emec.org.uk/marine-energy/tidal-devices/ (accessed on 30 September 2016).
4. Clarke, Joseph Andrew and Connor, Gary and Grant, Andrew and Johnstone, Cameron and Ordonez Sanchez, Stephanie. Contra-rotating marine current turbines: Single point tethered floating system—Stabilty and performance. In Proceedings of the 8rd European Wave and Tidal Energy Conference, Uppsala, Sweden, 7–10 September 2009.
5. Huang, S. Dynamic analysis of three-dimensional marine cables. *Ocean Eng.* **1994**, *21*, 587–605. [CrossRef]
6. Johanning, L.; Smith, G.H. Mooring design approach for Wave Energy Converter. *J. Eng. Marit. Environ.* **2006**, *220*, 159–174. [CrossRef]
7. Smith, J.L. Station keeping study for wec devices including compliant chain, compliant hybrid and taut arrangement. In Proceedings of the 27th Offshore and Arctic Marine Engineering Symposium, Estoril, Portugal, 15–20 June 2008.
8. Xu, S.; Wang, S.; Guedes Soares, C. Review of mooring design for floating wave energy converters. *Renew. Sustain. Energy Rev.* **2019**, *111*, 595–621. [CrossRef]
9. Coiro, D.P.; Troise, G.; Bizzarrini, N. Experiences in Developing Tidal Current and Wave Energy Devices for Mediterranean Sea. *Front. Energy Res.* **2018**, *6*, 136. [CrossRef]
10. Martynyuk, A.A.; Nikitina, N.V. The Theory of Motion of a Double Mathematical Pendulum. *Int. Appl. Mech.* **2000**, *36*, 1252–1258. [CrossRef]

11. Nevalainen, T.; Johnstone, C.; Grant, A. An Unsteady Blade Element Momentum Theory for Tidal Stream Turbines. In Proceedings of the 11th European Wave and Tidal Energy Conference, Plymouth, UK, 5–9 September 2015.
12. Maria Przybylska, W.S. Non-integrability of flail triple pendulum. *Chaos Solitons Fractals* **2013**, *53*, 60–74. [CrossRef]
13. Nevalainen, T.; Johnstone, C.; Grant, A. A Sensitivity Analysis on Tidal Stream Turbine Loads Caused by Operational, Geometric Design and Inflow Parameters. *Int. J. Mar. Energy* **2016**, *16*, 51–64. [CrossRef]
14. Elmas Anli, I.O. Classical and Fractional-Order Analysis of the Free and Forced Double Pendulum. *Engineering* **2010**, *2*, 935–949. [CrossRef]
15. Ohlhoff, A.P.R. Forces in the Double Pendulum. *ZAMM J. Appl. Math. Mech./Z. Angew. Math. Mech.* **2000**, *80*, 517–534. [CrossRef]
16. Wu, G.; Witz, J.; Ma, Q.; Brown, D. Analysis of wave induced drift forces acting on a submerged sphere in finite water depth. *Appl. Ocean Res.* **1994**, *16*, 353–361. [CrossRef]
17. Bossanyi, E. *BLADED for Windows Theory Manual*; Garrad Hassan and Partners Limited: Bristol, UK, 1997.
18. Gaonkar, G.; Peters, D. Review of dynamic inflow modeling for rotorcraft flight dynamics. In Proceedings of the 27th Structures, Structural Dynamics and Materials Conference, Structures, Structural Dynamics, and Materials and Co-Located Conferences, San Antonio, TX, USA, 19–21 May 1986. [CrossRef]
19. Schepers, J.G.; Snel, H. *Final Results of the EU JOULE Projects Dynamic Inflow*; Technical Report; US Government Printing Office: Washington, DC, USA, 1926.
20. Tuckerman, L.B. *Inertia Factors of Ellipsoids for Use in Airship Design*; Technical Report; US Government Printing Office, 1926.
21. Buckland, H.; Masters, I.; Chapman, J.; Orme, J. Blade Element Momentum Theory in Modelling Tidal Stream Turbines. In Proceedings of the 18th UK Conference on Computational Mechanics, Southampton, UK, 8–10 April 2010.
22. Chapman, J.C. Tidal Energy Device Hydrodynamics in Non-uniform Transient Flows. Ph.D. Thesis, Swansea University, Swansea, UK, 2008.
23. Whelan, J.M.R.; Graham, J.P. Inertia Effects on Horizontal Axis Tidal-Stream Turbines. In Proceedings of the 9th European Wave and Tidal Energy Conference, Lincoln, NE, USA, 24–26 June 2009.
24. Nevalainen, T. The Effect of Unsteady Sea Conditions on Tidal Stream Turbine Loads and Durability. Ph.D. Thesis, University of Strathclyde, Glasgow, UK, 2016.
25. British Oceanographic Data Centre. Wave Data. Available online: https://www.bodc.ac.uk/data/online_delivery/waves/search/ (accessed on 30 September 2016).
26. Simecatlantis. Simec Atlantis Energy Unveils World'S Largest Single Rotor Tidal Turbine, The AR2000. Available online: https://simecatlantis.com/2018/09/13/simec-atlantis-energy-unveils-worlds-largest-single-rotor-tidal-turbine-the-ar2000/ (accessed on 13 June 2016).
27. Airfoil Tool. NREL's S814 Airfoil (s814-nr). Available online: http://airfoiltools.com/airfoil/details?airfoil=s814-nr (accessed on 9 October 2021).
28. ABP Marine Environmental Research Ltd. Atlas of UK Marine Renewable Energy. Available online: https://www.renewables-atlas.info/ (accessed on 15 August 2017).
29. Fenton, J.D. A fifth-order Stokes theory for steady waves. *J. Waterw. Port Coastal Ocean Eng.* **1985**, *111*, 216–234. [CrossRef]

energies

MDPI

Article

Wave Energy in the Pacific Island Countries: A New Integrative Conceptual Framework for Potential Challenges in Harnessing Wave Energy

Jessica Borges Posterari [1,*] and Takuji Waseda [2]

1 Graduate Program in Sustainability Science, The University of Tokyo, 515 Kashiwanoha, Kashiwa 2778563, Chiba, Japan
2 Department of Ocean Technology Policy and Environment, The University of Tokyo, 515 Kashiwanoha, Kashiwa 2778563, Chiba, Japan; waseda@k.u-tokyo.ac.jp
* Correspondence: jessica.posterari@s.k.u-tokyo.ac.jp

Abstract: The Central and South Pacific have significant wave energy resources distributed through the region that are currently not being explored. Even though the wave energy resource in the Pacific has been studied, there is limited knowledge on the potential obstacles when inserting this new energy source into a unique and unexplored environment. Pacific Island countries (PICs) have distinctive characteristics that can become barriers to this technology, especially considering that local coastal and marine systems are fundamental for subsistence and local development. Thus, the success of a project relies on local acceptance. The current study developed an integrative conceptual framework for the PICs (ICFPICs) that derived from the integration of the elements of a political, economic, social, technological, environmental and legal (PESTEL) structured approach and further combined with a strengths, weaknesses, opportunities and threats (SWOT) approach to create a matrix that included relevant internal and external factors influencing a project. Four islands were analyzed through the ICFPICs to demonstrate the varying characteristics and challenges in the Pacific environment; the islands were Tubuai (French Polynesia), Viti Levu (Fiji), Rarotonga (Cook Islands), and 'Eua (Tonga). Applying the ICFPICs to each island shows that Tubuai has significant technological issues, Rarotonga has mostly economic issues, Viti Levu is the most developed island but also has several potential issues in the social sphere, while 'Eua has the fewest issues and is a viable candidate for further analysis. The ICFPICs can be used by decision makers, project developers, and stakeholders to recognize probable barriers when bringing wave energy technologies to the PICs and make informed decisions during the pre-feasibility stage.

Keywords: Pacific; wave energy; island environment; PESTEL; framework; SWOT

Citation: Borges Posterari, J.; Waseda, T. Wave Energy in the Pacific Island Countries: A New Integrative Conceptual Framework for Potential Challenges in Harnessing Wave Energy. *Energies* **2022**, *15*, 2606. https://doi.org/10.3390/en15072606

Academic Editors: Eugen Rusu, Kostas Belibassakis, George Lavidas and Peter V. Schaeffer

Received: 18 October 2021
Accepted: 29 March 2022
Published: 2 April 2022

Publisher's Note: MDPI stays neutral with regard to jurisdictional claims in published maps and institutional affiliations.

Copyright: © 2022 by the authors. Licensee MDPI, Basel, Switzerland. This article is an open access article distributed under the terms and conditions of the Creative Commons Attribution (CC BY) license (https://creativecommons.org/licenses/by/4.0/).

1. Introduction

The ocean is an abundant resource for island countries, and can be used as a source of energy in areas where natural resources are limited. The Pacific Island countries have a history of relying on importing diesel for energy purposes, which not only causes environmental concerns but also social and economic concerns. Diesel-based electricity in these countries is often associated with fluctuating prices and high electricity tariffs. Harnessing energy from the ocean is one of the alternatives to establishing energy security and strengthening resilient development, and includes wave energy, ocean current energy, tidal energy, and ocean thermal energy. There is limited literature available on resource assessment for currents, tidal, and thermal energy for the Pacific Ocean; however, there are recent studies on wave energy that show several potential sites within selected Pacific Island countries.

When it comes to wave energy in the Pacific Islands, so far two main studies have assessed and quantified the resource in different locations. The variables analyzed, data source, and methods used were distinct for both studies, nonetheless, they have shown that wave energy is a possibility for different locations inside the Pacific. In 1996, a study by Barstow and Falnes [1], using buoy measurements and GEOSAT satellite altimeter observations, analyzed the wave climate and wave energy resource for seven PICs (Cook Islands, Tonga, Fiji, Samoa, Vanuatu, Tuvalu, and Kiribati). They also included a state of the art review of wave energy technologies, environmental and political considerations, demography data, and energy needs in the selected countries.

The last study on this field was conducted in 2015, by Bosserelle et al. [2], and expanded the domain for wave energy resource assessment and analyzed 33 islands from 12 different countries. This assessment was carried out using the Centre for Australian Weather and Climate Research (CAWCR) wave hindcast, containing data from 1979 to 2010 with a resolution of 7 km around the Pacific Islands. This hindcast was validated by measurements from Pacific buoys and used to calculate wave statistics for the Pacific domain (Figure 1).

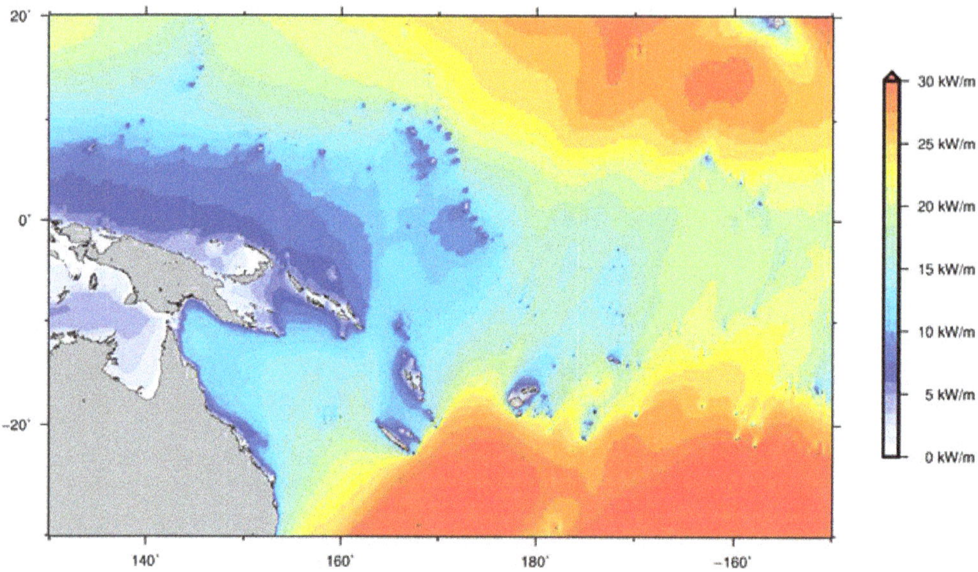

Figure 1. Average wave power in the Pacific (kW/m): reproduced from [2], Waves and Coasts in the Pacific, published by Pacific Community (SPC) 2015.

The South Pacific area shows significant potential for the deployment of wave energy converters based on the resource analysis and could be further explored to meet the renewable energy goals of the Pacific countries. Nevertheless, characterizing the resource itself does not provide enough information on the suitability of potential sites or the feasibility of wave energy at a specific location. As part of the site screening stage of a wave energy project, it is required to identify any potential challenges that could hinder the feasibility of installing and operating a wave energy converter. The project development process should assess the available data and their suitability to inform the initial feasibility of the project. Any data gaps and their relevance should be highlighted to identify, define, and prioritize the requirements for further and more detailed surveys [3]. This is done through a pre-feasibility study, which includes a preliminary resource assessment, as well as an overview of any potential obstacles for a project.

The nature of these obstacles can vary and, so far, there has been no predefined guideline on how to identify them. Combining all relevant information without a defined structure is a challenge since wave energy has direct and indirect relationships with different fields. Even though there is a lack of structured frameworks for the proposed scope, there is a significant amount of literature on different stages of a wave energy project, such as environmental impact assessment (EIA). An example would be the study by Mendonza et al. [4] that proposed a framework for environmental impacts on ocean energy devices, suggesting interactions between devices and the environment that should be considered for different devices. Their study mostly focused on the environmental dimension, which is in accordance with EIAs, while the present study focuses on the six PESTEL dimensions.

It is essential to gather pertinent information in the early stages of a project to increase the chances of success, especially in a new environment where wave energy has not been previously introduced. Two traditional concepts from the marketing and business fields that could be adapted to wave energy pre-feasibility studies are PESTEL and SWOT. These are two well-established decision- making tools that can structure information from a holistic perspective and assess unknown variables; however, there is limited literature on how to adapt them into the marine energy field. In this study we demonstrated how PESTEL and SWOT approaches can assist ocean wave energy projects in the Pacific to move forward. This has never been discussed in the literature and the presented framework can also be adapted to tidal, current, and thermal energy technologies in other regions.

1.1. PESTEL

PESTEL analysis has been evolving through the years as an integrative approach and can be found in different forms of literature. The origin of this approach is considered to be the book "Scanning the Business Environment" by Francis J. Aguilar [5], where the concept of economic, political, social and technological analysis (EPST) was first introduced. The acronym was later changed by Arnold Brown, becoming STEP, which was a rearrangement of the order of the words. The environmental factor was added subsequently becoming then social, economic, technical, political, and environmental (STEPE) factors. In general terms, the concept of environment involves far, near, and internal environments, including all types of factors related to the activities of the company [6]. The last addition to the acronym was the legal dimension, solidifying the concept of PESTEL that we have today. Even though the acronym can vary in the literature, the main idea behind it remains the same: provide the underlying structure for macro external environment analysis. PESTEL is commonly used for business and marketing research, being particularly useful to structure data and information that enables the company to predict situations and circumstances that it might encounter in the future [6].

Throughout the years, PESTEL has been adapted into different fields and its use has been expanded, even reaching the renewable energy sector. When it comes to the marine energy sector, the ocean energy converters have not fully reached market viability and are mostly in the research & development (R&D) stage. Thus, we are still not fully aware of the potential obstacles that are expected in new environments where ocean energy research does not exist. For this reason, there have been studies in the marine renewable energy field that incorporate the structure of PESTEL to conduct resource and feasibility analysis. Examples include the analysis done by Sandberg et al. [7] in 2016 regarding critical factors for wave energy converters in off-grid luxury resorts and small utilities, as well as a study on risk identification for the tidal industry using PESTEL by Kolios and Read [8] in 2013. Both papers utilize the PESTEL approach, however, the topics of interest are specific parts of the marine energy industry, and the outputs also differ. Sandberg et al. focused on the scenario of luxury resorts and small utilities and how viable wave energy would be; by contrast, Kolios and Read chose to focus on a literature review and a case study-oriented approach on the risks for the tidal industry.

1.2. SWOT and the PESTEL-SWOT Approaches

SWOT Analysis is a simple but powerful tool for sizing up an organization's resource capabilities and deficiencies, its market opportunities, and the external threats to its future [9]. It is a strategic planning framework that stands for strengths, weaknesses, opportunities, and threats. Strengths and weaknesses are internal factors and attributes of the organization, while opportunities and threats are external factors and attributes of the environment [10]. Heinz Weihrich [11] first introduced the concept of a matrix that identifies threats, opportunities, weaknesses, and strengths, naming it the TOWS matrix. Even though the matrix name varies in the literature, SWOT is a very well-known strategic and flexible framework that identifies key issues affecting business. Its adaption to different fields has been broadened, and it is possible to find studies applying SWOT to healthcare, agriculture, and tourism, among others. Within the marine energy field, Stingheru et al. [12] conducted a SWOT analysis of the European marine energy sector; this study highlighted positive and negative influences of harnessing marine energy at a European level to promote marine renewable energy.

It is also possible to combine PESTEL and SWOT to create a matrix that finds the strengths, weaknesses, opportunities, and threats by going through the external factors associated with PESTEL elements. Since the SWOT itself is a general matrix, the structure provided by PESTEL is useful for finding relevant external factors. The combined analyses of PESTEL and SWOT have not been adapted to marine energy technologies, nevertheless, they have been used for other renewable energy studies in the past. Damasceno and Abreu [13] created a PESTEL-SWOT evaluation method for the wind energy sector that assessed the favorable conditions and challenges for the wind energy market to expand in Brazil. Moreover, there is a study from Shadman et al. [14] utilizing stakeholder engagement, PESTEL, and SWOT analysis to assess the role of renewable energy for energy security in Malaysia. Finally, Islam and Mamun [15] have researched the possibilities and challenges of implementing renewable energy in island countries by utilizing both the PESTEL and SWOT approaches; their study did not focus on any particular region or technology, instead providing a broad view on the strengths, weaknesses, opportunities, and threats for renewable energy in developing and developed nations.

1.3. Novelty and Objectives

Wave energy is approached in several disciplines with a focus on different project stages, creating clusters of information that are interconnected, yet seldom integrated into the literature. This study applies concepts of interdisciplinary research into wave energy in the context of Pacific Island countries, combining a diverse range of elements from the PESTEL dimensions into a framework. Inserting wave energy into a new location will not only have technological and economic repercussions but will also affect the local environment and society. An integrative framework enables us to represent the diversity of issues and delineate the important variables that can turn a project unfeasible or create fundamental dissents between decision makers and stakeholders.

The options for the economic development of the PICs are restricted by limited natural resources, remoteness, and small land size. Local communities rely on sustainable use of their local resources for subsistence and income, which makes coastal and marine resources paramount to the local economy, society, and culture. Marine energy development is highly susceptible to local acceptance in this region, thus, identifying conflicts of use through a general framework will be crucial. There are four main objectives behind this study:

1. Review the literature on wave energy harnessing and the Pacific Islands environment;
2. Identify the potential challenges and important factors for wave energy in the Pacific;
3. Structure the information found using the PESTEL approach by combining elements into the relevant clusters—the ICFPICs;
4. Create a SWOT matrix using the identified factors from the ICFPICs as external factors.

The outcomes include a cluster diagram that represents the ICFPICs, a decision tree for the process of utilizing the framework, and a SWOT diagram constructed based on the information gathered. Section 2 describes the materials and methods for this study, including the resources used to construct the ICFPICs and the methodologies applied. Section 3 presents the results, which incorporates the ICFPICs diagram, a user-friendly decision tree, and a combined SWOT-PESTEL matrix. Discussions and case studies for

Pacific Island countries can be found in Section 4, while conclusions are summarized in Section 5.

2. Materials and Methods

2.1. Materials

2.1.1. Wave Energy Guidelines

With regards to project development on wave energy, the main guideline is "Guidelines for Project Development in the Marine Energy Industry" by EMEC [3], which also includes important information regarding feasibility assessment. Additional supporting documents are: "Protocols for wave and tidal resource assessment" [16], and "Impacts upon marine energy stakeholders" [17] by EquiMar Project; "Wave Energy Technology Brief" [18], "Renewable Energy Technology Innovation Policy" [19], "A Path to Prosperity: Renewable Energy for Islands" [20], and "Renewable Energy Opportunities For Island Tourism" [21] by IRENA; "Documentary summary of the environmental impact of renewable marine energy" [22], and "Civil society involvement and social acceptability of marine energy projects" [23] by MERiFIC.

2.1.2. Pacific Island Countries Reports

All the reports used here came from the two main organizations in the Pacific, which are the Pacific Community (SPC) and the Secretariat of the Pacific Regional Environment Programme(SPREP). These included: SPREP Annual Report [24], Status and Trends of Coral Reefs of the Pacific [25], Pacific Regional Energy Assessment [26], Pacific Marine Climate Change Report Card [27], and Pacific Community Results Reports [28]. Each document provided information on the current or latest status of the environmental, social, and economic sectors. They were the basis of understanding the local vulnerabilities and intricacies that should be considered during a project.

2.2. Methods

2.2.1. Integrated Conceptual Framework for the Pacific Island Countries

By studying the concepts behind wave energy harnessing and the Pacific environment, it was possible to construct an integrated conceptual framework for the Pacific Island countries (ICFPICs). The ICFPICs is integrative since it integrates knowledge and concepts from different fields and organize variables connected to a central idea; it is also a conceptual framework considering that all the information here gathered stems from literature reviews and methodological assumptions. After reviewing the available documents, guidelines, and reports, every item considered to be a potential challenge was categorized using the PESTEL approach and fit into a cluster diagram. Each PESTEL dimension was structured into separate clusters that revolved around a core concept, and where elements shared similar characteristics.

2.2.2. PESTEL-SWOT Combined Analysis

The ICFPICs provided an overview of the relevant elements for wave energy in the Pacific, which were be combined with the SWOT analysis to point towards the related internal and external factors. Each item from the ICFPICs was analyzed from the perspective of strengths, weaknesses, opportunities, and threats, resulting in a PESTEL-SWOT combined matrix. In this case, strengths and weaknesses were related to wave energy, representing the internal factors; opportunities and threats were external factors, being related to the macro-environmental variables that could justify or hinder a project. The resulting matrix shows the possible factors for each category and should be further adapted to individual case studies.

3. Results

Each division of the framework is explained below and further divided into important factors. Further information on each PESTEL element can be found in Appendix A, where the reasoning behind their connection to wave energy in the Pacific is explained.

The decision to invest in renewable energy, regardless of the source, is still connected to the political sphere of a country. This is particularly important for wave energy or marine technologies as a whole, considering their long lifecycle and higher investment rate of return. For the political sector, the main factors are political stability, renewable energy targets, and government support.

How affordable the energy is, the risks, and the benefits involved are all essential parts of the feasibility study of a wave energy project. The main concepts to be analyzed here are the cost of energy, the risks of the project, and the feed-in tariff. Therefore, for the economic sector, the main factors are economic stability, cost of energy, feed-in tariff, risk assessment, and access to funds.

Social aspects of an island nation are fairly complex; they are usually associated with the local environment and local economy making it an intricate task to classify which factors only belong to the social sphere. For instance, exploring natural resources, such as sand, can be for home construction or for export. For this study, social factors are all the activities that involve the local society, either to make profits or for subsistence purposes. For the social sector, the main factors are offshore mining, tourism, navigation, fishery, aquaculture, recreation sites, cultural and world heritage sites, and local acceptance.

The processes of building, maintaining, and connecting a WEC to the grid require additional infrastructure and resources. Bringing the device from the supplier to the potential site is a long process that relies on a port for logistic purposes, and also on specialists and qualified workers. In addition to that, if the main objective is to supply electricity to the main grid, suitable grid infrastructure is required to minimize installation costs. Therefore, for the technological sector, the main factors are electricity supply and demand, electricity grid, seaports, expertise, and logistics.

The island environment provides essential services to the local communities and therefore is a crucial element when analyzing the feasibility of any type of development. Wave energy converters are placed in the ocean, where there might be located important marine species or coral reefs. These devices might also be subjected to extreme weather scenarios and this factor will be accentuated if the WEC is situated in a hazard-prone area. For the environmental sector, the main factors are natural hazards, biodiversity, and coral reefs.

Lastly, there are mainly two forms of legal concerns regarding wave energy, which are related to the physical space and energy regulations. There are offshore zones that are either prohibited from being accessed or under protection, including areas being utilized as military bases, conservation areas, or the boundaries of the EEZ. Each country also has its own regulations when it comes to the generation of electricity and these should be accounted for. For the legal sector, the main factors are regulations, maritime zones, marine protected areas, military zones, and dependent territories.

3.1. Integrative Conceptual Framework for the PICs (ICFPICs)

Figure 2 shows an illustration of all the PESTEL components combined and each of their related factors. There is a total of 29 potential obstacles that were identified through the previous stages, and these are grouped by categories on the resulting visual representation. Each item should be further analyzed for a particular location by gathering relevant information and by mapping barriers when applicable.

To facilitate the process of utilizing the ICFPICs, a decision tree was created (Figure 3). The purpose is to assist the user in identifying potential challenges and determining which actions to take afterward. As an example, the tree starts with the ICFPICs item aquaculture, which is shown here as "Presence of aquaculture activities"; for this item, decision makers need to verify if it will present a challenge to the project and follow the necessary steps. In brief, for each challenge encountered the solutions will vary between five main options: changing the location of the site, conducting further analysis, consulting stakeholders, including additional costs to the project, and canceling or postponing the project. This is a generic tool that gives an overview of the process; however, it should be adapted to each case study and could potentially include additional steps or solutions.

Figure 2. Integrative conceptual framework for harnessing wave energy in the PICs (ICFPICs).

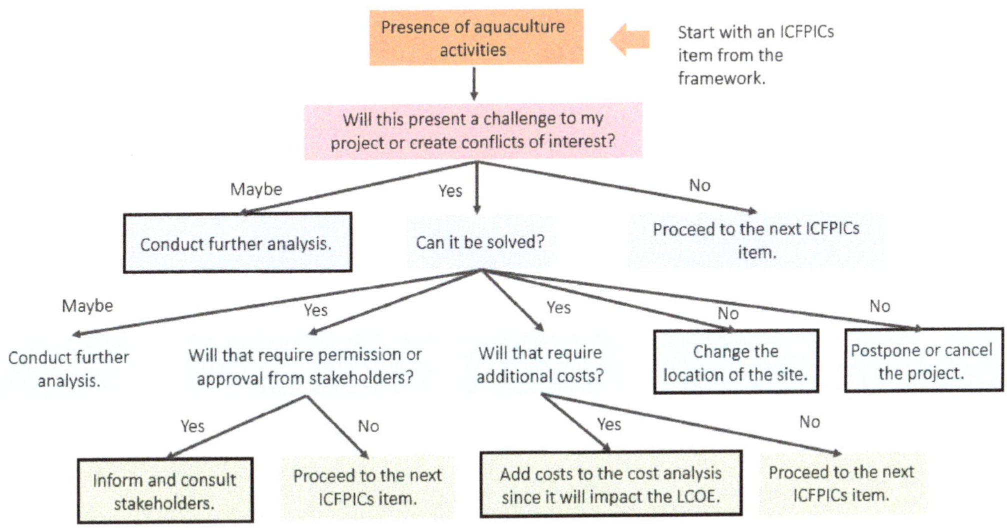

Figure 3. Decision tree exemplifying the process of utilizing the ICFPICs for a wave energy project in the Pacific environment; highlighted boxes are the main outcomes from the process.

3.2. The Combined PESTEL-SWOT Analysis

To demonstrate how PESTEL-SWOT analysis can be utilized, every item from the ICFPICs was studied as an external factor and fitted into a SWOT matrix (Table 1). Wave energy was surveyed in the context of the energy resource to be harnessed, and how it would contribute to the local society. The resulting matrix should be adapted to different case studies and provides a run-through of the feasibility of a wave energy project. Strengths and weaknesses are related to the process of installing, operating, and maintaining a wave energy converter device, which is mostly technological. Opportunities and threats are covered by the political, environmental, social, technological, economic, and legal categories.

Table 1. PESTEL-SWOT analysis combined; elements in the ICFPICs were used to create a SWOT matrix.

Strengths	**Weaknesses**
Alternative to importing fossil fuels. Robust structures that can survive harsh environments. Possibility of having a competitive cost of energy. Increases resilience through low-carbon development. Increases energy security by diversifying sources of energy. Island nations have ample ocean resources. High annual energy production that can cover energy demands of small islands. Several WEC devices are available to suit the local wave climate and geophysical conditions.	High discount rates. Uncertainty and risks are bound to the project. The wave energy sector is at the development stage. Significant distance from suppliers to the Pacific Island countries. Expertise is required to install, operate, and maintain the device. Lack of electricity grid in remote islands. A seaport is required to handle the equipment shipping.
Opportunities	**Threats**
Feed-in tariff scheme in practice. Local government support. Funding opportunities. Renewable energy targets. High wave energy resources distribution. Low seasonal, annual, and inter-annual variability. Multiple suitable locations for wave energy harnessing. Energy output can be used to power desalination plants. Job creation for different fields of expertise and training opportunities. Progress in the Sustainable Development Goals achievement through Goal 7.	Lack of regulation of wave energy. Marine protected areas, maritime zones, and military zones limiting the location of a WEC. Frequency of natural hazards such as hurricanes. Potential dangers to local biodiversity and coral reefs. Lack of sovereignty and additional bureaucracy. Political instability and economic instability. Presence of fishing, aquaculture, touristic, recreational, and offshore mining sites. Cultural and World Heritage sites with natural and cultural values. Lack of approval from the local communities.

4. Discussion

PICs face unique challenges and have distinct traits, such as narrow-based economies, limited natural resources, fragile ecosystems, reliance on subsistence activities, and remoteness from major markets. Thus, having a framework that encompasses regional characteristics is important when identifying challenges. The ICFPICs created here focused on wave energy on a regional level and is unique in the sense that such a concept has not been attempted yet.

There are different prospective applications for the ICFPICs; it can serve as a tool to identify potential challenges to a project, it can be used to identify key stakeholders, and lastly, it can be combined with the SWOT approach, giving an overview of both the barriers and opportunities for wave energy in the Pacific Island countries. The information gathered during the process of utilizing the framework is useful to create a SWOT matrix, which in turn will give an overview of the feasibility of a wave energy project by further categorizing the data into strengths, weaknesses, opportunities, and threats.

4.1. The ICFPICS Demonstrated for Four Case Studies

Four distinct Pacific Islands were selected to demonstrate how the ICFPICs can be utilized, and the potential challenges identified for every island are shown in Figures 4–7. The process included researching each item presented in the ICFPICs cluster diagram (Figure 2) and verifying if it was considered as a potential challenge. The analysis was based on country reports, regional reports, and official statements. Every item in Table 2 either posed a threat to a wave energy project on the island or could not be further verified; items that are not included were found to be non-threatening.

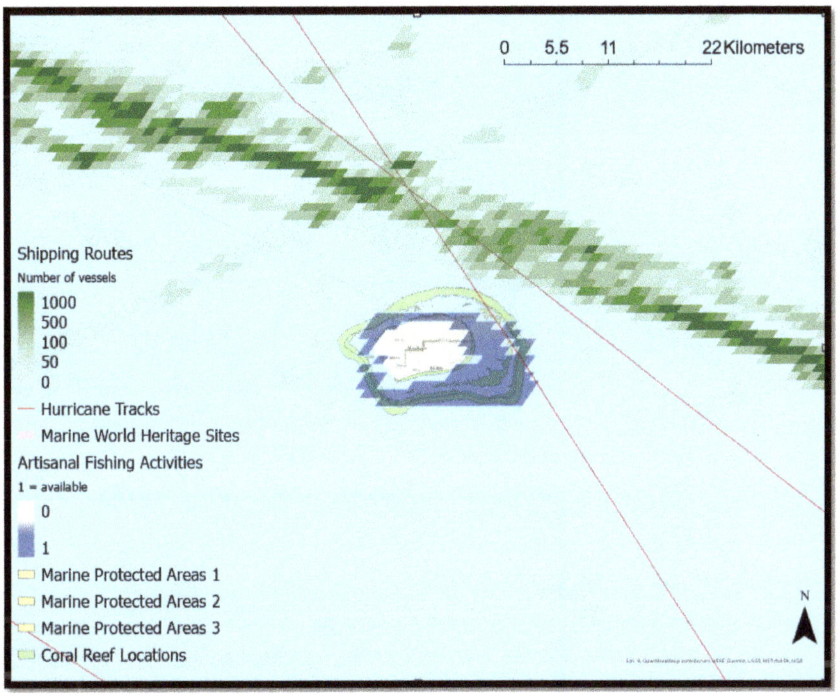

Figure 4. Map for Tubuai demonstrating challenges found through spatial analysis. The shipping routes are shown using a green color scale to represent the number of vessels that traveled through this area for one year (each raster has a 920 m^2 area). Artisanal fishing activities are present in the blue shaded areas. The remaining layers are colored separately, red for hurricane tracks, yellow for marine protected areas, light green for coral reefs, and pink for marine world heritage sites.

Tubuai is part of the Austral island group in French Polynesia, Viti Levu is the main island of Fiji, Rarotonga is the main island of the Cook Islands, and 'Eua is part of the Kingdom of Tonga. The analysis results demonstrate how each island has different characteristics and, consequently, different challenges. All four islands have established renewable energy targets, no signs of political instability, and no electricity grid available; however, since marine energy is relatively new, there is no information regarding government support. Cost of energy and risk are both unknown factors since resource assessment, risk assessment, and further analysis are needed. Additional common challenges were found to be the distance from the main markets, causing logistics issues, a lack of marine energy regulations, the presence of coral reefs, potential natural hazards, and a rich marine biodiversity environment. It is also important to note that access to funds and local acceptance will rely on the outputs from resource assessment and cost analysis, thereby being a potential barrier. The maps presented here were created using databases for protected areas [29], marine World Heritage Sites [30], coral reefs [31], hurricane tracks [32], and marine activities [33].

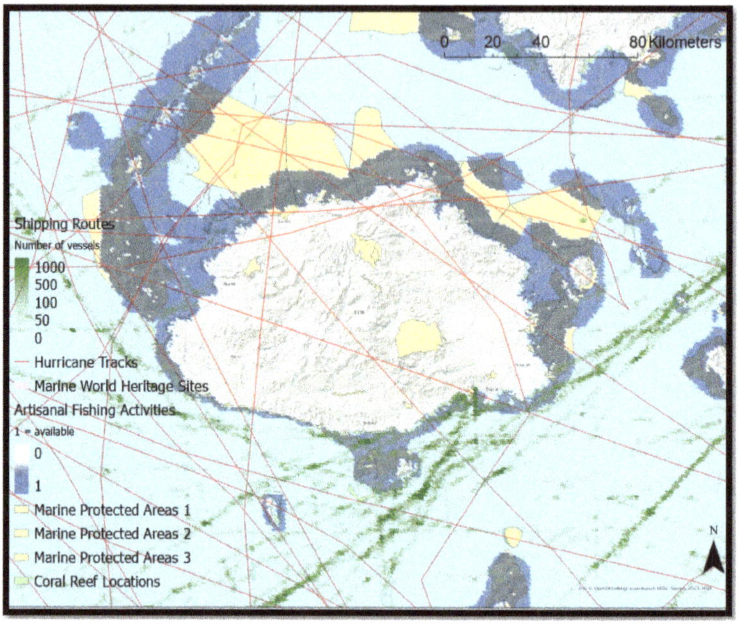

Figure 5. Map for Viti Levu demonstrating challenges found through spatial analysis. Color layers as per explained in Figure 4's caption.

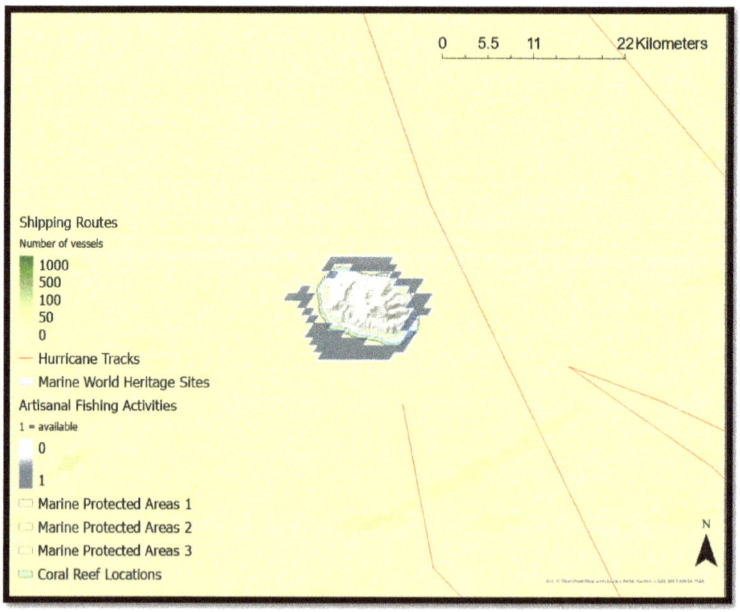

Figure 6. Map for Rarotonga demonstrating challenges found through spatial analysis. Color layers as per explained in Figure 4's caption.

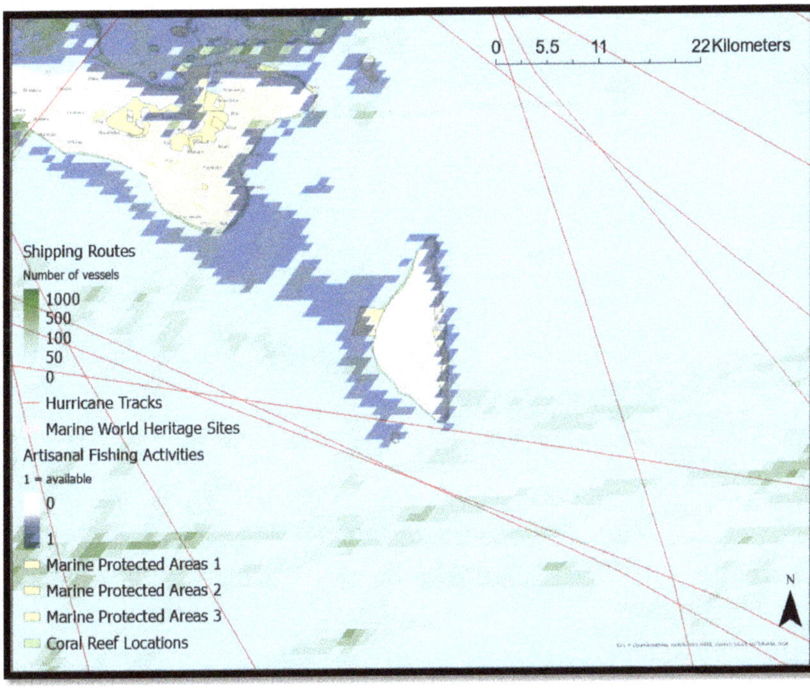

Figure 7. Map for 'Eua demonstrating challenges found through spatial analysis. Color layers as per explained in Figure 4's caption.

Tubuai is the most remote island among the selected sites, it is located in the south of French Polynesia with a population of approximately 2200 [34]. Due to the small population size, the energy demand is also low, and the technological aspects of wave energy might be a challenge. The Tubuai harbor has been expanded in 2014, yet, it has limited uses and might not accommodate large shipments. Even though there are touristic and recreational sites on the island, the influx of tourists is low; in 2013 it was recorded a total of 1899 visitors [34] for all Austral islands combined. The Historical Hurricane Tracks by the National Oceanic and Atmospheric Administration shows that there have been 11 storms in a 100 km radius around the island since 1970; only three so far have been classified as hurricanes. Additional challenges are the fishing activities, a large coral reef area surrounding the island, a proposed Marine Reserve zone for the Austral islands [35], artisanal fishing activities nearshore, high influx of vessels to the North, and the fact that French Polynesia is a Dependent Territory.

Tubuai is part of the Austral Island group in French Polynesia, Viti Levu is the main island of Fiji, Rarotonga is the main island of the Cook Islands, and 'Eua is part of the Kingdom of Tonga. The results of the analysis demonstrate how each island has different characteristics and, consequently, different challenges. All four islands have established renewable energy targets, no signs of political instability, and no electricity grid available; however, since marine energy is relatively new, there is no information regarding government support. Cost of energy and risk are both unknown factors since resource assessment, risk assessment, and further analysis are needed. Additional common challenges were found to be the distance from the main markets, causing logistics issues, a lack of marine energy regulations, the presence of coral reefs, potential natural hazards, and a rich marine biodiversity environment. It is also important to note that access to funds and local acceptance relies on the outputs from resource assessment and cost analysis, thereby being a potential barrier. The maps presented here were created using databases for protected areas [29], marine World Heritage Sites [30], coral reefs [31], hurricane tracks [32], and marine activities [33].

Table 2. Identified challenges for each island based on the ICFPICs.

	Tubuai (French Polynesia)	Viti Levu (Fiji)	Rarotonga (Cook Islands)	'Eua (Tonga)
Political	government support	government support	government support	government support
Economic	cost of energy risk assessment access to funds	cost of energy risk assessment access to funds	cost of energy risk assessment access to funds economic stability feed-in tariff	cost of energy risk assessment access to funds
Social	tourism fishery recreational sites local acceptance	tourism fishery recreational sites local acceptance offshore mining navigation aquaculture cultural and World Heritage Sites	tourism fishery recreational sites local acceptance offshore mining navigation aquaculture	tourism fishery recreational sites local acceptance
Technological	logistics expertise electricity supply and demand seaport electricity grid	logistics	logistics expertise	logistics expertise electricity supply and demand
Environmental	natural hazards biodiversity coral reef	natural hazards biodiversity coral reef	natural hazards biodiversity coral reef	natural hazards biodiversity coral reef
Legal	regulations maritime zones dependent territories	regulations marine protected areas military zones	regulations marine protected areas	regulations marine protected areas

Tubuai is the most remote island among the selected sites; it is located in the south of French Polynesia with a population of approximately 2200 [34]. Due to the small population size, the energy demand is also low, and the technological aspects of wave energy might be a challenge. The Tubuai harbor was expanded in 2014, yet it has limited uses and might not accommodate large shipments. Even though there are touristic and recreational sites on the island, the influx of tourists is low; in 2013 a total of 1899 visitors were recorded [34] for all Austral Islands combined. The Historical Hurricane Tracks by the National Oceanic and Atmospheric Administration shows that there have been 11 storms in a 100 km radius around the island since 1970; only three so far have been classified as hurricanes. Additional challenges are the fishing activities, a large coral reef area surrounding the island, a proposed marine reserve zone for the Austral Islands [35], artisanal fishing activities nearshore, a high influx of vessels to the North, and the fact that French Polynesia is a dependent territory.

Viti Levu is the largest island in Fiji and, therefore, an important place for social and economic activities. As of 2017, all eight provinces of Viti Levu combined (Ba, Ra, Nadroga-Navosa, Naitasiri, Tailevu, Namosi, Rewa, and Serua) accounted for 662,205 inhabitants [36]. In addition to the tourism and fishing industries, there is a possibility of aquaculture and deep-sea mining activities. Due to its development and high population status, Viti Levu has seaports, an electricity grid, high demand, and expertise available. Nevertheless, Fiji's geographical location makes it susceptible to environmental hazards; cyclones are a common occurrence, and there have been 45 storms in a 100 km radius around the island since 1956, including two category 5 hurricanes. Further challenges include the presence of a cultural World Heritage Site named Levuka Historical Port Town [37], a significant coral reef zone, several active shipping routes surrounding the island, and several marine protected areas. Additionally, the presence of the Fijian Navy and the location of its fleet should be considered for any offshore development. Despite the technological and economic advantages, Viti Levu is bound by environmental, legal, and social challenges.

Even though Rarotonga is the main island of the Cook Islands group, it has a small population size of approximately 10,572, according to the Cook Islands Demographic Profile 2006–2011 [38]. The economic stability is undefined, as are any feed-in tariff schemes or government programs to subside renewable energy. For the social aspect, there are potential sites for deep-sea mining in the exclusive economic zone, fishing activities, touristic attractions, recreational sites, nearshore corals, and navigation routes in the west, east, and south. Unlike Viti Levu, Rarotonga is not at the risk of intense hurricane events; nevertheless, there have been 21 storms tracked in a 100 km radius that should be considered, even though they were mostly tropical storms. Lastly, the Cook Islands are dedicating their entire exclusive economic zone, Marae Moana, an area of 1.9 million square kilometers (550,000 square nautical miles) to protection, conservation, and integrated management [39], which might lead to conflicts with the legal, environmental, and social sectors.

The last site, 'Eua, stood out among the selected islands due to having fewer potential challenges. 'Eua has small tourism, fishery, and maritime sectors; it also has a well-maintained electricity grid, a seaport, and the possibility of a feed-in tariff scheme. When it comes to the electricity supply factor, the small population size of fewer than 5000 residents might become an issue. Thus, one of the key elements to bring wave energy to 'Eua is balancing the energy output around the local demand to justify costs. The ICFPICs also identified possible challenges related to the presence of protected areas on the island, the need for expertise, and annual hurricane occurrences. Since 1958, there have been 47 category 1 or above hurricanes, which also included 3 category 4 hurricane events. Lastly, artisanal fishing activities are present in the west and east of 'Eua, as well as a high influx of vessels towards the west and south of the island. However, there are viable sites nearshore in the south that could be explored, especially considering that higher wave energy resources are found within this area.

In these examples, each island presented different challenges in the economic, social, technological, environmental, and legal categories, demonstrating the diversity of the Pacific Island countries. For instance, it would be particularly difficult to bring wave energy to Tubuai due to technological and environmental constraints being prevalent. Nevertheless, larger islands such as Viti Levu and Rarotonga can still present challenges, such as environmental hazards and conflicts of use of the offshore area. If we consider the minimum distance to avoid the obstacles presented in Figures 4–7, 'Eua has the lowest at 0.1 km, Rarotonga has the highest at 378.35 km, while Tubuai and Viti Levu have, respectively, 2.17 km and 1.39 km minimum distance. This means that 'Eua could potentially have onshore, nearshore, and offshore wave energy converter devices, increasing the diversity of options.

4.2. Further Analysis of 'Eua Island

The ICFPICs was applied for an in-depth analysis of the island of 'Eua. Furthermore, data obtained from Tongan governmental agencies and local institutions were used for this analysis to increase representation reliability. Data related to the fisheries, biodiversity, and tourism sectors were taken from Ministry of Fisheries [40], Marine and Coastal Biodiversity Management in Pacific Island Countries (MACBIO) [41], and Ministry of Tourism [42] sources, respectively. Each category from the ICFPICs is further explained below.

Political: Tonga experienced serious rioting in the capital Nuku'alofa in 2006 but adopted a democratic constitution in 2010 and appears to have returned to its earlier pattern of long-term political stability [43]. In addition to that, Tonga has already committed to achieving 50% renewable energy generation by 2020 and 70% by 2030, which is a motivator for wave energy to be further studied.

Economic: The cost of energy and the qualitative and quantitative risks need to be assessed in order to analyze if economic factors will be an obstacle for wave energy. Nevertheless, the final results will also depend on how the cost range compares to the current energy sources and to the average costs in the world; the cost alone cannot convey enough information, which is why the cost and risk analysis are necessary steps to identify any potential concerns.

Social: There are no documented offshore mining and aquaculture activities or heritage sites for the island of 'Eua. There are, however, villages that date back thousands of years and have cultural value, including Ohonua, Tufuvai, Pangai, Houma and Ha'atua. The island has only one port, which is situated at 'Ohonua and connects 'Eua to the Tongatapu island, the main route of the local ferries. When it comes to fisheries, 'Eua does not have a significant export rate and according to the latest statistics [44], only 12% of the households practice fishing for consumption or for sale. Nevertheless, there are sites near 'Eua using fish aggregating devices that have been deployed to increase fish production. Lastly, because of the presence of humpback whales, 'Eua is a fairly touristic island, with touristic activities being mainly whale watching, cave diving, and snorkeling.

Technological: The electrification rate for the Kingdom of Tonga is high and close to 100%, which also includes 'Eua. The network in 'Eua was also rebuilt in 2017 and was able to withstand Cyclone Gita, according to the TPL Annual Report from 2018 [45]. Therefore, grid-related issues are not a main concern. They also have the 'Ohonua port that serves as a navigation route between islands and for cargo transportation, which can be used for WEC shipments and related services. When it comes to expertise, however, there might be a lack of qualified professionals to work on the installation and maintenance of a WEC, considering that offshore development is non-existent in 'Eua. This would require additional funds but would open job opportunities and motivate public acceptance.

Environmental: 'Eua is situated in an area where hurricanes are relatively common occurrences. The cyclone season in Tonga is from November to April, however, the peak time for tropical cyclones in Tonga is from January to March with most events occurring in February [46]. The presence of cyclones should be accounted for in risk quantification and the cyclone season should be avoided in installation and maintenance procedures. Biodiversity is also another potential issue, with the humpback whales pathing around Tonga once a year from July to October, including around 'Eua. Pelagic sharks are also present on this island and are protected by the Kingdom of Tonga National Plan of Action (NPOA) Shark Plan [47]. In regard to corals, 'Eua has low coral reef resources, which are limited to the shallow areas around the island.

Legal: There are few regulations for marine energy in Tonga considering the lack of projects in this field. There are, however, important policies related to renewable energy, including: Renewable Energy Act 2009, Electricity Act 2007, Environment Impact Assessment Act 2003, Spatial Planning and Management Act 2012, and Petroleum Act. There are no documented maritime zones for 'Eua and there is also no military base in this area, which should not pose any risk to the project in terms of conflicts of use. The EEZ of 'Eua is relatively large, and it is unlikely that any wave energy development would trespass this area. There are, however, two marine protected areas in the island: 'Eua National Park and Tufuvai.

To better understand the boundaries for wave energy on the island of 'Eua, a map was created to add important locations found through the ICFPICs. The factors that could be mapped are present at Figure 8, which includes locations for fishing spots, touristic and recreational spots, important villages, the 'Ohonua port, and the ferry route between Tongatapu and 'Eua. Areas where the presence of humpback whales and pelagic sharks have been observed were also added, as well as the coral reef sites. Since the 'Euan population is mostly concentrated on the west coast, while a large part of the east coast is within the 'Eua National Park limits, the east coast was not considered for site selection.

Four sites representing the north, northwest, southwest and south were selected as potential WEC sites and are shown as "Analysis Points" on the map. The selection process included bathymetry analysis as well as proximity to the main populated districts. Table 3 provides an overview of the local wave climate and wave energy resource for each point, using annual climatology data from the CAWCR wave hindcast [48].

Table 3. Wave climate and wave energy parameters for each point.

	Point 1	Point 2	Point 3	Point 4
Mean Wave Height (m)	1.50	1.19	1.28	1.78
Mean Wave Period (s)	8.65	9.19	9.23	8.76
Mean Wave Energy Flux (kW/m)	10.94	11.02	12.96	19.66
Inter-Annual Variability (%)	7.75	9.52	9.13	5.87
Seasonal Variability (%)	14.67	29.99	30.89	24.77

Points 1 and 2 have the most constraints, being surrounded by fishing areas, beaches, and pelagic sharks. Furthermore, point 2 also has the port nearby, being in close proximity to the ferry route. The only obstacle identified for points 3 and 4 is the possible presence of pelagic sharks, which makes these sites the most suitable for wave energy harnessing in terms of feasibility factors. Presence of coral reefs should only be a concern for shallow areas nearshore, until approximately 5 m depth in most parts. Nonetheless, there is also the distance from grid-connected areas factor to be considered; point 4 is the furthest from any residential area and, therefore, requires longer transmission lines. When it comes to the physical resource, point 4 has the highest mean wave energy flux and the lowest inter-annual variability, being a strong candidate for a WEC, followed by point 3. Seasonal variability seems to be an issue for the most of 'Eua, including points 2–4.

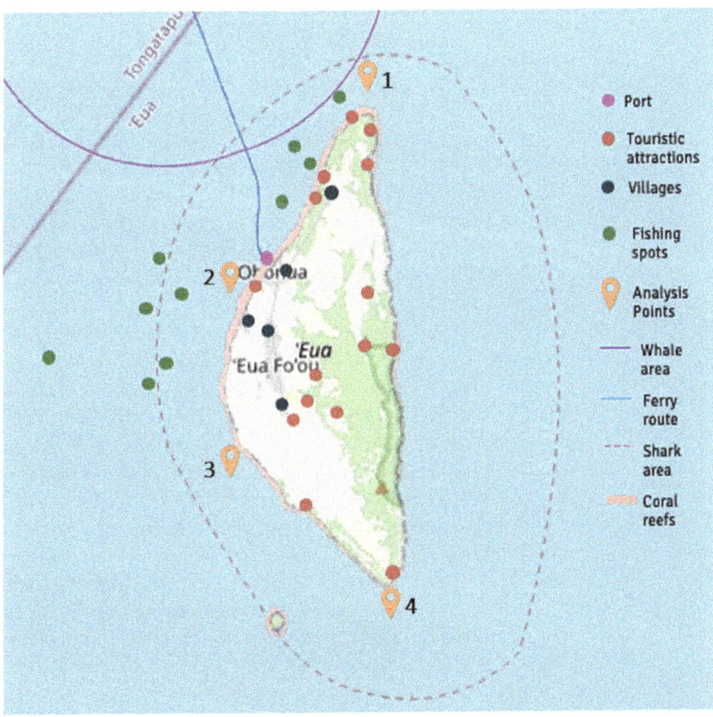

Figure 8. Map of 'Eua containing important sites and possible constraints. The map graphically presents the result of the PESTEL analysis and the potential sites for a WEC.

5. Conclusions

The Pacific region has already been shown to be promising in terms of wave energy resources even while the external factors remain unknown. This study proposed a framework that has an essential role in identifying scenarios where wave energy is considered feasible, making it a useful tool for project developers and decision makers. Figure 9 summarizes the functions of the ICFPICs and how it relates to its final goal. The framework allows the user to identify potential challenges and external factors through the ICFPICs elements, facilitating the process of defining suitable sites based on the results found before committing resources for site-specific feasibility studies and technical assessments.

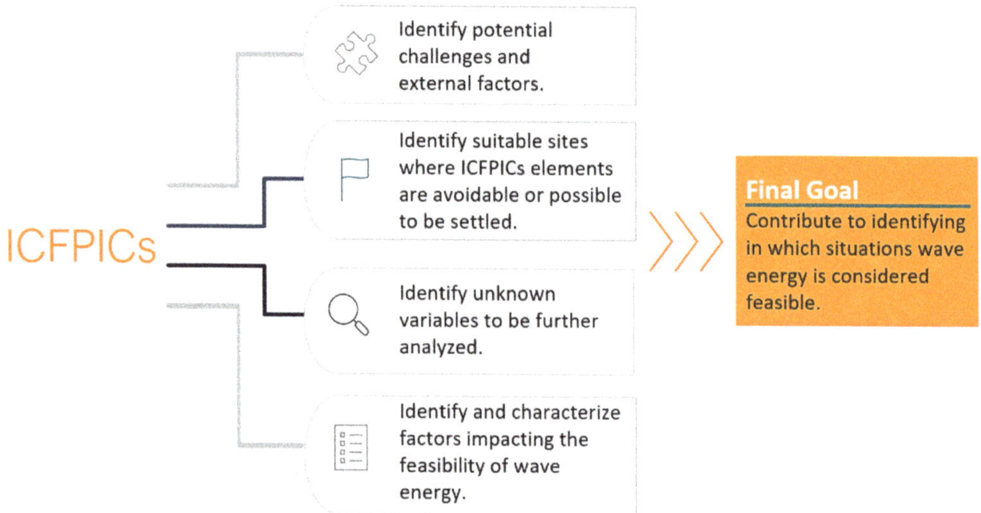

Figure 9. Diagram summarizing the functions of the ICFPICs.

It is recommended to conduct resource and cost analysis alongside the ICFPICs to assess the overall feasibility of a project. While the framework created here is useful for early assessments to provide a general overview of relevant factors, it should not replace further project stages such as environmental impact assessment, for instance. Given that the ICFPICs is flexible in its use and can tackle different issues, the user is also encouraged to adapt it to new locations and different marine energy technologies, such as tidal, current, and ocean thermal energy.

The four study cases presented earlier exemplified how the ICFPICs can find suitable locations and compare different scenarios for wave energy. All four islands are located in areas with high wave energy resources, nevertheless, several issues that could deem wave energy unfeasible were found. As a future step, there is value in 'Eua for wave energy harnessing, considering its favorable conditions found through a first assessment. Following this study, resource assessment, risk assessment, and cost analysis studies are suggested for 'Eua.

Possible obstacles found for wave energy in 'Eua are related to the local biodiversity, tourism, natural hazards, cost of energy, and economic risk. According to preliminary results, the coastal areas near Ha'atu'a (point 3) and Li 'Anga Huo 'A Maui (point 4) are recommended as potential WEC sites. Moreover, the levelized cost of energy (LCOE) should be quantified considering the risks of unplanned maintenance due to natural hazards, overhaul, wave climate variability, and uncertain shipping costs for the WEC infrastructure. The cost analysis also needs to include possible variability for the costs of a singular WEC device, as well as conversion rates and discount rates. Even though further quantitate analysis is necessary, with the ICFPICs it was possible to identify feasible scenarios for wave energy in different areas of 'Eua, as well as to characterize potential obstacles.

Author Contributions: Conceptualization, J.B.P.; funding acquisition, T.W.; methodology, J.B.P.; supervision, T.W.; writing—original draft, J.B.P.; writing—review & editing, T.W. All authors have read and agreed to the published version of the manuscript.

Funding: A part of this study was funded through the Science and Technology Research Partnership for Sustainable Development (SATREPS) program between Japan and Malaysia, titled "Development of Advanced Hybrid Ocean Thermal Energy Conversion (OTEC) Technology for Low Carbon Society and Sustainable Energy System: First Experimental OTEC Plant of Malaysia".

Data Availability Statement: Publicly available datasets were analyzed in this study. CAWCR Wave Hindcast can be found here: https://doi.org/10.4225/08/523168703DCC5 (accessed on 30 December 2021). The World Database on Protected Areas can be found here: https://www.protectedplanet.net/en/thematic-areas/wdpa?tab=WDPA (accessed on 14 December 2021). The World Marine Heritage Sites can be found here: https://www.marineregions.org/ (accessed on 13 December 2021). The Global distribution of warm-water coral reefs can be found here: https://doi.org/10.34892/t2wk-5t34 (accessed on 15 March 2022). The NOAA's International Best Track Archive for Climate Stewardship can be found here: https://www.ncei.noaa.gov/products/international-best-track-archive?name=ib-v4-access (accessed on 25 January 2022). Marine Activities data can be found here: https://knb.ecoinformatics.org/view/doi:10.5063/F1S180FS (accessed on 14 December 2021).

Acknowledgments: The authors would like to acknowledge SATREPS and its members, as well as The University of Tokyo and our research group "Applied Physical Oceanography" for their support during this study. Lastly, we would like to acknowledge the Kingdom of Tonga for recognizing this study and providing a research permit.

Conflicts of Interest: The authors declare no conflict of interest.

Appendix A

Appendix A.1. Potential Challenges

Appendix A.1.1. Political

Political Stability: Strong institutions and political leadership with the capacity to plan and manage policies and investments that support climate smart development are an essential building block of low carbon, climate resilient societies [49]. Therefore, the political stability of a country becomes an important variable when looking into wave energy for the Pacific Islands.

Renewable Energy Targets: Nations that have already established renewable energy targets are more likely to invest in wave energy technologies. Since most of the PICs have already committed to transitioning to renewable energy, this factor is going to serve as a motivation for wave energy to be utilized. However, wave energy has higher costs when compared to more traditional technologies and the absence of renewable energy targets might hamper a project.

Government Support: Wave energy currently needs government support for research and development (R&D) to compete with more mature technologies, such as wind and solar. However, this factor is also important for countries that have the potential for wave energy, since having support from the local government can facilitate the project. A cooperative agreement between government and developers will enable information sharing, which is essential to the pre-feasibility and feasibility stages where technical information is needed on a local scale. Having government approval is mandatory, nevertheless, different levels of support will either ease the process of project development or create additional hindrances.

Appendix A.1.2. Economic

Economic Stability: The economic stability of a country has a direct impact on the cost of energy, which is an important metric to analyze the viability of wave energy. Unstable economies can lead to high and unpredictable inflation, depreciation of the currency, lower investment opportunities, and low economic growth. The two main outcomes that can affect the viability of wave energy are the high exchange rates and unstable markets. Technology suppliers are outside the Pacific Island countries territory; therefore, the long-term wave energy project will be conducted using foreign currencies that might change in the future, altering projections. Furthermore, if there is negative economic growth, investors might refuse to start a project; given that the life cycle of a device is 25 years on average, any market uncertainties will reduce investor confidence and long-term stability is favorable.

Cost of Energy: This is the total costs for the generation of energy during the life-cycle of a device, including the capital, operation, maintenance, and decommissioning expenditures. Cost of energy (CoE) is one of the main indices to assess the economic feasibility of a project; another important index is the levelized cost of energy (LCOE), which can be calculated assuming different discount rates to levelize the costs for present value. Economic indices show how the costs for wave energy can compare to different sources of energy, including diesel generation, which is the main source for the PICs. If the costs are high, it will be more difficult to justify a wave energy project. Furthermore, there are factors within the CoE that should be included in the analysis, such as the distance to the shore, distance from the source point to the electricity grid, and water depth. Each one of the aforementioned factors can significantly increase the initial costs and is crucial when choosing a wave energy converter device. Additional important variables would be the discount rate and the conversion rate—considering that wave energy is still under development discount rates are expected to be high, moreover, the PICs will be importing the device, and conversion rates can fluctuate and generate a loss.

Feed-In Tariff: A FIT is a governmental incentive that ensures a premium fixed price for energy generated to the grid, making calculations of viability more predictable [7]. The FIT can make a project more appealing to investors and end-users by reducing price volatility and creating more opportunities for the renewable energy sector. On the same note, a lack of FIT schemes can hinder the chances of receiving funds and outside investment.

Risk Assessment: Any long-term investment will be bound to have risks related to different stages of the project. Risk assessment provides an understanding of risks, their causes, consequences, and their probabilities [50]. For wave energy projects, the risk is an important factor since it will influence economic parameters, such as Cost of energy (CoE) and levelized cost of energy (LCOE). Risks can take different forms, such as political and regulatory risk; counterparty, grid, and transmission link risk; currency, liquidity, and refinancing risk; as well as resource risk [51]. A qualitative and quantitative risk assessment should be performed to understand the source of the risks as well as their impact on the economic viability of the project.

Access to Funds: Ocean energy technologies demand high long-term investments, which are mostly due to the equipment costs, installation process, and discount rates. The Pacific Islands have a specific environment that creates drivers for ocean energy, such as high diesel costs and high ocean resources. Nevertheless, the high capital costs associated with these technologies preclude these island nations from constructing ocean energy facilities; financial and technical assistance must come from developed nations [52]. Access to funds is an important factor to justify a project, and lack thereof could impede or postpone the process significantly.

Appendix A.1.3. Social

Offshore Mining: According to Inniss et al. [53], "Marine mining has occurred for many years, with most commercial ventures focusing on aggregates, diamonds, tin, magnesium, salt, sulfur, gold, and heavy minerals. Activities have generally been confined to the shallow nearshore (less than 50 m water depth), but the industry is evolving and mining in deeper water looks set to proceed, with phosphate, massive sulfide deposits, manganese nodules, and cobalt-rich crusts regarded as potential future prospects". Pacific Island countries (PICs) are heavily dependent on natural resources and likely to remain so for the near future, making resource management an issue of critical importance for economic development [54]. Any project development must consider the presence of natural resources offshore and verify if there are any ocean policies regarding the use of these sites.

Tourism: Tourism represents a key driver of global economic growth and is a crucial component of the effort to alleviate poverty and achieve the other development objectives in many developing countries [21]. In the Pacific, tourism is a key sector of the local economy and is one of the main contributors to the gross domestic product. Tourism sector development offers Pacific Island countries a path to economic security that dovetails with broader development goals around infrastructure and employment [55]. For this reason, wave energy should not interfere with the tourism industry of any potential site and should seek the approval of possible stakeholders. Areas that are being used for diving, snorkeling, swimming, or that have any touristic purposes, ought to be mapped and circumvented.

Navigation: Islands are more reliant on marine transportation for commercial and non-commercial shipping due to lack of resources, dependence on international trade, and remoteness factors. There is the possibility of utilizing a WEC in the port structure since the port is an ideal location for a wave energy converter based on the overtopping principle as it can be easily integrated into the mound ruble without compromising the success of the project [56]. Nevertheless, if the WEC being studied will not be used as such, the proposed development should account for effects on navigation channels and seek to not interfere with main shipping routes.

Fishery: Fishery is one of the most crucial sectors of several islands inside the Pacific, considering its importance to the local economy and the subsistence of the local communities. Much of the region's nutrition, welfare, culture, employment, and recreation is based on the living resources in the zone between the shoreline and the outer reefs. The continuation of current lifestyles, the opportunities for future development, and food security are all highly dependent on coastal fisheries resources [57]. Considering the key role of the fishery sector, it will be imperative to map fishing areas and choose a site that does not coincide with this activity.

Aquaculture: According to Adams et al. [58], "Profitable aquaculture of penaeid shrimps and blacklip pearl oysters has now been established in some areas of the Pacific by commercial interests. Stand-alone enterprises producing penaeid shrimps for export markets are firmly established in New Caledonia, Fiji, and the Solomon Islands". The aquaculture sector in these countries is significant to their economy, nevertheless, there are still other examples of aquaculture activities being developed at different PICs which should also be accounted for. When choosing a suitable site for wave energy it is important to identify any aquaculture farms to avoid conflicts of use.

Recreation Sites: Recreation sites include the presence of beach areas or sites that are being used for sports, leisure, and additional activities that do not suit the previous factors. These can be used by both locals and tourists and could further be categorized as tourism; nevertheless, considering that several Pacific Islands do not have a high flux of tourists this will be then defined as a category of its own and will serve to identify important recreation sites.

Cultural and World Heritage Sites: Mixed cultural and natural World Heritage sites have both outstanding natural and cultural values and so are included on the World Heritage List according to a combination of cultural and natural heritage criteria [59]. These sites require tourism management and have regulations regarding the types of activities that are allowed, which means that using heritage sites for energy purposes or interfering with its lands is unviable. It is also important to note that aside from UNESCO Heritage Sites, any area with cultural value will create obstacles for project development.

Local Acceptance: People tend to accept renewable energy due to environmental issues (reduction of pollution by producing clean energy), but questions arise about environmental impacts, mainly those related to marine mammals, landscape/seascape changes, and noise [60]. It is important to consult key stakeholders, including members of the community, to share the benefits and potential impacts of the project and allow them to voice their opinions. A consensus between the local population and decision-makers can be achieved through stakeholder engagement plans to avoid any future conflicts of interest.

Appendix A.1.4. Technological

Electricity Supply and Demand: Each wave energy converter is capable of supplying a limited annual energy output; the actual output will vary depending on the local climate characteristics and if it is a singular device or an array of devices. To compensate for the high initial costs, it is common to establish a wave energy farm with high energy outputs. Nevertheless, Pacific Island countries encompass thousands of islands with varying population sizes, including remote islands with less than 1000 inhabitants. For a wave energy project to be viable, the chosen device needs to account for the relationship between energy output and energy demand, whilst keeping the costs competitive. For this reason, islands with higher population densities are more suitable locations.

Electricity Grid: Since islands with small-scale grid systems are more affected by fluctuations in renewable energy power supply than other areas connected to larger-scale grids, grid stability is a particularly important issue when increasing the renewable energy penetration rate in these areas [20]. Nevertheless, several islands still lack the basic infrastructure for grid connection and there are communities that are not yet connected to the grid. For this reason, not only is the stability of a grid an important aspect when studying the possibility of bringing wave energy to a site, but the presence of grid infrastructure is also crucial. The building of or improvement of an electricity grid will add costs to the installation process and can be detrimental to the feasibility of a project.

Seaports: Ports have an essential role in maritime logistic chains as they are the places where the cargoes are handled [57]. They also play an important role when it comes to wave energy since WECs are relatively large structures and might need several weeks for transportation and large-scale vessels. The process of receiving materials and supplies for a WEC will therefore require a port for operation; in case there is no infrastructure available, additional investment costs might be required.

Expertise: Even though having expertise available is preferred, the offshore industry in the Pacific is not yet developed and it is most likely that training programs will be necessary to conduct and maintain a wave energy project. The expertise factor is relatively complex, and even though the lack of expertise creates extra initial costs, it also benefits the local economy by creating job opportunities. Therefore, costs will increase but the chances of achieving public acceptance will be higher.

Logistics: Wave energy technologies are currently being developed in a limited number of countries, which might create supply-related drawbacks. Dedicated suppliers are not yet abundant due to the relatively small scale of the industry but suppliers in related applications may have the capacity to modify their existing products/services to supply the marine energy sector [61]. Due to the remoteness factor of Pacific Island countries, there will likely be additional costs in the process of importing, installing, and maintaining a wave energy converter device; the need for special vessels that will travel long distances will increase the initial, maintenance, and operation costs.

Appendix A.1.5. Environmental

Natural Hazards: Even though there is a lack of studies on the relationship between natural hazards and wave energy, it is well-known that storms can cause extreme wave events which might have an adverse impact on offshore structures. During storms and other extreme events, the stresses induced on the foundations, moorings, pylons, and sub-structures, etc., can exceed the design stress-causing failure of the device [62]. It is important to identify hazard-prone areas as well as the type and frequency of natural hazards to have a better understanding of risks. If a site has frequent storms this might add to unplanned maintenance costs and therefore may increase the overall costs of a project. For this factor, important variables include the number of past disaster events, frequency of natural hazards, as well as their intensity and proximity to the island. In the case of hurricanes, for instance, it is necessary to analyze the hurricane tracks, hurricane categories, frequency of events, wind speed, distance from the center of the storm to the island, and the number of events that caused damages and turned into disasters.

Biodiversity: Marine biodiversity plays an important role in the livelihood of the local population as well as in the environmental cycle. It is at present not clear what the scaling-up from the limited observations on individual or small clusters of devices to commercial-scale arrays will mean in terms of environmental effects and whether or not the effects observed to date are directly applicable [63]. Thus, since the effects on the local environment are still uncertain, areas that are rich in biodiversity should be avoided for wave energy projects to minimize the chance of negative impacts. It is also important to identify local protected species, endangered species, and key species during this stage to avoid incorporating areas of their natural habitat. This information will be used for early assessments: however, an EIA study will be required for further analysis and for identifying interactions between a WEC and the local environment.

Coral Reef: According to Moritz et al. [25], "The tropical Pacific region holds approximately 25% (about 66,000 km^2) of the global coral reef area. Spread across such a large area, these reefs vary considerably in terms of proximity to continents, reef structure, and biodiversity, as well as frequency and intensity of natural disturbances". Thus, the PICs hold a significant percentage of the global coral reef resources, which are also extremely valuable for the local environments and provide essential services. As to prevent any possible harmful interaction with the corals, a wave energy project should avoid utilizing areas with such environments for potential sites.

Appendix A.1.6. Legal

Regulations: Even though countries are expected to have regulations concerning the energy sector, the lack of specific regulations for marine energy might bring additional barriers or bureaucratic procedures. Since marine energy is not yet consolidated in the Pacific, there will be a high chance of encountering a lack of regulations for this market, and consultations with local government will be necessary to establish boundaries and define associated fees.

Marine Protected Areas: A marine protected area (MPA) is an area of intertidal or subtidal terrain, together with its overlying water and associated flora, fauna, and historical and cultural features, which has been reserved by law or other effective means to protect part or all of the enclosed environment [64]. There are different levels of protection, which result in different regulations regarding marine activities. Usually, marine exploration is prohibited in MPAs, while tourism and shipping activities might be limited. It is advisable to keep renewable energy generation outside the MPA boundaries to avoid any impact during the construction, operation, and decommissioning stages.

Maritime Zones: According to Goodall [65], "maritime zones are areas of ocean or sea which are or will be subject to national or international authority. They are delimited as parts of the seabed, water column and sea surface, the subdivision being on the grounds of political jurisdiction relating to the use and ownership of marine resources". These areas can include resource exploration, protected areas for marine species, disputed territories, and the exclusive economic zone (EEZ) boundary. To avoid any project constraints, maritime zones should be avoided and the WEC should remain inside the EEZ.

Military Zones: The Pacific Islands are strategically positioned between Eastern Asia and North America, which has sparked interest from different nations through the last decades. Due to their importance in terms of geographical position, it is possible to find military zones in the Pacific or agreements for future bases. For instance, The United States of America has air and naval bases in Guam and an intercontinental ballistic missile test site in Kwajalein Atoll in the Marshall Islands, which also supports space surveillance activities [66]. Utilizing these areas might either be prohibited or require an agreement between developers, local government, and responding authorities for the military zone.

Dependent Territories: There are still several Pacific Island Territories whose government does not hold full sovereignty and any developments on those areas will need to respond to different legislations. Those territories can be associated with the United States of America, France, Australia, the United Kingdom, and New Zealand; their levels of sovereignty might differ as well as their federal relationships. Bringing wave energy to these areas will require public acceptance from the local communities as well as from different governments, which might create additional difficulties.

References

1. Barstow, S.F.; Falnes, J. *Ocean Wave Energy in the South Pacific—The Resource and its Utilization*; South Pacific Applied Geoscience Commission (SOPAC) in conjunction with the Government of Norway: Suva, Fiji, 1996.
2. Bosserelle, C.; Reddy, S.; Kruger, J. *Waves and Coasts in the Pacific: Cost Analysis of Wave Energy in the Pacific*; Secretariat of the Pacific Community: Suva, Fiji, 2015.
3. European Marine Energy Centre (EMEC). *Guidelines for Project Development in the Marine Energy Industry*; BSI: London, UK, 2009.
4. Mendoza, E.; Lithgow, D.; Flores, P.; Felix, A.; Simas, T.; Silva, R. A framework to evaluate the environmental impact of OCEAN energy devices. *Renew. Sustain. Energy Rev.* **2019**, *112*, 440–449. [CrossRef]
5. Aguilar, F.J. *Scanning the Business Environment*; Macmillan Publishers: New York, NY, USA, 1967.
6. Yüksel, I. Developing a Multi-Criteria Decision Making Model for PESTEL Analysis. *Int. J. Bus. Manag.* **2012**, *7*, 52. [CrossRef]
7. Sandberg, A.B.; Klementsen, E.; Muller, G.; de Andres, A.; Maillet, J. Critical Factors Influencing Viability of Wave Energy Converters in Off-Grid Luxury Resorts and Small Utilities. *Sustainability* **2016**, *8*, 1274. [CrossRef]
8. Kolios, A.J.; Read, G. A political, economic, social, technology, legal and environmental (PESTLE) approach for risk identification of the tidal industry in the United Kingdom. *Energies* **2013**, *6*, 5023–5045. [CrossRef]
9. Thompson, A.A.; Strickland, A.J.; Gamble, J.E. *Crafting and Executing Strategy-Concepts and Cases*, 15th ed.; McGrawHill/Irwin: New York, NY, USA, 2007.
10. Gürel, E. Swot Analysis: A Theoretical Review. *J. Int. Soc. Res.* **2017**, *10*, 994–1006. [CrossRef]
11. Weihrich, H. The TOWS matrix—A tool for situational analysis. *Long Range Plan.* **1982**, *15*, 54–66. [CrossRef]
12. Stingheru, C.; Gasparotti, C.; Raileanu, A.; Rusu, E. A SWOT Analysis of the Marine Energy Sector at the European Level. *Acta Univ. Danub.* **2018**, *14*, 213–237.
13. Damasceno, V.S.; Abreu, Y.V. Avaliação Da Energia Eólica No Brasil Utilizando a Análise SWOT e PESTEL. *Interações* **2018**, *29*, 503–514. [CrossRef]
14. Shadman, S.; Chin, C.M.; Yap, E.H.; Sakundarini, N.; Velautham, S. The role of current and future renewable energy policies in fortifying Malaysia's energy security: PESTLE and SWOT analysis through stakeholder engagement. *Prog. Energy Environ.* **2021**, *16*, 1–17.
15. Islam, F.R.; Mamun, K.A. Possibilities and Challenges of Implementing Renewable Energy in the Light of PESTLE & SWOT Analyses for Island Countries. In *Smart Energy Grid Design for Island Countries*; Rabiul Islam, F.M., Al Mamun, K., Oo Amanullah, M.T., Eds.; Springer International Publishing: New York, NY, USA, 2017; pp. 1–19. [CrossRef]
16. Davey, T.; Venugopal, V.; Smith, H.; Smith, G.; Lawrence, J.; Cavaleri, L.; Bertotti, L.; Prevosto, M.; Girard, F.; Holmes, B. Protocols for wave and tidal resource assessment (Report No. D2.7). Report by University of Southampton. Report for EquiMar, Southampton, England, 2010. Available online: https://www.wiki.ed.ac.uk/download/attachments/9142387/EquiMar+D2.7+Resource+Assessment+Protocol.pdf?version=1 (accessed on 4 July 2020).
17. Stagonas, D.; Myers, L.E.; Bahaj, A.S. Impacts upon marine energy stakeholders (Report No. D5.8). Report by University of Southampton. Report for EquiMar, Southampton, England, 2011. Available online: https://www.wiki.ed.ac.uk/download/attachments/9142387/WP5_d5.8_final.pdf?version=1 (accessed on 4 July 2020).
18. International Renewable Energy Agency (IRENA). *Wave Energy Technology Brief*; IRENA Ocean Energy Technology Brief 4: Abu Dhabi, United Arab Emirates, 2014; Available online: https://www.irena.org/-/media/Files/IRENA/Agency/Publication/2014/Wave-Energy_V4_web.pdf (accessed on 4 July 2020).
19. International Renewable Energy Agency (IRENA). *Renewable Energy Technology Innovation Policy: A Process Development Guide*; IRENA: Abu Dhabi, United Arab Emirates, 2015.
20. International Renewable Energy Agency (IRENA). A Path to Prosperity: Renewable Energy for Islands. Third Edition. In Proceedings of the 22nd session of the Conference of the Parties (COP 22) to the United Nations Framework Convention on Climate Change (UNFCCC), Marrakech, Morocco, 7–18 November 2016.
21. International Renewable Energy Agency (IRENA). *Renewable Energy Opportunities for Island Tourism*; IRENA: Abu Dhabi, United Arab Emirates, 2014; Available online: www.irena.org (accessed on 15 August 2021).
22. Sotta, C. *Documentary summary of the environmental impact of renewable marine energy*; Task 3,2 of the MERIFIC Project. Report prepared within the framework of the MERIFIC Project "Marine Energy in Far Peripheral Island Communities"; MERIFIC Marine Energy in Far Peripheral and Island Communities: Manche, France, 2012.
23. Delvaux, P.A.G.; Rabuteau, Y.; Stanley, K. *Civil Society Involvement and Social Acceptability of Marine Energy Projects: Best Practices of the Marine Energy Sector*; Chapter 6.1.2; MERiFIC: Manche, France, 2013.
24. *Secretariat of the Pacific Regional Environment Programme (SPREP)*; SPREP: Apia, Samoa, 2019.
25. Moritz, C.; Vii, J.; Lee Long, W.; Tamelander, J.; Thomassin, A.; Planes, S. Status and Trends of Coral Reefs of the Pacific; Global Coral Reef Monitoring Network; 2018. Available online: https://solomonislands-data.sprep.org/system/files/status-coral-reefs-pacific.pdf (accessed on 2 November 2021).

26. Secretariat of the Pacific Regional Environment Programme (SPREP). *Pacific Regional Energy Assessment 2004: An assessment of the Key Energy Issues, Barriers to the Development of Renewable Energy to Mitigate Climate Change, and Capacity Development Needs to Removing the Barriers*; Wade, H., Johnston, P., Vos, J., Eds.; SPREP: Apia, Samoa, 2005.
27. Commonwealth Marine Economics Programme (CMEP). *Pacific Marine Climate Change Report Card 2018*; CMEP: Washington DC, WA, USA, 2018; p. 12.
28. Pacific Community (SPC). *Pacific Community Results Report 2018*; Pacific Community: New Caledonia, France, 2019.
29. UNEP-WCMC and IUCN. *Protected Planet: The World Database on Protected Areas (WDPA) and World Database on Other Effective Area-Based Conservation Measures (WD-OECM) [Online]*; UNEP-WCMC and IUCN: Cambridge, UK, 2021; Available online: www.protectedplanet.net (accessed on 26 December 2021).
30. Flanders Marine Institute. World Marine Heritage Sites (Version 1). 2013. Available online: https://www.marineregions.org/ (accessed on 3 November 2021).
31. UNEP-WCMC, WorldFish Centre, WRI, TNC (2021). *Global Distribution of Warm-Water Coral Reefs, Compiled from Multiple Sources including the Millennium Coral Reef Mapping Project. Version 4.1*; Includes contributions from IMaRS-USF and IRD (2005), IMaRS-USF (2005) and Spalding et al. (2001); UN Environment World Conservation Monitoring Centre: Cambridge, UK, 2010. [CrossRef]
32. NOAA's International Best Track Archive for Climate Stewardship (IBTrACS) Data. Available online: https://www.ncei.noaa.gov/products/international-best-track-archive?name=ib-v4-access (accessed on 5 November 2021).
33. Halpern, B.; Frazier, M.; Potapenko, J.; Casey, K.; Koenig, K.; Longo, C.; Lowndes, J.; Rockwood, C.; Selig, E.; Selkoe, K.; et al. Cumulative human impacts: Raw stressor data (2008 and 2013). *Knowl. Netw. Biocomplex.* **2015**. [CrossRef]
34. Institut De La Statistique De La Polynésie Française. *Fiche Communale*; Institut De La Statistique De La Polynésie Française: Papeete, French Polynesia, 2017.
35. Petit, N.J.; Tanret, D.; Lagouy, E.; et Maxime Chan, P.D.P. *Rāhui Nui nō Tuha'a Pae, Projet de grande réserve marine aux Iles Australes*; The Pew charitable Trusts, FAPE: Tahiti, French Polynesia, 2016; 66p.
36. Fiji Bureau of Statistics. *2017 Population and Housing Census*; Fiji Bureau of Statistics: Suva, Fiji, 2018; p. 13.
37. United Nations Educational, Scientific and Cultural Organization (UNESCO). Levuka Historical Port Town. Available online: https://whc.unesco.org/en/list/1399 (accessed on 14 April 2021).
38. Ministry of Finance and Economic Management. *Cook Islands Demographic Profile 2006–2011*; Government of the Cook Islands, Ministry of Finance and Economic Management: Rarotonga, Cook Islands, 2011.
39. Cook Islands Government. Marae Moana—Cook Islands Marine Park. United Nations, Ocean Conference, 2018. Available online: https://oceanconference.un.org/commitments/?id=15701 (accessed on 11 April 2021).
40. Ministry of Fisheries. *Personal Communication*; Ministry of Fisheries: Nuku'alofa, Tonga, 2020.
41. Ceccarelli, D.M.; Wendt, H.; Matoto, A.L.; Fonua, E.; Fernandes, L. *Biophysically Special, Unique Marine Areas of Tonga*; MACBIO: Suva, Fiji, 2017.
42. Ministry of Tourism. *Personal Communication*; Ministry of Tourism: Nuku'alofa, Tonga, 2020.
43. Firth, S. Instability in The Pacific Islands: A Status Report. Lowy Institute. 2018. Available online: https://www.lowyinstitute.org/publications/instability-pacific-islands-status-report (accessed on 30 June 2020).
44. Tonga Statistics Department. *Tonga 2016 Census of Population and Housing Volume 1: Basic Tables and Administrative Report*; Tonga Statistics Department: Nuku'alofa, Tonga, 2017.
45. Tonga Power Limited 1a. 2019. Annual Report 2019. Available online: www.tongapower.to (accessed on 23 May 2020).
46. Tonga Meteorological Service. Facts About Natural Hazards in Tonga. Available online: http://www.met.gov.to/index_files/Disasterfacts.htm (accessed on 30 June 2020).
47. Tonga Fisheries Division and Pacific Islands Forum Fisheries Agency. *Kingdom Of Tonga National Plan of Action (NPOA) Shark-Plan (2014–2016)*; Department of Environment: Nuku'alofa, Tonga, 2013.
48. Durrant, T.; Hemer, M.; Trenham, C.; Greenslade, D. CAWCR Wave Hindcast 1979–2010. *CSIRO Serv. Collection* **2013**, *8*. [CrossRef]
49. Foresight Africa. *Top Priorities for the Continent in 2016*; Brookings Institute: Washington, WA, USA, 2017.
50. International Electrotechnical Commission. *IEC/FDIS 31010 Risk Management—Risk Assessment Techniques Gestion des Risques—Techniques D'évaluation des Risques (IEC/FDIS 31010:2009(E))*; IEC: Geneva, Switzerland, 2009.
51. International Renewable Energy Agency (IRENA). *Unlocking Renewable Energy Investment: The Role of Risk Mitigation and Structured Finance*; IRENA: Abu Dhabi, United Arab Emirates, 2016.
52. Bruch, V.L. *An Assessment of Research and Development Leadership in Ocean Energy Technologies*; Sandia National Laboratories (SNL): Albuquerque, NM, USA, 1994. [CrossRef]
53. Inniss, L.; Simcock, A.; Ajawin, A.Y.A.; Alcala, A.C.; Bernal, P.; Calumpong, H.P.; Araghi, P.E.; Green, S.O.; Harris, P.; Kamara, O.K.; et al. *The First Global Integrated Marine Assessment World Ocean Assessment I*; United Nations: Cambridge, England, 2016; Available online: www.un.org/Depts/los/woa (accessed on 18 June 2020).
54. Chand, S. Managing natural resources in the Pacific Islands. In *International and Development Economics Working Papers idec01-6*; International and Development Economics: Suva, Fiji, 2001.
55. Everett, H.; Simpson, D.; Wayne, S. *Tourism as A Driver of Growth: A Pathway to Growth and Prosperity for Pacific Island Countries*; Asian Development Bank: Mandaluyong, Philippines, 2018.

56. Cascajo, R.; García, E.; Quiles, E.; Correcher, A.; Morant, F. Integration of Marine Wave Energy Converters into Seaports: A Case Study in the Port of Valencia. *Energies* **2019**, *12*, 787. [CrossRef]
57. Gillett, R. *Fisheries of the Pacific Islands: Regional and National Information*; FAO Regional Office for Asia and the Pacific: Bangkok, Thailand, 2011; 279p.
58. Adams, T.; Bell, J.; Labrosse, P. Current Status of Aquaculture in the Pacific Islands. In Proceedings of the Conference on Aquaculture in the Third Millennium, Bangkok, Thailand, 20–25 February 2000; pp. 295–305.
59. Wingham, E.; Cooper, B. *Our Pacific Heritage: The Future in Young Hands*; New Zealand National Commission for the United Nations Educational, Scientific, and Cultural Organisation (UNESCO): Wellington, New Zealand, 2004.
60. Simas, T.; Moura, A.; Batty, R.; Wilson, B.; Thompson, D.; Lonergan, M.; Norris, J. Uncertainties and Road Map (Report No. D6.3.2). Report by University of Southampton. Report for EquiMar, Southampton, England, 2010. Available online: https://www.wiki.ed.ac.uk/download/attachments/9142387/Equimar_D6-3-2.pdf?version=1 (accessed on 4 July 2020).
61. Myers, L.; Bahaj, A.; Retzler, C.; Bittencourt, C.; Flinn, J.; Sorenson, H.; Gardner, F. Assessment of the present status and future scenarios of the supply chain for marine energy arrays (Report No. D5.7). Report by University of Southampton. Report for EquiMar, Southampton, England, 2011. Available online: https://www.wiki.ed.ac.uk/download/attachments/9142387/WP5+deliverable+5.7+_final.pdf?version=1 (accessed on 4 July 2020).
62. Agarwal, A.; Venugopal, V.; Harrison, G.P. The assessment of extreme wave analysis methods applied to potential marine energy sites using numerical model data. *Renew. Sustain. Energy Rev.* **2013**, *27*, 244–257. [CrossRef]
63. Frid, C.; Andonegi, E.; Depestele, J.; Judd, A.; Rihan, D.; Rogers, S.I.; Kenchington, E. The environmental interactions of tidal and wave energy generation devices. *Environ. Impact Assess. Rev.* **2012**, *32*, 133–139. [CrossRef]
64. Kelleher, G.; Kenchington, R. *Guidelines for Establishing Marine Protected Areas (pp.-U, p.79)*; A marine conservation and development report; IUCN: Gland, Switzerland, 1992.
65. Goodall, B. Maritime zones. In *Encyclopedia of Hydrology and Lakes*; Springer: Dordrecht, The Netherlands, 1998.
66. Lum, T.; Vaughn, B. The Pacific Islands. Congressional Research Service, IF11208 Version 1. 2019. Available online: https://fas.org/sgp/crs/row/IF11208.pdf (accessed on 2 July 2020).

Journal of
Marine Science and Engineering

MDPI

Article

Design and Dynamic Stability Analysis of a Submersible Ocean Current Generator Platform Mooring System under Typhoon Irregular Wave

Shueei-Muh Lin [1,2,*], Chihng-Tsung Liauh [1,2] and Didi-Widya Utama [2]

1 Green Energy Technology Research Centre (GETRC), Kun Shan University, Tainan 71070, Taiwan; liauhct@mail.ksu.edu.tw
2 Department of Mechanical Engineering, Kun Shan University, Tainan 710, Taiwan; didiu@ft.untar.ac.id
* Correspondence: smlin45@gmail.com

Abstract: This research proposes a mooring system for an ocean current generator that is working under the impact of typhoon waves. The turbine and the platform are kept stable at a designed water depth to ensure that the generator remains undamaged and continuously generates electricity under excessive water pressure. In this design, the turbine generator is mounted in front of the floating platform by ropes and withstands the force of ocean currents, while the platform is anchored to the deep seabed with lightweight, high-strength PE ropes. In addition, two pontoons are used to connect the generator and the platform with ropes. When the balance is reached, the depth of the generator and the depth of the platform's dive can be determined by the length of the ropes. In this study, typhoon irregular wave is represented by the Jonswap wave spectrum. The irregular wave is simulated by six regular waves. The equation of motion of the mooring system is derived. The theoretical solution of the dynamic system is presented to determine the dynamic displacements of the platform, pontoon, turbine and the dynamic tensions of the ropes. The dynamic tensions of the ropes increase with the cross-sectional area of pontoon. The natural frequency of the mooring system depends on the parameters, including the mases of elements, the lengths of ropes and the cross-sectional area of pontoons. In the proposed mooring configuration, the dynamic tension of the rope is far less than the breaking strength of the rope; thus, the ocean turbine is stable, and no water that flows through will be disturbed by the floating platform.

Keywords: dynamic tension; displacement; ocean current; floating platform; turbine; pontoon; buffer spring

Citation: Lin, S.-M.; Liauh, C.-T.; Utama, D.-W. Design and Dynamic Stability Analysis of a Submersible Ocean Current Generator Platform Mooring System under Typhoon Irregular Wave. *J. Mar. Sci. Eng.* **2022**, *10*, 538. https://doi.org/10.3390/jmse10040538

Academic Editor: Eugen Rusu

Received: 13 February 2022
Accepted: 10 April 2022
Published: 14 April 2022

Publisher's Note: MDPI stays neutral with regard to jurisdictional claims in published maps and institutional affiliations.

Copyright: © 2022 by the authors. Licensee MDPI, Basel, Switzerland. This article is an open access article distributed under the terms and conditions of the Creative Commons Attribution (CC BY) license (https://creativecommons.org/licenses/by/4.0/).

1. Introduction

An excellent, natural energy resource is the Kuroshio strong current flowing along the east of Taiwan, which has an estimated electricity capacity of 4 GW [1]. Ocean current is one of the potential energy sources to be developed. However, the seabed beneath the Kuroshio current is almost over 1000 m in the area mentioned above. Moreover, several typhoons strike Taiwan every year. These two disadvantages must be solved before a current power generation system is constructed.

Investigation of fluid–structure interaction (FSI) is important for marine engineering, aircraft, engines, bridges and biotechnology. FSI is the interaction of some movable or deformable structures with an internal or surrounding fluid flow. Fluid–structure coupling can be simply divided into one-way coupling and two-way coupling. One-way coupling ignores the change of flow field space caused by structural deformation, so the calculation is more simplified. Anagnostopoulos [2] investigated the dynamic response analyses of offshore platforms under wave loadings and predicted the wave forces by means of Morison's equation. Therefore, the equation of motion for the lumped mass idealization of the platform was presented. The system was one-way coupling. It was found that the importance of fluid–structure interaction increased with higher dynamic amplifications. The effect of viscous damping due to the relative velocity between fluid and structure significantly decreased the resonant response. Istrati and Buckle [3] investigated the effect of FSI on connection forces in bridges caused by tsunami loads by using LS-DYNA software. It was found that the flexibility and the dynamic characteristics of the bridge structure significantly influenced the external tsunami loads on the bridge and the connection forces.

Xiang and Istrati [4] investigated the solitary wave–structure interaction of complex coastal deck geometries by using the Lagrangian–Eulerian (ALE) method with a multi-phase compressible formulation. It was found that, for small wave heights, the horizontal and uplift forces increased with the number of girders (Ng), while, for large waves, the opposite happened. Moreover, if the Ng was small, the wave particles accelerated after the initial impact on the offshore girder, leading to more violent slamming and larger pressures and forces on the deck. Conversely, if the Ng was large, unsynchronized eddies were formed in each chamber, which dissipated energy and resulted in weaker impacts on the deck. Obviously, if the surfaced structure is too large, the two-way coupling effect of wave–structure interaction needs to be considered. In addition, the multi-phase flow simulation needs to be considered in the numerical analysis, which is a very challenging problem and important in marine engineering. Some literature [4–6] is devoted to this research. Peregrine et al. [5] found that the breaking/broken waves and bores were dominated by significant turbulence effects and air entrapment. The hydrodynamic loads caused by the breaking wave on the marine decks were totally different from unbroken waves [6].

Firouz-Abadi et al. [7] investigated the stability analysis of shells conveying fluid. The boundary element method was applied to model the potential flow. It was found that the eigenvalues and mode shapes of the flow in the shell were strongly related to the unsteady pressure that induced the shell vibration. Bose et al. [8] investigated the flow-induced dynamic stability of a fluid–structure interaction (FSI) system comprising of a symmetrical NACA 0012 airfoil supported by non-linear springs. Lin et al. [9] investigated the wave propagation of an artery. A mathematical model was proposed to describe the wave propagation through an isotropic, elastic, thick tube filled with viscous and incompressible fluid. The tube is supported by the elastic muscle and simulated as the viscoelastic foundation. The flexural Young and Lamb wave modes through a tube wall are presented simultaneously. The dispersion curves and the energy transmissions of the three modes were investigated. It was found that the effect of the viscoelastic foundation constant on the wave speed and the transmission was significant.

The numerical method is usually used to investigate the dynamic behavior of the two-way-coupled FSI. In general, the numerical methods include the boundary element method [10], the finite volume method [11], the finite-element-based, arbitrary Lagrangian–Eulerian method [12], particle-based methods, such as smoothed particle hydrodynamics [13], and hybrid methods, such as coupled SPH-DEM [14] and coupled SPH-FEM [15].

There have been two cases where the performance of ocean current turbines was tested in seas: (1) One 50 kW ocean current turbine, developed by the Wanchi company (Kaohsiung City, Taiwan), was successfully moored to the 850 m deep seabed near the offshore of Pingtung County, Taiwan, by Chen et al. [1]. The current turbine generated about 26 kW under the current speed of 1.0 m/s; (2) An experimental 100-kW-class ocean current turbine was located off the coast of Kuchinoshima Island, Kagoshima Prefecture, and demonstrated by IHI and NEDO [16]. The current turbine generated about 30 kW under the current speed of 1.0 m/s. The turbine system 50 m below the sea surface was connected to the mooring foundation on the seabed at the depth of 100 m. The above experiments were conducted under the condition of small waves, and the influence of waves on the dynamic stability of the mooring system was not studied.

Zwieten et al. [17] investigated the C-plane prototype of an ocean current turbine with a hydrodynamic platform that was connected to the seafloor with a rope. This turbine, using its wingtips and canard to manipulate its depth and orientation in a temporally and spatially varying current, could generate maximum energy production. This study did not take the problem of turbine damage due to excessive water pressure when diving too deep into consideration. It also did not consider the disadvantages of the deeper ocean current, the lower flow rate and the smaller power generation. The effect of waves was also not considered.

One of the most challenging tasks for the ocean current turbine system is to develop a deep mooring technology because the targeted seabed is at a depth of almost 1000 m, as mentioned. To monitor the performance of the ocean turbine, the dynamic stability of the mooring system under the coupled effect of the ocean current and wave is needed [18–24]. Lin et al. [25] used the ocean current turbine system developed by the Wanchi company to investigate the dynamic stability of the system subjected to regular wave and current forces. The mooring system was composed of a turbine, a floating platform, traction ropes and a mooring foundation. Results showed that the effects of several parameters of the system on the dynamical stability of the ocean current turbine system were significant. However, the dynamic tension of the rope was not investigated in the study.

As the mooring foundation is set on the seabed over 1000 m deep, a long mooring rope is required. Consider the strength of the rope: lightweight, high-strength PE mooring ropes are more beneficial than chain and steel ropes. Lin and Chen [26] found that, when the ocean current velocity was 1 m/s and the rope length was about 2900 m, the drag force was 15 tons, and the rope was almost straight. In other words, the bending deformation of the PE rope was negligible. The deformation of the rope was longitudinal only. Accordingly, the mooring system is simulated in the linear elastic model to analyze the problem of dynamic stability. Consider an ocean current power generation system composed of a surfaced turbine, a floating platform, a towing rope and a mooring foundation [25]: whenever a typhoon hits, the turbine generator is towed back to the shore to avoid any possible damage, leaving the mooring system in the sea. Lin and Chen [26] proposed a protection method to protect the mooring system that avoids the damage caused by typhoon wave current. The principle of the design is that the platform generates a negative buoyancy to dive by letting water flow into its inner tank, and the pontoon is used to create a positive buoyancy. When the two elements are connected by a rope to achieve static equilibrium, the floating platform is submerged at a fixed depth determined by the rope length. Furthermore, the linear elastic model is used to construct the coupled motion equation of the system under a regular wave. The analytical solutions of the coupled equations are derived. It is theoretically verified that the proposed protection procedure can avoid the damage of the floating platform and the mooring line due to typhoon wave impact.

Lin et al. [27] simulated a mooring system for ocean current generation during non-typhoon periods and proposed a system that keeps the turbine statically stable at a designed underwater depth to ensure that the ocean current generator can generate electricity effectively. In their design, the turbine generator is connected to a surfaced platform, the platform is anchored to the deep mooring foundation by lightweight, high-strength polyethylene ropes and a pontoon is connected to ocean current turbines with rope. The static balance of the ocean current turbine is formed. Therefore, the depth of the current turbine can be determined by the length of the rope. Additionally, the linear elastic model is used to simulate the motion equation of the overall mooring system under a regular wave. The theoretical solution of the static and dynamic stability analysis of the mooring system is proposed. The dynamic displacements of the components and the dynamic tensions of ropes under the regular wave and ocean current are investigated. It is found that the effect of the wave phase on the dynamic response of the system is significant. The length of the rope can be adjusted to avoid resonance and reduce the tension of the rope. In addition, a buffer spring is used to reduce the dynamic tension of the rope to increase the safety and lifespan of the rope significantly.

To simplify the actual ocean waves, which are irregular, three approaches are commonly adopted: (1) the approximation of the wave field by a single, sinusoidal component with a given height, period and direction (regular waves); (2) the use of a limited number of harmonics of a primary wave to approximate non-sinusoidal properties (irregular waves); and (3) the representation of the water surface by an infinite summation of Fourier components (wave spectrum) [28]. Pierson and Moskowitz [29] presented the Pierson and Moskowitz wave spectrum. The assumption was that, if the wind blows steadily over a large area for a long time, the waves will reach equilibrium with the wind. This is the concept of a fully developed sea, which requires winds of a sea that continuously blow over hundreds of miles for several days to reach full development. Hasselmann et al. [30] experimentally found that the wave spectrum can never be fully developed. The wave spectrum continues to develop due to wave-to-wave interactions, even over long periods of time and distances. Therefore, the Pierson–Moskowitz spectrum is modified to add an additional and somewhat artificial factor to it to make the wave spectrum and experimental measurements more closely matched. The Jonswap wave spectrum is presented.

This study proposes a mooring design in which the ocean current generator still generates electricity when typhoon waves hit without interruption. To prevent the typhoon waves from invading the ocean current generator set, a process is adopted whereby the system dives below 60 m to avoid the damage of the typhoon waves. At the same time, to prevent diving too deep from damaging the turbine, the turbine is in a static balance at a predetermined depth underwater, and it must be able to maintain the a not-too-large dynamic displacement. The surface velocity of the Kuroshio in eastern Taiwan is relatively fast, and the deeper the water depth, the smaller the velocity. Therefore, the ocean current generator group should not be placed too deep. This study proposes a safe and efficient mooring system design and a linear elastic model to simulate the motion of the entire mooring system. Results for analyzing the static and dynamic stability of mooring systems, the dynamic displacements of turbines, floating platforms, pontoons and the dynamic tension of ropes under the action of typhoon waves and ocean currents are studied. The effects of several parameters on the dynamic behavior of the system are presented.

2. Mathematical Model

As shown in Figures 1 and 2, to prevent the damage of the typhoon waves, the turbine and the floating platform are submerged to a depth of less than 60 m. Therefore, the influence of the wave impact is almost negligible. In general, the dynamic response of a large-surfaced structure subjected to wave impact force, which is non-uniform and transient, is generally in the coupled translational–rotational (pitching, rolling and yawing) motion. The good conditions for ocean current power generation are high flow rate and stable flow direction, so the site is often a considerable distance from the shore: less affected by the coast and less likely to produce breaking waves. The impact of breaking waves is not considered in this manuscript.

When ocean currents flow through the blades of the ocean turbine, the turbine rotates and drives the power generator to generate electricity. Meanwhile, the turbine unit is subjected to the force of the ocean current; to fix the turbine unit, it is pulled by the floating platform connected by rope B. The floating platform provides buoyancy and is anchored to the deep seabed with lightweight, high-strength PE ropes. In addition, the ocean current turbine is connected to pontoon 4 via rope D, and the balance between the current generator and the pontoon is reached so that the depth of the turbine when the current is not affected can be determined by the length L_C of rope C. On one side of the floating platform, rope B is used to pull the ocean current generator, and the other side of the floating platform is pulled down and anchored on the deep seabed. The buoyancy of the floating platform can be adjusted to be smaller than that of static balance so that, when ropes A and B are pulled, the floating platform has negative buoyancy and pontoon 3 has positive buoyancy, and rope C is used to connect the floating platform and pontoon 3 to achieve a balance of positive and negative buoyancy. In this way, the depth of the floating platform can be calculated by the length L_C of rope C.

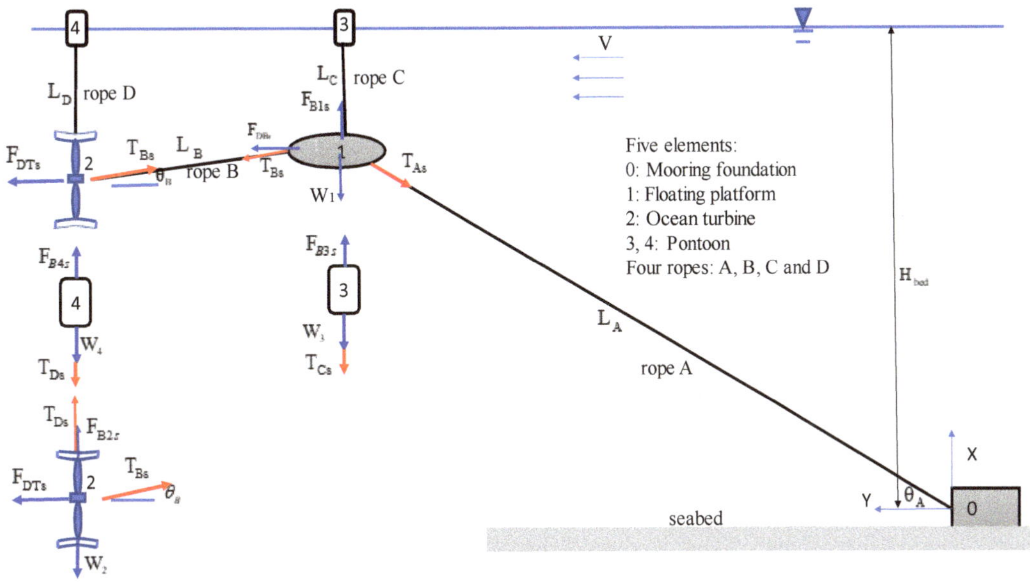

Figure 1. Coordinates of the current energy system composed of submarined ocean turbine, pontoons, floating platform, traction ropes and mooring foundation in the static state under steady ocean current.

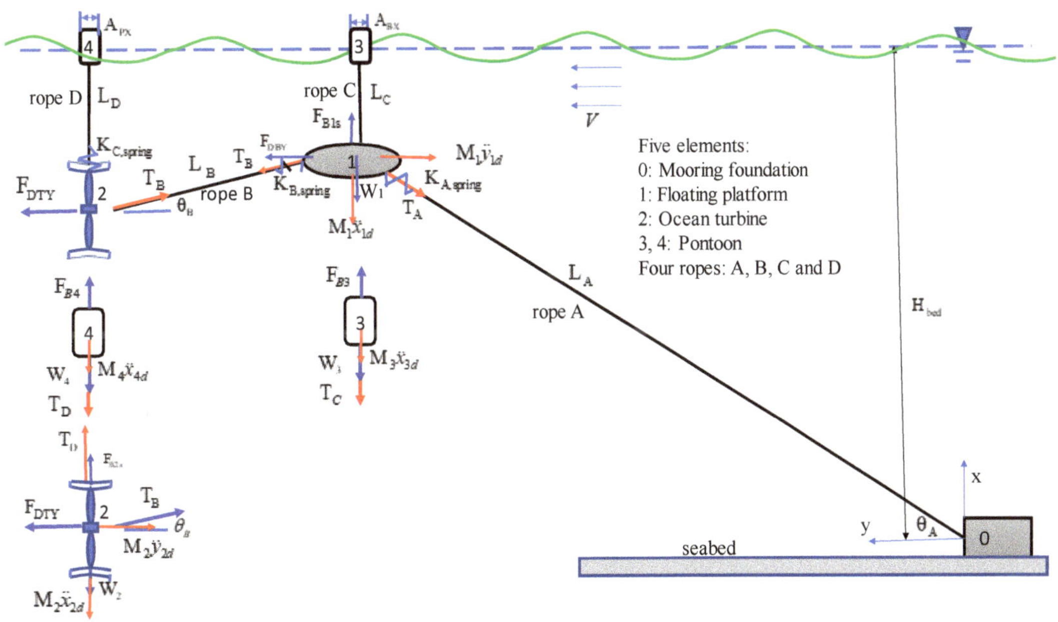

Figure 2. Coordinates of the current energy system composed of submarined ocean turbine, pontoon, floating platform, traction rope and mooring foundation in the dynamic state under steady ocean current and wave.

Lin and Chen [26] showed that a PE rope can be assumed to be a straight line under a certain amount of ocean current drag force because the force deformation of the PE rope is negligible. The linear elastic model presented by Lin and Chen [26] is used to analyze the motion equation of the overall mooring system.

Several assumptions are made based on these facts about ocean current energy converters (OCEC):

- The current flow is steady;
- The masses of the turbine, floating platform and the pontoon are concentrated;
- Lightweight and high-strength PE mooring ropes are used;
- Under the ocean velocity, the deformed configuration of PE rope is nearly straight;
- The elongation strain of the ropes is small;
- The tension of the rope is considered uniform.

According to these assumptions, the motion of the mooring system is translational. The coupled translational–rotational motion of a system subjected to non-uniform and impulsive force from the wave current will be discussed in future research. The coupled, linear, ordinary differential equations of the system are derived based on the assumptions. Due to the wave fluctuation, the buoyance forces of the pontoons stimulate the mooring system to vibrate. The coupled vibration motion of the system includes horizontal and vertical oscillations.

The global displacements (x_i, y_i) for the i-th element shown in Figures 1 and 2 are the sum of two parts: (1) the static one subjected to the steady current and (2) the dynamic one subjected to the wave, as follows:

$$x_i = x_{is} + x_{id}, \quad y_i = y_{is} + y_{id}, \; i = 1,\ 2,\ 3,\ 4 \tag{1}$$

where x and y are the vertical and horizontal displacements, respectively. Because of the pontoon buoyancy and the short length of rope between the turbine and the pontoon, the horizontal dynamic displacements of the turbine and pontoon 4 are almost the same: $y_{2d} \approx y_{4d}$. In a similar way, the horizontal dynamic displacements of the floating platform and pontoon 3 are almost the same: $y_{1d} \approx y_{3d}$. In addition, the total tensions of ropes A, B, C and D are also composed of two parts: (1) the static one and (2) the dynamic one, as follows:

$$T_i = T_{is} + T_{id}, \; i = \mathrm{A},\ \mathrm{B},\ \mathrm{C},\ \mathrm{D} \tag{2}$$

2.1. Static Displacements and Equilibrium under the Steady Current and without the Wave Effect

The static displacements of the five elements are:

$$\begin{gathered} x_0 = 0,\ y_0 = 0 \\ x_{1s} = H_{bed} - L_C = L_A \sin\theta_{As},\ y_{1s} = L_A \cos\theta_{As}; \\ x_{2s} = H_{bed} - L_D = x_{1s} - L_B \sin\theta_{Bs},\ y_{2s} = y_{1s} + L_B \cos\theta_{Bs} \\ x_{3s} = x_{1s} + L_C = H_{bed},\ y_{3s} = y_{1s} \\ x_{4s} = x_{3s} = x_{2s} + L_D = H_{bed},\ y_{4s} = y_{2s} \end{gathered} \tag{3}$$

Due to $x_{1s} >> x_{1d}$, the global inclined angle q_A can be expressed as:

$$\sin\theta_A = \frac{x_1}{L_\mathrm{A}} = \frac{x_{1s} + x_{1d}}{L_\mathrm{A}} \approx \frac{x_{1s}}{L_\mathrm{A}} = \sin\theta_{As} \tag{4}$$

Due to $x_{is} >> x_{id}$, the global inclined angle q_B can be expressed as:

$$\sin\theta_B = \frac{x_1 - x_2}{L_\mathrm{B}} = \frac{(x_{1s} + x_{1d}) - (x_{2s} + x_{2d})}{L_\mathrm{B}} \approx \frac{x_{1s} - x_{2s}}{L_\mathrm{B}} = \sin\theta_{Bs} \tag{5}$$

Under the steady current and without the wave effect, the static horizontal and vertical equilibriums of the floating platform are written, respectively, as shown in Figure 1.

$$T_{\mathrm{B}s}\cos\theta_{Bs} + F_{DBs} = T_{\mathrm{A}s}\cos\theta_{As} \tag{6}$$

$$F_{B1s} = T_{\mathrm{A}s}\sin\theta_{As} + T_{\mathrm{B}s}\sin\theta_{Bs} + W_1 \tag{7}$$

where T_{As}, T_{Bs}, F_{B1s} and W_1 are the static tensions of ropes A and B, the buoyancy of the floating platform and the weight of the floating platform, respectively. The steady drag of the floating platform under current $F_{DFs} = \frac{1}{2}C_{DFy}\rho A_{FY}V^2$.

The static horizontal and vertical equilibriums of the turbine are expressed, respectively, as:

$$T_{Bs}\cos\theta_{Bs} = F_{DTs} \tag{8}$$

where the steady drag of the turbine $F_{DTs} = C_{DTy}\frac{1}{2}\rho A_{Ty}V^2$.

$$F_{B2s} = W_2 - T_{Ds} - T_{Bs}\sin\theta_{Bs} \tag{9}$$

where T_{Ds}, F_{B2s} and W_2 are the static tensions of rope D, the static buoyancy and the weight of the turbine, respectively. The static vertical equilibrium of pontoon 3 is expressed as:

$$F_{B3s} = W_3 + T_{Cs} \tag{10}$$

where F_{B3s} and W_3 are the static buoyancy and the weight of pontoon 3, respectively. The static vertical equilibrium of the pontoon 4 is expressed as:

$$F_{B4s} = W_4 + T_{Ds} \tag{11}$$

where F_{B4s} and W_4 are the static buoyancy and the weight of pontoon 4, respectively.

2.2. Simulation of Irregular Wave

The irregular wave is represented by the Jonswap wave spectrum. The Jonswap wave spectrum is given as a modification of the Pierson–Moskowitz spectrum in accordance with DNV [28,31].

The wave energy spectrum is:

$$S_J(f) = B_J H_s^2 f_p^4 f^{-5}\exp\left[\frac{-5}{4}\left(\frac{f}{f_p}\right)^{-4}\right]\gamma^b \tag{12}$$

where f is the wave frequency, f_p is the peak frequency and H_s is the significant wave height.

$$B_J = \frac{0.06238\times(1.094-0.01915ln\gamma)}{0.230+0.0336\gamma-\frac{0.0185}{1.9+\gamma}},\ b = \exp\left[-0.5\left(\frac{f-f_p}{\sigma f_p}\right)^2\right],$$
$$\sigma = \begin{cases} 0.07, \text{ for } f \le f_p \\ 0.09, \text{ for } f > f_p \end{cases},\ \gamma = 3.3 \tag{13}$$

Referring to the information from the Central Meteorological Bureau Library of Taiwan about the typhoons that have invaded Taiwan from 1897 to 2019 [26,32] and selecting 150 typhoons that greatly affected Taiwan's Green Island, the significant wave height H_s during the 50-year regression period H_s =15.4 m, and the peak period P_w = 16.5 s.

Substituting the significant wave height H_s and the peak period P_w into Equations (12) and (13), the Jonswap wave spectrum is determined, as shown in Figure 3.

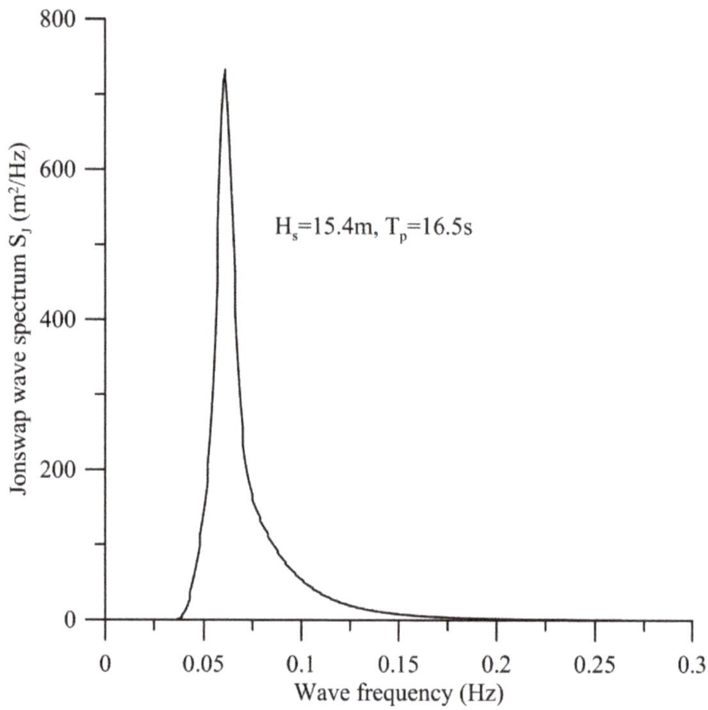

Figure 3. Jonswap wave spectrum.

The frequency domain of the wave spectrum is divided into N subdomains of $(\omega_0, \omega_1), \ldots, (\omega_{N-1}, \omega_N)$. The sea surface elevation of an irregular wave can be generated by the superposition of the regular wave components:

$$x_w = \sum_{i=1}^{N} a_i \sin\left(\Omega_i t - \vec{K}_i \cdot \vec{R} + \varphi_i\right) \tag{14}$$

where a_i, Ω_i, φ_i and $\vec{K}_i$ are the amplitude, angular frequency, phase angle and wave vector of the i-th regular wave, respectively. The angular frequency is $\Omega_i \in (\omega_{i-1}, \omega_i)$. The amplitude can be determined by:

$$\frac{1}{2}a_i^2(\Omega_i) = \int_{\omega_{i-1}}^{\omega_i} S(\omega)d\omega \tag{15}$$

The linear dispersion relation is considered [33]:

$$\Omega_i^2 = g\widetilde{k}_i \tanh \widetilde{k}_i H_{bed} \tag{16}$$

where g is gravity. The wave number $\widetilde{k}_i = \left|\vec{K}_i\right|$. Based on Equation (16), the wave number is obtained. Further, the wave length λ_i can be determined via the relation between the wave number $\widetilde{k}_i$ and the wave length $\widetilde{k}_i = 2\pi/\lambda_i$. Based on Equations (14)–(16) and Figure 4, and letting n = 6, the irregular wave is simulated by regular waves and listed in Table 1. It is assumed that the total wave energy flow rate of the regular waves is equal to that of the Jonswap wave spectrum. In cases 1~4, the numbers of regular waves range from 3 to 6. Every energy flow rate of every regular wave is assumed to be the same. According to Equation (15), the amplitude of each regular wave is the same, and there is no distinction between the dominated wave and the secondary wave. Moreover, the frequency of the regular wave and the given peak frequency are significantly different. In case 5, when six regular waves are used to simulate an irregular wave, every energy flow rate of every regular wave is different. It is obtained that the dominated wave frequency is consistent with the given peak frequency, and the amplitude of the dominated wave is significantly larger than that of other waves. The simulated results of case 6 are used later.

Table 1. Irregular wave simulated by regular waves [H_s = 15.4 m, P_w = 16.5 s, n = 6, H_{bed} = 1300 m].

Case	Number of Regular Waves		1	2	3	4	5	6
1	3	a_i (m)	2.603	2.603	2.603	-	-	-
		f_i (Hz)	0.0369	0.0390	0.1893	-	-	-
2	4	a_i (m)	2.255	2.255	2.255	2.255	-	-
		f_i (Hz)	0.0365	0.0382	0.0390	0.0398	-	-
3	5	a_i (m)	2.017	2.017	2.017	2.017	2.017	-
		f_i (Hz)	0.0365	0.0382	0.0390	0.0398	0.0406	-
4	6	a_i (m)	1.841	1.841	1.841	1.841	1.841	1.841
		f_i (Hz)	0.0365	0.0382	0.0390	0.0398	0.0406	0.1901
5	6	a_i (m)	1.142	4.208	2.630	1.364	0.843	0.605
		f_i (Hz)	0.0425	0.0600	0.0850	0.1150	0.1500	0.2664
		$\widetilde{k}_i$(1/m)	0.0073	0.0145	0.0291	0.0533	0.0906	0.2859
		λ_i (m)	861.5	433.3	215.9	117.9	69.3	22.0

2.3. Dynamic Equilibrium with the Effects of the Steady Current and Irregular Wave

The dynamic equilibrium in the vertical direction for pontoon 3 is:

$$M_3\ddot{x}_{3d} - F_{B3} + W_3 + T_C = 0 \quad (17)$$

where M_3 is the mass of pontoon 3. T_C is the tension of rope C. Substituting Equations (2) and (10) into Equation (17), one obtains:

$$M_3\ddot{x}_{3d} + T_{Cd} - F_{B3d} = 0 \quad (18)$$

where the dynamic tension of the rope C is:

$$T_{Cd} = K_{Cd}(x_{3d} - x_{1d}) \quad (19)$$

in which K_{Cd} is the effective spring constant. $x_{3d} - x_{1d}$ is the dynamic elongation between floating platform 1 and pontoon 3. Considering the safety of the rope, some buffer springs are used to serially connect the rope between elements 1 and 3. The effective spring constant of the rope–buffer spring connection is obtained:

$$K_{Cd} = \frac{K_{\text{rope C}}}{1 + K_{\text{rope C}}/K_{C,spring}} \quad (20)$$

where $K_{C,spring}$ is the constant of the spring connecting with rope C. The effective spring constant of the rope C, $K_{\text{rope C}} = E_C A_C / L_C$, where E_C, A_C, and L_C are the Young's modulus, cross-sectional area and length of rope C, respectively.

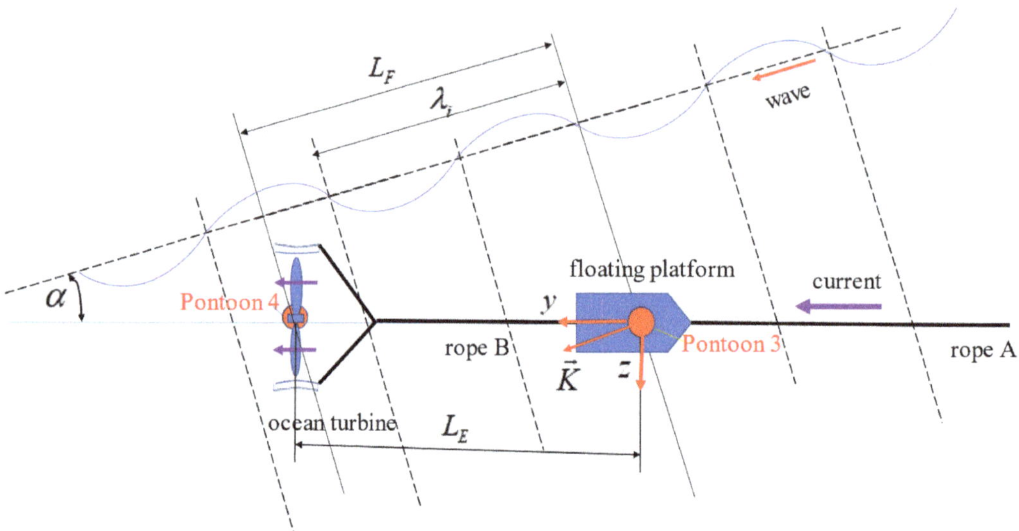

Figure 4. Relation among phase f_i, wave length λ_i and relative direction α of wave and current and distance L_E between the two pontoons.

Assuming the coordinates at pontoon 3 are as shown in Figure 3:

$$\overrightarrow{R}_{\text{pontoon 3}} = 0, \tag{21}$$

The sea surface elevation at pontoon 3 is:

$$x_{w,\text{pontoon 3}} = \sum_{i=1}^{N=6} a_i \sin(\Omega_i t + \varphi_i) \tag{22}$$

The coordinates at pontoon 4 are as shown in Figure 4:

$$\overrightarrow{R}_{\text{pontoon 4}} = L_E \overrightarrow{j} \tag{23}$$

The sea surface elevation at pontoon 4 is:

$$x_{w,\text{pontoon 4}} = \sum_{i=1}^{N=6} a_i \sin(\Omega_i t + \varphi_i + \phi_i) \tag{24}$$

where the phase angle $\phi_i = \frac{2\pi L_E}{\lambda_i} \cos\alpha$ and $L_E = \sqrt{L_B^2 - (L_C - L_D)^2}$. The values of the relative angle α and the wave length λ_i are naturally determined. Nevertheless, the length L_E can be changed to obtain the desired phase angle ϕ_i.

The wave force on the pontoons should include horizontal force and vertical force. Because the length of the ropes connecting the pontoon to the turbine and the carrier is more than 60 m, and the rope can only transmit the axial force and cannot transmit the lateral force, the effect of horizontal force on the dynamic stability of system can be ignored. The volume of the surfaced pontoon should be reduced as much as possible to reduce the wave force, which can increase the dynamic stability and safety of the system and can also be analyzed by the one-way-coupled FSI method. Because the volume of the pontoon is considered small, the horizontal wave force to the pontoon is small. In addition, the length of the ropes connecting the pontoon to the turbine and the carrier is more than 60 m, and the rope can only transmit the axial force and cannot transmit the lateral force; the effect of horizontal force on the dynamic stability of system can be ignored. The corresponding dynamic vertical buoyance of pontoon 3 can be expressed as:

$$F_{B3d}(t) = \sum_{i=1}^{N}[f_{Bs,i} \sin \Omega_i t + f_{Bc,i} \cos \Omega_i t] - A_{Bx}\rho g x_{3d} \tag{25}$$

where $f_{Bs,i} = A_{Bx}\rho g a_i \cos \varphi_i$, $f_{Bc,i} = A_{Bx}\rho g a_i \sin \varphi_i$. Substituting Equations (19) and (22) into Equation (18), one obtains:

$$M_3\ddot{x}_{3d} - K_{Cd}x_{1d} + (K_{Cd} + A_{Bx}\rho g)x_{3d} = \sum_{i=1}^{N}[f_{Bs,i} \sin \Omega_i t + f_{Bc,i} \cos \Omega_i t] \tag{26}$$

The pontoon is composed of two parts: (1) the floating section on the water surface and (2) the underwater container. The floating section on the water surface is cylindrical with equal diameters, and the dynamic buoyancy of the pontoon is related to its dynamic displacement. The mass of the pontoon can be controlled by the water in the underwater container. Therefore, the mass and cross-sectional area of the pontoon can be considered, and their independent, individual effects can be studied.

The dynamic equilibrium in the vertical direction for pontoon 4 is:

$$M_4\ddot{x}_{4d} - F_{B4} + W_4 + T_D = 0 \tag{27}$$

where M_4 is the mass of the pontoon 4, and T_D is the tension of the rope D. Substituting Equations (2) and (11) into Equation (29), one obtains:

$$M_4\ddot{x}_{4d} - F_{B4d} + T_{Dd} = 0 \tag{28}$$

where the dynamic tension of rope D is:

$$T_{Dd} = K_{Dd}(x_{4d} - x_{2d}) \tag{29}$$

in which K_{Dd} is the effective spring constant. $x_{4d} - x_{2d}$ is the dynamic elongation between floating platform 2 and pontoon 4. Considering the safety of the rope, some buffer springs are used to serially connect the rope between elements 2 and 4. The effective spring constant of the rope–buffer spring connection is obtained:

$$K_{Dd} = \frac{K_{\text{rope D}}}{1 + K_{\text{rope D}}/K_{D,spring}} \tag{30}$$

where $K_{D,spring}$ is the constant of the spring connecting with rope D. The effective spring constant of rope D, $K_{\text{rope D}} = E_D A_D / L_D$, where E_D, A_D, and L_D are the Young's modulus, cross-sectional area and length of the rope D, respectively.

According to Equation (24), the dynamic buoyance of pontoon 4 is:

$$F_{B4d}(t) = \sum_{i=1}^{N}[f_{Ts,i} \sin \Omega_i t + f_{Tc,i} \cos \Omega_i t] - A_{BT}\rho g x_{4d} \tag{31}$$

where $f_{Ts,i} = A_{BT}\rho g a_i \cos(\varphi_i + \phi_i)$, $f_{Tc,i} = A_{BT}\rho g a_i \sin(\varphi_i + \phi_i)$. Substituting Equations (29) and (31) into Equation (28), one obtains:

$$M_4\ddot{x}_{4d} - K_{Dd}x_{2d} + (K_{Dd} + A_{BT}\rho g)x_{4d} = \sum_{i=1}^{N}[f_{Ts,i} \sin \Omega_i t + f_{Tc,i} \cos \Omega_i t] \tag{32}$$

The dynamic equilibrium in the vertical direction for the floating platform is:

$$\left(M_1 + m_{eff,x}\right)\ddot{x}_{1d} - F_{B1s} + W_1 - T_C + T_A \sin \theta_A + T_B \sin \theta_B = 0 \tag{33}$$

where M_1 is the mass of the platform. The dynamic, effective mass of rope 1 in the x-direction, $m_{eff,x} = \frac{4f_g L_{As} \sin\theta_1}{\pi^2}$, was derived by Lin and Chen [15]. Substituting Equations (2) and (7) into Equation (33), one obtains:

$$\left(M_1 + m_{eff,x}\right)\ddot{x}_{1d} + T_{Cd} + T_{Ad}\sin\theta_A + T_{Bd}\sin\theta_B = 0 \tag{34}$$

where T_A is the dynamic tension of rope A.

$$T_{Ad} = K_{Ad}\delta_{Ad} \tag{35}$$

The dynamic elongation is $\delta_{Ad} = L_{Ad} - L_A$ where L_A and L_{Ad} are the static and dynamic lengths of rope A, respectively. The effective spring constant of the rope–buffer spring connection is:

$$K_{Ad} = \frac{K_{\text{rope A}}}{1 + K_{\text{rope A}}/K_{A,spring}} \tag{36}$$

where $K_{A,spring}$ is the constant of the spring connecting to rope A. The effective spring constant of rope A is $K_{\text{rope A}} = E_A A_A / L_A$, where E_A and A_A are the Young's modulus and the cross-sectional area of rope A, respectively. The static and dynamic lengths are:

$$L_A = \sqrt{x_{1s}^2 + y_{1s}^2},\ L_{Ad} = \sqrt{(x_{1s} + x_{1d})^2 + (y_{1s} + y_{1d})^2} \tag{37}$$

The approximated dynamic elongation is proposed by using the Taylor formula:

$$\delta_{Ad} = \frac{x_{1s}}{L_A}x_{1d} + \frac{y_{1s}}{L_A}y_{1d} \tag{38}$$

where the dynamic tension of rope B is:

$$T_{Bd} = K_{Bd}\delta_{Bd} \tag{39}$$

and where the dynamic elongation $\delta_{Bd} = L_{Bd} - L_{Bs}$. L_{Bs} and L_{Bd} are the static and dynamic lengths of rope B. The effective spring constant of the rope–buffer spring connection is:

$$K_{Bd} = \frac{K_{\text{rope B}}}{1 + K_{\text{rope B}}/K_{B,spring}} \tag{40}$$

where $K_{B,spring}$ is the constant of the spring connecting with rope B. The effective spring constant of rope B is $K_{\text{rope B}} = E_B A_B / L_B$, in which E_B and A_B are the Young's modulus and the cross-sectional area of rope B. The static and dynamic lengths are:

$$L_B = \sqrt{(x_{1s} - x_{2s})^2 + (y_{1s} - y_{2s})^2},\ L_{Bd} = \sqrt{(x_1 - x_2)^2 + (y_1 - y_2)^2} \tag{41}$$

Using the Tylor formula, one can obtain the approximated dynamic elongation:

$$\delta_{Bd} = \frac{x_{1s} - x_{2s}}{L_B}(x_{1d} - x_{2d}) + \frac{y_{1s} - y_{2s}}{L_B}(y_{1d} - y_{2d}) \tag{42}$$

Substituting Equations (19), (35), (38), (39) and (42) into Equation (34), one obtains:

$$\begin{aligned}&\left(M_1 + m_{eff,x}\right)\ddot{x}_{1d} + \left(K_{Ad}\tfrac{x_{1s}}{L_A}\sin\theta_A + K_{Bd}\tfrac{x_{1s}-x_{2s}}{L_B}\sin\theta_B - K_{Cd}\right)x_{1d} - \left(K_{Bd}\tfrac{x_{1s}-x_{2s}}{L_B}\sin\theta_B\right)x_{2d}\\ &+K_{Cd}x_{3d} + \left(K_{Ad}\tfrac{y_{1s}}{L_A}\sin\theta_A + K_{Bd}\tfrac{y_{1s}-y_{2s}}{L_B}\sin\theta_B\right)y_{1d} - \left(K_{Bd}\tfrac{y_{1s}-y_{2s}}{L_B}\sin\theta_B\right)y_{2d} = 0\end{aligned} \tag{43}$$

The dynamic equilibrium in the vertical direction for the turbine is:

$$-M_2\ddot{x}_{2d} - W_2 + F_{B2s} + T_D + T_B\sin\theta_B = 0 \tag{44}$$

Substituting Equations (2) and (9) into Equation (44), one obtains:

$$-M_2\ddot{x}_{2d} + T_{Dd} + T_{Bd}\sin\theta_B = 0 \tag{45}$$

Substituting Equations (29), (39) and (42) into Equation (45), one obtains:

$$\begin{aligned} M_2\ddot{x}_{2d} - K_{Bd}\tfrac{x_{1s}-x_{2s}}{L_B}\sin\theta_B x_{1d} + \left(K_{Dd} + K_{Bd}\tfrac{x_{1s}-x_{2s}}{L_B}\sin\theta_B\right)x_{2d} - K_{Dd}x_{4d} \\ -K_{Bd}\tfrac{y_{1s}-y_{2s}}{L_B}\sin\theta_B y_{1d} + K_{Bd}\tfrac{y_{1s}-y_{2s}}{L_B}\sin\theta_B y_{2d} = 0 \end{aligned} \tag{46}$$

The dynamic equilibrium in the horizontal direction for the floating platform is:

$$-\left(M_1 + m_{eff,y}\right)\ddot{y}_{1d} + F_{DFy} - T_A\cos\theta_A + T_B\cos\theta_B = 0 \tag{47}$$

where y_{1d} is the dynamic horizontal displacement of the floating platform. The dynamic effective mass of rope A in the y-direction is $m_{eff,y} = \frac{4f_g L_A\cos\theta_A}{\pi^2}$ [26]. The horizontal force on the platform due to the current velocity V and the horizontal velocity $\dot{y}_{1d}$ of the platform are expressed as [34]:

$$F_{DFY} = \frac{1}{2}C_{DFy}\rho A_{FY}(V - \dot{y}_{1d})^2 = \frac{1}{2}C_{DFy}\rho A_{FY}\left(V^2 - 2V\dot{y}_{1d} + \dot{y}_{1d}^2\right) \approx F_{DFs} - C_{DFy}\rho A_{FY}V\dot{y}_{1d} \tag{48}$$

Because $\dot{y}_{1d} << V$, the term $\dot{y}_{1d}^2$ is negligible. The drag coefficient of the floating platform is considered close to that of a bullet, i.e., $C_{DFy} \approx 0.3$, [26].

Substituting Equations (2), (7), (35), (38), (39), (42) and (48) into Equation (47), one obtains:

$$\begin{aligned} &\left(M_1 + m_{eff,y}\right)\ddot{y}_{1d} + C_{DFy}\rho A_{Fy}V\dot{y}_{1d} \\ &+\left(K_{Ad}\tfrac{x_{1s}}{L_A}\cos\theta_A - K_{Bd}\tfrac{x_{1s}-x_{2s}}{L_B}\cos\theta_B\right)x_{1d} + K_{Bd}\tfrac{x_{1s}-x_{2s}}{L_B}\cos\theta_B x_{2d} \\ &+\left(K_{Ad}\tfrac{y_{1s}}{L_A}\cos\theta_A - K_{Bd}\tfrac{y_{1s}-y_{2s}}{L_B}\cos\theta_B\right)y_{1d} + K_{Bd}\tfrac{y_{1s}-y_{2s}}{L_B}\cos\theta_B y_{2d} = 0 \end{aligned} \tag{49}$$

It is discovered from Equation (49) that the second term is the damping effect for vibration of the system. The damping effect depends on the parameters: (1) the damping coefficient C_{DyF}, (2) the damping area A_{BY} and (3) the current velocity V.

The dynamic equilibrium in the horizontal direction for the turbine is:

$$-M_2\ddot{y}_{2d} + F_{DTy} - T_B\cos\theta_B = 0 \tag{50}$$

where y_{2d} is the dynamic, horizontal displacement of the turbine. The horizontal force on the platform caused by the current velocity V and the horizontal velocity $\dot{y}_{2d}$ of the turbine is expressed as [34]:

$$F_{DTy} = C_{DTy}\frac{1}{2}\rho A_{Ty}(V - \dot{y}_{2d})^2 = C_{DTy}\frac{1}{2}\rho A_{Ty}\left(V^2 - 2V\dot{y}_{2d} + \dot{y}_{2d}^2\right) \approx F_{DTs} - C_{DyT}\rho A_{Ty}V\dot{y}_{2d} \tag{51}$$

where A_{Ty} is the effective operating area of the turbine. The theoretical effective drag coefficient of optimum efficiency is $C_{DTy} = 8/9$, [27]. Considering $\dot{y}_{2d} << V$, the term $\dot{y}_{2d}^2$ is negligible. Substituting Equations (2), (8), (39), (42) and (51) into Equation (50), one obtains:

$$\begin{aligned} M_2\ddot{y}_{2d} + C_{DTy}\rho A_{Ty}V\dot{y}_{2d} + K_{Bd}\tfrac{x_{1s}-x_{2s}}{L_B}\cos\theta_B x_{1d} \\ -K_{Bd}\tfrac{x_{1s}-x_{2s}}{L_B}\cos\theta_B x_{2d} + K_{Bd}\tfrac{y_{1s}-y_{2s}}{L_B}\cos\theta_B y_{1d} - K_{Bd}\tfrac{y_{1s}-y_{2s}}{L_B}\cos\theta_B y_{2d} = 0 \end{aligned} \tag{52}$$

It is discovered from Equation (52) that the second term is the damping effect for vibration of the system. The damping effect depends on the parameters: (1) the damping coefficient C_{DTy}, (2) the damping area A_{TY} and (3) the current velocity V.

Finally, the coupled equations of motion in terms of the dynamic displacements x_{1d}, x_{2d}, x_{3d}, x_{4d}, y_{1d}, and y_{2d} are discovered as Equations (26), (32), (34), (49), (52) and (56).

2.4. Solution Method

2.4.1. Free Vibration

Without the excitation of wave and under steady ocean current, the coupled motion of the system is in free vibration. According to Equations (26), (32), (34), (49), (52) and (56), the coupled equations of free vibration can be expressed as:

$$\mathbf{M}\ddot{\mathbf{Z}}_d + \mathbf{C}\dot{\mathbf{Z}}_d + \mathbf{K}\mathbf{Z}_d = 0 \tag{53}$$

where

$$\mathbf{Z}_d = \begin{bmatrix} x_{1d} \\ x_{2d} \\ x_{3d} \\ x_{4d} \\ y_{1d} \\ y_{2d} \end{bmatrix}, \mathbf{M} = \begin{bmatrix} \left(M_1 + m_{eff,x}\right) & 0 & 0 & 0 & 0 & 0 \\ 0 & M_2 & 0 & 0 & 0 & 0 \\ 0 & 0 & M_3 & 0 & 0 & 0 \\ 0 & 0 & 0 & M_4 & 0 & 0 \\ 0 & 0 & 0 & 0 & \left(M_1 + m_{eff,y}\right) & 0 \\ 0 & 0 & 0 & 0 & 0 & M_2 \end{bmatrix}$$
$$\mathbf{C} = \begin{bmatrix} C_1 & 0 & 0 & 0 & 0 & 0 \\ 0 & C_2 & 0 & 0 & 0 & 0 \\ 0 & 0 & C_3 & 0 & 0 & 0 \\ 0 & 0 & 0 & C_4 & 0 & 0 \\ 0 & 0 & 0 & 0 & C_5 & 0 \\ 0 & 0 & 0 & 0 & 0 & C_6 \end{bmatrix}, \mathbf{K} = \begin{bmatrix} K_{11} & K_{12} & K_{13} & 0 & K_{15} & K_{16} \\ K_{21} & K_{22} & 0 & K_{24} & K_{25} & K_{26} \\ K_{31} & 0 & K_{33} & 0 & 0 & 0 \\ 0 & K_{42} & 0 & K_{44} & 0 & 0 \\ K_{51} & K_{52} & 0 & 0 & K_{55} & K_{56} \\ K_{61} & K_{62} & 0 & 0 & K_{65} & K_{66} \end{bmatrix} \tag{54}$$

$$\begin{aligned}
&C_1 = C_2 = C_3 = C_4 = 0,\ C_5 = C_{DFy}\rho A_{Fy} V,\ C_6 = C_{DTy}\rho A_{Ty} V \\
&K_{11} = \left(K_{Ad}\tfrac{x_{1s}}{L_A}\sin\theta_A + K_{Bd}\tfrac{x_{1s}-x_{2s}}{L_B}\sin\theta_B + K_{Cd}\right) \\
&K_{12} = -\left(K_{Bd}\tfrac{x_{1s}-x_{2s}}{L_B}\sin\theta_B\right),\ K_{13} = -K_{Cd} \\
&K_{15} = \left(K_{Ad}\tfrac{y_{1s}}{L_A}\sin\theta_A + K_{Bd}\tfrac{y_{1s}-y_{2s}}{L_B}\sin\theta_B\right),\ K_{16} = -\left(K_{Bd}\tfrac{y_{1s}-y_{2s}}{L_B}\sin\theta_B\right) \\
&K_{21} = -K_{Bd}\tfrac{x_{1s}-x_{2s}}{L_B}\sin\theta_B,\ K_{22} = \left(K_{Dd} + K_{Bd}\tfrac{x_{1s}-x_{2s}}{L_B}\sin\theta_B\right), \\
&K_{24} = -K_{Dd},\ K_{25} = -K_{Bd}\tfrac{y_{1s}-y_{2s}}{L_B}\sin\theta_B,\ K_{26} = K_{Bd}\tfrac{y_{1s}-y_{2s}}{L_B}\sin\theta_B \\
&K_{31} = -K_{Cd} \qquad K_{42} = -K_{Dd} \\
&K_{33} = (K_{Cd} + A_{Bx}\rho g) \qquad K_{44} = (K_{Dd} + A_{BT}\rho g) \\
&K_{51} = \left(K_{Ad}\tfrac{x_{1s}}{L_A}\cos\theta_A - K_{Bd}\tfrac{x_{1s}-x_{2s}}{L_B}\cos\theta_B\right),\ K_{52} = K_{Bd}\tfrac{x_{1s}-x_{2s}}{L_B}\cos\theta_B \\
&K_{55} = \left(K_{Ad}\tfrac{y_{1s}}{L_A}\cos\theta_A - K_{Bd}\tfrac{y_{1s}-y_{2s}}{L_B}\cos\theta_B\right),\ K_{56} = K_{Bd}\tfrac{y_{1s}-y_{2s}}{L_B}\cos\theta_B \\
&K_{61} = K_{Bd}\tfrac{x_{1s}-x_{2s}}{L_B}\cos\theta_B,\ K_{62} = -K_{Bd}\tfrac{x_{1s}-x_{2s}}{L_B}\cos\theta_B, \\
&K_{65} = K_{Bd}\tfrac{y_{1s}-y_{2s}}{L_B}\cos\theta_B,\ K_{66} = -K_{Bd}\tfrac{y_{1s}-y_{2s}}{L_B}\cos\theta_B;
\end{aligned} \tag{55}$$

The solution of Equation (53) is assumed to be:

$$\mathbf{Z}_d = \begin{bmatrix} \overline{x}_{1d} & \overline{x}_{2d} & \overline{x}_{3d} & \overline{x}_{4d} & \overline{y}_{1d} & \overline{y}_{2d} \end{bmatrix}^{\mathrm{T}} = \left(\overline{\mathbf{z}}_{\mathbf{dc}}\cos\Omega t + \overline{\mathbf{z}}_{\mathbf{ds}}\sin\Omega t\right) \tag{56}$$

where $\overline{\mathbf{z}}_{\mathbf{dc}} = \begin{bmatrix} \overline{x}_{1d,c} & \overline{x}_{2d,c} & \overline{x}_{3d,c} & \overline{x}_{4d,c} & \overline{y}_{1d,c} & \overline{y}_{2d,c} \end{bmatrix}^{\mathrm{T}}$, $\overline{\mathbf{z}}_{\mathbf{ds}} = \begin{bmatrix} \overline{x}_{1d,s} & \overline{x}_{2d,s} & \overline{x}_{3d,s} & \overline{x}_{4d,s} & \overline{y}_{1d,s} & \overline{y}_{2d,s} \end{bmatrix}^{\mathrm{T}}$. Substituting Equation (56) into Equation (53), one obtains:

$$\begin{aligned}
&\left(\left(\mathbf{M}^{-1}\mathbf{K} - \Omega^2 \mathrm{I}\right)\overline{\mathbf{z}}_{\mathbf{dc}} + \Omega\mathbf{M}^{-1}\mathbf{C}\overline{\mathbf{z}}_{\mathbf{ds}}\right)\cos\Omega t \\
&+\left(\left(\mathbf{M}^{-1}\mathbf{K} - \Omega^2 \mathrm{I}\right)\overline{\mathbf{z}}_{\mathbf{ds}} - \Omega\mathbf{M}^{-1}\mathbf{C}\overline{\mathbf{z}}_{\mathbf{dc}}\right)\sin\Omega t = 0
\end{aligned} \tag{57}$$

Due to the orthogonality of $\sin\Omega t$ and $\cos\Omega t$, Equation (57) becomes:

$$\left(\mathbf{M}^{-1}\mathbf{K} - \Omega^2 \mathrm{I}\right)\overline{\mathbf{z}}_{\mathbf{dc}} + \Omega\mathbf{M}^{-1}\mathbf{C}\overline{\mathbf{z}}_{\mathbf{ds}} = 0 \tag{58}$$

$$-\Omega\mathbf{M}^{-1}\mathbf{C}\bar{\mathbf{z}}_{\mathbf{d}c}+\left(\mathbf{M}^{-1}\mathbf{K}-\Omega^2\mathrm{I}\right)\bar{\mathbf{z}}_{\mathbf{d}s}=0 \tag{59}$$

Further, Equation (58) can be expressed as:

$$\bar{\mathbf{z}}_{\mathbf{d}c}=-\left(\mathbf{M}^{-1}\mathbf{K}-\Omega^2\mathrm{I}\right)^{-1}\Omega\mathbf{M}^{-1}\mathbf{C}\bar{\mathbf{z}}_{\mathbf{d}s} \tag{60}$$

Substituting Equation (60) into Equation (59), one obtains:

$$\mathbf{Q}\bar{\mathbf{z}}_{\mathbf{d}s}=0 \tag{61}$$

where $\mathbf{Q}=\Omega^2\mathbf{M}^{-1}\mathbf{C}\left(\mathbf{M}^{-1}\mathbf{K}-\Omega^2\mathrm{I}\right)^{-1}\mathbf{M}^{-1}\mathbf{C}+\left(\mathbf{M}^{-1}\mathbf{K}-\Omega^2\mathrm{I}\right)$. The frequency equation is:

$$|\mathbf{Q}|=0 \tag{62}$$

The natural frequencies of the system can be determined via Equation (62).

2.4.2. Forced Vibration

Considering the excitation of the wave, the coupled Equations (26), (32), (34), (49), (52) and (56) can be rewritten in the matrix format as follows:

$$\mathbf{M}\ddot{\mathbf{Z}}_d+\mathbf{C}\dot{\mathbf{Z}}_d+\mathbf{K}\mathbf{Z}_d=\mathbf{F}_d \tag{63}$$

where

$$\begin{array}{c}\mathbf{F}_d=\left[\begin{array}{cccccc}0 & 0 & \sum\limits_{i=1}^{N}[f_{Bs,i}\sin\Omega_i t+f_{Bc,i}\cos\Omega_i t] & \sum\limits_{i=1}^{N}[f_{Ts,i}\sin\Omega_i t+f_{Tc,i}\cos\Omega_i t] & 0 & 0\end{array}\right]^T, \\ f_{Bs,i}=A_{Bx}\rho g a_i\cos\varphi_i,\ f_{Bc,i}=A_{Bx}\rho g a_i\sin\varphi_i \\ f_{Ts,i}=A_{BT}\rho g a_i\cos(\varphi_i+\phi_i),\ f_{Tc,i}=A_{BT}\rho g a_i\sin(\varphi_i+\phi_i)\end{array} \tag{64}$$

The solution of Equation (63) is assumed:

$$\mathbf{Z}_d=\left[\begin{array}{cccccc}x_{1d} & x_{2d} & x_{3d} & x_{4d} & y_{1d} & y_{2d}\end{array}\right]^{\mathrm{T}}=\sum_{i=1}^{N}(\mathbf{z}_{\mathbf{d}c,i}\cos\Omega_i t+\mathbf{z}_{\mathbf{d}s,i}\sin\Omega_i t), \tag{65}$$

where $\mathbf{z}_{\mathbf{d}c,i}=\left[\begin{array}{cccccc}x_{1d,c} & x_{2d,c} & x_{3d,c} & x_{4d,c} & y_{1d,c} & y_{2d,c}\end{array}\right]^{\mathrm{T}}$, $\mathbf{z}_{\mathbf{d}s,i}=\left[\begin{array}{cccccc}x_{1d,s} & x_{2d,s} & x_{3d,s} & x_{4d,s} & y_{1d,s} & y_{2d,s}\end{array}\right]^{\mathrm{T}}$. Substituting Equation (65) into Equation (63), one obtains:

$$\begin{array}{c}-\sum\limits_{i=1}^{N}\Omega_i^2(\mathbf{z}_{\mathbf{d}c,i}\cos\Omega_i t+\mathbf{z}_{\mathbf{d}s,i}\sin\Omega_i t)+\mathbf{M}^{-1}\mathbf{C}\sum\limits_{i=1}^{N}(-\Omega_i\mathbf{z}_{\mathbf{d}c,i}\sin\Omega_i t+\Omega_i\mathbf{z}_{\mathbf{d}s,i}\cos\Omega_i t) \\ +\mathbf{M}^{-1}\mathbf{K}\sum\limits_{i=1}^{N}(\mathbf{z}_{\mathbf{d}c,i}\cos\Omega_i t+\mathbf{z}_{\mathbf{d}s,i}\sin\Omega_i t)=\sum\limits_{i=1}^{N}(\mathbf{F}_{s,i}\sin\Omega_i t+\mathbf{F}_{c,i}\cos\Omega_i t)\end{array} \tag{66}$$

Multiplying Equation (66) by cos $\Omega_m t$ and integrating it from 0 to the period T_m, $2\pi/\Omega_m$, Equation (66) becomes:

$$\sum_{i=1}^{N}\mathbf{a}_{im}\mathbf{z}_{\mathbf{d}c,i}+\sum_{i=1}^{N}\mathbf{b}_{im}\mathbf{z}_{\mathbf{d}s,i}=\chi_{cm},\ m=1,2,\ldots,N \tag{67}$$

where

$$\begin{array}{l}\mathbf{a}_{im}=\left[\alpha_{im}\left(\mathbf{M}^{-1}\mathbf{K}-\Omega_i^2\mathrm{I}\right)-\beta_{im}\Omega_i\mathbf{M}^{-1}\mathbf{C}\right],\mathbf{b}_{im}=\left[\beta_{im}\left(\mathbf{M}^{-1}\mathbf{K}-\Omega_i^2\mathrm{I}\right)-\alpha_{im}\Omega_i\mathbf{M}^{-1}\mathbf{C}\right]\\ \chi_{cm}=\sum\limits_{i=1}^{N}\left(\mathbf{F}_{s,i}\beta_{im}+\mathbf{F}_{c,i}\alpha_{im}\right)\\ \alpha_{im}=\begin{cases}\frac{T_m}{2}, & i=m\\ \frac{\Omega_i\sin(\Omega_i T_m)}{(\Omega_i+\Omega_m)(\Omega_i-\Omega_m)}, & i\neq m\end{cases},\ \beta_{im}=\begin{cases}0, & i=m\\ \frac{\Omega_i(1-\cos(\Omega_i T_m))}{(\Omega_i+\Omega_m)(\Omega_i-\Omega_m)}, & i\neq m\end{cases}\end{array} \tag{68}$$

Multiplying Equation (66) by $\sin\Omega_m t$ and integrating it from 0 to the period T_m, $2\pi/\Omega_m$, Equation (66) becomes:

$$\sum_{i=1}^{N}\mathbf{c}_{im}\mathbf{z}_{\mathbf{d}c,i}+\sum_{i=1}^{N}\mathbf{d}_{im}\mathbf{z}_{\mathbf{d}s,i}=\chi_{sm},\ m=1,2,\ldots,N \tag{69}$$

where

$$\begin{array}{l}\mathbf{c}_{im}=\left[\beta_{mi}\left(\mathbf{M}^{-1}\mathbf{K}-\Omega_i^2\mathrm{I}\right)-\gamma_{im}\Omega_i\mathbf{M}^{-1}\mathbf{C}\right],\ \mathbf{d}_{im}=\left[\gamma_{im}\left(\mathbf{M}^{-1}\mathbf{K}-\Omega_i^2\mathrm{I}\right)-\beta_{mi}\Omega_i\mathbf{M}^{-1}\mathbf{C}\right]\\ \chi_{sm}=\sum\limits_{i=1}^{N}\left(\mathbf{F}_{s,i}\gamma_{im}+\mathbf{F}_{c,i}\beta_{mi}\right)\\ \gamma_{im}=\begin{cases}\frac{T_m}{2}, & i=m\\ \frac{\Omega_m\sin(\Omega_i T_m)}{(\Omega_i+\Omega_m)(\Omega_i-\Omega_m)}, & i\neq m\end{cases}\end{array} \tag{70}$$

Equations (67) and (69) can be written as:

$$\mathbf{BZ{=}F} \tag{71}$$

where

$$\mathbf{B}=\begin{bmatrix}\begin{bmatrix}\mathbf{a}_{11} & \mathbf{a}_{21} & \cdots & \mathbf{a}_{N1}\\ \mathbf{a}_{12} & \mathbf{a}_{22} & \cdots & \mathbf{a}_{N2}\\ \vdots & \vdots & \cdots & \vdots\\ \mathbf{a}_{1N} & \mathbf{a}_{2N} & \cdots & \mathbf{a}_{NN}\end{bmatrix}_{6N\times 6N} & \begin{bmatrix}\mathbf{b}_{11} & \mathbf{b}_{21} & \cdots & \mathbf{b}_{N1}\\ \mathbf{b}_{12} & \mathbf{b}_{22} & \cdots & \mathbf{b}_{N2}\\ \vdots & \vdots & \cdots & \vdots\\ \mathbf{b}_{1N} & \mathbf{b}_{2N} & \cdots & \mathbf{b}_{NN}\end{bmatrix}_{6N\times 6N}\\ \begin{bmatrix}\mathbf{c}_{11} & \mathbf{c}_{21} & \cdots & \mathbf{c}_{N1}\\ \mathbf{c}_{12} & \mathbf{c}_{22} & \cdots & \mathbf{c}_{N2}\\ \vdots & \vdots & \cdots & \vdots\\ \mathbf{c}_{1N} & \mathbf{c}_{2N} & \cdots & \mathbf{c}_{NN}\end{bmatrix}_{6N\times 6N} & \begin{bmatrix}\mathbf{d}_{11} & \mathbf{d}_{21} & \cdots & \mathbf{d}_{N2}\\ \mathbf{d}_{12} & \mathbf{d}_{22} & \cdots & \mathbf{d}_{N2}\\ \vdots & \vdots & \cdots & \vdots\\ \mathbf{d}_{1N} & \mathbf{d}_{2N} & \cdots & \mathbf{d}_{NN}\end{bmatrix}_{6N\times 6N}\end{bmatrix}_{12N\times 12N},$$

$$\mathbf{Z}{=}\begin{bmatrix}\begin{bmatrix}\mathbf{z}_{\mathbf{d}c,1}\\ \mathbf{z}_{\mathbf{d}c,2}\\ \vdots\\ \mathbf{z}_{\mathbf{d}c,N}\end{bmatrix}_{6N\times 1}\\ \begin{bmatrix}\mathbf{z}_{\mathbf{d}s,1}\\ \mathbf{z}_{\mathbf{d}s,2}\\ \vdots\\ \mathbf{z}_{\mathbf{d}s,N}\end{bmatrix}_{6N\times 1}\end{bmatrix}_{12N\times 1},\quad \mathbf{F}=\begin{bmatrix}\begin{bmatrix}\chi_{c1}\\ \chi_{c2}\\ \vdots\\ \chi_{cN}\end{bmatrix}_{6N\times 1}\\ \begin{bmatrix}\chi_{s1}\\ \chi_{s2}\\ \vdots\\ \chi_{sN}\end{bmatrix}_{6N\times 1}\end{bmatrix}_{12N\times 1} \tag{72}$$

The solution of Equation (65) is:

$$\mathbf{Z{=}B}^{-1}\mathbf{F} \tag{73}$$

Further, one can derive the dynamic tensions of ropes under irregular wave as follows: The dynamic tension of rope A is:

$$T_{Ad}=\sum_{i=1}^{N}T_{Adc,i}\cos\Omega_i t+T_{Ads,i}\sin\Omega_i t \tag{74}$$

where $T_{Adc,i} = K_{Ad}\left(\frac{x_{1s}}{L_A}x_{1dc,i} + \frac{y_{1s}}{L_A}y_{1dc,i}\right)$, $T_{Ads,i} = K_{Ad}\left(\frac{x_{1s}}{L_A}x_{1ds,i} + \frac{y_{1s}}{L_A}y_{1ds,i}\right)$.

The dynamic tension of rope B is:

$$T_{Bd} = \sum_{i=1}^{N} T_{Bdc,i}\cos\Omega_i t + T_{Bds,i}\sin\Omega_i t \tag{75}$$

where

$$T_{Bdc,i} = K_{Bd}\left[\frac{x_{2s}-x_{1s}}{L_B}(x_{2dc,i} - x_{1dc,i}) + \frac{y_{2s}-y_{1s}}{L_B}(y_{2dc,i} - y_{1dc,i})\right],$$
$$T_{Bds,i} = K_{Bd}\left[\frac{x_{2s}-x_{1s}}{L_B}(x_{2ds,i} - x_{1ds,i}) + \frac{y_{2s}-y_{1s}}{L_B}(y_{2ds,i} - y_{1ds,i})\right].$$

The dynamic tension of rope C is:

$$T_{Cd} = \sum_{i=1}^{N} T_{Cdc,i}\cos\Omega_i t + T_{Cds,i}\sin\Omega_i t \tag{76}$$

where $T_{Cdc,i} = K_{Cd}(x_{3\mathbf{dc},i} - x_{1dc,i})$, $T_{Cds,i} = K_{Cd}(x_{3\mathbf{ds},i} - x_{1ds,i})$.

The dynamic tension of rope D is:

$$T_{Dd} = \sum_{i=1}^{N} T_{Ddc,i}\cos\Omega_i t + T_{Dds,i}\sin\Omega_i t \tag{77}$$

where $T_{Ddc,i} = K_{Dd}(x_{4\mathbf{dc},i} - x_{2dc,i})$, $T_{Dds,i} = K_{Dd}(x_{4\mathbf{ds},i} - x_{2ds,i})$.

3. Numerical Results and Discussion

This study investigates the dynamic response of two kinds of mooring system under the typhoon irregular wave: (1) the diving depth of the turbine L_D = 60 m, the diving depth of the floating platform $L_C \geq 60$ m and (2) the diving depth of the floating platform L_C = 60 m, the diving depth of the turbine $L_D \geq 60$ m. Meanwhile, the effects of several parameters on the dynamic response are investigated.

Firstly, the first kind of mooring system is investigated. Consider the conditions in Figure 5a,b: (1) the depth of seabed H_{bed} = 1300 m, (2) the cross-sectional area of pontoon 3 connecting to floating platform A_{BX} = 2.12 m^2, (3) the cross-sectional area of pontoon 4 connecting to turbine A_{BT} = 2.12 m^2, (4) no buffer spring, (5) the ropes A, B, C and D are made of some commercial, high-strength PE dyneema; Young's modulus E_{PE} = 100 GPa, weight per unit length $f_{g,PE}$ = 16.22 kg/m, diameter D_{PE} = 154 mm, cross-sectional area A_{PE} = 0.0186 m^2, fracture strength $T_{fracture}$ = 759 tons, (6) the static diving depth of the turbine L_D = 60 m, (7) the horizontal distance between the turbine and floating platform L_E = 100 m, (8) the inclined angle of the rope A, $\theta_A = 30^\circ$, (9) the current velocity V = 1 m/s, (10) the irregular wave is simulated by six regular waves which are listed in Table 1, (11) the wave phase angles φ_i, $i = 1, 2, \ldots, 6$ are assumed as $\{30^\circ, 60^\circ, 90^\circ, 120^\circ, 170^\circ, 270^\circ\}$, (12) the masses of turbine, floating platform and pontoons $M_1 = 300$ *tons*, $M_2 = 838$ *tons*, $M_3 = M_4 = 250$ *tons*, (13) the cross-sectional area of the floating platform and turbine $A_{FY} = 23$ m^2 and $A_{TY} = 500$ m^2, (14) the effective damping coefficients $C_{DFy} = 0.3$ and $C_{DTy} = 8/9$, (15) the static axial force to turbine F_{DTs} = 180 tons and (16) the relative orientation between current and wave $\alpha = 60^\circ$.

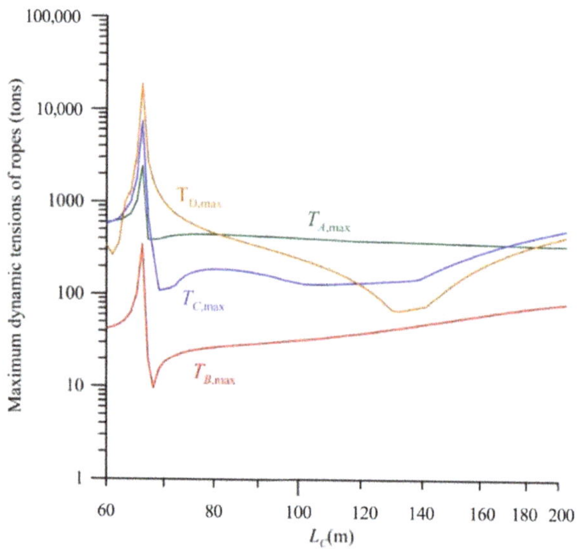

(a)

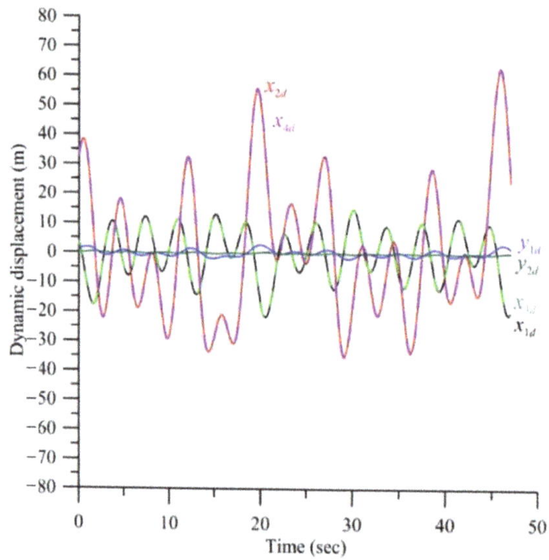

(b)

Figure 5. Dynamic tensions of ropes and Dynamic displacements of elements under the typhoon irregular wave. (**a**) Dynamic tension of the four ropes under the typhoon irregular wave as a function of the diving length for L_E = 100 m; (**b**) Dynamic displacements of the four elements at resonance for L_C = 66 m.

Figure 5a demonstrates the effect of the diving depth of the floating platform L_C on the maximum dynamic tensions of the four ropes, $T_{A,max}$, $T_{B,max}$, $T_{C,max}$ and $T_{D,max}$ under the typhoon irregular wave when the diving depth of the turbine L_D = 60 m. The irregular wave is simulated by six regular waves which are listed in Table 1. When the depth L_C increases from 60 m, the dynamic tensions of the ropes increase significantly. If $L_{C,res}$ = 66 m, the resonance happens, and the maximum dynamic tensions $T_{A,max}$ = 2422 tons, $T_{C,max}$ = 7329 tons and $T_{D,max}$ = 18,835 tons, which is over that of the fracture strength of rope, $T_{fracture}$ = 759 tons. Figure 5b demonstrates the vibration mode at the resonance. It is found that the displacements x_{2d} and x_{4d} of turbine 2 and pontoon 4 are largest. Therefore, the maximum dynamic tension is that of rope D, $T_{D,max}$.

When L_C increases further, the dynamic tension decreases sharply. If L_C > 80 m, all the dynamic tensions are significantly less than the fracture strength of rope, $T_{fracture}$ = 759 tons. If L_C = 80 m, $T_{A,max}$ = 442 tons, $T_{B,max}$ = 27 tons, $T_{C,max}$ = 187 tons, then$T_{D,max}$ = 478 tons. The maximum one among the four dynamic tensions is $T_{max} = T_{D,max}$ = 478 tons. If L_C = 150 m, the maximum dynamic tension $T_{max} = T_{A,max}$ = 367 tons. This is because the natural frequency changes with the length L_C. The excitation frequencies of the irregular wave are different to the natural frequency of the mooring system. Therefore, the resonance does not exist. It is found that the greater the diving depth of the floating platform L_C, the smaller the maximum dynamic tension. In other words, the mooring system of the diving depth of the floating platform L_C = 150 m is better than that of L_C = 80 m. Because the diving depth of the floating platform is different to that of turbine, the water flowing through the floating platform does not interfere with the flow field of the turbine. Moreover, for L_C > 80 m, the dynamic tension $T_{A,max}$ of rope A decreases with the diving depth L_C. This is because the angle θ_A of rope A decreases with the diving depth L_C. The towing force is horizontal due to the ocean velocity. Meanwhile, the dynamic tension $T_{B,max}$ of rope B increases with the diving depth L_C. It is because the angle θ_B of rope B increases with the diving depth L_C.

Figure 6 presents the relation between the diving depth L_C of the floating platform and the dynamic tensions of ropes under the typhoon irregular wave for the distance L_E = 200 m. Aside from the distance L_E =200 m, all other parameters are the same as those of Figure 5. It can be observed in Figure 6 that, when L_E = 200 m, the maximum resonant position $L_{C,res}$ = 80 m is different to $L_{C,res}$ = 66 m for L_E = 100 m in Figure 5. The effect of the horizontal distance between the turbine and floating platform L_E on the dynamic tension with L_C = 150 m is negligible.

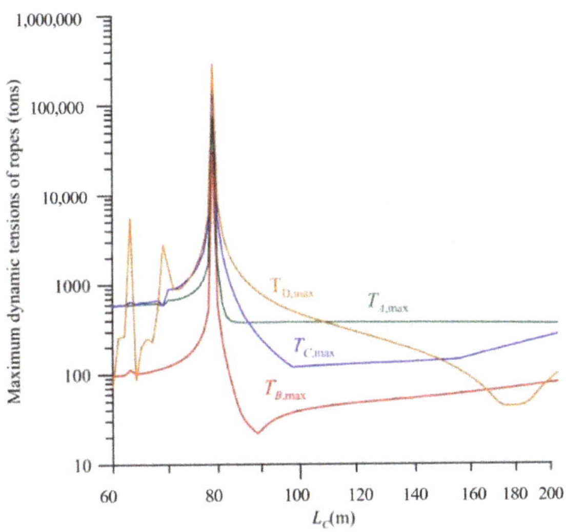

Figure 6. Dynamic tension of the four ropes under the typhoon irregular wave as a function of the diving length L_C and the horizontal distance between the turbine and floating platform L_E.

Figure 7 shows the effects of the diving depth of the floating platform L_C and the mass of pontoons M_3 and M_4 on the dynamic tensions of the four ropes, $T_{A,max}$, $T_{B,max}$, $T_{C,max}$ and $T_{D,max}$ under the typhoon irregular wave. In this case, the mass of pontoons $M_3 = M_4$ = 150 tons; other parameters are the same as those of Figure 6. It is found that, if the mass of pontoons $M_3 = M_4$ = 150 tons, the resonance occurs at several diving depths of the floating platform L_C, and the maximum dynamic tensions are over that of the fracture strength of rope, $T_{fracture}$ = 759 tons. In other words, if the weight of the pontoon is too low, the dynamic displacement of the system is too intense, resulting in the excessive dynamic tension of the rope.

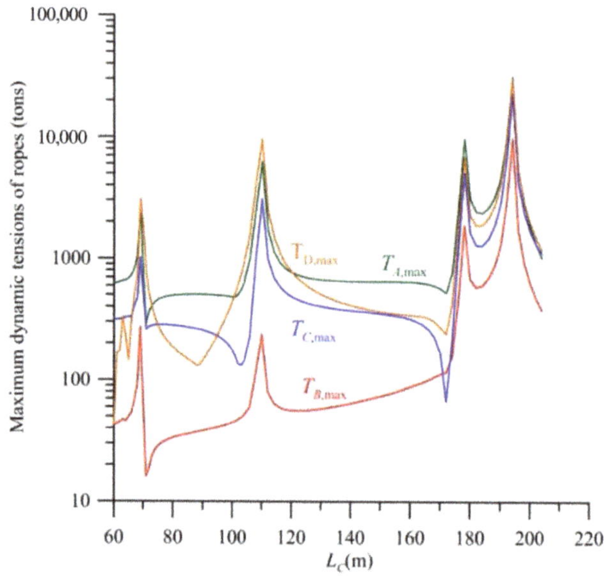

Figure 7. Dynamic tension of the four ropes under the typhoon irregular wave as a function of the diving length L_C and the mass of pontoons M_3 and M_4.

Further, the second kind of mooring system is investigated. Figure 8 demonstrates the effect of the diving depth of the floating platform L_D and the mass of pontoons M_3 and M_4 on the maximum dynamic tensions of the four ropes, $T_{A,max}$, $T_{B,max}$, $T_{C,max}$ and $T_{D,max}$ under the typhoon irregular wave when the diving depth of the turbine L_C = 60 m and the horizontal distance between the turbine and floating platform L_E = 100 m. All the other parameters are the same as those of Figure 5. It is found that there is no resonance. The dynamic tension increases with the diving depth of the floating platform L_D, especially in the case where M_3 = M_4 = 150 tons. The maximum tension is that of rope A, $T_{A,max}$, which is close or over that of the fracture strength of rope, $T_{fracture}$ = 759 tons. It is concluded that this mooring system should not be proposed.

Figure 9 demonstrates the effect of the diving depth of the turbine L_D and the mass of pontoons M_3 and M_4 on the maximum dynamic tensions of the four ropes, $T_{A,max}$, $T_{B,max}$, $T_{C,max}$ and $T_{D,max}$ under the typhoon irregular wave when the diving depth of the floating platform L_C = 60 m and the horizontal distance between the turbine and floating platform L_E = 200 m. All the other parameters are the same as those in Figure 8. It is found that the maximum tension of the four ropes is the dynamic tension of rope A, $T_{A,max}$. If the mass of pontoons M_3 = M_4 =150 tons, the maximum tension $T_{A,max}$ decreases with the diving depth of the turbine L_D. However, it is the reverse for the case of the mass of pontoons M_3 = M_4 = 250 tons. Moreover, the dynamic tension $T_{A,max}$, with the mass of pontoons M_3 = M_4 = 150 tons, is obviously less than that of the mass of pontoons M_3 = M_4 = 250 tons.

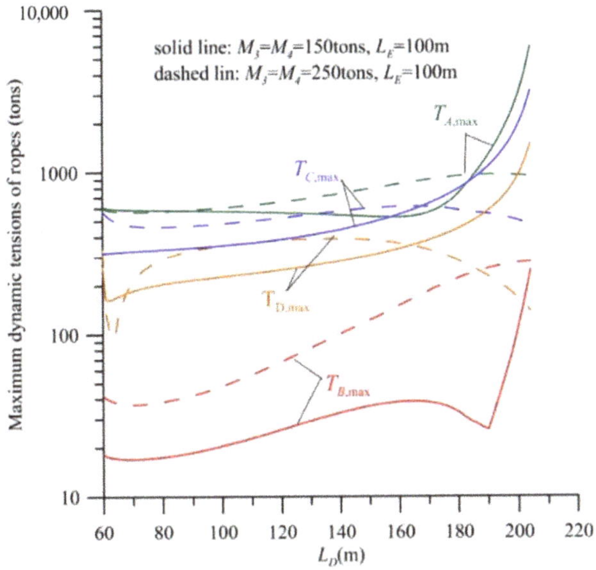

Figure 8. Dynamic tension of the four ropes under the typhoon irregular wave as a function of the diving length L_D and the mass of pontoons M_3 and M_4 for L_E = 100 m.

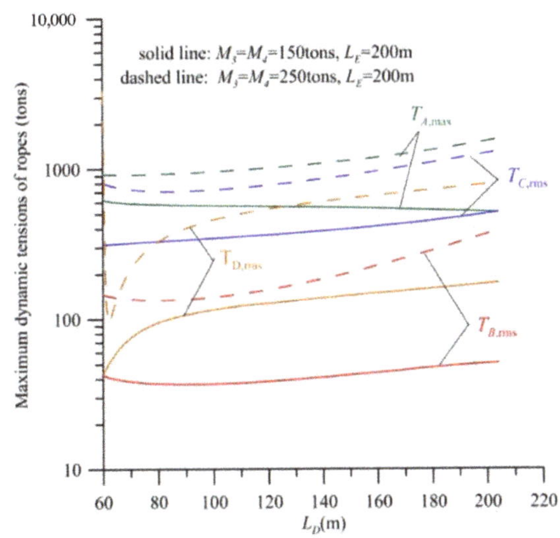

Figure 9. Dynamic tension of the four ropes under the typhoon irregular wave as a function of the diving length L_D and the mass of pontoons M_3 and M_4 for L_E = 200 m.

On the eastern coast of Taiwan, the average velocity of the Kuroshio at a depth of 150 m is 0.65 m/s, and that at a depth of 30 m is 1.1 m/s [35]. It is well known that the potential energy of ocean current can be estimated by using the formula $\eta\frac{1}{2}\rho AV^3$, where h is the efficiency, r is the density, A is the operating area and V is the flow velocity. Based on the formula, the ratio of the potential power generation of the diving depth of the turbine L_D = 30 m to that of L_D = 150 m is about 4.85. In other words, the deeper the diving depth of the turbine L_D, the smaller the power generation.

Figure 10a demonstrates the dynamic displacements of the turbine, floating platform and pontoons.

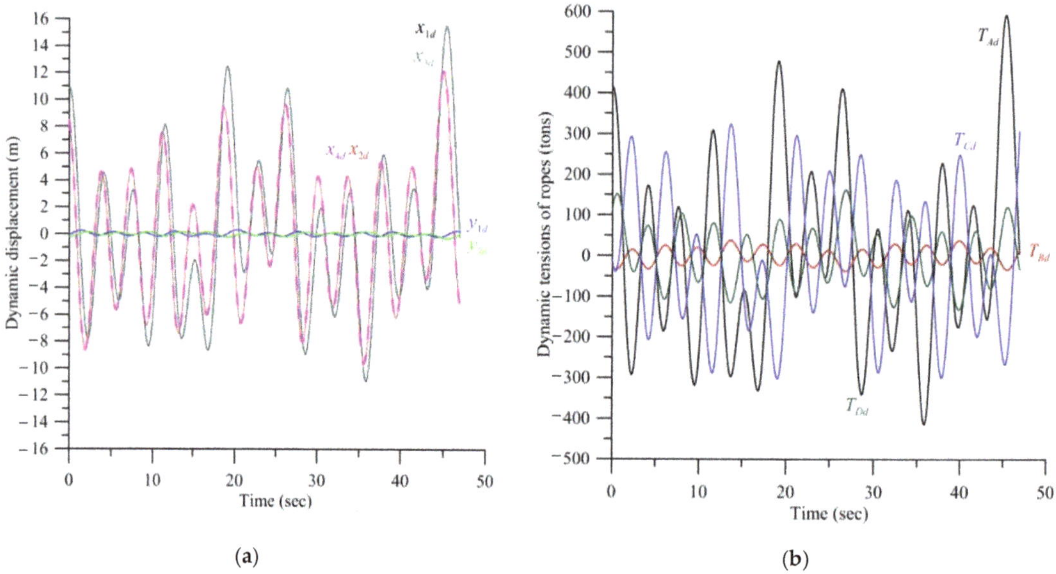

Figure 10. (**a**) Dynamic displacements of the four elements and (**b**) dynamic tensions of ropes under typhoon irregular wave for $L_C = 60$ m, $L_D = 70$ m.

The diving depth of the floating platform L_C = 60 m, the diving depth of the turbine L_D = 70 m and the mass of pontoons $M_3 = M_4$ = 150 tons. The other parameters are the same as those in Figure 9. Dynamic displacements are multi-frequency coupled. The horizontal displacements of the turbine and the floating platform y_{1d} and y_{2d} are very small, the amplitude is about 0.30 m, the vertical displacements x_{1d} and x_{3d} are large and the amplitude is about 15.5 m, which is close to the significant wave H_S = 15.4 m. The amplitudes of vertical displacements x_{2d} and x_{4d} are about 15.5 m. The amplitudes of vertical displacements x_{1d} and x_{3d} are about 9.69 m. The vertical displacements of pontoon 3 and the floating platform directly connected by using rope C are synchronized and similar. The vertical displacements of pontoon 4 and the turbine directly connected by using rope D are synchronized and similar.

Figure 10b demonstrates the dynamic tension of the rope. The maximum dynamic tension $T_{A,max}$ of rope A connecting the floating platform and the mooring foundation is about 589 tons. The maximum dynamic tension $T_{B,max}$ of rope B connecting the turbine and the floating platform is about 38 tons. The maximum dynamic tension $T_{C,max}$ of rope C connecting pontoon 3 and the floating platform is about 322 tons. The maximum dynamic tension $T_{D,max}$ of rope D connecting pontoon 4 and the turbine is about 75 tons.

Figure 11a demonstrates the dynamic displacements of the turbine, floating platform and pontoons. The diving depth of the floating platform L_C = 150 m, the diving depth of the turbine L_D = 60 m. The other parameters are the same as those in Figure 6. Dynamic displacements are multi-frequency coupled. The horizontal displacements of the turbine and the floating platform y_{1d} and y_{2d} are very small, the amplitude is about 0.14 m, the amplitudes of vertical displacements x_{1d} and x_{3d} are about 8.6 m and the amplitudes of vertical displacements x_{2d} and x_{4d} are about 9.6 m, which are significantly lower than the significant wave H_S = 15.4 m. The vertical displacements of pontoon 3 and the floating platform directly connected by using rope C are synchronized and similar. The vertical displacements of pontoon 4 and the turbine directly connected by using rope D are synchronized and similar.

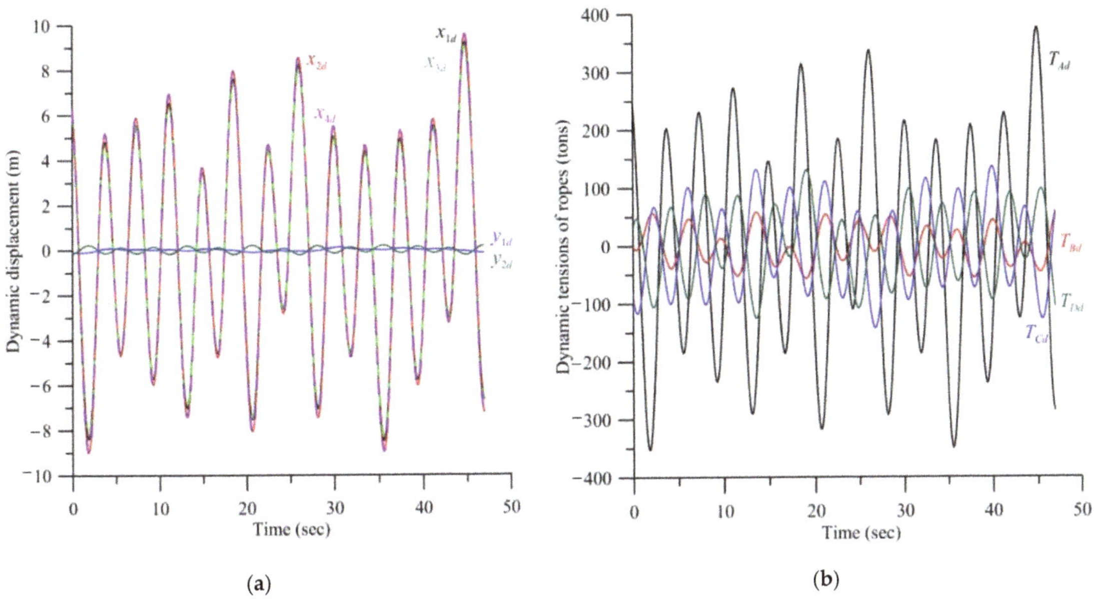

Figure 11. (**a**) Dynamic displacements of the four elements and (**b**) dynamic tensions of ropes under typhoon irregular wave for $L_C = 150$ m, $L_D = 60$ m.

Figure 11b demonstrates the dynamic tension of rope. The maximum dynamic tension $T_{A,max}$ of rope A connecting the floating platform and the mooring foundation is about 375 tons. The maximum dynamic tension $T_{B,max}$ of rope B connecting the turbine and the floating platform is about 58 tons. The maximum dynamic tension $T_{C,max}$ of rope C connecting pontoon 3 and the floating platform is about 143 tons. The maximum dynamic tension $T_{D,max}$ of rope D connecting pontoon 4 and the turbine is about 131 tons.

Figure 12 demonstrates the effects of the diving depth of the floating platform L_C and the buffer spring connected in series with ropes C and D on the dynamic tension of the rope. The diving depth of the turbine L_D = 60 m. The effective spring constants of the two buffer springs are $K_{C,spring} = K_{D,spring} = K_{\text{rope A}}$. The other parameters are the same as those in Figure 5. Compared with Figure 5, it is found that the dynamic tensions $T_{A,max}$, $T_{B,max}$ and $T_{C,max}$ of the ropes A, B and C are significantly reduced at the resonance point, but the effect on $T_{D,max}$ is not obvious and is still over the fracture strength $T_{fracture}$. If the diving depth of the floating platform $L_C > 72$ m, the effect of the buffer springs on the dynamic tensions is negligible. It is concluded that the effect of the buffer springs on the dynamic tensions of this mooring system is slight.

Figure 13 demonstrates the effects of the cross-sectional area of pontoon A_{BX}, A_{PX} and the diving depth of the floating platform L_C on the dynamic tensions of the four ropes. The cross-sectional area of the two pontoons is $A_{BX} = A_{PX} = 4$ m^2. The other parameters are the same as those in Figure 5. Compared with Figure 5, it is found that the dynamic tensions are significantly increased. At the resonance point, the dynamic tension is over the fracture strength $T_{fracture}$. If the diving depth of the floating platform $L_C > 85$ m, the dynamic tension is close to the fracture strength $T_{fracture}$. It is concluded that the larger the cross-sectional area of the pontoon, the larger the dynamic tension.

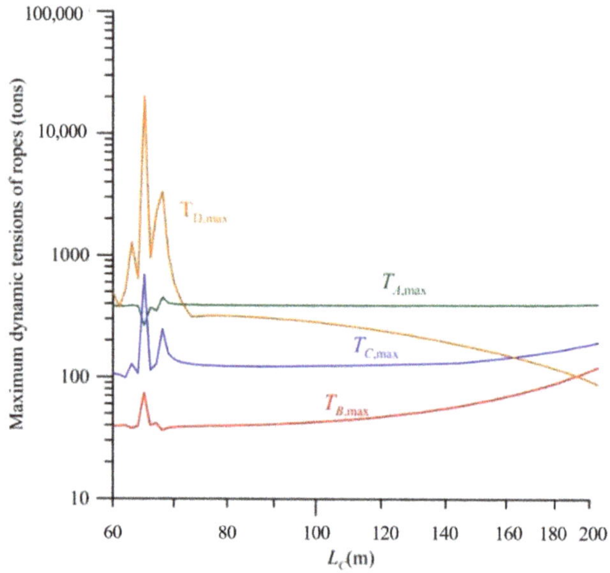

Figure 12. Dynamic tension of the four ropes under the typhoon irregular wave as a function of the diving length L_C and the buffer springs $K_{C,spring}$, $K_{D,spring}$.

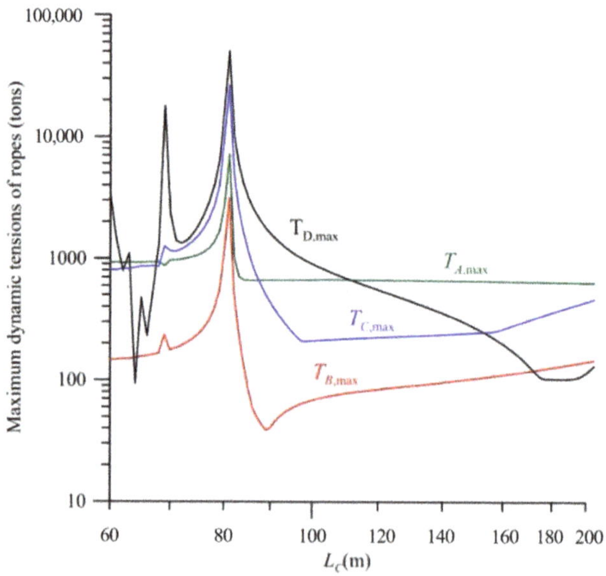

Figure 13. Dynamic tension of the four ropes under the typhoon irregular wave as a function of the cross-sectional area of the two pontoons A_{BX}, A_{PX} and the diving depth of the floating platform L_C.

Figure 14 demonstrates the effects of the significant wave height H_s and the peak period P_w on the dynamic tension $T_{A,max}$. Based on Equations (14)–(16), the irregular wave is simulated by six regular waves, i.e., $n = 6$. The six regular waves share according to the ratio of energy {2,35,8,4,3,1}. The amplitude a_i, frequency f_i, the wave number k_i and wave length l_i can be determined. The diving depths $L_C = 60$ m and $L_D = 150$ m. The horizontal distance between the turbine and floating platform $L_E = 200$ m. Two buffer springs are connected in series with ropes C and D. The effective spring constants of the two buffer springs are $K_{C,spring} = K_{D,spring} = K_{\text{rope A}}$. The other parameters are the same as those of Figure 5. It is found that the more the significant wave height H_s, the larger the dynamic tension $T_{A,max}$. For the peak period T_p = 13.5 s, the dynamic tension $T_{A,max}$ increases dramatically with the significant wave height Hs. With the increase of the peak period T_p, the increase rate of the dynamic tension $T_{A,max}$ becomes low.

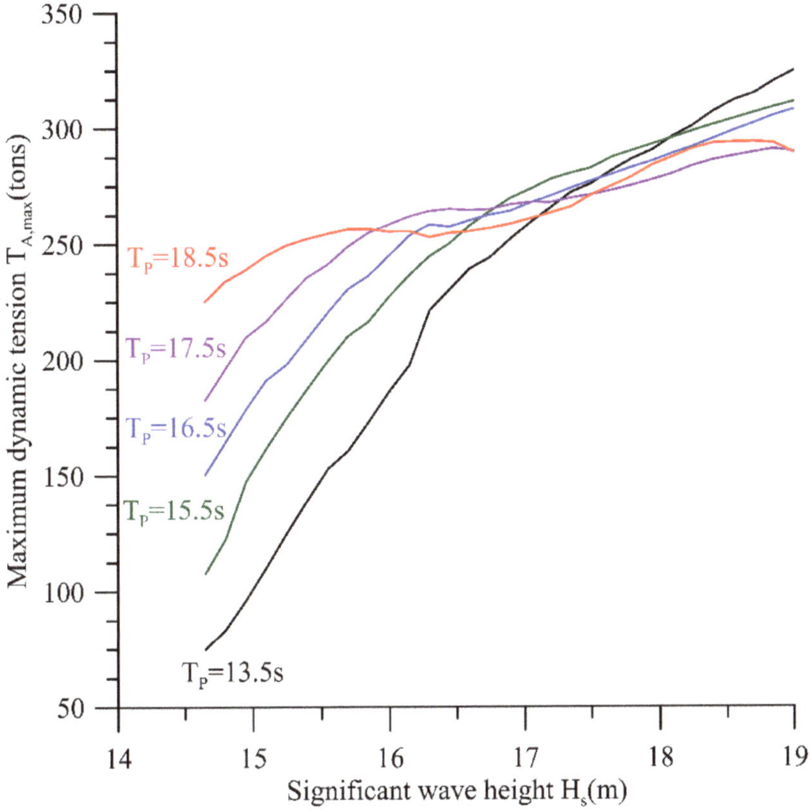

Figure 14. Dynamic tension $T_{A,max}$ under the typhoon irregular wave as a function of the significant wave height H_s and the peak period T_p.

Figure 15 demonstrates the effect of the relative angle a of the wave and ocean current on the dynamic tension. The significant wave height H_s = 15 m, and the peak period P_w = 16.5 s. The other parameters are the same as those of Figure 14. It is observed that the effect of the relative angle a of the wave and ocean current on the dynamic tension is significant. $T_{A,max}(\alpha = 180°)$ and $T_{C,max}(\alpha = 180°)$ are much larger than $T_{A,max}(\alpha = 0°)$ and $T_{C,max}(\alpha = 0°)$, respectively. Moreover, $T_{D,max}(\alpha = 90°)$ is significantly larger than $T_{A,max}(\alpha = 0°)$ and $T_{A,max}(\alpha = 180°)$.

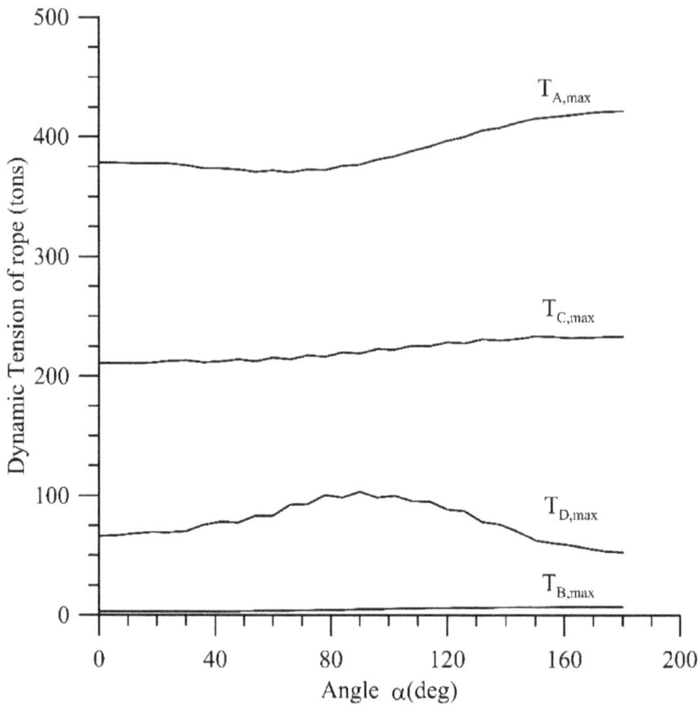

Figure 15. Dynamic tension of the four ropes under the typhoon irregular wave as a function of the relative angle α.

Based on the frequency Equation (62), the effects of the masses M_3, M_4 and M_2 and the distance L_E and the areas on the natural frequencies are investigated and listed in Table 2. It is found that the larger the cross-sectional areas of pontoon A_{Bx} and A_{BT}, the higher the natural frequencies of the system. The larger the masses of pontoon M_3 and M_4, the lower the natural frequencies of the system. The larger the mass of turbine M_2, the lower the first natural frequency of the system. However, the effect of the mass of turbine M_2 on the second natural frequency of the system is negligible. The larger the distance between the turbine and the floating platform L_E, the higher the second natural frequency of the system. However, the effect of the distance L_E on the first natural frequency of the system is negligible.

Table 2. The first two natural frequencies f_{n1} and f_{n2} as a function of the masses M_3, M_4 and M_2, the distance L_E and the areas A_{Bx}, A_{BT} for M_1 = 300 tons.

L_E (m)	M_3, M_4 (tons)	M_2 (tons)	$A_{Bx} = A_{BT}$ = 2.12 m^2		$A_{Bx} = A_{BT}$ = 4 m^2	
			f_{n1} (Hz)	f_{n2} (Hz)	f_{n1} (Hz)	f_{n2} (Hz)
130	250	838	0.0220	0.0703	0.0302	0.0761
		535	0.0258	0.0703	0.0355	0.0761
200		838	0.0220	0.0806	0.0302	0.0857
		535	0.0258	0.0806	0.0355	0.0857
130	150	535	0.0277	0.0776	0.0379	0.0839
200			0.0277	0.0890	0.0380	0.0946

4. Conclusions

This paper studies the safe design of a mooring system for an ocean current generator that is working under the impact of typhoon waves. Two mooring designs are investigated, and one safe and feasible mooring system is proposed. The proposed mooring design can stabilize the turbine and platform around a certain predetermined water depth, thereby, maintaining the stability and safety of the ocean current generator. The effects of several parameters on the dynamic response under irregular wave impact were discovered as follows:

(1) Considering the first mooring configuration, the diving depth L_D of the turbine is fixed at 60 m. When the diving depth of the floating platform L_C =150 m, the dynamic tension is significantly less than the fracture strength $T_{Fracture}$ of rope, and it is far from the resonance. Moreover, because the diving depth L_D of the turbine is far from the depth L_C of the floating platform, the floating platform does not interrupt the turbine water flow. Because the floating platform is a structure without a rotating mechanism in it, such as the rotating blade of a turbine, the water-proof at the depth of 150 m under sea surface is easily constructed. Therefore, this mooring configuration is safe and feasible;

(2) Considering the second mooring configuration, the diving depth L_C of the floating platform is fixed at 60 m. When the diving depth L_D of the floating platform is larger than the diving depth L_C, there is no resonance point, but the dynamic tension $T_{Ad,max}$ of rope A is obviously larger than that of the first method and close to the fracture strength $T_{Fracture}$. It is found [32] that, for the Kuroshio current on the eastern coast of Taiwan, the greater the depth under the sea surface, the lower the current flow rate. The ratio of the potential power generation of the diving depth of the turbine L_D = 30 m to that of L_D = 150 m is about 4.85. Moreover, because there are the rotating blades of the turbine, the water-proof at the higher pressure under sea surface is difficult to construct. Therefore, the second mooring configuration is not recommended;

(3) The larger the area of pontoons A_{BX} and A_{TX}, the larger the maximum dynamic tensions, especially for $T_{Ad,max}$;

(4) For the first mooring configuration, if the weight of the pontoon is too low, the dynamic displacement of the system is too intense, resulting in the excessive dynamic tension of the rope;

(5) The effect of the buffer springs on the dynamic tensions of the first mooring configuration is slight.

The coupled translational and rotational motions will be studied in another manuscript. Moreover, the transient response of the system subjected to impact force will be investigated in the future.

Author Contributions: Conceptualization, S.-M.L. and C.-T.L.; methodology, S.-M.L.; software, S.-M.L. and D.-W.U.; validation, S.-M.L.; formal analysis, S.-M.L.; investigation, S.-M.L. and D.-W.U.; resources, S.-M.L. and C.-T.L.; data curation, D.-W.U.; writing—original draft preparation, S.-M.L.; writing—review and editing, C.-T.L.; visualization, S.-M.L.; supervision, S.-M.L.; funding acquisition, S.-M.L. and C.-T.L. All authors have read and agreed to the published version of the manuscript.

Funding: This work was financially supported by the Green Energy Technology Research Center from The Featured Areas Research Center Program within the framework of the Higher Education Sprout Project by the Ministry of Education (MOE) in Taiwan and the National Academy of Marine Research of Taiwan, R. O. C. (NAMR110051).

Institutional Review Board Statement: Not applicable.

Informed Consent Statement: Not applicable.

Data Availability Statement: The figures and the tables in this manuscript have clearly described all the data of this study.

Acknowledgments: This work was financially supported by the Green Energy Technology Research Center from The Featured Areas Research Center Program within the framework of the Higher Education Sprout Project by the Ministry of Education (MOE) in Taiwan and the National Academy of Marine Research of Taiwan (NAMR110051).

Conflicts of Interest: The authors declare no conflict of interest. The funders had no role in the design of the study; in the collection, analyses, or interpretation of data; in the writing of the manuscript, or in the decision to publish the results.

Nomenclature

A_{BX}, A_{BT}	cross-sectional area of pontoons 3 and 4, respectively
A_{BY}, A_{TY}	damping area of platform and turbine under current, respectively
a_i	amplitude of the i-th regular wave
C_{DFy}, C_{DTy}	damping coefficient of floating platform and turbine
F_B	buoyance
F_D	drag under current
f	wave frequency
H_{bed}	depth of seabed
H_s	significant wave height
g	gravity
K	effective spring constant
$\vec{K}_i$	wave vector of the i-th regular wave
$\widetilde{k}_i$	wave number of the i-th regular wave
L_i, i = A,B,C,D	length of rope i
L_i	length of rope i
M_i	mass of element i
$m_{eff,x}, m_{eff,y}$	vertical and horizontal effective mass of rope A, respectively
P_w	peak period of wave
$\vec{R}$	coordinate
T_i	tension force of rope i
t	time variable
V	ocean current velocity
x_i, i = 1~4	vertical displacements of the floating platform, the turbine and the pontoons, respectively
x_w	sea surface elevation
y_1, y_2	horizontal displacements of the floating platform and the turbine, respectively
α	relative angle between the directions of wave and current
ρ	density of sea water
Ω_i	angular frequency of the i-th regular wave
ω	angular frequency
φ_i	phase angle of the i-th regular wave
ϕ_i	phase delay of the i-th regular wave
θ_i	angles of rope i
λ_i	length of the i-th regular wave
δ_i	elongation of rope i

Subscript

0~4	mooring foundation, floating platform, turbine and two pontoons, respectively
A, B, C, D	Ropes A, B, C and D, respectively
s, d	static and dynamic, respectively
PE	high-strength PE dyneema rope
p	peak

References

1. Chen, Y.Y.; Hsu, H.C.; Bai, C.Y.; Yang, Y.; Lee, C.W.; Cheng, H.K.; Shyue, S.W.; Li, M.S. Evaluation of test platform in the open sea and mounting test of KW Kuroshio power-generating pilot facilities. In Proceedings of the 2016 Taiwan Wind Energy Conference, Keelung, Taiwan, 24–25 November 2016.
2. Anagnostopoulos, S.A. Dynamic response of offshore platforms to extreme waves including fluid-structure interaction. *Eng. Struct.* **1982**, *4*, 179–185. [CrossRef]
3. Istrati, D.; Buckle, I.G. Effect of fluid-structure interaction on connection forces in bridges due to tsunami loads. In Proceedings of the 30th US-Japan Bridge Engineering Workshop, Washington, DC, USA, 21–23 October 2014.
4. Xiang, T.; Istrati, D. Assessment of Extreme Wave Impact on Coastal Decks with Different Geometries via the Arbitrary Lagrangian-Eulerian Method. *J. Mar. Sci. Eng.* **2021**, *9*, 1342. [CrossRef]
5. Peregrine, D.H.; Bredmose, H.; Bullock, G.; Obhrai, C.; Muller, G.; Wolters, G. Violent water wave impact on walls and the role of air. In Proceedings of the 29th International Conference on Coastal Engineering, Lisbon, Portugal, 19–24 September 2004; Volume 1, pp. 4005–4017.
6. Istrati, D.; Buckle, I.; Lomonaco, P.; Yim, S. Deciphering the Tsunami Wave Impact and Associated Connection Forces in Open-Girder Coastal Bridges. *J. Mar. Sci. Eng.* **2018**, *6*, 148. [CrossRef]
7. Firouz-Abadi, R.; Noorian, M.; Haddadpour, H. A fluid–structure interaction model for stability analysis of shells conveying fluid. *J. Fluids Struct.* **2010**, *26*, 747–763. [CrossRef]
8. Bose, C.; Badrinath, S.; Gupta, S.; Sarkar, S. Dynamical Stability Analysis of a Fluid Structure Interaction System Using a High Fidelity Navier-stokes Solver. *Procedia Eng.* **2016**, *144*, 883–890. [CrossRef]
9. Lin, S.-M.; Wang, W.-R.; Lee, S.-Y.; Chen, C.-W.; Hsiao, Y.-C.; Teng, M.-J. Wave modes of a pre-stressed thick tube conveying blood on the viscoelastic foundation. *Appl. Math. Model.* **2015**, *39*, 466–482. [CrossRef]
10. Belibassakis, K. A boundary element method for the hydrodynamic analysis of floating bodies in variable bathymetry regions. *Eng. Anal. Bound. Elements* **2008**, *32*, 796–810. [CrossRef]
11. Tsui, Y.-Y.; Huang, Y.-C.; Huang, C.-L.; Lin, S.-W. A Finite-Volume-Based Approach for Dynamic Fluid-Structure Interaction. *Numer. Heat Transf. Part B Fundam.* **2013**, *64*, 326–349. [CrossRef]
12. Xiang, T.; Istrati, D.; Yim, S.C.; Buckle, I.G.; Lomonaco, P. Tsunami Loads on a Representative Coastal Bridge Deck: Experimental Study and Validation of Design Equations. *J. Waterw. Port Coast. Ocean Eng.* **2020**, *146*, 04020022. [CrossRef]
13. Westphalen, J.; Greaves, D.M.; Raby, A.; Hu, Z.Z.; Causon, D.M.; Mingham, C.G.; Omidvar, P.; Stansby, P.K.; Rogers, B.D. Investigation of Wave-Structure Interaction Using State of the Art CFD Techniques. *Open J. Fluid Dyn.* **2014**, *4*, 18–43. [CrossRef]
14. Ren, B.; Jin, Z.; Gao, R.; Wang, Y.-X.; Xu, Z.-L. SPH-DEM Modeling of the Hydraulic Stability of 2D Blocks on a Slope. *J. Waterw. Port Coast. Ocean Eng.* **2014**, *140*, 04014022. [CrossRef]
15. Hasanpour, A.; Istrati, D.; Buckle, I. Coupled SPH–FEM Modeling of Tsunami-Borne Large Debris Flow and Impact on Coastal Structures. *J. Mar. Sci. Eng.* **2021**, *9*, 1068. [CrossRef]
16. IHI; NEDO. The Demonstration Experiment of the IHI Ocean Current Turbine Located off the Coast of Kuchinoshima Island, Kagoshima Prefecture, Japan. 14 August 2017. Available online: https://tethys.pnnl.gov/project-sites/ihi-ocean-current-turbine (accessed on 28 August 2021).
17. VanZwieten, J.; Driscoll, F.; Leonessa, A.; Deane, G. Design of a prototype ocean current turbine—Part II: Flight control system. *Ocean Eng.* **2006**, *33*, 1522–1551. [CrossRef]
18. Xiang, G.; Xu, S.; Wang, S.; Soares, C.G. *Advances in Renewable Energies Offshore: Comparative Study on Two Different Mooring Systems for a Buoy*; Taylor & Francis Group: London, UK, 2019; pp. 829–835.
19. Paduano, B.; Giorgi, G.; Gomes, R.P.F.; Pasta, E.; Henriques, J.C.C.; Gato, L.M.C.; Mattiazzo, G. Experimental Validation and Comparison of Numerical Models for the Mooring System of a Floating Wave Energy Converter. *J. Mar. Sci. Eng.* **2020**, *8*, 565. [CrossRef]
20. Chen, D.; Nagata, S.; Imai, Y. Modelling wave-induced motions of a floating WEC with mooring lines using the SPH method. In Proceedings of the 3rd Asian Wave and Tidal Energy Conference, Marina Bay Sands, Singapore, 25–27 October 2016; pp. 24–28.
21. Davidson, J.; Ringwood, J.V. Mathematical Modelling of Mooring Systems for Wave Energy Converters—A Review. *Energies* **2017**, *10*, 666. [CrossRef]
22. Touzon, I.; Nava, V.; De Miguel, B.; Petuya, V. A Comparison of Numerical Approaches for the Design of Mooring Systems for Wave Energy Converters. *J. Mar. Sci. Eng.* **2020**, *8*, 523. [CrossRef]
23. Lin, S.M. Nonlinear vibration control of a tall structure with tuned liquid column damper. *Mech. Adv. Mater. Struct.* **2016**, *23*, 146–155. [CrossRef]
24. Bek, M.; Amer, T.; Sirwah, M.A.; Awrejcewicz, J.; Arab, A.A. The vibrational motion of a spring pendulum in a fluid flow. *Results Phys.* **2020**, *19*, 103465. [CrossRef]
25. Lin, S.M.; Chen, Y.Y.; Hsu, H.C.; Li, M.S. Dynamic Stability of an Ocean Current Turbine System. *J. Mar. Sci. Eng.* **2020**, *8*, 687. [CrossRef]
26. Lin, S.-M.; Chen, Y.-Y. Dynamic Stability and Protection Design of a Submarined Floater Platform Avoiding Typhoon Wave Impact. *J. Mar. Sci. Eng.* **2021**, *9*, 977. [CrossRef]
27. Lin, S.-M.; Chen, Y.-Y.; Liauh, C.-T. Dynamic Stability of the Coupled Pontoon-Ocean Turbine-Floater Platform-Rope System under Harmonic Wave Excitation and Steady Ocean Current. *J. Mar. Sci. Eng.* **2021**, *9*, 1425. [CrossRef]

28. Young, I.R. Regular, Irregular Waves and the Wave Spectrum. *Encycl. Maritime Offshore Eng.* **2017**, *2017*, 1–10. [CrossRef]
29. Pierson, W.J., Jr.; Moskowitz, L. A proposed spectral form for fully developed wind seas based on the similarity theory of S. A. Kitaigorodskii. *J. Geophys. Res. Space Phys.* **1964**, *69*, 5181–5190. [CrossRef]
30. Hasselmann, K.; Barnett, T.P.; Bouws, E.; Carlson, H.; Cartwright, D.E.; Enke, K.; Ewing, J.A.; Gienapp, H.; Hasselmann, D.E.; Kruseman, P.; et al. Measurements of wind-wave growth and swell decay during the Joint North Sea Wave Project (JONSWAP). *Ergnzungsheft zur Deutschen Hydrogr. Z. Reihe* **1973**, *A(8)*, 1–95. Available online: https://www.researchgate.net/publication/256197895 (accessed on 12 February 2022).
31. DNVGL-CG-0130-Wave Loads. 2018. Available online: http://www.dnvgl.com (accessed on 3 February 2022).
32. Shyue, S.W. *Development and Promotion of Key Technologies of Ocean Current Energy*; OAC108001; Ocean Affairs Council: Kaohsiung, Taiwan, 2019; pp. 4–103. (In Chinese)
33. Aggarwal, A.; Pákozdi, C.; Bihs, H.; Myrhaug, D.; Chella, M.A. Free Surface Reconstruction for Phase Accurate Irregular Wave Generation. *J. Mar. Sci. Eng.* **2018**, *6*, 105. [CrossRef]
34. Lin, S.-M. Effective dampings and frequency shifts of several modes of an inclined cantilever vibrating in viscous fluid. *Precis. Eng.* **2010**, *34*, 320–326. [CrossRef]
35. Chen, Y.Y. *Research on the Facility Configuration and Executing Procedure of the Ocean Current Energy Test Field in Taiwan*; NAMR110051; Ocean Affairs Council: Kaohsiung, Taiwan, 2021; pp. 3.36–3.45. (In Chinese)

energies

MDPI

Article

Site Selection for Ocean Thermal Energy Conversion Plants (OTEC): A Case Study in Panama

Guillermo Lopez [1,2,3], Maria de los Angeles Ortega Del Rosario [1,2,*], Arthur James [1,2,3] and Humberto Alvarez [4]

1 Department of Mechanical Engineering, Universidad Tecnológica de Panamá, Panama City 0819-07289, Panama; guillermo.lopez2@utp.ac.pa (G.L.); arthur.james@utp.ac.pa (A.J.)
2 Research Group—Iniciativa de Integración de Tecnologías para el Desarrollo de Soluciones Ingenieriles (I2TEDSI), Universidad Tecnológica de Panamá, Panama City 0819-07289, Panama
3 Sistema Nacional de Investigación (SNI), Clayton, City of Knowledge Edf. 205, Panama City 0819-10280, Panama
4 Department of Industrial Engineering, Universidad Tecnológica de Panamá, Panama City 0819-07289, Panama; humberto.alvarez@utp.ac.pa
* Correspondence: maria.ortega@utp.ac.pa

Abstract: This research addressed a need for technical evaluation of the oceanic scenario of Panama for the use of Ocean Thermal Energy Conversion (OTEC). Its bathymetry and location can potentially lead to the exploitation of OTEC, diversifying the energy matrix and helping achieve sustainability. Nevertheless, site selection for OTEC can be a complex task since it involves various alternatives, with different quantitative and qualitative criteria, which may conflict in some cases. Optimization and multiple criteria (MCD) methods have been used lately to address these issues; however, their use is still limited. Here, Analytic Hierarchical Analysis (AHP) is proposed as a MCD method for site selection. Six sites of interest were considered as the alternatives for a plant installment. These sites were chosen, excluding the environmentally and aboriginal protected areas. The quantitative criteria considered were surface and deep-water temperatures, coastline distance, gross and net efficiency. Those variables related to the efficiency, such as the water temperatures, can be considered the most influential, leading to Punta Burica, located on Panama's Pacific coast, as the best option (96.17%).

Keywords: Ocean Thermal Energy Conversion (OTEC); site selection; multi-criteria decision (MCD); hierarchy analysis process (AHP); renewable energy

Citation: Lopez, G.; Ortega Del Rosario, M.d.l.A.; James, A.; Alvarez, H. Site Selection for Ocean Thermal Energy Conversion Plants (OTEC): A Case Study in Panama. *Energies* **2022**, *15*, 3077. https://doi.org/10.3390/en15093077

Academic Editors: Kostas Belibassakis, Eugen Rusu and George Lavidas

Received: 18 March 2022
Accepted: 13 April 2022
Published: 22 April 2022

Publisher's Note: MDPI stays neutral with regard to jurisdictional claims in published maps and institutional affiliations.

Copyright: © 2022 by the authors. Licensee MDPI, Basel, Switzerland. This article is an open access article distributed under the terms and conditions of the Creative Commons Attribution (CC BY) license (https://creativecommons.org/licenses/by/4.0/).

1. Introduction

The sun serves as the main source of energy on the planet. The incident radiation is absorbed by the ozone in the stratosphere and much by the clouds. About 35% of it is reflected into space, and then 66% of the remaining energy is absorbed by the Earth's surface. Considering that more than two-thirds of the Earth's surface is hosted by water, the oceans represent the most prominent solar energy collectors and, thus, the largest global energy reservoir [1,2].

The most considerable ocean thermal energy is hosted near the Equator [3–5]. Geographically, Panama is located in Central America, bordering the Caribbean Sea and the Pacific Ocean, between 7° and 10° North latitude and 77° and 83° West longitude. Therefore, Panama's ocean scenario displays thermal energy resource potential [6–8]. However, since Panama has a broad oceanic scenario, knowing its location is not enough to guarantee the sustainability of implementing marine energy exploitation, such as OTEC (ocean thermal energy conversion).

OTEC technology uses the temperature difference between warm ocean surface waters and deeper cold waters. The zones that achieve adequate thermal differences to take advantage of the oceanic thermal resource are generally close to tropical areas, near the

Equator, as shown in Figure 1. However, because temperature jumps are around 20 °C, a low efficiency can be reached. Despite this, the global theoretical potential for OTEC has been estimated at 44,000 TWh/year, making OTEC an attractive alternative to cater to particular needs of society [9–11].

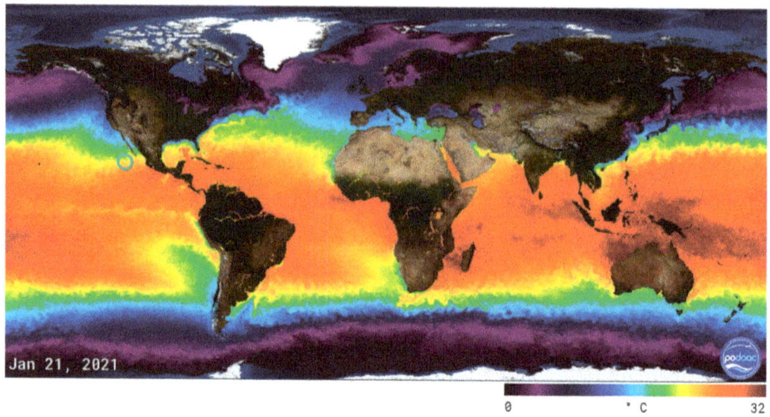

Figure 1. Global oceanic thermal resource and location of Panama (enclosed in the light blue circle) [12].

A good site selection is a key factor in implementing a sustainable and successful project that aims to exploit ocean thermal energy, regardless of the targeted objective or benefit [13]. Therefore, the proper site selection for the implementation and viable installation involves the analysis of various variables, qualitative and quantitative.

Many criteria can help assess the potential of the thermal resource for a site that can later be used by OTEC technology addressing sustainability. For instance, this technology's efficiency depends on variables such as depth, the temperature of deep cold seawater, warm sea surface temperatures, and anomalies of warm sea surface temperatures [13,14]. Thus, site selection can quickly become a multi-criteria decision-making (MCD) problem, and it generates a conflictive scenario that houses technical, social, economic, environmental, and legislative criteria [13]. Moreover, studies on this topic implementing the classic methods of MCD are limited because many of the decisions for selecting OTEC sites are subjective concerning the criteria of each decision-maker (DM). The accurate decision regarding the site selection for OTEC systems represents the fundamental basis in the planning process for the use of ocean thermal energy and sustainability of this technology [15].

The difficulty for a site selection process lies in the number of indicators or the amount of data available and comparing each set of indicators corresponding to each alternative of interest, besides granting an assessment that justifies a degree of importance among the criteria considered. Then, it is paramount first to identify the aim to use the local oceanic thermal resource. OTEC allows obtaining various benefits from the use of the oceanic thermal resource [14,16], including different classifications for OTEC plant (On-Shore or Off-Shore). OTEC technology is used for: electric power generation [17], air conditioning [11], industrial refrigeration [6], aquaculture [18], liquid hydrogen production [19] and desalinated water [19]. Then, the alternatives with the most significant potential within the oceanography can be selected considering the environmental legislative regulations, for instance, whether the planned lands belong to the group of areas or regions environmentally protected. After this, appropriate criteria set can be drawn by means that match the project's goals of interest.

This problem addresses various variables, some of which are not directly related to the nature of the oceanic scenario. For instance, some qualitative aspects generate added value to the implementation of this technology, such as environmental, political, and social impact [14]. Most studies that have been conducted are related to the analysis of one or more of the criteria influencing decision-making for the selection of OTEC sites. These criteria include temperature gradient, bathymetric characteristics, environmental, social, and economic impact [16].

Addressing sustainability according to environmental, social, and economic impacts can be a rather difficult task since the lack of absolute instruments for evaluations or available data can endanger the precision and reliability of the results. Even though some of them have been assessed previously in the literature, such as environmental impact assessment and sustainability [14,16], sustainability of society [13,14], energy sustainability [3,20], economic sustainability [14], sustainability of construction and maintenance [13,21], and sustainability of the auxiliary condition [14].

These considerations and several variables that categorize the site selection for OTEC render the site selection a complex study scenario. Furthermore, one must consider the dependence between the relevance given to each of these aspects by specialists or researchers during the decision-making process and the discrepancy that this can project due to the degree of knowledge and experience each possesses.

In Panama's scenario, the lack of legislation and regulations and the lack of information and specialists can render a complex task for implementing OTEC technology. This research only includes variables or quantitative indicators justified and validated with software that use oceanographic data in real-time, such as the World Ocean Atlas and the NOAA database, compared with the corresponding literature.

The qualitative variables depend on the evaluator's degree of knowledge, experience, and specialty. Therefore, it carries the uncertainty corresponding to the sensitivity of human perception, which generates a more complex scenario in the validation and justification of the weighting for assessing the importance of these indicators.

2. Materials and Methods

This study aimed to analyze and select potential sites for OTEC exploitation within the Republic of Panama. A Hierarchical Analysis Method (AHP) was used as a MCD to analyze the criteria to identify the advantages, disadvantages, applicability, and reliability of a site within the oceanic territory of Panama. Subsequently, the analysis of this problem was approached as a particular case of MCD, using the initial process proposed by Thomas L. Saaty (The Analytic Hierarchy Process, 1980); for the formulation of any case of AHP-MCD [22,23].

All the possible criteria and sub-criteria and the indicators that address the subject under study were identified. An exclusion–inclusion criteria model was performed to address the qualitative criteria. Then, a model was proposed for this case, presenting a hierarchical structure where all aspects (indicators) relevant to the justification of the problem under study were considered. Each selected indicator was evaluated and quantified by considering their direct effects on the net production of the system, besides on guaranteeing the potential of the oceanic thermal resource housed in each alternative; for this valorization, the Saaty fundamental scale was used.

A comparison of the matrices was performed to justify the assessment or priority of each indicator at the time of being compared with itself and with the others. These comparisons represent the importance of the criteria, establishing the most significant importance to the criterion with the greatest relevance to the subject under study.

Next, the prioritization and synthesis were carried out. In this stage of the AHP, the different priorities considered for resolving the problem are provided. The priority represents an abstract unit valid for any scale in which the decision-maker's preferences consider appropriate when integrating tangible, intangible, quantitative, and qualitative aspects.

Subsequently, the consistency of the randomly generated matrix of paired comparisons was justified by using the method provided by the AHP to estimate the degree of consistency between the paired opinions provided by decision-makers. Therefore, the consistency radius, consistency index, and random index are calculated to justify whether the judgments are inconsistent or have a reasonable level of consistency.

The radius of consistency is a ratio or quotient in which if its values exceeds 0.10, it indicates that the judgments are inconsistent; therefore, the original values of the matrix of paired comparisons should be reconsidered and edited. For the case where CR is less than 0.10, this represents a good and reasonable indication in paired comparisons.

Finally, the prioritization matrix of alternatives is presented. In this, it is possible to justify and identify the best alternative within the oceanic territory where the OTEC plant must be installed. In Figure 2, we can see a flowchart corresponding to the methodology addressed in this research.

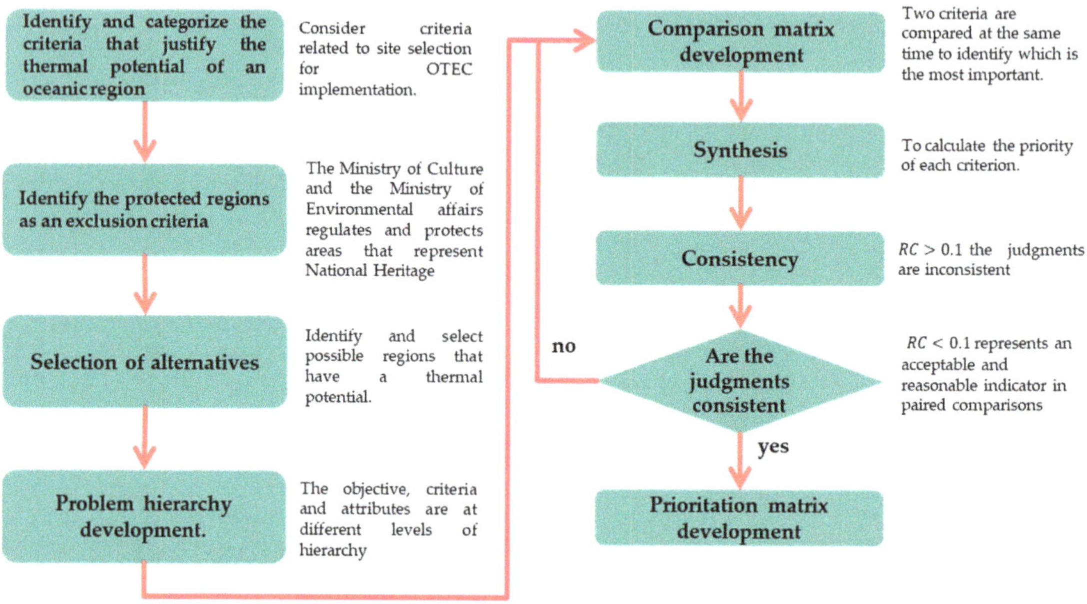

Figure 2. The methodology proposed for developing and implementing the AHP-MCD method in site selection for OTEC.

2.1. The Analytic Hierarchy Process Method (AHP)

The AHP method analyzes and develops complex decision-making problems of multiple MCD criteria [22]. The AHP is based on identifying all the variables involved in a problem, linking them according to all possible solutions, and concluding [23]. The AHP method efficiently and graphically organizes the respective information for any problem under study [24]. The AHP is a hierarchy with priorities, where these show the overall preference corresponding to one of the decision alternatives [23].

The AHP method uses direct quantitative allocation scales to prioritize the criteria and make comparisons between criteria pairs [25]. This step aims to build a vector of priorities or weights that evaluates the relative importance of the decision-making unit given to each criterion. Table 1 shows the fundamental scale proposed by Saaty for this process [26].

Table 1. Saaty's scale for the absolute numbers corresponding to the priority and importance of these considering their respective definitions and degree of contribution [22,23,27].

Intensity of Importance	Definition	Description
1	Both criteria and elements are of equal importance.	Two activities also contribute to the objective.
2	Between equal and moderately preferable.	
3	Weak or moderate importance of one over the other.	Experience and judgment slightly favor one activity over another.
4	Between moderate and strongly preferable.	
5	Essential or strong importance of one criterion over the other.	Experience and judgment strongly favor one activity over another.
6	Between strong and extremely preferable.	
7	Demonstrated importance of one criterion over another.	One activity is greatly favored over another; their proven mastery in practice.
8	Between very strong and extremely preferable.	
9	Extreme importance.	Evidence that favors one activity over another it is of the highest possible order of affirmation.
Higher reciprocals		
If the activity i have assigned one of the above non-zero numbers compared to the activity j, then j has the reciprocal value compared to i		
If the activities are very close		
It can be difficult to assign the best value, but compared to other contrasting activities, the size of the small numbers would not be too noticeable, but they can still indicate the relative importance of the activities.		

2.2. Application of the AHP Method

2.2.1. Prioritization

The AHP requires a preference or priority to each alternative decision, considering the extent to which each criterion contributes. A mathematical procedure called synthesis summarizes the information and provides a hierarchy of criteria for alternatives in terms of global preference [23].

2.2.2. Paired Comparisons and Paired Comparison Matrix

The fundamental bases of the AHP are the paired comparisons [23]. The AHP uses the Saaty table to assess relative preferences when comparing two elements. The matrix of paired comparisons contains alternative comparisons or criteria. Therefore, if one considers a matrix A of dimensions $n \times n$, with the relative judgments about the criteria, and a_{ij} is the element (i, j) of A, for $i = 1, 2, \ldots, n$, and $j = 1, 2, \ldots, n$.

Then, one can state that A is a matrix of paired comparisons of n criteria if a_{ij} is the measure of the preference of the criterion in the row i when compared to the criterion in the column j. When $i = j$, the value of a_{ij} will be equal to 1 since the criterion is being compared with itself.

$$A = \begin{bmatrix} 1 & a_{12} & \cdots & a_{1n} \\ a_{21} & 1 & \cdots & a_{2n} \\ \vdots & \vdots & \vdots & \vdots \\ a_{n1} & a_{n2} & \cdots & 1 \end{bmatrix} \quad (1)$$

In addition, it is fulfilled that: $a_{ij} = a_{ji} = 1$ that is:

$$A = \begin{bmatrix} 1 & a_{12} & \cdots & a_{1n} \\ 1/a_{12} & 1 & \cdots & a_{2n} \\ \vdots & \vdots & \vdots & \vdots \\ 1/a_{1n} & 1/a_{2n} & \cdots & 1 \end{bmatrix} \quad (2)$$

The complexity of these decisions becomes more significant the greater the scope of the problem is. This can be addressed by using methodologies that allow structuring the problem, modeling it, efficiently weighing the criteria relevant to that decision, and then defining the alternative that best suits them.

2.2.3. Synthesis

The synthesis consists of calculating the priority corresponding to each criterion being compared. The following is the procedure corresponding to three steps, which provide an approximation of the synthesized priorities [23]:

- Add the values in each column of the matrix of paired comparisons.
- Divide each element of such a matrix by the total of its column; the resulting matrix is called the normalized paired comparison matrix.
- Calculate the average of the elements of each row of the relative priorities of the elements being compared.

2.2.4. Priority Matrix

It summarizes the priorities for each alternative to each criterion. For m criteria and n alternatives [23]:

$$\begin{array}{c} \\ \text{Alternative 1} \\ \text{Alternative 2} \\ \cdots \\ \text{Alternative} \end{array} \begin{array}{c} \begin{array}{cccc} \text{Criterion 1} & \text{Criterion 2} & \cdots & \text{Criterion m} \end{array} \\ \begin{bmatrix} P_{11} & P_{12} & \cdots & P_{13} \\ P_{21} & P_{22} & \cdots & P_{23} \\ \cdots & \cdots & \cdots & \cdots \\ P_{n1} & P_{n2} & \cdots & P_{n3} \end{bmatrix} \end{array} \tag{3}$$

where Pij is the priority of alternative i over criterion j, for $i = 1, 2, \ldots, n$; $y\, j = 1, 2, \ldots, m$.

The column vector results from the product of the priority matrix with the priority vector of the criteria.

$$\begin{bmatrix} P_{11} & P_{12} & \cdots & P_{1m} \\ P_{21} & P_{22} & \cdots & P_{2m} \\ \cdots & \cdots & \cdots & \cdots \\ P_{n1} & P_{n2} & \cdots & P_{nm} \end{bmatrix} \begin{bmatrix} P'_1 \\ P'_2 \\ \cdots \\ P'_m \end{bmatrix} = \begin{bmatrix} P_{g1} \\ P_{g2} \\ \cdots \\ P_{g3} \end{bmatrix} \tag{4}$$

where P_{gi} is the overall priority (over the overall goal) of the alternative i $(i = 1, 2, \ldots, n)$.

2.2.5. Consistency

It refers to the consistency of the judgments expressed by each DM during the stage of paired comparisons. It is necessary to remember that perfect consistency is tough to achieve. Therefore, some inconsistency is estimated for any set of paired comparisons after these judgments. The AHP provides a method for estimating the degree of consistency between the paired opinions provided by those responsible. If the degree of consistency is acceptable, the decision process is continued; although, if this is unacceptable, the DM must reconsider its judgments corresponding to the paired comparisons and then continue with the analysis.

To determine whether a level of consistency is "reasonable," it is necessary to establish a quantifiable measure for the comparison matrix A $n \times n$ where n represents the number of alternatives to compare. If the matrix A is perfectly consistent, it generates a normalized matrix A $n \times n$ of elements w_{ij} (for $i,\ j = 1,\ 2,\ \ldots,\ n$). Such, all columns are identical, that is [23]:

$$w_{12} = w_{13} = \ldots = w_{1n} = w_1;\ w_{21} = w_{23} = \ldots = w_{2n} = w_2;$$
$$w_{n1} = w_{n2} = \ldots = w_{nn} = w_n$$
$$\mathrm{N} = \begin{bmatrix} W_1 & W_1 & \cdots & W_1 \\ W_2 & W_2 & \cdots & W_2 \\ \vdots & \vdots & \vdots & \vdots \\ W_n & W_n & \cdots & W_n \end{bmatrix} \tag{5}$$

From the given definition of A, we have:

$$\begin{bmatrix} 1 & W_1/W_2 & \cdots & W_1/W_n \\ W_2/W_1 & 1 & \cdots & W_2/W_n \\ \vdots & \vdots & \vdots & \vdots \\ W_n/W_1 & W_n/W_2 & \cdots & 1 \end{bmatrix} \begin{bmatrix} W_1 \\ W_2 \\ \vdots \\ W_n \end{bmatrix} = \begin{bmatrix} nW_1 \\ nW_2 \\ \vdots \\ nW_n \end{bmatrix} = n \begin{bmatrix} W_1 \\ W_2 \\ \vdots \\ W_n \end{bmatrix} \tag{6}$$

More compactly, we say that A is consistent if and only if:

$$AW = nW \tag{7}$$

where W is a column vector of relative weights W_j $(j = 1,\ 2,\ \ldots,\ n)$ approximated with the average of the n elements of the line in the normalized matrix N. The AHP quantifies the consistency ratio (CR) employing Equation (8), being this the quotient between the consistency index (CI) of A and the random consistency index.

$$CR = \frac{CI}{IA} \tag{8}$$

where CI is the consistency index of A and is calculated using Equation (9).

$$CI = \frac{n_{max} - 1}{N - 1} \tag{9}$$

IA represents the random consistency index of A; it is the consistency index of a matrix of paired comparisons generated randomly. This depends on the number of elements that are compared. However, there are other alternatives to estimate the random consistency index; some specialists suggest the Equation (10).

$$IA = \frac{1.98(n - 2)}{2} \tag{10}$$

The Consistency Ratio (CR): represents an indicator of the acceptability of the consistency of the analysis. If $CR > 0.10$ indicates that judgments are inconsistent; therefore, it recommends reconsidering and modifying the original values of the matrix of paired comparisons. However, for the case where $RC \leq 0.10$ it represents a reasonable level of consistency in paired comparisons. The consistency ratio can be calculated using Equation (11).

$$CR = \frac{IC}{IA} \tag{11}$$

2.3. Location of the Study Area

Panama is in the intertropical region near the terrestrial Equator, with coastal limits in the Caribbean Sea and Pacific oceans (Figure 1). Therefore, Panama's oceanic territory reflects significant ocean thermal potential; since its geographical location could allow benefiting from the use of this resource [5].

The corresponding process to justify the selection of a suitable site objectively and reliably for the installation of OTEC technology represents a process and analysis of the ocean scenario of each region. This process is linked to the norms or legislations of each nation. Even within the same nation, this analysis procedure could show certain discrepancies between the variables corresponding to the potential of each site and the viability of the technology [16,28]. To assess these matters, a set of inclusion-exclusion criteria was defined based on subjects such as environmentally protected areas, defined by the Environmental Affairs Ministry of Panama [29], aboriginal protected areas, distance to a potential population that can make a profit from these technologies, available data, among others. This process led to the selection of six potential sites. Table 2 below mentions the six sites of interest within Panama's ocean scene.

Table 2. Sites of interest to evaluate the ocean thermal energy potential in the Republic of Panama.

Name	Coordinate	Province	Ocean
Punta Burica	82.875° W, 7.875° N	Chiriqui	Pacific
Punta Mariato	80.875° W, 6.875° N	Veraguas	
Chepigana	78.375° W, 7.125° N	Darien	
Colon Island	81.875° W, 9.625° N	Bocas del toro	Caribbean Sea (Atlantic side)
Calovébora	81.125° W, 9.375° N	Veraguas	
El porvenir	79.125° W, 9.875° N	Guna Yala	

2.4. Criteria for Estimating the Potential and Sustainability of the Oceanic Thermal Resource

Currently, criteria have been identified that justify the potential of the thermal resource hosted in any site of interest. However, in this research, problems were addressed considering certain criteria or indicators that justify the ocean thermal potential identified as the variables that govern decision-making in selecting sites for OTEC from the perspective of sustainability and environmental impact.

From the literature, it was possible to identify specific global criteria that address the problem from the literature. This study's criteria to be considered are warm water temperatures (T_{cw}), depth, thermal efficiency, and absolute efficiency. The data was taken from the monthly statistical mean (2016–2018) using the Ocean Data View (ODV) software version 5.5.1. In addition, the variables were evaluated regarding anomalies in surface water temperatures during annual periods; for each case, their respective temperature jumps, or thermoclines are presented.

Next, Figure 3 shows the diagram corresponding to the problem under study. The main objective is to guarantee the operationality of OTEC in the oceanic territory of Panama, sustainably responsible for the environment, and that complies with Panama's legislation. In addition, the alternatives of interest and the criteria used in the analysis and development for the selection were identified.

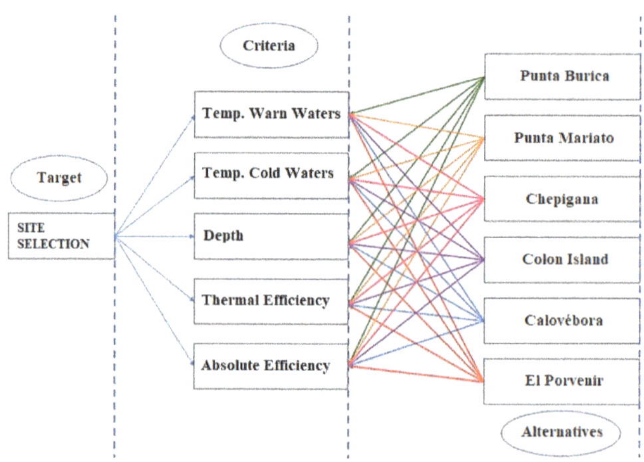

Figure 3. Hierarchical site selection analysis tree for OTEC technology.

2.5. Sea Surface Temperatures (SST)

The SST is closely related to the operationality of OTEC systems. Within the Rankine cycle, the evaporating agent of this technology will be the warm seawater in which a temperature of about 26 °C is necessary to reach the expected efficiency [3,30]. Therefore, it is necessary to evaluate whether the alternatives of interest record these readings.

These data were considered from the World Ocean Atlas 2018 (WOA18) database. This is a set of climatological fields of SST, salinity, density, cold water temperatures, distances to coasts, and in situ bathymetry. For this investigation, the scale of 0.25 × 0.25 degrees of pressure was used. The coordinates corresponding to the meteorological stations have been identified for each site. Later, these data were synchronized with the ODV software, and the respective quantitative values were obtained for each criterion to be considered.

In Figure 4, the annual behavior of the surface temperatures corresponding to the three selected sites in the Panamanian Pacific can be observed, SST ≈ 26 °C was considered, this being a reference as the adequate SST for OTEC systems to show an efficiency actual acceptable [18,31].

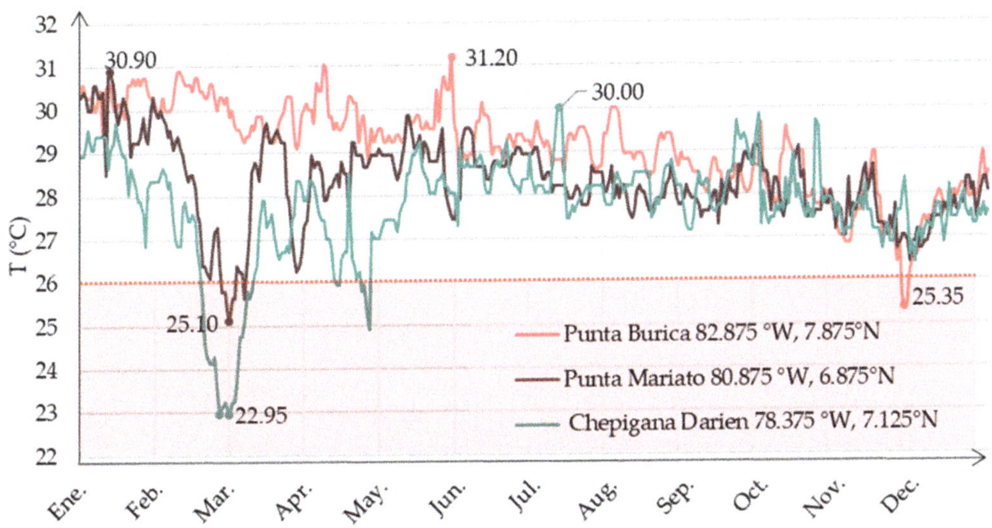

Figure 4. Panama's Pacific SST for 2018 obtained from World Ocean Atlas in 2018.

The most negligible temperature anomalies for shorter periods were recorded in Punta Burica. During this period the SST reached values approximately 31 °C being higher al recommended in his research [3,32]. The SST of Punta Mariato did not reach the maximum temperatures observed in Punta Burica (31 °C), but they were very close 30.90 ± 0.01. This site shows disturbances that are manifested during March, as shown in Figure 5, corresponding to the annual anomalies of the SST of Panama's Pacific coast. The Gulf of Panama experienced changes in its SST of approximately 3.0 °C, corresponding to the purple color and the PODAAC scale.

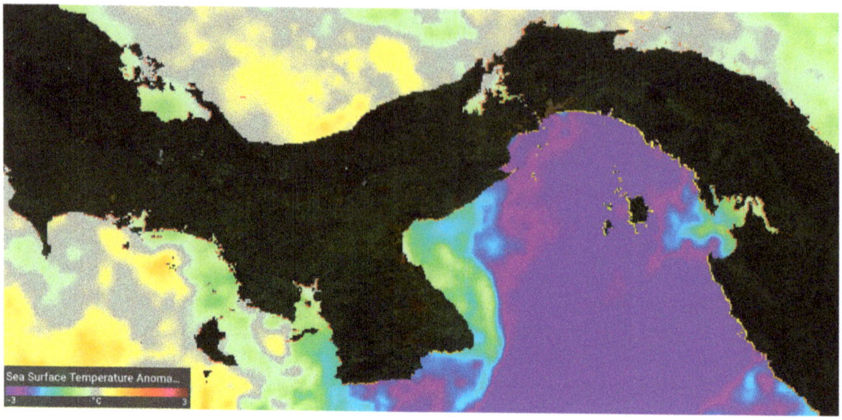

Figure 5. Surface temperature anomalies of the oceanic territory in the Pacific of Panama in March 2019.

The Sea Surface Temperature Anomalies (SSTA) represent the changes experienced by the SST greater than the temperature unit in a period. These oscillations experienced by warm surface waters are directly proportional to the operationality of the OTEC technology in terms of efficiency and net production [33,34].

In the case of Chepigana, greater differences can be observed; this behavior can be related to the warm current that emerges from the Gulf of Panama through the process called EKMAN transport [8,35,36], in which due to the displacement of the water masses product of the winds and the Coriolis effect, the cold deep waters rise and emerge on the surface. These phenomena manifest themselves during the March reflecting SST $\approx 23 \pm 0.001$ °C.

Since this value was lower than the recommended in the literature [3,15,29], this site was considered less attractive. Figure 6 shows the anomalies corresponding to the SST of the Panamanian Caribbean. This scenario shows better stability in STT anomalies than those observed in the Pacific.

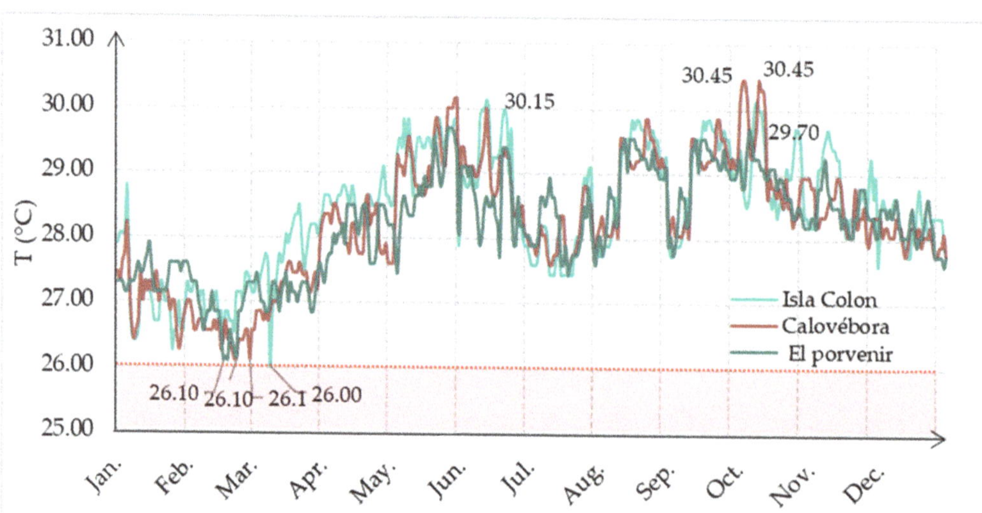

Figure 6. Anomalies of the surface temperature of the Oceanic territory in the Caribbean of Panama in the year 2018. Obtained from World Ocean Atlas in 2018.

For this case, other indicators or variables should be considered that allow justifying which of the three sites hold the most significant thermal potential considering the operationality of the OTEC system since this Panamanian Caribbean scenario projects insignificant discrepancies. Figure 7 shows the SST of the sites with the greatest potential in the Caribbean and Pacific region of Panama and their respective disturbances and shows the minimum 25.35 ± 1.07 °C and maximum 31.20 ± 1.07 °C annual SST values.

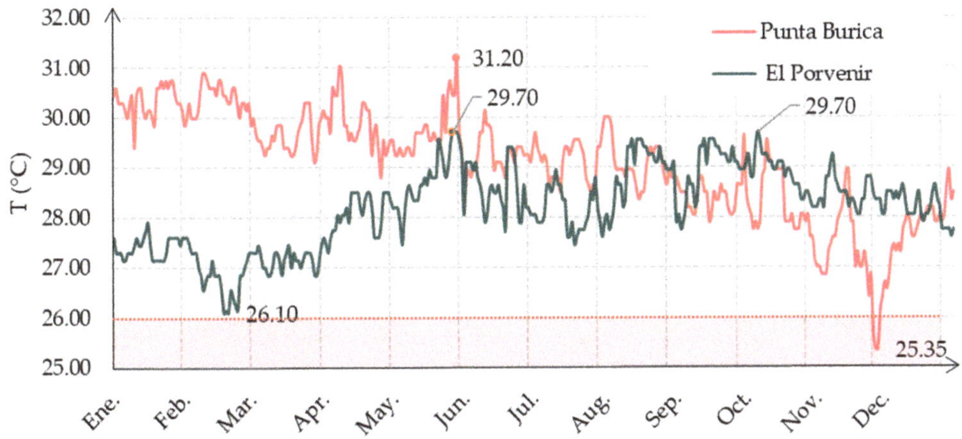

Figure 7. Surface temperature anomalies of the territory of Panama in 2018.

As shown in Figure 4, Punta Burica reaches less than 26 °C during a short period during December; however, the temperature remained above this reading from that month. The difference in these variables between these two sites is not considered from a quantitative point of view. Therefore, it would be necessary to consider other variables to select a site with the best conditions to implement OTEC technology.

Figure 8 presents the disturbances corresponding to the SST of Punta Burica in the Pacific of Panama during the years 2016–2018. When comparing the anomalies shown by the SST between these years, during the period between November 2016 and the beginning of December 2016, a cooling below 26 °C was identified. When evaluating the remaining disturbances, these reflected temperatures seemed suitable for implementing OTEC technology.

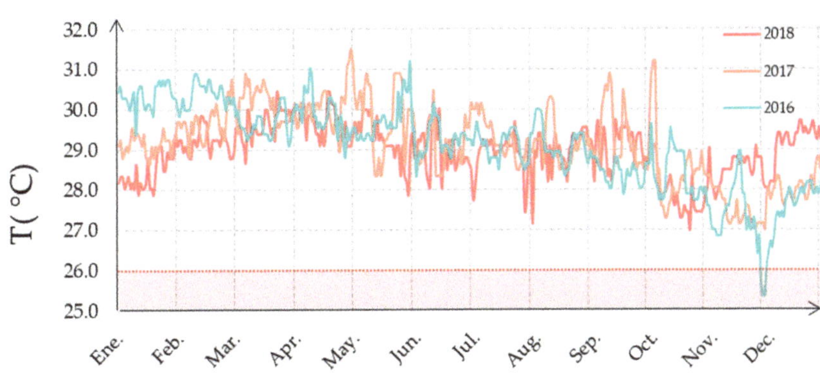

Figure 8. Disturbances of surface water temperatures (SST) in Punta Burica during the years 2016–2018.

2.6. Temperatures of Cold Sea Waters (SST) and Depth

The variable temperatures of cold waters of the deep sea are directly linked to depth. For the operationality of OTEC systems, waters with temperatures of approximately 4–5 °C needed and these are housed at depths on average between 700–1000 m [1,11,37,38]. In previous research such as OTEC Alternative for the Electric Power Generation in Panama, we can observe the bathymetric map of Panama and identify where these waters are housed [5].

In the particular case of implementing this technology in order to generate electricity, pumping systems consume approximately three-quarters of the production of the design [15,30,39]. Therefore, if the site of interest hosts cold waters at depths greater than those recommended [1,31], the system will have to overcome a greater hydrostatic force in the suction line; thus, the consumption by the pumping system will increase, and the net production of the system will be reduced, which could reduce the sustainability of the system [1,15,40,41]. Figures 9 and 10 show that the thermoclines for Punta Burica in the Pacific and El Porvenir in the Caribbean Sea, respectively, can be observed for five annual periods.

After considering the respective measurements in Figures 9 and 10, it is possible to justify if there is a directly proportional relationship between the criterion cold water temperatures and depth. In addition, this relationship is linked to the net production of the system [15,42].

Figure 10 shows how in the oceanic scenario in Punta Burica, cold waters can be found at 1200 m. However, these temperatures are housed at a depth greater than 1500 m for the case corresponding to El Porvenir. For both scenarios, the results projected during the five annual periods are observed, depicting similar values, which indicates the stability of the resource during each period that could guarantee the continuous and stable operationality of OTEC.

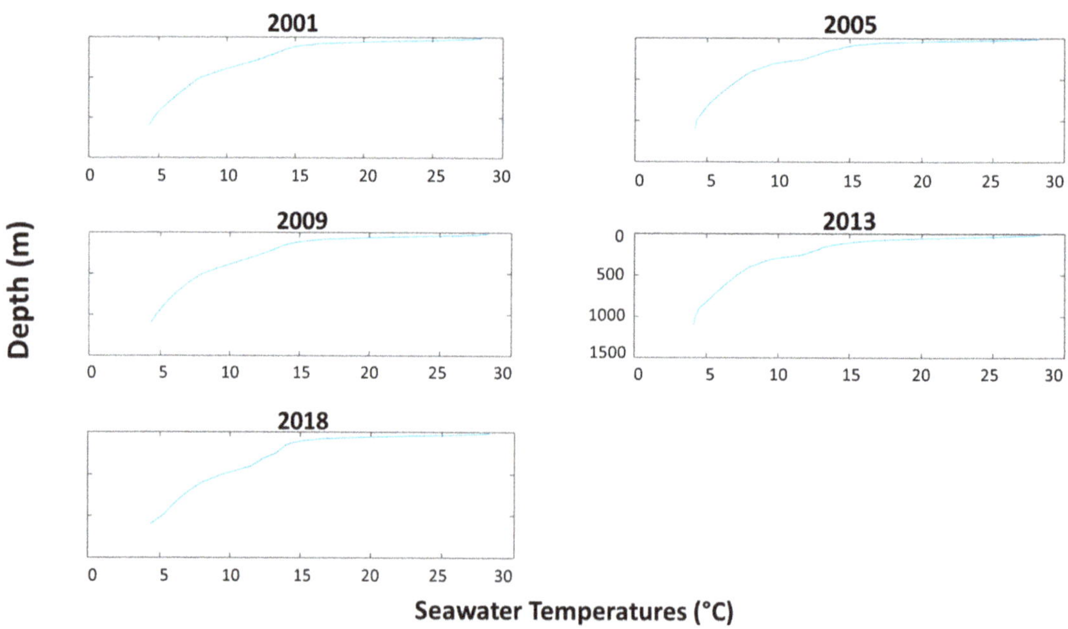

Figure 9. Punta Burica Thermocline 82.875° W, 7.875° N.

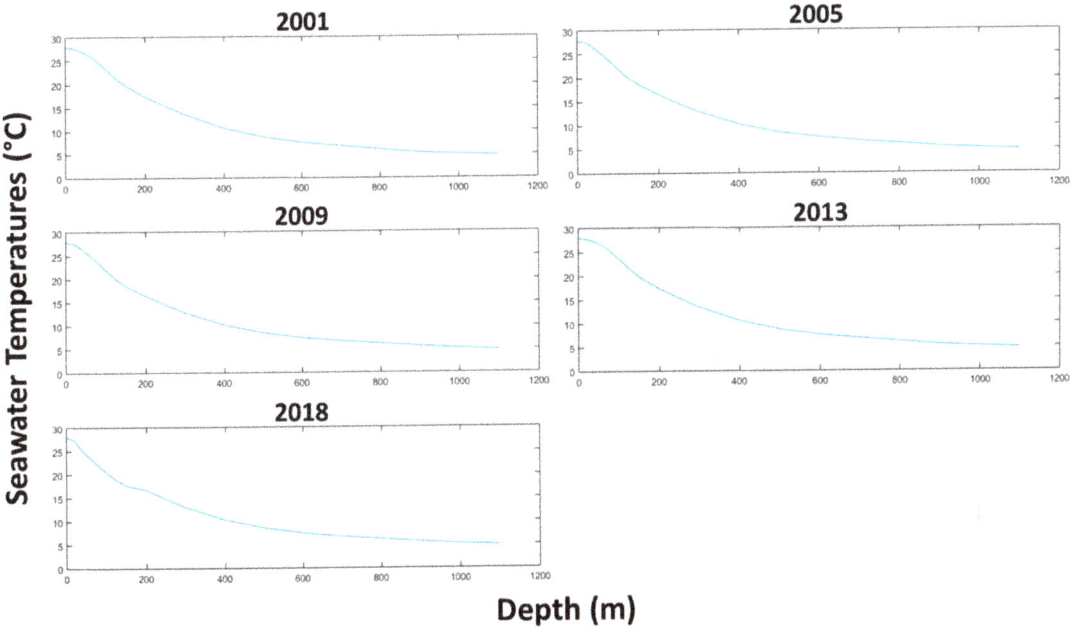

Figure 10. El Porvenir Thermocline 79.125° W, 9.875° N.

In addition, it can be observed that after reaching 4 °C in deep-sea waters achieving a unit of increase of this criterion per unit length of depth becomes linear; therefore, it is more complex and less attractive from a production point of view.

These criteria are related to technical, environmental, structural, operational, and economic aspects, raising the complexity of ensuring the proper selection of a site for OTEC. Figure 11 reflects the relationship between temperature differential and depth, where the cold waters of the deep sea are housed. These are used by OTEC technology in its condensation system.

El Porvenir hosts cold waters at 1000 m, where these waters have measurements of 5.1 °C, considering that the lowest sea surface water temperature recorded in the 2018 period was 20.97 °C; this site showed a temperature difference of 21.77 °C. If the variables corresponding to Punta Burica are considered and then compared with El Porvenir, it is observed that proper cold waters that ensure OTEC's functionality are reached at greater depths in this site. However, this site shows greater potential in its surface scenario.

Identifying which site is the most suitable for OTEC involves considering other variables such as the depth of cold water. In El Porvenir it is greater than in Punta Burica, which affects the amount of energy that the cold-water pump must be supplied to the condensation system. Both sites comply with the recommendations [3,32].

In Figure 12, the bathymetric variables corresponding to Punta Burica are depicted. The figures also depict the location of the data collection point, and the corresponding thermocline can also be observed. The temperature profiles under study are projected and their variation along with the distance to the coast. A stable scenario is projected for this site with favorable temperatures for OTEC technology.

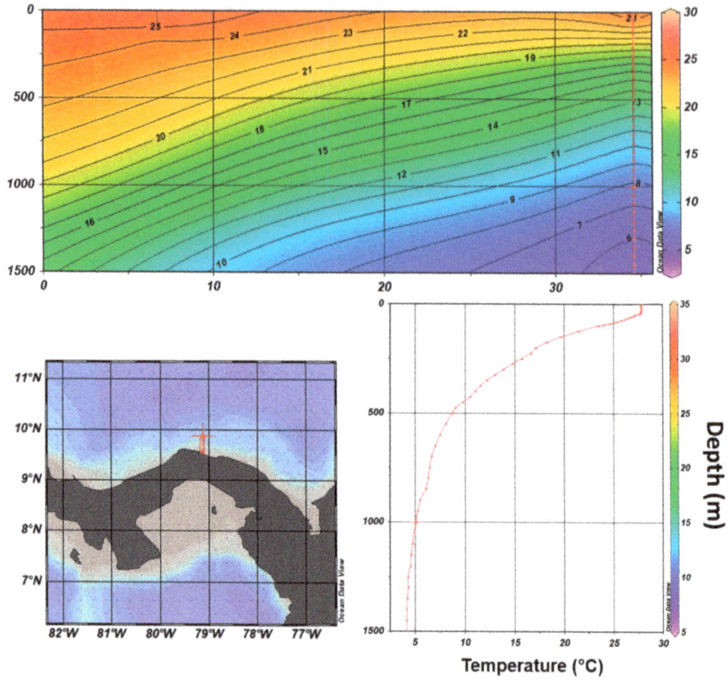

Figure 11. Bathymetric variables corresponding to El Porvenir 79.125° W, 9.875° N.

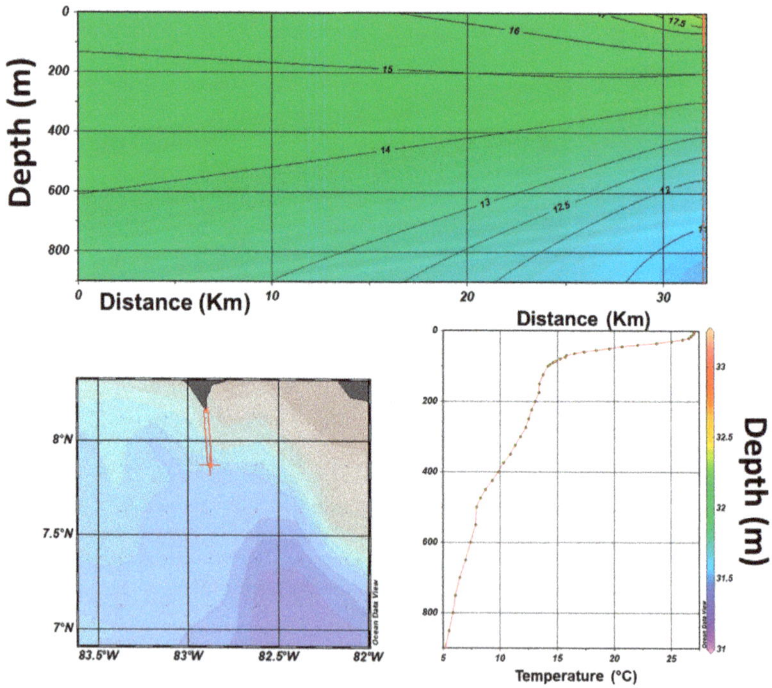

Figure 12. Bathymetric variables corresponding to Punta Burica.

2.7. Thermal and Absolute Efficiency

For thermal performance, two criteria are considered in this study: thermal efficiency and absolute efficiency. In the case of OTEC efficiency, whether the energy resource can be used or not depends largely on the efficiency of OTEC, which can be estimated by the ideal Rankine cycle, by the Equation (12) [30,43]:

$$\eta = \frac{T_W - T_C}{2(T_W)} \tag{12}$$

where η: ORC efficiency. T_W: Absolute warm water temperatures at a depth of 20 m. T_C: Absolute coldwater temperature at a depth of 700–1000 m. It is considered that certain losses induce the reduction of the production of electrical energy. This a gross OTEC efficiency (∂, in%) [14,44] can be described as:

$$\partial = \eta_t \frac{\Delta T}{2(T_W)} = \eta_t \frac{(T_w - T_c)}{2(T_W)} \tag{13}$$

where ∂: gross efficiency of the OTEC system. η_t: efficiency of the turbogenerator or turbine.

3. Application of the AHP Method

3.1. Thermal and Absolute Efficiency

Implementing OTEC technology currently faces various challenges in its low efficiencies and absence of legislation that regulates its implementation. However, there is another issue of interest for the sustainable implementation of OTEC. The appropriate selection for the installation of an OTEC plant aims to guarantee its sustainable operationality. This theme addresses environmental aspects, operational sustainability, legislation, and the Sustainable Development Goals (SDGs), which represent quantitative and qualitative aspects that directly affect the justification for selecting the OTEC site.

After evaluating the oceanic scenario of Panama, considering three sites of the Panamanian Caribbean and the Panamanian Pacific, the following research question arises:

What is the most appropriate site for the installation of the OTEC plant?

This topic was categorized as a particular case study MCD-AHP. Therefore, the considerations involving the problem under study were identified:

- Technical analysis of the thermal potential that is hosted in each site of interest was proposed; these sites are the alternatives in our AHP.
- Only quantitative variables were considered.
- The quantitative variables were evaluated in three annual periods corresponding by using ODV software.
- Different depths shall be considered for the above measurements.
- Decision-making was carried out considering state of the art.
- The SAATY table was used to establish the quantitative importance of each indicator and alternative.
- Qualitative variables for this case were not considered.
- Only the criteria were considered; the sub-criterion for this case is not considered due to the lack of information in our panorama.

In Table 3, the alternatives for OTEC installation were considered with their respective criteria. These were identified, accounting for the previous review where specific global criteria that address the problem were identified; in addition, the values were justified using ODV.

Table 3. Corresponding criteria for each site are considered as an alternative for the installation of OTEC in Panama.

Alternatives.	Criteria or Indicators				
	Sea Surface Temperatures (°C)	Cold Water Temperatures (°C @ 900 m)	Distance to Coastline (km)	Thermal Efficiency (%)	Gross Efficiency (%)
Punta Burica (S1)	29.09 ± 1.07	5.18 ± 0.10	32.069 ± 0.10	0.0396	0.3699
Punta Mariato(S2)	28.07 ± 0.99	5.16 ± 0.10	30.900 ± 0.10	0.0384	0.3679
Chepigana (S3)	27.71 ± 1.20	5.02 ± 0.10	30.000 ± 0.10	0.0377	0.3685
Colon Island (S4)	28.42 ± 0.94	5.48 ± 0.10	30.150 ± 0.10	0.0381	0.3632
Colovébora (S5)	28.22 ± 0.97	6.55 ± 0.10	30.450 ± 0.10	0.0360	0.3456
El Porvenir (S6)	28.20 ± 0.80	5.54 ± 0.10	29.700 ± 0.10	0.0376	0.3616

3.2. Prioritization

Table 4 shows the priorities corresponding to each criterion identified and considered in the MCD-AHP for the appropriate selection of the site. The quantitative value assigned to each criterion was justified using the Saaty table, and the priority of each of these is based on its relevance in the operationality of the system addressed in state-of-the-art [45].

Table 4. Hierarchy of the criteria is considered an alternative for installing OTEC in Panama for each site.

Warm Water Temperature (°C)	Cold Water Temperatures (°C @ 900 m)	Distance to Coast (km)	Thermal Efficiency (%)	Gross Efficiency (%)
1	3	4	2	3

The first criterion considered represents the TSS. This criterion evaluates the consistency of energy generation taking advantage of the oceanic thermal resource since it allows measuring the temperature difference between warm surface waters and cold waters of the deep sea. In Table 4, this criterion was assigned the value of 1 because during the process of developing the matrix of comparison of criteria, regardless of the value assigned for it at the time of being compared, it will be compared with itself, therefore, it represents equality [13,46].

Coldwater temperatures represent the second criterion; this allows us to assess the temperature differentials between sea waters. It was therefore considered moderately preferable (3). When analyzing ec.1, it is evident that the value of the OTEC efficiency is proportional to the increase in the temperature difference between surface and deep waters [47]. In addition, there is a sub-criterion linked to this, representing the distance or depth where the cold waters of the deep sea are housed. This criterion depends on the increase in the hydrostatic load in the suction line and, therefore, the power needed for the condensation circuit pump, consequently the OTEC plant's net production.

The third indicator or criterion is the distances to coastlines, the magnitude of this criterion is directly proportional to the thermal energy housed per unit area. Therefore, it was considered between moderate and strongly preferable (4). This indicator allows estimating the operational consistency of OTEC [14,16]. In addition, this criterion represents a cost in the design of an Offshore flat, since it demands cables to transport the energy resource produced to the supply network located on land, it is important to mention that for this type of installation, it is necessary to contemplate the energy losses per unit length in the energy transport cable and the costs generated by the transport of personnel.

Thermal efficiency represents the fourth indicator. It is related to the thermal potential housed in the area of interest, as OTEC shows low efficiencies for this indicator, it was assigned a value of (2) that represents between equal and moderately preferable. In addition, this indicator allows to evaluate the consistency of the operationality of the OTEC plant [16,21].

Gross efficiency represents the fifth indicator. This allows us to consider the losses in OTEC production since it contemplates for this analysis the efficiency of the turbine [31,44]. For this case, it was assigned a weighting of 3, which is moderately preferable with respect to the other criteria.

Note that this process arises from quantitative values of each variable of interest in each alternative. After this, we proceed to carry out a priority establishment analysis, where this process is clearly subjective to the knowledge and experience of decision-makers and finally identify the importance of each criterion.

3.3. *Paired Comparisons and Paired Comparison Matrix*

In Table 5 the matrix of the results of the paired comparison can be observed.

Table 5. Paired criteria comparison matrix.

CRITERIA	Warm Water Temperature (°C)	Cold Water Temperatures (°C @ 900 m)	Distance to Coast (km)	Thermal Efficiency (%)	Gross Efficiency (%)
Water temperature quality (°C)	1	3	4	2	3
Cold water temperatures (°C @ 900 mts)	1/3	1	4	1/2	3
Distance to coast (km)	1/4	1/4	1	1/4	1/4
Thermal efficiency (%)	1/2	2	4	1	2
Gross efficiency (%)	1/3	1/3	4	1/2	1

3.4. *Synthesis*

In Table 6 are the results corresponding to the synthesis or normalized matrix. These data represent the result corresponding to the procedure for establishing the approximation of the priorities of each alternative.

Table 6. Synthesis matrix used to calculate the weighting vector from the criteria comparison matrix.

Standardized Matrix					Weighting
0.41379	0.45570	0.23529	0.47059	0.32432	0.3799
0.13793	0.15190	0.23529	0.11765	0.32432	0.1934
0.10345	0.03797	0.05882	0.05882	0.02703	0.0572
0.20690	0.30380	0.23529	0.23529	0.21622	0.2395
0.13793	0.05063	0.23529	0.11765	0.10811	0.1299

3.5. *Priority Matrix and Consistency*

Table 7 presents the overall priority corresponding to each decision alternative that is summarized in the column vector.

Table 7. Random consistency indices corresponding to the sample size.

Number of Items to be Compared	1	2	3	4	5	6	7	8	9	10
Random Consistency Index (IA)	0	0	0.58	0.89	1.11	1.24	1.32	1.40	1.45	1.49

4. Results

4.1. *Prioritization Warm Water Temperatures (°C)*

Considering the TSS criterion, Figure 13 shows the prioritization of this indicator with respect to each alternative under study. From the results obtained, we can say that the best alternative identified under this consideration is site number one (Punta Burica).

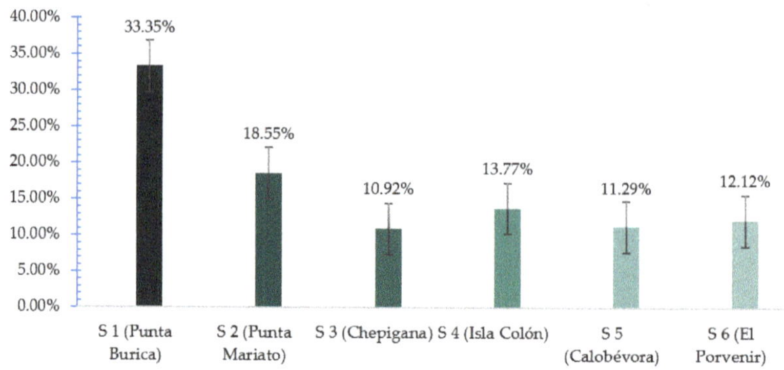

Figure 13. Prioritization percentages for the criterion: Temperatures of warm sea waters corresponding to each alternative.

The TSS is directly linked to the thermal efficiency of the OTEC system since the efficiency of the OTEC system is dependent on the temperature of warm waters, as shown in Equation (1). Therefore, the sustainability in operation of an OTEC Plant is intimately linked to the disturbances experienced by TSS.

Punta Burica showed the highest percentage of prioritization, 33.35% with respect to the other sites or alternatives, whereas site three (Chepigana) achieved the minimum value of 10.92% for this comparison. These data attained an error rate of approximately 8%, which provided reliability in the results for the prioritization of SST.

4.2. Prioritization Distance to the Coastline (km)

The distance to coast criterion denotes a determining variable in site selection since considerable lengths of power lines are needed to transport the energy produced by OTEC to the grid on land.

This criterion can also be categorized as one of the techno-economic aspects of OTEC since it considers maintenance costs, energy transport lines, and continuous transport of labor personnel, the higher this variable, the greater these preserves will be directly linked.

Figure 14 depicts the results corresponding to the distance to coast criterion priority, site one (Punta Burica). This shows the minimum percentage of 10.19% concerning the other five study sites because it is more distant from the site where the cold waters of the deep sea are housed. For this case, site six (El Porvenir) shows the highest hierarchy with a 32.88% prioritization, which tells us that only considering this criterion, our best alternative to implement OTEC would be site six.

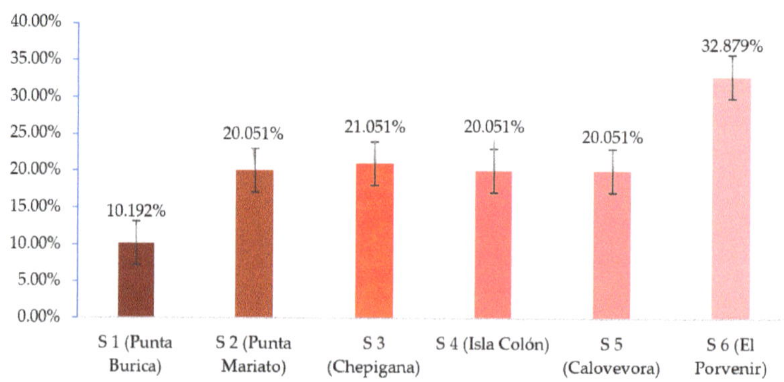

Figure 14. Prioritization percentages for the criterion: Distance to the coastline (km), corresponding for each alternative.

4.3. Thermal Efficiency Prioritization (%)

OTEC efficiency represents an indicator closely linked to the potential of the oceanic thermal resource as an energy resource. This can be estimated using the ideal Rankine cycle using Equation (12).

When considering the criterion or thermal efficiency indicator, site one (Punta Burica) shows the highest hierarchy with 23.91% regarding the other remaining sites under study. Although, the difference of this indicator is minimally considered to the rest of the alternatives. Figure 15 shows that site six (El Porvenir) is the one that presents the minimum percentage of 11.17% of hierarchy for this criterion.

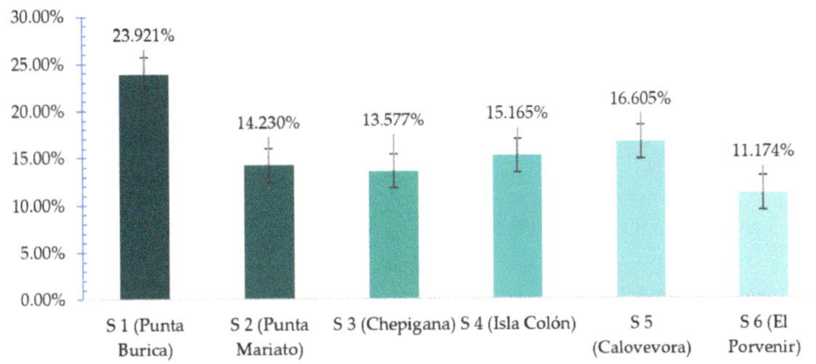

Figure 15. Prioritization percentages for the criterion: Thermal Efficiency (%), corresponding for each alternative.

When considering only this criterion, our less favorable alternative to installing an efficient OTEC system would be El Porvenir, and the most propitious site would be Punta Burica, with a prioritization percentage of 23.92%, note that for this case, the error percentage is approximately 4%, which projects us a greater reliability in the results obtained.

4.4. Prioritization Gross Efficiency (%)

This indicator considers the turbine's efficiency, which induces certain losses due to the continuous use of this element, reducing the production of electrical energy. The analysis corresponding to this can be seen in Figure 16. As for the evaluation corresponding to the gross efficiency criterion, considering all the alternatives, the picture is similar for the case of thermal efficiency, where site one (Punta Burica) presents the highest hierarchy with a percentage of 25.083% and site three with the minimum registered of 7.067%. These results reflect an error rate of approximately 4%, giving us reliability of 96% when considering this criterion individually.

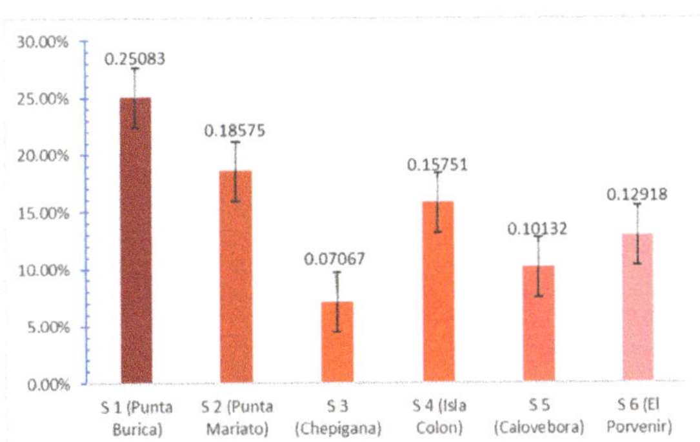

Figure 16. Prioritization percentages for the criterion: Gross efficiency (%), corresponding for each alternative.

4.5. Prioritization Deep Sea Cold Water Temperatures (°C)

Finally, considering the temperature criterion of cold water of deep-sea being these primordial for the condensation system of the OTEC plant. The temperatures of these waters are intimately linked to thermal efficiency, gross system efficiency and OTEC net power.

Achieving cold water masses from the deep sea directly affects the condensation system. If these are housed at greater depths, the greater the losses to be overcome in the suction line. Therefore, the power and consumption corresponding to this pumping system will increase proportionally.

Figure 17 shows this panorama corresponding to the prioritization of this indicator, where Punta Burica and site two (Punta Mariato) present the highest percentages of hierarchy, 24.55%, and 22.76%, respectively, both located in the Pacific Oceanic scenario. Site three (Chepigana), which is located in the Gulf of Panama, registers a minimum of 6.38%. For this case, the results show the reliability of approximately 95%.

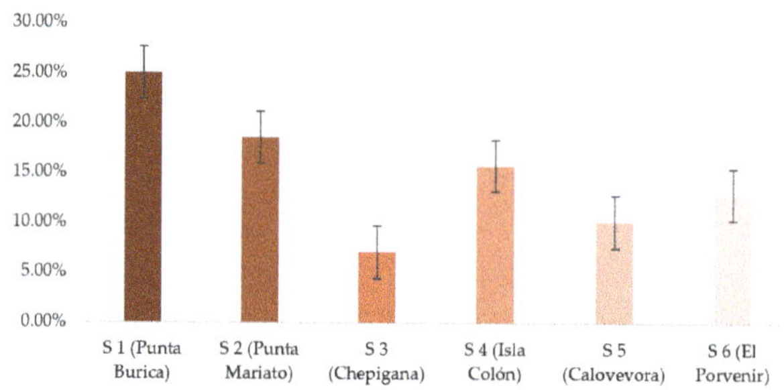

Figure 17. Prioritization percentages for the criterion: Coldwater temperatures (°C) corresponding for each alternative.

The Caribbean sites: Isla Colón, Calovébora, and El Provenir, have hierarchies of 15.751%, 10.285%, and 12.018%, respectively. The Caribbean scenario has warm and coldwater temperatures more stable during annual periods; however, the necessary cold waters are more distant from the coasts or mainland.

4.6. Hierarchical Prioritization

Figure 18 presents a summary of the prioritization percentages for each of the criteria studied in decision-making regarding the selection of the most favorable site for the use of the oceanic thermal resource in Panama. Here, it can be identified the prioritization of each corresponding criterion for each alternative; for some of these, the distance to the coastline criterion has greater prioritization. However, in other alternatives, warm water temperatures adopt the prioritization to the others.

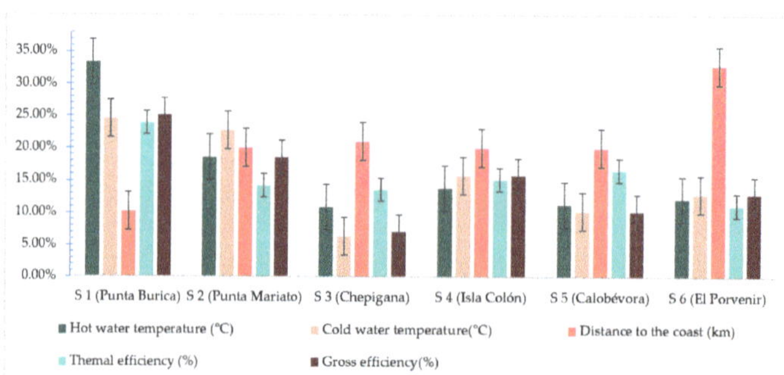

Figure 18. Prioritization percentages corresponding to each respective criterion of each alternative for the use of Panama's oceanic thermal resource.

Then, when using the MCD-AHP method to evaluate this issue, Punta Burica performs as the best alternative for using the oceanic thermal resource through the sustainable implementation of OTEC technologies.

It is important to mention that this alternative turns out to be the most appropriate of the six without considering the added value induced by the qualitative variables. That is because there are currently no legislations that allow us to estimate this value in terms of social and environmental impact.

4.7. Prioritization of Alternatives (Standardized Matrix)

Figure 19 shows the percentages corresponding to the prioritization of our alternatives with a Critical Radius of 0.054, which is satisfactory, considering the AHP model, which also justifies the consistency in the paired comparisons made during our analysis. Punta Burica represents the best technical alternative for installing and executing an OTEC plant in the open sea.

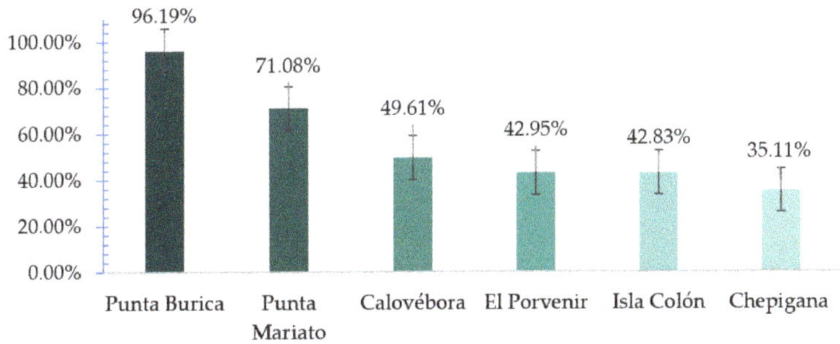

Figure 19. Alternatives according to the prioritization or importance of each of the categories with respect to the others.

5. Discussion

After addressing the AHP and identifying the multiple criteria for decision making (MCD), Punta Burica in the Gulf of Chiriquí in the Pacific of Panama is justified as the best alternative (96.19%) for the sustainable installation of an OTEC plant. According to the criteria selected as indicators, this site performs the best in four of the five, apart from the distance to the coastline.

Each criterion selected was justified under considerations of standards, legislation, and research corresponding to other countries where this technology is currently implemented to use the oceanic thermal resource. Therefore, the use of this resource will urgently demand the development of legislation that regulate the implementation of these technologies in Panamanian territory; in addition, these would allow the development of a new MCD analysis for site selection where the subjective variables corresponding to the environmental and social impact that was not included for this analysis will be included since this information in Panama does not exist. When considering the techno-economic, social, scientific, commercial, and environmental panorama corresponding to this theme, the analysis should be carried out considering the variables of a qualitative nature. These represent an added cost; thus, the site selected in this study as the most appropriate for the installation and execution of this technology could be affected.

Analyzing TSS in this study proved that the Panamanian oceanic territory presents a stable annual resource, in which the minimum temperature stated in the literature (26 °C), was achieved for most of the period analyzed. The latter is of great importance because it provides a notion of sustainability and continuity in energy production. Furthermore, one can note that the criteria selected are well linked to the actual efficiency and production; thus, their effects on each site are pretty similar.

Panama has an oceanic territory with great thermal potential. Its location in the intertropical zone near the terrestrial Equator, with coastal limits in the Caribbean Sea and the Pacific Ocean, categorizes it as part of the 98 nations with the greatest potential for thermal energy in its oceanography. Its location in the intertropical zone near the terrestrial Equator, with coastal limits in the Caribbean Sea and the Pacific Ocean, categorizes it as part of the 98 nations with the greatest potential for thermal energy in its oceanography. However, the lack of data, research and specialists in this scientific area limits the implementation of this technology within its territory. In addition, there is currently no legislation that promotes and regulate the implementation of this technology.

As future work, it is proposed to use census data to measure the impact on the nearest population to each potential site as an indicator of performance and the effect it could have on marine life. In addition, with the site selection made, it is intended to carry out a model and development of a 100 MW system in Punta Burica based on its meteorological data. Although for this case, some issues such as turbine efficiency, pipes roughness, heat losses, pressure drops, and other phenomena must be considered. This analysis could give us an estimate of the performance of an OTEC plant under these conditions, considering the inherent restrictions.

Author Contributions: Conceptualization, A.J., M.d.l.A.O.D.R. and G.L.; methodology, G.L., M.d.l.A.O.D.R. and H.A.; software, G.L.; validation, H.A.; formal analysis, G.L. and H.A.; writing—original draft preparation, G.L.; writing—review and editing, M.d.l.A.O.D.R.; visualization, G.L. and M.d.l.A.O.D.R.; supervision, A.J. and M.d.l.A.O.D.R.; funding acquisition, G.L., A.J. and M.d.l.A.O.D.R. All authors have read and agreed to the published version of the manuscript.

Funding: This research was funded by the Secretaría Nacional de Ciencia, Tecnología e Innovación (SENACYT) (Convenio de Colaboración Educativa 62-2017), as well by Sistema Nacional de Investigación (SNI).

Institutional Review Board Statement: Not applicable.

Informed Consent Statement: Not applicable.

Data Availability Statement: Not applicable.

Acknowledgments: We thank the Secretaría Nacional de Ciencia, Tecnología e Innovación (SENACYT) of the Republic of Panama for its commitment and financial support to the the Master of Science in Mechanical Engineering of the Faculty of Mechanical Engineering of the Universidad Tecnológica de Panamá. Additionally, we want to acknowledge the Sistema Nacional de Investigación (SNI) for its partial financial support of this work.

Conflicts of Interest: The authors declare no conflict of interest.

References

1. Avery, W.H.; Wu, C. Renewable energy from the ocean. *Mar. Policy* **2002**, *26*, 471–479. [CrossRef]
2. Ruiz, G.; Paola, E.; Huante, A.G.; Cueto, Y.R.; Graniel, J.F.B.; Mendieta, M.Á.A.; Acevedo, E.C.; Cupul, J.G.T.; Medina, V.M.R.; Casarín, y.R.S. *Conversión de Energía Térmica Oceánica (otec) Estado del Arte*; CEMIE-Oceano. Energías Renovables del Océano; Universidad de Campeche Mexico: Universidad Autónoma de Campeche 2017 Instituto de Ecología, Pesquerías y Oceanografía del Golfo de México (EPOMEX): Campeche, Mexico, 2017.
3. Vega, L.A. Ocean Thermal Energy Conversion. *Struct. Eng. Int.* **2007**, *4*, 85–88. [CrossRef]
4. Syamsuddin, M.L.; Attamimi, A.; Nugraha, A.P.; Gibran, S.; Afifah, A.Q.; Oriana, N. OTEC Potential in the Indonesian Seas. *Energy Procedia* **2015**, *65*, 215–222. [CrossRef]
5. López, G.; James, A.; Ortega, M.; Jurado, F. OTEC Alternative for the Electric Power Generation in Panama. In Proceedings of the 2019 7th International Engineering, Sciences and Technology Conference (IESTEC), Panama, Panama, 9–11 October 2019.
6. Devis-Morales, A.; Montoya-Sánchez, R.A.; Osorio, A.F.; Otero-Díaz, L.J. Ocean thermal energy resources in Colombia. *Renew. Energy* **2014**, *66*, 759–769. [CrossRef]
7. Matthew, C. *Development of a Marine Energy Roadmap for Panama*; The University of Edinburgh: Edinburgh, UK, 2016.
8. Palka, E.J. A geographic overview of panama: Pathway to the Continents and Link between the Seas. *Río Chagres Panama* **2005**, *52*, 3–18.
9. Rajagopalan, K.; Nihous, G.C. Estimates of global Ocean Thermal Energy Conversion (OTEC) resources using an ocean general circulation model. *Renew. Energy* **2013**, *50*, 532–540. [CrossRef]

10. NASA. State of the Ocean. 2021. Available online: https://podaac-tools.jpl.nasa.gov/soto/#b=BlueMarble_ShadedRelief_Bathymetry&l=GHRSST_L4_MUR_Sea_Surface_Temperature(la=true),MODIS_Aqua_CorrectedReflectance_TrueColor&ve=-96.12456736007292,2.1504913711860656,-72.36761727890408,12.873147621186066&pl=false&pb=false&d=2019-04-25&ao=false&as=2019-04-18&ae=2019-04-25&asz=1/day&afr=500&tlr=days (accessed on 5 April 2021).
11. International Renewable Energy Agency. Ocean Thermal Energy Conversion Technology Brief. United Arab Emirated: IRENA. 2014. Available online: https://www.irena.org/-/media/Files/IRENA/Agency/Publication/2014/Ocean_Thermal_Energy_V4_web.pdf (accessed on 5 April 2021).
12. The Physical Oceanography Distributed Active Archive Center. State of the Ocean. 2019. Available online: https://podaac-tools.jpl.nasa.gov/soto/#b=BlueMarble_ShadedRelief_Bathymetry&l=GHRSST_L4_MUR_Sea_Surface_Temperature(la=true),MODIS_Aqua_CorrectedReflectance_TrueColor&ve=-95.4960923194885,2.0186554336860656,-72.9960923194885,13.004983558686066&pl=false&pb=false&d=2019-04-25&ao=false&as=2019-04-18&ae=2019-04-25&asz=1/day&afr=500&tlr=days (accessed on 5 September 2021).
13. Garduño-Ruiz, E.P.; Silva, R.; Rodríguez-Cueto, Y.; García-Huante, A.; Olmedo-González, J.; Martínez, M.L.; Wojtarowski, A.; Martell-Dubois, R.; Cerdeira-Estrada, S. Criteria for optimal site selection for ocean thermal energy conversion (Otec) plants in Mexico. *Energies* **2021**, *14*, 2121. [CrossRef]
14. Zhang, J.; Xu, C.; Song, Z.; Huang, Y.; Wu, Y. Decision framework for ocean thermal energy plant site selection from a sustainability perspective: The case of China. *J. Clean. Prod.* **2019**, *225*, 771–784. [CrossRef]
15. Chen, F.; Liu, L.; Peng, J.; Ge, Y.; Wu, H.; Liu, W. Theoretical and Experimental Research on the Thermal Performance of Ocean Thermal Energy conversion System Using the Rankine Cycle Mode. *Energy* **2019**, *183*, 497–503. [CrossRef]
16. García, A.; Rodríguez, Y.; Garduño, E.; Hernández, R. General Criteria for Optimal Site Selection for the Installation of Ocean Thermal Energy Conversion (OTEC) Plants in the Mexican Pacific. In *Ocean Thermal Energy Conversion (OTEC)—Past, Present, and Progress*; IntechOpen: London, UK, 2016; p. 13.
17. Negara, R.; Koto, J. Potential of 100 KW of Ocean Thermal Energy Conversion in Karangkelong, Sulawesi Utara, Indonesia. *Int. J. Environ. Res. Clean Energy* **2017**, *305*, 10.
18. Garduño Ruiz, E.P.; Huante, G.; Cueto, A.R.Y.; Bárcenas Graniel, J.F.; Mendieta Alatorre, M.Á.; Cerezo Acevedo, E.; Tobal Cupul, J.G.; Romero Medina, V.M.; Silva Casarín, R. *Conversión de Energía Térmica Oceánica (OTEC)*; Cemie-Océano, Universidad Autónoma de Campeche: Campeche, México, 2017.
19. Garduño, E.; Garcia, A. De Energía Térmica Oceánica. 2017. Available online: https://cemieoceano.mx/downloads/libros/CEMIEOceano_Gradiente_Termico.pdf (accessed on 5 April 2021).
20. Lockheed Martin Mission Systems & Sensors (MS2). *Ocean Thermal Extractable Energy Visualization: Final Technical Report*; Ocean Thermal Energy Resource Assessment; Department of Energy: Washington, DC, USA, 2012; 88p.
21. Doorga, J.R.S.; Gooroochurn, O.; Motah, B.A.; Ramchandur, V.; Sunassee, S. A novel modelling approach to the identification of optimum sites for the placement of ocean thermal energy conversion (OTEC) power plant: Application to the tropical island climate of Mauritius. *Int. J. Energy Environ. Eng.* **2018**, *9*, 363–382. [CrossRef]
22. Saaty, T.L. *Proceso de Análisis Jerárquico (AHP)*; J. Wiley: New York, NY, USA, 1980.
23. White, G.P. Decision making with the analytic hierarchy process. *Omega* **1987**, *15*, 283–290. [CrossRef]
24. Sergio, A.; Redondo, L. Utilidad de metodos como AHP. *Cuad. Adm.* **2007**, *20*, 65–84.
25. Mendez, M.; Álamo, A.L.; Frutos, M.; Ascua, R. Fundamentos de Ayuda Multicriterio a la Decisión. 2014. Available online: https://www.researchgate.net/publication/317882380_FUNDAMENTOS_DE_AYUDA_MULTI-CRITERIO_A_LA_DECISION (accessed on 5 April 2021).
26. Mellinas Fernandez, M. Análisis Comparativo de Técnicas de Generación Eléctrica; AHP y TOPSIS Fuzzificado. 2012. Available online: https://repositorio.upct.es/bitstream/handle/10317/2963/pfc4430.pdf?sequence=1 (accessed on 5 April 2021).
27. José María, M.J. El Proceso Analítico Jerárquico (ahp). Fundamentos, Metodología y Aplicaciones. 2010. Available online: https://users.dcc.uchile.cl/~{}nbaloian/DSS-DCC/ExplicacionMetodoAHP(ve%20rpaginas11-16).pdf (accessed on 5 April 2021).
28. Chen, T.Y. Multiple criteria decision analysis using a likelihood-based outranking method based on interval-valued intuitionistic fuzzy sets. *Inf. Sci.* **2014**, *286*, 188–208. [CrossRef]
29. López, G.; James, A.; Ortega, M.D.L.A. Evaluation of the Oceanic Thermal Potential on the Coasts of Panama. Available online: https://tethys.pnnl.gov/sites/default/files/publications/PAMEC-2020-E-BOOK.pdf (accessed on 5 April 2021).
30. Vera, D.; Baccioli, A.; Jurado, F.; Desideri, U. Modeling and optimization of an ocean thermal energy conversion system for remote islands electrification. *Renew. Energy* **2020**, *162*, 1399–1414. [CrossRef]
31. Vega, L.A.; Michaelis, D. First Generation 50 MW OTEC Plantship for the Production of Electricity and Desalinated Water. In Proceedings of the Offshore Technology Conference, Houston, TX, USA, 3–6 May 2010. [CrossRef]
32. Wu, C.-x.; Wu, B.-j.; Yin, Y. A Review of Ocean Thermal Energy Utilization. *Adv. N&R Energy* **2014**, *2*, 275–280.
33. Langer, J.; Quist, J.; Blok, K. Recent progress in the economics of ocean thermal energy conversion: Critical review and research agenda. *Renew. Sustain. Energy Rev.* **2020**, *130*, 109960. [CrossRef]
34. Yang, M.H.; Yeh, R.H. Analysis of optimization in an OTEC plant using organic Rankine cycle. *Renew. Energy* **2014**, *68*, 25–34. [CrossRef]
35. International Renewable Energy Agency. *Renewables Readiness Assessment: Panama*; IRENA: Abu Dhabi, United Arab Emirates, 2018.

36. British Embassy. *An Examiniation of Ocean Energy Potential in Panama: Identifying the Next Steps*; British Embassy: Hong Kong, China, 2016.
37. Masutani, S.M.; Takahashi, P.K. Ocean Thermal Energy Conversion (OTEC). *Encycl. Ocean Sci.* **2010**, 167–173. [CrossRef]
38. Díez, P.F. *Energía Maremotérmica*, 12th ed.; Universidad De Cantabria: Cantabria, Spain, 2003; Available online: http://www.ingenieroambiental.com/3008/ENERG%CDA%20mareotermica.pdf (accessed on 5 April 2021).
39. Sang, Y.; Karayaka, H.B.; Yan, Y.; Yilmaz, N.; Souders, D. Ocean (Marine) Energy. In *Comprehensive Energy Systems*; Elsevier: Amsterdam, The Netherlands, 2018; Volume 1. [CrossRef]
40. Vega, L.A. Economics of Ocean Thermal Energy Conversion (OTEC). In *Ocean Energy Recovery: The State of the Art*; ASCE: Reston, VA, USA, 1992; Chapter 7; pp. 152–181.
41. Deethayat, T.; Kiatsiriroat, T.; Thawonngamyingsakul, C. Performance analysis of an organic Rankine cycle with internal heat exchanger having zeotropic working fluid. *Case Stud. Therm. Eng.* **2015**, *6*, 155–161. [CrossRef]
42. Jung, J.Y.; Lee, H.S.; Kim, H.J.; Yoo, Y.; Choi, W.Y.; Kwak, H.Y. Thermoeconomic analysis of an ocean thermal energy conversion plant. *Renew. Energy* **2016**, *86*, 1086–1094. [CrossRef]
43. Mohd Idrus, N.H.; Musa, M.N.; Yahya, W.J.; Ithnin, A.M. Geo-Ocean Thermal Energy Conversion (GeOTEC) power cycle/plant. *Renew. Energy* **2017**, *111*, 372–380. [CrossRef]
44. Wang, M.; Jing, R.; Zhang, H.; Meng, C.; Li, N.; Zhao, Y. An Innovative Organic Rankine Cycle (ORC) based Ocean Thermal Energy Conversion (OTEC) System with Performance Simulation and Multi-Objective Optimization. *Appl. Therm. Eng.* **2018**, *145*, 743–754. [CrossRef]
45. Fetanat, A.; Khorasaninejad, E. A novel hybrid MCDM approach for offshore wind farm site selection: A case study of Iran. *Ocean Coast. Manag.* **2015**, *109*, 17–28. [CrossRef]
46. Avery, W.H. OTEC—Ocean Thermal Energy Conversion. 1993. Available online: https://www.worldcat.org/title/ocean-thermal-energy-conversion/oclc/232665992&referer=brief_results (accessed on 5 April 2021).
47. Meegahapola, L.; Udawatta, L.; Witharana, S. The Ocean thermal energy conversion strategies and analysis of current challenges. In Proceedings of the 2007 International Conference on Industrial and Information Systems, Peradeniya, Sri Lanka, 9–11 August 2007; pp. 123–128.

energies

MDPI

Article

GIS-MCDM-Based Approach to Site Selection of Wave Power Plants for Islands in China

Meng Shao [1], Shulei Zhang [1], Jinwei Sun [1,*], Zhixin Han [2], Zhuxiao Shao [1] and Chuanxiu Yi [1]

1 College of Engineering, Ocean University of China, Qingdao 266100, China; mengya@ouc.edu.cn (M.S.); zhangshulei@stu.ouc.edu.cn (S.Z.); szx0617@163.com (Z.S.); yichuanxiu@stu.ouc.edu.cn (C.Y.)
2 Economics and Management School, Wuhan University, Wuhan 430072, China; hanzhixin30@whu.edu.cn
* Correspondence: sjw@ouc.edu.cn; Tel.: +86-18661819603

Abstract: On-site development of wave energy resources is a promising way to overcome power-shortage problems on islands. It is necessary to select suitable islands to deploy wave power plants, which are influenced by multiple factors related to resources, technology, economy, society, and environment. This study develops a two-stage decision framework to identify feasible islands and determine priority order based on geographic information systems (GIS) and multicriteria decision-making (MCDM). In the exclusion stage, unfeasible marine areas are excluded based on exclusion criteria and feasible island alternatives are identified. In the evaluation stage, alternatives are evaluated by evaluation criteria using the combined weighting method and the technique for order of preference by similarity to ideal solution (TOPSIS)-grey relation analysis (GRA) method. As the combined weighting method is based on the fuzzy group decision-making (GDM)-analytic hierarchy process (AHP) and the entropy method, it can effectively reduce subjective deviation. The proposed framework is applied in Shandong Province. It identifies 13 inhabited islands feasible for constructing wave power plants, among which Daguan, South Changshan, and Xiaoguan are the optimal ones. Sensitivity analysis is performed to verify the feasibility of the proposed framework. The results show that it is effective and could provide reference for practical engineering.

Keywords: site selection; GIS; MCDM; wave power plants; island alternatives

Citation: Shao, M.; Zhang, S.; Sun, J.; Han, Z.; Shao, Z.; Yi, C. GIS-MCDM-Based Approach to Site Selection of Wave Power Plants for Islands in China. *Energies* **2022**, *15*, 4118. https://doi.org/10.3390/en15114118

Academic Editors: Eugen Rusu, Kostas Belibassakis and George Lavidas

Received: 21 April 2022
Accepted: 1 June 2022
Published: 3 June 2022

Publisher's Note: MDPI stays neutral with regard to jurisdictional claims in published maps and institutional affiliations.

Copyright: © 2022 by the authors. Licensee MDPI, Basel, Switzerland. This article is an open access article distributed under the terms and conditions of the Creative Commons Attribution (CC BY) license (https://creativecommons.org/licenses/by/4.0/).

1. Introduction

Against the background of rising environmental concerns and the depletion of fossil energy reserves, renewable energy resources are expected to be an important part of the world's future energy supply [1–3]. Marine energy, as a type of renewable energy with wide distribution, abundant reserves, and broad development prospects, has received considerable attention in many coastal countries around the world [4,5]. Wave energy is one of the major forms of marine energy, with strong predictability, high stability, and significantly higher density than other marine-energy sources [6,7]. Research has shown that the world's available wave energy could reach 2 billion kW, equivalent to twice the current total power generation [8]. At present, the harnessing and exploitation of wave energy in China is still in the research and development stage. To promote wave energy development, there has been an urgent push for research on selecting satisfactory wave power plant sites. Appropriate site selection is the prerequisite for wave energy industrialization, and it directly affects electricity-generation capacity and future socioeconomic benefits [9,10].

At present, the harnessing and exploitation of wave energy is often applied on islands that are far from the shore [11,12]. Because of the limitations of power-grid access, a large number of inhabited islands currently face power-shortage problems, which has placed great constraints on local economic development and population growth [13,14]. With island development becoming more and more important, the construction of reliable and affordable island power systems has become an urgent task [15,16]. The local development and utilization of wave energy resources around inhabited islands will be a promising path [17]. Considering the high construction costs of wave power plants, it is necessary to select feasible islands and determine their prioritization for the deployment of wave power plants.

Selecting a wave power plant site involves multiple factors related to resources, technology, economy, society, and environment; it is usually regarded as a complex multi-criteria decision-making (MCDM) problem. Applying MCDM to site-selection decision-making can support dealing with multiple, often conflicting criteria in a structured way, allowing different preferences to be considered. Another excellent tool, geographic information systems (GIS), can help decision-makers carry out the collection, storage, management, calculation, analysis, and visualization of geo-referenced data [18]. In previous studies of renewable-energy site selection, GIS has been frequently combined with MCDM to form decision-support tools to exclude unsuitable sites based on restrictions or to calculate site-suitability indexes based on the established criteria system [19–22].

Moreover, some scholars try to carry out research from multi-objective planning, and the most widely used method is data envelopment analysis (DEA) [23]. DEA is a methodology based on linear programming to measure the relative efficiency of homogenous decision-making units (DMUs) with multiple inputs and multiple outputs [24,25]. Wang et al. (2022) proposed a combined method based on DEA, Grey Analytic Hierarchy Process (G-AHP), and Grey Technique for Order Preference by Similarity to Ideal Solution (G-TOPSIS) for solar PV power plants site selection, in which DEA was used in the first phase to select high-efficiency locations based on various measurable criteria [26]. Pambudi et al. (2019) presented a hierarchical fuzzy data envelopment analysis model for identifying suitable locations for the construction of wind farms in the Indonesian archipelago [27]. However, in the evaluation process, DEA focuses on economic cost and power generation efficiency, and can only perform quantitative analysis [28,29]. On the other hand, each DMU obtains the weights from the most favorable aspect, and it will cause these weights to be different with different DMUs so that the characteristics of each DMU lack comparability, and the results obtained in this way may be not reliable.

In terms of energy sources, previous studies of site selection have mainly focused on solar, onshore wind, and offshore wind power; few, however, have investigated wave power. To the best knowledge of the authors, there has been no research on site selection for wave power plants for islands in the existing literature. This is likely because the harnessing and exploitation of solar and wind energy have entered the development stage in terms of industrialization and practical use, and wave energy is still in the research and development stage, or the early stage of industrialization. Therefore, research on site selection for wave energy is of great significance for making progress in the industrialization of renewable energy.

A few studies investigating wave power plant site selection have been conducted in various areas. Ghosh et al. (2016) employed analytic hierarchy process (AHP) to obtain evaluation criteria weights and then used an artificial neural network to determine a suitability index for wave-energy-conversion device site selection in the UK and Jamaica [30]. Abaei et al. (2017) developed a new site-selection decision method to estimate the expected utility of different sites for wave power plants in Tasmania; the approach was based on a Bayesian network model and could be extended to influence diagrams [31]. Vasileiou et al. (2017) used GIS and AHP to obtain evaluation criteria weights and employed weighted linear combination (WLC) to determine suitable areas for hybrid offshore wind and wave energy systems in Greece [32]. Gradden et al. (2016) proposed a GIS-based approach for the site selection of hybrid wind and wave energy platforms along the Atlantic-facing coasts of Europe [33]. Shao et al. (2020) employed GIS, AHP, and WLC methods to calculate a suitability index and drew a suitability map for constructing wave energy power stations in Qingdao, China [34]. Nobre et al. (2009) proposed a framework based on a combination of reclassification and weighting procedures in a GIS environment. In that framework, expert experience and WLC were applied to determine suitability for wave farm deployment in an area off the southwest coast of Portugal [35]. Flocard et al. (2016) determined criteria weights based on expert experience and obtained a suitability index for wave energy converter site selection using WLC [36].

The literature review reveals that a few studies have undertaken large-scale site selection for wave power plants based on a combination of GIS and MCDM methods. Those studies have primarily been limited to obtaining regional suitability indexes or classes for wave development. It appears, however, that no studies have considered small-scale site selection to determine the priority order of feasible site alternatives. Yet, large-scale and small-scale site selection are both essential components of research on site-selection decision-making. Moreover, in previous studies, criteria weighting and alternatives evaluation have been regarded as two core and troublesome stages that affect the decision results. Some researchers have done effective work on these two stages; nevertheless, certain problems remain to be solved, as outlined below.

(1) In published studies, AHP is the most common weighting method [37]. As a subjective weighting method, AHP relies on experts' subjective judgment to give a comparative matrix and determine criteria weights. However, the process has difficulty on avoiding subjective deviations caused by factors such as insufficient expert knowledge or experience, which affect the reliability of the weighting results.

To reduce subjective deviations and improve reliability, some researchers have attempted to improve traditional AHP by integrating it with other approaches. Integrating fuzzy theory with AHP can determine criteria weights by considering fuzzy linguistic variables from decision-makers. Sánchez-Lozano et al. expressed expert-group opinions using triangular fuzzy numbers (TFNs) and then used them in the AHP method [38]. Ayodele et al. proposed interval type-2 fuzzy AHP, which reduces uncertainty in decision-making processes [39,40]. A few researchers have combined subjective and objective weighting methods to determine criteria weights; this approach can not only consider subjective expert judgment but also reflect information in the data itself [41].

(2) Currently, the alternative evaluation methods in wave energy site selection research are mainly limited to WLC. WLC is a classical, simple, straightforward MCDM method. In recent years, it has been popularized and applied to many decision-making problems. However, it has also been criticized because its mathematical model is not sufficiently clear. Accordingly, many MCDM methods have been proposed and employed. Among them, TOPSIS (technique for order preference by similarity to an ideal solution), which has a clear mathematical model, is generally considered to be the most scientific and convenient one. Currently, this method has been used in several studies of solar energy site selection and wind energy site selection, but never been used in the field of wave energy site selection.

Sánchez-Lozano et al., used traditional TOPSIS to assess alternatives for solar power plants [42]. Several researchers used fuzzy TOPSIS to evaluate alternatives [43–47]. Fang et al. proposed an extended TOPSIS method to rank the order of photovoltaic power plant sites [43]. Sánchez-Lozano employed two different MCDM methods, TOPSIS and Elimination and Choice Expressing the Reality TRI (ELECTRE-TRI), to evaluate and classify suitable locations for solar farms; that study also examined the differences and similarities between the two methods [44].

Despite the popularity and application of TOPSIS, it still has some limitations and needs to be improved. In alternative evaluation, the classical TOPSIS method only considers the distances to the best and worst ideals while ignoring other dimensions.

This study develops a two-stage decision framework based on GIS and MCDM for wave power plant site selection for islands, and applies it in Shandong, China. The framework solves the aforementioned problems and its innovation lies in the following aspects:

(1) It includes both large-scale site selection and small-scale site selection. The first stage aims to exclude unfeasible marine areas and identity island alternatives for constructing wave power plants. The second stage aims to evaluate island alternatives to determine priority order.

(2) A combined weighting method is proposed to determine criteria weights, based on a combination of subjective and objective weighting methods. The subjective weighting method consists of fuzzy theory, AHP, and group decision-making (GDM). The objective weighting method employs the entropy method, which handles information from an objective standpoint. The proposed weighting method avoids subjective bias and improves the accuracy of the results; meanwhile, it overcomes the shortcomings of single subjective or objective weighting methods.
(3) An integrated TOPSIS-GRA (grey relation analysis) approach is proposed for alternative evaluation of wave power plant locations. In this approach, the distance used in the TOPSIS method is employed to represent the position similarity between alternatives. The grey relational grade used in GRA is mainly employed to describe the shape similarity between alternatives.

The rest of this paper is organized as follows. Section 2 establishes a criteria system for site selection, including exclusion and evaluation criteria. Section 3 presents the decision framework for site selection and introduces GIS, the combined weighting method, and TOPSIS-GRA. Section 4 presents a case study of Shandong Province to identify the suitable islands for the siting of wave power plants; sensitivity analysis is performed as well. Finally, the conclusions and outlook are presented in Section 5.

2. Criteria System

The site selection of wave power plants for islands is influenced by plenty of factors. After studying much research on site selection decision making and consulting experts, exclusion and evaluation criteria for wave power plant site selection are confirmed as follows; detailed data sources are provided in Section 4.1.

2.1. Exclusion Criteria

To exclude islands unsuitable for wave energy development, exclusion criteria are established based on the existing literature, the actual situation of the islands, and expert opinions. These criteria take technical, economic, social, and environmental factors into consideration.

2.1.1. Marine Ecological Red Line

To maintain marine ecological health and security, China has designated important marine ecological function areas as key control areas; these are called "marine ecological red line" (MERL). All development activities that might affect ecosystems are prohibited or restricted in MERL areas [48]. Considering these restrictions and environmental factors, areas covered by MERL are excluded for wave energy development.

2.1.2. Wave Power Density

Wave power density (WPD) is an important technical criterion for evaluating wave energy resources considering both wave height and wave period. As an exclusion criterion, the lowest WPD value should meet the technical feasibility requirements of wave-energy-generation devices.

2.1.3. Water Depth

For economic and technical reasons, water depth (WD) imposes many space restrictions on wave power plants site selection. The type and installation method of wave energy generation devices and cost-related issues (e.g., wiring, operating, and maintenance costs) are all affected by WD [49,50].

2.2. Evaluation Criteria

To determine the preference order of islands, 14 evaluation criteria related to resource, natural, economic, social and environmental factors are identified to evaluate island alternatives. Table 1 presents the classifications of evaluation criteria.

Table 1. Evaluation criteria.

First-Level Criteria	Second-Level Criteria	Benefit/Cost	Qualitative/Quantitative
Resource criteria	Wave power density (WPD) (kW/m)	B	Quantitative
	Wave height (WH) (m)	B	Quantitative
Natural criteria	Seabed geology (SG)	B	Qualitative
	Water depth (WD) (m)	C	Quantitative
	Coastal erosion (CE)	C	Qualitative
	Geological disaster (GD)	C	Qualitative
Economic criteria	Distance from the shore (DS) (m)	C	Quantitative
	Distance from the port (DP) (m)	C	Quantitative
	Population served (PS)	B	Quantitative
Social/environmental criteria	Fishing potential (FP)	C	Qualitative
	Tourism potential (TP)	C	Qualitative
	Shipping density (SD)	C	Qualitative
	Policy encouragement (PE)	B	Qualitative
	Electricity demand (ED)	B	Qualitative

2.2.1. Resource Criteria

Resource conditions are critical for the economic viability and technical feasibility of wave power plants. WPD and wave height are both benefit criteria, and they are important resource criteria for wave energy resource evaluation [32–36]. These two criteria are used to describe how much wave energy is available for wave-energy generation at a site. The greater the WPD and wave height, the more suitable it is for wave energy development.

2.2.2. Natural Criteria

Natural criteria affect the construction and operation of wave power plants for islands, including WD, seabed geology, coastal erosion, and geological disaster. Seabed geology is a benefit criterion, while WD, coastal erosion, and geological disaster are cost criteria. WD limits the type and placement of wave energy generation devices. Installing wave energy generation devices in areas with a large WD will increase foundation costs and technical difficulties [30,32,33]. Seabed geology affects the installation of energy generation devices and submarine cables [49,51,52]. It is very valuable for selecting a suitable seabed geology for installing wave energy generation devices. After that, a developer can determine the appropriate slope, installation location, and connection route to the coast for installation. Coastal erosion indicates the soil erosion of a near-shore beach zone [30,52,53]. Installing wave energy generation devices around areas with severe coastal erosion will increase the difficulty of construction and reduce the stability of power generation. Geological disaster refers to the frequency of geological disasters around the island; areas with a high frequency of geological disaster are not suitable for constructing wave-energy power plants [52].

2.2.3. Economic Criteria

Economic criteria affect the construction and operation costs of wave power plants. They include distance from the shore, distance from the port, and population served. Distance from the shore and distance from the port are cost criteria, and population served is a benefit criterion. Distance from the shore is related to operation and maintenance costs; being far from the shore increases the cost of maintaining wave energy generation devices [32–34]. Distance from the port affects construction and installation costs; areas close to ports are better for constructing wave power plants because the related costs will be comparatively low [32,33,35,36]. Population served refers to the population of the island served by the wave power plant [32]. It reflects the number of potential energy consumers; the larger the population served, the more urgent the power demand of the island.

2.2.4. Social/Environmental Criteria

Wave power plants may affect social benefits and environmental conditions around islands. Social/environmental criteria include fishing potential, tourism potential, shipping density, policy encouragement, and electricity demand. Fishing potential, tourism potential, and shipping density are cost criteria; policy encouragement and electricity demand are benefit criteria. Areas with high fishing potential and tourism potential are less suitable for wave energy development. Constructing wave power plants in areas with good fishing potential will affect the normal economic activities of island residents [36,54] Meanwhile, wave-energy-generation devices also cause visual and noise disturbances for tourists, which will affect the economic benefits of local tourism [30,55,56]. The deployment of wave-energy-generation devices should not disturb primary shipping routes since the probability of collision with the devices will increase [30,32,33,36]. Policy encouragement is important for achieving a successful, long-lasting wave power plant since a reliable institutional policy framework can promote constructing wave power plants [47,57,58]. The island population, infrastructure construction, distance from the shore, and current energy supply situation determine the island's power demand [59–61]. The greater the demand for power, the more urgent the need for construction.

After quantifying the above qualitative criteria through reclassification, vector normalization (VN) is further employed to normalize all criteria values. The purpose of normalization is to eliminate differences between attributes in dimensionality and order of magnitude. Normalization can affect the decision result by affecting the diversity of attribute data (DAD) [62]. VN does not change DAD and is considered to be the best normalization method for TOPSIS. The formula for VN is

$$x_{ij}^* = \frac{x_{ij}}{\sqrt{\sum_{i=1}^{m} x_{ij}^2}}, (i = 1, \ldots, m, j = 1, \ldots, n). \tag{1}$$

where x_{ij} represents the attribute value of ith alternative against jth criterion, x_{ij}^* represents the normalized attribute value, m represents the number of alternatives, and n represents the number of criteria.

3. Methodology

3.1. Decision Framework for Site Selection

The decision-making process for site selection consists of two stages: exclusion stage and evaluation stage. Figure 1 shows the decision-making framework.

Stage 1: Exclusion stage

This stage aims to exclude unfeasible marine areas and identify feasible islands through exclusion criteria sets. GIS is introduced in this stage to handle spatial data. In this study, GIS datasets for MERL, WPD, WD are established, each dataset can generate a thematic map. By superimposing these maps, unfeasible marine areas and feasible island alternatives are identified.

Stage 2: Evaluation stage

To rank island alternatives, MCDM methods are employed in this stage. After identifying feasible islands in the study area, the island alternatives are evaluated based on 14 evaluation criteria using the combined weighting method and TOPSIS-GRA. Criteria weights are obtained by the combined weighting method, including fuzzy GDM-AHP and entropy method. After obtaining the criteria weights, TOPSIS-GRA is used to rank the islands.

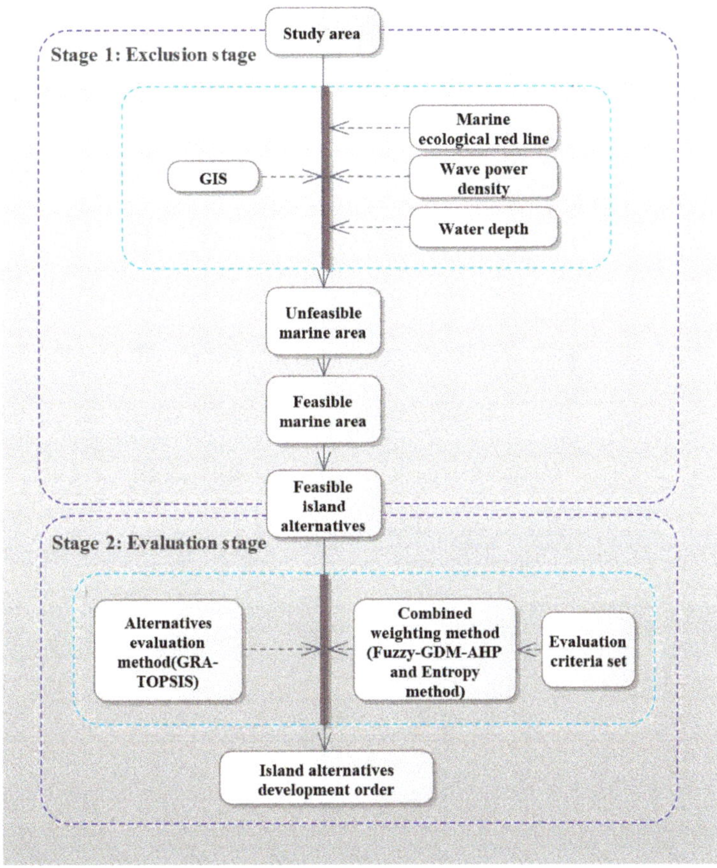

Figure 1. Decision-making framework for site selection.

3.2. GIS

GIS is an information system used to deal with data, maps, and spatial information [63]. GIS tools can support planning and decision making in site-selection problems [64–67]. In this study, GIS is used to exclude unfeasible marine areas by three exclusion criteria related to economic, technical, and social constraints. Inverse distance weighted (IDW) interpolation and Euclidean distance in GIS are used to obtain 100 m × 100 m raster datasets of WPD, WD. After obtaining the raster data of the exclusion criteria, the Boolean overlay operation is employed in the overlay analysis in GIS to exclude unfeasible areas. After that, islands feasible for constructing wave power plants are identified for subsequent study.

3.3. Combined Weighting Method

In total, 4 first-level criteria and 14 s-level criteria are set for site selection. Fuzzy GDM-AHP is used for the determination of first-level criteria weights. The entropy method is utilized to determine second-level criteria weights.

3.3.1. Fuzzy GDM-AHP

AHP is a well-known MCDM method invented by Saaty as a decision-making tool; it is widely used for its simple calculation process and straightforward theory [68]. Traditional AHP has some shortcomings, such as subjective deviations, insufficient reliability, and an inability to reflect human thinking processes. To overcome these shortcomings, fuzzy set theory and GDM theory are introduced into AHP to determine the first-level criteria weights.

Fuzzy set theory was introduced by Zadeh to deal with vague, imprecise, and uncertain problems [69]. Fuzzy decision-making is a rational decision-making method that considers human subjectivity. In a fuzzy environment, linguistic variables are transformed into TFNs, which take the ambiguity and uncertainty of expert judgment into account [70]. By integrating TFNs into AHP, decision-making processes can be described more accurately.

Expert judgment is the most important part of the AHP method. To reduce the bias of individual expert evaluation and make the evaluation results more objective, GDM theory is introduced into the calculation of evaluation criteria weights [71]. By selecting experts from different fields and empowering them according to their importance, the advantages of expert judgment can be maximized, and the accuracy and reliability of results can be improved [72].

The process of calculating evaluation criteria weights through fuzzy GDM-AHP is as follows [73]:

(1) Establish a fuzzy pairwise comparison matrix:

Let $F = [c_{kv}]_{n\times n}$ be the matrix for n criteria against the goal. c_{kv} is a fuzzy set representing the relative importance of criterion k over v. Then, assume $c_{kv} = \frac{1}{c_{vk}}$.

Figure 2 shows the possible assessment values of c_{kv} in the pairwise comparison matrix, represented as TFNs [74].

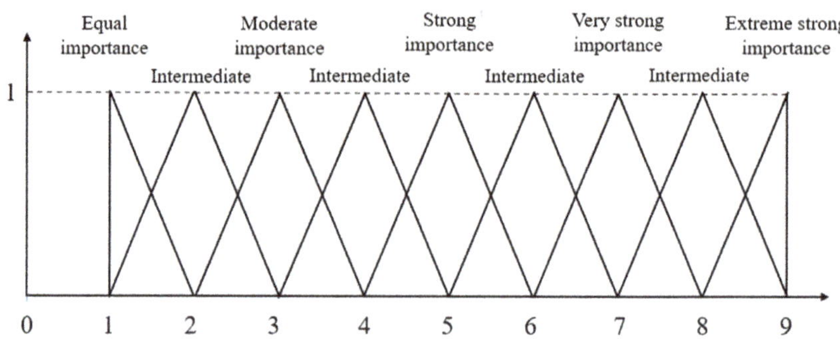

Figure 2. Degree of influence and corresponding TFNs for expert assessment.

(2) Synthesize judgements using GDM:

There are t experts forming an expert set $E = \{E_1, E_2, \cdots, E_t\}$; the weights of experts are $\{e_1, e_2, \cdots, e_t\}$. Let $c_{kv}^{(t)} = \left(l_{kv}^{(t)}, m_{kv}^{(t)}, u_{kv}^{(t)}\right)$ be a TFN representing the relative importance of c_k over c_v judged by DM_t. After GDM, let $c_{kv} = (l_{kv}, m_{kv}, u_{kv})$ be the aggregated relative importance of c_k over c_v judged by all experts. c_{kv} can be calculated as follows:

$$l_{kv} = l_{kv}^{(1)e_1} l_{kv}^{(2)e_2} \cdots l_{kv}^{(t)e_t}; \tag{2}$$

$$m_{kv} = m_{kv}^{(1)e_1} m_{kv}^{(2)e_2} \cdots m_{kv}^{(t)e_t}; \tag{3}$$

$$u_{kv} = u_{kv}^{(1)e_1} u_{kv}^{(2)e_2} \cdots u_{kv}^{(t)e_t}. \tag{4}$$

(3) Calculate the fuzzy weights of the criteria:

The geometric normalized average method can be used to calculate the fuzzy weights of criteria. Where the values are fuzzy, not crisp, the weight vector will be achieved through the following formula:

$$(w_{lk}, w_{mk}, w_{uk}) = \frac{\left(\prod_{v=1}^{n}(l_{kv}, m_{kv}, u_{kv})\right)^{1/n}}{\sum_{k=1}^{n}\left(\prod_{v=1}^{n}(l_{kv}, m_{kv}, u_{kv})\right)^{1/n}}, \tag{5}$$

where (w_{lk}, w_{mk}, w_{uk}) is the fuzzy weight of the k-th criterion.

(4) Defuzzify the fuzzy weights:

Fuzzy sets are difficult to compare directly because they are partially ordered rather than linear or strictly ordered crisp values. So, we defuzzify the obtained weights to calculate the crisp value of each criterion weight as follows:

$$w_{FCk} = \frac{w_{lk} + 4w_{mk} + w_{uk}}{6}. \tag{6}$$

where w_{FCk} is the crisp weight of the k-th first-level criterion.

3.3.2. Entropy Method

The entropy method is used to calculate criteria weights according to the size and difference degree of the value of the sample data [75,76]. The larger the entropy, the smaller the influence of the evaluation criterion on decision-making; that is, the weight of the criterion is smaller. The process for calculating the weight of the evaluation criteria by the entropy method is as follows:

(1) Normalize the decision matrix:

Different criteria can be of different scales. A given decision matrix should first be transformed into a dimensionless space via

$$p_{ij} = \frac{x_{ij}}{\sum_{i=1}^{m} x_{ij}}(i = 1, 2, \cdots, m; j = 1, 2, \cdots, n), \tag{7}$$

where p_{ij} is the probability of the j-th criteria in the i-th alternative.

(2) Calculate the entropy of the j-th criteria:

$$E_j = -\mathrm{K}\sum_{i=1}^{m} p_{ij} \ln p_{ij}, \tag{8}$$

$$\mathrm{K} = \frac{1}{\ln m}, \tag{9}$$

where E_j is the entropy of the j-th criteria, and K is the coefficient.

(3) Calculation of objective weights:

$$w_{SCj} = \frac{1 - E_j}{\sum_{j=1}^{n}(1 - E_j)}. \tag{10}$$

where w_{SCj} is the weight of the j-th second-level criterion.

3.3.3. Combined Algorithm

To obtain the criteria weights for site selection, a combined weighting algorithm is proposed, which is to solve the weights of the first- and second-level criteria respectively. The first-level criteria weights are calculated by fuzzy GDM-AHP, the second-level criteria weights are calculated by the entropy method, and the combined weight is calculated by the following:

$$w_{Cj}^{*} = w_{FCk} \cdot w_{SCj(k)}. \tag{11}$$

where w^*_{Cj} is the combined weight of the j-th criterion, w_{FCk} is the weight of the k-th first-level criterion, and $w_{SCj(k)}$ is the weight of the j-th second-level criterion under the k-th first-level criterion.

3.4. TOPSIS-GRA

This study proposes a novel hybrid method integrating TOPSIS and GRA to obtain the optimal site for a wave power plant.

TOPSIS method, first developed by Hwang and Yoon [77], is commonly used for addressing the rank issue. The basic idea of TOPSIS is that the best decision is the one that is closest to the ideal situation and farthest from the non-ideal situation. Although TOPSIS is widely used in many fields, it has some shortcomings. TOPSIS introduces two reference points and ranks alternatives by comparing the distances from alternatives to these points. It can express the position similarity between alternatives, but it does not consider the shape similarity between the alternatives. The GRA method was originally developed by Deng and is suitable for making decisions in multiple-attribute situations [78]. The limitation of TOPSIS can be overcome by the grey relation coefficient of the GRA model [79,80]. The combination of TOPSIS and GRA measures the relations among alternatives based on the degree of similarity or difference in both the position and shape of the alternatives.

The process for TOPSIS-GRA is as follows:

(1) Calculate the weighted normalized decision matrix:

$$v_{ij} = w^*_{Cj} \cdot x^*_{ij} (i = 1, 2, \cdots, m; j = 1, 2, \cdots, n), \tag{12}$$

$$V = \left(v_{ij}\right)_{m \times n} = \begin{bmatrix} v_{11} & v_{12} & \cdots & v_{1n} \\ v_{21} & v_{22} & \cdots & v_{2n} \\ \vdots & \vdots & \ddots & \vdots \\ v_{m1} & v_{m2} & \cdots & v_{mn} \end{bmatrix}, \tag{13}$$

where v_{ij} denotes the weighted normalized criterion value of the j-th criterion in the i-th alternative.

(2) Determine the positive ideal solutions (A^+) and negative ideal solutions (A^-):

$$A^+ = \{v_1^+, \ldots, v_n^+\} = \left\{\left(\max_i v_{ij}, j \in J\right)\left(\min_i v_{ij}, j \in J'\right)\right\}, \ (i = 1, \ldots, m), \tag{14}$$

$$A^- = \{v_1^-, \ldots, v_n^-\} = \left\{\left(\min_i v_{ij}, j \in J\right)\left(\max_i v_{ij}, j \in J'\right)\right\}, \ \left(i = 1, \ldots, m\right), \tag{15}$$

where J and J' refer to the benefit criteria set and cost criteria set, respectively.

(3) Calculate the Euclidean distance d_i^+ and d_i^- of each alternative from positive ideal solutions (PIS) and negative ideal solutions (NIS):

$$d_i^+ = \sqrt{\sum_{j=1}^n \left(v_{ij} - v_j^+\right)^2}, \ (i = 1, \ldots, m), \tag{16}$$

$$d_i^- = \sqrt{\sum_{j=1}^n \left(v_{ij} - v_j^-\right)^2}, \ (i = 1, \ldots, m), \tag{17}$$

where d_i^+ is the distance from alternative i to PIS, and d_i^- is the distance from alternative i to NIS.

(4) Calculate the grey relational coefficients:

Based on the weighted normalized decision matrix, the grey relational coefficient between the i-th alternative and the PIS with respect to the j-th criterion is calculated as follows:

$$r_{ij}^{+} = \frac{\underset{i}{min}\underset{j}{min}\left|v_{ij} - v_j^{+}\right| + \rho \underset{i}{max}\underset{j}{max}\left|v_{ij} - v_j^{+}\right|}{\left|v_{ij} - v_j^{+}\right| + \rho \underset{i}{max}\underset{j}{max}\left|v_{ij} - v_j^{+}\right|}, \tag{18}$$

$$R^{+} = \left[r_{ij}^{+}\right]_{m \times n}, \tag{19}$$

where ρ is the distinguishing coefficient, and R^{+} is the grey relational coefficient matrix with PIS. In this study, the distinguishing coefficient is set as 0.5.

Similarly, the grey relational coefficient between the i-th alternative and the NIS with respect to the j-th criterion can be obtained as follows:

$$r_{ij}^{-} = \frac{\underset{i}{min}\underset{j}{min}\left|v_{ij} - v_j^{-}\right| + \rho \underset{i}{max}\underset{j}{max}\left|v_{ij} - v_j^{-}\right|}{\left|v_{ij} - v_j^{-}\right| + \rho \underset{i}{max}\underset{j}{max}\left|v_{ij} - v_j^{-}\right|}, \tag{20}$$

$$R^{-} = \left[r_{ij}^{-}\right]_{m \times n}, \tag{21}$$

where R^{-} is the grey relational coefficient matrix with NIS.

(5) Calculate the grey relational grade:

The grey relational grade is used for the overall evaluation of alternatives depending on all criteria. It is defined as the average value of relational coefficients at different criteria. For the i-th alternative, the grey relational grades from PIS and NIS are given as follows:

$$g_i^{+} = \frac{1}{n}\sum_{j=1}^{n} r_{ij}^{+}; \tag{22}$$

$$g_i^{-} = \frac{1}{n}\sum_{j=1}^{n} r_{ij}^{-}. \tag{23}$$

(6) Calculate a new relational grade:

Normalize the Euclidean distances and grey relational grades obtained from Equations (5) and (7), as follows:

$$D_i^{+} = \frac{d_i^{+}}{max d_i^{+}},\ D_i^{-} = \frac{d_i^{-}}{max d_i^{-}},\ G_i^{+} = \frac{g_i^{+}}{max g_i^{+}},\ G_i^{-} = \frac{g_i^{-}}{max g_i^{-}}; \tag{24}$$

$$S_i^{+} = \alpha D_i^{-} + \beta G_i^{+}; \tag{25}$$

$$S_i^{-} = \alpha D_i^{+} + \beta G_i^{-}. \tag{26}$$

Among them, the larger the values of D_i^{-} and G_i^{+}, the closer the alternative is to the positive ideal solution in position and shape. The larger the values of D_i^{+} and G_i^{-}, the closer the alternative is to the negative ideal solution in position and shape. In the above formulas, α and β are the weights of position and shape, respectively, in the calculation of the similarity degree of the alternative and ideal solutions, reflecting the decision-maker's preference for position and shape factors. In this study, α and β are both set as 0.5.

The new relational grade is as follows:

$$Z_i = \frac{S_i^{+}}{S_i^{+} + S_i^{-}}. \tag{27}$$

(7) Rank alternatives according to the values of Z_i:

The order of alternatives is ranked according to the value of relative closeness to each of the alternatives. A greater value of Z_i indicates a higher priority in the alternatives.

4. Case Study

4.1. Study Area

In the "Marine Renewable Energy Development Plan" in China, Shandong Province is positioned as a key area for marine renewable energy development [51,58,81]. It borders the Bohai Sea and the Yellow Sea, with a coastline of approximately 3345.41 km, rich in wave energy resources. It has a developed marine economy, and there is a huge demand for energy due to busy marine activities along the coast. At the same time, Shandong Province has gathered many powerful marine science research institutes and related enterprises in China, which is an important condition for the development and utilization of wave energy [82].

Shandong Province has jurisdiction over 589 islands, among which 32 are inhabited. Given the distance from the mainland, the economic activities of the inhabited islands are severely restricted by power dilemmas [83]. In addition, traditional power-generation modes are costly and cause serious pollution. Clean wave energy can be easily obtained around islands, which will not cause pollution and can greatly alleviate the power-shortage problems in the islands [84].

This study investigated site selection for wave power plants for the inhabited islands of Shandong Province. Based on locations and development conditions, the latitude and longitude of the study area (Figure 3) are selected from 34°24′ N to 38°58′ N and 117°34′ E to 123°37′ E. Considering the requirements for the accuracy of the results, the evaluation units in the study area are divided into 100 m × 100 m grids. Table 2 shows the data description and source of each criterion.

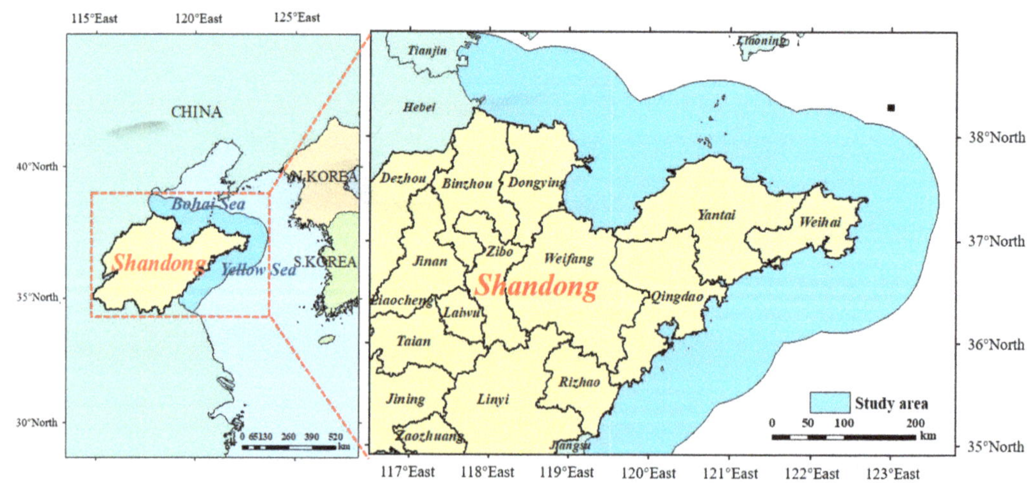

Figure 3. Study area.

Table 2. Data descriptions and sources of criteria.

Criterion	Data Description	Data Resource
Marine ecological red line	Vector data	Yellow Sea and Bohai Sea Marine Ecological Red Line Delineation Plan, released by Shandong Provincial People's Government
Wave power density	Average wave power density of Shandong Province in 39 years; 100 m ×100 m grid data (kW/m)	General Bathymetric Chart of the Oceans (GEBCO) and European Center for Medium Weather Forecast (ECMWF)
Wave height	Average wave height of Shandong Province in 39 years; 100 m ×100 m grid data (m)	General Bathymetric Chart of the Oceans (GEBCO) and European Center for Medium Weather Forecast (ECMWF)
Seabed geology	Score 1–9; the higher the score, the better the seabed geology	China Offshore Ocean Atlas (submarine topography and landforms)
Water depth	100 m ×100 m grid data (m)	General Bathymetric Chart of the Oceans (GEBCO)
Coastal erosion	Score 1–9; the higher the score, the greater the degree of coastal erosion	China Island History (Shandong Volume)
Geological disaster	Score 1–9; the higher the score, the greater the frequency of geological disaster	China Island History (Shandong Volume)
Distance from the shore	100 m × 100 m grid data (m)	Shortest Euclidean distance to the land coastline in ArcGIS
Distance from the port	(m)	Shortest Euclidean distance to the major ports (obtained from the Transportation Department of Shandong Province) in ArcGIS
Population served	Amount of island population	Statistical Yearbooks released by the government
Fishing potential	Score 1–9; the higher the score, the greater the fishing potential	Marine Ecological Environmental Protection Plan, released by the Ecological Environment Department of Shandong Province
Tourism potential	Score 1–9; the higher the score, the greater the tourism potential	Marine Ecological Environmental Protection Plan, released by the Ecological Environment Department of Shandong Province
Shipping density	Score 1–9; the higher the score, the higher the shipping density	Marine Ecological Environmental Protection Plan, released by the Ecological Environment Department of Shandong Province
Policy encouragement	Score 1–9; the higher the score, the greater the policy encouragement	Shandong Province Island Protection Plan, released by the Department of Oceans and Fisheries of Shandong Province
Electricity demand	Score 1–9; the higher the score, the greater the electricity demand	China Island History (Shandong Volume) and Statistical Yearbook released by the government

4.2. Exclusion of Unfeasible Areas

In this study, unfeasible marine areas are excluded by three exclusion criteria. The MERL of Shandong Province includes 10 types of areas: marine nature reserves; special marine protected areas; important estuarine ecosystems; important coastal wetlands; important fishery waters; special protected islands; natural landscape and historical and cultural heritage areas; important coastal tourist areas; important sandy shorelines; and sand source protected sea area [85,86]. All of these areas should be excluded. The Simulating Waves Nearshore (SWAN) model is used to simulate the wave field, and the 39-year average WPD distribution in the study area could be obtained by calculation [87,88]. WPD data are point feature data with an accuracy of $1' \times 1'$. Considering the existing wave-energy-generation devices and the data for Shandong Province, marine areas with a WPD lower than 1 kW/m are regarded as undeveloped sea areas [87]. WD data are point-feature data with an accuracy of 0.1° × 0.1°. Considering currently available technology and installation types, areas with a WD greater than 50 m are excluded. The exclusion range of each criterion is shown in Table 3.

Table 3. Exclusion range of each criterion.

Exclusion Criteria	Exclusion Range
Marine ecological red line	All
Wave power density	<1 kW/m
Water depth	>50 m

ArcGIS software is used for overlay analysis. The thematic map of unfeasible areas is obtained by superimposing the respective maps of these three criteria. Figure 4 shows a different thematic map for each exclusion criterion. Figure 5 shows the unfeasible and feasible marine areas determined by the combination of the three maps.

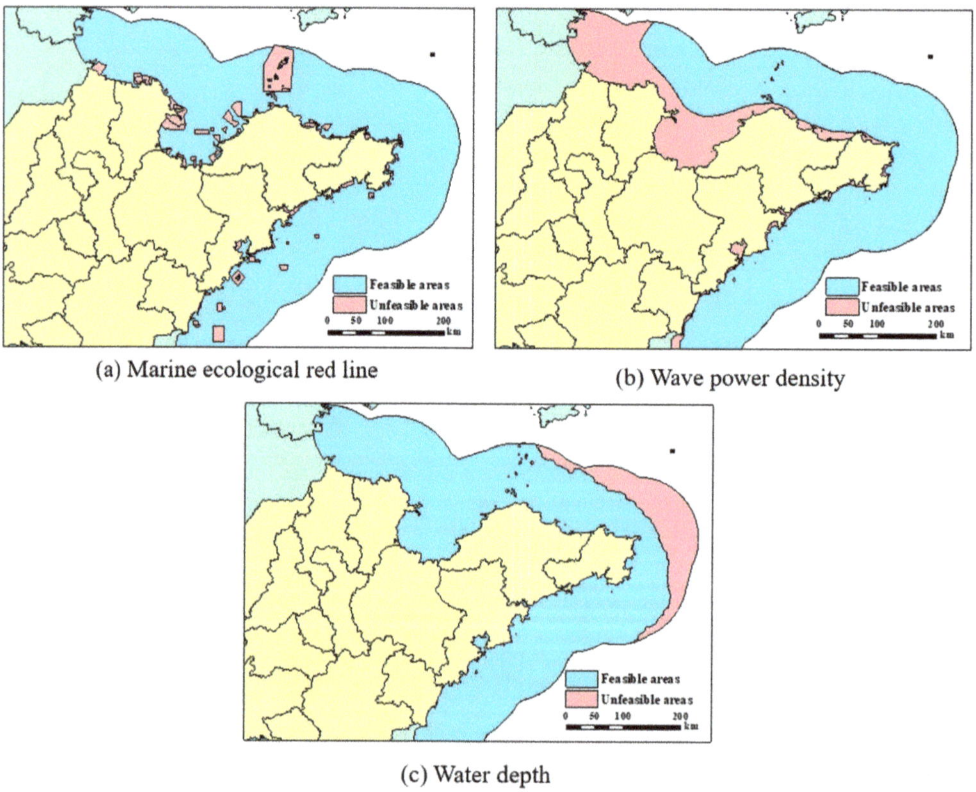

Figure 4. Unfeasible areas based on the three exclusion criteria.

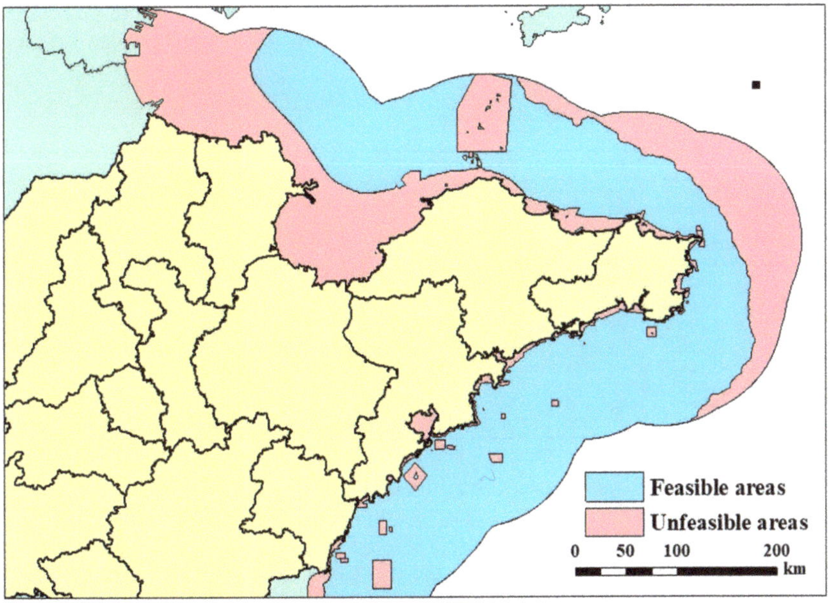

Figure 5. Unfeasible areas and feasible areas.

4.3. Feasible Islands Identification and Data Acquisition

After excluding the unfeasible parts of the study area, thirteen inhabited islands that can feasibly develop wave energy are identified: South Changshan, North Changshan, Temple, Daheishan, Xiaoheishan, Jiming, Nanhuang, East Little Qingdao, Zhucha, Muguan, Daguan, Xiaoguan, and Zhaitang, which constitute alternative set $A = \{A_1, A_2, \cdots A_{13}\}$. A_1, A_2, A_3, A_4, A_5 are located in the northern part of Yantai. A_6, A_7, A_8 belong to Weihai, and the other five islands are located in the east and south of Qingdao. Figure 6 shows the distribution of the thirteen island alternatives for constructing wave power plants.

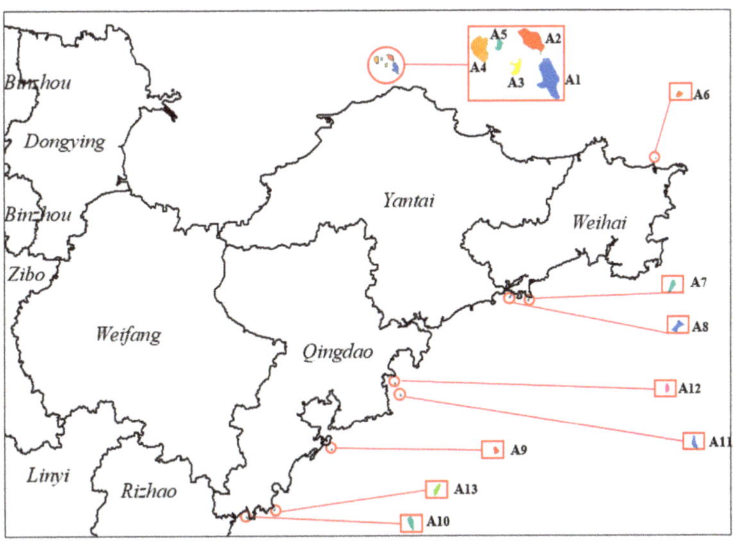

Figure 6. Thirteen island alternatives for constructing wave power plants.

Through data investigation, on-site observation, and numerical simulation, the attribute values of the evaluation criteria of each alternative are obtained, as shown in Table 4.

Table 4. Attribute value matrix.

Alternatives	Island	WPD	WH	SG	WD	CE	GD	DS	DP	PS	FP	TP	SD	PE	ED
A_1	South Changshan	1.0402	0.7734	4	11.1637	5	6	5884.52	9428.9	24400	8	8	8	8	7
A_2	North Changshan	1.1703	0.8282	4	12.4547	5	7	6288.24	1839.7	3300	7	7	6	8	7
A_3	Temple	1.0698	0.7727	4	11.33	4	3	5637.07	1561.9	1700	8	7	7	6	6
A_4	Daheishan	1.1477	0.8261	3	12.1661	6	5	5440.56	2064.25	1500	7	7	6	6	6
A_5	Xiaoheishan	1.1491	0.8213	3	12.4229	7	3	5801.7	2144.13	270	7	7	6	8	5
A_6	Jiming	1.2674	0.8411	3	24.3263	4	3	3161.31	412.32	200	4	5	7	4	3
A_7	Nanhuang	1.3127	0.6818	6	6.9741	3	6	9395.83	98.77	520	6	4	5	4	2
A_8	East Little Qingdao	1.2463	0.5634	6	7.4281	3	6	10,623.75	211.82	300	5	4	4	4	3
A_9	Zhucha	1.7929	0.7446	5	18.0511	5	4	1979.26	400.12	560	4	4	2	7	4
A_{10}	Muguan	1.2549	0.5589	4	11.2361	5	3	3434.31	138.52	180	7	4	8	5	2
A_{11}	Daguan	2.2626	0.7583	4	11.1103	6	4	5200.64	975.35	120	3	5	5	7	3
A_{12}	Xiaoguan	2.0068	0.6483	4	5.8069	6	5	5179.35	936.13	100	3	6	5	7	2
A_{13}	Zhaitang	1.6374	0.7475	6	15.8693	4	3	5216.56	130.14	1100	4	4	4	8	4

4.4. Determination of Criteria Weights

The evaluation criteria weights are determined by the combined weighting method. First-level criteria weights can be solved based on the fuzzy GDM-AHP method. Second-level criteria weights are calculated based on the entropy method.

(1) Determination of first-level criteria weights

The first-level criteria weights are calculated by Equations (2)–(4). Matrices of pairwise comparisons are created based on five experts in the fields of economics, marine energy technology, and the social sciences, using a fuzzy scale from (1,1,1) to (8,9,9). Expert weights are specified as (0.3, 0.3, 0.2, 0.1, 0.1). Appendix A shows the fuzzy pairwise comparison matrix generated by the five experts. Table 5 shows the fuzzy values of the first-level criteria weights. Through defuzzification and normalization, the weights calculated by Equation (6) are (0.4896, 0.1779, 0.1286, 0.2038).

Table 5. Fuzzy values of the first-level criteria weights.

w_1	(0.2772, 0.4966, 0.8208)
w_2	(0.1102, 0.1740, 0.3148)
w_3	(0.0735, 0.1305, 0.2146)
w_4	(0.1217, 0.1990, 0.3663)

The calculation results show that resource criteria account for almost 50% of the weight. It means resource criteria are the most important and should be considered more in the site-selection process. Resource criteria have always been the most important criteria in decision making for renewable energy power plant site selection [23,25,28,29]. The weight of social/environmental criteria is the second largest at 20.38%, indicating that the external conditions of social/environmental criteria can restrict or promote site selection to a certain extent. The weights of natural criteria and economic criteria are 17.79% and 12.86%, respectively, indicating slightly less importance.

(2) Determination of second-level criteria weights

The second-level criteria weights are calculated by Equations (7)–(10). Figure 7 shows the calculation results. From the calculation results, it can be seen that the weight of WPD under resource criteria is much larger than that of WH, indicating that WPD has a greater impact on site selection. The weight of PS under economic criteria accounts for 70.69%, indicating its high importance among economic criteria.

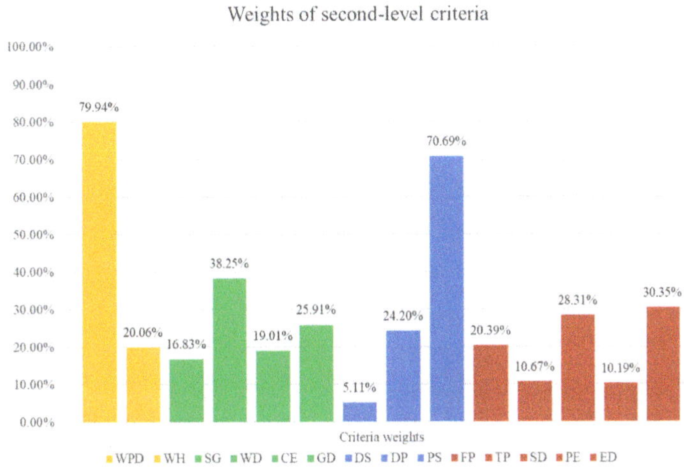

Figure 7. Weights of second-level criteria.

(3) Determination of combined criteria weights

Figure 8 shows the combined weights based on Equation (11). According to the calculation results, the weight of WPD is the largest at 39.14%. As a resource criterion, WPD plays a vital role in the process of site selection. The criteria weights of WD and population served are close to 10%, indicating that these two criteria also have a relatively large impact on site selection. At the same time, the weights of other criteria are relatively small, and the impact on overall decision making is relatively small, but their role in the process of site selection should not be ignored.

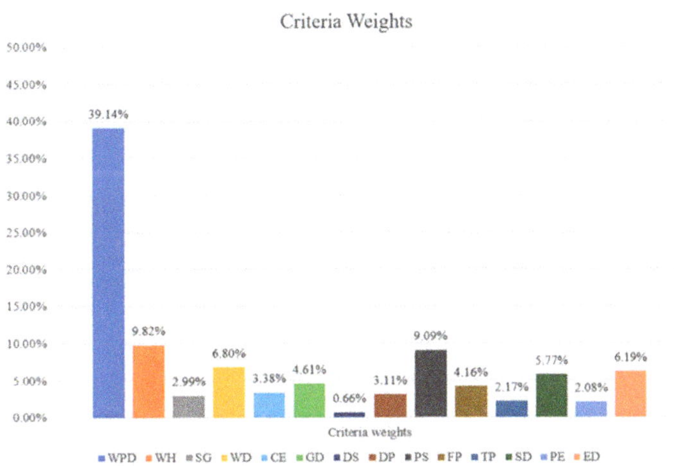

Figure 8. Criteria weights.

4.5. Evaluation of Feasible Islands

The 13 identified inhabited islands of Shandong Province are evaluated and ranked using TOPSIS-GRA to determine the precedence sequences for development. Table 6 shows the final results and rankings of the 13 islands, obtained on the basis of Section 3.4.

Based on the complete assessment results obtained by the proposed decision framework, the top five optimal islands are Daguan, South Changshan, Xiaoguan, Zhucha, and

Zhaitang, respectively. Daguan is found to be the best site for establishing a wave power plant owing to its optimal wave energy conditions and good other features. The National Ocean Technology Center established a hybrid solar–wind–wave independent power system on Daguan in 2010 [89]. To some extent, this also shows that the resources and social environmental conditions of Daguan are suitable for wave energy development.

Table 6. Ranking of site alternatives.

Rank	d_i^+	d_i^-	r_i^+	r_i^-	Z_i	Alternatives	Island
1	0.7638	1.0000	1.0000	0.9409	0.5399	A11	Daguan
2	0.7596	1.0000	1.0000	0.9779	0.5351	A1	South Changshan
3	0.7866	0.8272	0.9727	0.9560	0.5081	A12	Xiaoguan
4	0.7809	0.6613	0.9820	0.8871	0.4963	A9	Zhucha
5	0.8273	0.5627	0.9852	0.9256	0.4689	A13	Zhaitang
6	0.9291	0.4041	0.9554	0.9696	0.4173	A7	Nanhuang
7	0.9278	0.3442	0.9495	0.9247	0.4112	A2	North Changshan
8	0.9859	0.3803	0.9553	0.9738	0.4053	A8	East Little Qingdao
9	0.9688	0.3311	0.9313	0.9982	0.3909	A10	Muguan
10	0.9989	0.3096	0.9409	0.9780	0.3875	A3	Temple
11	0.9729	0.2994	0.9387	0.9940	0.3863	A4	Daheishan
12	0.9732	0.2990	0.9373	0.9921	0.3861	A6	Jiming
13	1.0000	0.2905	0.9401	1.0000	0.3809	A5	Xiaoheishan

South Changshan ranks second. It has the largest population served and the best social and environmental conditions. Given the large number of residents, the island is in urgent need of developing wave power plants to alleviate power pressures. Xiaoguan has the second-largest WPD and the smallest WD, leading it to the third place. Ranking fourth, Zhucha has the smallest distance to ports, and it performs relatively well for WPD and wave height. Finally, Zhaitang ranks fifth, performing best for distance to the shore and performing relatively well for WPD and population served.

4.6. Sensitivity Analysis

In decision making, various uncertain issues affect decision accuracy, such as the different risk attitudes of DMs, different weights of evaluation criteria, and different MCDM methods for the final ranking. Hence, it is necessary to test the sensitivity of the ranking results.

4.6.1. Varying Expert Weights

A sensitivity analysis based on equal expert weights is performed, as shown in Figure 9. The results obtained from equal expert weights are very similar to the original results. It is worth noting that the top nine islands remain unchanged, and only two islands have changed in development order. Therefore, the ranking results remain stable for variable expert weights.

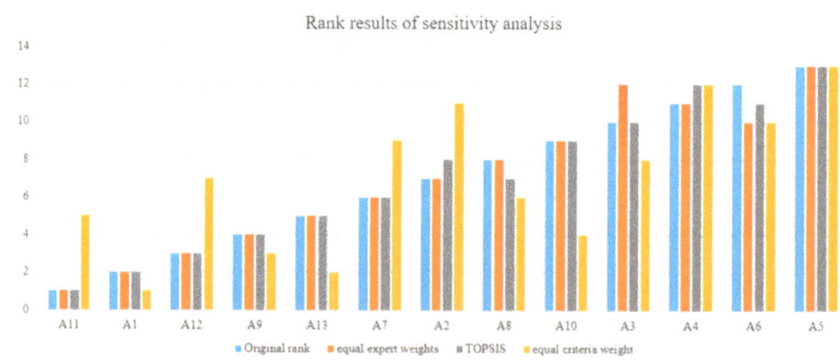

Figure 9. Ranking results of the sensitivity analysis.

4.6.2. Varying Criteria Weights

Because the criteria weights affect the final results, equal criteria weights are set to test its impact on the decision results. Figure 9 shows the final rankings. With the adjustment of the criteria weights, the ranking results change accordingly. The rankings of all alternatives fluctuate within five ranks. A1 performed best in population served; when the criteria weights are equal, it ranks first. A11, A1, A9, and A13 still perform fairly well, ranking among the top five. A5 is still last with equal criteria weights. When criteria weights are equal, the order of islands will inevitably change since resource conditions are the decisive criteria for site selection. A significant reduction in resource condition criteria will inevitably change the ranking results, reflecting the characteristics of sensitivity. Therefore, when the importance of criteria is quite different, it is necessary to find a suitable algorithm to solve the criteria weights.

4.6.3. Varying the Ranking Method

Different MCDM methods have different calculation principles, and the obtained ranking results might also be different. TOPSIS is used to rank islands to test the universality of the results. Figure 9 shows that the ranking of islands is generally stable, and the top six optimal islands remain unchanged. The results under TOPSIS change only four alternatives; A2, A8, A4, and A6 are changed in the development order. This comparative analysis demonstrates the practicability of the proposed model.

5. Conclusions

To address the problems of wave-power-plant site selection for islands in China, this study proposed a two-stage decision framework, including both large- and small-scale site selection, based on a combination of GIS, fuzzy GDM-AHP, entropy method and GRA-TOPSIS. This approach enabled us to identify feasible islands and determine priority order. The main contributions of this study were as follows:

- While the combined weighting method was used to obtain criteria weights, the subjective bias, which was the shortcoming of single subjective weighting method, was largely reduced. The loss of decision information was reduced by employing TFNs to represent the attitudes of experts; in addition, a combination of GDM theory and entropy method made decision making more reliable and reduced the ambiguity in actual problems.
- TOPSIS and GRA were combined to rank island alternatives, considering both position similarity and shape similarity between alternatives. As TOPSIS-GRA used the grey correlation degree, as well as distances from the alternatives to PIS and NIS, to construct a new relative closeness for ranking alternatives, the decision-making accuracy was improved.
- The proposed framework was applied in Shandong Province. A total of 13 feasible inhabited islands were identified for constructing wave power plants. The top five optimal islands were Daguan, South Changshan, Xiaoguan, Zhucha, and Zhaitang, in order. These results could provide a reference for decision-makers to build wave power plants. Sensitivity analysis was employed by varying the expert weights, criteria weights, and ranking methods. The results demonstrated that the proposed framework was effective and feasible.

The proposed methodology framework can be generally applied to other energy sources by changing the criteria system. Future research on wave power plants site selection can be conducted as follows: first, attribute values can be used in the fuzzy environment to improve the precision of the results. Second, while the fuzzy sets in this paper are TFNs, the trapezoidal fuzzy numbers, intuitionistic fuzzy sets and interval hesitant fuzzy sets can be used in subsequent research to improve the flexibility of fuzzy sets in dealing with fuzzy and uncertain problems.

Author Contributions: Conceptualization, M.S. and S.Z.; methodology, M.S.; software, S.Z.; validation, S.Z., J.S. and Z.H.; formal analysis, J.S.; investigation, S.Z. and J.S.; resources, M.S., Z.H. and Z.S.; data curation, Z.S.; writing—original draft preparation, S.Z.; writing—review and editing, M.S. and C.Y.; visualization, C.Y; funding acquisition, M.S. and J.S. All authors have read and agreed to the published version of the manuscript.

Funding: This research was funded by National Natural Science Foundation of China, grant number 51609224; Shandong Provincial Natural Science Foundation of China, grant numbers ZR2020QE297 and Qingdao Postdoctoral Application Research Project.

Institutional Review Board Statement: Not applicable.

Informed Consent Statement: Not applicable.

Data Availability Statement: Not applicable.

Acknowledgments: The research was financially supported by the National Natural Science Foundation of China (Grant No. 51609224), Shandong Provincial Natural Science Foundation of China (ZR2020QE297) and Qingdao Postdoctoral Application Research Project.

Conflicts of Interest: The authors declare no conflict of interest.

Appendix A

Table A1. Fuzzy pairwise comparison matrix of expert 1 (Weight: 0.3).

	Resource Criteria (C_1)	Natural Criteria (C_2)	Economic Criteria (C_3)	Social/Environmental Criteria (C_4)
C_1	(1, 1, 1)	(1, 2, 3)	(3, 4, 5)	(1, 2, 3)
C_2	(1/3, 1/2, 1)	(1, 1, 1)	(1, 1, 2)	(1, 1, 2)
C_3	(1/5, 1/4, 1/3)	(1/2, 1, 1)	(1, 1, 1)	(1/3, 1/2, 1)
C_4	(1/3, 1/2, 1)	(1/2, 1, 1)	(1, 2, 3)	(1, 1, 1)

Table A2. Fuzzy pairwise comparison matrix of expert 2 (Weight: 0.3).

	Resource Criteria (C_1)	Natural Criteria (C_2)	Economic Criteria (C_3)	Social/Environmental Criteria (C_4)
C_1	(1, 1, 1)	(2, 3, 4)	(2, 3, 4)	(2, 3, 4)
C_2	(1/4, 1/3, 1/2)	(1, 1, 1)	(1, 1, 2)	(1/2, 1, 1)
C_3	(1/4, 1/3, 1/2)	(1/2, 1, 1)	(1, 1, 1)	(1/4, 1/3 1)
C_4	(1/4, 1/3, 1/2)	(1, 1, 2)	(2, 3, 4)	(1, 1, 1)

Table A3. Fuzzy pairwise comparison matrix of expert 3 (Weight: 0.2).

	Resource Criteria (C_1)	Natural Criteria (C_2)	Economic Criteria (C_3)	Social/Environmental Criteria (C_4)
C_1	(1, 1, 1)	(2, 3, 4)	(3, 4, 5)	(2, 3, 4)
C_2	(1/4, 1/3, 1/2)	(1, 1, 1)	(1, 2, 3)	(1, 1, 1)
C_3	(1/5, 1/4, 1/3)	(1/3, 1/2, 1)	(1, 1, 1)	(1/2, 1, 1)
C_4	(1/4, 1/3, 1/2)	(1, 1, 1)	(1, 1, 2)	(1, 1, 1)

Table A4. Fuzzy pairwise comparison matrix of expert 4 (Weight: 0.1).

	Resource Criteria (C_1)	Natural Criteria (C_2)	Economic Criteria (C_3)	Social/Environmental Criteria (C_4)
C_1	(1, 1, 1)	(3, 4, 5)	(3, 4, 5)	(1, 2, 3)
C_2	(1/5, 1/4, 1/3)	(1, 1, 1)	(1, 1, 1)	(1/2, 1, 1)
C_3	(1/5, 1/4, 1/3)	(1, 1, 1)	(1, 1, 1)	(1/2, 1, 1)
C_4	(1/3, 1/2, 1)	(1, 1, 2)	(1, 1, 2)	(1, 1, 1)

Table A5. Fuzzy pairwise comparison matrix of expert 5 (Weight: 0.1).

	Resource Criteria (C_1)	Natural Criteria (C_2)	Economic Criteria (C_3)	Social/Environmental Criteria (C_4)
C_1	(1, 1, 1)	(3, 4, 5)	(3, 4, 5)	(3, 4, 5)
C_2	(1/5, 1/4, 1/3)	(1, 1, 1)	(1, 1, 2)	(1/2, 1, 1)
C_3	(1/5, 1/4, 1/3)	(1/2, 1, 1)	(1, 1, 1)	(1/3, 1/2, 1)
C_4	(1/5, 1/4, 1/3)	(1, 1, 2)	(1, 2, 3)	(1, 1, 1)

Table A6. Fuzzy pairwise comparison matrix by GDM.

	Resource Criteria (C_1)	Natural Criteria (C_2)	Economic Criteria (C_3)	Social/Environmental Criteria (C_4)
C_1	(1.0000, 1.0000, 1.0000)	(1.7617, 2.8137, 3.8367)	(2.6564, 3.6693, 4.6762)	(1.5784, 2.6253, 3.6457)
C_2	(0.2606, 0.3554, 0.5676)	(1.0000, 1.0000, 1.0000)	(1.0000, 1.1487, 2.0237)	(0.7071, 1.0000, 1.2311)
C_3	(0.2138, 0.2725, 0.3764)	(0.4941, 0.8706, 1.0000)	(1.0000, 1.0000, 1.0000)	(0.3453, 0.5451, 0.8123)
C_4	(0.2743, 0.3809, 0.6335)	(0.8123, 1.0000, 1.4142)	(1.2311, 1.8346, 2.8958)	(1.0000, 1.0000, 1.0000)

References

1. Aleixandre-Tudó, J.; Castelló-Cogollos, L.; Aleixandre, J.L.; Aleixandre-Benavent, R. Renewable energies: Worldwide trends in research, funding and international collaboration. *Renew. Energy* **2019**, *139*, 268–278. [CrossRef]
2. Arens, M.; Åhman, M.; Vogl, V. Which countries are prepared to green their coal-based steel industry with electricity?—Reviewing climate and energy policy as well as the implementation of renewable electricity. *Renew. Sustain. Energy Rev.* **2021**, *143*, 110938. [CrossRef]
3. Rahman, S.M.; Miah, M.D. The impact of sources of energy production on globalization: Evidence from panel data analysis. *Renew. Sustain. Energy Rev.* **2017**, *74*, 110–115. [CrossRef]
4. Dolman, S.; Simmonds, M. Towards best environmental practice for cetacean conservation in developing Scotland's marine renewable energy. *Mar. Policy* **2010**, *34*, 1021–1027. [CrossRef]
5. Quero García, P.; García Sanabria, J.; Chica Ruiz, J. Marine renewable energy and maritime spatial planning in Spain: Main challenges and recommendations. *Mar. Policy* **2021**, *127*, 104444. [CrossRef]
6. Liang, B.; Shao, Z.; Wu, Y.; Shi, H.; Liu, Z. Numerical study to estimate the wave energy under Wave-Current Interaction in the Qingdao coast, China. *Renew. Energy* **2021**, *101*, 845–855.
7. Liang, B.; Shao, Z.; Wu, G.; Shao, M.; Sun, J. New equations of wave energy assessment accounting for the water depth. *Appl. Energy* **2017**, *188*, 130–139. [CrossRef]
8. International Energy Agency. World Energy Balances 2018 Overview. 2018. Available online: https://www.iea.org/events/statistics-world-energy-balances-2018-overview (accessed on 1 March 2022).
9. Kim, S.; Lee, H.; Kim, H.; Jang, D.H.; Kim, H.J.; Hur, J.; Cho, Y.S.; Hur, K. Improvement in policy and proactive interconnection procedure for renewable energy expansion in South Korea. *Renew. Sustain. Energy Rev.* **2018**, *98*, 150–162. [CrossRef]
10. Robertson, B.; Dunkle, G.; Gadasi, J.; Garcia-Medina, G.; Yang, Z.Q. Holistic marine energy resource assessments: A wave and offshore wind perspective of metocean conditions. *Renew. Energy* **2021**, *170*, 286–301. [CrossRef]
11. State Oceanic Administration. The 13th Five-Year Plan for National Marine Economic Development. 2017. Available online: https://www.ndrc.gov.cn/fzggw/jgsj/dqs/sjdt/201705/P020190909487471217145.pdf (accessed on 1 March 2022).
12. Javed, M.S.; Ma, T.; Jurasz, J.; Canales, F.A.; Lin, S.Q.; Ahmed, S.; Zhang, Y.J. Economic analysis and optimization of a renewable energy based power supply system with different energy storages for a remote island. *Renew. Energy* **2021**, *164*, 1376–1394. [CrossRef]
13. Blechinger, P.; Cader, C.; Bertheau, P.; Huyskens, H.; Seguin, R.; Breyer, C. Global analysis of the techno-economic potential of renewable energy hybrid systems on small islands. *Energy Policy* **2016**, *98*, 674–687. [CrossRef]
14. Mehrjerdi, H. Modeling and optimization of an island water-energy nexus powered by a hybrid solar-wind renewable system. *Energy* **2020**, *197*, 117217. [CrossRef]
15. Cross, S.; Padfield, D.; Ant-Wuorinen, R.; Phillip, K.; Sanna, S. Benchmarking island power systems: Results, challenges, and solutions for long term sustainability. *Renew. Sustain. Energy Rev.* **2017**, *80*, 1269–1291. [CrossRef]
16. Cabrera, P.; Lund, H.; Carta, J.A. Smart renewable energy penetration strategies on islands: The case of Gran Canaria. *Energy* **2018**, *162*, 421–443. [CrossRef]
17. Sun, Z.; Zhang, H.C.; Liu, X.L.; Ding, J.; Xu, D.L.; Cai, Z.W. Wave energy assessment of the Xisha Group Islands zone for the period 2010–2019. *Energy* **2021**, *220*, 119721. [CrossRef]
18. Sánchez-Lozano, J.M.; García-Cascales, M.S.; Lamata, M.T. Identification and selection of potential sites for onshore wind farms development in region of Murcia Spain. *Energy* **2014**, *73*, 311–324. [CrossRef]
19. Al Garni, H.Z.; Awasthi, A. Solar PV power plant site selection using a GIS-AHP based approach with application in Saudi Arabia. *Appl. Energy* **2017**, *206*, 1225–1240. [CrossRef]

20. Baseer, M.A.; Rehman, S.; Meyer, J.P.; Alam, M.M. GIS-based site suitability analysis for wind farm development in Saudi Arabia. *Energy* **2017**, *141*, 1166–1176. [CrossRef]
21. Giamalaki, M.; Tsoutsos, T. Sustainable siting of solar power installations in Mediterranean using a GIS/AHP approach. *Renew. Energy* **2019**, *141*, 64–75. [CrossRef]
22. Shao, M.; Han, Z.X.; Sun, J.W.; Xiao, C.S.; Zhang, S.L.; Zhao, Y.X. A review of multi-criteria decision making applications for renewable energy site selection. *Renew. Energy* **2020**, *157*, 377–403. [CrossRef]
23. Zografidou, E.; Petridis, K.; Arabatzis, G.; Dey, P. Optimal design of the renewable energy map of Greece using weighted goal-programming and data envelopment analysis. *Comput. Oper. Res.* **2016**, *66*, 313–326. [CrossRef]
24. Cavone, G.; Dotoli, M.; Epicoco, N.; Seatzu, C. Intermodal terminal planning by Petri Nets and Data Envelopment Analysis. *Control Eng. Pract.* **2017**, *69*, 9–22. [CrossRef]
25. Mahmoudi, R.; Emrouznejad, A.; Khosroshahi, H.; Khashei, M.; Rajabi, P. Performance evaluation of thermal power plants considering CO_2 emission: A multistage PCA, clustering, game theory and data envelopment analysis. *J. Clean. Prod.* **2019**, *223*, 641–650. [CrossRef]
26. Wang, C.; Dang, T.; Nguyen, N.; Wang, J. A combined Data Envelopment Analysis (DEA) and Grey Based Multiple Criteria Decision Making (G-MCDM) for solar PV power plants site selection: A case study in Vietnam. *Energy Rep.* **2022**, *8*, 1124–1142. [CrossRef]
27. Pambudi, G.; Nananukul, N. A hierarchical fuzzy data envelopment analysis for wind turbine site selection in Indonesia. *Energy Rep.* **2019**, *5*, 1041–1047. [CrossRef]
28. Tavana, M.; Khalili-Damghani, K.; Santos Arteaga, F.; Mahmoudi, R.; Hafezalkotob, A. Efficiency decomposition and measurement in two-stage fuzzy DEA models using a bargaining game approach. *Comput. Ind. Eng.* **2018**, *118*, 394–408. [CrossRef]
29. Cavone, G.; Dotoli, M.; Epicoco, N.; Morelli, D.; Seatzu, C. Design of Modern Supply Chain Networks Using Fuzzy Bargaining Game and Data Envelopment Analysis. *IEEE Trans. Autom. Sci. Eng.* **2020**, *17*, 1221–1236. [CrossRef]
30. Ghosh, S.; Chakraborty, T.; Saha, S.; Majumder, M.; Pal, M. Development of the location suitability index for wave energy production by ANN and MCDM techniques. *Renew. Sustain. Energy Rev.* **2016**, *59*, 1017–1028. [CrossRef]
31. Abaei, M.M.; Arzaghi, E.; Abbassi, R.; Garaniya, V.; Penesis, I. Developing a novel risk-based methodology for multi-criteria decision making in marine renewable energy applications. *Renew. Energy* **2017**, *102*, 341–348. [CrossRef]
32. Vasileiou, M.; Loukogeorgaki, E.; Vagiona, D.G. GIS-based multi-criteria decision analysis for site selection of hybrid offshore wind and wave energy systems in Greece. *Renew. Sustain. Energy Rev.* **2017**, *73*, 745–757. [CrossRef]
33. Cradden, L.; Kalogeri, C.; Barrios, I.M.; Galanis, G.; Ingram, D.; Kallos, G. Multi-criteria site selection for offshore renewable energy platforms. *Renew. Energy* **2016**, *87*, 791–806. [CrossRef]
34. Shao, M.; Han, Z.X.; Sun, J.W.; Zhang, S.L.; Zhao, Y.X.; Shao, Z.X. A GIS-MCDM Combination for the Site Assessment of Wave Energy Power Station in Qingdao, Eastern China. In Proceedings of the 30st ISOPE Conference, Shanghai, China, 14–19 June 2020; Volume 1, pp. 137–143.
35. Nobre, A.; Pacheco, M.; Jorge, R.; Lopes, M.F.P.; Gato, L.M.C. Geo-spatial multi-criteria analysis for wave energy conversion system deployment. *Renew. Energy* **2009**, *34*, 97–111. [CrossRef]
36. Flocard, F.; Ierodiaconou, D.; Coghlan, I.R. Multi-criteria evaluation of wave energy projects on the south-east Australian coast. *Renew. Energy* **2016**, *99*, 80–94. [CrossRef]
37. Shahabi, H.; Keihanfard, S.; Ahmad, B.B.; Amiri, M.J.T. Evaluating Boolean, AHP and WLC methods for the selection of waste landfill sites using GIS and satellite images. *Environ. Earth Sci.* **2014**, *71*, 4221–4233. [CrossRef]
38. Sánchez-Lozano, J.M.; García-Cascales, M.S.; Lamata, M.T. Evaluation of suitable locations for the installation of solar thermoelectric power plants. *Comput. Ind. Eng.* **2015**, *87*, 343–355. [CrossRef]
39. Ayodele, T.R.; Ogunjuyigbe, A.S.O.; Odigie, O.; Munda, J.L. A multi-criteria GIS based model for wind farm site selection using interval type-2 fuzzy analytic hierarchy process: The case study of Nigeria. *Appl. Energy* **2018**, *228*, 1853–1869. [CrossRef]
40. Ayodele, T.R.; Ogunjuyigbe, A.S.O.; Odigie, O.; Jimoh, A.A. On the most suitable sites for wind farm development in Nigeria. *Data Brief* **2018**, *19*, 29–41. [CrossRef]
41. Wu, Y.; Zhang, J.; Yuan, J.; Geng, S.; Zhang, H. Study of decision framework of offshore wind power station site selection based on ELECTRE-III under intuitionistic fuzzy environment: A case of China. *Energy Convers. Manag.* **2016**, *113*, 66–81. [CrossRef]
42. Sánchez-Lozano, J.M.; Teruel-Solano, J.; Soto-Elvira, P.L.; García-Cascales, M.S. Geographical Information Systems (GIS) and Multi-Criteria Decision Making (MCDM) methods for the evaluation of solar farms locations: Case study in south-eastern Spain. *Renew. Sustain. Energy Rev.* **2013**, *24*, 544–556.
43. Fang, H.; Li, J.; Song, W. Sustainable site selection for photovoltaic power plant: An integrated approach based on prospect theory. *Energy Convers. Manag.* **2018**, *174*, 755–768. [CrossRef]
44. Sánchez-Lozano, J.M.; García-Cascales, M.S.; Lamata, M.T. Comparative TOPSIS-ELECTRE TRI methods for optimal sites for photovoltaic solar farms. Case study in Spain. *J. Clean. Prod.* **2016**, *127*, 387–398. [CrossRef]
45. Rezaei, M.; Mostafaeipour, A.; Qolipour, M.; Tavakkoli-Moghaddam, R. Investigation of the optimal location design of a hybrid wind-solar plant: A case study. *Int. J. Hydrogen Energy* **2018**, *43*, 100–114. [CrossRef]
46. Sánchez-Lozano, J.M.; García-Cascales, M.S.; Lamata, M.T. GIS-based onshore wind farm site selection using Fuzzy Multi-Criteria Decision Making methods. Evaluating the case of Southeastern Spain. *Appl. Energy* **2016**, *171*, 86–102. [CrossRef]

47. Sindhu, S.; Nehra, V.; Luthra, S. Investigation of feasibility study of solar farms deployment using hybrid AHP-TOPSIS analysis: Case study of India. *Renew. Sustain. Energy Rev.* **2017**, *73*, 496–511. [CrossRef]
48. Lu, W.; Liu, J.; Xiang, X.; Song, W.; Mcilgorm, A. A comparison of marine spatial planning approaches in China: Marine functional zoning and the marine ecological red line. *Mar. Policy* **2015**, *62*, 94–101. [CrossRef]
49. Chaouachi, A.; Covrig, C.F.; Ardelean, M. Multi-criteria selection of offshore wind farms: Case study for the Baltic States. *Energy Policy* **2017**, *103*, 179–192. [CrossRef]
50. Lin, Y.F.; Dong, S.; Wang, Z.F.; Guedes Soares, C. Wave energy assessment in the China adjacent seas on the basis of a 20-year SWAN simulation with unstructured grids. *Renew. Energy* **2019**, *136*, 275–295. [CrossRef]
51. State Oceanic Administration. *China Offshore Ocean Atlas (Submarine Topography and Landforms)*, 1st ed.; Ocean Press: Beijing, China, 2016; pp. 6–28.
52. Compilation Committee of "Chinese Islands History". *China Island History (Shandong Volume)*, 1st ed.; Ocean Press: Beijing, China, 2013; pp. 27–368.
53. Rangel-Buitrago, N.; Neal, W.J.; de Jonge, V.N. Risk assessment as tool for coastal erosion management. *Ocean Coast. Manag.* **2020**, *186*, 105099. [CrossRef]
54. Wu, B.; Yip, T.L.; Xie, L.; Wang, Y. A fuzzy-MADM based approach for site selection of offshore wind farm in busy waterways in China. *Ocean Eng.* **2018**, *168*, 121–132. [CrossRef]
55. Simão, A.; Densham, P.J.; Haklay, M. Web-based GIS for collaborative planning and public participation: An application to the strategic planning of wind farm sites. *J. Environ. Manag.* **2009**, *90*, 2027–2040. [CrossRef]
56. Wu, Y.; Geng, S. Multi-criteria decision making on selection of solar-wind hybrid power station location: A case of China. *Energy Convers. Manag.* **2014**, *81*, 527–533.
57. Lee, A.H.I.; Chen, H.H.; Kang, H.Y. Multi-criteria decision making on strategic selection of wind farms. *Renew. Energy* **2009**, *34*, 120–126. [CrossRef]
58. Chen, H.H.; Kang, H.Y.; Lee, A.H.I. Strategic selection of suitable projects for hybrid solar-wind power generation systems. *Renew. Sustain. Energy Rev.* **2010**, *14*, 413–421. [CrossRef]
59. Maleki, A.; Pourfayaz, F.; Hafeznia, H.; Rosen, M.A. A novel framework for optimal photovoltaic size and location in remote areas using a hybrid method: A case study of eastern Iran. *Energy Convers. Manag.* **2017**, *153*, 129–143. [CrossRef]
60. Vafaeipour, M.; Hashemkhani, S.; Hossein, M.; Varzandeh, M. Assessment of regions priority for implementation of solar projects in Iran: New application of a hybrid multi- criteria decision making approach. *Energy Convers. Manag.* **2014**, *86*, 653–663. [CrossRef]
61. Dong, J.; Feng, T.; Yang, Y.; Ma, Y. Macro-site selection of wind/solar hybrid power station based on ELECTRE-II. *Renew. Sustain. Energy Rev.* **2014**, *35*, 194–204.
62. Chen, P.Y. Effects of normalization on the entropy-based TOPSIS method. *Expert Syst. Appl.* **2019**, *136*, 33–41. [CrossRef]
63. Villacreses, G.; Gaona, G.; Martínez-Gómez, J.; Jijón, D.J. Wind farms suitability location using geographical information system (GIS), based on multi-criteria decision making (MCDM) methods: The case of continental Ecuador. *Renew. Energy* **2017**, *109*, 275–286. [CrossRef]
64. Charabi, Y.; Gastli, A. PV site suitability analysis using GIS-based spatial fuzzy multi-criteria evaluation. *Renew. Energy* **2011**, *36*, 2554–2561. [CrossRef]
65. Doorga, J.R.S.; Rughooputh, S.D.D.V.; Boojhawon, R. Multi-criteria GIS-based modelling technique for identifying potential solar farm sites: A case study in Mauritius. *Renew. Energy* **2019**, *133*, 1201–1219. [CrossRef]
66. Gigović, L.; Pamučar, D.; Božanić, D.; Ljubojević, S. Application of the GIS-DANP-MABAC multi-criteria model for selecting the location of wind farms: A case study of Vojvodina, Serbia. *Renew. Energy* **2017**, *103*, 501–521. [CrossRef]
67. Ali, S.; Taweekun, J.; Techato, K.; Waewsak, J.; Gyawali, S. GIS based site suitability assessment for wind and solar farms in Songkhla, Thailand. *Renew. Energy* **2019**, *132*, 1360–1372. [CrossRef]
68. Saaty, T.L. *The Analytic Hierarchy Process*, 1st ed.; McGraw-Hill: New York, NY, USA, 1980; pp. 25–78.
69. Zadeh, L.A. Fuzzy sets. *Inf. Control* **1965**, *8*, 338–353. [CrossRef]
70. Yazdani-Chamzini, A.; Yakhchali, S.H. Tunnel Boring Machine (TBM) selection using fuzzy multicriteria decision making methods. *Tunn. Undergr. Space Technol.* **2012**, *30*, 194–204. [CrossRef]
71. Ertay, T.; Kahveci, A.; Tabanli, R.M. An integrated multi-criteria group decision-making approach to efficient supplier selection and clustering using fuzzy preference relations. *Int. J. Comput. Integr. Manuf.* **2011**, *24*, 1152–1167. [CrossRef]
72. Kar, A.K. A hybrid group decision support system for supplier selection using analytic hierarchy process, fuzzy set theory and neural network. *J. Comput. Sci.* **2015**, *6*, 23–33. [CrossRef]
73. Zimmer, K.; Fröhling, M.; Breun, P.; Schultmann, F. Assessing social risks of global supply chains: A quantitative analytical approach and its application to supplier selection in the German automotive industry. *J. Clean. Prod.* **2017**, *149*, 96–109. [CrossRef]
74. Yan, L.; Claudia, M.; Eckert, C.E. A review of fuzzy AHP methods for decision-making with subjective judgements. *Expert Syst. Appl.* **2020**, *161*, 1137.
75. Zhao, X.; Qi, Q.; Li, R. The establishment and application of fuzzy comprehensive model with weight based on entropy technology for air quality assessment. *Symp. Secur. Detect. Inf. Process* **2010**, *7*, 217–222. [CrossRef]
76. Zhao, D.F.; Li, C.B.; Wang, Q.Q.; Yuan, J.J. Comprehensive evaluation of national electric power development based on cloud model and entropy method and TOPSIS: A case study in 11 countries. *J. Clean. Prod.* **2020**, *277*, 123190. [CrossRef]

77. Hwang, C.L.; Yoon, K. *Multiple Attribute Decision Making: Methods and Applications*, 1st ed.; Springer: New York, NY, USA, 1981; pp. 173–184.
78. Deng, J.L. Introduction to grey system. *J. Grey Syst.* **1989**, *1*, 1–24.
79. Zhang, M.; Li, G.X. Combining TOPSIS and GRA for supplier selection problem with interval numbers. *J. Cent. South Univ.* **2018**, *25*, 1116–1128. [CrossRef]
80. Wang, P.; Zhu, Z.Q.; Wang, Y.H. A novel hybrid MCDM model combining the SAW, TOPSIS and GRA methods based on experimental design. *Inform. Sci.* **2016**, *345*, 27–45. [CrossRef]
81. National Development and Reform Commission. The 13th Five-Year Plan for Renewable Energy Development. 2016. Available online: http://www.nea.gov.cn/135916140_14821175123931n.pdf (accessed on 18 March 2022).
82. Qiu, S.; Liu, K.; Wang, D.; Ye, J.; Liang, F. A comprehensive review of ocean wave energy research and development in China. *Renew. Sustain. Energy Rev.* **2019**, *113*, 109271. [CrossRef]
83. Department of Oceans and Fisheries of Shandong Province. Shandong Province Island Protection Plan. 2013. Available online: http://hyj.shandong.gov.cn/xwzx/sjdt/201312/t20131219_507515.html (accessed on 18 March 2022).
84. Moschos, E.; Manou, G.; Dimitriadis, P.; Afentoulis, V.; Koutsoyiannis, D.; Tsoukala, V.K. Harnessing wind and wave resources for a Hybrid Renewable Energy System in remote islands: A combined stochastic and deterministic approach. *Energy Procedia* **2017**, *125*, 415–424. [CrossRef]
85. Shandong Provincial Government. Regulations on the Management of Bohai Sea Ecological Protection Red Line in Shandong Province. 2013. Available online: http://www.weifang.gov.cn/xxgk/shyj/201812/P020181203518669818345.pdf (accessed on 20 March 2022).
86. Shandong Provincial Government. Regulations on the Management of Yellow Sea Ecological Protection Red Line in Shandong Province. 2016. Available online: http://hyj.yantai.gov.cn/module/download/downfile.jsp?classid=0&filename=1703201647326397572.pdf (accessed on 20 March 2022).
87. Liang, B.; Gao, H.; Shao, Z. Characteristics of global waves based on the third-generation wave model SWAN. *Mar. Struct.* **2019**, *64*, 35–53. [CrossRef]
88. Gao, H.J.; Liang, B.C.; Shao, Z.X. A global climate analysis of wave parameters with a focus on wave period from 1979 to 2018. *Appl. Ocean Res.* **2021**, *111*, 102652. [CrossRef]
89. Wang, X.G.; Guo, Y.; Zhang, Y.F. Island-Based Ocean Renewable Energy Multi-Energy Complementary Power Supply Technology. In Proceedings of the 2016 Chinese Society for Environmental Sciences Annual Conference Proceedings (Volume IV), Haikou, China, 14 October 2016; pp. 923–931.

processes

MDPI

Article

Emerging Floating Photovoltaic System—Case Studies High Dam and Aswan Reservoir in Egypt

Nagananthini Ravichandran [1], Hady H. Fayek [2] and Eugen Rusu [3,*]

1 Independent Researcher, Power Electronics and Drives, Tamil Nadu 624204, India; naganantthini.eee@gmail.com
2 Electromechanics Engineering Department, Faculty of Engineering, Heliopolis University, Cairo 11785, Egypt; hadyhabib@hotmail.com or hady.habib@hu.edu.eg
3 Department of Mechanical Engineering, Faculty of Engineering, 'Dunarea de Jos' University of Galati, 47 Domneasca Street, 800008 Galati, Romania
* Correspondence: eugen.rusu@ugal.ro

Abstract: The world has a target of achieving 100% renewable energy by the end of the century. This paper presents a case study to establish a new floating photovoltaic park (FPV) in Egyptian dams. In Egypt, two hydroelectric dams, namely High Dam and Aswan Reservoir, together produce 2.65 GW in the Upper-Egypt region. The addition of 5 MW FPV for each dam is simulated using the Helioscope software application. A comparison between the performance of the dams with and without adding the FPV is presented in terms of the evaporation rate and total produced energy. A comparison between different types of FPV, namely polycrystalline, thin film and mono-crystalline in the two dams are also presented. The results show that installing FPV in the Egyptian dams will drive the dams to better performance in terms of carbon dioxide reduction, water-saving from reducing evaporation and increasing hydropower generation.

Keywords: water–energy nexus; floating photovoltaic; high dam; Aswan Reservoir

Citation: Ravichandran, N.; Fayek, H.H.; Rusu, E. Emerging Floating Photovoltaic System—Case Studies High Dam and Aswan Reservoir in Egypt. *Processes* **2021**, *9*, 1005. https://doi.org/10.3390/pr9061005

Academic Editor: Masoud Soroush

Received: 8 May 2021
Accepted: 4 June 2021
Published: 6 June 2021

Publisher's Note: MDPI stays neutral with regard to jurisdictional claims in published maps and institutional affiliations.

Copyright: © 2021 by the authors. Licensee MDPI, Basel, Switzerland. This article is an open access article distributed under the terms and conditions of the Creative Commons Attribution (CC BY) license (https://creativecommons.org/licenses/by/4.0/).

1. Introduction

The penetration level of renewable energies worldwide is continually increasing, and by the end of 2020, the net cumulative capacity added by the renewable energy system was 261 GW [1]. Due to the stochastic nature of sustainable energy technologies in terms of variability and reliability, not enough installed capacity of renewables is yet established worldwide to measure the impacts of each renewable energy technology in terms of construction footprints, environmental consequences, socio-economic studies and the reliability of those technologies over the years [2]. With the increase in renewable energy shares in different locations, all the previous impacts will be clearer until the world reaches fully sustainable power generation. Hydropower, wind turbines and photovoltaic technology are the three highest renewable energy technologies in terms of installed capacity. In the last 10 years, the installed capacity of photovoltaic technology has had a noticeable continuous increase and now, the cumulative photovoltaic power generation capacity is 707.5 GW. The solar PV system held the highest newly added renewable energy sources in 2020 with a capacity of 127 GW, which is 14.41% more than the wind energy system. The installed capacity of hydro power plants worldwide is 1150 GW, which represents 16% of the world generated electricity [3]. Particularly in Egypt, the government is focusing on increasing the penetration level of renewable energies to face the continuous increase in electricity demand [4]. Egypt has two hydro power stations with a total installed capacity of 2.65 GW, which was representative of all the renewable energy technologies of the Egyptian grid until the end of the last century. Egypt is rich with solar energy potential; the amount of solar energy incidence per square meter varies between 5 and 8 kWh per day with a duration of 3000–4000 h per year [5]. The Egyptian maximum demand in 2019 reached 32 GW, while the maximum penetration level of renewable energies in Egypt in 2020 was 10% [3].

The Upper-Egypt region has a target of 100% renewable energy operation with its two dams and the world's largest photovoltaic park. The total generation of the region is 4.45 GW, in which 59% of the renewable power is generated from the hydropower plants of High Dam and Aswan Reservoir. However, several factors obstruct the maintenance of the capacity of the hydropower generation source. In 2011, Ethiopia started the construction of the Grand Ethiopian Renaissance Dam (GERD) on the Blue Nile in a place named Guba, approximately 60 km from Sudan. GERD will badly affect the High Dam (HD) and Aswan Reservoir (AR) in Egypt in terms of hydropower generation and the Nile water level. The hydropower is projected to decrease by 20–30% [6]. The Nile water level is anticipated to decrease by 0.4–0.75 m [7]. Additionally, the irradiation periods in the location of the reservoir increases the evaporation rate drastically, which results in poor power generation during high power demand periods. Therefore, to maintain the level of the reservoir by mitigating the evaporation rate, various countries experimented with a potential covering system. This covering system will reduce the action of wind and irradiation on the water surface, thereby reducing the rate of evaporation using monomolecular films, floating devices, suspended shading covers, and wind retarding devices [8,9]. One such covering system using PV panels can effectively shadow the reservoir from sunlight while generating power. This newly emerged solar technology, named a floating PV system, gained widespread global attention due to higher efficiency than a ground and roof-mounted PV system [10–12]. A water evaporative cooling mechanism and lower soiling loss improves the PV efficiency up to 30% according to a study in Indonesia, which has a climate similar to that of Egypt [13]. Several numerical and experimental analyses on the FPV system in minimizing water loss from evaporation while generating power have been carried out, with impressive results [14–17]. Apart from mounting FPV systems on lakes and ponds with the purpose of drinking and irrigation, it is efficient for installation on reservoirs with hydroelectric power plants [18]. The major advantage in hybridizing FPV on HEPP is the already available grid connection and water saved from evaporation can be effectively directed to hydropower generation [19,20]. Experimental results show the efficient yield from solar and hydropower in implementing this hybrid system [19–23]

Thus, FPV-HEPP integration has high potential, especially in countries with high temperatures such as Egypt, in addressing the water–energy demand. The use of FPV will secure the operation of this region as a 100% sustainable region and importing sustainable energy to other regions in the Egyptian grid.

The main contributions of this paper are:

- The numerical analysis of the FPV system and its benefits if added to Aswan Reservoir and High Dam.
- The comparison between the three types of PV modules, mono-crystalline, polycrystalline, and thin film, to be used in an FPV installed above Aswan Reservoir and High Dam.
- A water-saving effect study when the FPV is added to Aswan Reservoir and High Dam, considering what the hydropower can generate from the saved water in the presence of FPV.
- The environmental impacts of adding an FPV to both Dams.

The present study analyzes the potential of the FPV system upon implementation in the major hydropower reserves of Egypt: High Dam and Aswan Reservoir. The paper is organized as follows: Section 2 is an overview of the FPV, while Section 3 illustrates the need to reduce evaporation in High Dam and Aswan Reservoir. Section 4 illustrates the electrical performance of the two dams, while Section 5 presents the numerical analysis of the FPV in High Dam and Section 6 presents the numerical analysis of the FPV in Aswan Reservoir. Section 7 illustrates the FPV from the view of cost, carbon dioxide emissions and water–energy nexus. Sections 8 and 9 illustrate the results, discussion and conclusion, respectively.

2. Overview of FPV System

Photovoltaics are now the fastest growing source of generating electricity worldwide. There are different methods to install PV systems in different infrastructures. The most famous mounting methods are:

1. Roof-top photovoltaic;
2. Canal photovoltaic;
3. Off-shore photovoltaic;
4. On-shore photovoltaic;
5. Agri-photovoltaics;
6. Floating photovoltaic (FPV).

In 2007, the first FPV system was installed with 20 kW capacity in Japan [24]. At present, they are installed in many locations worldwide, with a total installed capacity of 1.10 GW according to the World Bank report in 2019 [25]. Compared to ground or roof-mounted PV systems, water-mounted PV systems are proved to have very good performance among different countries and conditions. The main advantages of FPV systems are the absence of land acquisition, increased efficiency, less soiling and shading losses, the maintenance of water quality, the prevention of algal blooms, automatic water evaporative cooling and saving water from evaporation [26–31]. The most famous FPV stations worldwide are:

1. Coal mining subsidence area of Huainan City, China with of 60 MW capacity;
2. Yamakura solar power plant in Japan has an installed capacity of 13.70 MW;
3. Pei County in China has an installed capacity of 9.98 MW;
4. Umenoki in Japan has an installed capacity of 7.55 MW;
5. Jining GCL in China has an installed capacity of 6.78 MW;
6. Hirotani Ike Floating Solar Plant in Japan has an installed capacity of 6.80 MW;
7. Queen Elizabeth II Reservoir in the UK has an installed capacity of 6.34 MW;
8. Cheongpung Lake in South Korea has an installed capacity of 3 MW;
9. Otae Province in South Korea has an installed capacity of 3 MW.

There are remarkable benefits from floating PV technologies, which are:

- Inverting underutilized bodies of water into money-making ponds.
- Zero costs when it comes to land acquisition.
- Lower grid connection costs and enhanced accessibility to the existing infrastructure.
- Increased output and operational efficiency as a result of the water's inherent cooling nature when evaporated.
- Reduction in algae growth.
- Improved plant load factor (PLF).

The structural components involved in the FPV system are floating platforms, cables, mooring and an anchoring mechanism, as shown in Figure 1. Floating platforms are the supporting structure with enough buoyancy to float itself with the installed PV array. Pontoon-based floating structures made up of high density polyethylene (HDPE), is the most commonly used floating structure [27]. Apart from that, metal-based structures with steel pipes are also used to support the PV array on water [28,29]. In order to ensure the stability of the floating structure from the action of waves and wind currents, a proper anchoring mechanism should be implemented. Based on the soil type and water level of the reservoir, the floating structure is anchored. Anchors made up of concrete blocks are placed on the bottom of the water body and connected to the edges of the PV array through mooring lines. [27,29]. Based on the position of the FPV array, anchoring on the embankment of the reservoir or on the nearby land area is possible [13,20,21,32]. Mooring lines ensure the flexibility and stability of the FPV system during severe wind and waves. Elastic mooring lines are used to make the FPV structure more flexible during a drift in water level during monsoon and empty reservoir conditions [30]. The power generated from the PV array installed on the floating structure is connected to the substation through underwater cables. Based on the distance of the substation from the FPV array, the inverter station is either placed on the ground or on a separate floating platform near the PV array to reduce the resistive losses [20–22,29,33]. Consequently, to increase the overall efficiency of the system, a cleaning and tracking mechanism can be implemented. In the present study, a pontoon-based floating PV, and a bottom anchoring system with elastic mooring lines is considered.

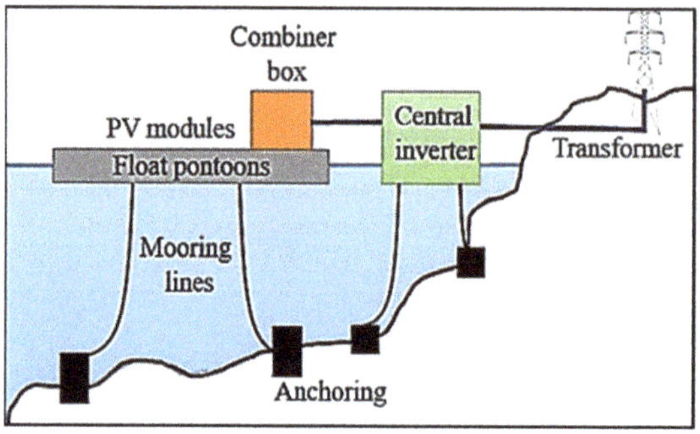

Figure 1. Key design elements for power generation through FPV system.

3. Need for Reducing Evaporation Rate

Egypt is situated in the north-eastern part of Africa that experiences hot desert climatic conditions all year round. The daytime temperatures in the geographical location are extremely hot, with temperature levels greater than 45 °C, and the annual average sunshine hours is more than 3500 h. With very good irradiation levels, Egypt experiences a high number of patent evolutions in the energy generation, transmission and distribution sector, in which solar PV technology from the renewable energy sources play a major role [7,8]. Additionally, these high radiation levels increase the rate of evaporation, which leads to water scarcity conditions.

The entire population of Egypt depends on the Nile River as the major water source to satisfy the irrigation, domestic and industrial water needs. A significant number of dams and canals were constructed in the early 1950s to preserve and consume the water during dry summer seasons. As mentioned above, this clear, bright and sunny year-round season elevates the evaporation rate of open water storage surfaces. Thus, an increase in evaporation rate increases the water loss, and the purpose of the water storage system is affected. Additionally, the drift in the water head level of the reservoir affects the continuous power generation through hydroelectric power plants. Various covering systems are practiced, reducing the impacts of radiation and temperature on a reservoir with a large surface area. Floating PV panels are an effective covering system in mitigating the evaporation rate while generating power. Additionally, with high radiation and sunlight hours of the region, the conventional PV system experiences issues of overheating of the panels by the formation of hotspots which experience the over proportional heating of solar cells in comparison with other cells of the module. These issues are eliminated by the water evaporative cooling of panels by placing PV panels on the water surface; thus, the generated power is more highly efficient than the rooftop and ground-mounted PV systems. Further, the FPV system with an equivalent capacity of the hydroelectric power plant has the potential to provide intermittent operation, ensuring a continuous power supply.

Potential Evapotranspiration Estimation

Initially, in order to calculate the amount of water saved by the FPV as it covers surface, it is necessary to estimate the annual water loss (liters/year) through potential evapotranspiration (PET) from the reservoir [30]. Many conventional methodologies are in practice to estimate the PET at different times of year based on the geographical conditions of the location [30]. The calculation needs the data of radiation with duration, temperature (surface, air, wet-bulb and dew point), wind velocity, latitude, and latent heat of vaporization, humidity, pressure and albedo over a period of time. Evaporation estimation techniques are mainly categorized into pan evaporation, the water budget method, the mass transfer method, and the water and energy balance method [30,33].

The Penman–Monteith method of estimating the evaporation rate is the most commonly used methodology, as it is highly recommended for its accuracy. The calculation needs the daily meteorological data of temperature, relative humidity, wind speed and irradiation incident on the horizontal surface with the geographical information including latitude and altitude above the sea level of the particular location. This meteorological data for Aswan High Dam and Aswan Reservoir for the period of 10 years (2010–2019) were obtained from the NASA website to calculate the annual water loss through PET [18].

These two dams also have a specific installed capacity and currently operate as the source of hydroelectric power projects. Since water is the direct source of power generation in HEPP, the rate of evaporation directly affects the power production, which is the major reason for selecting these dams for the location of the FPV system to balance the water–energy nexus. The evaporation rate is calculated using the Penman method by Equation (1) [33].

$$ET_o = \frac{0.408\,\Delta\,(R_n - G) + \gamma\,\frac{900}{T+273}\,u_2\,(e_s - e_a)}{\Delta + \gamma(1 + 0.34)u_2}\left(\text{mm day}^{-1}\right) \quad (1)$$

where ET_o is the reference evapotranspiration $\left(\text{mm day}^{-1}\right)$, Δ is the slope of the vapor pressure curve $\left(\text{kPa }^{\circ}\text{C}^{-1}\right)$, u_2 is the wind speed at 2-m height $\left(\text{ms}^{-1}\right)$, R_n is the net radiation $\left(\text{MJ m}^{-2}\text{day}^{-1}\right)$. G is the soil heat flux density $\left(\text{MJ m}^{-2}\text{day}^{-1}\right)$, e_s is the saturation vapor pressure (kPa), e_a is the actual pressure (kPa), γ is the psychometric constant $\left(\text{kPa}^{\circ}\text{C}^{-1}\right)$ and T is the mean daily air temperature at 2 m height (°C).

From the calculated results, Figure 2 shows the sum of the monthly potential evaporation rate (mm), which is calculated from 2000 to 2019, that has a higher level of PET half of the year, with July being the peak [34]. Additionally, from the trend line of Figure 3a, it is observed that the cumulative rate of PET is increasing every year and Figure 3b shows the annual water loss of the reservoir through evaporation. This gives a major concern in regard to mitigating the loss of available freshwater resources through evaporation. Over the selected period, Figure 4 shows the month-wise sum of PET values. In the months of June and July, the water loss through evaporation is higher than others, with values reaching more than 200 mm.

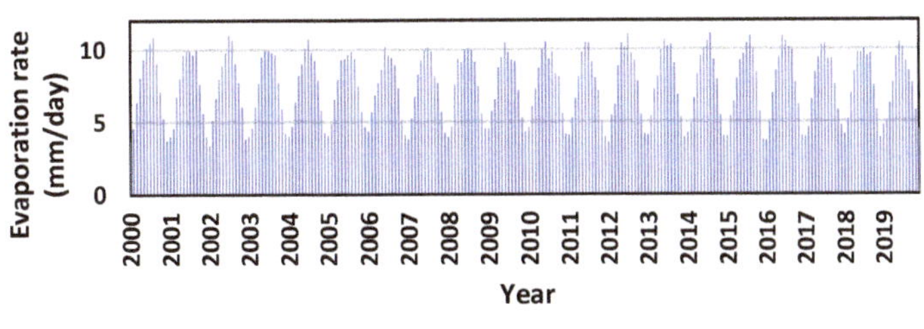

Figure 2. Evaporation rate in Egyptian dams among years in mm/day.

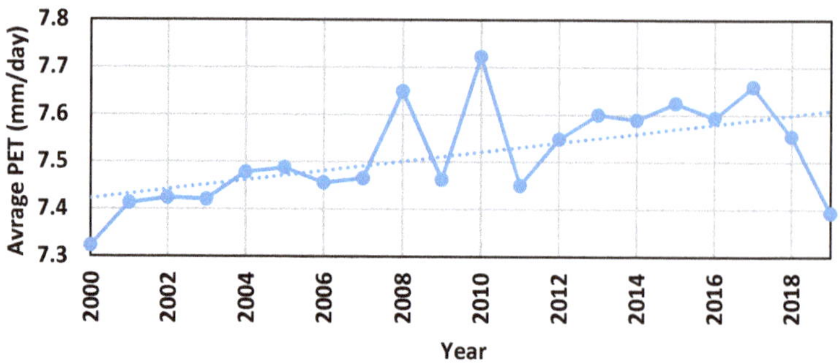

(**a**)

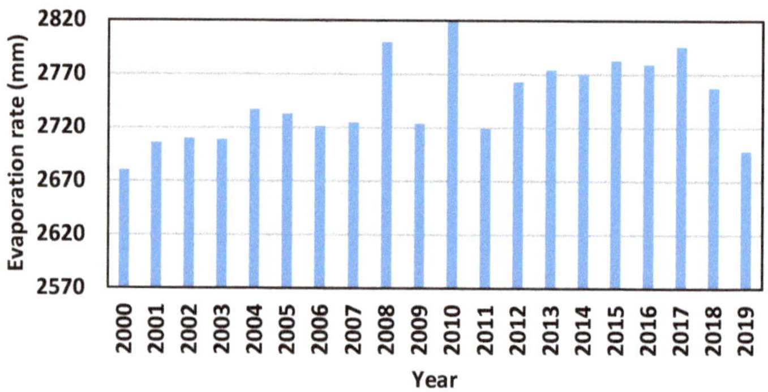

(**b**)

Figure 3. (**a**) Annual average of potential evaporation rate per day (**b**) annual water loss through evaporation (mm).

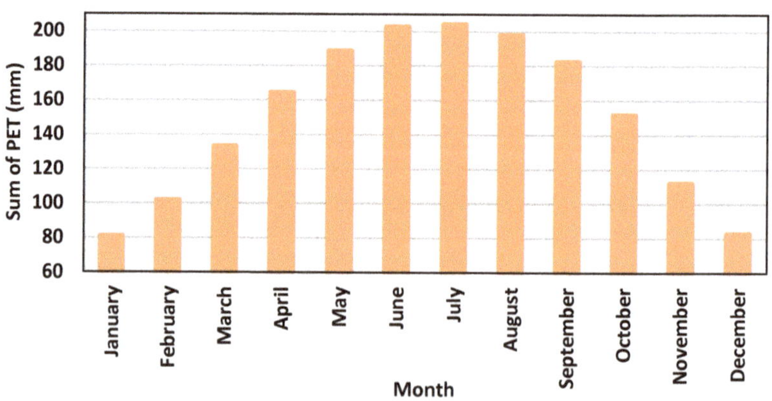

Figure 4. Sum of monthly potential evaporation rate over the period of (2000–2019).

4. Electrical Performance of FPV System in Egyptian Reservoirs

High Dam construction work was started in 1960, finished in 1971 and the operational head storage was reached 1976, and was the world's largest dam at that time. The dam has 12 generating units with equal capacities; the total capacity of the power plant is 2.10 GW, and it was built to ensure water supply during dry seasons, as stated by its engineer. Its secondary and tertiary goals were to increase power production in Egypt and with it, drive industrialization. Now, it represents 5–6% of the total generated electricity of Egypt. Aswan Reservoir was built at the beginning of the 20th century and now it has a full capacity of 0.55 GW; it was built to feed the Aswan governorate with electricity before Egypt established a national power system [4,7].

The electrical performance of the floating photovoltaic system (FPV) is analyzed on the water surface of Aswan High Dam and Aswan Reservoir using monocrystalline, polycrystalline and thin film type PV panels. The model, type, dimensions, standard test conditions (STCs) nominal operating cell temperature (NOCT), and electrical performance of these PV panels used in the study is presented in Appendix A. Further, the floating platform is simulated in two different orientation and tracking mechanisms. To avoid variation in the electrical output of the system, the power output capacities of the three types of PV panels are selected in the same range. For all these analyses, the area covered by the FPV is kept constant in Aswan High Dam, while in Aswan Reservoir, the installed capacity is considered as constant. The floating platform for the present study is assumed to be a pontoon based on the dimension mentioned in the previous experiments [33]. Each pontoon occupies a pair of PV panels tilted at an angle of 25 degrees and placed at a distance of 0.15 m apart from each other. The pontoons are interconnected in a way to form the FPV array, providing 0.50 m distance between each row in an FPV array to provide catwalks. These catwalks are necessary for maintenance and the water evaporative cooling of the PV panels. The edges of the complete floating desk are coupled to the anchors, which are either placed on the embankment of the reservoir or in the bottom of the water in the form of concrete blocks surface through mooring lines [27,29]. Previous studies have focused in detail on this developing solar technology [27,33–35]. The present study will focus more on hybrid power production and its associated environmental and economic benefits.

5. High Dam

The reservoir area of Aswan High Dam receives an average annual global horizontal irradiance (GHI) of 2301.50 kWh/m^2, and the plane of array (POA) irradiance differs according to the type of panel used. Figure 5 shows the FPV location on the Aswan High Dam that is decided based on the ease of grid connection and undulated water surface for effective mooring. In this study, 50,000 m^2 of the total reservoir area is covered by the floating platform to install PV panels. The experiment to analyze the variation in the power generation capacity for portrait and landscape orientation in the presence and absence of a tracking mechanism is conducted and listed in Tables 1 and 2.

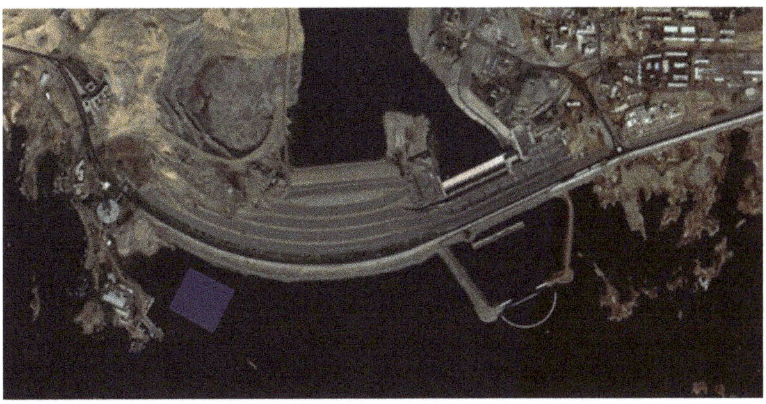

Figure 5. Location FPV system on Aswan High Dam covering 50,000 m^2.

Table 1. Electrical performance of FPV system for various panel types and orientation of High Dam (constant area 50,000 m^2).

S.No	Mounting	Type of PV Panel	Orientation	POA Irradiance (kWh/m^2)	PV Power (MWp)	PV Energy (GWh/year)	PR (%)	Specific Yield (kWhp/kWp)
1.	Fixed Mount (β = 25°)	Poly crystalline	Portrait	2491.1	5.18	9.77	75.7	1884.8
			Landscape	2491.1	4.75	9.34	78.8	1963.2
		Mono crystalline	Portrait	2491.1	4.90	9.08	74.3	1850.5
			Landscape	2491.1	4.50	8.72	77.8	1937.1
		Thin Film	Portrait	2494.9	6.30	11.93	75.8	1892.0
			Landscape	2494.9	5.87	11.71	80.0	1995.9
2.	Single-axis tracking	Poly crystalline	Portrait	2616.6	4.94	10.28	79.6	2081.5
			Landscape	2732.2	4.55	9.87	79.3	2166.4
		Mono crystalline	Portrait	2612.1	4.67	9.58	78.5	2049.5
			Landscape	2727.6	4.31	9.18	78.1	2130.5
		Thin Film	Portrait	2582.8	6.00	12.70	82.0	2118.0
			Landscape	2683.8	5.60	12.30	81.8	2196.1

Table 2. Electrical performance of FPV system for various panel types and orientation of High Dam (constant area 50,000 m^2).

S.No	Mounting	Type of PV Panel	Orientation	Modules	TOF (%)	SAF (%)	Operating Temperature (°C)	Inverter AC Nameplate (MW)
1.	Fixed Mount (β = 25°)	Poly crystalline	Portrait	21,602	99.7	93.7	41.2	4.16
			Landscape	19,812	99.7	97.8	41.7	3.83
		Mono crystalline	Portrait	20,872	99.7	93.4	41.2	3.95
			Landscape	19,148	99.7	97.8	41.7	3.61
		Thin Film	Portrait	15,010	99.9	91.4	41.0	5.05
			Landscape	13,973	99.9	96.5	41.6	4.72
2.	Single-axis tracking	Poly crystalline	Portrait	18,339	104.7	98.6	42.5	3.97
			Landscape	18,975	107.4	98.4	43.0	3.66
		Mono crystalline	Portrait	19,890	104.6	98.7	42.5	3.75
			Landscape	18,339	109.2	98.3	43.0	3.46
		Thin Film	Portrait	14,279	103.4	98.8	42.3	4.81
			Landscape	13,339	107.4	98.4	42.8	4.50

5.1. FPV with Fixed Tracking System

For a fixed type tracking system, the tilt angle (β) of the PV panel is analyzed from 0 to 89 degrees and it is found that the maximum plane of array (POA) irradiation is obtained at an angle of 25 degrees. At this angle, the obtained POA irradiance is 2491.11 kWh/m^2 for crystalline type PV panels, and 2494.91 kWh/m^2 for thin film PV panels.

5.1.1. Crystalline PV Panels

By covering the water surface area of 50,000 m^2 with 21,602 polycrystalline modules in portrait orientation, results from the installed capacity are obtained as 5.18 MW with an annual yield of 9.72 GWh. Compared to portrait orientation, placing the PV panels in landscape orientation in the same area results in a decrease in the installed capacity and the energy yield of the FPV system by 9.05% and 4.14%, respectively, with a decrease in the number of modules and inverter requirement. For the given nameplate DC capacity, the performance of the FPV system is predicted by the performance ratio (*PR*), which is calculated using Equation (2) [16].

$$PR = \frac{E_G}{P_{OUT} \times \frac{I_{POA}}{I_{STC}}} \tag{2}$$

where, E_G represents the total energy supplied to the grid (kWh), P_{OUT} is the total power output from the FPV system (kW), I_{POA} is the plane of array irradiance (kW/m^2), and I_{STC} is the irradiance at standard test conditions (I_{STC} = 1000 W/m^2).

Similarly, the specific yield capacity is the ratio of the amount of energy generated to the nameplate DC capacity of the FPV system. These two values are found to be lower in the portrait orientation than the landscape-oriented panels. Meanwhile, using monocrystalline modules the nameplate DC capacity is obtained as 4.75 MW with 20,872 panels while placing them vertically and 19,812 panels in the horizontal direction. The total FPV system yields 9.08 GWh per year and 8.72 GWh per year in the vertical and horizontal direction, respectively. The energy yield, PR and specific yield of the system using monocrystalline panels is lesser than polycrystalline panels.

5.1.2. Thin Film PV Panel

The electrical and mechanical properties of the thin film solar structure are different from the mono- and poly-crystalline solar modules [33]. This change in properties is reflected in its performance while placing it on the reservoirs in the present study. The POA irradiance of the FPV array consisting of thin film solar cells is 2494.9 kW/m^2. In vertically oriented thin film cells, the installed capacity of the entire array is 22% and 29% higher than poly and mono crystalline panels, respectively. Similarly, compared to their counterparts, the energy yield is increased by 22 and 34%. Eventually, the performance ratio and the specific yield capacity of the system with thin film PV modules experience a higher rise than the crystalline panels in both orientations.

5.2. FPV with Single Axis Tracking System

Single- and dual-axis tracking systems are used to maximize the power output by focusing the PV panel normally to the direction of the irradiation in its availability. The dual-axis type of tracking system is slightly difficult to implement in FPV systems, which is still under analysis. In this study, single-axis tracking or east–west tracking of the FPV system is analyzed for crystalline and thin film type of PV panels placed horizontally and vertically. In the fixed tracking system, the modules are south-oriented, while in single axis tracking system the azimuth angle of the modules is changed to feasibly rotate the panels east to west. This change in the position of the PV panels decreases the number of panels occupied in the area and eventually, the installed capacity of the system. However, with the decreased installed capacity, it is observed that the output of the system in all six cases is considerably increased while implementing the tracking mechanism. Figure 6 shows the comparison POA irradiance level on the system with and without the tracking mechanism with reference to the global horizontal irradiation (GHI) levels in the reservoir. The irradiance levels are the data obtained from the polycrystalline module placed in portrait orientation.

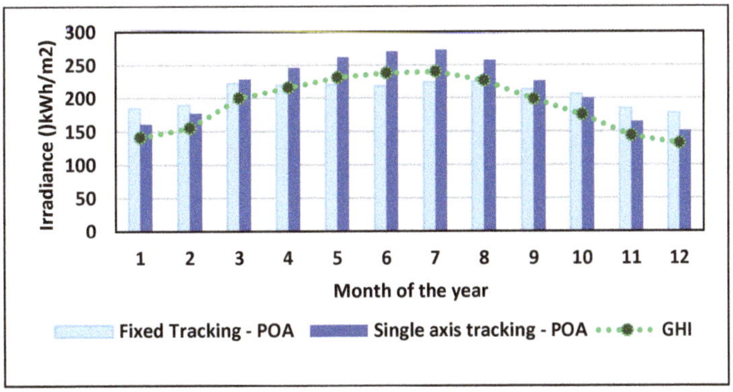

Figure 6. Plane of array (POA) irradiance variation with and without tracking system.

5.3. Losses and Irradiation Utilization Factor

Reducing losses in the PV system is a significant challenge to address. In the FPV system, the soiling losses due to dust particles ranges from 1% to 3%, while in the land-based PV system it is 2% to 5%. Additionally, the optimum temperature of the PV panel is maintained by the automatic water evaporative cooling mechanism, while a land-based PV system requires separate water or air-based cooling systems. In the present study, with both types of orientation as listed in Table 3, the temperature loss is less slightly lesser in polycrystalline PV modules.

Table 3. Losses in the FPV system.

S.No	Mounting	Type of PV Panel	Orientation	Inverter Loss (%)	Shading Loss (%)	Irradiance Loss (%)	Temperature Loss (%)
1.	Fixed Mount (β = 25°)	Poly crystalline	Portrait	1.9	6.3	0.2	10.2
			Landscape	2.0	2.2	0.2	10.2
		Mono crystalline	Portrait	2.5	6.6	0.2	11.2
			Landscape	2.3	2.2	0.1	11.3
		Thin Film	Portrait	2.0	8.6	0.7	8.4
			Landscape	2.1	3.5	0.6	8.5
2.	Single-axis tracking	Poly crystalline	Portrait	2.0	1.4	0.1	10.7
			Landscape	1.9	1.7	0.0	10.8
		Mono crystalline	Portrait	2.4	1.3	0.1	11.7
			Landscape	2.5	1.7	0.0	11.8
		Thin Film	Portrait	2.1	1.2	0.5	8.8
			Landscape	2.1	1.6	0.4	8.9

Eventually, shading losses due to near shading objects, including trees and buildings, and far shading objects including mountains and embankment of the reservoir can highly impact the total DC output of the PV system. In the present study, it is found to be higher in the portrait mode of panel arrangement than in the landscape condition. The increase in the shading losses of about 34% while placing the PV panels vertically is mainly due to shades from the top end of front row panels on the bottom edges of the subsequent row of panels. This increase in the shading loss is the key reason for the reduction in the tilt orientation factor (TOF) and solar access factor (SAF) [35].

6. Aswan Reservoir

A similar but slightly different numerical analysis of the floating PV system is performed on covering Aswan Reservoir. Figure 7 shows the FPV location on Aswan Reservoir that is decided based on the ease of grid connection and undulated water surface for effective mooring. In this study, the system set to have equal DC power capacity in portrait as well as landscape orientation with the percentage of area covering the total reservoir varied eventually. The electrical performance of the FPV system on Aswan Reservoir of 5 MW installed capacity with various orientations and tracking mechanism is simulated, and the results are listed in Tables 4 and 5. The annual GHI of Aswan Reservoir is the same as Aswan High Dam, while varied POA irradiance, annual energy generated from the installed FPV capacity with their PR and specific yield is listed. For a fixed FPV system, the optimal tilt angle of the module is kept as 25°, with the floating platform design similar to the aforementioned study.

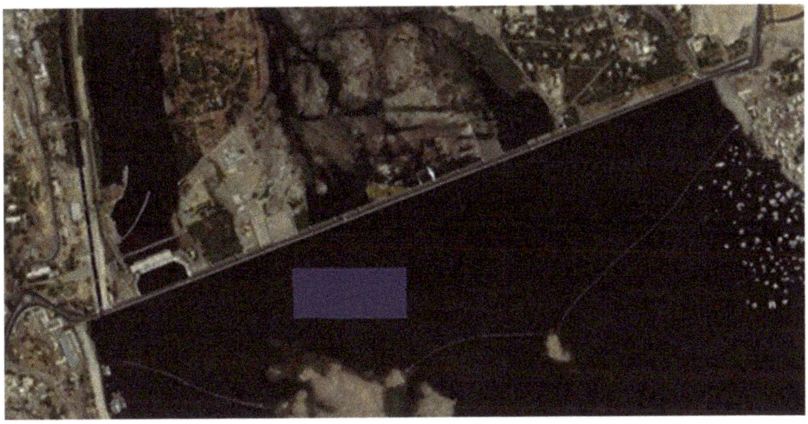

Figure 7. Location of FPV on the Aswan Reservoir of 5MW capacity.

Table 4. Electrical performance of FPV system for various panel types and orientation of Aswan Reservoir—5 MW capacity.

S.No	Mounting	Type of PV Panel	Orientation	Area (m^2)	POA Irradiance (kWh/m^2)	PV Energy (GWh/year)	PR (%)	Specific Yield (kWhp/kWp)
1.	Fixed Mount (β = 25°)	Poly crystalline	Portrait	45,103.62	2491.1	9.569	76.8	1913.8
			Landscape	52,142.54	2491.1	9.934	79.8	1968.8
		Mono crystalline	Portrait	54,677.37	2491.1	9.527	76.5	1905.4
			Landscape	61,594.18	2491.1	9.729	78.1	1945.9
		Thin Film	Portrait	42,288.17	2494.9	9.777	78.4	1955.5
			Landscape	46,725.92	2494.9	10.13	81.2	2026.7
2.	Single-axis tracking	Poly crystalline	Portrait	54,171.21	2678.4	10.65	79.5	2129.5
			Landscape	60,782.67	2800.7	11.09	79.2	2217.5
		Mono crystalline	Portrait	57,322.46	2632.6	10.5	78.3	2093.8
			Landscape	64,078.02	2796.3	10.91	78.0	2182.0
		Thin Film	Portrait	44,342.92	2639.5	10.81	81.9	2161.9
			Landscape	49,455.60	2753.1	1.26	81.8	2251.2

Table 5. Electrical performance of FPV system for various panel types and orientation of Aswan Reservoir—5 MW capacity.

S.No	Mounting	Type of PV Panel	Orientation	No.of Modules	TOF (%)	SAF (%)	Operating Temperature (°C)	Inverter Capacity (MW)	Water Saving (m^3)
1.	Fixed Mount (β = 25°)	Poly crystalline	Portrait	15,625	99.7	94.6	41.3	4.02	123,812.00
			Landscape	15,625	99.7	98.6	41.8	4.02	143,134.25
		Mono crystalline	Portrait	21,276	99.7	94.5	41.6	4.02	150,092.50
			Landscape	21,276	99.7	98.6	41.8	4.02	169,079.53
		Thin Film	Portrait	11,904	99.8	94.5	41.3	4.02	116,083.45
			Landscape	11,904	99.8	98.0	41.8	4.02	128,265.31
2.	Single-axis tracking	Poly crystalline	Portrait	20,833	107.2	98.5	42.8	4.02	148,703.07
			Landscape	20,833	112.1	98.0	43.4	4.02	166,851.90
		Mono crystalline	Portrait	21,276	106.9	98.6	42.8	4.02	157,353.57
			Landscape	21,276	111.9	98.1	43.3	4.02	175,897.84
		Thin Film	Portrait	11,904	105.6	98.6	42.6	4.02	121,723.84
			Landscape	11,904	110.2	98.2	43.1	4.02	135,758.44

In all types of PV panels, the area required to install the PV modules in landscape orientation is higher than the portrait orientation. The reservoir area covered by the FPV system results minimum in portrait orientated thin film PV modules and maximum in landscape-oriented mono-crystalline modules. However, with the equivalent number of modules in both orientation type, the annual energy generated from the horizontally oriented crystalline array results higher. This higher energy yield from the FPV system in landscape orientation irrespective of their panel type is mainly due to the reduced shading losses (See Table 6). The amount of water saved per year by the FPV covering system is calculated by the actual annual evaporation rate when the reservoir is uncovered. The volume of water prevented from evaporation varies with respect to the area covered in each type of FPV system, as given in Table 5. The inverter nameplate is selected by keeping the constant DC to AC ratio as 1.24 and the total number of inverters required, and their associated losses are calculated. The major losses in the FPV system affecting the energy generated are irradiance, shading and temperature loss. Shading losses in the PV modules are comparatively high in the fixed mount tracking system in comparison to the single axis tracking system. The position of the floating platform is fixed on the reservoir with zero or fewer impacts of short and long-distance shadows such as hills, trees, and dam walls. However, the shading losses due to the adjacent panels even at optimum tilt angle are unavoidable. In contrast, the temperature losses in the system with a tracking mechanism experiences higher levels than the system without a tracking mechanism. The continuous exposure to the irradiation from the sunlight in order to enhance the energy yield results in an increase in the operating temperature of the PV cell and the module. However, the water evaporative cooling of the FPV modules boosts the energy generated from the system.

Table 6. Losses in the FPV system in Aswan Reservoir.

Sl.No	Mounting	Type of PV Panel	Orientation	Inverter Loss (%)	Shading Loss (%)	Irradiance Loss (%)	Temperature Loss (%)
1.	Fixed Mount	Poly crystalline	Portrait	2.0	5.4	0.2	9.7
			Landscape	2.0	1.4	0.1	9.8
		Mono crystalline	Portrait	2.4	3.7	0.1	11.2
			Landscape	2.4	1.4	0.1	11.3
		Thin Film	Portrait	2.1	5.5	0.6	8.4
			Landscape	2.1	2.0	0.6	8.5
2.	Single-axis tracking	Poly crystalline	Portrait	2.0	1.5	0.1	10.7
			Landscape	2.0	2.0	0.0	10.8
		Mono crystalline	Portrait	2.5	1.5	0.0	11.8
			Landscape	2.5	1.9	0.0	11.9
		Thin Film	Portrait	2.1	1.4	0.4	8.9
			Landscape	2.1	1.8	0.3	9.0

7. Proposed System Analysis

7.1. Cost Analysis

In the past ten years, solar PV technology experienced a tremendous decrease in the levelized cost of energy (LCOE) every year which in turn increases its total installed capacity as well as the employment technology globally [36]. The cost breakup of the FPV system (see Table 7) is similar to the conventional PV system, with the additional cost requirement of complete transportation and installation of a floating structure which includes pontoons, mooring cables, anchors, screws and tensors.

In the year 2019, the global weighted average LCOE and the total installation cost of solar photovoltaic technology was 0.068 USD/kWh and 995 USD/kW, respectively, which is 13.1% and 17.6% lower compared to its value in the previous year [36]. This fall in the average LCOE as well as in the installation cost is experienced every year due to the newly commissioned utility-scale renewable energy power of solar PV technology. The capacity factor of the solar PV system is defined as the weighted average capacity factor (CF) of the solar PV system, which gained a step ahead every year: it was 18% in the year 2019, which is 23.3% higher in the last ten years range [36].

Table 7. Floating PV system cost component of different PV panels.

S.No	Category	Cost Component	Crystalline (2019 USD/kW)	Thin Film (2019 USD/kW)
1.	Module and inverter	Modules	557.01	360.57
		Inverters	89.8	89.8
2.	BoS	Racking and mounting	108.5	108.5
		Grid connection	53.4	53.4
		Cabling/wiring	47.8	47.8
		Safety and security	22.7	22.7
		Monitoring and control	25.3	25.3
3.	Installation	Mechanical installation	46.1	46.1
		Electrical installation	31.4	31.4
		Inspection	3.3	3.3
4.	Soft costs	Margin	104.0	104.0
		Financing costs	60.6	60.6
		System design	79.5	79.5
		Permitting	58.1	58.1
		Incentive application	18.8	18.8
		Costumer acquisition	14.8	14.8
5.	Floating structure	Pontoons	1583.35	-
		Moorings	178.76	178.76
		Installations	608.61	608.61
		Project Management	93.56	93.56
		Annual O & M	2.26	1.58

7.2. Carbon Dioxide Analysis

The prime consideration in switching to renewables from fossil fuel is to reduce carbon emission. From 1970 to 2017, Egypt experienced 422 Mt of CO_2 emission, which is a 72.95% increase, mainly through coal and oil. In Egypt, the replaced renewable energy generation system avoided 7.14 million tons of CO_2 emission, particularly through hydro and solar technology [36].

To assess the environmental benefits of the FPV system, the equivalent carbon dioxide emissions involved directly and indirectly in the manufacturing, installation and delivery of the entire FPV system is analyzed first. As per the report of climate transparency in the year 2018 on the country-specific electricity factors, the CO_2 emission per kWh of generated power in Africa was 0.9609 kg. Thus, the CO_2 emission avoided for power generating capacity of the present study is calculated and listed in Table 8. The CO_2 savings from the solar energy production from the FPV fixed mount system and FPV system with single-axis tracking in portrait orientation in all three types of PV panels are calculated for a service life of 20 years, considering the specific carbon emission rate to be 0.9609 kg CO_2/kWh for Aswan High Dam and Aswan Reservoir with HEPP of 2.1 GW cumulative power generation capacity.

Table 8. CO_2 saving from fixed mount and single axis tracking FPV systems.

FPV System	FPV Energy (MWh/Year)	CO_2 Saving from Solar Energy Production (Tons)	CO_2 Saving from Reduction in Evaporation (Tons)	Total CO_2 Saving (Tons)
Fixed mount	1934	38,680.00	410.61	39,090.61
Single-axis tracking	2193	43,860.00	410.61	44,270.61

As the FPV system also has the added advantage of the reduction in water loss through evaporation, the reduction in carbon emission due to evaporation mitigation is also calculated, and the results are listed in Table 8. Besides this, the potential net/loss CO_2 by reducing the water evaporation must also be taken into account. The specific energy (kWh/m^3) is a potential energy use indicator of the embodied energy associated with water provision and storage in the irrigation reservoir and it is expressed as a ratio of energy consumption to water volume supplied [37]. The average value of specific electricity capacity (SEC) varies for different water sources [38], for the recycled water stored in the reservoir closer to the field location is an average of 0.5 kWh/m^3 [38]. Therefore, the energy saving from the reduction in water evaporation is straightforwardly calculated for the lifetime of the project. The amount of water saved by covering Aswan High Dam and Aswan Reservoir is 42,731.56 m^3/year. The energy saving from the reduction in water evaporation over the lifetime of 20 years is 427315.6 kWh, with a reduction of 410.61 t CO_2. Taking this into account, the total potential CO_2 saving by the FPV systems with tracking are estimated as 44,270.61 t CO_2 and it is 11.70% higher than the fixed mount FPV system.

7.3. Water–Energy Nexus Analysis

The water–energy nexus regulation is the notorious advantage of FPV, i.e., mitigating the potential water loss through evaporation while generating highly efficient power from the renewable energy source. This saved water without being lost from evaporation can be effectively used for the purpose of hydroelectric generation, irrigation or drinking. By directing the amount of water saved from evaporation for hydropower generation, the FPV covering system acts as a virtual battery and increases the energy yield from the hydro plants of the installed reservoir [18,30,33]. Thus, the cumulative renewable energy generation of the nation increases.

The installed hydropower generation capacity of the Aswan High Dam is 2100 MW, which produces 10,042 GWh annually. In the present study, by covering the 0.5 km^2 area of Aswan High Dam water surface, the annual water saving is about 137,252.86 m^3, i.e., 0.1 million cubic meter (MCM) that increases by 63.56 MWh of hydropower production per year. Similarly, Aswan Reservoir has an installed hydropower plant of 592 MW capacity, and the average water saving by the FPV system increases by 21.76 MWh of hydropower annually.

8. Discussion on the Limitations of the Study

The present study analyzes the performance of the FPV system in reducing evaporation and hybrid power generation upon implementation in the HEPP reservoir. Being the first-ever study to analyze the FPV system in Egypt, the installed capacity of the system is limited to 5 MW to make it feasible for large scale analysis and real-time implementation. FPV projects covering entire reservoir areas in Agost, Spain and Silver Lake, USA results in a reduction in evaporation rate by 75% and 90%, respectively [21,39]. However, it is not advisable to cover the entire reservoir to completely eliminate the evaporation rate [40]. This is because the complete shading of the water surface with an FPV covering system affects the water quality and biodiversity [10,23,40]. As a trade-off, considering the water–energy demand, it is advisable to cover less than 40% of the entire surface of the water body [40]. In the present analysis, an average of 5000 m^2 is required for installing an FPV system of 5 MW capacity. In such a case, covering 4% of the HD reservoir area of (5250 km^2) by an FPV system tends to have an equal capacity as the hydroelectric power plant of 2.1 GW and it is possible to provide intermittent operation.

In the present study, the position of the FPV system is considered based on eliminating near and far shadows and area, which tends not to have the complete dry condition. However, proper mooring analysis on the water depth and soil type is essential to ensure the stability of the system. The FPV structure of the present study is considered as a single large structure and the cost of the mooring and anchoring system is calculated accordingly; however, for operational safety and maintenance, the existing system is divided into small capacities. The system is analyzed only for a single-axis tracking mechanism to obtain a high energy yield. This is because dual-axis tracking on an FPV system is still challenging to implement due to the continuous action of mild waves [10]. Besides focusing the PV panel toward a high irradiation point, the tracking mechanism has to endure the persistent disturbances of the floating from the action of waves. In both sensor-based and astronomically calculated tracking types, inappropriate controller designs fail to find the brightest spot in the sky. This hunting condition of the tracking device extracts more power from the motor. Considering this condition, the single-axis tracking mechanism is cost effective and efficient for the floating PV system [41].

The power generation capacity and environmental benefits such as carbon footprint and evaporation mitigation are analyzed and implemented worldwide. The progress of the present study is to mainly focus on the panel interconnection topologies and multilevel inverter connection. Being in the initial stage of development, research on the FPV system related to standards of the structural components used and their impacts on the water quality on large scale implementation is needed. Besides lakes, reservoirs and hydropower plants, the future of the FPV is on the offshore platform [27,28,42]. Thin film PV modules are the growing technology that is suitable for potential FPV implementation in marine regions. However, the harsh marine environment is still slowing down the offshore FPV implementation and experimental studies [10,28].

9. Conclusions

The paper shows the possibilities of establishing FPV in Egyptian dams in terms of High Dam and Aswan Reservoir. The study shows that installing FPV will lead to an increase in the power generation of the hybrid hydro PV system to an annual increase of up to 11.9 GWh for High Dam and 11.3 GWh for Aswan Reservoir. The hybrid system will lead to water saving by 0.1 MCM for both dams. Three alternatives are presented to select the type of the FPV in terms of mono-crystalline, polycrystalline, and thin film; the results show that polycrystalline is the best solution in Egypt for FPV. The study also proposed two different scenarios to mount FPV in terms of fixed mounting and single-axis tracking and the results show that single-axis tracking achieved a higher energy rate by 4.96%. Using FPV and a hydro hybrid system achieved carbon dioxide reduction by 44,270.61 tons.

Author Contributions: Conceptualization, H.H.F. and N.R.; methodology, H.H.F.; validation, N.R., H.H.F. and E.R.; formal analysis, N.R.; investigation, H.H.F.; resources, H.H.F.; data curation, N.R.; writing—original draft preparation, H.H.F.; writing—review and editing, H.H.F. and E.R.; visualization, N.R.; supervision, E.R.; project administration, H.H.F.; funding acquisition, E.R. All authors have read and agreed to the published version of the manuscript.

Funding: This work was carried out in the framework of the research project DREAM (Dynamics of the REsources and technological Advance in harvesting Marine renewable energy), supported by the Romanian Executive Agency for Higher Education, Research, Development and Innovation Funding—UEFISCDI, grant number PN-III-P4-ID-PCE-2020–0008.

Institutional Review Board Statement: Not the case this study does not involve humans or animals.

Informed Consent Statement: Not the case this study does not involve humans or animals.

Data Availability Statement: The data that support the findings of this study are available in NASA Prediction of Worldwide Energy Resource at https://power.larc.nasa.gov (accessed on 7 May 2021).

Conflicts of Interest: The authors declare no conflict of interest.

Nomenclature

AC	Alternating current
AR	Aswan Reservoir
CF	Capacity factor
DC	Direct current
FPV	Floating photovoltaic
HD	High Dam
HEPP	Hydroelectric power plant
GERD	Grand Ethiopian Renaissance Dam
GHI	Global Horizontal Irradiance
LCOE	Levelized Cost of Energy
MCM	Million Cubic Meter
Mt	Metric Ton
NOCT	Nominal operating cell temperature
PET	Potential Evapotranspiration
POA	Plane of Array
PR	Performance ratio
PV	photovoltaic
RES	Renewable Energy Sources
SAF	solar access factor
SEC	Specific electricity consumption
STC	Standard test conditions
TOF	tilt orientation factor
β	Tilt angle

Appendix A. Electrical Specifications of PV Panels

Table A1. Panel types and description.

S.No	Parameters	Poly Crystalline	Mono Crystalline	Thin Film
1.	Manufacturer	Canadian Solar	Solar World	First Solar
2.	Model Name	CS6P-240P	SW 235	FS-6420 December 2017
3.	Power output capacity (W)	240	235	420
4.	Dimensions (mm)	1638 × 982 × 40	1001 × 1675 × 31	2009 × 1232 × 49
5.	Weight (kg)	20	21.2	36
6.	Temperature coefficient at maximum power (/°C)	−0.43	−0.45	−0.35
7.	Temperature coefficient open circuit voltage (/°C)	−0.34	−0.33	−0.28
8.	Temperature coefficient short circuit voltage (/°C)	+0.065	+0.042	+0.04
9.	Operating temperature (°C)	−40 to +85	−40 to +90	−40 to +85
10.	Module efficiency (%)	14.61	13.12	17
11.	Power tolerance (W)	±5	±3	±5
12.	Standard Test Conditions (STC)—Irradiance of 1000 W/m^2, Temperature 25 °C & air mass AM1.5			
13.	peak power of PV module (W)	240	235	420
14.	maximum power voltage (V)	29.9	30.3	180.4
15.	maximum power current (A)	8.03	7.77	2.33
16.	open circuit voltage (V)	37	37.5	218.5
17.	short circuit current (A)	8.53	8.19	2.54
18.	Nominal operating cell temperature (NOCT)—Irradiance of 800 W/m^2 and ambient temperature of 20 °C			
19.	Peak power of PV module (W)	170	170.9	317.2
20.	Voltage at maximum power (V)	27.0	27.5	168.7
21.	Current at maximum power (A)	6.18	6.22	1.88
22.	open circuit voltage (V)	33.8	34.0	206.3
23.	Short circuit current (A)	6.76	6.6	2.04

References

1. Global Energy Review in 2020. Available online: https://www.iea.org/reports/global-energy-review-2020/renewables (accessed on 14 April 2021).
2. United Nations. The Impact of Renewable Energy Technologies on Global Energy Efficiency. Available online: https://www.un.org/en/chronicle/article/impact-renewable-energy-technologies-global-energy-efficiency (accessed on 14 April 2021).
3. Hydropower Status Report. Available online: https://www.hydropower.org/resources/status-report (accessed on 14 April 2021).
4. Fayek, H.H.; Abdalla, O.H. Maximization of Renewable Power Generation for Optimal Operation of the Egyptian Grid. In Proceedings of the 2020 IEEE 29th International Symposium on Industrial Electronics (ISIE), Delft, The Netherlands, 17–19 June 2020; pp. 1033–1038. Available online: https://ieeexplore.ieee.org/document/9152450 (accessed on 14 April 2021). [CrossRef]
5. Abdalla, O.H.; Fayek, H.H.; Ghany, A.A. Steady-State and Transient Performances of the Egyptian Grid with Benban Photovoltaic Park. In Proceedings of the Cigre Egypt 2019 Conference, The Future of Electricity Grids—Challenges and Opportunities, Cairo, Egypt, 6–8 March 2019; p. 205. Available online: https://www.researchgate.net/publication/342420850_Steady-State_and_Transient_Performances_of_the_Egyptian_Grid_with_Benban_Photovoltaic_Park (accessed on 14 April 2021).
6. Abdelhaleem, F.S.; Helal, E.Y. Impacts of Grand Ethiopian Renaissance Dam on Different Water Usages in Upper Egypt. *Br. J. Appl. Sci. Technol.* **2015**, *8*, 461–483. [CrossRef]
7. Hamed, K.H. Stochastic Investigation of the GERD-AHD Interaction through First Impoundment and Beyond. In *Grand Ethiopian Renaissance Dam Versus Aswan High Dam*; Negm, A.M., Abdel-Fattah, S., Eds.; The Handbook of Environmental Chemistry; Springer International Publishing AG: Berlin/Heidelberg, Germany, 2018. [CrossRef]
8. Assouline, S.; Narkis, K.; Or, D. Evaporation suppression from water reservoirs: Efficiency considerations of partial covers. *Water Resour. Res.* **2011**, *47*. [CrossRef]
9. Wurbs, R.A.; Ayala, R.A. Reservoir evaporation in Texas, USA. *J. Hydrol.* **2014**, *510*, 1–9. [CrossRef]
10. Acharya, M.; Devraj, S. *Floating Solar Photovoltaic (FSPV): A Third Pillar to Solar PV Sector?* TERI Discussion Paper: Output of the ETC India Project; The Energy and Resources Institute: New Delhi, India, 2019.
11. Sahu, A.; Yadav, N.; Sudhakar, K. Floating photovoltaic power plant: A review. *Renew. Sustain. Energy Rev.* **2016**, *66*, 815–824. [CrossRef]
12. Ranjbaran, P.; Yousefi, H.; Gharehpetian, G.; Astaraei, F.R. A review on floating photovoltaic (FPV) power generation units. *Renew. Sustain. Energy Rev.* **2019**, *110*, 332–347. [CrossRef]
13. Sukarso, A.P.; Kim, K.N. Cooling Effect on the Floating Solar PV: Performance and Economic Analysis on the Case of West Java Province in Indonesia. *Energies* **2020**, *13*, 2126. [CrossRef]
14. Farfan, J.; Breyer, C. Combining floating solar photovoltaic power plants and hydropower reservoirs: A virtual battery of great global potential. *Energy Procedia* **2018**, *155*, 403–411. [CrossRef]
15. Liu, L.; Sun, Q.; Li, H.; Yin, H.; Ren, X.; Wennersten, R. Evaluating the benefits of integrating floating photovoltaic and pumped storage power system. *Energy Convers. Manag.* **2019**, *194*, 173–185. [CrossRef]
16. Nazififard, M.; Taheri, S.M.; Nazififardarani, K. Novel Floating Photovoltaic Cover Systems Generating Electricity and Prevent Evaporative Losses for Agriculture Industry in Iran. In Proceedings of the 3rd International Conference of IEA, Tehran, Iran, 28 February 2017; pp. 1–5.
17. Lee, N.; Grunwald, U.; Rosenlieb, E.; Mirletz, H.; Aznar, A.; Spencer, R.; Cox, S. Hybrid floating solar photo-voltaics-hydropower systems: Benefits and global assessment of technical potential. *Renew. Energy* **2020**, *162*, 1415–1427. [CrossRef]
18. Nagananthini, R.; Nagavinothini, R. Investigation on floating photovoltaic covering system in rural Indian reservoir to minimize evaporation loss. *Int. J. Sustain. Energy* **2021**, 1–25. [CrossRef]
19. Rosa-Clot, M.; Tina, G.M. Integration of PV Floating with Hydroelectric Power Plants (HPPs). In *Floating PV Plants*; Academic Press: Cambridge, MA, USA, 2020; pp. 89–100.
20. Cazzaniga, R.; Rosa-Clot, M.; Rosa-Clot, P.; Tina, G.M. Integration of PV floating with hydroelectric power plants. *Heliyon* **2019**, *5*, e01918. [CrossRef] [PubMed]
21. Mckay, A. Floatovoltaics: Quantifying the Benefits of a Hydro fits of a Hydro of a Hydro ydro f PV floating with hydroel. Bachelor's Theses, Pomona College, Claremont, CA, USA, 2013; p. 74. Available online: https://scholarship.claremont.edu/pomona_theses/74 (accessed on 14 April 2021).
22. Lopes, M.P.C.; Neto, S.D.A.; Branco, D.A.C.; de Freitas, M.A.V.; Fidelis, N.D.S. Water-energy nexus: Floating photovoltaic systems promoting water security and energy generation in the semiarid region of Brazil. *J. Clean. Prod.* **2020**, *273*, 122010. [CrossRef]
23. Gorjian, S.; Sharon, H.; Ebadi, H.; Kant, K.; Scavo, F.B.; Tina, G.M. Recent technical advancements, economics and environmental impacts of floating photovoltaic solar energy conversion systems. *J. Clean. Prod.* **2021**, *278*, 124285. [CrossRef]
24. Trapani, K.; Santafé, M.R. A review of floating photovoltaic installations: 2007–2013. *Prog. Photovolt. Res. Appl.* **2014**, *23*, 524–532. [CrossRef]
25. World Bank. Floating PV Power Plants. Available online: http://documents1.worldbank.org/curated/en/579941540407455831/pdf/Floating-Solar-Market-Report-Executive-Summary.pdf (accessed on 14 April 2021).
26. Grubišić-Čabo, F.; Nižetić, S.; Giuseppe Marco, T. Photovoltaic panels: A review of the cooling techniques. *Trans. FAMENA* **2016**, *40*, 63–74.

27. Trapani, K.; Millar, D.L. Proposing offshore photovoltaic (PV) technology to the energy mix of the Maltese islands. *Energy Convers. Manag.* **2013**, *67*, 18–26. [CrossRef]
28. Trapani, K.; Millar, D.L. The thin film flexible floating PV (T3F-PV) array: The concept and development of the prototype. *Renew. Energy* **2014**, *71*, 43–50. [CrossRef]
29. Kim, S.H.; Yoon, S.J.; Choi, W. Design and construction of 1 MW class floating PV generation structural system using FRP members. *Energies* **2017**, *10*, 1142. [CrossRef]
30. Rao, B.B.; Sandeep, V.M.; Rao, V.U.M.; Venkateswarlu, B. Potential Evapotranspiration Estimation for Indian Conditions: Improving Accuracy through Calibration Coefficients. 2012. Available online: http://nicra-icar.in/nicrarevised/images/Books/Potential%20Evapotranspiration%20estimation.pdf (accessed on 21 February 2021).
31. Perakis, C.; El Shenawy, E.; El Ghetany, H.; Kyriakarakos, G. Design of autonomous PV/RO desalination systems—Case studies for Egypt and Greece. *Desalination Water Treat.* **2017**, *74*, 12–20. [CrossRef]
32. Ferrer-Gisbert, C.; Ferrán-Gozálvez, J.J.; Redón-Santafé, M.; Gisbert, P.F.; Sánchez-Romero, F.J.; Torregrosa-Soler, J.B. A new photovoltaic floating cover system for water reservoirs. *Renew. Energy* **2013**, *60*, 63–70. [CrossRef]
33. Nagananthini, R.; Nagavinothini, R.; Balamurugan, P. Floating Photovoltaic Thin Film Technology—A Review. In *Intelligent Manufacturing and Energy Sustainability*; Springer: Singapore, 2020; pp. 329–338.
34. NASA Prediction of Worldwide Energy Resource. Available online: https://power.larc.nasa.gov (accessed on 2 February 2021).
35. Folsom Labs. Helioscope Solar Design Software. 2020. Available online: https://help.helioscope.com/ (accessed on 9 March 2021).
36. *Renewable Energy Capacity Statistics 2021*; The International Renewable Energy Agency (IRENA): Abu Dhabi, United Arab Emirates, 2021; ISBN 978-92-9260-342-7. Available online: https://www.irena.org/-/media/Files/IRENA/Agency/Publication/2021/Apr/IRENA_RE_Capacity_Statistics_2021.pdf (accessed on 14 April 2021).
37. Allen, R.G.; Pereira, L.S.; Raes, D.; Smith, M. *Crop Evapotranspiration-Guidelines for Computing Crop Water Requirements—FAO Irrigation and Drainage Paper 56*; FAO: Rome, Italy, 1998; Volume 300, p. D05109.
38. Wang, H.; Yang, Y.; Keller, A.A.; Li, X.; Feng, S.; Dong, Y.N.; Li, F. Comparative analysis of energy intensity and carbon emis-sions in wastewater treatment in USA, Germany, China and South Africa. *Appl. Energy* **2016**, *84*, 873–881. [CrossRef]
39. Taboada, M.; Caceres, L.; Graber, T.; Galleguillos, H.; Cabeza, L.; Rojas, R. Solar water heating system and photovoltaic floating cover to reduce evaporation: Experimental results and modeling. *Renew. Energy* **2017**, *105*, 601–615. [CrossRef]
40. Haas, J.; Khalighi, J.; de la Fuente, A.; Gerbersdorf, S.; Nowak, W.; Chen, P.-J. Floating photovoltaic plants: Ecological impacts versus hydropower operation flexibility. *Energy Convers. Manag.* **2020**, *206*, 112414. [CrossRef]
41. Cazzaniga, R.; Cicu, M.; Rosa-Clot, M.; Tina, G.; Ventura, C. Floating photovoltaic plants: Performance analysis and design solutions. *Renew. Sustain. Energy Rev.* **2018**, *81*, 1730–1741. [CrossRef]
42. Trapani, K.; Millar, D.L.; Smith, H.C. Novel offshore application of photovoltaics in comparison to conventional marine renewable energy technologies. *Renew. Energy* **2013**, *50*, 879–888. [CrossRef]

energies

MDPI

Article

Hydrodynamic Analysis of Twin-Hull Structures Supporting Floating PV Systems in Offshore and Coastal Regions

Alexandros Magkouris [1], Kostas Belibassakis [1] and Eugen Rusu [2,*]

1 School of Naval Architecture & Marine Engineering, National Technical University of Athens, 15780 Athens, Greece; alexmagouris@gmail.com (A.M.); kbel@fluid.mech.ntua.gr (K.B.)
2 Department of Mechanical Engineering, University Dunarea de Jos of Galati, 800008 Galati, Romania
* Correspondence: erusu@ugal.ro; Tel.: +40-740205534

Abstract: In this paper, a novel model based on the boundary element method (BEM) is presented for the hydrodynamic analysis of floating twin-hull structures carrying photovoltaic panels, supporting the study of wave responses and their effects on power performance in variable bathymetry regions. The analysis is restricted to two spatial dimensions for simplicity. The method is free of any mild-slope assumptions. A boundary integral representation is applied for the near field in the vicinity of the floating body, which involved simple (Rankine) sources, while the far field is modeled using complete (normal-mode) series expansions that are derived using separation of variables in the constant depth half-strips on either side of the middle, non-uniform domain, where the depth exhibited a general variation, overcoming a mild bottom-slope assumption. The numerical solution is obtained by means of a low-order panel method. Numerical results are presented concerning twin-hull floating bodies of simple geometry lying over uniform and sloping seabeds. With the aid of systematic comparisons, the effects of the bottom slope and curvature on the hydrodynamic characteristics (hydrodynamic coefficients and responses) of the floating bodies are illustrated and discussed. Finally, the effects of waves on the floating PV performance are presented, indicating significant variations of the performance index ranging from 0 to 15% depending on the sea state.

Keywords: hydrodynamic analysis; floating bodies; general bathymetry; BEM

Citation: Magkouris, A.; Belibassakis, K.; Rusu, E. Hydrodynamic Analysis of Twin-Hull Structures Supporting Floating PV Systems in Offshore and Coastal Regions. *Energies* **2021**, *14*, 5979. https://doi.org/10.3390/en14185979

Academic Editor: Athanasios Kolios

Received: 19 August 2021
Accepted: 16 September 2021
Published: 20 September 2021

Publisher's Note: MDPI stays neutral with regard to jurisdictional claims in published maps and institutional affiliations.

Copyright: © 2021 by the authors. Licensee MDPI, Basel, Switzerland. This article is an open access article distributed under the terms and conditions of the Creative Commons Attribution (CC BY) license (https://creativecommons.org/licenses/by/4.0/).

1. Introduction

The energy yield of floating photovoltaics (FPVs) is in the spotlight, as offshore photovoltaic (PV) installations present significant advantages over corresponding onshore ones (see [1,2]). These, among others, include the ample surface available for arrangements in farms, the nearshore/coastal regions and the open sea, including locations that are already licensed for offshore wind parks (in the area between wind turbines), as well as the potential of hybridization with offshore wind energy. The development of offshore FPV parks is particularly important for southern European regions, e.g., in the Mediterranean Sea, since solar radiation in southern latitudes is relatively high, nearly 150–200% greater than that of the Atlantic Sea Ocean, the North Sea and Baltic Sea regions [3], while wind and wave potentials are comparatively low. Furthermore, offshore PV installations present increased efficiency due to the cooling effects of water and wind, which are triggered by the interaction of airflow with the solar panels [4]. It is worth mentioning here that, according to a recent report from DNV GL, it is expected that offshore FPVs will reach maturity by 2030 (see also DNVGL-RP-0584-Edition 2021-03).

On the other hand, although several solar farms have been developed on closed water basins, such as lakes, reservoirs and dams, implementing installations in the open sea is a challenging task, as their interaction with several environmental factors is not yet fully understood [5]. In the offshore and nearshore region, safety and viability require the design and construction of resilient FPV structures that can withstand the harsh marine environment. Regarding the deployment of floating structures of relatively large dimensions in nearshore and coastal areas, it is also expected that bathymetric variations will have significant effects on their responses under wave loads, which also affect the performance of the power output due to oscillatory motions of the structure and the panels arranged on the deck. Stability requirements, in conjunction with a lightweight structure with a center of gravity at a relatively increased height above the keel, led to the consideration of a twin hull structure with a more complicated response pattern and resonance characteristics. In this study, a hydrodynamic model is developed to predict the dynamic responses of a floating structure supporting photovoltaic panels on a deck while being subject to wave loads. For the treatment of complicated resonance phenomena, as well as the effects of finite and possibly variable bathymetry, which characterizes nearshore and coastal regions, a general model based on boundary element methods (BEM) is developed, which is capable of modeling the involved phenomena. The model is then systematically applied in selected examples to produce preliminary results, which are illustrative of the effect of dynamic motions on the energy efficiency of a floating unit.

The interaction of free surface gravity waves with floating bodies at intermediate depths in areas that are characterized by non-uniform seabed topographies is a mathematically interesting problem, which can be used to analyze a wide range of applications, such as the design and performance evaluation of ships and other floating structures operating in nearshore areas. Theoretical aspects of the problem of small-amplitude water waves propagating in a finite water depth and their interaction with floating and/or submerged bodies have been presented under various geometric assumptions by many authors [6–8] regarding the existence of trapped modes in a channel with obstructions. Furthermore, shallow-water conditions are frequently encountered in marine applications. When floating structures or docks are moored in shallow-water areas, accurate predictions of the motions induced by the prevailing sea state are needed, not only for optimizing the mooring system, depending on the stability needs of each configuration, but also for ensuring that the under-keel clearance remains sufficient for the structure to avoid grounding in extreme (for the area under study) weather and sea conditions. In many applications, the water depth is assumed to be constant, which is practically valid in cases where there are small depth variations or the floating body's dimensions are small compared to the bottom variation length. However, in applications involving the utilization of floating bodies in coastal waters, the variations of bathymetry may cause significant effects on the hydrodynamic behavior of ships and structures, especially concerning the wave-induced responses. Under the assumption of slowly varying bathymetry, mild-slope model have been developed for the analysis of wave-induced floating body motion [9]. To treat environments that are characterized by steeper bathymetric variations, e.g., near the coast or the entrances of ports and harbors, extended models are required (see, e.g., Ohyama and Tsuchida [10]).

In the present work, a novel method based on BEM is used for the hydrodynamic analysis of twin-hull floating structures with PV systems; the method is capable of treating the effects of varying bathymetry without any mild-slope assumptions. In particular, a low-order panel based on linear wave theory is developed and verified. Following the hybrid formulation by Yeung [11], the present method utilizes the simplicity of Rankine sources, in conjunction with appropriate representations of the wavefield in the exterior semi-infinite domain, as presented by Nestegard and Sclavounos [12] for 2D radiation problems in deep water and by Drimer and Agnon [13] in the case of finite water depth. The far field is modeled using complete (normal-mode) series expansions, which are derived using separation of variables in the two constant-depth half-strips separating the variable bathymetry region from the regions of wave incidence and wave transmission (see Figure 1). Numerical results are presented concerning twin-hull floating bodies of simple geometry over uniform and sloping seabeds. With the aid of systematic comparisons, the effects of bottom slope on the hydrodynamic characteristics (hydrodynamic coefficients and responses) are presented and discussed. Finally, results are presented regarding the effect of wave-induced motions on floating PV performance, indicating significant variations in the performance index ranging from 0 to 15%, depending on the sea state.

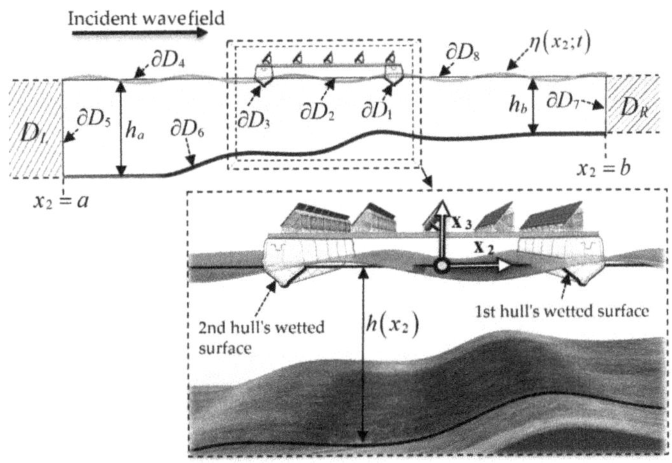

Figure 1. Geometric configuration of the 2D problem.

2. Mathematical Formulation

The 2D problem concerning the hydrodynamic behavior of a twin hull floating body of arbitrary cross-section in a coastal–marine environment is considered, as illustrated in Figure 1. A Cartesian coordinate system $\mathbf{x} = (x_1, x_2, x_3)$ is introduced, with the origin placed at the mean water level, coinciding with the structure's center of flotation, with the x_3-axis pointing upwards. The configuration is considered unchanged in the x_1-direction and, therefore, the analysis is limited to the x_2x_3 plane, modeling a two-dimensional cross-section.

The environment comprises a water layer bounded by the free surface at $x_3 = 0$ and the rigid bottom at depth $h = h(x_2)$. It is assumed that $h = h(x_2)$ exhibited a general variation, i.e., the corresponding bathymetry is defined by parallel, straight contours lying between two regions of constant but different water depths: $h = h_a$ in the region of wave incidence and $h = h_b$ in the region of transmission. The fluid is assumed to be homogeneous, inviscid and incompressible and its motion irrotational with a small width. The wavefield in the region is excited by a harmonic incident field, with propagation direction normal to the depth contours (along the x_2-axis). Without loss of generality, a left-incident wave field is assumed (see Figure 1). Thus, in the context of linearized wave theory, the fluid motion is fully described by the 2D wave potential $\Phi(x_2, x_3; t)$, with the velocity field being equal to $\mathbf{v}(\mathbf{x}, t) = \nabla\Phi(\mathbf{x}, t)$. Assuming that the free-surface elevation and the wave velocities are small, the potential function $\Phi(x_2, x_3; t)$ satisfies the linearized wave equations (see, e.g., [14,15]). Under these assumptions, the wavefield is time-harmonic and its potential function can be represented by the time-independent (normalized) complex potential function φ as:

$$\Phi(x_2, x_3; t) = \mathrm{Re}\left\{ -\frac{igA}{\omega} \varphi(x_2, x_3; \mu) \cdot \exp(-i\omega t) \right\}, \tag{1}$$

where $H = 2A$ is the incident wave height, g is the acceleration of gravity, $\mu = \omega^2/g$ is the frequency parameter and $i = \sqrt{-1}$. The free surface elevation is obtained in terms of the wave potential at $x_3 = 0$ as follows:

$$\eta(x_2; t) = -\frac{1}{g} \frac{\partial\Phi(x_2, 0; t)}{\partial t}. \tag{2}$$

In addition to the physical boundaries (floating body, free surface, seabed), we further introduce two vertical interfaces on either side of the body, serving as incidence/radiation/

transmission boundaries. Therefore, the boundary ∂D of the two-dimensional domain D, occupied by the fluid, is decomposed into eight subsections ∂D_i, $i = 1,2,\ldots,8$, as illustrated in Figure 1, so that D is enclosed by the curve $\partial D = \cup_{i=1}^{8} \partial D_i$, with ∂D_1 and ∂D_3 being the right- and left-hand sides, respectively, of the twin-hull's wetted surface. The sections of ∂D numbered 2, 4 and 8 correspond to the water-free surface, while ∂D_6 is the impermeable seabed. Finally, the wave incidence occurs via ∂D_5, which also serves, along with ∂D_7, as a radiation interface for the diffracted field due to the presence of the (fixed) body, as well as the radiation fields that develop due to the wave-induced body's motions.

Apart from the non-uniform domain D containing the floating body, the total flow field is considered to be of infinite length and, therefore, also comprises the uniform semi-infinite subdomains D_L and D_R, where the depth is constant and equal to h_a and h_b, respectively. Hence, the function $h = h(x_2)$ is of the form:

$$h(x_2) = \begin{cases} h_a, & x_2 \leq a \\ h(x_2), & a < x_2 < b \\ h_b, & x_2 \geq b \end{cases} \tag{3}$$

The function $\varphi = \varphi(x_2, x_3; \mu)$ appearing in Equation (1) is the normalized potential in the frequency domain, which will hereafter be simply written as $\varphi(x_2, x_3)$. Using standard floating body hydrodynamic theory [15,16], the potential is decomposed as follows:

$$\varphi(x_2, x_3) = \varphi_p(x_2, x_3) + \frac{\mu}{A} \sum_{k=2}^{4} \xi_k \varphi_k(x_2, x_3) \tag{4}$$

where $\varphi_p(x_2, x_3) = \varphi_I(x_2, x_3) + \varphi_D(x_2, x_3)$ is the propagating field, with $\varphi_I(x_2, x_3)$ being the incident field, which corresponds to the solution of the wave propagation problem across the non-uniform subdomain in the absence of the floating structure and $\varphi_D(x_2, x_3)$ being the diffraction potential, which accounts for the presence of the body, fixed in its mean position. Moreover, the functions $\varphi_k(x_2, x_3)$, $k = 2,3,4$, denote the radiation potentials associated with the motion of the twin-hull structure, corresponding to its three degrees of freedom (DOF), i.e., the linear transverse motion (sway: $k = 2$), the linear vertical motion (heave: $k = 3$) and the rotation about the longitudinal (x_1) axis (roll: $k = 4$). Finally, ξ_k, $k = 2,3,4$, stand for the complex amplitudes of the corresponding wave-induced motions.

The sub-problems, whose solutions provide the potential functions $\varphi_k(x_2, x_3)$, $k = p$, 2,3,4, in the variable bathymetry region, were formulated as radiation-type problems in the bounded subdomain D, with the aid of the following general representations of the wave potential $\varphi(x_2, x_3)$ in the left- and right-side semi-infinite strips D_L and D_R, which are obtained using separation of variables (see, e.g., [17]):

$$\begin{aligned}\varphi_p^{(L)}(\mathbf{x}) = {} & \left[\exp\left(ik_0^{(L)} x_2\right) + C_0^{(L)} \exp\left(-ik_0^{(L)} x_2\right)\right] Z_0^{(L)}(x_3) + \\ & + \sum_{n=1}^{\infty} C_n^{(L)} \exp\left[k_n^{(L)}(x_2 - a)\right] Z_n^{(L)}(x_3), \mathbf{x} \in D_L\end{aligned} \tag{5a}$$

$$\begin{aligned}\varphi_k^{(L)}(\mathbf{x}) = {} & C_0^{(L)} \exp\left(-ik_0^{(L)} x_2\right) Z_0^{(L)}(x_3) + \\ & + \sum_{n=1}^{\infty} C_n^{(L)} \exp\left[k_n^{(L)}(x_2 - a)\right] Z_n^{(L)}(x_3), \mathbf{x} \in D_L, k = 2,3,4\end{aligned} \tag{5b}$$

$$\begin{aligned}\varphi_k^{(R)}(\mathbf{x}) = {} & C_0^{(R)} \exp\left(ik_0^{(R)} x_2\right) Z_0^{(R)}(x_3) \\ & + \sum_{n=1}^{\infty} C_n^{(R)} Z_n^{(R)}(x_3) \exp\left[k_n^{(R)}(b - x_2)\right], \mathbf{x} \in D_R, k = p,2,3,4\end{aligned} \tag{5c}$$

The first term ($n = 0$) in the series Equations (5) is the propagating mode, while the remaining ones ($n = 1,2,\ldots$) are the evanescent modes with $C_n^{(i)}$ ($n = 0,1,2,\ldots$) being the corresponding coefficients. The first term of $\varphi_p^{(L)}(\mathbf{x})$ is further separated into a unit-amplitude mode propagating toward D, playing the role of the incident field, and the additional mode $C_0^{(L)}\exp(-ik_0^{(L)}x_2)Z_0^{(L)}(x_3)$, propagating toward $-\infty$ in the x_2-direction, which is the reflected field coming from the diffraction potential $\varphi_D(\mathbf{x})$. In the above expansions, the functions $\left\{Z_n^{(i)}\right\}_{n=0}^{\infty}$ are defined as $Z_n^{(i)} = \cosh\left[k_n^{(i)}\left(z+h^{(i)}\right)\right] / \cosh\left(k_n^{(i)}h^{(i)}\right)$ and are obtained using separation of variables via the vertical Sturm–Liouville problem, to which Laplace's equation reduces in the constant depth strips $\{\,D_L|-h_a<x_3<0,\ x_2<a\}$ and $\{\,D_R|-h_b<x_3<0,\ x_2>b\}$. The corresponding eigenvalues $k_0^{(i)}$ and $\left\{k_n^{(i)}\right\}_{n=1}^{\infty}$ are respectively obtained as the real root and the imaginary roots of the dispersion relation: $\omega^2 = k^{(i)}g\cdot\tanh\left(k^{(i)}h^{(i)}\right)$, $i = L, R$, where g denotes the acceleration due to gravity. The completeness of the expansions derives from the standard theory of regular eigenvalue problems (see, e.g., [18]). Based on the above representations, the hydrodynamic problems concerning the propagating and radiation potentials $\varphi_k(x_2,x_3)$ were formulated as radiation-type problems, satisfying the following systems of equations, boundary conditions and matching conditions for $k = p,2,3,4$:

$$\frac{\partial^2\varphi_k(x_2,x_3)}{\partial x_2{}^2}+\frac{\partial^2\varphi_k(x_2,x_3)}{\partial x_3{}^2}=0,\ \mathbf{x}\in D|_{(Domain\ of\ Transmission)} \tag{6a}$$

$$\frac{\partial\varphi_k(\mathbf{x})}{\partial n}-\mu\varphi_k(\mathbf{x})=0,\ \mu=\frac{\omega^2}{g},\ \mathbf{x}\in(\partial D_2\cup\partial D_4\cup\partial D_8)|_{(Free\ Surface)} \tag{6b}$$

$$\frac{\partial\varphi_k(\mathbf{x})}{\partial n}=0,\ \mathbf{x}\in\partial D_6|_{(Seabed)} \tag{6c}$$

$$\frac{\partial\varphi_k(\mathbf{x})}{\partial n}=N_k(\mathbf{x}),\ \mathbf{x}\in(\partial D_1\cup\partial D_3)|_{(Wetted\ Surface)} \tag{6d}$$

$$\frac{\partial\varphi_k(\mathbf{x})}{\partial n}-T_L\left[\varphi_k^{(L)}(\mathbf{x})\right]=Q_k,\ \mathbf{x}\in\partial D_5|_{(\text{Incidence /Reflection /Radiation})} \tag{6e}$$

$$\frac{\partial\varphi_k(\mathbf{x})}{\partial n}-T_R\left[\varphi_k^{(R)}(\mathbf{x})\right]=0,\ \mathbf{x}\in\partial D_7|_{(\text{Radiation})} \tag{6f}$$

The above boundary sections are also illustrated in Figure 1. Moreover, in Equations (6a)–(6f), $n = (0,n_2,n_3)$ denotes the unit normal vector to the boundary ∂D, directed to its exterior. The boundary data N_k, $k = 2,3,4$ appearing on the right-hand side of Equation (6d) are defined by the components of the generalized normal vector on the wetted surface boundary section $\partial D_1\cup\partial D_3$: $N_2 = n_2$, $N_3 = n_3$ and $N_4 = x_2n_3 - x_3n_2$, and constitute the (unit-amplitude) excitations of the system in Equation (6a)–(6f) for each k. N_p is set to 0 so that the solution of the propagating field is obtained by treating the floating body as an impermeable, immobile solid boundary. Finally, the operators $T_L\left[\varphi_k^{(L)}(\mathbf{x})\right]$ and $T_R\left[\varphi_k^{(R)}(\mathbf{x})\right]$ are appropriate Dirichlet-to-Neumann (DtN) maps (see, e.g., [19]), ensuring the complete matching of the fields $\varphi_k(\mathbf{x})$, $k = p,2,3,4$, on the vertical interfaces ∂D_5 and ∂D_7, respectively. These operators are derived from Equations (5a)–(5c), exploiting the completeness properties of the vertical bases $\left\{Z_n^{(i)}(z), n = 0,1,2,\ldots\right\}$, $i = L, R$. More details are provided in Appendix A.

3. The BEM for Floating Twin-ull Structures

3.1. The Incidence, Diffraction and Radiation Problems

The corresponding problems of the propagating and radiation potentials $\varphi_k(\mathbf{x})$, $k = p,2,3,4$, given as Equations (6) were treated by means of boundary integral equation formulations that are based on the single-layer potential (see, e.g., [20]). Accordingly, the following integral representations are introduced for $\varphi_k(\mathbf{x}),\ k = p, 2, 3, 4$, in the bounded subdomain D:

$$\varphi_k(\mathbf{x}) = \int_{\partial D} \sigma_k(\mathbf{x}')G(\mathbf{x}', \mathbf{x})d\updownarrow(\mathbf{x}'), \mathbf{x} = (x_2, x_3) \in D, \mathbf{x}' \in \partial D,\ k = p, 2, 3, 4 \tag{7}$$

where $G(\mathbf{x}', \mathbf{x}) = \ln|\mathbf{x}' - \mathbf{x}|/2\pi$ is the Green's function of the Laplace equation in 2D free-space; $\sigma_k(\mathbf{x}')$ is a source/sink strength distribution, defined on the boundary of the bounded subdomain D for each of the four subproblems; and $d\updownarrow(\mathbf{x}')$ denotes the differential element along the boundary ∂D (see, e.g., [21,22]). Based on the properties of the single-layer distributions, the corresponding normal derivatives of the functions $\varphi_k(\mathbf{x}),\ k = p, 2, 3, 4$, on the boundary ∂D are given by (see, e.g., [21]):

$$\frac{\partial \varphi_k(\mathbf{x})}{\partial n} = -\frac{\sigma_k(\mathbf{x})}{2} + \int_{\partial D} \sigma_k(\mathbf{x}')\frac{\partial G(\mathbf{x}', \mathbf{x})}{\partial n} d\updownarrow(\mathbf{x}'),\ (\mathbf{x}, \mathbf{x}') \in \partial D. \tag{8}$$

Using the above in Equations (6b)–(6f), we obtain a system of boundary integral equations, with support on the different sections of ∂D for the determination of the corresponding unknown source distribution $\sigma_k(\mathbf{x}),\ \mathbf{x} \in \partial D, k = p, 2, 3, 4$, for each of the potential functions $\varphi_k(\mathbf{x}),\ k = p, 2, 3, 4$. The final system read as follows for $k = p, 2, 3, 4$:

$$\begin{array}{l} -\frac{\sigma_k(\mathbf{x})}{2} + \int_{\partial D} \sigma_k(\mathbf{x}')\frac{\partial G(\mathbf{x}',\mathbf{x})}{\partial n} d\ell(\mathbf{x}') + \\ \quad -\mu \int_{\partial D} \sigma_k(\mathbf{x}')G(\mathbf{x}', \mathbf{x})d\ell(\mathbf{x}') = 0,\ \mathbf{x} \in (\partial D_2 \cup \partial D_4 \cup \partial D_8),\ \mathbf{x}' \in \partial D, \end{array} \tag{9a}$$

$$-\frac{\sigma_k(\mathbf{x})}{2} + \int_{\partial D} \sigma_k(\mathbf{x}')\frac{\partial G(\mathbf{x}', \mathbf{x})}{\partial n} d\ell(\mathbf{x}') = 0,\ \mathbf{x} \in \partial D_6,\ \mathbf{x}' \in \partial D, \tag{9b}$$

$$-\frac{\sigma_k(\mathbf{x})}{2} + \int_{\partial D} \sigma_k(\mathbf{x}')\frac{\partial G(\mathbf{x}', \mathbf{x})}{\partial n} d\ell(\mathbf{x}') = N_k(\mathbf{x}),\ \mathbf{x} \in (\partial D_1 \cup \partial D_3),\ \mathbf{x}' \in \partial D, \tag{9c}$$

$$\begin{array}{l} -\frac{\sigma_k(\mathbf{x})}{2} + \int_{\partial D} \sigma_k(\mathbf{x}')\frac{\partial G(\mathbf{x}',\mathbf{x})}{\partial n} d\ell(\mathbf{x}') + \\ \quad -T_L\left[\int_{\partial D} \sigma_k(\mathbf{x}')G(\mathbf{x}', \mathbf{x})d\ell(\mathbf{x}')\right] = Q_k,\ \mathbf{x} \in \partial D_5,\ \mathbf{x}' \in \partial D, \end{array} \tag{9d}$$

$$\begin{array}{l} -\frac{\sigma_k(\mathbf{x})}{2} + \int_{\partial D} \sigma_k(\mathbf{x}')\frac{\partial G(\mathbf{x}',\mathbf{x})}{\partial n} d\ell(\mathbf{x}') + \\ \quad -T_R\left[\int_{\partial D} \sigma_k(\mathbf{x}')G(\mathbf{x}', \mathbf{x})d\ell(\mathbf{x}')\right] = 0,\ \mathbf{x} \in \partial D_7,\ \mathbf{x}' \in \partial D, \end{array} \tag{9e}$$

From the above systems' solutions $\sigma_k,\ k = p, 2, 3, 4$, the corresponding potential functions $\varphi_k(\mathbf{x}),\ k = p, 2, 3, 4$ and all quantities associated with them were calculated using Equations (7) and (8) in the bounded subdomain D. The solutions of the system consisting of Equations (9a)–(9e) are obtained numerically by means of a low-order boundary element method based on simple (Rankine) sources (see also [23]). The geometry of the different sections of ∂D is approximated using linear segments on which the source distribution is taken to be piecewise constant. In this case, the boundary integrals in Equations (9a)–(9e) associated with each element's contribution can be analytically calculated (see, e.g., [24]) and the systems of boundary integral equations reduce to an equal number of algebraic systems, whose unknowns are the vectors $\left\{\sigma_{kj}\right\}_{j=1}^{M}$, $k = p, 2, 3, 4$ with M being the number of linear boundary elements used to approximate the geometry of ∂D.

3.2. Equations of Motion

The total hydrodynamic loads (forces and moment) on the twin-hull structure consist of the Froude-Krylov loads, which are solely due to the undisturbed incident field $\varphi_I(\mathbf{x})$, the diffraction loads caused by the pressure field generated by the presence of the fixed floating body and the radiation loads due to the pressure fields of the wavefields "radiated" by the oscillating body. Based on the calculated propagating potential $\varphi_p(\mathbf{x})$(consisting of the incident and diffraction potentials), the summation of the Froude-Krylov and the diffraction-induced hydrodynamic forces, as well as the corresponding moment (F_k, $k = 2,3,4$), are calculated using surface integration, as follows:

$$F_k = i\omega\rho \int_{\partial D_1 \cup \partial D_3} \varphi_P(\mathbf{x}) \cdot \mathbf{N}_k(\mathbf{x}) d\ell(\mathbf{x}),\ k = 2,3,4,\ \mathbf{x} \in (\partial D_1 \cup \partial D_3) \quad (10)$$

where ρ denotes the fluid (water) density and $\mathbf{N}_k$, $k = 2,3,4$, is the generalized normal vector on the wetted surface (also defined in Section 2). Moreover, from the radiation potentials $\varphi_k(\mathbf{x})$, $k = 2,3,4$, the hydrodynamic coefficients are calculated using:

$$\omega^2 A_{kl} + i\omega B_{kl} = i\omega\rho\Pi_{kl},\ l,k = 2,3,4, \text{ where} \quad (11a)$$

$$\Pi_{kl} = \int_{\partial D_1 \cup \partial D_3} \varphi_l(\mathbf{x}) \mathbf{N}_k(\mathbf{x}) d\ell(\mathbf{x}),\ l,k = 2,3,4,\ \mathbf{x} \in (\partial D_1 \cup \partial D_3) \quad (11b)$$

In the above expressions, $\mathbf{A}_{(3\times3)}$ is the (symmetric) matrix of the added inertial coefficients, which, for each frequency, corresponded to the proportion of the radiation loads in phase with the structure's acceleration (in the frequency domain). $\mathbf{B}_{(3\times3)}$ is the corresponding matrix of hydrodynamic damping coefficients, which consists of the part of the radiation loads in phase with the structure's velocity. Details about the definitions of the hydrodynamic forces and coefficients, as well as the system of equations of motion, can be found in [15] or in ship hydrodynamics textbooks (see, e.g., [16,24]). The latter quantities allow us to formulate and solve the equations of motion of the floating body in the inhomogeneous domain. The general form of the equations of motion in the frequency domain for the 2D twin-hull structure considered is:

$$\left\{-\omega^2[\mathbf{M} + \mathbf{A}(\omega)] - i\omega\,\mathbf{B}(\omega) + \mathbf{C}\right\}\xi = F \quad (12)$$

where $\mathbf{C}$ is the hydrostatic restoring forces and moments acting on the structure.

Due to the symmetry of the body with respect to the vertical axis x_3, the component N_3 of the generalized normal vector is symmetric, while the components $\mathbf{N}_k$, $k = 2,4$, are antisymmetric. Assuming that the seabed profile variations do not significantly alter the radiation potentials $\varphi_k(\mathbf{x})$, $k = 2,3,4$, near the floating structure, the potential function $\varphi_3(\mathbf{x})$ is also symmetric and the functions $\varphi_k(\mathbf{x})$, $k = 2,4$ are antisymmetric. This fact implies that $\Pi_{32} = \Pi_{34} = 0$ and $\Pi_{23} = \Pi_{42} = 0$. Therefore, the dynamic equations of motion relating to the oscillations of the body are simplified in the following form, where the heaving motion (ξ_3) is decoupled from the sway and roll motions (ξ_2, ξ_4) of the twin hull:

$$\left[-\omega^2(M + A_{22}) - i\omega B_{22}\right]\xi_2 - \left(\omega^2 A_{24} + i\omega B_{24}\right)\xi_4 = F_2, \quad (13a)$$

$$\left[-\omega^2(M + A_{33}) - i\omega B_{33} + 2\rho g B_{(H)}\right]\xi_3 = F_3, \quad (13b)$$

$$\left(-\omega^2 A_{42} - i\omega B_{42}\right)\xi_2 + \left[-\omega^2(I_{44} + A_{44}) - i\omega B_{44} + Mg \cdot GM\right]\xi_4 = F_4, \quad (13c)$$

where $B_{(H)}$ is the breadth of each individual hull and GM denotes the metacentric height. The total mass equals $M = \rho \cdot \nabla$, referring to unit length in the transverse direction (kg/m), where ρ denotes the fluid's density and ∇ is the displacement volume of the structure. Moreover, due to the symmetry of the floating structure, its center of buoyancy (B) is located on the vertical line $x_2 = 0$ and its x_3 coordinate is calculated as the center of area of the submerged volume's cross-section. The center of gravity (G) is also located at $x_2 = 0$ due to symmetry of the configuration and its x_3 coordinate is considered to be located at the waterplane $(x_3 = 0)$. A total radius of gyration per unit length in the transverse direction of $R_G = \left(B_{(T)} - B_{(H)}\right)/2$ is considered about the longitudinal axis (x_1), where $B_{(T)}$ is the total breadth of the twin-hull structure and, therefore, $I_{44} = M(R_G)^2$. The metacentric radius was evaluated as $BM = I/\nabla$, where I is the second moment of area of the waterplane, calculated using applying Steiner's theorem as:

$$I = 2\left[\left(\frac{B_{(H)}{}^3}{12}\right) + B_{(H)} \cdot \left(\frac{B_{(T)} - B_{(H)}}{2}\right)^2\right], \tag{14}$$

which also refers to the unit length in the transverse direction (x_1). Finally, the metacentric height was calculated as $GM = KB + BM - KG$, where K is a reference point with coordinates $(0, x_3)$. The above equations can also be modified to include other external forces, as e.g., mooring forces or spring terms (see, e.g., Section 3.5 of [25]). The solution of the above system (13) provides us with the complex amplitudes of the corresponding motions of the twin hull: ξ_k, $k = 2, 3, 4$. Then, the total wave potential is obtained using Equation (4), from which the hydrodynamic pressure is obtained using Bernoulli's theorem. The wave loads on the floating structure are calculated using pressure integration on the wetted surface $\partial D_1 \cup \partial D_3$.

4. Numerical Results

4.1. Comparison with Other Methods and Verification

The results obtained by the previously described numerical model are here compared to previous research for verification purposes. The results concern a twin-hull floating structure whose individual hulls are cylindrical, with a draft equal to the radius, which results in wetted surfaces whose cross-section shapes are semicircles. Numerical results regarding the above configuration were presented by Ohcusu [26] in 1969 and Rhee [27] in 1982 concerning the amplitude ratio of the radiated fields' wave height away from the body divided by the amplitude of the forced oscillation that excites the field itself in calm water.

Figure 2 illustrates the aforementioned ratio regarding the heave and sway motions in the case of unit amplitude of the twin-hull structure. The wetted surface of each hull is semicircle of radius R in the x_2x_3 plane, while each of the two semicircles' centers are at a distance P from the origin, following the notation of Rhee [27]. Therefore, the two centers are at a distance $2P$ apart and the configuration is defined so that $2P/R = 3$. The results concern the radiation fields that propagate in deep water, which is achieved in the present numerical model by setting the depth as a constant and equal to half the wavelength, for each simulated frequency, as calculated from the dispersion relation for deep-water $(\lambda = 2\pi g/\omega^2)$.

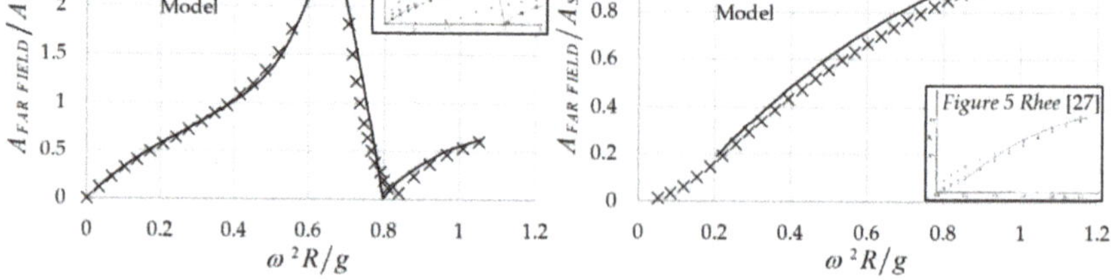

Figure 2. Amplitude ratios for heave and sway for a twin-hull floating structure with semicircle hull cross-sections $(2P/R = 3,\ h/R = \infty)$.

The domain extends to three wavelengths away from the floating body in both directions and the free surface elevation is evaluated by the discrete BEM model at the last free surface boundary element away from the structure (adjacent to the first boundary element of the radiation boundary). The amplitude ratios of Figure 2 are presented as functions of the non-dimensional frequency parameter $\omega^2 R/g$.

Indicative results are illustrated in Figure 3 concerning wave fields generated by unit-amplitude forced oscillations of the twin hull in sway and heave, with the non-dimensional frequency parameter $\omega^2 R/g$ set to 1. The amplitude ratios are equal to 0.992 and 0.520 for sway and heave, respectively, as also shown in Figure 2.

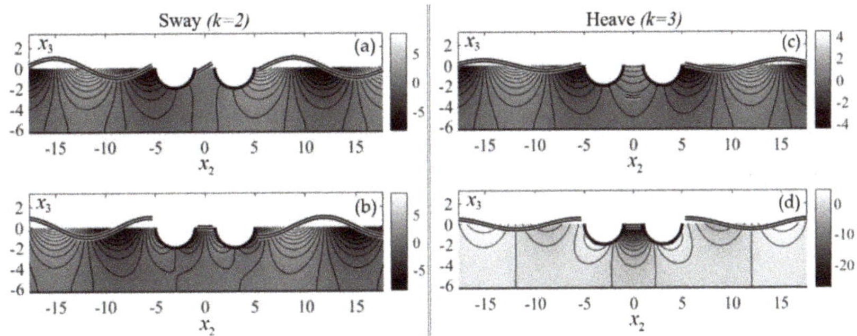

Figure 3. Sway and heave radiation fields for $2P/R = 3$, $h = \lambda/2$ and $\omega^2 R/g = 1$. (**a**,**b**) Real and imaginary part of the normalized sway field $\varphi_2(\mathbf{x})$ and corresponding free-surface elevation. (**c**,**d**) Real and imaginary part of the normalized heave field $\varphi_3(\mathbf{x})$ and corresponding free-surface elevation.

An identical twin-hull structure was studied by Dabssi et al. [28] in 2008 regarding its hydrodynamic coefficients. Figure 3 illustrates the added mass and damping of the floating structure in heave (ξ_3). The added mass (A_{33}) has been divided by the structure's mass, while the damping coefficient for heave (B_{33}) has been divided by the mass times the angular frequency ω so that all presented quantities are non-dimensional. It is noted that the displacement in this case does not need to be numerically calculated since it equals the sum of volumes of two half-cylinders of radius R that were considered to extend to unit length in the transverse direction (x_1) and therefore is equal to πR^2. The results of Figure 4 concern the heaving motion of the twin-hull in a finite water depth h, where $h/R = 2$. The calculated data sets are presented as functions of the non-dimensional wavenumber kR.

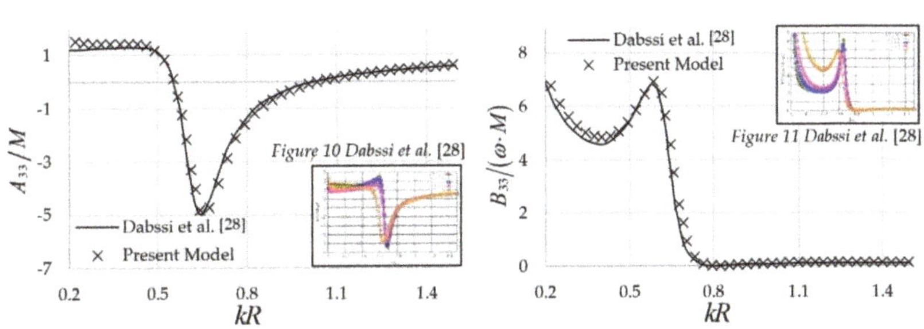

Figure 4. Added mass and damping coefficient of heave, for a twin-hull floating structure of semicircle hull cross sections ($2P/R = 3$, $h/R = 2$).

4.2. Hydrodynamic Analysis of Floating Twin-Hull Structure in Waves

The effect of sloping seabed environments on the hydrodynamic characteristics of a twin hull is here illustrated by considering the case of a structure of non-dimensional total breadth equal to $B_{(T)}/h = 2/3$, with the non-dimensional breadth and draft of each hull set to $B_{(H)}/h = T/h = 1/10$, where h denotes the mean water depth of the inhomogeneous domain D. The individual hulls that make up the twin-hull layout were modeled via the cross-section of a Wigley hull at $x_1 = 0$, which is given by the analytical relation:

$$x_2 = \mp \frac{B_{(H)}}{2} \cdot \left[1 - \left(\frac{x_3}{T}\right)^2\right]. \tag{15}$$

The configuration is considered to be located in an inhomogeneous region (see Figures 5 and 6). The center of gravity was selected to coincide with the center of flotation. The center of buoyancy (B), which is calculated as the center of area of the submerged volume's cross-section, is located at $(x_2 = 0, x_3 = -0.375 \cdot T)$ and, thus, the non-dimensional metacentric height of this layout is $GM/h = 1.179$.

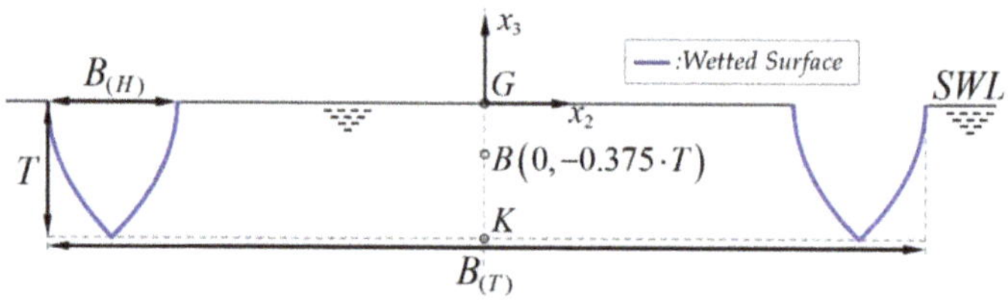

Figure 5. Outline of the modeled configuration and basic dimensions.

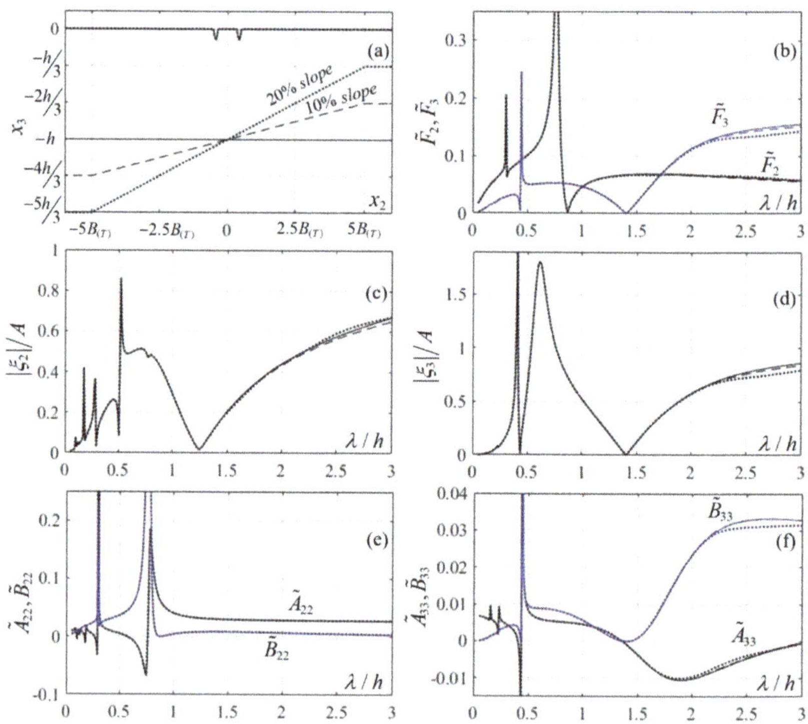

Figure 6. (**a**) Floating body and domains of transmission ($B_{(T)}/h = 2/3$, $B_{(H)}/h = T/h = 1/10$). (**b**) Hydrodynamic forces $\tilde{F}_k$, $k = 2, 3$. (**c**,**d**) RAOs in sway and heave motions, respectively. (**e**,**f**) Hydrodynamic coefficients A_{22}, B_{22} and A_{33}, B_{33}, respectively. All quantities were plotted vs. the non-dimensional wavelength λ/h, where h denotes the average water depth.

Numerical results are presented in Figures 6 and 7 concerning the hydrodynamic behavior of this floating structure in constant depth and over two linear shoals characterized by (constant) bottom slopes of 10 and 20%, respectively (see Figure 6a). The shoaling environments were achieved using a linear depth reduction of $2h/3$ and $4h/3$, respectively, over a depth variation distance of $10B_{(T)}$, with the mean water depth of all three domains of transmission being equal to h. The results concerning the homogeneous domain were plotted using solid lines, while the results concerning the inhomogeneous transmission domains with bottom slopes of 10 and 20% were plotted using dashed lines and dotted lines, respectively.

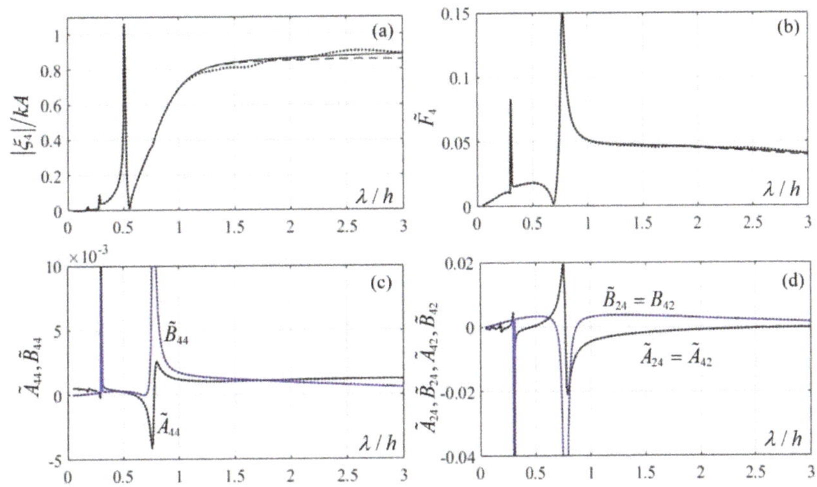

Figure 7. (**a**) RAO in roll motion. (**b**) Hydrodynamic moment $\tilde{F}_4$. (**c**,**d**) Hydrodynamic coefficients A_{44}, B_{44} and A_{24}, B_{24}, respectively. All quantities are plotted vs. the non-dimensional wavelength λ/h, where h denotes the average water depth.

In particular, Figure 6b illustrates the normalized hydrodynamic forces as functions of the non-dimensional wavelength λ/h for all three considered domains of transmission, where $\lambda = 2\pi/k_0$ is the wavelength corresponding to the mean water depth h, as obtained through application of the dispersion relation: $\omega^2 = k_0 g \cdot \tanh(k_0 h)$. The normalization used for the hydrodynamic forces is $\tilde{F}_k = F_k/\rho g h A,\ k = 2,3$, where A is the incident wave amplitude.

Figure 6c,d depict the twin hull's response amplitude operators (RAOs) associated with its two linear motions, i.e., sway (ξ_2) and heave (ξ_3). The body's linear responses were normalized as $RAO_k = \tilde{\xi}_k = |\xi_k|/A,\ k = 2,3$. Finally, in Figure 6e,f corresponding results concerning the hydrodynamic coefficients are presented. The matrix $\mathbf{A}_{(3\times3)}$ of added inertial coefficients and the matrix $\mathbf{B}_{(3\times3)}$ of hydrodynamic damping coefficients were normalized as:

$$\tilde{\mathbf{A}} = \frac{\mathbf{A}}{\rho}\begin{pmatrix} h^{-2} & h^{-2} & h^{-3} \\ h^{-2} & h^{-2} & h^{-3} \\ h^{-3} & h^{-3} & h^{-4} \end{pmatrix},\ \tilde{\mathbf{B}} = \sqrt{\frac{h}{g}} \cdot \frac{\mathbf{B}}{\rho}\begin{pmatrix} h^{-2} & h^{-2} & h^{-3} \\ h^{-2} & h^{-2} & h^{-3} \\ h^{-3} & h^{-3} & h^{-4} \end{pmatrix}. \quad (16)$$

Figure 7a illustrates the twin hull's RAO associated with the angular motion, i.e., roll (ξ_4). The body angular response is normalized as $RAO_4 = \tilde{\xi}_4 = |\xi_4|/kA$, with k being the wavenumber corresponding to the mean water depth h. The corresponding normalized hydrodynamic moment $\tilde{F}_4 = F_4/\rho g h^2 A$ is shown in Figure 7b. Figure 7c depicts the diagonal elements (A_{44}, B_{44}) of the added inertia and damping matrices. Moreover, the (non-diagonal) elements of the symmetric added inertia and hydrodynamic damping matrices are shown in Figure 7d. All results are plotted as functions of the non-dimensional wavelength λ/h.

Finally, indicative results regarding the total induced wavefields are depicted in Figure 8 for non-dimensional wavelength equal to $\lambda/h = 2.4$. In particular, Figure 8a illustrates the real part of the total potential $\varphi(\mathbf{x})$; see Equation (4) for the three considered cases of 0, 10 and 20% bottom slope (see Figure 6a) in the general vicinity of the floating twin-hull structure. Figure 8b depicts the imaginary parts of the corresponding potential functions. The configuration was made dimensional by setting $h = 30$ m and an incident field of amplitude $A = 1.5$ m has been considered. The alterations to the wavefields due to the non-uniform topography profiles are clearly seen, as the equipotential lines intersect each seabed boundary section perpendicularly, which implies the satisfaction of the impermeability boundary condition.

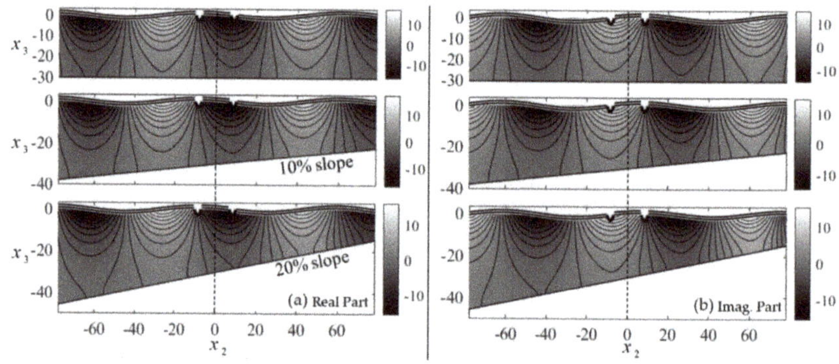

Figure 8. (**a**) Real and (**b**) Imaginary Part of the total complex wave potential and corresponding free surface elevation for a twin-hull floating structure of breadth $B = 20$ m and three considered cases of bottom slope 0, 10 and 20% in an environment with a mean depth of $h = 30$ m in the case of an incident wave of wavelength $\lambda/h = 2.4$ and amplitude $A = 1.5$ m.

5. Effects of Floating Structure Response in Waves on Floating PV Performance

The energy efficiency of a floating photovoltaic (FPV) unit is based on several parameters, many of which are the result of the surrounding marine environment. Some of the factors that affect the energy efficiency of FPV are common with corresponding land-based units, with similar power output levels, while others are absent in land installations.

In open seas, there is generally a higher level of humidity than inland, as well as lower ambient temperatures. The decreased temperatures are a result of various factors, which among others, include [29] the water's transparency, which results in the incoming solar radiation being transmitted to inner layers of the medium rather than just the surface layer, as well as the fraction of incident irradiation that is naturally used for evaporation. Furthermore, the wind speed is usually higher due to long fetch distances compared to land. The above parameters can help to maintain a low operating temperature of the solar cells, which, in turn, leads to close-to-optimal performance of the solar panel. The latter's efficiency decreases with increasing temperatures. More importantly, PV efficiency is strongly dependent on the angle of incidence (AOI) of solar irradiation, which, in offshore FPV installations, is directly affected by the dynamic wave-induced motions. In particular, the power output of photovoltaic cells is strongly affected by the angle of incidence (AOI) of solar irradiation and the plane of array (POA) irradiance, which is given by the following equation:

$$POA = DNI\cos(AOI) + DHI + RI, \tag{17}$$

where *DNI*, *DHI* and *RI* are the direct, diffuse and reflected irradiance components on a tilted surface, respectively. To provide indicative results regarding the effect of wave-induced motions of the floating structure on the power output, we considered an offshore installation in the geographical sea area of the southern Aegean Sea. For the latter area, the optimized values for tilt and azimuth angles of the photovoltaic installations, respectively, are $\theta_T = 30^o$ and $\theta_A = 135^o$ using data extracted from the Sandia Module Database, which is provided by the PV_LIB toolbox (https://pvpmc.sandia.gov/applications/pv_lib-toolbox/ (accessed on 12 August 2021)).

In this work, a preliminary assessment of a floating photovoltaic system's energy efficiency is made for twin-hull structures, taking into account data regarding the dynamic motions of the floating unit carrying the panels, as derived by the hydrodynamic model presented earlier, while the interesting effects of temperature and humidity will be studied in future work. The linear motions, i.e., sway (k = 2) and heave (k = 3), are considered to have no important effect on the tilt angle of the panels and, therefore, the angle of incidence. Hence the effect of the unit's mobility is limited to the angular oscillation i.e., the roll motion (k = 4) under excitation from the beam incident waves.

For this purpose, response data was simulated by assuming specific sea conditions. The latter are characterized by a frequency spectrum used to describe the incident waves. We considered the floating twin-hull structure of total breadth $B_{(T)} = 20$ m examined in the previous section in constant water depth $h = 30$ m. The sea state is described by a Brettschneider spectrum model (see [30], Chapter 2.3), as follows:

$$S(\omega; H_s, T_p) = \frac{5}{16} H_s^2 \frac{\omega_p^4}{\omega^5} \exp\left[-\frac{5}{4}\left(\frac{\omega}{\omega_p}\right)^{-4}\right] \tag{18}$$

where H_s is the significant wave height, $\omega_p = 2\pi/T_p$ is the peak frequency and T_p the corresponding peak period.

The roll responses calculated by the present model, as discussed in the previous section, were used to evaluate the fluctuations of the AOI and the effect on the power output performance of a PV system consisting of panels, with the aforementioned values of tilt (relative to the horizontal deck of the structure) and azimuth angles. Specifically, the roll spectrum was calculated using the RAO of the roll motion (see Figure 7a) of the twin-hull structure using:

$$S_4(\omega) = RAO^2(\omega) k^2 S(\omega) \tag{19}$$

where the wavenumber k is given by the dispersion relation of water waves for the water depth considered. Based on the calculated roll spectrum, time series of the roll motion $\xi_4(t; H_s, T_p)$ of the above floating twin-hull structure were simulated, for the considered configuration (structure and coastal environment) and incident waves, characterized by the parameters (H_s, T_p) using the random-phase model [30], Chapter 8.2, (see also [31]).

The results were normalized using the value corresponding to calm water (flat horizontal deck of the structure) in the same sea environment, which results in the following definition of the performance index:

$$PI(t) = \frac{a \ \cos(\alpha_m + \xi_4(t)) + b}{a \ \cos(\alpha_m) + b} \tag{20}$$

where $a = DNI$ and $b = DHI + RI$ for the geographical area and sea environment considered, respectively, and α_m is a representative value for the angle of incidence.

As an example, the numerical results concerning the calculated roll response of a floating twin-hull structure of breadth $B_{(T)} = 20$ m at depth $h = 30$ m with an incident wave spectrum (dashed line) and roll angle spectrum (solid line) of the structure in the case of incident waves of significant wave height $H_S = 0.5$ m and peak period $T_P = (2\pi/\omega_p) = 4$ s are presented in Figure 9. Based on the calculated roll spectrum, the simulated time series of the roll motion of the above floating twin-hull structure for the considered coastal environment and incident waves are shown in Figure 10 for a time interval of 1 h. Furthermore, in the same figure, a representative small interval of 3 min was obtained using the random-phase model [30,31] for a sea state that was characterized by significant wave height $H_S =$

0.5 m and peak period $T_P = 4$ s, as generated by winds corresponding to the Beaufort scale levels $BF = 1 - 2$. In this case, indicative results concerning the effect of waves and roll responses of the structure on the performance index are shown in Figure 11, as calculated by Equation (20) using a representative value of the mean angle of incidence $\alpha_m = 5°$ and omitting, as a first approximation, the effect of diffuse and reflected irradiance components $(b \approx 0)$. The value of the performance index in calm water was $PI_{CALM} = 0.9962$. In the considered case of incident waves, which were characterized by a very low energy content, the RMS value of the estimated performance index dropped to $PI_{RMS} = 0.9947$. The latter's mean value, as well as the corresponding calm-water value, are shown in Figure 11 using cyan and red lines, respectively.

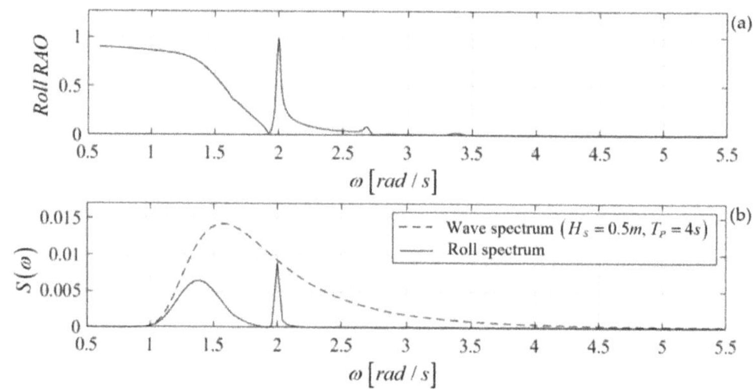

Figure 9. (**a**) Roll response of the floating twin-hull structure of breadth $B = 20$ m at a mean depth $h = 30$ m. (**b**) Incident wave spectrum (dashed line) and roll angle spectrum (solid line) of the structure in the case of incident waves with a significant wave height $H_S = 0.5$ m and a peak period $T_P = 4$ s.

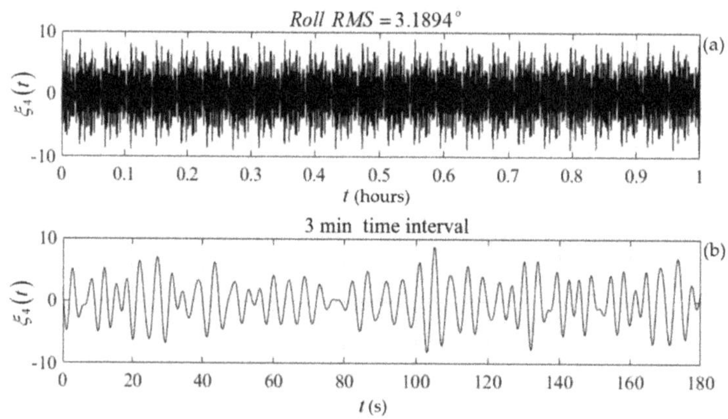

Figure 10. Simulated time series of the floating twin-hull structure's roll motion. Total breadth $B_{(T)} = 20\,m$ at a depth $h = 30\,m$ in the case of incident waves with a significant wave height equal to $H_S = 0.5\,m$ and a peak period $T_P = 4\,$s. (**a**) A 1 h long time series and (**b**) indicative roll motion over a 3 min long interval.

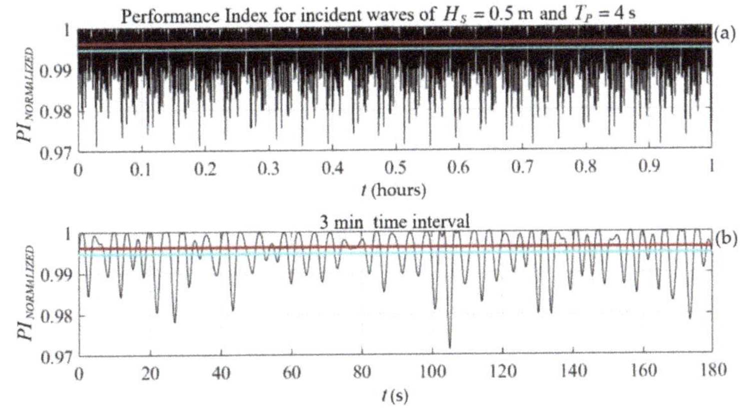

Figure 11. Performance Index of FPV on the floating twin hull structure of breadth $B_{(T)} = 20$ m at a depth $h = 30$ m in the case of incident waves with a significant wave height $H_S = 0.5$ m and a peak period $T_P = 4$ s. Roll motion time series (**a**) in a 1 h long time interval and (**b**) in an indicative 3 min long time interval.

6. Discussion

Following previous works [32,33], concerning the investigation and modeling of marine renewable energy systems, the present method focused on the estimation of the effect of wave-induced responses on the performance index of a twin-hull FPV structure in various sea conditions, as defined by the wave climatology of the offshore–coastal site where the system was deployed. As an example, the results concerning the performance index that is associated with the wave effects (Equation (20)) of the floating twin-hull structure of breadth $B_{(T)} = 20$ m at a water depth $h = 30$ m that are presented and discussed above are given in Table 1 for wind waves corresponding to the Beaufort scale from $BF = 1$ (relatively calm sea) to $BF = 5 - 6$ conditions.

Table 1. Performance indexes for different sea conditions.

BF	Sea Condition	$H_S(m)$	$T_P(s)$	PI_{RMS}
1–2	1–2	0.5	4	0.9947
3	3	1	6	0.9771
4–5	4	2	8	0.9203
5–6	4–5	3	9	0.8475

We observe that the effect of roll responses results in fluctuations of the AOI that could cause a significant drop in the performance index as the sea condition changed from calm to moderate and higher severities. A more complete picture of the sea state's effect on the FPV module's power output, as estimated using the present method, is shown in Figure 12, indicating its usefulness for supporting the systematic analysis and design of the system, including the offshore structure, as well as the electric production and storage subsystems. Although inevitable fluctuations of the AOI due to waves in offshore PV units reduce the power output, this negative effect could be balanced or even reversed by the cooling effect and other factors resulting from the marine environment, which is a subject that is left to be considered in future work.

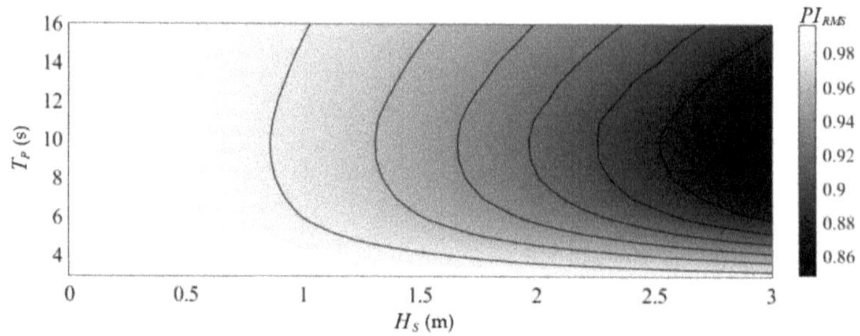

Figure 12. Contour map of the normalized performance index as a function of the prevailing sea state (significant wave height H_S and peak period T_P).

7. Conclusions

A BEM model was developed and applied to the hydrodynamic analysis of twin-hull structures in variable bathymetry regions and was used to predict their hydrodynamic responses and their effects concerning the power output of offshore FPV systems. The analysis was restricted to two spatial dimensions for simplicity. After verification of the method with comparisons against data from the literature, the method was systematically applied and the derived numerical results are presented for floating bodies of simple geometry, lying over uniform and sloping seabeds. With the aid of systematic comparisons, the effects of bottom slope on the hydrodynamic characteristics (hydrodynamic coefficients and responses) of the floating bodies were illustrated and discussed. Finally, response data that was simulated for specific sea conditions, characterized by frequency spectra, were considered to describe the incident waves interacting with a floating twin-hull structure, in order to evaluate the effect of wave-induced fluctuations on the power output performance of the floating PV system. The effects of waves on the floating PV performance are presented, indicating significant variations of the performance index ranging from 0 to 15% depending on the sea state. Future work will be directed toward (a) the detailed analysis of wave and wind environmental factors and their effects on the resulting system's performance, (b) the extension of the model to 3D including 6-DOF wave motion analysis of the floating structures over general bathymetry and evaluation of their performance and (c) a systematic application of the present method to realistic cases that support the optimized design of floating PV modules in specific marine environments.

Author Contributions: Conceptualization, K.B.; methodology, K.B. and A.M; software, A.M. and K.B.; validation, A.M.; writing—original draft preparation, A.M.; writing—review and editing, K.B. and E.R.; visualization, E.R.; supervision, K.B. All authors have read and agreed to the published version of the manuscript.

Funding: This work was supported by the research project DREAM (Dynamics of the Resources and Technological Advance in Harvesting Marine Renewable Energy), funded by the Romanian Executive Agency for Higher Education, Research, Development and Innovation Funding–UEFISCDI, grant number PNIII-P4-ID-PCE-2020-0008.

Acknowledgments: The doctoral studies of A. Magkouris are currently supported by the Special Account for Research Funding (E.L.K.E.) of the National Technical University of Athens (N.T.U.A.) operating in accordance with Law 4485/17 and National and European Community Law.

Conflicts of Interest: The authors declare no conflict of interest.

Abbreviations

AOI	Angle of incidence
BC	Boundary condition
BEM	Boundary element method
BM	Metacentric radius
DHI	Diffuse horizontal irradiance
DNI	Direct normal irradiance
DOF	Degree of freedom
DtN	Dirichlet-to-Neumann
FPV	Floating photovoltaic
GM	Metacentric height
KB	Distance from reference point K to center of buoyancy
KG	Distance from reference point K to center of gravity
PI	Performance index
POA	Plane of array
PV	Photovoltaic
RAO	Response amplitude operator
RI	Reflected irradiance
RMS	Root mean square (here of the time series data)
SWL	Still-water level

Appendix A. Dirichlet-to-Neumann Operators

By projecting the terms of Equation (5a) on the orthonormal basis, spanned by the normalized eigenfunctions $\widetilde{Z}_n^{(i)}(z) = Z_n^{(i)}(z)/\left\| Z_n^{(i)} \right\|$, with $\left\| Z_n^{(i)} \right\|$ standing for the L_2-norm of each vertical function:

$$\left\| Z_n^{(i)} \right\| = \left\{ \int_{z=-h_i}^{0} \left[Z_n^{(i)}(x_3) \right]^2 dx_3 \right\}^{1/2}, i = L, R, \tag{A1}$$

we obtain

$$\left\langle \varphi_p^{(L)}(\mathbf{x}) \cdot \widetilde{Z}_n^{(L)}(x_3) \right\rangle = \begin{cases} \exp\left(ik_0^{(L)} x_2\right) + C_0^{(L)} \exp\left(-ik_0^{(L)} x_2\right), & n = 0 \\ C_n^{(L)} \exp\left[k_n^{(L)} (x_2 - a)\right], & n \geq 1 \end{cases} \tag{A2}$$

where $\langle f(x_3), g(x_3) \rangle = \int_{x_3=-h_a}^{x_3=0} [f(x_3) \cdot g(x_3)]\, dx_3$. Therefore, the reflection coefficient in the left half-strip D_L is equal to

$$C_0^{(L)} = \frac{-\exp\left(ik_0^{(L)} x_2\right) + \left\langle \varphi_p^{(L)}(\mathbf{x}) \cdot \widetilde{Z}_0^{(L)}(x_3) \right\rangle}{\exp\left(-ik_0^{(L)} x_2\right)}, \ (x_2, x_3) \in D_L \tag{A3}$$

Moreover, by calculating the derivative of Equation (5a) with respect to the unit normal vector n (which is directed opposite to the x_2 direction on ∂D_5) and replacing in Equation (6e)

$$\begin{aligned} -\frac{\partial \varphi_p^{(L)}(\mathbf{x})}{\partial n} &= ik_0^{(L)} \left[\exp\left(ik_0^{(L)} x_2\right) - C_0^{(L)} \exp\left(-ik_0^{(L)} x_2\right) \right] Z_0^{(L)}(x_3) + \\ &+ \sum_{n=1}^{\infty} k_n^{(L)} C_n^{(L)} \exp\left[k_n^{(L)} (x_2 - a)\right] Z_n^{(L)}(x_3) = -T_L\left[\varphi_p^{(L)}(\mathbf{x})\right] - Q_p, \mathbf{x} \in D_L, \end{aligned} \tag{A4}$$

and by using Equation (A2), we conclude to

$$T_L\left[\varphi_p^{(L)}(\mathbf{x})\right] = ik_0^{(L)}\widetilde{Z}_0^{(L)}(x_3)\left\langle \varphi_p^{(L)}(\mathbf{x}) \cdot \widetilde{Z}_0^{(L)}(x_3)\right\rangle + \\ -\sum_{n=1}^{\infty} k_n^{(L)}\widetilde{Z}_n^{(L)}(x_3)\left\langle \varphi_p^{(L)}(\mathbf{x}) \cdot \widetilde{Z}_n^{(L)}(x_3)\right\rangle, \tag{A5}$$

where

$$Q_p = -2ik_0^{(L)} \exp\left(ik_0^{(L)}x_2\right)\widetilde{Z}_0^{(L)}(x_3). \tag{A6}$$

Similarly, for $k = 2,3,4$, we obtain

$$-\frac{\partial\varphi_k^{(L)}(\mathbf{x})}{\partial n} = -ik_0^{(L)}C_0^{(L)} \exp\left(-ik_0^{(L)}x_2\right)Z_0^{(L)}(x_3) + \\ +\sum_{n=1}^{\infty} k_n^{(L)}C_n^{(L)} \exp\left[k_n^{(L)}(x_2-a)\right]Z_n^{(L)}(x_3) = -T_L\left[\varphi_k^{(L)}(\mathbf{x})\right] - Q_k, \mathbf{x} \in D_L. \tag{A7}$$

Using the above results, we obtain $T_L\left[\varphi_k^{(L)}(\mathbf{x})\right] = T_L\left[\varphi_p^{(L)}(\mathbf{x})\right]$, $k = 2,3,4$, and $Q_k = 0$, $k = 2,3,4$. Similarly, for the wavefield in the domain D_R:

$$\frac{\partial\varphi_p^{(R)}(\mathbf{x})}{\partial n} = ik_0^{(R)}C_0^{(R)} \exp\left(ik_0^{(R)}x_2\right)Z_0^{(R)}(x_3) + \\ +\sum_{n=1}^{\infty} k_n^{(R)}C_n^{(R)} \exp\left[k_n^{(R)}(b-x_2)\right]Z_n^{(R)}(x_3) = T_R\left[\varphi_p^{(R)}(\mathbf{x})\right], \mathbf{x} \in D_R. \tag{A8}$$

and thus

$$T_R\left[\varphi_p^{(R)}(\mathbf{x})\right] = ik_0^{(R)}Z_0^{(R)}(x_3)\left\langle \varphi_k^{(R)}(\mathbf{x}) \cdot \widetilde{Z}_0^{(R)}(x_3)\right\rangle + \\ +\sum_{n=1}^{\infty} k_n^{(R)}Z_n^{(R)}(x_3)\left\langle \varphi_k^{(R)}(\mathbf{x}) \cdot \widetilde{Z}_n^{(R)}(x_3)\right\rangle. \tag{A9}$$

References

1. Trapani, K.; Santafé, M.R. A review of floating photovoltaic installations: 2007-13. *Prog. Photovolt.* **2014**, *23*, 524–532. [CrossRef]
2. Grech, M.; Stagno, M.L.; Aquilina, M.; Cadamuro, M.; Witzke, U. Floating Photovoltaic Installation in Maltese Sea Waters. In Proceedings of the 32nd European Photovoltaic Solar Energy Conference and Exhibition, Munich, Germany, 20–24 June 2016.
3. Diendorfer, C.; Haider, M.; Lauermann, M. Performance analysis of offshore solar power plants. *Energy Procedia* **2014**, *49*, 2462–2471. [CrossRef]
4. Sukarso, A.P.; Kim, K.N. Cooling Effect on the Floating Solar PV: Performance and Economic Analysis on the Case of West Java Province in Indonesia. *Energies* **2020**, *13*, 2126. [CrossRef]
5. Muscat, M. *A Study of Floating PV Module Efficiency*; Institute for Sustainable Energy, University of Malta: Msida, Malta, 2014.
6. Mei, C.C. *The applied Dynamics of Ocean Surface Waves*, 2nd ed.; World Scientific: Singapore, 1989.
7. Kuznetsov, N.; Maz'ya, V.; Vainberg, B. *Linear Water Waves*; Cambridge University Press: Cambridge, UK, 2002.
8. Evans, D.V.; Kuznetsov, C.M. Chapter 4. In *Gravity Waves in Water of Finite Depth*; Hunt, J.N., Ed.; Computational Mechanics Publications: Southampton, UK, 1997.
9. Takagi, K.; Naito, S.; Hirota, K. Hydrodynamic Forces Acting on a Floating Body in a Harbour of Arbitrary Geometry. In Proceedings of the Third International Offshore and Polar Engineering Conference, Singapore, 6–11 June 1993; Volume 3, pp. 192–199.
10. Ohyama, T.; Tsuchida, M. Expanded mild-slope equations for the analysis of wave-induced ship motion in a harbor. *Coast. Eng.* **1997**, *30*, 77–103. [CrossRef]
11. Yeung, R.W. A hybrid integral-equation method for time harmonic free surface flows. In Proceedings of the 1st International Conference on Numerical Ship Hydrodynamics, Gaithersburg, MD, USA, 20–22 October 1975.
12. Nestegard, A.; Sclavounos, P.D. A numerical solution of two-dimensional deep water wave body problems. *J. Ship Res.* **1984**, *28*, 48–54. [CrossRef]
13. Drimer, N.; Agnon, Y. A hybrid boundary element method for second-order wave-body interaction. *Appl. Ocean Res.* **1994**, *16*, 27–45. [CrossRef]
14. Wehausen, J.; Laitone, E. Surface Waves. In *Encyclopedia of Physics, FD III 9*; Springer: Berlin, Germany, 1960.
15. Wehausen, J.V. The motion of floating bodies. *Annual Rev. Fluid Mech.* **1971**, *3*, 237–268. [CrossRef]
16. Newman, J.N. *Marine Hydrodynamics*; MIT Press: Cambridge, MA, USA, 1977.
17. Massel, S. Extended refraction-diffraction equations for surface waves. *Coast. Eng.* **1993**, *19*, 97–126. [CrossRef]

18. Coddington, E.A.; Levinson, N. *Theory of Ordinary Differential Equations*; McGraw Hill: New York, NY, USA, 1955.
19. Givoli, D. Non-reflecting boundary conditions. *J. Comp. Phys.* **1991**, *94*, 1–29. [CrossRef]
20. Wehausen, J.V. *Methods for Boundary-Value Problems in Free-Surface Flows*; David Taylor Naval Ship Research and Development Center Rep.: Bethesda, MD, USA, 1974.
21. Kress, R. *Linear Integral Equations*; Springer: Berlin, Germany, 1989.
22. Katz, J.; Plotkin, A. *Low Speed Aerodynamics*; McGraw Hill: Singapore, 1991.
23. Belibassakis, K.A. A boundary element method for the hydrodynamic analysis of floating bodies in variable bathymetry regions. *Eng. Anal. Bound. Elem.* **2008**, *32*, 796–810. [CrossRef]
24. Lewis, E.V. (Ed.) *Principles of Naval Architecture*; Society of Naval Architects and Marine Engineers; SNAME: Jersey City, NJ, USA, 1989; Volume III.
25. Drimer, N.; Agnon, Y.; Stiassnie, M. A simplified analytical model for a floating breakwater in water of finite depth. *Appl. Ocean Res.* **1992**, *14*, 33–41. [CrossRef]
26. Ohkusu, M. *On the Heaving Motion of Two Circular Cylinders on the Surface of a Fluid*; Reports of Research Institute for Applied Mechanics; Kyushu University: Kyushu, Japan, 1969.
27. Rhee, K.P. Investigation of the Effect of Water Depths on Two-dimensional Hydrodynamic Coefficients for Twin-Hull Sections. *J. Soc. Nav. Archit. Korea* **1982**, *19*, 39–45.
28. Dabssi, N.; Chagdali, M.; Hémon, A. Hydrodynamic Coefficients and Forces on Multihulls in Shallow Water with Constant or Variable Depth. *Transport* **2008**, *23*, 245–252. [CrossRef]
29. Dörenkämper, M.; Wahed, A.; Kumar, A.; De Jong, M.; Kroon, J.; Reindl, T. The cooling effect of floating PV in two different climate zones: A comparison of field test data from the Netherlands and Singapore. *Sol. Energy* **2021**, *214*, 239–247. [CrossRef]
30. Goda, Y. *Random Seas and Design of Maritime Structures*; World Scientific: Singapore, 2000.
31. Ochi, M.K. *Ocean Waves*; The stochastic approach; Cambridge University Press: Cambridge, UK, 1998.
32. Belibassakis, K.; Bonovas, M.; Rusu, E. A Novel Method for Estimating Wave Energy Converter Performance in Variable Bathymetry Regions and Applications. *Energies* **2018**, *11*, 2092. [CrossRef]
33. Bonovas, M.; Belibassakis, K.; Rusu, E. Multi-DOF WEC Performance in Variable Bathymetry Regions Using a Hybrid 3D BEM and Optimization. *Energies* **2019**, *12*, 2108. [CrossRef]

Journal of
Marine Science and Engineering

MDPI

Article

Capture Power Prediction of the Frustum of a Cone Shaped Floating Body Based on BP Neural Network

Wei Wang [1], Yanjun Liu [1,2,3,*], Fagang Bai [1] and Gang Xue [1]

1 Institute of Marine Science and Technology, Shandong University, Qingdao 266200, China; wangwei_sdu@mail.sdu.edu.cn (W.W.); baifagang14@163.com (F.B.); xuegangzb@163.com (G.X.)
2 School of Mechanical Engineering, Shandong University, Jinan 250061, China
3 Key Laboratory of High Efficiency and Clean Mechanical Manufacture, Ministry of Education, Jinan 250061, China
* Correspondence: lyj111ky@163.com

Abstract: How to improve the power generation of wave energy converters (WEC) has become one of the main research objectives in wave energy field. This paper illustrates a framework on the use of back propagation (BP) neural network in predicting capture power of the frustum of a cone shaped floating body. Mathematical model of single floating body is derived, and radius, semi-vertical angle, mass, submergence depth, power take-off (PTO) damping coefficient, and stiffness coefficient are identified as key variables. Commercial software ANSYS-AQWA is used for numerical simulations to obtain hydrodynamic parameters, and then capture power is calculated by these parameters. A database containing 100 samples is established by Latin hypercube sampling (LHS) method, and a simple feature study is conducted. A BP neural network model with high accuracy is designed and trained for predictions based on built database. The results show that forecasting results and desired outputs are in great agreement with error percentage not greater than 4%, correlation coefficient (CC) greater than 0.9, P value close to 1, and root mean square error (RMSE) less than 139 W. The proposed method provides a guideline for designers to identify basic parameters of the floating body and system damping coefficient.

Keywords: structure parameters; ANSYS-AQWA simulations; feature study; BP neural network; power predictions

Citation: Wang, W.; Liu, Y.; Bai, F.; Xue, G. Capture Power Prediction of the Frustum of a Cone Shaped Floating Body Based on BP Neural Network. *J. Mar. Sci. Eng.* **2021**, *9*, 656. https://doi.org/10.3390/jmse9060656

Academic Editor: Spyros A. Mavrakos

Received: 7 May 2021
Accepted: 10 June 2021
Published: 13 June 2021

Publisher's Note: MDPI stays neutral with regard to jurisdictional claims in published maps and institutional affiliations.

Copyright: © 2021 by the authors. Licensee MDPI, Basel, Switzerland. This article is an open access article distributed under the terms and conditions of the Creative Commons Attribution (CC BY) license (https://creativecommons.org/licenses/by/4.0/).

1. Introduction

Wave energy converters (WEC), a new type of energy extractor with little pollution, are expected to be a reliable alternative to the current generation method. There are two stages for an oscillating body WEC transforming wave energy into other forms of energy like electricity. A floating body is firstly required to capture the wave energy induced by a wave's motion. Then the moving floating body drives a generator to generate power. An intact oscillating body WEC system is generally composed of a moving floating body, a power take-off (PTO) system, and an anchor chain, etc. At present, the conversion efficiency of WECs is relatively low, so the main research objective is to improve the power generation of a specific device.

One method is to design a different floating body's shape, and the shape is usually irregular curved surface. McCabe [1] researched the optimization of the shape of a wave energy collector to improve energy extraction by genetic algorithms, and a benchmark collector shape was identified. Colby [2] used evolutionary algorithms to optimize the ballast geometry and achieved 84% improvement in power output. Fang [3] designed a mass-adjustable float, and a new optimization calculation method was proposed. Multi-freedom buoys have been also proposed in [4–6]. They can translate or rotate in more than one freedom, so more wave energy can be captured. Another means is to design an innovative PTO system. Reabroy [7] proposed a novel floating device integrated with a fixed breakwater. The simulations and experiments proved that installing a breakwater can greatly improve the conversion efficiency. Liang [8] designed a novel PTO system which is inside the buoy with a mechanical motion rectifier (MMR). This mechanism can convert the bidirectional wave motion into unidirectional rotation of the generator by two one-way bearings. Li [9] improved this device by substituting one-way bearings for two one-way clutches. Chen [10] proposed a new point-absorber WEC with an outer-floater and a built-in power take-off mechanism. Besides, array-type WECs, integrated with many buoys and PTO systems, are also researched to achieve large scale power generation. The typical one is WaveNET [11], developed by Albatern in Scotland. Sun [12] proposed an array-type energy-capturing mechanism integrated with marine structures. Liu [13] proposed an array-type WEC combined with oscillating buoy.

The factors that affect the power generation have also been studied recently. Zou [14] analyzed the effects of spring force, mass force, and damping force on energy conversion efficiency based on a 3D wave tank model. Yu [15] and Wu [16] discussed the influence of the floating body's shape, PTO damping coefficient, system stiffness coefficient, and geometry parameters on power generation. Zheng [17] established an optimization model of energy conversion performance via genetic algorithm. Ma [18] researched the two-body floating point absorber and the results showed that stiffness coefficient had less effect on the power generation than damping coefficient. Ji [19] proved that PTO damping coefficient and submerged body volume were the most important parameters that affect the output power, and that the significant wave height had little influence on conversion efficiency. Tongphong [20] analyzed the effects of wave frequency, PTO damping coefficient, and structure form (floating or fixed) on capture factors.

Wave load and hydrodynamic parameters are vital factors in the analysis process of floating body's motions. Numerical simulations are widely used in hydrodynamic performance analysis to obtain these parameters. Ma [21] used ANSYS-AQWA software to assess the hydrodynamic performance and energy conversion of a pitching float WEC and analyzed key factors' influences on the performance. Amiri [22] presented a numerical simulation scheme for a point wave absorber and analyzed its performance. Yu [23] applied Reynolds-Averaged Navier-Stokes (RANS) computational method for analyzing the hydrodynamic heave response of a specific WEC device.

In addition to the traditional physical model [24,25], novel methods and models based on big data and machine learning have also been presented. Law [26] carried out wave prediction over a large distance downstream using artificial neural network, introducing machine learning algorithm into ocean engineering. Desouky [27] utilized non-linear autoregressive with exogenous input network to predict the surface elevation with the help of an ahead located sensor. Kumar [28] used the Minimal Resource Allocation Network (MRAN) and the Growing and Pruning Radial Basis Function (GAP-RBF) network to predict the daily wave heights based on real marine data. Some elevating measurements are also proposed to assess the performance of predictions in [29]. Avila [30] combined Fuzzy Inference Systems (FIS) and Artificial Neural Networks (ANN) to forecast wave energy in Canary Islands. Wang [31] predicted power outputs of a WEC in shallow water, taking bottom effects into accounts. Halliday [32] utilized Fast Fourier Transform (FFT) to predict wave behavior in short term based on real marine data. Davis [33] used a nonlinear Extended Kalman Filter to estimate the wave excitation force based on experimental wave tank data. Ni [34] combined the Long Short-Term Memory (LSTM) algorithm and the principal component analysis (PCA) together to predict the power generation of a WEC.

Different from traditional mathematical model, this paper presents an agent model using BP neural network to determine the complex non-linear reflection between design variables and power generation. Power predictions are the foundation of multi-objective optimizations of a floating body. Accurate power prediction can provide a guide for the electricity consumption, allocation, and distribution in power grid. Through the prediction, the unknown generation power becomes measurable, so reasonable manners can be arranged to increase the grid capacity.

The remainder of this paper is organized as follows: Section 2 develops the mathematical model of the oscillating float-type WEC. In Section 3, a sample database is established by LHS method, and a simple feature study is conducted. The geometric model and simulations of each sample are done in ANSYS-AQWA (developed by ANSYS company, based in Canonsburg, Pennsylvania, USA) in Section 4. Section 5 designs a BP neural network model and it is used to predict the capture power. Results and discussion are given in Section 6 and conclusions are presented in Section 7.

2. Mathematical Model

A schematic diagram of the oscillating body WEC is shown in Figure 1. To simplify the study, some assumptions are made as below:

1. linear wave theory and potential flow theory are suitable for this model, and they are used to describe wave motion;
2. only the heave motion of the floating body is considered;
3. the viscous force and mooring force acting on the floating body are ignored [10];
4. the PTO system is linear.

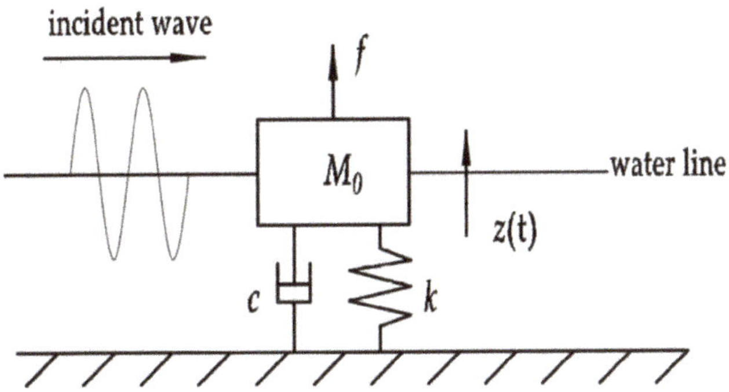

Figure 1. Mechanical model.

Under three assumptions, the following forces act on the floating body: hydrodynamic forces (excitation and radiation force); hydrostatic buoyancy; PTO damping force; rigid restoring force. According to the theory of fluid mechanics and Newton's law, the governing equation of the floating body can be expressed as follows:

$$M_0\ddot{z}(t) = f_E - f_R - f_S - f_{PTO} - f_K \tag{1}$$

where M_0 represents mass; $z(t)$ represents the heave displacement; f_E represents excitation force; f_R represents radiation force; f_S represents hydrostatic buoyancy; f_{PTO} represents PTO damping force; f_K represents rigid restoring force.

The excitation force imparts on the floating body by the incoming wave. It is the summation of the Froude-Krylov force f_{FK} and the diffraction force f_D, so it can be written as follows:

$$f_E = f_{FK} + f_D \tag{2}$$

The radiation force is induced by the floating body's motion and can be decomposed into an added mass term and a radiation damping term [25], so it can be expressed as follows:

$$f_R = A_M\ddot{z}(t) + B_C\dot{z}(t) \tag{3}$$

where A_M and B_C are the added mass and radiation damping in the vertical direction, respectively.

The hydrostatic buoyancy, induced by seawater static pressure, is the resultant force of gravity and buoyancy. It is a force that restores the structure to hydrostatic equilibrium and is linear with the heave displacement of the floating body. It can be written as:

$$f_S = \rho g A_W z(t) \tag{4}$$

where ρ is seawater density; g is acceleration of gravity; A_W is water cross area of the floating body. In this paper, the value of PTO damping is relatively large, and the heave displacement of the floating body is small. As a result, it is assumed that the water cross area of the floating body does not change. It is the section where the water line is located when the floating body is in still water. Therefore, the hydrostatic buoyancy can be expressed as:

$$f_S = \frac{1}{4}\pi\rho g D^2 z(t) \tag{5}$$

where D is the diameter of the floating body.

The energy conversion system can be simplified to a linear spring damping system, so the PTO damping force is

$$f_{PTO} = c\dot{z}(t) \tag{6}$$

where c is the damping coefficient of the PTO system.

The rigid restoring force is proportional to the heave displacement, and it can be written as follows:

$$f_k = kz(t) \tag{7}$$

where k is stiffness coefficient.

Reformulate Equation (1) through Equations (2), (3), (5)–(7):

$$[M_0 + A_M]\ddot{z}(t) + (B_C + c)\dot{z}(t) + (\rho g A_W + k)z(t) = f_E(t) \tag{8}$$

Apply Fourier transform to Equation (8) and obtain another governing equation in the frequency domain. It is

$$\left[(\rho g A_W + k) + j\omega(B_C + c) - \omega^2(M_0 + A_M)\right]Z(\omega) = F_E(\omega) \tag{9}$$

where ω is the wave frequency; j is imaginary unit; $Z(\omega)$ and $F_E(\omega)$ are functions of displacement and excitation force in the frequency domain, respectively.

In the frequency domain, the excitation force can be expressed by the product of the unit excitation force and the incident wave amplitude [35]. It is

$$F_E(\omega) = F_{unit}(\omega)A(\omega) \tag{10}$$

Equation (9) can be rewritten as follows:

$$\left[(\rho g A_W + k) + j\omega(B_C + c) - \omega^2(M_0 + A_M)\right]Z(\omega) = F_{unit}(\omega)A(\omega) \tag{11}$$

Formula (11) is a typical damped and forced vibration equation, so the natural frequency and damping factor can be expressed as below:

$$\omega_n = \sqrt{\frac{\rho g A_W + k}{M_0 + A_M}} \tag{12}$$

$$\beta_n = \frac{B_C + c}{2(M_0 + A_M)} \tag{13}$$

From Equations (12) and (13), the natural frequency and damping factor of a given WEC change over added mass and added damping.

According to Equation (11), the heave response in the frequency domain is

$$Z(\omega) = \frac{F_{unit}(\omega)A(\omega)}{(\rho g A_W + k) + j\omega(B_C + c) - \omega^2(M_0 + A_M)} \tag{14}$$

The average power in one wave period, captured by the floating body with heave motion, can be written as the product of damping force and vertical velocity. The work done by damping force is the energy absorbed by the floating body, so the mean capture power is

$$\begin{aligned} P_{\text{mean}} &= \frac{1}{T}\int_0^T f_{PTO}\dot{z}(t)dt \\ &= \frac{1}{T}\int_0^T c\dot{z}(t)^2 dt \\ &= \frac{1}{2}\omega^2 c|Z(\omega)|^2 \\ &= \frac{1}{2}c\frac{\omega^2|F_E|^2}{[-\omega^2(M_0+A_M)+k+\rho g A_W]^2+\omega^2(B_C+c)^2} \\ &= \frac{1}{2}c\frac{|F_E|^2}{[\frac{-\omega^2(M_0+A_M)+k+\rho g A_W}{\omega}]^2+(B_C+c)^2} \end{aligned} \tag{15}$$

The mean capture power reaches the maximum when the following conditions are met.

$$k = \omega^2(M_0 + A_M) - \rho g A_W \tag{16}$$

$$c = \begin{cases} B_C, B_C > 0 \\ -B_C, B_C < 0 \end{cases} \tag{17}$$

This stiffness and damping are called the best stiffness and the best damping, respectively. When $B_c > 0$, the natural frequency, damping factor, and displacement are

$$\omega_n = \omega \tag{18}$$

$$\beta_n = \frac{B_C}{M_0 + A_M} \tag{19}$$

$$Z(\omega) = -\frac{jF_{unit}(\omega)A(\omega)}{2\omega B_C} \tag{20}$$

The max capture power is

$$P_{\text{max}} = \frac{|F_E|^2}{8B_C} \tag{21}$$

3. Design of Experiments (DOE) Method

3.1. Latin Hypercube Sampling

The sampling method is of great importance in experimental design. A good sampling method can result in more reasonable sample distribution, leading to a better model with higher accuracy. In this paper, a Latin hypercube sampling (LHS) method is utilized to generate sample points. Different from random sampling, LHS has a high efficiency of space filling by maximizing the stratification of each edge distribution, which improves the uniformity.

According to Equation (15), the factors that determine the capture power under given wave conditions are PTO damping coefficient c, system stiffness coefficient k, wave exciting force F_E, float mass m, added mass A_M, and added damping B_C. Added mass, added damping, and wave exciting force are related to the geometry and submergence depth of the floating body. The geometric features of the floating body depend on radius R, semi-vertical angle α, and mass m. As a result, four main geometric parameters, including radius R, semi-vertical angle α, mass m, and submergence depth d, plus two system parameters, PTO damping coefficient c and stiffness coefficient k, are selected as key variables that affect the capture power.

The sample space of six key variables are defined as follows:

$$\begin{cases} d \in [2,3] \\ R \in [1.5,3] \\ m \in [7000,8000] \\ \alpha \in [5,25] \\ c \in [10,000,30,000] \\ k \in [3000,6000] \end{cases} \tag{22}$$

A database covering 100 sample points is established (see in Appendix A) and scatter diagrams of these samples are shown in Figure 2.

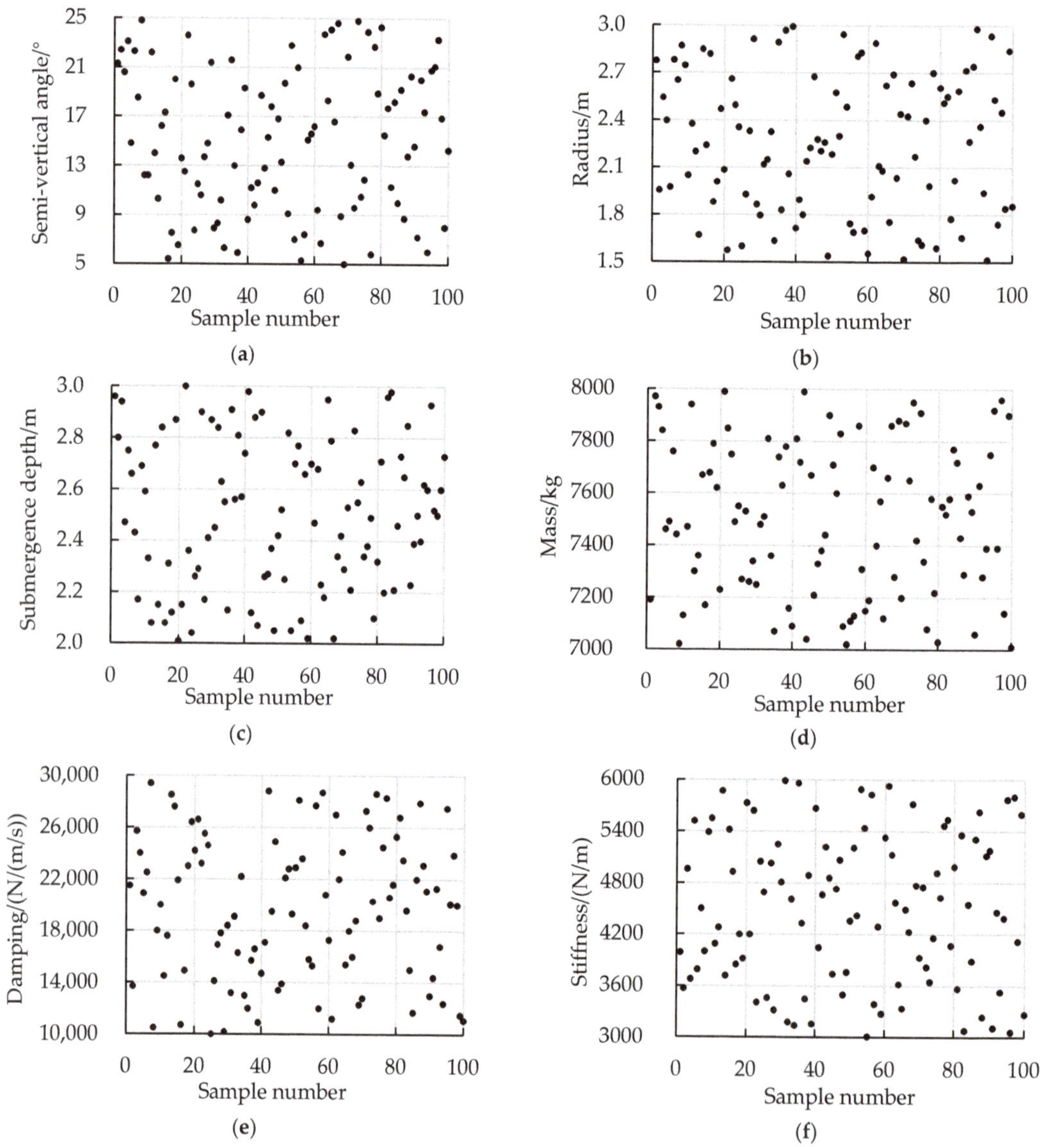

Figure 2. Sample scatter diagrams. (**a**) semi-vertical angle; (**b**) radius; (**c**) submergence depth; (**d**) mass; (**e**) damping; (**f**) stiffness.

In Figure 2, each variable fills the whole sample space and the standard deviation of the value is small. It can reflect the relationship between the factor and the response in the six spaces.

3.2. Feature Study

Suitable feature study on the data set can give an insight to the correlation between the inputs and output. Pearson correlation analysis is conducted in this section to identify the correlation between six key variables and the capture power. Figure 3 shows the correlation coefficients in different wave situations. In this heatmap, a negative value means a negative correlation, and a positive value means a positive correlation. A large absolute value means a strong correlation.

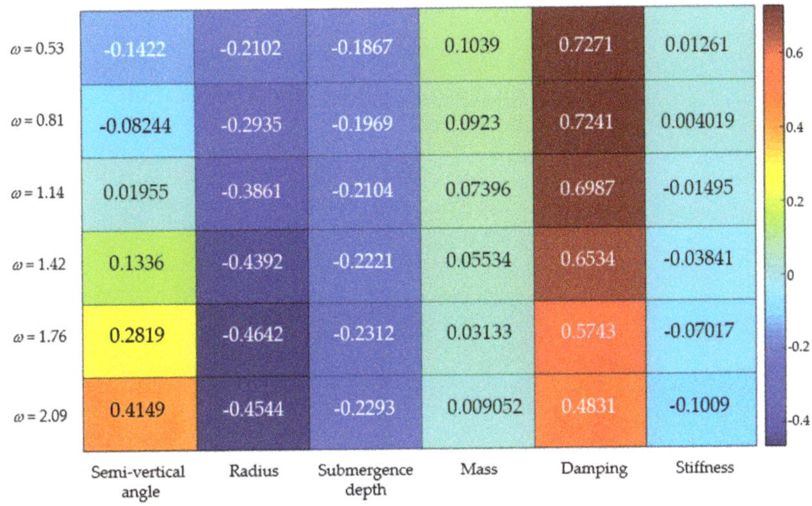

Figure 3. The correlation coefficients between the inputs and output.

In general, radius, submergence depth, and damping show a strong correlation, while semi-vertical angle, mass, and stiffness behave a weak correlation. Besides, the correlation is different at different wave frequency. When the wave frequency is 0.53 and 0.81 rad/s, semi-vertical angle shows a negative correlation, while a positive correlation comes up at other frequencies. The similar situation happens on stiffness. Mass and damping behave a positive correlation, while radius and submergence depth show a negative correlation all the time.

4. Numerical Simulations

4.1. Simulation Scheme

The structural schematic of the cone shaped floating body investigated in this study is shown in Figure 4.

The height of the cylinder part above the waterline is a constant, 0.5 m. In ANSYS Design Modeler, the 3D geometry with given parameters is created.

In this paper, ANSYS-AQWA, a commercial computation software based on potential flow theory, is utilized to calculate hydrodynamic parameters. The simulation process, including numerical modeling, parameters setting, mesh generation, and data post-processing, can be conducted in the graphical interface directly. The basic simulation steps for each sample are as follows:

1. The moment of inertia and center of mass of the floating body are calculated in Static Structural module;

2. Set solution environment in hydrodynamic diffusion module. The water line is at z = 0, the seawater depth is 200 m, and the surface area are 100 m × 100 m. Details of the point mass, additional damping, and additional hydrostatic stiffness are set according to the results obtained in Static Structural module and parameters in Table A1. In this study, the considered wave range is from 0 to 0.4 Hz, meaning that the wave circular frequency within 2.5 rad/s needs to be simulated. Therefore, the defeaturing tolerance and maximum element size are 0.5 m and 1 m, respectively. The maximum allowed wave frequency is 0.61 Hz in this scheme;
3. Solve the model in the frequency domain and obtain Diffraction and Froude-Krylov force F_e, added mass A_M, and radiation damping B_C.

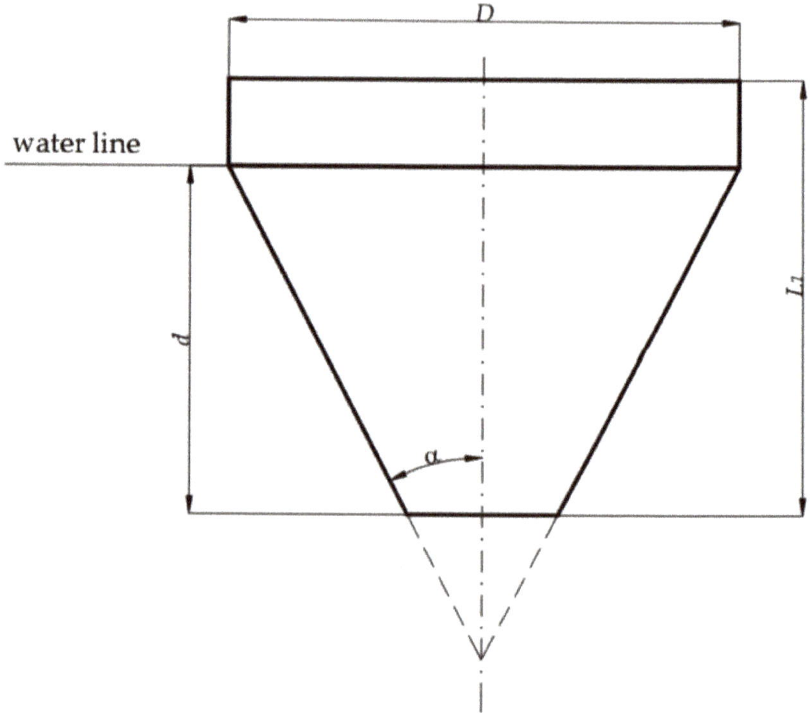

Figure 4. Schematic of the floating body's structure.

For each simulation, the given frequency range is divided into 52 frequency points. The mean power for each sample at each frequency is calculated. The results of sample 1 and sample 2 are shown in Figure 5.

With the increase in wave frequency, the capture power rises firstly and then drops steadily. For each sample, there is a unique optimal frequency in which the capture power can reach the maximum. The 100 samples' capture power are calculated so that they can be used as training set and test set for BP neutral network. Only two samples' results are presented in this figure.

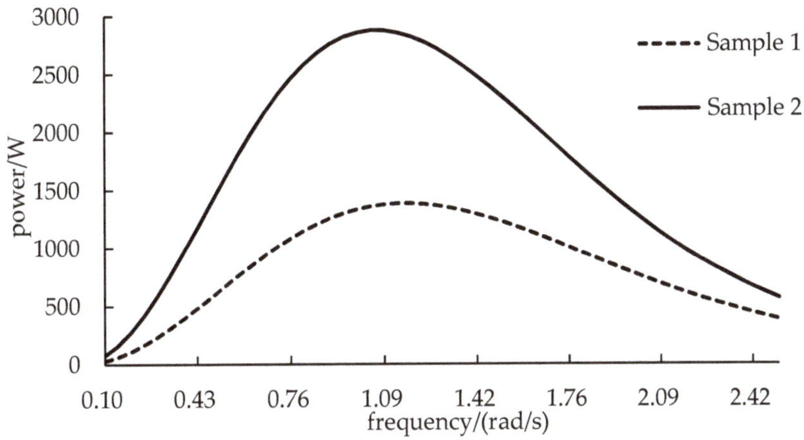

Figure 5. Mean capture power of sample 1 and sample 2.

4.2. Theoretical Verification of Simulations

Falnes [36] illustrated that the maximum power that a heaving axisymmetric body can absorb is

$$P_{\max} = \frac{J\lambda}{2\pi} \tag{23}$$

where J is the wave energy flux; λ is the wavelength. For deep-water waves, $\lambda = g/2\pi$. J is

$$J = \frac{\rho g^2 T H^2}{32\pi} \tag{24}$$

where T and H are wave period and height, respectively. Budal's upper bound [36] gave another upper limit power that a submerged body with given volume V can absorb. It is

$$P_u < \frac{\pi \rho g V H}{4T} \tag{25}$$

where V is the volume of the submerged part. The point of intersection of two theoretical curves can be defined as (Tc, Pc). P_c is

$$P_c = \frac{\rho g^2}{32\pi}\sqrt{2VH^3} \tag{26}$$

In this study, Equations (23) and (25) are used to verify the validity of simulations. To make comparisons, the results are normalized by dividing P_c. The three curves are shown in Figure 6.

It can be found that the capture power curves of two samples are in the area enclosed by curve P_{max}, curve P_u, and coordinate axes, which means the simulation scheme is accurate and reliable. All the samples are verified successfully and only two of them are demonstrated in this section.

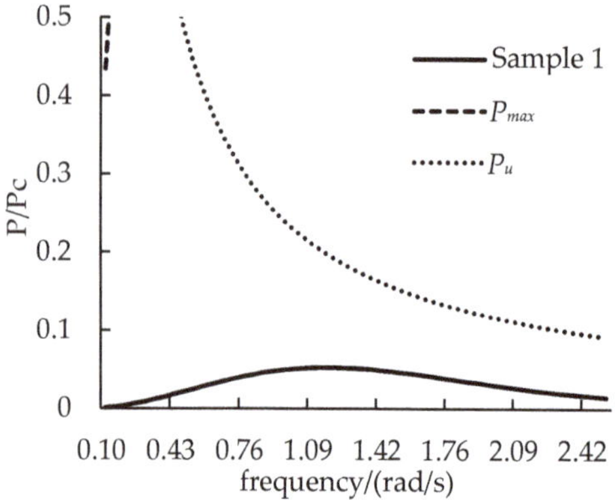

(a)

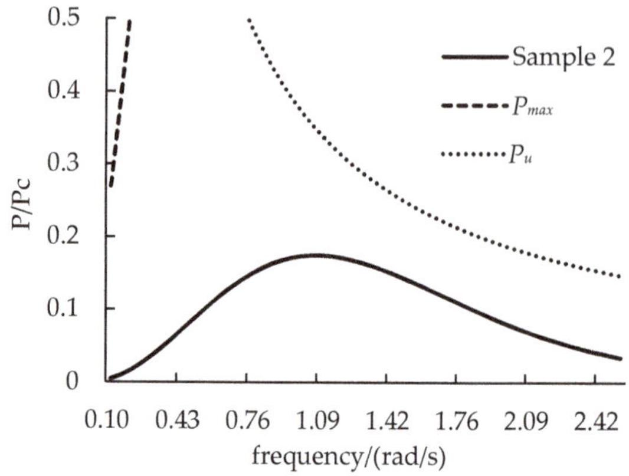

(b)

Figure 6. Power curves from simulations and theories. (**a**) results of sample 1; (**b**) results of sample 2.

5. BP Neural Network

The back propagation (BP) neural network is a kind of feedforward neural network trained by error back propagation algorithm. It is a most widely used form, and is composed of many nonlinear transformation units. This algorithm has a strong non-linear mapping ability and can simulate any nonlinear continuous functions with much higher accuracy theoretically. After the network is trained, the reflection between the inputs and outputs can be obtained and memorized. They are shown on the weights of each layer. BP neural network's structure is flexible, which means the number of layers and neurons can be changed according to research objectives. A BP neural network generally includes an input layer, one or two hidden layers, and an output layer. Full connections are applied between layers. More details about BP neural network can be seen in [37].

5.1. Neural Network Design

The first step to design a good neural network is to identify the number of hidden layers. A three layers neural network, which contains only one hidden layer, can simulate any reflection from n-dimensional inputs to m-dimensional outputs. Hence, a three-layer neural network with one hidden layer is selected in this study. Next, the nodes of each layer need to be identified. In this study, six key variables are selected, so the number of nodes in input layer is six. Only one parameter, capture power, needs to be predicted, so the number of nodes in output layer is one. Finally, the number of nodes in hidden layer needs to be identified. There is an empirical formula [38] that can be referred to identify the number of hidden nodes.

$$l = \sqrt{n+m} + a \tag{27}$$

where l, n, and m are the number of nodes in hidden layer, input layer, and output layer, respectively; a is an adjustment constant ranging from 1 to 10.

In this paper, the number of hidden nodes is tested from 3 to 12 to identify the most suitable value. MSE is used to elevate the performance, and the results are shown in Table 1.

Table 1. The number of hidden nodes and the values of MSE.

Number of Nodes	3	4	5	6	7	8	9	10	11	12
MSE	0.0215	0.0160	0.0236	0.0238	0.0274	0.0263	0.0371	0.0338	0.0364	0.0341

MSE reaches a minimum when the number of hidden nodes is 4, which is the optimal value for this case. The final BP neural network structure designed in this paper is shown in Figure 7.

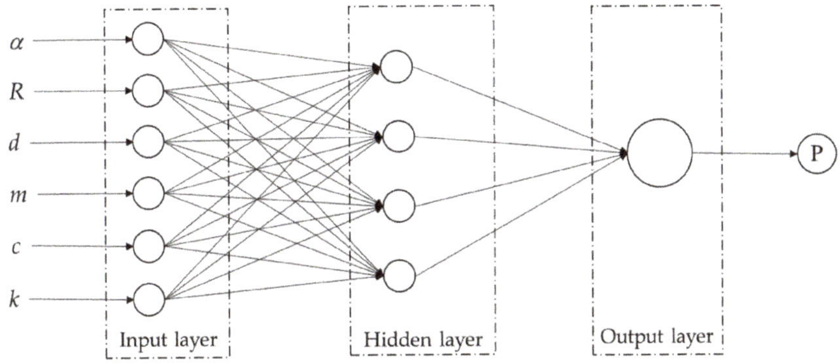

Figure 7. Structure of the designed BP neural network.

According to the structure, the output b_j of input layer can be expressed as follows:

$$b_j = f_1(\sum_{i=1}^{6} w_{ij}x_i + \theta_j),\ j = 1,2,3,4 \tag{28}$$

where w_{ij} is the weight from the input layer to the hidden layer; x_i is the input variable; θ_j is the threshold value of the hidden layer. The output y of the BP neural network is

$$y = f_2(\sum_{j=1}^{4} w_j b_j + \theta'),\ j = 1,2,3,4 \tag{29}$$

where w_j is the weight from the hidden layer to the output layer; θ' is the threshold value of the output layer.

5.2. Data Standardization and Neural Network Training

Before training, data standardization for individual features needs to be conducted to improve training speed. The standardization formula used in this paper is

$$x = \frac{x_i - x_{\min}}{x_{\max} - x_{\min}} \tag{30}$$

where x is the standardized result; $x_{\max}$ and $x_{\min}$ are the maximum and minimum values in the dataset, respectively. The standardized data have a distribution range between 0 and 1.

The network training process is to adjust the weights and thresholds so that the value of loss function reduces to a minimum. The training parameters for this model are shown in Table 2.

Table 2. Training parameters.

Weight Change Rate	Learning Rate	Training Epochs	Performance (Judged by Mean Square Error)	Minimum Gradient	Validation Checks
0.01	0.05	1000	10^{-5}	10^{-7}	6

Tangent sigmoid function (*tansig*) is adopted for the hidden layer, and the linear function (*purelin*) is adopted for the output layer. In the training process, mean squared error is used as loss function. It is defined as

$$MSE = \frac{1}{m}\sum_{i=1}^{m} (\hat{y} - y)^2 \tag{31}$$

where m is the number of samples; $\hat{y}$ is the observed value; y is the real value. In this paper, the top 80 samples are defined as training set. This model is trained in MATLAB R2019a, and the trendline of MSE for training set is shown as Figure 8.

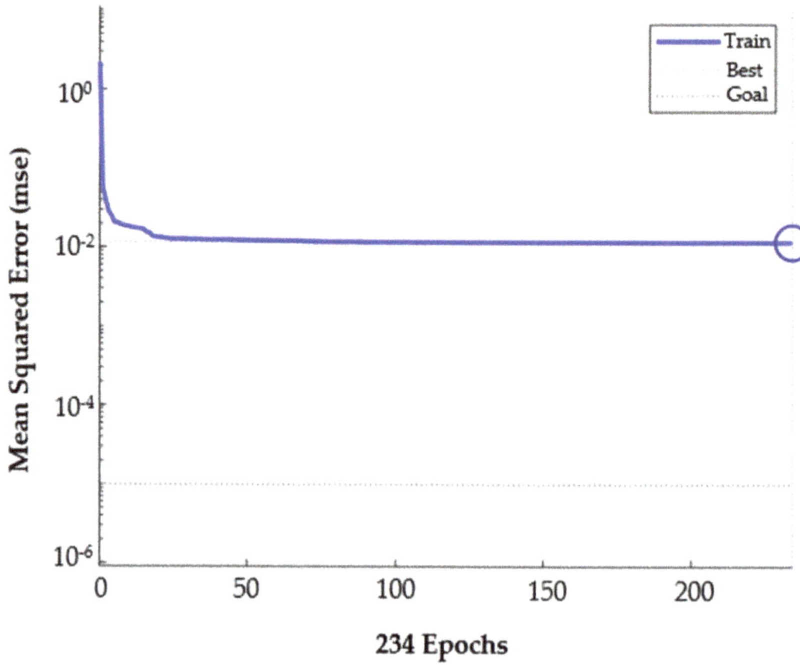

Figure 8. The trendline of training loss versus epochs.

The training process is terminated at 234 epochs because the gradient reaches the minimum (10^{-7}). The rest 20 samples are used to test, and the forecasting results after being de-standardized are shown in the next section.

6. Results and Discussion

In this section, six groups' forecasting data (ω = 0.53 rad/s, ω = 0.81 rad/s, ω = 1.14 rad/s, ω = 1.42 rad/s, ω = 1.76 rad/s, and ω = 2.09 rad/s) is given because they are the most common wave frequency. The desired outputs and forecasting results are presented in Figure 9 under different frequency. For each sample, the output power at 52 frequency points can be predicted.

In Figure 9a, the deviation of five samples (85, 90, 94, 99, and 100), which are at the lowest position of the graph, are relatively large, with mean error about 60 W. In Figure 9c, the error of sample 81 is the largest, with approximately 500 W. The forecasting results of sample 95 and 96 are rather larger than desired outputs in Figure 9d,f, and the error of sample 89 is around 200 W in Figure 9f. The highest accuracy is at ω = 0.81 rad/s and almost all the forecasting points fit the desired points. In contrast, the worst result is at ω = 2.09 rad/s and there are five forecasting results deviating the desired outputs.

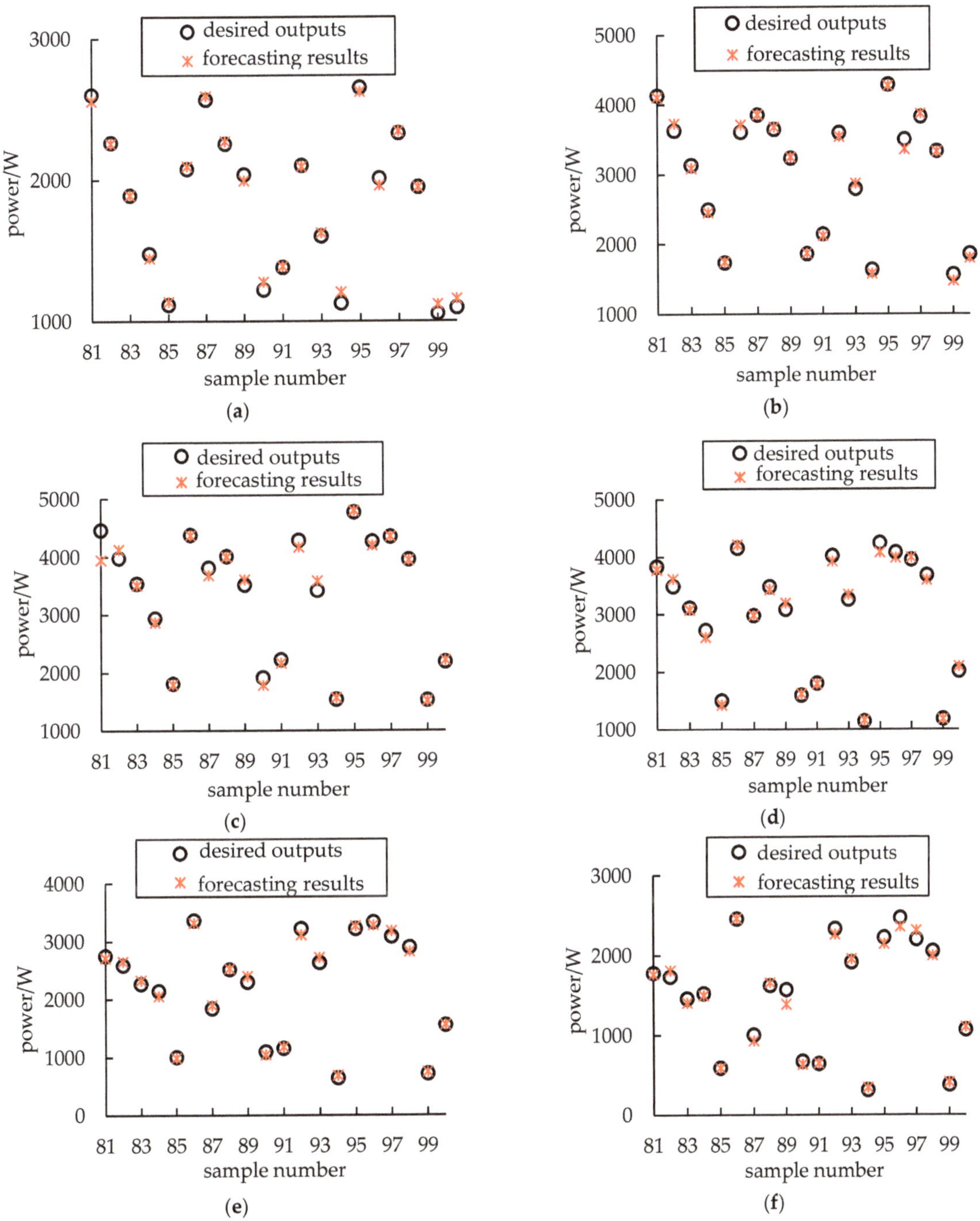

Figure 9. Comparisons between desired outputs and forecasting results. (**a**) ω = 0.53 rad/s; (**b**) ω = 0.81 rad/s; (**c**) ω = 1.14 rad/s; (**d**) ω = 1.42 rad/s; (**e**) ω = 1.76 rad/s; (**f**) ω = 2.09 rad/s.

To further verify the accuracy of the BP model, correlation coefficient (CC), root mean square error (RMSE), and error percentage are introduced in this section. They are defined as follows [29]

$$CC = \frac{\sum_{i=1}^{m}(t_i - \bar{t})(y_i - \bar{y})}{\sqrt{\sum_{i=1}^{m}(t_i - \bar{t})^2 \sum_{i=1}^{m}(y_i - \bar{y})^2}} \tag{32}$$

$$RMSE = \sqrt{\frac{1}{m}\sum_{i=1}^{m}(t_i - y_i)^2} \tag{33}$$

$$e = \frac{1}{m}\sum_{i=1}^{m}\left|\frac{y_i - t_i}{t_i}\right| \tag{34}$$

where m is the number of forecasting results; t_i is the desired value; y_i is the output of the network; $\bar{t}$ and $\bar{y}$ are average values of desired and forecasting results, respectively. The significance analysis of ANOVA is also conducted in MATLAB R2019a, and the statistical parameters (after de-standardization) are listed in Table 3.

Table 3. The statistical parameters between desired and forecasting results.

Wave Frequency (rad/s)	CC	RMSE (W)	Error Percentage	*p* Value
0.53	0.95931	37.5	2.03%	0.9588
0.81	0.90129	60.5	1.79%	0.9769
1.14	0.92105	138.7	2.45%	0.9206
1.42	0.91295	78.8	2.22%	0.9691
1.76	0.92558	61.6	2.7%	0.9789
2.09	0.94852	69.3	4%	0.9362

The values of CC are greater than 0.9, meaning that the correlations with each group are well fitted. The values of RMSE do not exceed 140 W, and the error percentage is no more than 4%, indicating that desired outputs and forecasting results are reasonably fitted. All P values are close to 1, which means there is no significant difference between desired and forecasting outputs. These validation factors indicate that this model has a good prediction accuracy and meets the engineering requirement.

7. Conclusions

In this paper, capture power predictions of a specific shape floating body are attempted based on mathematical model, ANSYS-AQWA simulations, and BP neural network. The key variables are identified and the simulation scheme is proposed. A sample database is built by LHS and the corresponding power of each sample is calculated. In the end, a BP neural network, of which training set is from simulation results, is designed to predict the capture power at different wave frequency. Its performance and accuracy are also evaluated through statistical parameters.

According to the results, the conclusions can be given as follows:

1. A mathematical model is constructed to identify the most important factors that affect the capture power. Four geometric parameters (radius, semi-vertical angle, mass, and submergence depth) and two system parameters (PTO damping coefficient and stiffness coefficient) are identified as key variables;
2. A BP neural network with high accuracy is designed and it is used to predict the capture power. The error percentage of top five groups is less than 2.5%, and that of the last group is 4%. The values of CC are greater than 0.9 and that of RMSE are less than 80 W except for the third group, of which the value of RMSE is 138.7 W. The P values are close to 1. However, due to the error of simulations caused by commercial software, this method needs experimental data to support.

Author Contributions: Methodology, W.W.; Software, F.B.; Validation, W.W. and F.B.; Formal analysis, W.W.; Investigation, Y.L.; Resources, Y.L. and G.X.; Writing—original draft preparation, W.W.; Writing—review and editing, G.X.; Funding acquisition, Y.L. All authors have read and agreed to the published version of the manuscript.

Funding: This research was funded by the Joint Research Fund under cooperative agreement between the National Natural Science Foundation of China and Shandong Provincial People's Government (grant number U1706230), Shandong Province Major Science and Technology Innovation Project (grant number 2018CXGC0104), Marine Renewable Energy Fund Project (grant number GHME2017YY01), and National Key Research and Development Plan (2017YFE0115000).

Institutional Review Board Statement: Not applicable.

Informed Consent Statement: Not applicable.

Data Availability Statement: Not applicable.

Acknowledgments: The authors gratefully acknowledge the financial support from the Joint Research Fund under cooperative agreement between the National Natural Science Foundation of China and Shandong Provincial People's Government (grant number U1706230), Shandong Province Major Science and Technology Innovation Project (grant number 2018CXGC0104), Marine Renewable Energy Fund Project (grant number GHME2017YY01), and National Key Research and Development Plan (2017YFE0115000).

Conflicts of Interest: The authors declare no conflict of interest.

Appendix A

Table A1. Details of 100 sample points.

No.	α	R	d	m	c	k
1	21.3	2.775	2.96	7190	21,500	3990
2	22.4	1.955	2.80	7970	13,700	3570
3	20.6	2.540	2.94	7930	25,700	4960
4	23.1	2.395	2.47	7840	24,000	3680
5	14.8	1.975	2.75	7460	20,900	5520
6	22.3	2.780	2.66	7490	22,500	3790
7	18.5	2.650	2.43	7760	29,400	4500
8	24.8	2.870	2.17	7440	10,500	4000
9	12.2	2.745	2.69	7020	18,000	5390
10	12.2	2.050	2.59	7130	20,000	5550
11	22.2	2.375	2.33	7470	14,500	4090
12	14.0	2.200	2.08	7940	17,600	4280
13	10.3	1.670	2.77	7300	28,500	5870
14	16.2	2.850	2.15	7360	27,600	3720
15	17.3	2.240	2.84	7670	21,900	5420
16	5.4	2.820	2.08	7170	10,700	4930
17	7.5	1.880	2.31	7680	14,900	3850
18	20.0	2.010	2.12	7790	23,000	4200
19	6.5	2.470	2.87	7620	26,400	3920
20	13.6	2.085	2.01	7230	24,200	5730
21	12.5	1.575	2.15	7990	26,600	4200
22	23.6	2.660	3.00	7850	23,200	5640
23	19.6	2.495	2.36	7750	25,500	3410
24	7.7	2.355	2.04	7490	24,600	5050
25	11.5	1.600	2.26	7550	10,000	4690
26	10.6	1.930	2.29	7270	14,100	3460
27	13.7	2.330	2.90	7530	16,900	5030
28	14.8	2.915	2.17	7260	17,800	3320

Table A1. *Cont.*

No.	α	R	d	m	c	k
29	21.4	1.865	2.41	7340	10,200	5250
30	7.9	1.795	2.87	7250	18,400	4810
31	8.3	2.120	2.45	7480	13,200	5990
32	10.2	2.150	2.84	7510	19,100	3180
33	6.3	2.325	2.63	7810	16,300	4610
34	17.1	1.635	2.55	7360	22,200	3140
35	21.6	2.895	2.13	7070	13,000	5960
36	13.0	1.830	2.91	7740	12,000	4330
37	5.9	2.970	2.56	7630	15,700	3450
38	15.9	2.060	2.81	7780	16,600	4890
39	19.3	2.995	2.57	7160	10,900	3160
40	8.6	1.715	2.74	7090	14,700	5670
41	11.2	1.895	2.98	7810	17,100	4050
42	9.8	1.800	2.12	7720	28,800	4660
43	11.6	2.140	2.88	7990	19,500	5220
44	18.7	2.225	2.07	7040	24,900	4860
45	12.8	2.675	2.90	7670	13,400	3740
46	15.3	2.280	2.26	7210	13,900	4730
47	17.8	2.205	2.27	7330	22,100	5070
48	11.0	2.260	2.37	7380	22,800	3500
49	16.8	1.540	2.05	7440	19,300	3760
50	13.3	2.185	2.42	7900	22,900	4360
51	19.7	2.575	2.52	7710	28,100	5210
52	9.1	2.300	2.25	7600	23,600	4420
53	22.8	2.945	2.82	7830	18,400	5890
54	7.0	2.485	2.05	7090	15,800	5440
55	21.0	1.745	2.70	7020	15,300	3010
56	5.3	1.690	2.77	7110	27,700	5830
57	7.4	2.805	2.09	7130	12,000	3390
58	15.1	2.830	2.66	7860	28,700	4290
59	15.6	1.700	2.02	7310	20,800	3280
60	16.2	1.555	2.70	7150	17,300	5330
61	9.4	1.915	2.47	7190	11,200	5930
62	6.7	2.890	2.68	7700	27,000	5130
63	23.7	2.110	2.23	7400	22,000	4570
64	18.3	2.080	2.18	7570	24,100	3620
65	24.1	2.620	2.95	7120	15,400	3340
66	16.6	1.755	2.79	7660	18,000	4490
67	24.6	2.690	2.02	7860	16,000	4230
68	8.9	2.035	2.34	7280	18,800	5720
69	5.0	2.440	2.42	7880	12,300	4770
70	21.9	1.520	2.29	7200	12,800	3930
71	13.1	2.425	2.53	7870	27,300	4750
72	9.6	2.635	2.21	7650	26,000	3820
73	24.8	2.170	2.83	7950	20,300	3650
74	10.5	1.640	2.55	7420	28,600	4160
75	11.9	1.610	2.63	7910	19,000	4920
76	23.9	2.400	2.34	7340	24,500	4630
77	5.8	1.985	2.38	7080	28,300	5470
78	22.7	2.700	2.49	7580	20,600	5540
79	18.9	1.590	2.10	7220	21,600	4070
80	24.3	2.605	2.32	7030	25,300	4990
81	15.5	2.510	2.71	7550	26,800	3570
82	17.7	2.550	2.20	7520	23,500	5360
83	11.3	1.775	2.96	7580	19,600	3080
84	18.2	2.020	2.98	7770	15,000	4550
85	10.0	2.585	2.21	7720	11,700	3890

Table A1. *Cont.*

No.	α	R	d	m	c	k
86	19.2	1.655	2.46	7430	22,000	5310
87	8.7	2.715	2.73	7290	27,900	5630
88	13.8	2.265	2.65	7590	23,100	3240
89	20.3	2.740	2.85	7530	21,100	5120
90	14.6	2.980	2.23	7060	13,000	5180
91	7.2	2.360	2.39	7630	14,400	3110
92	20.0	1.940	2.50	7280	21,300	4460
93	17.4	1.515	2.40	7390	16,800	3530
94	6.0	2.935	2.62	7750	12,400	4390
95	20.8	2.530	2.60	7920	27,500	5770
96	21.1	1.740	2.93	7390	20,100	3060
97	23.3	2.450	2.52	7960	23,900	5800
98	16.9	1.840	2.50	7140	20,000	4120
99	8.0	2.840	2.60	7900	11,500	5600
100	14.3	1.855	2.73	7010	11,100	3270

References

1. Mccabe, A.P.; Aggidis, G.A.; Widden, M.B. Optimizing the shape of a surge-and-pitch wave energy collector using a genetic algorithm. *Renew. Energy* **2010**, *35*, 2767–2775. [CrossRef]
2. Colby, M.K.; Nasroullahi, E.M.; Tumer, K. Optimizing ballast design of wave energy converters using evolutionary algorithms. In Proceedings of the 13th Annual Genetic and Evolutionary Computation Conference (GECCO 2011), Dublin, Ireland, 12–16 July 2011; pp. 1739–1746.
3. Fang, H.; Jin, L. Investigation on Resonance Response of Mass-Adjustable Float in Wave Energy Conversion System. *Energy Procedia* **2019**, *158*, 315–320. [CrossRef]
4. Giorgi, G.; Ringwood, J.V. Analytical representation of nonlinear Froude-Krylov forces for 3-DoF point absorbing wave energy devices. *Ocean Eng.* **2018**, *164*, 749–759. [CrossRef]
5. Le-Ngoc, L.; Gardiner, A.I.; Stuart, R.J.; Caughley, A.J.; Huckerby, J.A. Progress in the development of a multimode self-reacting wave energy converter. *OCEANS* **2010**, *2010*, 1–7.
6. Gao, H.; Yu, Y. The dynamics and power absorption of cone-cylinder wave energy converters with three degree of freedom in irregular waves. *Energy* **2018**, *143*, 833–845. [CrossRef]
7. Reabroy, R.; Zheng, X.; Zhang, L.; Zang, J.; Yuan, Z.; Liu, M.; Sun, K.; Tiaple, Y. Hydrodynamic response and power efficiency analysis of heaving wave energy converter integrated with breakwater. *Energy Convers. Manag.* **2019**, *195*, 1174–1186. [CrossRef]
8. Liang, C.; Ai, J.; Zuo, L. Design, fabrication, Simulation and testing of an ocean wave energy converter with mechanical motion rectifier. *Ocean Eng.* **2017**, *136*, 190–200. [CrossRef]
9. Li, X.; Martin, D.; Liang, C.; Chen, C.; Parker, R.G.; Zuo, L. Characterization and verification of a two-body wave energy converter with a novel power take-off. *Renew. Energy* **2020**, *163*, 910–920. [CrossRef]
10. Chen, Z.; Zhou, B.; Zhang, L.; Li, C.; Zang, J.; Zheng, X.; Xu, J.; Zhang, W. Experimental and numerical study on a novel dual-resonance wave energy converter with a built-in power take-off system. *Energy* **2018**, *165*, 1008–1020. [CrossRef]
11. WaveNET–the Floating, Flexible Wave Energy Generator. Available online: https://newatlas.com/albatern-wavenet-squid-floating-wave-energy/34942/ (accessed on 22 April 2021).
12. Sun, P.; Li, Q.; He, H.; Chen, H.; Zhang, J.; Li, H.; Liu, D. Design and optimization investigation on hydraulic transmission and energy storage system for a floating-array-buoys wave energy converter. *Energy Convers. Manag.* **2021**, *235*, 113998. [CrossRef]
13. Liu, Z.; Qu, N.; Han, Z.; Zhang, J.; Zhang, S.; Li, M.; Shi, H. Study on energy conversion and storage system for a prototype buoys-array wave energy converter. *Energy Sustain. Dev.* **2016**, *34*, 100–110. [CrossRef]
14. Zou, J. Numerical Simulation of Wave Energy Use Efficiency of Oscillating Buoy and Its Influence Parameters. Master's Dissertation, Harbin Institute of Technology, Harbin, China, December 2014.
15. Yu, L. The Performance Research on the Oscillating Float Type Wave Energy Converter. Master's Dissertation, Harbin Engineering University, Harbin, China, March 2016.
16. Wu, L. Study on Hydrodynamic Performance and Energy Conversion Efficiency of a Three-Dimensional Two-Body WEC. Master's Dissertation, Harbin Engineering University, Harbin, China, March 2017.
17. Zheng, X. Study on the Hydrodynamic Performance for Two Kinds of Point Absorbing Wave Energy Device. Doctor's Dissertation, Harbin Engineering University, Harbin, China, June 2016.
18. Ma, Y.; Zhang, A.; Yang, L.; Li, H.; Zhai, Z.; Zhou, H. Motion simulation and performance analysis of two-body floating point absorber wave energy converter. *Renew. Energy* **2020**, *157*, 353–367. [CrossRef]
19. Ji, X.; Shami, E.A.; Monty, J.; Wang, X. Modelling of linear and non-linear two-body wave energy converters under regular and irregular wave conditions. *Renew. Energy* **2020**, *147*, 487–501. [CrossRef]

20. Tongphong, W.; Kim, B.H.; Kim, I.C.; Lee, Y.H. A study on the design and performance of ModuleRaft wave energy converter. *Renew. Energy* **2021**, *163*, 649–673. [CrossRef]
21. Ma, Y.; Ai, S.; Yang, L.; Zhang, A.; Liu, S.; Zhou, B. Research on design and optimization of the pitching float wave energy converter. *Energy Sci. Eng.* **2020**, *8*, 3866–3882. [CrossRef]
22. Amiri, A.; Panahi, R.; Radfar, S. Parametric study of two-body floating-point wave absorber. *J. Mar. Sci. Appl.* **2016**, *15*, 41–49. [CrossRef]
23. Yu, Y.; Li, Y. A RANS Simulation of the Heave Response of a Two-Body Floating-Point Wave Absorber. In Proceedings of the 21st International Offshore and Polar Engineering Conference, Maui, HI, USA, 19–24 June 2011.
24. Bosma, B.; Zhang, Z.; Brekken, T.K.A.; Özkan-Haller, H.T.; McNatt, C.; Yim, S.C. Wave energy converter modeling in the frequency domain: A design guide. In Proceedings of the 2012 IEEE Energy Conversion Congress and Exposition (ECCE), Raleigh, NC, USA, 15–20 September 2012; pp. 2099–2106.
25. Bosma, B.; Brekken, T.K.A.; Özkan-Haller, H.T.; Yim, S.C. Wave energy converter modeling in the time domain: A design guide. In Proceedings of the 1st IEEE Conference on Technologies for Sustainability, Portland, OR, USA, 1–2 August 2013; pp. 103–108.
26. Law, Y.Z.; Santo, H.; Lim, K.Y.; Chan, E.S. Deterministic wave prediction for unidirectional sea-states in real-time using Artificial Neural Network. *Ocean Eng.* **2020**, *195*, 106722. [CrossRef]
27. Desouky, M.A.A.; Abdelkhalik, O. Wave prediction using wave rider position measurements and NARX network in wave energy conversion. *Appl. Ocean Res.* **2018**, *82*, 10–21. [CrossRef]
28. Kumar, N.K.; Savitha, R.; Al Mamun, A. Regional ocean wave height prediction using sequential learning neural networks. *Ocean Eng.* **2017**, *129*, 605–612. [CrossRef]
29. Zheng, C.; Li, C. Variation of the wave energy and significant wave height in the China Sea and adjacent waters. *Renew. Sustain. Energy Rev.* **2015**, *43*, 381–387. [CrossRef]
30. Avila, D.; Marichal, G.N.; Padron, I.; Quiza, R.; Hernandez, A. Forecasting of wave energy in Canary Islands based on Artificial Intelligence. *Appl. Ocean Res.* **2020**, *101*, 102189. [CrossRef]
31. Wang, Y. Efficient prediction of wave energy converters power output considering bottom effects. *Ocean Eng.* **2019**, *181*, 89–97. [CrossRef]
32. Halliday, J.R.; Dorrell, D.G.; Wood, A.R. An application of the Fast Fourier Transform to the short-term prediction of sea wave behaviour. *Renew. Energy* **2011**, *36*, 1685–1692. [CrossRef]
33. Davis, A.F.; Fabien, B.C. Wave excitation force estimation of wave energy floats using extended Kalman filters. *Ocean Eng.* **2020**, *198*, 106970. [CrossRef]
34. Ni, C.; Ma, X.; Wang, J. Integrated deep learning model for predicting electrical power generation from wave energy converter. In Proceedings of the 25th International Conference on Automation and Computing (ICAC), Lancaster University, Lancaster, UK, 5–7 September 2019; pp. 192–197.
35. Peng, J. Study on Hydrodynamic Performance for Oscillating Float Buoy Wave Energy Converter. Doctor's Dissertation, Shandong University, Jinan, China, May 2014.
36. Falnes, J. A review of wave-energy extraction. *Mar. Struct.* **2007**, *20*, 185–201. [CrossRef]
37. Hornik, K. Approximation capabilities of multilayer feedforward networks. *Neural Netw.* **1991**, *4*, 251–257. [CrossRef]
38. Ma, X.; Guan, Y.; Mao, R.; Zheng, S.; Wei, Q. Modeling of lead removal by living Scenedesmus obliquus using backpropagation (BP) neural network algorithm. *Environ. Technol. Innov.* **2021**, *22*, 101410. [CrossRef]

Journal of
Marine Science and Engineering

Article

Experimental and Numerical Study on the Characteristics of Motion and Load for a Floating Solar Power Farm under Regular Waves

Jun-Hee Lee [1], Kwang-Jun Paik [1,*], Soon-Hyun Lee [1], Jun Hwangbo [1] and Tae-Hyu Ha [2]

[1] Department of Naval Architecture and Ocean Engineering, Inha University, Incheon 22201, Korea; harry049@naver.com (J.-H.L.); vori93@naver.com (S.-H.L.); bearjun87@naver.com (J.H.)
[2] Steel Structure Research Group, POSCO, Incheon 21998, Korea; hath@posco.com
* Correspondence: kwnagjun.paik@inha.ac.kr; Tel.: +82-32-860-7331

Abstract: Recently, the demand for floating solar power farms in lakes and coasts (rather than on land) has been increasing rapidly. It is important to develop a numerical analysis technique that considers environmental conditions to predict structural stability and accurate motion response while designing a floating solar power farm. In this study, we performed a comparison under conditions similar to those of the Inha University towing tank (IUTT) model test to verify the numerical analysis method. The results revealed that heave and pitch movements were dominant under head sea conditions. Relative behavior occurred because of the hinge connection of each unit, and complex motion characteristics appeared depending on the wave conditions. The numerical method was verified based on the motion response and load of the floating solar farm. The validity of the results was also confirmed.

Keywords: floating solar power farm; wave-induced motion; RAO (response amplitude operator); motion; CFD; model test

Citation: Lee, J.-H.; Paik, K.-J.; Lee, S.-H.; Hwangbo, J.; Ha, T.-H. Experimental and Numerical Study on the Characteristics of Motion and Load for a Floating Solar Power Farm under Regular Waves. *J. Mar. Sci. Eng.* **2022**, *10*, 565. https://doi.org/10.3390/jmse10050565

Academic Editor: Eugen Rusu

Received: 31 March 2022
Accepted: 19 April 2022
Published: 21 April 2022

Publisher's Note: MDPI stays neutral with regard to jurisdictional claims in published maps and institutional affiliations.

Copyright: © 2022 by the authors. Licensee MDPI, Basel, Switzerland. This article is an open access article distributed under the terms and conditions of the Creative Commons Attribution (CC BY) license (https://creativecommons.org/licenses/by/4.0/).

1. Introduction

With a new climate system that replaces the Kyoto Protocol, which expired in 2020, the Paris Climate Agreement has been applied since January 2021, which includes strengthening global warming control targets and reducing greenhouse gas emissions. As follow, various policies have been enacted to encourage the development of new renewable energy sources to reduce greenhouse gas emissions.

In addition to international treaties, South Korea is promoting a "Renewable Energy 3020" policy to increase the ratio of renewable-energy power generation by 2030. The goal of the policy is to increase the renewable energy power generation ratio by 20% or more, and to supply over 95% of the capacity of the newly installed platform with clean energy such as solar and wind power [1]. Since 2012, the Renewable Portfolio Standard (RPS), which mandates a certain percentage of power generation as new renewable energy for industrialization such as solar power, wind power, and hydrogen, has also been enforced [2].

Recently, in the case of solar power generation, the demand for installation on water (e.g., lakes, coasts, and on the sea) is increasing rapidly due to the lack of installation area on land. It is more convenient to secure space on water and in the sea than on land, minimizes the effect of natural disasters, and eases the development large-scale power generation complexes [3].A floating solar power farm system is frequently used as a unified platform considering its installation convenience, expandability, and mobility.

On the other hand, since the coast and sea are exposed directly to natural disturbance (e.g., seawater, wind, and waves), technologies that can minimize the effects of waves and actively reduce fluctuations are needed [4–6]. Owing to the periodic motion of waves, fatigue loads are generated on structures (e.g., floating bodies, supporting frames, panels, and coupling devices), and a design is needed for supporting structures considering structural stability. To consider the stability of a floating solar power generation system, experiments and numerical studies on the motion performance and connections between adjacent platforms are still underway.

Oliveira-Pinto et al. [7] provided a literature review on the applicability of floating solar power farms and introduced the technologies currently available, as well as the technical challenges and risk factors when designing studies in the marine environment.

Sahu et al. [8] illustrated the concept and advantages of floating solar power systems and studied each component of solar power systems. The environmental loads can cause deformation and stress in the module, causing microcracks and reducing productivity and durability. The technology of thin-film research was proposed to withstand harsh environmental conditions. It was noted that remote sensing and GIS (geographic information system)-based technologies can be utilized to determine the potential of solar panels.

The floating structure uses individual HDPE (high-density polyethylene), and the floating bodies were connected to a pin. It was said that the connection part was weak due to environmental factors. When this connector is exposed to high wave heights, stress concentration could occur and cause system problems. It was shown that the behavior of floating systems is complex [9,10].

Shi et al. [11] introduced a network modeling method for the dynamic prediction of multi-module floating structures and conducted it experimentally in the wave. The connector was made on three-axis motion, and RAOs (response amplitude operator) were compared at various frequencies through the experiment. Recently, the SCOTRA company developed a connector system that allows floating bodies to have their own motility in all directions.

When waves interact with a floating solar power farm (FSP), the motions of the FSP generate inertial forces and dynamic loads on the structures. Non-linear effects make it difficult to predict what will happen on the platforms. It could increase the motion response complexity exponentially of floating bodies. Therefore, recent studies are trying to predict the motion of the system using the using the computational fluid dynamics (CFD) [12,13].

All previous studies have stated that it is important to predict the load at the design stage due to the non-linear environmental factors that affect the FSP depending on the wave conditions. Also, it is necessary to consider motion response with different FSP components (e.g., connecting method, the type of mooring system, the array of floating bodies, and the position and angle of the solar panel).

In this study, we focused on the motion response and tried to understand the relative motion that may occur in a unit platform. Owing to the establishment of numerical techniques, the simulations were performed for a floating solar power farm in the form of a unit under wave conditions. Uniaxial hinge and catenary techniques were applied to the connector and mooring system to determine the motion performance of the floating solar power farm for six degrees of freedom (6DOF). This was verified by an experiment conducted in the Inha University towing tank (IUTT). It investigated fluid forces such as load and pressure distribution, which are difficult to determine using the experiment.

2. Experimental Method

CFD is essential for evaluating the wave load and motion performance of an FSP in a real sea area. In addition, a model test is performed to verify the numerical analysis method.

2.1. Geometry of a Floating Solar Power Farm

In general, an FSP consists of a floating body, a frame bar that supports the floating body, and a solar panel. In this study, the FSP components for the model test were designed as shown in Figure 1. The floating body has a length, breadth, and depth of 0.12 m, 0.105 m, and 0.12 m, respectively.

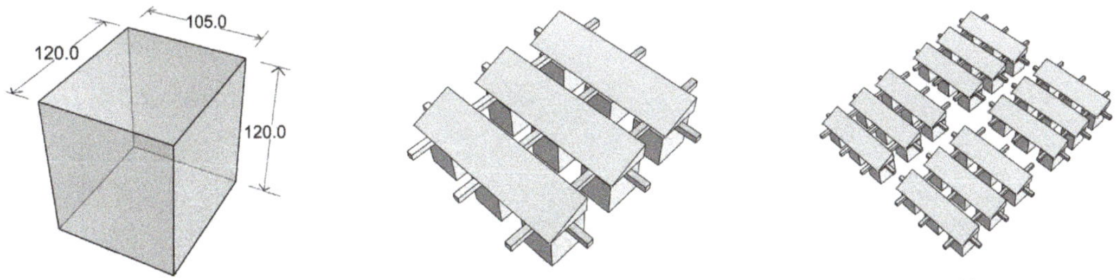

(**a**) (**b**) (**c**)

Figure 1. Geometry of a floating solar power farm used in the experiment: (**a**) floating body; (**b**) unit; (**c**) block.

The unit consists of nine floating bodies. These were designed as cuboids for convenient production. To understand the relative behavior between units under wave conditions, two-row and two-column matrix forms were combined and called a block. Table 1 lists the specifications of the FSP components. In addition, a uniaxial hinge system with a length of 50 mm was constructed so that relative behavior between units could be investigated. The position of the connector was installed on the parallel line where the frame bars were located. Figure 2 shows the overall geometry of the model tested in this study.

Table 1. Main particulars of a floating solar power farm.

Title 1	Floating Body	Unit	Block
Length, L (m)	0.120	0.738	1.526
Breadth, B (m)	0.105	0.640	1.330
Depth, D (m)	0.120	0.160	0.160
Draft, T (m)	0.060	0.060	0.060

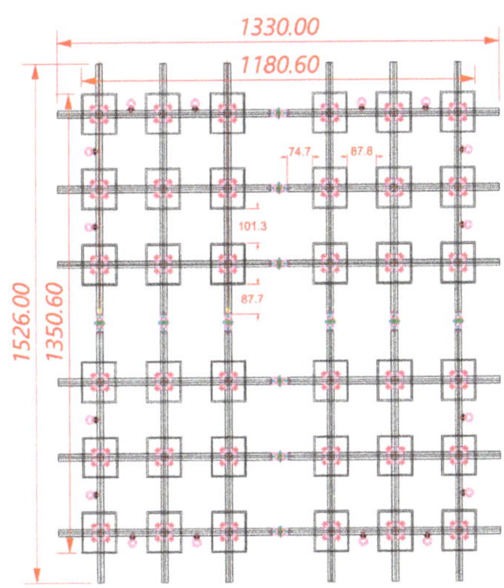

Figure 2. Schematic of the floating solar power farm in the experiment.

2.2. Experimental Setup and Procedure

The experiment in this study was conducted in the IUTT (see Figure 3). The IUTT consists of a wave maker that can generate a maximum wavelength of 2.0 m and wave height of 0.2 m; a wave absorber; and a concrete tank that has a length, breadth, and depth of 50 m, 3.5 m, and 1.5 m, respectively. The breadth of the FSP at the waterline is about 1.1806 m (see Figure 2). The model size was selected to minimize the effect of blockage.

Figure 3. The towing tank in Inha University.

The movement of the FSP was extracted as time-series data using a camera (OptiTrack Prime X13, Motion Technologies, Inc., Seoul, Republic of Korea) considering its high frequency, accuracy, and convenient installation. Four cameras were used in this study to capture motion based on the principle that infrared light emitted from a camera is reflected by a marker and sensed by the camera. The camera tracks the marker with positional errors less than ± 0.2 mm and rotational errors less than 0.5°. To keep the position of the FSP, weights and wires were used as anchors and mooring lines. The 7 × 7 stainless-steel wires had a diameter of 1.2 mm, an axial stiffness of 565.2 N/m, and a weight of 0.0063 kg/m. The mooring line connection in the FSP was placed in the center of the frame bar connection the floating bodies. The anchor was fixed at a parallel position 1.5 m away from those positions, and the length of the mooring line was set to 1.9 m.

In general, the mooring system of the FSP is a taut spread to restrain its motion. However, since the water depth could be changed according to the tide condition, the mooring system in this study was assumed to be a catenary mooring considering a loosened state at low tide.

A one-component force meter has been manufactured to measure tension up to 50 N with a linearity, hysteresis, and reproducibility of 0.7%, −0.4%, and 0.6%, respectively. A tension gauge was installed on the upstream mooring lines of the FSP.

As shown in Table 2, the experimental conditions were selected within the range that can be experimented with using the IUTT. L_{unit} refers to the waterline length of a unit in the x-direction. Two incident wave angles (139°, 180°), two wave steepnesses (H/λ = 0.03, 0.05), and wavelength ratios (λ/L_{unit} = 1.6–3.2) were considered in the experiment. Here, λ is the wavelength and H is the wave height. Only regular waves were used.

Schematic diagrams of the model test of the FSP under wave conditions are shown in Figures 4 and 5. The direction of wave travel, direction from the FSP toward its right, and direction opposite to that of gravity were set as the positive x-, y-, and z-directions. An ultrasonic wave height meter was installed 2.0 L_{unit} in front of the FSP to ensure the reliability of the waves generated by real-time wave height information. The first and second rows were defined as the first and second groups, respectively. The experimental setup of the test model of the FSP under wave conditions is shown in Figure 6.

Table 2. Test conditions of the experiment.

No	Incident Wave Angle	H/λ	λ/L_{unit}
1	180	0.03	1.6
2	180	0.03	2.0
3	180	0.03	2.4
4	180	0.03	2.8
5	180	0.03	3.2
6	180	0.05	1.6
7	180	0.05	2.0
8	180	0.05	2.4
9	139	0.03	1.6
10	139	0.03	2.0
11	139	0.03	2.4
12	139	0.03	2.8
13	139	0.03	3.2

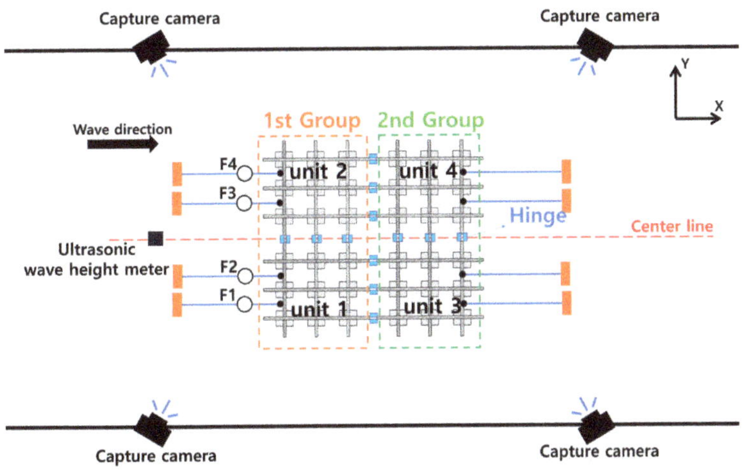

Figure 4. Diagram of the experimental setup for head sea conditions.

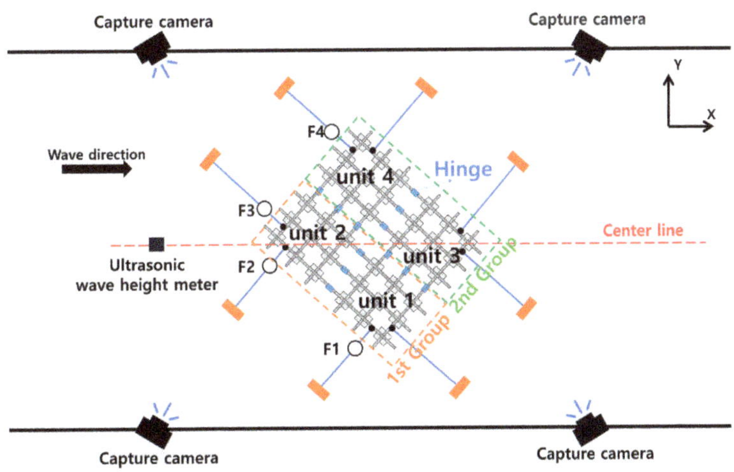

Figure 5. Diagram of the experimental setup for oblique sea conditions.

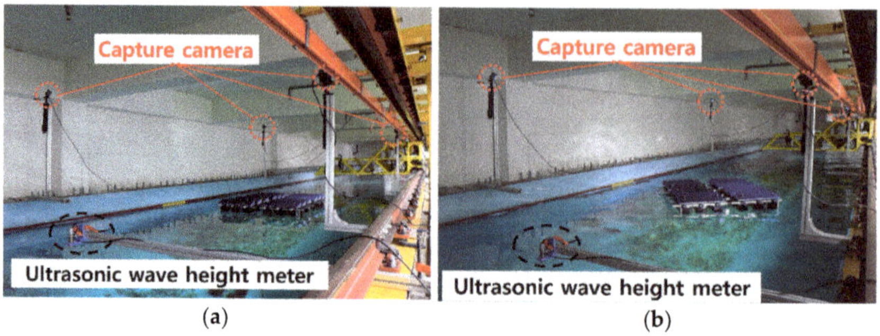

Figure 6. Experimental setup: (**a**) head sea conditions; (**b**) oblique sea condition.

3. Numerical Modeling

3.1. Governing Equations

In this study, the flow around the FPS was assumed to be an incompressible turbulent flow. The continuity equation and Reynolds average Navier–Stokes (RANS) equations were used as the governing equations. These can be expressed as shown in Equations (1) and (2):

$$\frac{\partial \rho}{\partial t} + \nabla \cdot (\rho \overline{u_i}) = 0 \quad (1)$$

$$\rho \overline{u_j} \frac{\partial \overline{u_i}}{\partial x_j} = \rho \overline{f}_i + \frac{\partial}{\partial x_j}\left[-\overline{p}\delta_{ij} + \mu \left(\frac{\partial \overline{u_i}}{\partial x_j} + \frac{\partial \overline{u_j}}{\partial x_i} \right) - \rho \overline{u_i' u_j'} \right] \quad (2)$$

where u_i is the three-dimensional velocity vector in the x, y, and z directions. $\overline{u}$ is the mean component and p, ρ, μ, and f are the pressure, density, dynamic viscosity, and body-force per mass, respectively. The left-hand side of this equation represents the variation in the mean momentum of a fluid element owing to the unsteadiness in the mean flow and convection by the mean flow $-\rho \overline{u_i' u_j'}$ owing to the fluctuating velocity field. It is generally referred to as the Reynolds stress. The nonlinear Reynolds stress term requires additional modeling to close the RANS equation for solving. In this study, the Realizable $k - \epsilon$ turbulence model was used.

3.2. Numerical Method and Setup

The commercial CFD program STAR-CCM + v15.06 [14] was used to determine the motion and load characteristics of the FSP under wave conditions. The Dynamic fluid–body interaction (DFBI) model was used for the movement of the FSP. Furthermore, an overset mesh was used considering the deformation of the grid system owing to the wave, which leads to larger movement of the FSP. Six-DOF analyses were performed, and the free surface was generated by a volume-of-fluid (VOF) model showing non-mixed fluid. The wave was generated as the fifth-order Stokes waves [15].

The domain and boundary conditions used in the computation under head sea and oblique sea conditions are shown in Figures 7 and 8, respectively. The boundary conditions were set to a velocity inlet, and the forcing method supported by STAR-CCM + was utilized to minimize the dissipation of generated waves. The size of the forcing zone is shown in the red area.

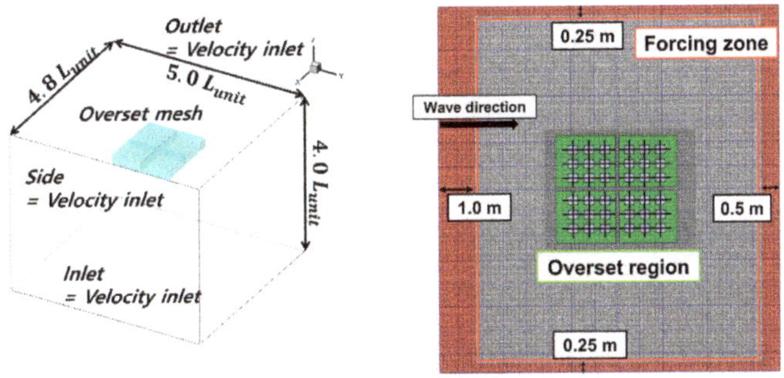

Figure 7. Computation domain and boundary conditions for head sea conditions.

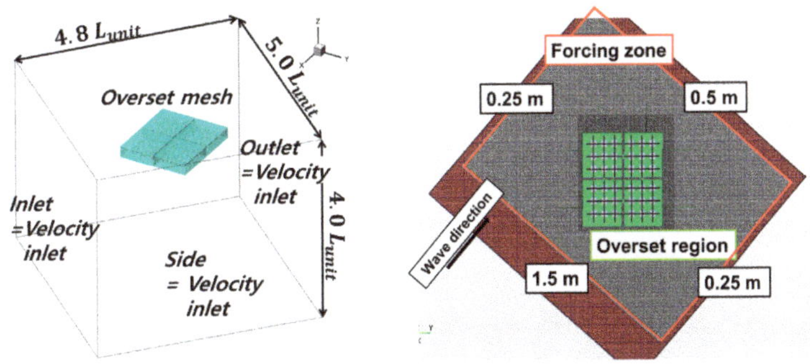

Figure 8. Computation domain and boundary conditions for oblique sea conditions.

In a previous study, the effects of the wave (e.g., wave run-up, reflections, and diffractions) occurred owing to the blunt geometry and spacing of the floating bodies [16,17]. Due to the forcing boundary absorbed the effects of the wave, the domain size was set as shown in Figures 7 and 8, respectively. The computation was an implicit unsteady condition, and the time step was set approximately $T_e/250$ s. Here, T_e is the wave period. Approximately 5 million and 6 million grids were used in the computations for the head sea and oblique sea conditions, respectively.

A catenary coupling system was used to express a mooring line in the computation. Catenary coupling models an elastic, quasi-stationary catenary such as a chain or towing rope, which hangs between two endpoints and is subject to its weight in the gravity field. In the local cartesian coordinate system, the catenary shape is given by Equations (3) and (4):

$$x = au + b\sinh + \alpha$$

$$y = a\cosh(u) + \frac{b}{2}\sinh^2(u) + \beta \tag{3}$$

$$a = \frac{c}{\lambda_0 g},\ b = \frac{ca}{DL_{eq}},\ c = \frac{\lambda_0 L_{eq} g}{\sinh(u_2) - \sinh(u_1)}$$

$$\tan\phi = \sinh(u) \tag{4}$$

where the curve parameter u_i is related to the inclination angle ϕ of the catenary curve by Equation (4). g is the gravitational acceleration. λ_0 and L_{eq} are the mass per unit length and relaxation length, respectively, of the catenary under force-free conditions. D is the stiffness of the catenary, and α and β are integration constants that depend on the positions of the two endpoints and the total mass of the catenary.

Also, a revolute joint coupling system was applied to express a hinge system in the computation. A revolute joint connects two 6-DOF bodies, each of which has its own local-body coordinate system. The position and axis direction of each body must coincide. However, the position and axis of the revolute joint can vary over time.

The overall forms in which the mooring lines and hinges were applied to the FSP are shown in Figure 9. The same geometry and physical property values of the model used in the experiment were applied in the computation simulations. The test conditions used to verify the numerical method are listed in Table 3.

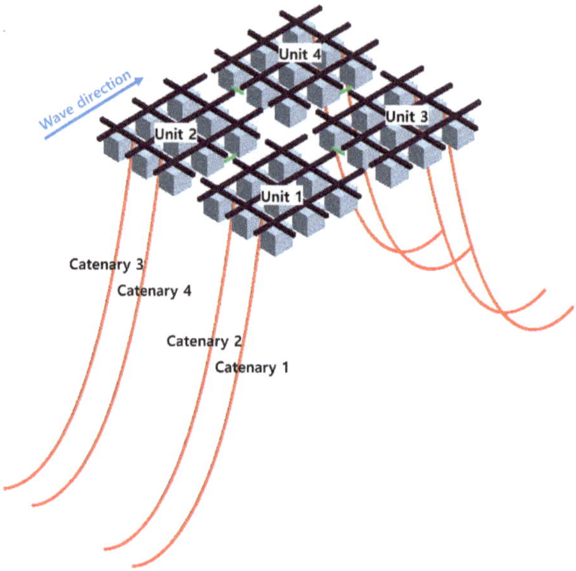

Figure 9. Computation setup of mooring lines (catenary) and the hinge system.

Table 3. Test conditions of computation.

No	Incident Wave Angle	H/λ	λ/L_{unit}
1	180	0.03	1.6
2	180	0.03	2.0
3	180	0.03	2.4
4	180	0.03	2.8
5	180	0.03	3.2
6	180	0.05	1.6
7	180	0.05	2.0
8	180	0.05	2.4
9	139	0.03	2.4
10	139	0.03	2.8

4. Validation and Verification

The accuracy of the free surface is an important factor that needs to be considered before the CFD interpretation of the motion properties and flow of an FSP subjected to waves.

Previously, it was found that the importance of the variables such as time step, damping, and forcing constants can prevent wave reflection under boundary conditions, and can minimize waves [18,19]. In general, the computations that apply to the wave take a long computational time. Kim et al. [20] performed a numerical sensitivity simulation for two-dimensional waves and completed a numerical error distribution map based on the grid and CFL (Courant–Friedrichs–Lewy) conditions. Through those results obtained from the two-dimensional problem, a FPSO (floating production storage and offloading) motion simulation was conducted under wave conditions. The tested two-dimensional wave conditions showed a relatively good agreement compared to potential results, which shows that the verification of the waves through the two-dimensional test is valid. In this study, two-dimensional sensitivity analysis for waves was performed, and this was applied to the FSP to perform grid convergence analysis.

4.1. Two-Dimensional Wave Generation Sensitivity

This section presents the results of an examination of the computational conditions to accurately simulate waves under the conditions given in the experiment. Considered as a 2D problem, the target waves are two types of stokes waves (λ = 1.35 m, H = 0.0405 m; and λ = 1.80 m, H = 0.054 m). The total simulation time corresponded to eight cycles of the wave. The data were extracted at three points (x = +1.650 m, +0.675 m, and −0.675 m from the origin) to verify the correctness of the implementation for the target wave (see Figure 10).

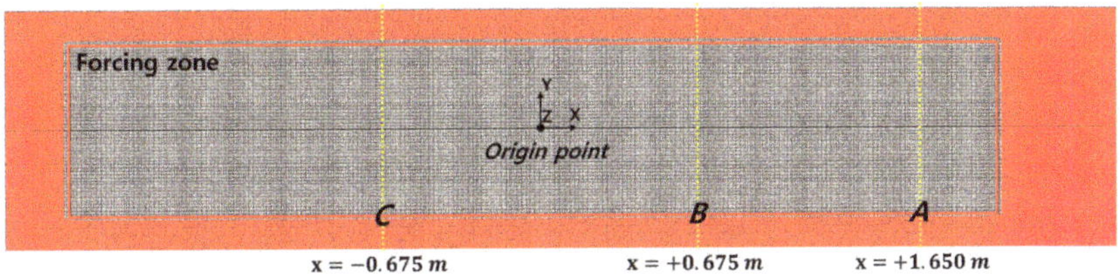

Figure 10. Computation domain in waves through grid sensitivity analysis.

The medium grid was set to approximately 135 cells per wavelength and 20 cells per wave height with λ = 1.35 m and H = 0.0405 m. The basic grid size was varied by $\sqrt{2}$ times. In total, three types of grid systems were constructed with coarse and fine grids. The coarse grid was set to approximately 95 cells per wavelength and 12 cells per wave height. Meanwhile, the fine grid was set to approximately 190 cells per wavelength and 26 cells per wave height. In all the grid systems, the time step was set to a value at which the CFL was lower than one.

The wave profiles of the numerical and theoretical solutions obtained at each point by the grid systems were compared as shown in Figures 11 and 12, where the black solid line is the theoretical solution. The coarse, medium, and fine grids are the green-triangle, blue-square, and red-circle symbols, respectively.

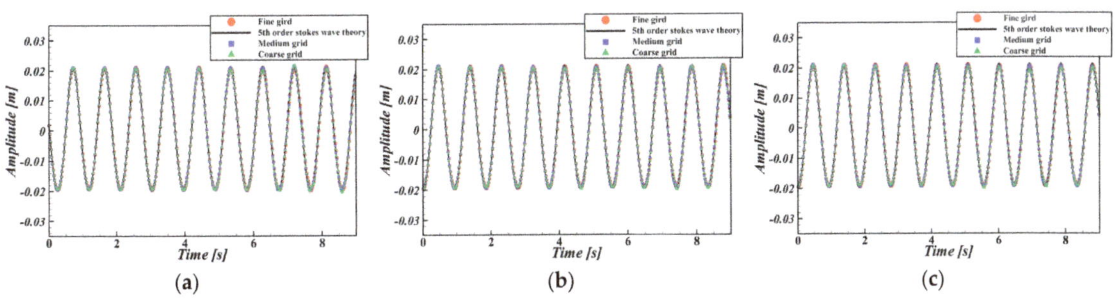

Figure 11. Comparison of wave profiles at λ = 1.350 m: (**a**) x = +1.650 m from the origin; (**b**) x = +0.675 m from the origin; (**c**) x = −0.675 m from the origin.

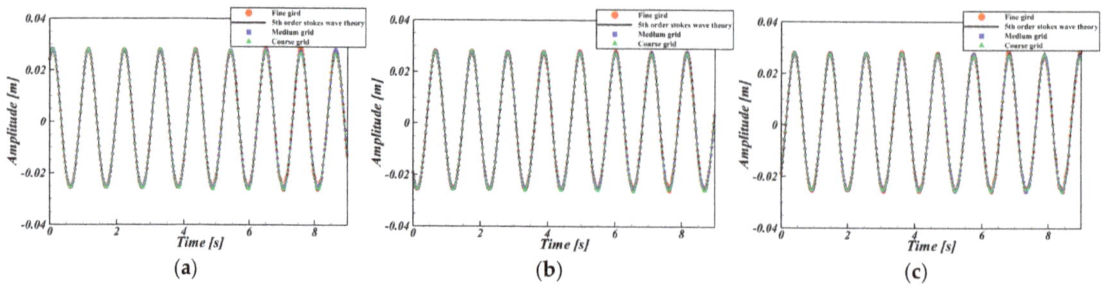

Figure 12. Comparison of wave profiles at λ = 1.800 m: (**a**) x = +1.650 m from the origin; (**b**) x = +0.675 m from the origin; (**c**) x = −0.675 m from the origin.

Table 4 shows the wave amplitude difference between the theoretical and numerical solutions according to the variation in the grid system at each position. The fine grid shows the best results. However, it requires an excessively long calculation time because of the number of grids and time steps. The medium grid appears to be suitable considering efficiency. In this study, we selected a medium-grid system to express the free surface.

Table 4. Comparison of wave amplitude for each position according to the grid system.

Wavelength [m]	Position	5th-Order Stokes Waves Theory [mm]	Coarse Grid [%diff]	Medium Grid [%diff]	Fine Grid [%diff]
1.35	A	40.86	−1.25%	−0.64%	−0.41%
1.35	B	40.86	−1.96%	−1.16%	−0.76%
1.35	C	40.86	−3.01%	−1.37%	−1.27%
1.80	A	53.82	−0.27%	−0.34%	−0.41%
1.80	B	53.82	−1.27%	−1.11%	−1.02%
1.80	C	53.82	−1.76%	−1.51%	−1.05%

4.2. Grid Convergence Test for the Floating Solar Power Farm under Wave Conditions

In this section, the grid convergence tests for the FSP computation are presented. The test subject was set to one unit, and the free surface was set to three stages, as described in Section 4.1. The grid convergence index (GCI) [21–23] of the test conditions are listed in Table 5. The GCI in this study represents the difference in the result value due to the grid variation represented by Grid-1, Grid-2, and Grid-3. The grid numbers of Grid-1, Grid-2, and Grid-3 were 0.96 million, 1.36 million, and 1.92 million, respectively. The number of grids was changed by changing the basic grid. In the grid convergence test, the difference of motion RAOs was reduced with the increase of grid size. The heave RAO GCI is 0.00048, 0.00031, and 0.00055 at wavelength ratios of 1.6, 2.4, and 3.2, and the pitch RAO GCI is 0.00057, 0.00183, and 0.00180 at the same wavelength ratios.

Table 5. Comparison of the motions under the wave conditions (H/λ = 0.03).

Variables	λ/L_{unit}	Time Step (s)	Heave RAO GCI (%)	Pitch RAO GCI (%)
	1.6		0.00048	0.00057
Grid	2.4	CFL = 1.0	0.00031	0.00183
	3.2		0.00055	0.00180

The results of FSP RAOs according to the variation in the grid systems are shown in Figure 13. The coarse, medium, and fine grids and experiment results are the green-triangle, blue-square, red-circular, and black-circular symbols, respectively. Unlike other grid systems, Grid-1 indicates lower RAOs in heave and pitch motions. At short wavelengths, the number of grids was insufficient for the wave height to express the free surface.

Even when Grid-3 was expressed effectively for motion RAOs of the FSP and free surface, it required a computation-time approximately two times that for Grid-2. In addition, Grid-2 indicated that the accuracy of the free surface and the results of motion RAOs did not differ significantly from those for Grid-3. The heave and pitch RAOs show GCI values below 1.0% for Grid-3 and Grid-2. In addition, the motion RAOs for Grid-2 show satisfactory agreement with the experiment results with an error of about 3.0%. The computations in this study were performed using Grid-2.

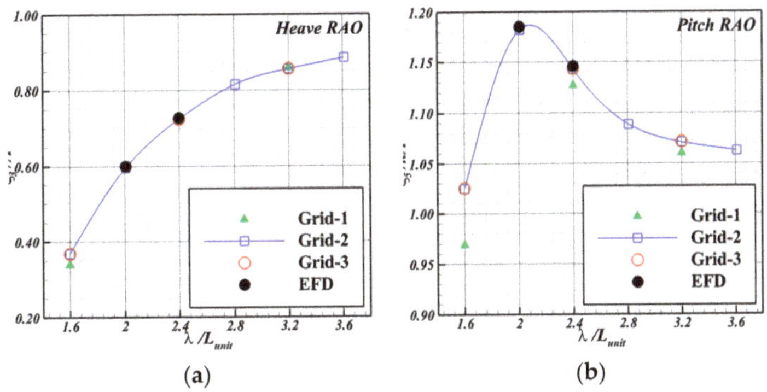

Figure 13. Comparison of motion RAOs according to grid conditions: (**a**) heave; (**b**) pitch.

5. Results

A model test was conducted using the IUTT to investigate the motion and load characteristics of the FSP under wave conditions. The results were compared with the computation results obtained with identical conditions. It was attempted to ensure the reliability and validity of the calculation based on the motions generated in each unit by the connection and the load applied to the mooring lines.

5.1. Head Sea Conditions at H/λ = 0.03

The experimental results of the motion RAOs of each unit of the FSP under waves at H/λ = 0.03 according to the wavelength are shown in Figure 14. Here, the circular and square symbols represent an odd-numbered and even-numbered unit, respectively. Black and blue denote the first and second groups, respectively.

In the case of surge motion, the surge motion generated in the wave direction in each unit was expressed. The difference between the groups was large at λ/L_{unit} = 2.0. Subsequently, the units tended to converge as the wavelength increased.

In the case of heave motion, the heave RAO increased linearly with the increase in wavelength, and relative behavior occurred between the first and second groups at λ/L_{unit} = 2.0 and 2.4. It appeared that when the incident wave lost energy while contacting the first group, this effect was reduced by the motion of RAOs of the second group. In addition, it was considered that a similar heave RAO occurred at λ/L_{unit} = 1.6 owing to a small wave height and short period.

In the case of roll motion, the hinge motion parallel to the wave travel direction was predominant because of the head sea conditions, the FSP was bilaterally symmetric, and almost no roll RAO was generated.

In the case of pitch motion, because the wavelength was longer than the FSP length for λ/L_{unit} = 2.8, a motion that rides on a wave appeared. Furthermore, the pitch RAO tended to converge as the wavelength increased. In addition, at λ/L_{unit} = 2.4, the length of the FSP became equal to the wavelength, and the maximum pitch RAO occurred.

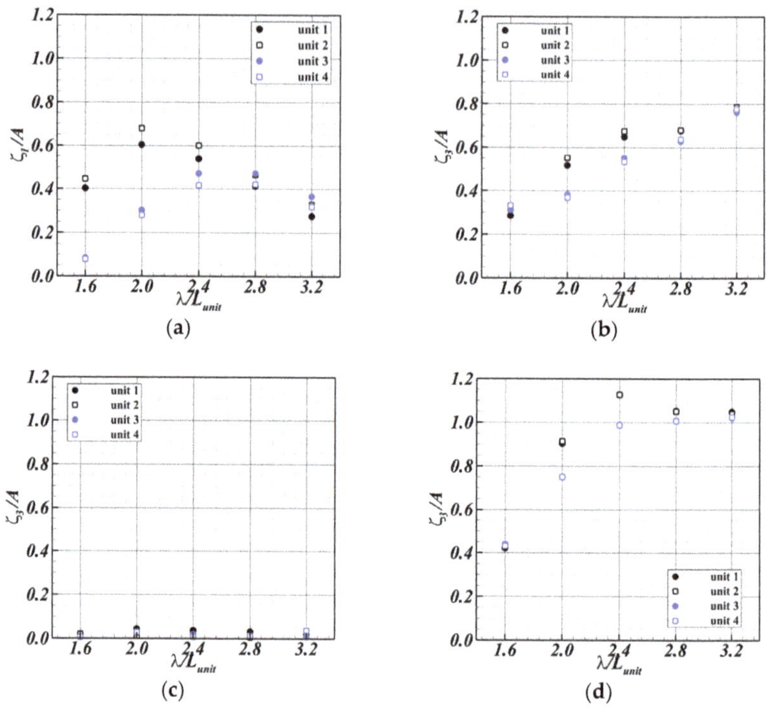

Figure 14. Motion RAOs of the experiment according to the wavelength at H/λ = 0.03: (**a**) surge; (**b**) heave; (**c**) roll; (**d**) pitch.

The experimental and computation results were compared to verify the wave pattern around the FSP according to the wavelength (see Figure 15). Here, the images on the left and right represent the experiment and computation, respectively. It was observed that the waves were diffracted and reflected by floating bodies, and wave run-up was also observed. When the waves passed through the floating bodies, the bodies were significantly affected by the periodic variation in the incident wave, which especially appeared at a relatively short period condition (λ/L_{unit} = 1.6). Then, the nonlinear wave profile reveled a non-uniform load on the support structure coupled to the suspended bodies [24]. At λ/L_{unit} = 2.4, the deformation of the first and second groups were similar to the experiment. Also, the motions of riding on the wave were observed at long wavelengths.

The computation results of motion RAOs according to wavelength were compared with the experimental results (see Figures 16 and 17). Here, the black-square and red-circular symbols represent the experimental and computation results, respectively. The heave RAOs are shown in Figure 16. Similar results were observed overall. The heave RAOs increased linearly with the wavelength. The pitch RAOs are shown in Figure 17. The computation results matched a maximum value at λ/L_{unit} = 2.4, which is the same as the experiment, which was influenced by the wave conditions. However, the heave and pitch RAOs showed a difference from motion RAOs.

At H/λ = 0.03, λ/L_{unit} = 2.4, the time series data of the total tension was measured on the mooring lines connected to the FSP (see Figure 18). Here, the black-solid and red-dotted lines represent the experimental and computation results, respectively. The tension time series shows two points. One is the maximum load point where the FSP was farthest from the initial position. The second point is the average load point where the position of the FSP converged owing to the restoration of the mooring lines. Compared to the experiment, it showed a similar trend that the period in which the load was generated and the two tension change points. The total tension of the mooring lines is expressed as the maximum and average loads (see Table 6). The maximum tension was observed at λ/L_{unit} = 2.0, where the difference of surge RAO between the units was large. We observed a tendency to converge even at tensions with $\lambda/L_{unit} \geq 2.4$.

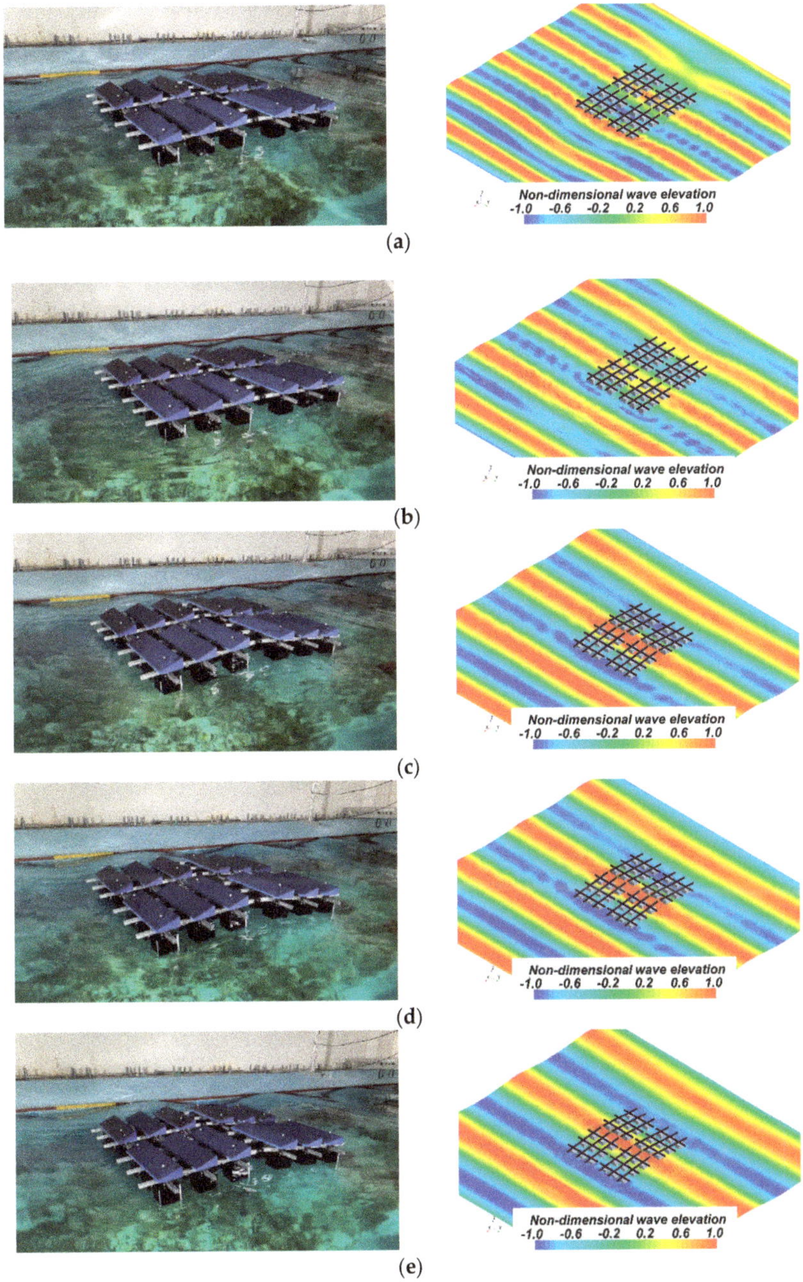

Figure 15. Wave pattern of experiment (**left**) and computation (**right**) according to the wavelength for H/λ = 0.03 under head sea: (**a**) λ/L_{unit} = 1.6; (**b**) λ/L_{unit} = 2.0; (**c**) λ/L_{unit} = 2.4; (**d**) λ/L_{unit} = 2.8; (**e**) λ/L_{unit} = 3.2.

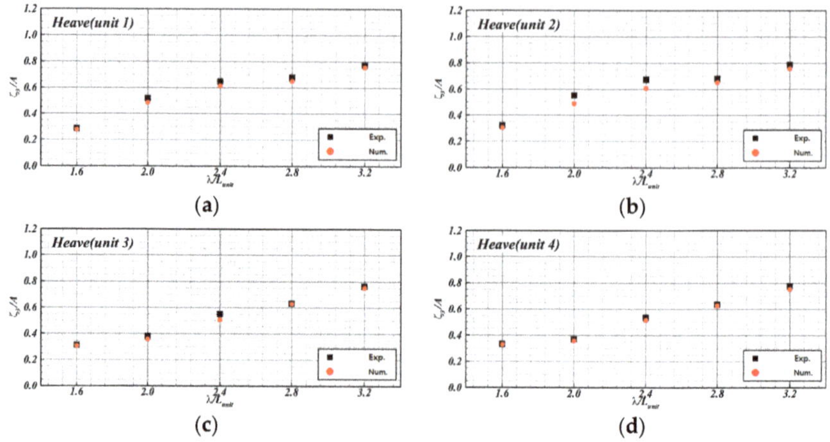

Figure 16. Heave RAOs for H/λ = 0.03 under head sea: (**a**) Unit 1; (**b**) Unit 2; (**c**) Unit 3; (**d**) Unit 4.

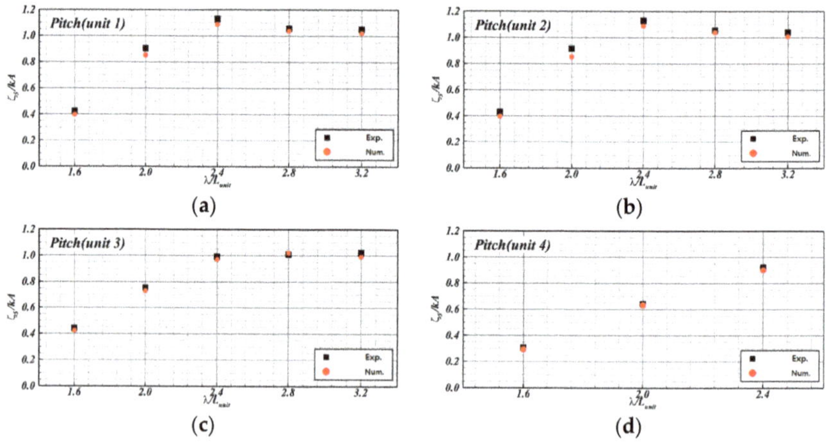

Figure 17. Pitch RAOs for H/λ = 0.03 under head sea: (**a**) Unit 1; (**b**) Unit 2; (**c**) Unit 3; (**d**) Unit 4.

Table 6. Comparison of total tension determined by experiment and computation at H/λ = 0.03.

Incident Wave Angle	H/λ	λ/L_{unit}	Max. Tension [N] (Exp.)	Max. Tension [N] (Num.)	Mean. Tension [N] (Exp.)	Mean. Tension [N] (Num.)
180	0.03	1.6	15.23	13.79	9.30	8.83
180	0.03	2.0	37.52	35.00	21.11	19.22
180	0.03	2.4	20.28	21.64	12.92	13.77
180	0.03	2.8	10.29	11.94	6.01	6.55
180	0.03	3.2	11.99	12.23	6.95	6.69

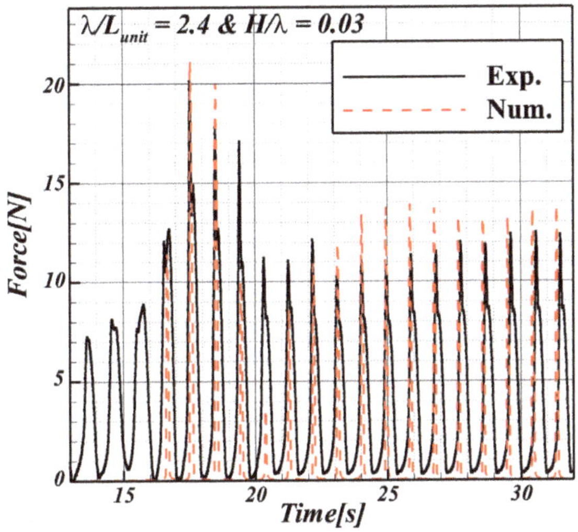

Figure 18. Time series data of total tension at λ/L_{unit} = 2.4 and H/λ = 0.03.

5.2. Head Sea Condition at Different Wave Steepness

The time series motion data for the heave and pitch of Unit 1 in the experiment according to the wave steepness for λ/L_{unit} = 1.6 and 2.4 are shown in Figure 19. Here, the pink and blue colors represent λ/L_{unit} = 1.6 and 2.4, respectively. The solid and dotted lines represent H/λ = 0.03 and 0.05, respectively. When the wave steepness is constant, the variation at λ/L_{unit} = 2.4 is larger than that for 1.6, because the height of the wave varies significantly.

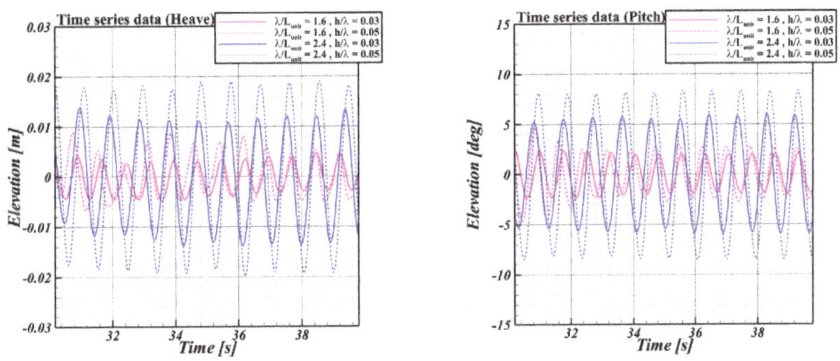

(a) (b)

Figure 19. Time series motion data according to wave steepness at Unit 1: (**a**) heave; (**b**) pitch.

The experimental results of the motion RAOs of each unit of the FSP under waves according to wave steepness are shown in Figure 20. In the motion RAOs, the results for H/λ = 0.03 were higher than those for 0.05. This tendency was converse to that of the time series motion data. It was thought that the nonlinearity of the motion increased with respect to the rate of increase in wave steepens. In addition, the difference appeared to be smaller in the second group than in the first. The wave lost energy while contacting the first group, and the second group in contact with the losing wave had reduced motion.

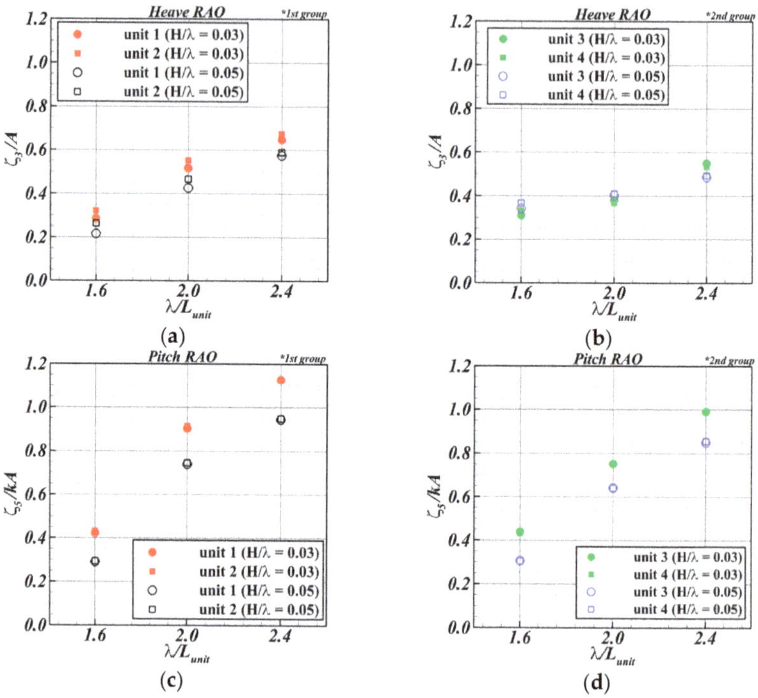

Figure 20. Motion RAOs of the experiment according to wave steepness: (**a**) heave (first group); (**b**) heave (second group); (**c**) pitch (first group); (**d**) pitch (second group).

The experiment and computation results were compared to verify the wave pattern around the FSP according to the wavelength (see Figure 21). In $H/\lambda = 0.03$, a large motion characteristic appeared at the center of the hinge. The bigger wave run-up and diffraction were observed, which showed similar results in computation. Also, the green water observed at $\lambda/L_{unit} = 2.4$ in the experiment was also observed in the computation results.

The computation results of motion RAOs according to the wavelength at $H/\lambda = 0.03$ were compared with the experimental results (see Figures 22 and 23). Similarly, as in the previous section, the amplitude of the RAOs increased as the wavelength increased. The total tension of the mooring lines is expressed as the maximum and average loads (see Table 7).

The tension was larger at $\lambda/L_{unit} = 2.0$, and the maximum load occurred at a wavelength ratio equal to that mentioned in Section 5.2. In addition, as the wave height increased, the wave energy increased with the load. It is expressed non-dimensionally to analyze the tension by excluding the wave conditions (see Figure 24). The tension coefficient ($C_{tension}$) is identical to that in Equation (5):

$$C_{Tension} = \frac{F_{Total\ tenstion}}{\frac{\rho g A^2 B_T^2}{L_{unit}}} \quad (5)$$

where ρ is the density, g is the gravitational acceleration, A is the wave amplitude, and B_T is the total breadth of the floating bodies at one unit. Regarding the tension coefficient, $H/\lambda = 0.03$ appeared to be larger than 0.05. This was because the tension increased with the increase in wave height. However, the non-dimensional tension coefficient tended to decrease because of the nonlinear decrease in the pitch RAO.

(**a**)

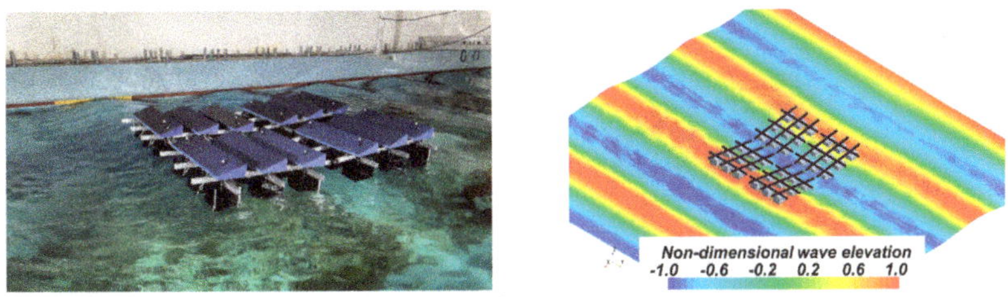

(**b**)

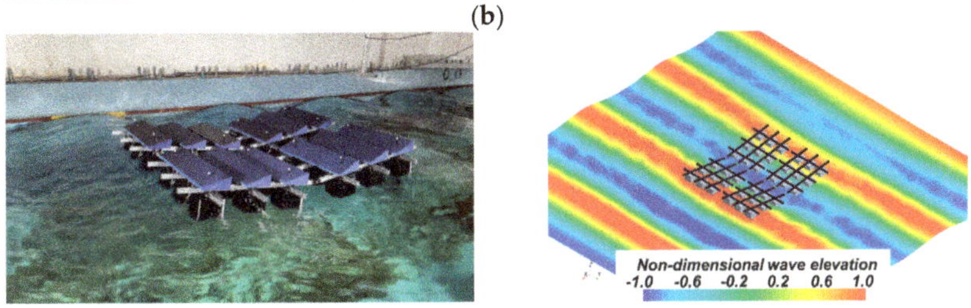

(**c**)

Figure 21. Wave pattern of experiment (**left**) and computation (**right**) according to the wavelength for H/λ = 0.05 under head sea: (**a**) λ/L_{unit} = 1.6; (**b**) λ/L_{unit} = 2.0; (**c**) λ/L_{unit} = 2.4.

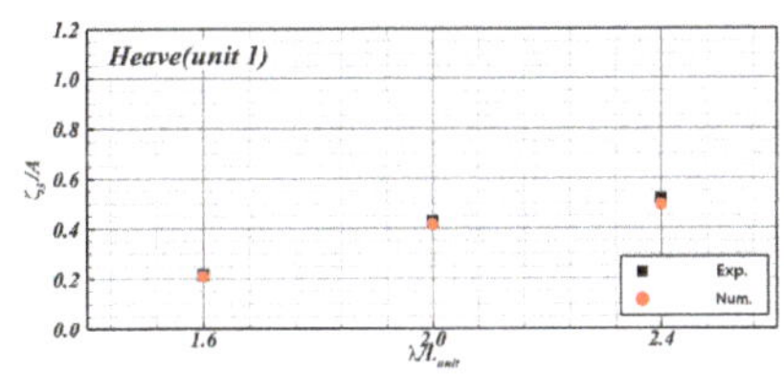

(**a**)

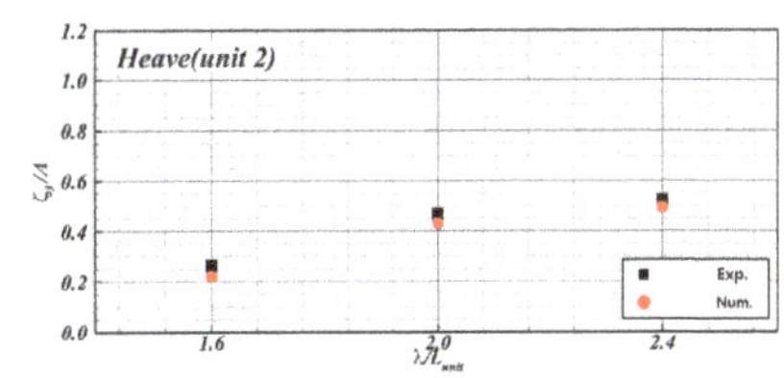

(**b**)

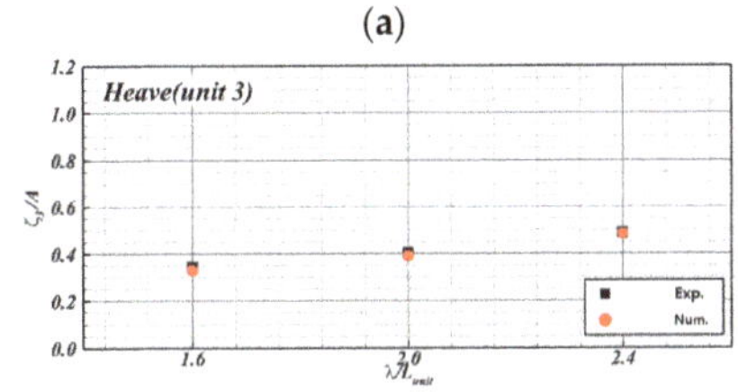

(**c**)

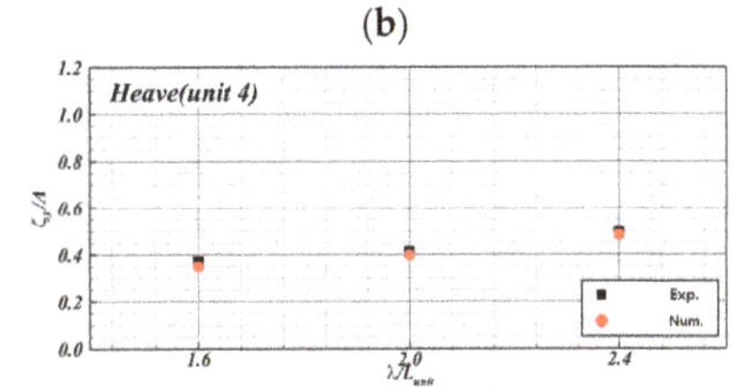

(**d**)

Figure 22. Heave RAOs for H/λ = 0.05 under head sea: (**a**) Unit 1; (**b**) Unit 2; (**c**) Unit 3; (**d**) Unit 4.

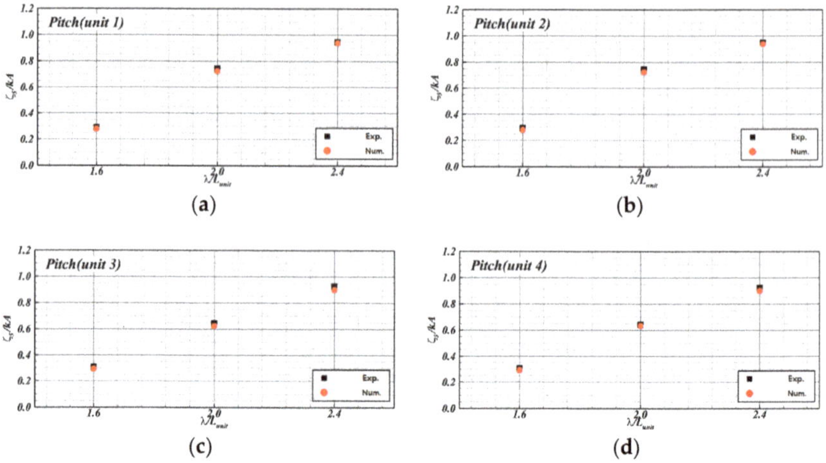

Figure 23. Pitch RAOs for $H/\lambda = 0.05$ under head sea: (**a**) Unit 1; (**b**) Unit 2; (**c**) Unit 3; (**d**) Unit 4.

Table 7. Comparison of total tension determined by experiment and computation at $H/\lambda = 0.05$.

Incident Wave Angle	H/λ	λ/L_{unit}	Max. Tension [N] (Exp.)	Max. Tension [N] (Num.)	Mean. Tension [N] (Exp.)	Mean. Tension [N] (Num.)
180	0.05	1.6	17.07	20.41	9.47	11.02
180	0.05	2.4	44.76	48.96	25.29	27.33
180	0.05	3.2	32.93	35.66	17.27	20.23

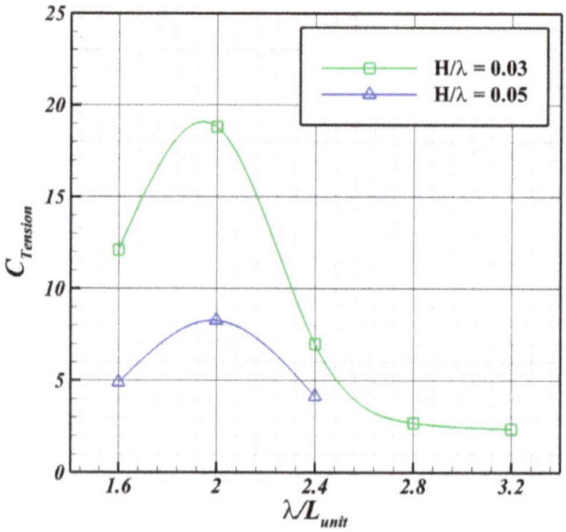

Figure 24. Comparison of tension coefficient according to wave steepness.

5.3. Oblique Sea Conditions at $H/\lambda = 0.03$

In this study, we conducted a model test under oblique wave conditions by rotating the FSP, which leads the waves to enter diagonally. Units 2 and 4 of the FSP were arranged in a straight line, and the FSP was rotated by 41°. The experimental results of the motion RAOs of each unit of the FSP under oblique waves are shown in Figure 25.

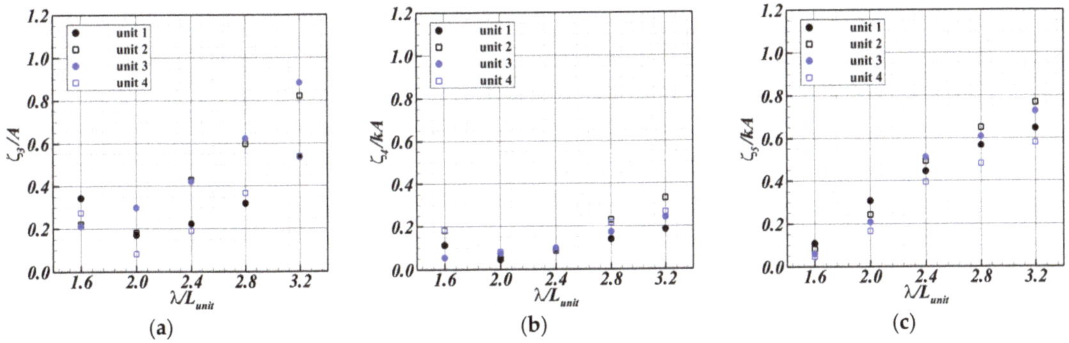

Figure 25. Motion RAOs of the experiement according to wavelength under oblique sea: (**a**) heave; (**b**) roll; (**c**) pitch.

In the case of heave motion, similar motion characteristics occurred between units (Units 1 and 4, and Units 2 and 3) located at the same position. In Unit 3, the waves that passed between Units 1 and 4 overlapped, thereby causing excessive motion.

Roll motion varies significantly depending on the wave direction, and this increases gradually according to the wavelength. In the uniaxial hinge system, the units in a column exhibited similar movements. In the case of pitch motion, the motion was lower than that under head sea conditions. When the wavelength was increased, the relative movements between the units were getting bigger. On the other hand, in the experiment, it was observed that the diffracted waves spread widely, and the waves would re-enter the FSP through the walls. In addition, the overlapping and dissipation of waves between floating bodies (which were difficult to observe in the experiment) were inspected.

The computation under oblique sea conditions was performed using the numerical method verified for head sea conditions. The results were compared with the experimental results for λ/L_{unit} = 2.4 and 2.8. The experimental and computation results were compared to verify the wave pattern around the FSP according to the wavelength (see Figure 26). The generated wave encountered the floating body at the front of Unit 2. The effect of wave (e.g., diffraction, upwelling, run-up, and reflection) appeared to be larger than those under the head sea conditions because the geometry of the FSP.

The computation results of the motion RAOs according to wavelength were compared with the experimental results (see Figure 27). The oblique sea conditions showed similarities to head sea conditions. The translational heave motion showed a tendency to match. However, the roll and pitch motions differed, and it is thought that the difference of the hinge causes errors between experiments and simulations. This was determined to be an error that occurred in the hinge gap, as mentioned in the previous section. Theoretically, the uniaxial hinge does not cause simultaneous rotational motion of the horizontal and vertical axes. However, in the experiment, it was determined that the rotational motion occurred simultaneously owing to the hinge gap and that additional rotational motion was generated during overlapping rotation. However, the motion occurred only in the axial direction in the computation. This was observed a difference in the rotational components.

The time series data of the total tension measured on the mooring lines connected to the FSP under oblique sea are shown in Figure 28. The total tension of the mooring lines is expressed as the maximum and average loads in the oblique sea (see Table 8). The periods in which the load was generated, and the load exhibited similar tendencies. There was a difference in the load pattern after the maximum tension. This is unlike the head sea condition. Furthermore, it was determined that the oblique sea condition did not achieve precise geometrical similarity for the mooring line system. It was then determined that the external force applied to the FSP appeared to be relatively low tension because the waves were dispersed and escaped from the configuration of FSP and the movements were smaller than those under the head sea.

(a)

(b)

Figure 26. Wave pattern of experiment (**left**) and computation (**right**) according to the wavelength under oblique sea: (**a**) λ/L_{unit} = 2.4; (**b**) λ/L_{unit} = 2.8.

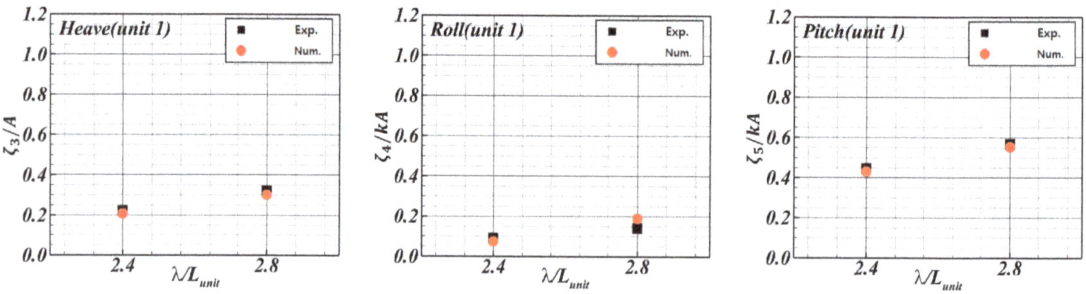

(a)

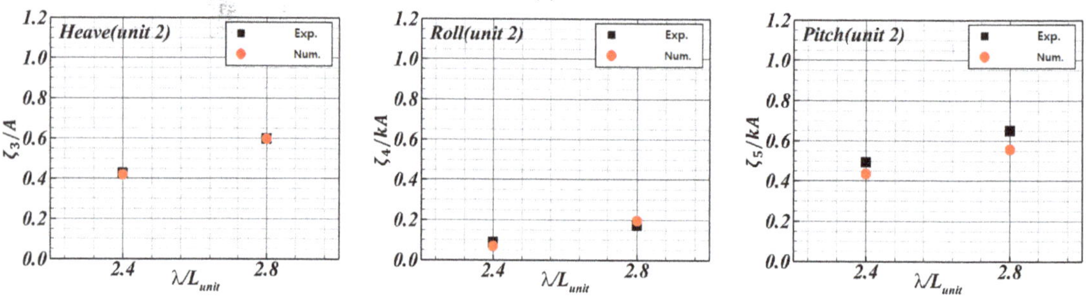

(b)

Figure 27. *Cont.*

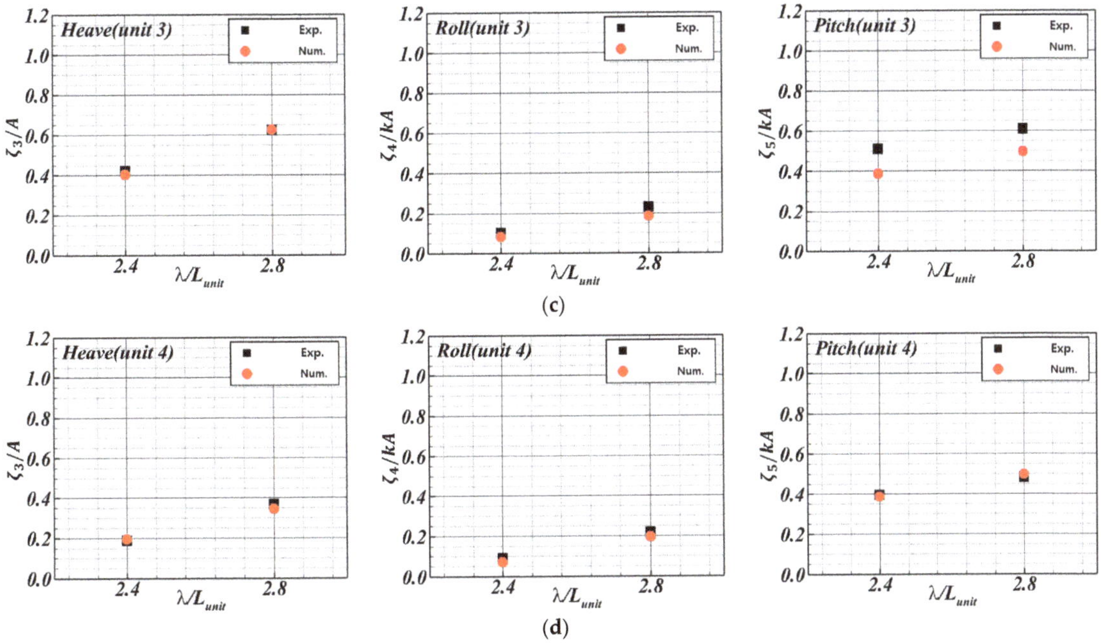

Figure 27. Motion RAOs of heave (**left**), roll (**center**) and pitch (**right**) under oblique sea: (**a**) Unit 1; (**b**) Unit 2; (**c**) Unit 3; (**d**) Unit 4.

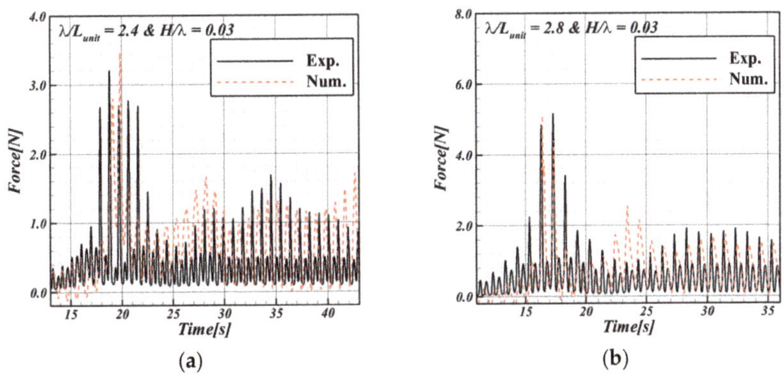

Figure 28. Time series data of total tension under oblique sea: (**a**) λ/L_{unit} = 2.4; (**b**) λ/L_{unit} = 2.8.

Table 8. Comparison of total tension determined by experiment and computation under oblique sea.

Incident Wave Angle	H/λ	λ/L_{unit}	Max. Tension [N] (Exp.)	Max. Tension [N] (Num.)	Mean. Tension [N] (Exp.)	Mean. Tension [N] (Num.)
139	0.05	2.4	3.345	3.206	1.342	1.631
139	0.05	2.8	5.117	5.129	1.484	1.717

According to these results, it is necessary to develop an accurate computational technique for a connection similar to that in the experiment. When applying multi-directional articulations to the FSP, substantial research needs to be carried out on articulations for a more accurate interpretation.

6. Conclusions

In this study, a physical method (experiment) and numerical method (CFD) were used to observe an accurate motion response of an FSP. The validity and reliability of the numerical method was confirmed through a validation process. Both results were compared, and these showed good agreement.

1. For the experiment, a model of a two-row and two-column unit matrix-type floating solar power farm was manufactured. The connecting part was composed of uniaxial hinges. Motion capture cameras and tension meters were used to measure the motion characteristics and tension, respectively. We established experimental techniques, e.g., the construction of a measurement platform for 3D object motion and the construction of a floating solar power farm (FSP) experiment under wave conditions. The results were used to verify the numerical analysis.
2. The conditions of the three waves were calculated and compared with the experimental results to verify the numerical technique. The wave generated in the computation was identical to that generated in the experiment. Although there were differences in the tension and mechanical error generated in the experiment, similar trend of results was obtained for the wave pattern generated around the floating solar power farm and the motion responses to each motion component. The tension also verified the similarity.
3. Through the experiment and CFD, each unit group of the FSP showed identical motion under head sea conditions. The longitudinal motion appeared to be larger than the other motion responses. As the wavelength increased, the heave and pitch RAOs increased linearly, and the pitch RAOs tended to converge after a wavelength ratio of 2.4. The largest total load applied to the mooring lines was generated at a wavelength ratio of 2.0, which yielded a large difference in surge motion for each unit.
4. The motion response of the floating solar power farm was compared according to the wave height in the head sea condition. Although the motion increased when the wave height increases, the motion RAOs showed the converse tendency. This showed that the motions are proportional to the wave height. On the other hand, the non-dimensional tension with a wave steepness of 0.03 was larger than that for 0.05. It was thought that although the tension increased, the tension coefficient tended to decrease owing to the nonlinear decrease in pitch RAOs.
5. The waves approached the floating solar power farm at an angle (oblique waves, 41 degrees). This revealed various motion characteristics. One of these was roll motion, which was not observed in the head sea conditions. It had a highly complex motion response. In addition, the waves dispersed due the configuration of the FSP. The motion appeared to be smaller than that under the head sea condition. The load applied to the tension also appeared to be low.

The verified numerical technique can become the basis for research on developing a mooring system that can determine various environmental conditions and motion responses and maximize the efficiency of solar panels and floating solar power farms considering the wave load. The design can be achieved by considering the load distribution of the support structure. In a future study, various floating solar farms with over two rows and columns would be simulated to investigate the complex interactions among floating units in detail.

Author Contributions: Conceptualization, computation, data analysis, post processing, and writing—original draft preparation, review and editing, J.-H.L.; conceptualization, data analysis, writing—review and editing, funding, and supervision, K.-J.P.; conceptualization and data analysis and review, S.-H.L.; conceptualization and data analysis, J.H.; visualization and investigation, T.-H.H. All authors have read and agreed to the published version of the manuscript.

Funding: The Evaluation of Wave Load and Motion Performance of the Support Structure for the Floating Solar Power Plant (2021X009), by POSCO, Republic of Korea, and the Competency Development Program for Industry Specialists of the Korean Ministry of Trade, Industry and Energy (MOTIE), operated by the Korean Institute for Advancement of Technology (KIAT). (No. P0012646, HRD program for Global Advanced Engineer Education Program for Future Ocean Structures).

Institutional Review Board Statement: Not applicable.

Informed Consent Statement: Not applicable.

Data Availability Statement: Not applicable.

Acknowledgments: This research was supported by the Evaluation of Wave Load and Motion Performance of the Support Structure for the Floating Solar Power Plant (2021X009), by POSCO, Republic of Korea, and was funded and conducted under the Competency Development Program for Industry Specialists of the Korean Ministry of Trade, Industry and Energy (MOTIE), operated by Korean Institute for Advancement of Technology (KIAT). (No. P0012646, HRD program for Global Advanced Engineer Education Program for Future Ocean Structures).

Conflicts of Interest: The authors declare no conflict of interest.

References

1. Renewable Energy 3020 Implementation Plan (Ministry of Trade, Industry and Energy, Republic of Korea). Available online: http://www.motie.go.kr/motie/ne/presse/press2/bbs/bbsView.do?bbs_cd_n=81&bbs_seq_n=159996 (accessed on 20 March 2021).
2. Alsharif, M.H.; Kim, J.; Kim, J.H. Opportunities and Challenges of Solar and Wind Energy in South Korea: A Review. *Sustainability* **2018**, *10*, 1822. [CrossRef]
3. Park, M.; Park, S.; Seong, B.; Choi, Y.; Jung, S.P. Current Status and Prospective of Offshore Wind Power to Achieve Korean Renewable Energy 3020 Plan. *J. Korean Soc. Environ. Eng.* **2021**, *43*, 195–205. [CrossRef]
4. Hu, C.; Sueyoshi, M.; Liu, C.; Kyozuka, Y.; Ohya, Y. Numerical and experimental study on a floating platform for offshore renewable energy. In Proceedings of the International Conference on Ocean, Offshore and Arctic Engineering, Nantes, France, 9–14 June 2013.
5. Kim, K.H.; Hong, J.P.; Park, S.; Lee, K.; Hong, K. An Experimental Study on Dynamic Performance of Large Floating Wave-Offshore Hybrid Power Generation Platform in Extreme Conditions. *J. Korean Soc. Mar. Environ. Energy* **2016**, *19*, 7–17. [CrossRef]
6. Kwon, Y.J.; Nam, B.W.; Kim, N.; Jung, D.H.; Hong, S.Y.; Kim, H.J. Numerical and Experimental Study on Motion Response of 1MW OTEC Platform. *J. Ocean Eng. Technol.* **2017**, *31*, 81–90. [CrossRef]
7. Oliveira-Pinto, S.; Stokkermans, J. Marine floating solar plants: An overview of potential, challenges and feasibility. *Proc. Inst. Civ. Eng. Marit. Eng.* **2020**, *173*, 120–135. [CrossRef]
8. Sahu, A.; Yadav, N.; Sudhakar, K. Floating photovoltaic power plant: A review. *Renew. Sustain. Energy Rev.* **2016**, *66*, 815–824. [CrossRef]
9. Terre, C. Hydrelio®Technolo. 2018. Available online: https://issuu.com/cieletterre/docs/c_t_catalog_hydrelio__technology_20 (accessed on 13 November 2021).
10. Abyn, H.; Maimun, A.; Koto, J.; Rafiqul Islam, M.; Magee, A.; Bodaghi, B.; Pauzi, M.; Siow, C.L.; Mobassher Tofa, M. Hydrodynamic interaction of floating structure in regular waves. *J. Teknol. Sci. Eng.* **2014**, *66*, 91–96. [CrossRef]
11. Shi, Q.J.; Zhang, H.C.; Xu, D.L.; Qi, E.R.; Tian, G.; Ding, J.; Wu, Y.S.; Lu, Y.; Li, Z.W. Experimental validation of network modeling method on a three-modular floating platform model. *Coast. Eng.* **2018**, *137*, 92–102. [CrossRef]
12. Li, P.; Faltinsen, O.M.; Lugni, C. Nonlinear vertical accelerations of a floating torus in regular waves. *J. Fluids Struct.* **2016**, *66*, 589–608. [CrossRef]
13. Al-Yacouby, A.M.; Halim, E.R.B.A.; Liew, M.S. Hydrodynamic analysis of floating offshore solar farms subjected to regular waves. *Lect. Notes Mech. Eng.* **2020**, *20*, 375–390.
14. Siemens. Simcenter STAR-CCM+ User Guide, Version 2020.2. Available online: https://support.sw.siemens.com/2020 (accessed on 1 January 2020).
15. Fenton, J.D. A Fifth-Order Stokes Theory for Steady Waves. *J. Waterw. Port Coast. Ocean Eng.* **1985**, *111*, 216–234. [CrossRef]
16. Lee, S.B.; Han, S.Y.; Choi, Y.M.; Kwon, S.H.; Jung, D.W.; Park, J.S. Study on wave run-up phenomenon over vertical cylinder. *J. Ocean Eng. Technol.* **2013**, *27*, 62–67. [CrossRef]
17. Cheng, L.; Lin, P. The Numerical Modeling of Coupled Motions of a Moored Floating Body in Waves. *Water* **2018**, *10*, 1748. [CrossRef]
18. Perić, R.; Abdel-Maksoud, M. Reliable damping of free-surface waves in numerical simulations. *Ship Technol. Res.* **2016**, *63*, 1–13. [CrossRef]
19. Perić, R.; Abdel-Maksoud, M. Analytical prediction of reflection coefficients for wave absorbing layers in flow simulations of regular free-surface waves. *Ocean Eng.* **2018**, *147*, 132–147. [CrossRef]

20. Kim, J.; Park, I.R.; Suh, S.B.; Kang, Y.D.; Hong, S.Y.; Nam, B.W. Motion Simulation of FPSO in Waves through Numerical Sensitivity Analysis. *J. Ocean Eng. Technol.* **2018**, *32*, 166–176. [CrossRef]
21. Celik, I.B.; Ghia, U.; Roache, R.J.; Freitas, C.J.; Coleman, H.; Raad, P.E. Procedure for estimation and reporting of uncertainty due to discretization in CFD applications. *J. Fluids Eng. Trans.* **2008**, *130*, 0780011–0780014.
22. Kim, S.J.; Choi, Y.S.; Cho, Y.; Choi, J.W.; Hyun, J.J.; Joo, W.G.; Kim, J.H. Analysis of the Numerical Grids of a Francis Turbine Model through Grid Convergence Index Method. *KSFM J. Fluid Mach.* **2020**, *23*, 16–22. [CrossRef]
23. Tezdogan, T.; Demirel, Y.K.; Kellett, P.; Khorasanchi, M.; Incecik, A.; Turan, O. Full-scale unsteady RANS CFD simulations of ship behaviour and performance in head seas due to slow steaming. *Ocean Eng.* **2015**, *97*, 186–206. [CrossRef]
24. Chen, W.; Gao, Y.; Wang, L.; Guo, Z.; Wang, B. Numerical simulation of wave run-up on four cylinders in a square configuration. *Appl. Ocean Res.* **2021**, *108*, 102519. [CrossRef]

Journal of
Marine Science and Engineering

MDPI

Article

A Decision Support Tool for Long-Term Planning of Marine Operations in Ocean Energy Projects

Francisco X. Correia da Fonseca *, Luís Amaral and Paulo Chainho

WavEC-Offshore Renewables, Edifício Diogo Cão, Doca de Alcântara Norte, 1350-352 Lisboa, Portugal; luis.amaral@wavec.org (L.A.); paulo@wavec.org (P.C.)
* Correspondence: francisco.fonseca@wavec.org

Abstract: Ocean energy is a relevant source of clean renewable energy, and as it is still facing challenges related to its above grid-parity costs, tariffs intended to support in a structured and coherent way are of great relevance and potential impact. The logistics and marine operations required for installing and maintaining these systems are major cost drivers of marine renewable energy projects. Planning the logistics of marine energy projects is a highly complex and intertwined process, and to date, limited advances have been made in the development of decision support tools suitable for ocean energy farm design. The present paper describes the methodology of a novel, open-source, logistic and marine operation planning tool, integrated within DTOceanPlus suite of design tools, and responsible for producing logistic solutions comprised of optimal selections of vessels, port terminals, equipment, as well as operation plans, for ocean energy projects. Infrastructure selection logistic functions were developed to select vessels, ports, and equipment for specific projects. A statistical weather window model was developed to estimate operation delays due to weather. A vessel charter rate modeling approach, based on an in-house vessel database and industry experience, is described in detail. The overall operation assumptions and underlying operating principles of the statistical weather window model, maritime infrastructure selection algorithms, and cost modeling strategies are presented. Tests performed for a case study based a theoretical floating wave energy converter produced results in good agreement with reality.

Keywords: logistics; marine operation planning; dtoceanplus; decision support; ocean energy

Citation: Correia da Fonseca, F.X.; Amaral, L; Chainho, P. A Decision Support Tool for Long-Term Planning of Marine Operations in Ocean Energy Projects. *J. Mar. Sci. Eng.* **2021**, 9, 810. https://doi.org/10.3390/jmse9080810

Academic Editors: Eugen Rusu, Kostas Belibassakis and George Lavidas

Received: 9 June 2021
Accepted: 22 July 2021
Published: 27 July 2021

Publisher's Note: MDPI stays neutral with regard to jurisdictional claims in published maps and institutional affiliations.

Copyright: © 2021 by the authors. Licensee MDPI, Basel, Switzerland. This article is an open access article distributed under the terms and conditions of the Creative Commons Attribution (CC BY) license (https://creativecommons.org/licenses/by/4.0/).

1. Introduction

The decarbonization of the energy sector is urgent, requiring global action to achieve our long-term climate goals and to mitigate the impacts of climate change [1]. To meet our ambitious emission cuts, innovation in low-carbon technologies and mass deployment of renewable energy generation will be fundamental [2,3].

As a renewable energy resource, ocean energy is clean, abundant, and powerful. Wave and tidal energy are attractive sources of renewable energy, as they have low variability when compared to wind, can be accurately forecast, and are fit to respond to the electricity demand during night-time [4]. Additionally, the production profile of wave and tidal energy systems is complementary to wind and solar, smoothing the otherwise peaking nature of renewables in the production mix [5]. It is estimated that about 100 GW of wave and tidal energy capacity can be deployed in Europe by 2050 [6], creating significant carbon emission reductions as well as economic growth opportunities. Europe's seas and oceans could therefore play a fundamental role in the decarbonization of the energy sector, contributing to the transition from a power system based on imported fossil fuels, to a flexible and interconnected system based on clean, renewable, and infinite domestic resources [7]. However, the ocean energy sector is still facing challenges related to performance, reliability, and survivability, which ultimately translates into above grid-parity costs.

Logistics and marine operations are major cost drivers of marine renewable energy projects. Even though researchers have made significant progress over the last years in what concerns the installation and operation and maintenance (O&M) planning of offshore wind farms [8–11], advances for ocean energy farms have been more modest, attesting for the lower maturity of the sector [12]. For offshore wind, the installation costs typically represent one fifth to one third of the project's CapEX [13–15], while O&M activities represent about one-fourth to one-half of the total lifetime costs of the project [16]. However, for less mature sectors such as wave and tidal energy, slightly larger percentages may be expected [17]. These costs are typically amplified when deploying projects in further offshore waters, as the marine operations related to the construction, installation, maintenance, and decommissioning of such farms become increasingly challenging. Even though deploying farms further offshore is expected to improve resource availability and consequently increase the expected power output of the farm (while also minimizing competition for space and visual disturbance [18]), more severe weather climates and larger distances to shore translate into lower farm accessibility, higher risks of work delays, and ultimately larger project costs. As a significant fraction of the marine operation costs can be attributed to vessel charter (according to Dalgic et al. [19], approximately 73% of the total O&M costs are related to vessel hiring [19]), even modest reductions in operation duration may result in significant cost-reductions [20].

Planning the logistics of offshore renewable projects is a highly intertwined process, with multiple conflicting objectives and alternatives, and a large optimization potential. Given the complexity associated with planning such logistics, computational tools have been developed to support decision-making at different project stages. Computational tools can be distinguished according to (i) decision-making time-scale (long-term strategic planning based on historical weather data, or short-term daily operational planning based on weather forecasts), (ii) project phase (installation, O&M, and decommissioning), (iii) target sector (offshore wind, ocean energy, or both), (iv) licensing type (open-source, private, or commercial), and (v) software functionalities (e.g., weather window modeling, operation planning, infrastructure selection, failure/degradation modeling, revenue modeling, and techno-economic assessments).

Table 1 shows a list of the main logistic support tools developed to date with the goal of supporting offshore projects. It can be seen that most development efforts have focused on producing O&M simulation tools to estimate the OPEX of offshore wind projects. Some of these tools were developed to simulate the degradation of farm components and the occurrence of failures, replicating real-world decisions in respect to the scheduling of preventive and corrective maintenance activities [21]. This is the case of ECN O&M Tool, the ECN O&M Calculator (formerly OMCE) and the O2M model of DNV-GL [22]. As most operations carried out at sea are significantly weather dependent, computational tools generally include weather window models to estimate the potential waiting on weather contingencies. Some commercial and sector-agnostic tools, such as Mermaid [23] and ForeCoast Marine [24] (marked as "Agnostic" in Table 1), have focused almost exclusively on this type of service, quantifying weather risks for different operation types and target sectors. Another similar commercial product worth mentioning is StormGeo [25], which also provides short-term decision support based on near-future weather forecasts.

Table 1. Examples of logistic support tools for marine renewable energy projects, featuring their functionalities. Tools were labeled according to their capabilities, namely, to select suitable infrastructure solutions in respect to ports (*P*), vessels (*V*), and equipment (*E*).

Organisation	Product Name	Open Source	Applicable to Ocean Energy	Weather Window Analysis	Infra-Structure Selection	Optimal Operation Plan	Inst	O&M	Decom.
DNV-GL	O2M	No	No	Hindcast	No	No	No	Yes	No
DNV-GL	OMCAM	No	No	Hindcast	No	No	No	Yes	No
ECN	ECN O&M Access	No	No	Forecast	No	No	No	Yes	No
DTOcean 1.0 /2.0	DTO Logistics module [26,27]	Yes	Yes	Hindcast	P, V, E	Yes	Yes	Yes	No
DTOceanPlus	**DTO+ LMO module**	**Yes**	**Yes**	**Hindcast**	**P, V, E**	**Yes**	**Yes**	**Yes**	**Yes**
ECN (TNO)	ECN O&M Calculator (OMCE) [28]	No	No	Hindcast	No	No	No	Yes	No
Fraunhofer IWES	Multi-Agent-System	No	No	Hindcast	No	No	No	Yes	No
ROMEO	ROMEO O&M Tool [29]	N/A	No	Forecast	No	No	No	Yes	No
SINTEF Energy Research	NoWIcob [30]	No	No	Hindcast	V	No	No	Yes	No
SINTEF Ocean (MARINTEK)	Vessel fleet optimization models [31,32]	No	No	Hindcast	V	No	No	Yes	No
Shoreline	Shoreline Design [33]	No	No	Hindcast	No	No	Yes	Yes	No
Strathclyde University	StrathOW-OM [34,35]	No	No	Hindcast	No	No	No	Yes	No
EDF Group	ECUME-I [36,37]	No	No	Hindcast	No	No	No	Yes	No
Wave Energy Scotland	WES O&M Tool [38]	Yes	Yes	Hindcast	No	No	No	Yes	No
James Fisher and Sons	Mermaid [23]	No	Agnostic	Hindcast	No	No	-	-	-
JBA Consulting	ForeCoast Marine [24]	No	Agnostic	Hindcast	No	No	-	-	-
StormGEO	StormGEO S-Planner [25]	No	Agnostic	Forecast	No	No	-	-	-

However, it is possible to observe that very limited advances have been made in the development of logistic support tools suitable for ocean energy farm design. Existing computational tools either focus exclusively on the O&M phase (e.g., WES O&M Tool [38]) or are limited in functionality. A reduced number of tools has been developed to address vessel selection in offshore wind projects (e.g., NOWICOB [30] and StrathOW-OM [35]). Still, these tools do not consider the selection of ports and equipment (nor their impacts on optimal vessel selection), and most importantly, are not easily adaptable to ocean energy projects [39]. Moreover, most existent tools are notably user input-intensive, and thus unsuitable for project and technology developers at early development stages where uncertainties and unknowns are large. Finally, despite the growing number of open-source initiatives, which have been found to contribute significantly to sector innovations and to the European Union's economy [40,41], most computational tools were developed under private or commercial licenses with limited published research. As such, these tools miss out on the key benefits of open-source projects, related to higher transparency, robustness and scrutiny, as well as continuous improvements through community collaborations.

At a project development phase, integrating preliminary plans for the installation, maintenance, and decommissioning in the design process has the potential to reveal unexpected impacts of certain component design decisions on logistic costs. This is a particularly important step for ocean energy projects as less mature technologies have higher cost-optimization potential, frequently achievable with simple concept adjustments. In order to address the identified research gaps and market needs, the Logistics and Marine Operations (LMO) module was developed and integrated within DTOceanPlus software, an open-source suite of design tools for ocean energy projects [42]. The LMO module is responsible for designing and planning the project life-cycle phases (i.e., installation, maintenance, and decommissioning) of ocean energy projects. Reflecting the most recent experiences and best practices of the offshore wind sector, the LMO module produces logistic solutions comprised of optimal selections of vessels, port terminals, equipment, as well as operation plans, for ocean energy projects. An innovative methodology to optimize the selection of vessels, port terminals, and equipment was developed. A novel vessel cost modeling methodology was implemented in order to take into consideration the impacts of vessel capabilities on charter price, and reveal cost reduction pathways. Comprehensive, purpose-built databases of offshore operations, vessels, ports, and equipment were generated to support the main functionalities of the tool, even when unknowns are large and data availability is limited. These databases will be made freely available upon the final release of the DTOceanPlus software. Leveraging on its main functionalities, the Logistics and Marine Operations module proposes optimal logistic solutions that minimize total project costs, guiding project design and strategic investment decisions in ocean energy projects at different stages of technological and project maturity.

The present paper describes in detail the novel Logistics and Marine Operations tool, one of the seven design modules of the DTOceanPlus software. In Section 2, the DTOceanPlus software is briefly presented. The underlying operating principle, main functionalities, and methodology of the Logistic and Marine Operations tool are described in detail in Section 3. A brief test case showcasing the functionalities of the DTOceanPlus Logistics module is described in Section 4. The most important outcomes of the work are summarized in Section 5.

2. DTOceanPlus Suite of Design Tools

DTOceanPlus is an open source, integrated suite of design tools, developed to support the selection, development, deployment, and assessment of ocean energy systems at different stages and levels of aggregation (component, sub-system, and array). Building on the results from the original DTOcean project [43–45], at the time of writing, DTOceanplus software is being developed within a 3-years long EU-funded H2020 project with the same name [42], aimed at accelerating the commercialization of the ocean energy sector.

As illustrated in Figure 1, DTOceanPlus was developed in a modular fashion, with a set of independent but integrated tools:

- **Structured Innovation tool**, to support innovation and the selection of technology concepts;
- **Stage Gate tool**, to guide technology developers in their technology development process;
- **Deployment Design design tools**, to support optimal device and array deployment:
  - *Site Characterization (SC)*, to characterize the deployment site in respect to environmental (e.g., met-ocean) and geotechnical conditions;
  - *Machine Characterization (MC)*, to characterize the device's prime mover;
  - *Energy Capture (EC)*, to design and optimize the device hydrodynamics at an array level;
  - *Energy Transformation (ET)*, to design the Power Take-off (PTO) unit and control strategies;
  - *Energy Delivery (ED)*, to design the electrical power transmission system of the farm;
  - *Station Keeping (SK)*, to produce foundations and mooring designs;
  - *Logistics and Marine Operations (LMO)*, to design logistical solutions related to the installation, operation, maintenance, and dismantling operations.
- **Assessment design tools**, devised to evaluate ocean energy projects in respect to key metrics:
  - *System Performance and Energy Yield (SPEY)*, to assess projects in respect to their energy performance;
  - *System Lifetime Costs (SLC)*, to assess projects from the economic and financial investment perspectives;
  - *System Reliability, Availability, Maintainability, Survivability (RAMS)*, to evaluate different reliability aspects of ocean energy projects;
  - *Environmental and Social Acceptance (ESA)*, to evaluate ocean energy projects in respect to their environmental and social impacts.
- Underlying **digital models** to provide a standard framework for the description of ocean energy sub-systems, as well as a **global database** with reference data from various sources.

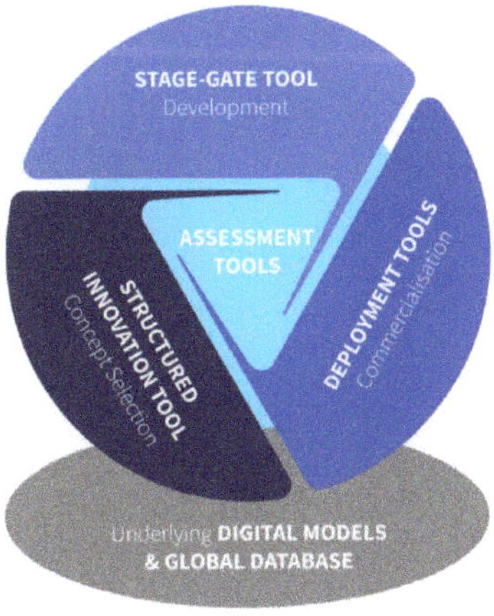

Figure 1. Representation of DTOceanPlus tools.

3. Logistics and Marine Operations Module

3.1. Operating Principle

The Logistics and Marine Operation (LMO) module was designed to generate optimal logistical solutions for the installation, O&M, and decommissioning phases of ocean energy projects. These logistic solutions consist of an operation plan (featuring dates and sequence of activities) and an optimal combination of vessels, equipment and ports that minimize the costs of each operation individually, reducing the capital and operational expenditures (CAPEX and OPEX) of the project.

The operating principle behind the LMO module is similar for all three life cycle stages of the project, and can be described as a sequence of different steps, as schematized in Figure 2. First, the LMO module collects design inputs from the user and previously run Deployment design modules (listed in Section 2), data related to the devices and subsystems that must be installed, maintained, and dismantled throughout the lifetime of the project (see Figure 3. More information about the inputs available in [46]). These attributes are subsequently converted into project logistic requirements (e.g., *N* monopiles with specific dimensions and a given weight need to be lifted, transported, and installed). Second, based on the specified components and identified requirements, the corresponding marine operations that must be carried out are identified (e.g., pile installation). In respect to O&M, two maintenance types are simultaneously considered in the LMO module: preventive (time-based maintenance) and corrective, based on failure events generated by the RAMS module [47] taking into consideration component failure distributions.

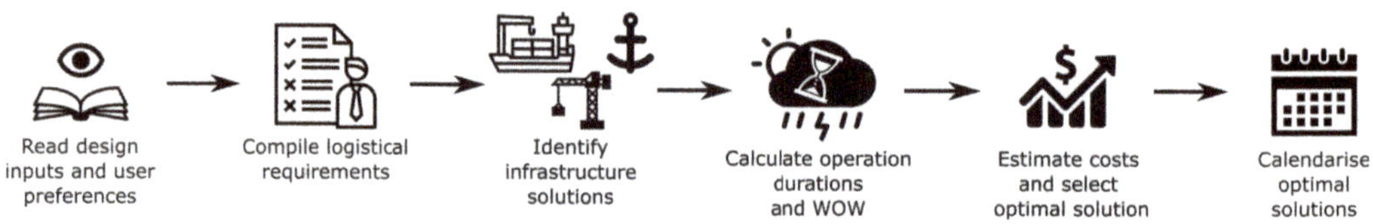

Figure 2. Working principle of the Logistics module.

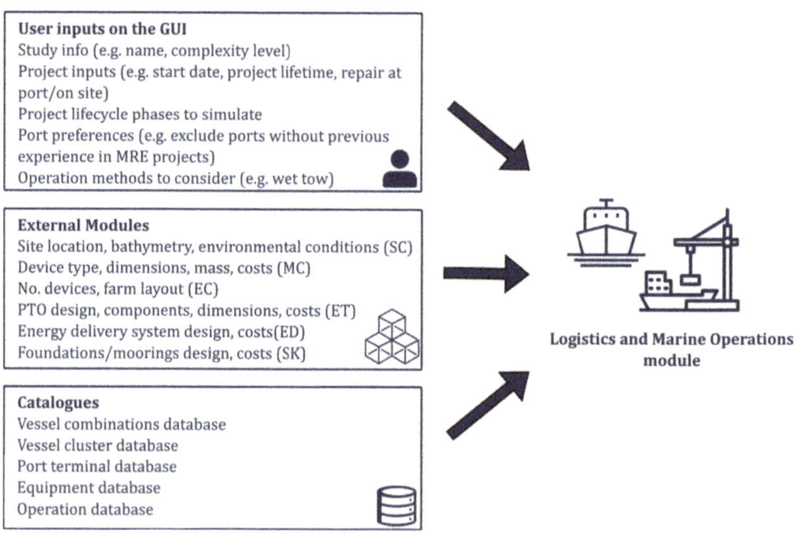

Figure 3. Schematic representation of the inputs to the LMO module.

In a third step, the process of identifying feasible infrastructure solutions begins. Vessels, ports, and equipment must not only meet their minimum individual prerequisites imposed by the project, but also be compatible between each other. Then, for each potential logistic solution, defined by a specific operation plan and infrastructure setup, the operation net durations and expected waiting on weather (WOW) are computed based on historical weather data and a operation catalog, featuring reference operation durations and weather restrictions (e.g., maximum wave height and maximum wind speed). Following the calculation of the total operation durations (including weather delays), for each logistic solution, the operation total costs can be calculated by considering the daily costs of the infrastructure setup. Finally, the logistic solution that presents the lowest operation total costs can be chosen as the optimal solution.

In the DTOceanPlus software suite, each module was developed with three levels of complexity (Cpx1, Cpx2, and Cpx3) to accommodate different stages of project maturity, and different amounts of data availability and uncertainties. In LMO, the main differences between complexity Cpx2 and Cpx3 are the certainty of the inputs and whether default values are assumed instead of requesting these from the user. Alternatively, the simplified mode (Cpx1) can be used for early stage technologies, at lower technology readiness levels (TRL 1-3), or whenever limited information is available about the technology design and project specifics. The simplified mode may also be used to provide a quick and rough estimate for higher TRL projects. Finally, a "study comparison" feature was implemented in DTOceanPlus to enable the user to evaluate and compare different inputs, strategies, and scenarios, in respect to their impacts on the logistic solutions.

In order to carry out the design of the installation, maintenance, and decommissioning phases, the LMO module employs databases of vessels, port terminals and operations, that will be publicly available once the final version of the DTOceanPlus suite of tools is released.

3.2. *Compilation of Operations and Logistic Requirements*

In a first step, the Logistics and Marine Operations module reads component design inputs generated by previously run DTOceanPlus Deployment Design modules and/or introduced by the user. Based on the specified component designs, number of components, and user preferences, the LMO module identifies and proposes to the user a list and sequence of operations to install, service and/or dismantle a given ocean energy farm (see Table 2 for the list of operations considered in LMO). For each operation, relevant operational methods (e.g., transportation method and piling method) are read from upstream modules (e.g., cable burial layout is generated by the Energy Delivery module, using specific cable burial methods), or requested to the user, as described in Table 3.

Based on the identified operations and specified operation methods, infrastructure requirements are defined (e.g., the vessel's deck area must be sufficient to transport at least one device or system). Infrastructure requirements are compiled in Table 4.

Table 2. List of operations for each project phase included in the Logistics and Marine Operations module.

Installation Operations	Maintenance Operations	Decommissioning Operations
Foundations installation	Topside inspection	Device removal
Moorings installation	Underwater inspection	Collection point removal
Support structures installation	Mooring inspection	Moorings removal
Collection point installation	Array cable inspection	Foundations removal
Export cable installation	Export cable inspection	
Array cable installation	Device retrieval	
Post-lay cable burial	Device repair at port	
External protection installation	Device redeployment	
Device installation	Device repair on site	
	Mooring line replacement	
	Array cable replacement[1]	
	Export cable repair	

Table 3. Operation methods considered in the LMO module.

Method List	Defined by	Options			
Transportation	User	Dry	Wet	-	-
Port load-out	User/Default	Lift	Float	Skidded	Railed
Cable burial	User/ED module	Ploughing	Jetting	Cutting	-
Burial sequence	User/Default	Simultaneous	Post-lay	-	-
Cable landfall	User/Default	OCT	HDD	-	-
Piling	User/Default	Hammering	Vibro-piling	Drilling	-

Table 4. Infrastructure requirements considered in the LMO module.

#	Port Terminal	Vessel	Piling Equipment	Cable Burial Equipment	ROV Equipment	Divers
1	Terminal type	Crane capacity	Depth rating	Depth rating	Depth rating	Depth rating
2	Terminal draught	Vessel draft	Pile sleeve diameter	Burial depth	ROV class	-
3	Terminal area	Bollard pull	Penetration depth	Cable diameter	-	-
4	Onshore crane capacity	DP class	Soil type	Cable MBR	-	-
5	Quay soil strength	Deck area	-	-	-	-
6	Max. distance to site	Deck strength	-	-	-	-
7	Past experience	Max. cargo	-	-	-	-
8	-	Turntable storage	-	-	-	-
9	-	Turntable capacity	-	-	-	-
10	-	Max. no. passengers	-	-	-	-

3.3. *Infrastructure Pre-Selection*

3.3.1. Feasible Infrastructure

On a first stage, feasibility functions are used to assess whether vessels, ports and equipment, listed in the DTOceanPlus catalogs, meet the absolute minimum requirements for the job (e.g., sufficient vessel deck area, adequate terminal dry dock dimensions, acceptable ROV depth rating, etc.). These functions are simple mathematical Boolean formulations that filter out the maritime infrastructure noncomplying with the previously defined logistic requirements (see Section 3.2). A default safety factor of 10% is implemented in the feasibility functions to reflect uncertainties and account for potential margins of of error, although this value may be modified by the user.

The identification of feasible ports and equipment follows a simple elimination process, where instances of the port terminals and equipment databases are discarded based on the operation requirements. The port terminal database consists of 203 terminals, from 12 different EU countries, with 21 parameters, including name, type, country, location, terminal entrance width, draught, maximum load, and terminal area, to name a few (see Table 5). Similarly, six main types of equipment (e.g., cable burial tools, piling hammers, and ROVs, to name a few) are considered in DTOceanPlus and included in the equipment catalogs. However, the fleet selection process is slightly more complex.

3.3.2. Fleet Selection Methodology

There are numerous approaches to conduct a given offshore operation. Devices may be transported from port to site, on deck of a large crane vessel with adequate cranes to carry out the installation procedures. Alternatively, they may be loaded at port onto a transport barge, which would in turn be towed a set of tugs or a Anchor Handling Tug and Supply (AHTS) vessel. In case devices were structurally designed for the purpose, they may be individually wet-towed directly to site. Low draft floating converters may be transported using a semi-submersible vessel, with the capability of ballasting down and submerging its deck to unload the converter in the water. Based on the existing experience in the offshore renewable energy sector, a vessel combination (VC) database was developed, namely, for the device installation operation (see Table 6), featuring combinations of different vessel types in different quantities and under different roles. For simplicity, for each vessel combination, three major vessel roles were defined, with different evaluation criteria:

1. Main vessels: vessels that play a central role in the operation, being responsible for key activities such as lifting and transporting components, driving a monopile, and installing a cable, to name a few. These vessels are thus assessed in respect to their main attributes (e.g., deck area, crane capabilities, etc.) depending on the vessel type and operation plan.
2. Tow vessels: vessels that are responsible for towing a device/structure (wet tow), or a non-propelled barge (dry tow). Tow vessels are assessed in terms of their bollard pull capabilities, which must be sufficiently large to meet the estimated bollard pull requirements for safely towing an object or barge.
3. Support vessels: vessels that can be used to support lifts, control marine traffic, and assist device positioning, but do not have a central role in the operation itself.

The fleet selection algorithm follows a two-stage elimination process. It starts by discarding unsuitable vessel combinations that do not meet project requirements or user preferences (e.g., wet tow is not allowed). Then, for each feasible vessel combination, the fleet selection algorithm searches in a vessel cluster database for vessels that meet the technical requirements associated with the attributed roles (e.g., sufficient deck area, sufficient bollard pull to wet tow the device). A given vessel is deemed "feasible" if capable of performing the minimum work criteria (e.g., in case of on deck transportation, vessel must have sufficient deck area to accommodate at least one device per trip).

In DTOceanPlus, a vessel cluster database was compiled based on the statistical analysis of an original database with 14,847 vessels and 46 technical parameters. The very large size of the original database ensured the representativeness of the considered vessel list, although prohibited directly using it in DTOceanPlus due to data privacy issues and to keep computational requirements manageable. In the vessel cluster database, vessels sharing a large number of similar characteristics were grouped into clusters, using the K-Means unsupervised machine learning algorithm [48]. For each technical parameter of each vessel cluster, the p25, p50, and p75 statistical values were computed and stored. Deliverable D5.8 [46] provides more information about the vessel clustering process.

Table 5. Example port terminal entry of the DTOceanPlus port catalog.

Terminal Parameter	Value
Terminal id	T114
Name of Port	Viana do Castelo
Terminal name	Dry dock #1
Country	Portugal
Terminal coordinates (lat, lon)	(41.675, −8.8383)
Experience in MRE projects	TRUE
Storage area [m^2]	100,000
Slipway	TRUE
Terminal type	Dry-dock
Terminal width [m]	32
Terminal length [m]	203
Quay load bearing capacity [t/m^2]	0.8
Terminal draught [m]	3.5
Terminal area [m^2]	6500
Terminal hinterland area [m^2]	3300
Gantry crane lift capacity [t]	80
Tower crane lift capacity [t]	200
Jack-up suitability	TRUE

Table 6. Example vessel combinations for the device installation operation, stored in the DTOceanPlus Vessel Combinations catalog.

Id	Type	Transportation	Qty	Main Vessel	Qty	Tow Vessel	Qty	Support Vessel
VC_001	Device Installation	On deck	1	Propelled crane vessel	-	-	-	-
VC_002	Device Installation	On deck	1	Jack-up Vessel	-	-	-	-
VC_003	Device Installation	On deck	1	SOV Gangway	-	-	-	-
VC_004	Device Installation	On deck	1	AHTS	-	-	-	-
VC_005	Device Installation	Dry tow	1	Non propelled crane Vessel	1	Tug	-	-
VC_006	Device Installation	Dry tow	1	Transport Barge	1	Tug	-	-
VC_007	Device Installation	On deck	1	Semi-submersible	-	-	1	Multicat
VC_008	Device Installation	Wet tow	-	-	1	AHTS	-	-
VC_009	Device Installation	Wet tow	-	-	2	AHTS / Tug	-	-
VC_010	Device Installation	Wet tow	-	-	1	AHTS / Tug	1	Multicat
VC_011	Device Installation	Wet tow	-	-	2	AHTS / Tug	1	Multicat
VC_012	Device Installation	Wet tow	-	-	3	AHTS / Tug	1	Multicat

3.3.3. Infrastructure Matching

Once feasible infrastructure have been identified, infrastructure-matching functions assess the compatibility between each infrastructure type in the context of an integrated solution. In this step, independently feasible but incompatible infrastructure solutions are discarded (e.g., port entrance width must be larger then vessel beam, port draught must be compatible with vessel draft, etc.). Once the infrastructure matching algorithm has been run, suitable infrastructure combinations are produced to be further analyzed in respect to total operation duration and ultimately, costs.

3.4. *Definition of Activity Sequence*

In the LMO module, operations (e.g., foundation installation) are broken down into smaller, uninterruptible tasks that must be carried out, referred to as activities (e.g., mobilization, transit, and positioning), with specific durations and weather restrictions. For each operation, activity flowcharts were developed, featuring the activity blocks, precedence rules (i.e., which activity comes next), and condition nodes which define multiple potential paths that an operation may take. Condition nodes were defined as static, when based on previously defined component types and operation methods (e.g., foundation type is a pile, transportation method is dry), or dynamic, when dependent on the considered infrastructure solution and operation stage (e.g., vessel is already full or not). In the flowcharts, activities may have a constant duration, or a dynamic duration when the activity length depends on external criteria such as distance and vessel transit speeds (for transits and tows) or soil conditions (e.g., for piling activities such a "Seafloor drilling"). In Table 7, the activity flowchart of the foundation installation operation is presented as an example.

In Tables 8 and 9, the cable burial and piling speeds are presented, respectively, for different soil conditions. Activities, such as piling and cable burial, have specific speeds that are highly dependent on the seabed geology. In the Logistics and Marine Operations module, the piling and cable burial speeds were compiled and adapted from the literature review carried in the original DTOcean project [45,49], for the considered soil types, as defined by Kervella, Y. [50].

Based on the operation flowcharts, specified project characteristics, and infrastructure solution, a sequencing algorithm computes the full sequence of activities, from start to finish, that must be carried out to perform a given operation. Flowcharts are stored as tables in the operation catalogs, allowing for modifications to the durations, weather limits, and sequencing, by advanced users. This activity sequence is then fed into the weather window model described in Section 3.5.

Table 7. Flowchart of the foundation installation operation, featuring activity blocks, precedence rules, and condition nodes (shown in italic).

ID	Activity Name	Next Activity	Duration [h]	Decision Paths
OP01_A0	Mobilization	OP01_A1	48	-
OP01_A1	Vessel preparation & loading	T_C0	48	-
T_C0	*cond_stat_methods:transport*	T_C1_1;T_C1_2	-	0-dry;1-wet
T_C1_1	*cond_stat_methods:load_out*	T_A0;T_A0;T_A1	-	0-float away;1-lift away;2-skidded/trailer
T_C1_2	*cond_stat_methods:load_out*	T_A7;T_A4;T_A5	-	0-float away;1-lift away;2-skidded/trailer
T_A0	Lift item onto vessel deck	T_D0	3	-
T_A1	Place item on steel rail/trailer	T_A2	2	-
T_A2	Translate item onto vessel deck	T_D0	2	-
T_D0	*cond_dynam_deck full*	T_D1;T_A3	-	0-false;1-true
T_D1	*cond_dynam_quay empty*	T_C1_1;T_A3	-	0-false;1-true
T_A3	Seafastening	T_A8	1	-
T_A4	Lift item from the quay to the water	T_A9	2	-
T_A5	Place item on marine slipway	T_A6	2	-
T_A6	Push/pull item to the water	T_A9	2	-
T_A7	Flood Dry-dock	T_A9	6	-
T_A8	Item transportation on vessel deck	OP01_A2	transit_site	-
T_A9	Item towed to site	OP01_A2	tow	-
OP01_A2	Positioning	OP01_C0	1	-
OP01_C0	*cond_stat_object:type*	OP01_A3;OP01_A6;OP01_A13		0-pile;1-suction caisson;2-gravity base
OP01_A3	Leveling and positioning of guiding template	P_C0	2	-
P_C0	*cond_stat_methods:piling*	P_A0;P_A4;P_A6	-	0-drilling;1-hammering;2-vibro-piling
P_A0	Rig and pile leveling and positioning	P_A1	2	-
P_A1	Seafloor drilling	P_A2	drill	-
P_A2	Pile lowering into aperture	P_A3	1	-
P_A3	Flushing and grouting	OP01_A5	1	-
P_A4	Pile levelling and positioning	P_A5	1	-
P_A5	Hammering pile into seafloor	OP01_A5	hammer	-
P_A6	Pile levelling and positioning	P_A7	1	-
P_A7	Vibro-piling into seafloor	OP01_A5	vibro	-
OP01_A5	Removal of guiding template	OP01_D0	1	-
OP01_A6	Hosting	OP01_A7	2	-
OP01_A7	Lower caisson to seabed	OP01_A8	1	-
OP01_A8	Penetration into seabed due to weight	OP01_C1	2	-
OP01_C1	*cond_stat_requirements:rov*	OP01_A9;OP01_A11	-	0-inspection;1-work
OP01_A9	Pump water from caisson	OP01_A10	2	-
OP01_A10	Undock suction pump	OP01_D0	1	-
OP01_A11	Pump water from caisson with ROV	OP01_A12	2	-
OP01_A12	Undock suction pump ROV	OP01_D0	0.2	-
OP01_A13	Hosting	OP01_A14	2	-
OP01_A14	Lowering GBA to seabed	OP01_D0	1	-
OP01_D0	*cond_dynam_deck empty*	OP01_A15;OP01_A16	-	0-false;1-true
OP01_A15	Transit to next site	OP01_A2	0.2	-
OP01_A16	Transit from site to port	OP01_D1	transit_port	-
OP01_D1	*cond_dynam_quay empty*	OP01_A1;OP01_A17	-	0-false;1-true
OP01_A17	Demobilization		48	-

Table 8. Cable burial speeds (m/h), for different soil types and cable burial methods, stored in the DTOceanPlus operations catalog.

Installation Method	Very Soft Clay	Soft Clay	Firm Clay	Stiff Clay	Very Stiff Clay	Hard Clay	Very Loose Sand	Loose Sand	Medium Dense Sand	Dense Sand	Very Dense Sand	Gravels & Pebbles
Jetting	450	450	250	0	0	0	300	300	200	0	0	0
Ploughing	0	375	500	550	550	300	100	100	350	100	100	300
Cutting	0	325	325	75	75	75	0	0	275	275	275	0
Dredging	150	100	75	50	50	50	150	150	100	75	75	75
Surface lay	700	700	700	700	700	700	700	700	700	700	700	700

Table 9. Pile installation speeds (m/h), for different soil types and piling methods, stored in the DTOceanPlus operations catalog.

Installation Method	Very Soft Clay	Soft Clay	Firm Clay	Stiff Clay	Very Stiff Clay	Hard Clay	Very Loose Sand	Loose Sand	Medium Dense Sand	Dense Sand	Very Dense Sand	Gravels & Pebbles
Drilling	0	0	0.65	0.5	0.5	0.25	0	0	0	0	0	0
Hammering	15	12.5	7.5	4.5	4.5	0	20	20	15	5	5	5
Vibro-driving	175	75	0	0	0	0	375	375	250	75	75	75
Suction pump	200	100	0	0	0	0	375	375	250	100	100	0
ROV with jetting	475	475	250	0	0	0	250	250	250	0	0	0

3.5. Weather Window Model

Weather window analysis is a crucial step in the strategic planning of marine operations in order to estimate potential weather delays and operation costs. The most common approach is to simulate a project subject to several years of historical environmental conditions, being commonly referred to as hindcast analysis [51]. Given the random nature of the met-ocean conditions at a given site, sample size must be sufficiently large to appropriately capture the potentially large annual variability. Even though more is better, 20-years of continuous weather data is a commonly accepted reference. As maritime operations are typically planned on hourly basis, DNV standards recommend linearly interpolating the raw met-ocean conditions when necessary to generate an hourly time series [52]. Subsequently, the time series of met-ocean conditions can be analyzed as a single continuous record.

The environmental conditions observed at a given offshore location can be understood as a multivariate stochastic process [53–55], whereas each environmental parameter (wind speed, significant wave height, peak wave period, and current speed) is interdependent and can be described by statistical distributions with specific joint probabilities but clear ensemble seasonal trends [56,57]. Even though cyclic patterns may be observed throughout the year (e.g., the summer season is typically calmer than winter, even though summer storms should not be overlooked), it may be reasonable to assume data stationarity for smaller time periods [58]. This method is known as piecewise stationarity and consists of grouping the entire met-ocean time series by seasons or months and carrying out separate calculations. The assumption weather data stationarity implies that the statistical properties (e.g., mean, variance, and autocorrelation) of the historical dataset are constant, and is typically assumed reasonable for fixed monthly blocks.

Following a hindcast simulation approach, the underlying principle of the LMO's weather window algorithm consists of attempting to schedule the specified operations in the past. Once the sequence of activities, durations, and weather restrictions have been specified for each operation (see Section 3.2), the algorithm attempts to iteratively initiate the operation in different time-steps of the historical time-series of met-ocean conditions, each iteration corresponding to a different simulation. Initial time-steps are randomly selected using a Monte Carlo approach, taking as user input the percentage of time-steps to analyze in each month (where 100% corresponds to analyzing the entire time-series). For each simulation, in case both the first and subsequent time-steps are deemed workable (i.e., OLCs are met) for a period that is equal or longer than the entire operation duration, then the operation can be carried out without any delays. Otherwise, waiting on weather is required, which may include waiting at port (WAP), and/or waiting on site (WOS), i.e., between consecutive activity blocks. The waiting on site is defined with a maximum duration and weather limit, which may not be exceeded.

For an operation with n activities, starting in time-step t, the total operation duration $d_{op.total}$ would be defined as shown in Equation (1) below, where $d_{net,i}$ refers to the net duration of activity i of the operation.

$$d_{op.total}(t) = \mathrm{WOW}(t) + \sum_{i=1}^{n} d_{net,i} = \sum \mathrm{WAP} + \sum \mathrm{WOS} + \sum_{i=1}^{n} d_{net,i} \quad (1)$$

Assuming monthly stationarity for the weather conditions, the waiting times calculated for each initial time-step can be grouped and statistically analyzed in monthly blocks. Given that the monthly waiting on weather values do not follow a normal distribution, the statistical parameters such as median (p50) and the interquartile ranges (p25–p75) can be used to estimate the expected value and quantifying statistical dispersion. As an example, Figure 4 shows a hypothetical non-exceedance distribution plotted for a given operation "op.A", considering all WOW values that occurred in every month of February of the entire 20-years long time series. As shown in Figure 4, there is a 50% probability that the waiting time for the specified operation will be equal or lower than about 28 h, whereas the p25 and p75 values are equal to 22 and 38 h, respectively. According to the estimated interquartile range, there is a 50% probability that the waiting time for this operation will be in the range of 22–38 h, for the month of February.

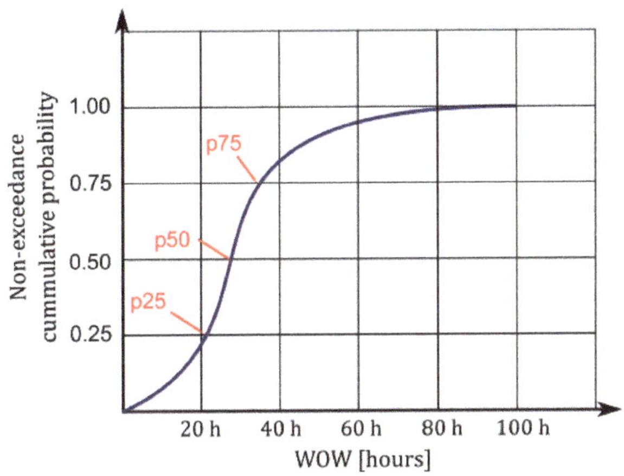

Figure 4. Illustrative representation of the non-exceedance probability of waiting on weather for an example operation in a given month.

For each operation, the weather window model thus computes monthly weather window statistics featuring the expected weather delay (p50) and resulting total operation duration for the different months of the year. The advantage of calculating the weather delays for each month of the year is that potential cost-reduction approaches, such as changing the starting month or optimizing the sequence of operations, may be unveiled to the user. The monthly weather window statistics are illustrated in Table 10 for a given operation with a net duration of 30 h.

Table 10. Monthly weather window statistics, in hours, for a given operation with a net duration of 30 h when scheduled in different months.

	Jan	Feb	Mar	Apr	May	Jun	Jul	Aug	Sep	Oct	Nov	Dec
WOW (p50, in h)	43	38	22	15	14	10	8	8	15	39	55	64
Total duration (h)	73	68	52	45	44	40	38	38	45	69	85	94

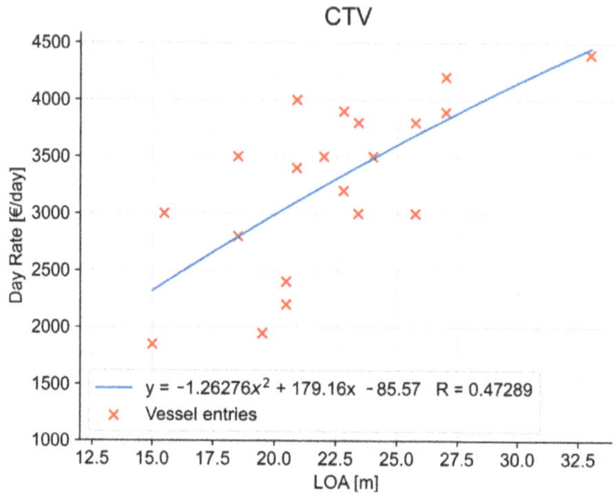

Figure 5. Example regression of the daily charter rates for crew transfer vessels (CTVs) as a function of the vessel's length overall (LOA), based on existing database.

3.6. Calculation of Operation Costs

3.6.1. Vessel Costs

Vessel costs play a large role in the total costs of an offshore renewable energy project. Total vessel costs can be broken down into vessel chartering and fuel expenditures. The daily operating costs per day can be calculated as shown in Equation (2).

$$c_{vessel} = c_{charter} + c_{fuel} \tag{2}$$

3.6.2. Daily Vessel Charter Rates

The cost of chartering a given vessel depends on several factors, such as vessel characteristics and capabilities, as well as surrounding market conditions. Contract duration and contractual frameworks typically also play a role. Smaller tonnage vessels such as CTVs, tugs, and survey vessels are commonly chartered out on a time charter basis (e.g., BIMCO Supply time [59]) with a clearly defined vessel day rate. However, larger vessels such as jack-up vessels, crane vessels, and cable laying vessels are mostly hired as part of comprehensive service agreements such as EPCI[2] or T&I[3] contracts (e.g., FIDIC or Logic [60]). In order to be able to compare different types of contracts, average daily charter costs that exclude consumables such as fuel and harbour costs, were used.

Vessel characteristics such as age, size, crane capabilities, deck area, dynamic positioning (DP) equipment, and engine power are known to have an impact on the total vessel costs. Based on guidance from Global Renewable Shipbrokers (GRS) [61], a offshore vessel broker, major cost drivers for the vessel charter rates were identified for each vessel type. Even though the vessel charter rates are dependent on a large number of variables, for simplicity and to provide a first cost estimate, vessel charter costs were modeled as a function of a single parameter for each vessel type. As shown in Figure 5, cost functions that model charter day rates for the different vessel types were then derived, based on a curve fitting applied to database points gathered from: (i) DTOcean 1.0 vessel database, (ii) WavEC's in-house vessel database, (iii) cost figures provided by *ECN* [62] and *GRS*, (iv) from industry expert experience. Different regression models, including linear, polynomial, exponential, logarithmic and piece-wise regressions, were adjusted to find a best fit based on the R-squared coefficient, while eliminating fits that result in cost inflections within the analyzed domain. The resulting cost functions were compiled in Table 11. It can be seen that even though charter price variability is not fully explained by a single parameter, important relationships were obtained, with the potential to guide vessel selection decisions.

Table 11. Daily charter rate regression curves, for different vessel types, in Euros, as a function of their input parameters (x).

Vessel Type	Input Parameter	Domain Validity	Function	R^2
Tug	Bollard Pull (tonnes)	$13 \leq x < 25$ $25 \leq x < 70$ $70 \leq x \leq 80$	$c_{chart} = 151.34x - 467.47$ $c_{chart} = 2.18x + 3261.61$ $c_{chart} = 508.57x - 32186$	0.9611
Multicat	LOA (m)	$21 \leq x < 28$ $28 \leq x < 35$ $35 \leq x \leq 42$	$c_{chart} = 63.23x + 1812.4$ $c_{chart} = 916.74x - 22086$ $c_{chart} = 10,000$	0.96626
AHTS[4]	Bollard Pull (tonnes)	$70 < x \leq 338$	$c_{chart} = -8.3 \times 10^{-3}x^2 + 114.90x - 261.87$	0.6857
CLV[5]	Total cable storage (ton)	$565 \leq x \leq 10,000$	$c_{chart} = 2.46 \times 10^{-4}x^2 + 7.25x + 53,090$	0.4716
CTV[6]	LOA (m)	$15 \leq x \leq 33$	$c_{chart} = -1.26x^2 + 179.16x - 85.57$	0.4729
DSV[7]	LOA (m)	$35 \leq x \leq 150$	$c_{chart} = 4308.81 \exp(0.02x)$	0.96580
Guard Vessel	Service speed (knots)	$7 \leq x \leq 24$	$c_{chart} = 77.11x + 1345.48$	0.99948
Non-propelled Barge	Barge dimensions	$1557 \leq x \leq 19,950$	$x = ve.LOA * ve.beam * ve.draft$ $c_{chart} = 953.92log(x) - 6761.18$	0.87829
Jack-up vessel	Crane lift capacity (tonnes)	$50 \leq x < 755$ $755 \leq x < 896$ $896 \leq x \leq 4400$	$c_{chart} = 64.71x + 21,448.41$ $c_{chart} = 586.18x - 372,275$ $c_{chart} = 26.83x + 128,892$	0.77216
Propelled crane vessel	Crane lift capacity (tonnes)	$4 \leq x \leq 3300$	$c_{chart} = -5.44 \times 10^{-3}x^2 + 88.91x + 12,714.58$	0.99548
Non-propelled crane vessel	Crane lift capacity (tonnes)	$4 \leq x < 2108$ $2108 \leq x < 3300$	$c_{chart} = 60.62x - 9075.34$ $c_{chart} = 15.74x + 85,536.50$	0.55486
PSV[8]	Free Deck Space (m^2)	$30 \leq x \leq 5005$	$c_{chart} = 1.005x + 8969.85$	1.00
Rock Dumper	Stone cargo capacity (tonnes)	$5400 \leq x \leq 69,212$	$c_{chart} = 3.99x + 69,212.41$	0.26059
SOV[9] with gangway	No. Passengers	$x < 60$ $x \geq 60$	$c_{chart} = 24,000$ $c_{chart} = 50,000$	N.D.
SOV gangway relevant	No. Passengers	$x < 60$ $x \geq 60$	$c_{chart} = 24,000$ $c_{chart} = 42,000$	N.D.
Survey vessel	LOA (m)	$23 \leq x \leq 56$	$c_{chart} = 333.33x - 4166.67$	0.66484

3.6.3. Daily Vessel Fuel Costs

Given that the considered vessel charter rates excluded fuel costs, vessel fuel consumption had to be estimated. Fuel consumption contributes significantly to the total operation costs, but also to the emissions and carbon footprint of the project. Total vessel fuel consumption depends on several different factors, namely number of engines (main and auxiliary), engine power, engine efficiency, operation duration, mobilized ancillary equipment, transit speed and distance, as well as weather conditions. In order to provide a first fuel consumption estimate, the LMO module calculates the average vessel fuel consumption per day as

$$f_{cs} = \text{TIP} \cdot \text{ALF} \cdot \text{SFOC} \cdot 24 \cdot \frac{1}{1000^2} \quad (3)$$

In Equation (3), TIP is the vessel's total installed power (in kW), ALF is the average load factor, and SFOC is specific fuel oil consumption (in g/kWh) [63]. According to the ship broker's experience in vessel chartering for offshore wind projects, an average load factor of 80%, and a specific fuel oil consumption of 210 g/kWh were indicated as reference values. However, these values may be modified by the user. The daily fuel costs can thus be estimated by multiplying the daily fuel consumption f_{cs} by a reference price of fuel, as shown in Equation (4). In respect to the fuel price p_{fuel}, the marine diesel oil (MDO) price in the port of Rotterdam, 515 €/ton, was taken as a reference [64]. However, when running the LMO module, this value may be modified by the user to reflect different fuel prices or even other fuel types such as heavy fuel oil (HFO).

$$c_{fuel} = f_{cs} \cdot p_{fuel} \quad (4)$$

3.6.4. Equipment Costs

For a given operations, the equipment costs can be simply calculated as the product of the operation duration and the sum of the daily (and/or half-day) renting cost of the selected equipment for that operation. Daily and half-day renting costs figures are available in the equipment databases.

3.6.5. Spare Part Costs

For O&M operations, in case of component failure, the cost of the spare components are calculated using the costs of a new component, as designed by other modules (or introduced by the user) and compiled in the Bill of Materials (BOM).

3.6.6. Port Terminal Costs

Port expenditures are generally port-specific, varying greatly with the type of contract and duration, leased storage area, the size of the vessels, and need for ancillary equipment such as cranes. However, according to the literature, total port expenditures amount on average to about 0.5% of the total costs of offshore wind projects [65]. The port terminal costs were thus included by adding an extra 0.5% to the total operation costs.

3.6.7. Total Operation Costs

For each logistic solution, the total operation costs based on the operation duration d_{op}, selected equipment, vessel fleet, port costs, and spare components (for O&M operations), are calculated as described in Equation (5).

$$C_{op} = d_{op} \cdot (c_{ports} + c_{vessels} + c_{equip}) + C_{spare} \quad (5)$$

Based on the total operation costs calculated for each logistic solution candidate, the optimal operation solution that minimizes total costs can be selected for each operation.

3.7. Operation Calendarization

For the installation and decommissioning project life-cycle phases, the operation calendarization functionality is responsible for scheduling the previously identified optimal operations on the project calendar, taking into consideration the project start date, operation net duration, expected weather delays in the considered month, as well as predefined operation sequence. For maintenance operations, the operation calendarization functionality schedules corrective maintenance operations in the aftermath of component failure, and preventive maintenance activities following the predefined preventive maintenance frequency. The periodicity of preventive maintenance operations, as well as the device shutdown requirement, which expresses whether device shutdown is assumed when carrying out the operation, are compiled in the maintenance catalog, as presented in Table 12. The preventive maintenance frequency values may be modified by the user to fit project specific requirements.

In case of component failure, or device shutdown requirement during preventive maintenance, resulting downtime per device is stored. The interrelationships between farm components and the ability of each device to produce and deliver its energy to the grid were represented in a hierarchy structure. This automatically generated tree-like structure, described in detail in [47], allows to evaluate the impacts of a given component failure (e.g., array cable) in the farm energy production (i.e., downtime of respective devices).

Table 12. Catalog of preventive maintenance operations, featuring operation name, annual periodicity, and device shutdown requirements.

ID	Name	Periodicity (Years)	Device Shutdown
1	Topside inspection	1	Yes
2	Underwater inspection	2	Yes
3	Mooring inspection	1	Yes
4	Array cable inspection	2	No
5	Export cable inspection	4	No

4. Case Study

In order to demonstrate the functionalities of the Logistics and Marine Operations module, a case study was developed for the installation of a floating wave energy converter, inspired on Sandia's Reference Model 3 (RM3) [66]. Sandia's RM3 device consists of a 260 kW heaving point absorber. The overall design and dimensions are represented in Figure 6.

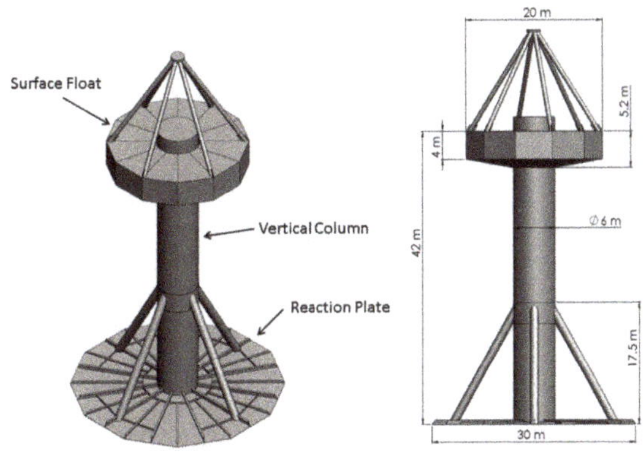

Figure 6. Sandia's RM3 floating wave energy reference model design and dimensions.

In the present case study, a deployment location in Europe was selected with similar characteristics (bathymetry and wave energy resource) to the RM3 reference site (Eureka, in Humboldt County, California). The mean reference site wave energy density is 33.5 kW/m. In Figure 7, the selected project site is depicted, as well as relevant nearby ports (stored in the terminal catalog) that were considered during the algorithm's port selection process. August 1st was specified as the installation starting date due to being the month with best weather conditions.

As a floating device, it was considered that the converter would be transported from port to site by wet towage. Drag-anchors were considered for station keeping. In order to export the generated power to shore, a 3.3 kV export cable with a total length of 6680 m, mostly buried at 0.5 m depth, was considered. The dimensions and characteristics of the subsystems were compiled in Table 13.

Based on the introduced list of components, the LMO module recognized that three operations would be required, in the recommended sequence: (i) installation of the mooring system, (ii) installation of the export cable, and (iii) installation of the device. It is suggested by the algorithm that the mooring system is pre-laid, an increasingly common practice in floating wind projects, followed by the cable installation operation to reduce the risks of cable damage during the mooring installation activities. Finally, the installation of the device consists of wet-towing the converter to site and connecting the pre-laid moorings and umbilical cable.

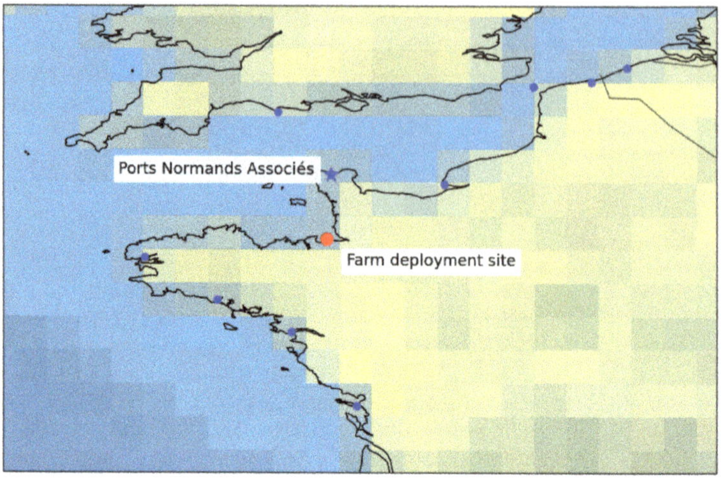

Figure 7. Project site in the North sea, including the farm deployment location (in red), relevant ports (blue circles), and optimal port identified by the port selection algorithm (blue star). Map generated in Python using Cartopy library [67].

Table 13. Dimensions and characteristics of the floating wave energy converter and sub-systems.

Component	**No.**	**Type**	**Mass**	**Length**	**Width**	**Height**	**Draft**	**Tow draft**
Device	1	Floating WEC	680,000 kg	30 m	30 m	42 m	35 m	15 m
Component	**No.**	**Type**	**Mass**	**Length**	**Width**	**Height**		
Anchor	3	Drag-anchor	9,535 kg	5.472 m	5.898 m	3.291 m		
Component	**No.**	**Material**	**Mass**	**Length**	**Diameter**			
Mooring line	3	Nylon	4,703 kg	340.7 m	0.146 m			
Component	**No.**	**Type**	**Mass**	**Length**	**Diameter**	**Voltage**	**MBR**	**Burial depth**
Power cable	1	Export	8,700 kg	6,680 m	0.079 m	3.3 kV	1.15 m	0.5 m

The results of the LMO module, featuring the selected vessels, durations, and costs for each operation, are shown in Table 14. Leveraging on the port terminal and vessel databases, as the algorithm identified the optimal port-fleet combination in respect on project costs. The *Ports Normands Associés*, in the north coast of France was selected for all three installation operations. An Anchor Handling Tug Support (AHTS) vessel was recommended for the mooring installation, a Cable Laying Vessel (CLV) for the cable installation (including cable burial), and two tugs for the device installation. Despite being the least energetic month, results suggest that the expected waiting on weather in August is not negligible, representing 41% and 27% of the total expected operation duration for the mooring and cable installation operations, respectively. It can be observed that for a single device, the total installation costs amounts to approximately 1.8 M€. Given that the installation of the cable and moorings are the largest contributors to the total project commissioning costs, significant economies of scale can be expected for projects with a higher number of devices, as multiple components would be installed per trip, avoiding unnecessary transits. The obtained installation cost figures showed good agreement with the RM3 installation cost breakdown, presented in Sandia's in-depth study [66]. It was found that differences in the results were mainly caused by the mobilization and demobilization assumptions in Sandia's study, which were not reproduced in LMO. A screenshot of the results page of the LMO module is shown in Figure 8.

Table 14. Results of the LMO module for the case study.

Operation	Mooring Installation	Cable Installation	Device Installation
Operation sequence	1st	2nd	3rd
Number of vessels	1	1	2
Selected vessels	AHTS	Cable Laying Vessel	Tug, Tug
Selected terminal		Ports Normands Associés-Flamands Sud	
Selected equipment	ROV	ROV, plough	ROV
Mobilisation	96.0 h	96.0 h	96.0 h
Total transit	15.6 h	15.2 h	38.1 h
Work at port	67.0 h	24.0 h	2.0 h
Work on site (at sea)	21.0 h	54.0 h	3.0 h
Waiting on weather	163.0 h	85.4 h	0.0 h
Total operation duration	362.6 h	274.7 h	139.1 h
Vessel fuel consumption	548.28 ton	469.13 ton	87.86 ton
Terminal costs	€3345	€5829	€652
Vessel costs	€550,945	€1,035,114	€85,223
Equipment costs	€118,020	€130,614	€45,240
Total operation costs	€672,310	€1,171,557	€131,115

Figure 8. Screenshot of the Logistics and Marine Operations module: Installation results. Darker bars in the Gantt chart represent the estimated weather delays for each operation.

5. Conclusions

The present work describes the development of a novel methodology for designing the installation, maintenance, and decommissioning phases of ocean energy projects. Given the sensitivity of given marine operations to weather conditions and its impacts on project costs, a statistical weather window model was developed to estimate potential weather delays. Based on a database of vessels relevant to offshore renewable energy projects, simplified cost functions were produced for each vessel type to estimate the daily charter rates. Employing a systematic approach to infrastructure selection, and leveraging on comprehensive and user-modifiable databases of operations, vessels, port terminals, and equipment, the Logistics and Marine Operations module produces operation plans and optimal infrastructure solutions that satisfy project requirements and minimize total project costs. Tests performed for a case study based on a theoretical floating wave energy converter produced results in good agreement with the detailed study conclusions.

Given its open-source licensing and its community collaborative environment, continuous improvements of the Logistics and Marine Operations module are foreseen. Future research plans include improving functionalities and further demonstrating the developed methodology using data from real ocean energy projects, benchmarking against the outputs of other logistic support tools.

Author Contributions: Investigation and Methodology, F.X.C.d.F. and P.C.; software and coding, L.A. and F.X.C.d.F.; writing, F.X.C.d.F., review, F.X.C.d.F., P.C., and L.A. All authors have read and agreed to the published version of the manuscript.

Funding: This work has been partially supported by European Union's Horizon 2020 research and innovation programme under grant agreement No 785921, project DTOceanPlus (Advanced Design Tools for Ocean Energy Systems Innovation, Development and Deployment).

Institutional Review Board Statement: Not applicable.

Informed Consent Statement: Not applicable.

Data Availability Statement: Data will be made available in Zenodo, a publicly accessible repository, at the end of the DTOceanPlus H2020 project. The authors may be contacted for further information.

Acknowledgments: In addition to EU funding and partner contributions, the present research was partially supported by Global Renewables Shipbrokers (GRS Offshore) by providing valuable insights and expertise in the topics of offshore vessel data modeling, charting cost estimates and vessel fuel consumption. A special thank you goes to Manuel Rentschler for his contribution to the coding of the maintenance algorithm, António Maximiano for his contribution in debugging the weather window tool, and António Sarmento to his review work.

Conflicts of Interest: The authors declare no conflicts of interest.

Notes

1 In DTOceanPlus, in the occurrence of array cable failure, it is assumed that the entire array cable is replaced. However, for export cables, it is assumed that the damaged section is repaired instead.
2 Engineering, Procurement, Construction and Installation
3 Transport and Installation
4 Anchor Handling Tug Supply vessel
5 Cable Laying Vessel
6 Crew Transport Vessel
7 Dive Support Vessel
8 Platform Supply Vessel
9 Service Offshore Vessel

References

1. United Nations Framework Convention on Climate Change. The Paris Agreement. 2015. Available online: https://unfccc.int/sites/default/files/english_paris_agreement.pdf (accessed on 21 January 2020).
2. Gerbaulet, C.; von Hirschhausen, C.; Kemfert, C.; Lorenz, C.; Oei, P.Y. European electricity sector decarbonization under different levels of foresight. *Renew. Energy* **2019**, *141*, 973–987. [CrossRef]
3. Zeyringer, M.; Fais, B.; Keppo, I.; Price, J. The potential of marine energy technologies in the UK—Evaluation from a systems perspective. *Renew. Energy* **2018**, *115*, 1281–1293. [CrossRef]
4. Widén, J.; Carpman, N.; Castellucci, V.; Lingfors, D.; Olauson, J.; Remouit, F.; Bergkvist, M.; Grabbe, M.; Waters, R. Variability assessment and forecasting of renewables: A review for solar, wind, wave and tidal resources. *Renew. Sustain. Energy Rev.* **2015**, *44*, 356–375. [CrossRef]
5. Dias, F.; Renzi, E.; Gallagher, S.; Sarkar, D.; Wei, Y.; Abadie, T.; Cummins, C.; Rafiee, A. Analytical and computational modelling for wave energy systems: The example of oscillating wave surge converters. *Acta Mech. Sin.* **2017**, *33*, 647–662. [CrossRef]
6. Ocean Energy Forum. Ocean Energy Strategic Roadmap 2016, Building Ocean Energy for Europe. 2016. Available online: https://www.oceanenergy-europe.eu/wp-content/uploads/2017/10/OEF-final-strategic-roadmap.pdf (accessed on 6 November 2020).
7. European Commission. Blue Energy—Action Needed to Deliver on the Potential of Ocean Energy in European Seas and Oceans by 2020 and Beyond. Communication from the Commission to the European Parliament, the Council, the European Economic and Social Committee and the Committee of the Regions. 2014. Available online: https://op.europa.eu/en/publication-detail/-/publication/e4aea330-19b7-42d8-9d5c-0da648e76792 (accessed on 26 July 2021).
8. Shafiee, M. Maintenance logistics organization for offshore wind energy: Current progress and future perspectives. *Renew. Energy* **2015**, *77*, 182–193. [CrossRef]
9. Thomsen, K.E. *Offshore Wind a Comprehensive Guide to Successful Offshore Wind Farm Installation*, 2nd ed.; Academic Press: Cambridge, MA, USA, 2014; p. 404.
10. Irawan, C.A.; Ouelhadj, D.; Jones, D.; Stålhane, M.; Sperstad, I.B. Optimisation of maintenance routing and scheduling for offshore wind farms. *Eur. J. Oper. Res.* **2017**, *256*, 76–89. [CrossRef]

11. Carroll, J.; McDonald, A.; Dinwoodie, I.; McMillan, D.; Revie, M.; Lazakis, I. Availability, Operation & Maintenance Costs of Offshore Wind Turbines with Different Drive Train Configurations. *Wind Energy* **2017**, *20*, 361–378. [CrossRef]
12. Gray, A. *Modelling Operations and Maintenance Strategies for Wave Energy Arrays*; University of Exeter: Exeter, UK, 2017; p. 315.
13. Ahn, D.; Shin, S.C.; Kim, S.Y.; Kharoufi, H.; Kim, H.C. Comparative evaluation of different offshore wind turbine installation vessels for Korean west–south wind farm. *Int. J. Nav. Archit. Ocean Eng.* **2017**, *9*, 45–54. [CrossRef]
14. Morandeau, M.; Walker, R.T.; Argall, R.; Nicholls-Lee, R.F. Optimisation of marine energy installation operations. *Int. J. Mar. Energy* **2013**, *3–4*, 14–26. [CrossRef]
15. BVG Associates. Value Breakdown for the Offshore Wind Sector. In *Commissioned for Renewables Advisory Board*; February 2010; pp. 1–20. https://assets.publishing.service.gov.uk/government/uploads/system/uploads/attachment_data/file/48171/2806-value-breakdown-offshore-wind-sector.pdf (accessed on 26 July 2021).
16. Poulsen, T.; Hasager, C.; Jensen, C. The Role of Logistics in Practical Levelized Cost of Energy Reduction Implementation and Government Sponsored Cost Reduction Studies: Day and Night in Offshore Wind Operations and Maintenance Logistics. *Energies* **2017**, *10*, 464. [CrossRef]
17. Low Carbon Innovation Coordination Group. *Carbon Innovation Coordination Group Technology Innovation Needs Assessment (TINA) Marine Energy Summary Report*; Technical Report; August 2012. Available online: https://assets.publishing.service.gov.uk/government/uploads/system/uploads/attachment_data/file/593459/Technology_Innovation_Needs_Assessment___Marine.pdf (accessed on 26 July 2021)
18. Soukissian, T.; Denaxa, D.; Karathanasi, F.; Prospathopoulos, A.; Sarantakos, K.; Iona, A.; Georgantas, K.; Mavrakos, S. Marine Renewable Energy in the Mediterranean Sea: Status and Perspectives. *Energies* **2017**, *10*, 1512. [CrossRef]
19. Dalgic, Y.; Lazakis, I.; Turan, O. Vessel charter rate estimation for offshore wind O&M activities. In Proceedings of the 15th International Congress of the International Maritime Association of the Mediterranean IMAM 2013. Developments in Maritime Transportation and Exploitation of Sea Resources, A Coruna, Spain, 14–17 October 2013; pp. 899–907. [CrossRef]
20. Walker, R.T.; Van Nieuwkoop-Mccall, J.; Johanning, L.; Parkinson, R.J. Calculating weather windows: Application to transit, installation and the implications on deployment success. *Ocean Eng.* **2013**, *68*, 88–101. [CrossRef]
21. Gray, A. *What Are Operations and Maintenance Simulation Tools?—An Explanation of O&M Models in the Offshore Renewable Energy Sector*. August 2020. Available online: https://ore.catapult.org.uk/app/uploads/2020/08/OM_Model_Review_Paper_FINAL.pdf (accessed on 26 July 2021)
22. Welte, T.M.; Sperstad, I.B.; Halvorsen-Weare, E.E.; Netland, Ø.; Nonås, L.M.; Stålhane, M., Operation and Maintenance Modelling. In *Offshore Wind Energy Technology*; John Wiley & Sons, Ltd.: Hoboken, NJ, USA, 2018; pp. 269–303. [CrossRef]
23. James Fisher. Mojo Mermaid. Available online: http://www.mojomermaid.com/ (accessed on 6 November 2020).
24. JBA Consulting. ForeCoast Marine. Available online: https://www.forecoastmarine.com/ (accessed on 6 November 2020).
25. StormGEO Ltd. StormGEO S-Planner. Available online: https://www.stormgeo.com/products/s-suite/s-planner/ (accessed on 6 November 2020).
26. Teillant, B.; Chainho, P.; Raventos, A.; Nava, V.; Jeffrey, H. A Decision Supporting Tool for the Lifecycle Logistics of Ocean Energy Arrays. In Proceedings of the 5th International Conference on Ocean Energy, Halifax, Canada, 4–6 November 2014.
27. Teillant, B.; Chainho, P.; Raventos, A.; Sarmento, A.; Jeffrey, H. Characterization of the logistic requirements for the marine renewable energy sector. In Proceedings of the 1st International Conference on Renewable Energies Offshore, Lisbon, Portugal, 24–26 November 2014.
28. Rademakers, L.W.M.M.; Braam, H.; Obdam, T.S. Estimating Costs of Operation & Maintenance for Offshore Wind Farms. 2008. Available online: https://repository.tudelft.nl/islandora/object/uuid:ff4a94c7-5f57-4872-aba2-aad622656c16 (accessed on 26 July 2021).
29. Athanasios, K.; Feargal, B. ROMEO-Review of Existing Cost and O&M Models, and Development of a High-Fidelity Cost/revenue Model for Impact Assessment. 2018. Available online: https://www.romeoproject.eu/wp-content/uploads/2018/12/D8.1_ROMEO_Report-reviewing-exsiting-cost-and-OM-support-models.pdf (accessed on 25 January 2021).
30. Hofmann, M.; Sperstad, I.B. NOWIcob—A Tool for Reducing the Maintenance Costs of Offshore Wind Farms. *Energy Procedia* **2013**, *35*, 177–186. [CrossRef]
31. SINTEF. Decision Support Tools. Available online: https://www.sintef.no/projectweb/marwind/dst/ (accessed on 25 January 2021).
32. Gutierrez-Alcoba, A.; Ortega, G.; Hendrix, E.M.; Halvorsen-Weare, E.E.; Haugland, D. A model for optimal fleet composition of vessels for offshore wind farm maintenance. *Procedia Comput. Sci.* **2017**, *108*, 1512–1521. [CrossRef]
33. Shoreline. Design Entire Wind Farms by Simulating, Modelling and Analysing Scenarios in a Risk-Free Virtual Environment. Available online: https://www.shoreline.no/solutions/design/ (accessed on 24 January 2021).
34. Dalgic, Y.; Dinwoodie, I.; Lazakis, I.; McMillan, D.; Revie, M. Optimum CTV Fleet Selection for Offshore Wind Farm O&M Activities. In Proceedings of the ESREL 2014, Wroclaw, Poland, 14–18 September 2014; p. 9.
35. Dalgic, Y.; Lazakis, I.; Dinwoodie, I.; McMillan, D.; Revie, M.; Majumder, J. The influence of multiple working shifts for offshore wind farm O&M activities—STRATHOW-OM Tool. In Proceedings of the Design and Operation of Offshore Wind Farm Support Vessels, London, UK, 28–29 January 2015; p. 9. Available online: https://www.researchgate.net/publication/271525141_The_influence_of_multiple_working_shifts_for_offshore_wind_farm_OM_activities_-_StrathOW-OM_Tool (accessed on 26 July 2021)

36. Douard, F.; Domecq, C.; Lair, W. A Probabilistic Approach to Introduce Risk Measurement Indicators to an Offshore Wind Project Evaluation—Improvement to an Existing Tool Ecume. *Energy Procedia* **2012**, *24*, 255–262. [CrossRef]
37. Paterson, J. Metocean Risk Analysis in Offshore Wind Installation. Ph.D. Thesis, University of Edinburgh, Scotland, UK, 2018. Available online: https://era.ed.ac.uk/handle/1842/35981/ (accessed on 15 July 2021).
38. Wave Energy Scotland. Operations and Maintenance Simulation Tool: User Guide. 2017. Available online: https://www.waveenergyscotland.co.uk/media/1182/wes-om-tool-user-guide_rev1.pdf (accessed on 10 December 2020).
39. Sperstad, I.B.; Stålhane, M.; Dinwoodie, I.; Endrerud, O.E.V.; Martin, R.; Warner, E. Testing the robustness of optimal access vessel fleet selection for operation and maintenance of offshore wind farms. *Ocean Eng.* **2017**, *145*, 334–343. [CrossRef]
40. Grzegorzewska, P. First Results of the Study on the Impact of Open Source | Joinup. Available online: https://joinup.ec.europa.eu/collection/open-source-observatory-osor/news/first-results-study-impact-open-source (accessed on 26 July 2021).
41. Fraunhofer ISI. European Commission: Open Source Study (First Results). Available online: https://openforumeurope.org/wp-content/uploads/2020/11/OFE_Fraunhofer_OS_impact_study_5_Nov.pdf (accessed on 26 July 2021).
42. European Commission. Advanced Design Tools for Ocean Energy Systems Innovation, Development and Deployment | Projects | H2020 | CORDIS. 2018. Available online: https://cordis.europa.eu/project/rcn/214811_en.html (accessed on 21 December 2020).
43. The University of Edinburgh. DTOcean FP7-ENERGY EU Project: Optimal Design Tools for Ocean Energy Arrays. Available online: https://cordis.europa.eu/project/id/608597 (accessed on 26 July 2021).
44. The DTOcean Developers. DTOcean: Optimal Design Tools for Ocean Energy Arrays. Available online: https://dtocean.github.io/ (accessed on 11 December 2020).
45. Teillant, B.; Chainho, P.; Vrousos, C.; Charbonier, K.; Ybert, S.; Giebhardt, J. Deliverable 5.6: Report on Logistical Model for Ocean Energy and Considerations. 2016. Available online: https://www.dtoceanplus.eu/content/download/2541/file/DTO_WP5_ECD_D5.6.pdf (accessed on 26 July 2021).
46. Correia da Fonseca, F.X.; Amaral, L.M.B.; Rentschler, M.U.T.; Arede, F.; Chainho, P.; Yang, Y.; Noble, D.R.; Petrov, A.; Nava, V.; Germain, N.; et al. Logistics and Marine Operations Tools—Alpha version. Public Deliverable D5.7, DTOceanPlus, May. Available online: https://www.dtoceanplus.eu/Publications/Deliverables/Deliverable-D5.7-Logistics-and-Marine-Operations-Tools-alpha-version (accessed on 16 July 2021).
47. Yi, Y.; Nambiar, A.; Luxcey, N.; Correia da Fonseca, F.X.; Amaral, L. Reliability, Availability, Maintainability and Survivability Assessment Tool—Alpha Version. Public Deliverable D6.3, DTOceanPlus. 2020. Available online: https://www.dtoceanplus.eu/Publications/Deliverables/Deliverable-D6.3-Reliability-Availability-Maintainability-and-Survivability-Assessment-Tool-alpha-version (accessed on 16 July 2021).
48. Raval, U.R.; Chaita, J. Implementing & Improvisation of K-means Clustering Algorithm. *IJCSMC* **2016**, *5*, 191–203. Available online: https://www.ijcsmc.com/docs/papers/May2016/V5I5201647.pdf
49. Teillant, B.; Raventos, A.; Chainho, P.; Victor, L.; Goormachtigh, J.; Collin, A.; Hardwick, J.; Weller, S.; Guerrini, M.; Giebhardt, J.; et al. Deliverable 5.2: Characterization of Logistic Requirements. 2014.
50. Youen, K. Site Characterisation—Alpha Version. Public Deliverable D5.2, DTOceanPlus. 2020. Available online: https://www.dtoceanplus.eu/Publications/Deliverables/Deliverable-D5.2-Site-Characterisation-alpha-version (accessed on 16 July 2021).
51. Chen, Y.; Cao, P.; Mukerji, P. Weather Window Statistical Analysis for Offshore Marine Operations. In Proceedings of the Eighteenth (2008) International Offshore and Polar Engineering Conference, Vancouver, BC, Canada, 6–11 July 2008; The International Society of Offshore and Polar Engineers (ISOPE): Vancouver, BC, Canada, 2008; p. 1.
52. Det Norske Veritas (DNV). *Modelling and Analysis of Marine Operations*; Recommended Practices ed.; RP-H103. 2011. Available online https://rules.dnv.com/docs/pdf/DNVPM/codes/docs/2011-04/RP-H103.pdf (accessed on 26 July 2021)
53. Ochi, M.K. *Ocean Waves—The Stochastic Approach*; Cambridge University Press: Cambridge, UK, 1998; doi:10.1017/CBO9780511529559. [CrossRef]
54. Goda, Y. *Random Seas and Design of Maritime Structures*; World Scientific Publishing: Singapore, 2000; doi:10.1142/9789812385444_0001. [CrossRef]
55. Holthuijsen, L.H. *Waves in Oceanic and Coastal Waters*; Cambridge University Press: Cambridge, UK, 2007.
56. Sparks, N.J. IMAGE: A multivariate multi-site stochastic weather generator for European weather and climate. *Stoch. Environ. Res. Risk Assess.* **2018**, *32*, 771–784. [CrossRef]
57. Yang, X.C.; Zhang, Q.H. Joint probability distribution of winds and waves from wave simulation of 20 years (1989–2008) in Bohai Bay. *Water Sci. Eng.* **2018**, *6*, 296–307. [CrossRef]
58. Sasaki, W.; Iwasaki, S.I.; Matsuura, T.; Iizuka, S. Recent increase in summertime extreme wave heights in the western North Pacific. *Geophys. Res. Lett.* **2005**, *32*. [CrossRef]
59. Kennedys. Notes from the Bar: BIMCO SUPPLYTIME 2017. Available online: https://kennedyslaw.com/thought-leadership/article/notes-from-the-bar-bimco-supplytime-2017/ (accessed on 30 December 2020).
60. Harling, M.; Gard, W.; James, S. Future Trends in Procurement Strategy: The Influence of New Nuclear and Offshore Energy Projects. 2013. Available online: https://uk.practicallaw.thomsonreuters.com/3-549-1845?transitionType=Default&contextData=(sc.Default)&firstPage=true (accessed on 26 July 2021).
61. Global Renewables Shipbrokers. Available online: https://www.grs-offshore.com/ (accessed on 10 January 2021).

62. ECN. Energy Research Centre of the Netherlands. Available online: https://www.ecn.nl/energy-research/index.html (accessed on 4 January 2021).
63. Lundh, M.; Garcia-Gabin, W.; Tervo, K.; Lindkvist, R. Estimation and Optimization of Vessel Fuel Consumption. *IFAC PapersOnLine* **2016**, *49*, 394–399. [CrossRef]
64. Ship & Bunker. World Bunker Prices. Available online: https://shipandbunker.com/prices (accessed on 16 December 2020).
65. Scottish Enterprise. Oil and Gas 'Seize the Opportunity' Guides: Offshore Wind. Technical Report SE/4504/May16, Scottish Enterprise. 2016. Available online: https://www.offshorewindscotland.org.uk/media/1116/sesdi-oil-and-gas-div-guide-offshore-wind.pdf (accessed on 26 July 2021).
66. Neary, V.S.; Previsic, M.; Jepsen, R.A.; Lawson, M.J.; Yu, Y.H.; Copping, A.E.; Fontaine, A.A.; Hallett, K.C.; Murray, D.K. Methodology for Design and Economic Analysis of Marine Energy Conversion (MEC) Technologies. 2014. Available online: https://energy.sandia.gov/wp-content/gallery/uploads/SAND2014-9040-RMP-REPORT.pdf (accessed on 21 January 2021).
67. UK Met Office. Cartopy, a Cartographic Python Library with Matplotlib Support for Visualisation. Available online: https://pypi.org/project/Cartopy/ (accessed on 27 April 2021).

energies MDPI

Article

A Proposed Guidance for the Economic Assessment of Wave Energy Converters at Early Development Stages

Amélie Têtu [1,*,†] and Julia Fernandez Chozas [2,†]

[1] Department of the Built Environment, Aalborg University, 9220 Aalborg, Denmark
[2] Julia F. Chozas, Consulting Engineer, 1360 Copenhagen, Denmark; info@juliafchozas.com
* Correspondence: amt@build.aau.dk; Tel.: +45-9940-2924
† These authors contributed equally to this work.

Citation: Têtu, A.; Fernandez Chozas, J. A Proposed Guidance for the Economic Assessment of Wave Energy Converters at Early Development Stages. *Energies* **2021**, *14*, 4699. https://doi.org/10.3390/en14154699

Academic Editor: Eugen Rusu

Received: 27 May 2021
Accepted: 21 July 2021
Published: 3 August 2021

Publisher's Note: MDPI stays neutral with regard to jurisdictional claims in published maps and institutional affiliations.

Copyright: © 2021 by the authors. Licensee MDPI, Basel, Switzerland. This article is an open access article distributed under the terms and conditions of the Creative Commons Attribution (CC BY) license (https://creativecommons.org/licenses/by/4.0/).

Abstract: Wave energy is one of the most promising renewable energies available with its very large resource. The waves generated by the wind field are steadier than the wind field itself, rendering wave energy more consistent than wind energy. It is also more predictable than wind and solar. Wave energy is making continuous progress towards commercialisation, and thanks to an increasing number of deployments at sea, the sector is increasing the understanding of the costs and economies of these projects. No wave energy converter has been demonstrated to be commercially viable, and it is yet to be proven that wave energy can contribute to the renewable energy mix. In this context, and in order to find an economically viable solution for exploiting wave energy, it is important to assess the economic potential of a particular concept throughout the entire technological development process. At early development stages, this assessment can be challenging and present large uncertainties. Notwithstanding, it is important to perform the economic assessment already at the early stages in order to identify possible bottlenecks or potential improvements or modifications of a concept. This work presents guidance for the economic evaluation of a wave energy concept at an early development stage by setting up the economic frame based on a target LCoE. It involves the understanding of the entry cost to be achieved for a specific target market and evaluating the breakdown of costs based on a detailed technology agnostic database of costs. The guidance is then applied to a new type of wave energy converter, in which the primary coupling with the waves is through hydrodynamic lift forces.

Keywords: wave energy; wave energy converter; LCoE; cost breakdown; capital expenditure; operating expenditure; commercialisation; lift-based wave energy converter

1. Introduction

As it is detailed in the Offshore Renewable Energy Strategy released by the European Commission in November 2020 [1], the European Union (EU) is raising its climate targets for 2030 and is committed to becoming climate-neutral by 2050. To achieve this, the EU is setting ambitious targets, including the generation of more energy at sea and from the sea. The goals for 2030 include an EU offshore wind energy capacity of 60 GW and an ocean energy capacity (including wave and tidal energy) of 1 GW. The targets are even more ambitious for 2050, where the aim is at installing 300 GW of offshore wind and 40 GW of wave and tidal energy.

The European strategic energy technology plan (SET-Plan) declaration of intent for ocean energy [2] has also set ambitious economic targets for wave and tidal energy technologies. Wave energy technologies are expected to reach a levelised cost of energy (LCoE) of 200 EUR/MWh in 2025, of 150 EUR/MWh in 2030 and of 100 EUR/MWh in 2035 (export infrastructure costs or the costs for delivering the electricity to onshore substations are taken into account within the LCoE). These numbers show that the economic and market potential of wave energy in Europe is large and reaching the SET-Plan target numbers is an ambitious goal.

When a new type of wave energy converter starts its development path (the point in time when the initial idea or its working principle is conceived), it is a challenge to estimate its potential economic value when reaching the commercialisation stage. It is also commonly agreed that the primary metric for judging the economic potential of energy technologies is the LCoE. However, it is also commonly agreed that estimates of the LCoE for wave energy technologies are affected by the lack of a dominating technology as well as uncertainties caused by unproven technologies in terms of electricity generation [3–6]. In an attempt to answer these limitations, other metrics have been introduced to compare technologies at low Technology Readiness Levels (TRL) [7–9]. For example, the ACE metric, which is the ratio of the average climate capture width to the Characteristic Capital Expenditure (ACE), was introduced in [8] to assess the economic capabilities of wave energy concepts at early stages of technology development. In [10], the ACE is calculated as an optional cost metric for assessing technologies at early stages when not sufficient or reliable data for calculating the LCoE is available. Together with the Hydrodynamic Quality Factor (HQF), the ACE can be useful to compare different WEC concepts. This method is convenient when comparing concepts at low TRLs. However, when setting up the development pathway for a wave energy project, the target market indicator is the LCoE. When looking solely at one technology throughout its development lifetime, choosing a single metric, such as the LCoE, can therefore facilitate tracking the economic performances of the technology so that it follows the preferred pathway for development [11].

Assessing the LCoE of a technology at the early stages of technology development is associated with some uncertainties. Instead of a bottom-up approach, a top-down approach can be used, where the LCoE of a technology is defined by the entry LCoE value for the target market. There will still be some uncertainties in the costs found using this approach, but it will give a range of target costs for a technology to reach in order to achieve the end goal of commercialisation in the specific market.

In this context, guidance for the economic assessment of a wave energy technology at an early development stage or TRL (TRL1 to TRL4) [12] is proposed in this work with a focus on a target market while considering the uncertainties associated with the calculations. A methodology that can be applied to any project deployed at any location around the globe is presented in the following section, and this methodology is afterwards applied to a specific wave energy technology at an early stage of development. Limitations of the methodology and uncertainties on the calculations are also discussed. The methodology can eventually enable the identification of possible improvements for the particular concept studied.

2. Proposed Methodology

The proposed methodology to assess the economic potential of a wave energy converter at early development stages is presented in this section. The first step is to define the target market for the commercialisation of a specific technology. Afterwards, all assumptions and relevant relations are introduced to enable the reverse calculation of costs associated with a particular project. Then, a detailed breakdown of costs is presented, which can enable identifying bottlenecks and possibilities for the improvement of the technology. Finally, a review of estimates for economic indicators, such as Capital Expenditures (CAPEX), Operational Expenditures (OPEX), capacity factors (C_f) and availability, is presented in order to aid the economic evaluation.

2.1. *Target Market for Commercialisation*

In order to setup the right frame for the economic assessment of a wave energy concept, it is important to identify the target market for commercialisation. For a wave energy technology aiming at the utility-scale market, the SET-Plan has defined an LCoE target in the range of 100 to 200 EUR/MWh [2]. These numbers can be compared to the target LCoE for offshore wind energy for 2030 of 65 EUR/MWh, including grid connection [13]. An exemption to this could be islands that are not connected to the mainland grid and have their own power supply based on traditional diesel generators. In this case, a competitive LCoE to the diesel alternative depends on oil prices. These are variable and will fluctuate throughout the project lifetime, but something in the range of 300 to 400 EUR/MWh, as reported for Kiritimati Island, in the Pacific Ocean [14], seems reasonable. Similar values were reported in [15] for a diesel engine farm in the United Kingdom. As a comparison, in 2018, a value for diesel generation in the range of 560 to 730 EUR/MWh was reported for Christmas Island, Australia [16]. Overall, identifying the target market has a significant impact in the boundary conditions for the economic assessment. (Even though this work focuses on the target market, existing frameworks for accelerating the development of wave energy is not to be neglected when developing a new concept. See for example [17] for existing European frameworks.)

2.2. Assumptions and Economic Computations

In order to perform an economic assessment at the early stages of the concept development (TRLs 1–4), some assumptions are made. Through reverse calculations, the initial estimates on the CAPEX and OPEX of the project can be obtained. The goal is to get an indication of the values that would allow achieving a given target LCoE. To perform the reverse calculation, some values need to be assumed for the following parameters:

- Discount rate, r, which is usually assumed constant over the project lifetime. The discount rate is the rate used in the discounted cash flow analysis to calculate the present value of future cash flows [18].
- Project lifetime, n. The project lifetime is the life expectancy for the deployed project. Typical values for ocean renewable energy projects are 20–25 years [19].
- Capacity factor, C_f. The capacity factor is defined as the power output of a plant divided by its maximal power capability. As an example, for offshore wind energy, this can go above 45%; a life capacity factor of 49% has been reported for the Anholt offshore wind farm in Denmark [20].
- Availability factor, $a_{\%}$. The availability factor is the fraction of time for which a plant is producing electricity over the project lifetime. It is often expressed in percentage.

Based on the project interest rate and the project lifetime, it is possible to extend the discount rate on an annual basis by calculating the annualisation factor ($A_f(n,r)$):

$$A_f(n,r) = \frac{r}{1-(1+r)^{n+1}}. \tag{1}$$

The Normalised Annual Energy Production (NAEP), in MWh per installed kW per year, can be calculated using the following relation:

$$\text{NAEP} = 8765 \cdot C_f \cdot a_{\%} \tag{2}$$

where 8765 corresponds to the number of hours in a year.

The LCoE (in EUR/MWh) relates the parameters, CAPEX, OPEX, NAEP, and the annualisation factor through the following simplified equation:

$$\text{LCoE} = \frac{A_f \cdot \text{CAPEX} + \text{OPEX}}{\text{NAEP}}. \tag{3}$$

By assuming a given LCoE, NAEP and annualisation factor, a relation between the CAPEX (in EUR per kW of installed power) and the OPEX (in EUR per kW of installed power per year) can be obtained. Because the OPEX is typically expressed as a percentage of the CAPEX (this relation is further described in Section 2.3.2) by choosing a given percentage, the CAPEX can be estimated; and from that, the OPEX can also be derived.

2.3. Breakdown of Costs

The costs related to a wave energy project are usually divided into CAPEX and OPEX. These two categories are further divided into cost centres. This categorisation is performed in order to quantify the influence of the different cost centres on the overall cost of a project. For wave energy projects, this exercise was performed in [4,21–25], but in all cases, either a single technology was considered or a limited number of technologies were considered.

A literature review was performed by Têtu and Fernandez Chozas in [26] to build a cost database for wave energy projects [27]. This database is the baseline for the technology agnostic breakdown of costs presented in this section, which can be used to assign ranges of cost for the different cost centres when target values for the CAPEX and OPEX are known.

2.3.1. CAPEX

CAPEX for a wave energy project can be summarised as all the expenditures associated with the project development, its deployment and commissioning until the operation of the WEC farm starts. It also includes decommissioning at the end of the project life. A thorough literature review of costs related to the CAPEX can be found in [26]. Costs found in this category include costs related to the multifaceted process of developing a WEC farm from inception through to the handing over of the farm to the customer. A cost breakdown for the CAPEX is presented in Figure 1 based on the work compiled in [26]. The different costs centres are explained further in the following subsections.

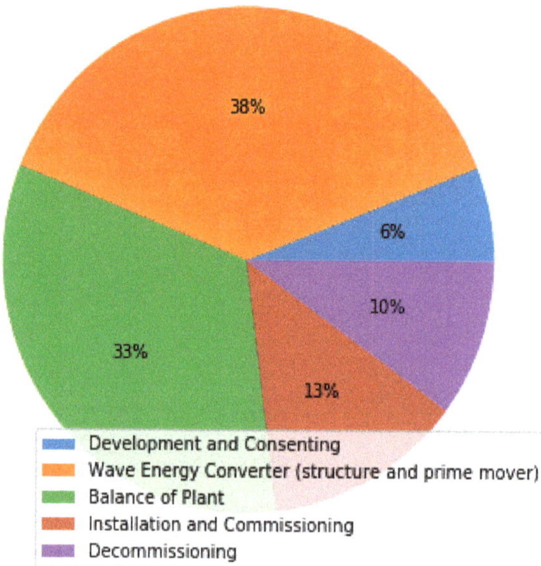

Figure 1. CAPEX breakdown of costs for the different cost centres.

Development and Consenting

The development and consenting services account for the following costs: project management, design engineering, planning and consenting. These costs are normally reported as percentage of CAPEX and this percentage is expected to decrease proportionally as the installed power capacity increases and standard procedures are developed [28]. Percentage values range from 2% up to 12% of the CAPEX [21,29–31], and this may well be due to the fact that these cost are very project (i.e., single prototype or wave energy farm) and site-specific. From a general point of view and by taking the experience gathered in the wind energy sector [32], development and consenting can be estimated at 6% of the CAPEX.

Wave Energy Converter (Structure and Prime Mover)

The wave energy converter, including its structure and prime mover, has been reported many times to have the most significant impact on CAPEX. The structure cost includes the detailed infrastructure design and supply of all components from the mooring attachment point, excluding the power take-off system. This also includes delivery to a port. Again, the costs reported in the literature are very project-specific, as they depend on the structure's main materials and volumes [21,24,25,33]. A value of 38% of the CAPEX is reasonable according to [26], where a range of 27% to 38% of the CAPEX has been found for this category.

Balance of Plant

The Balance of plant costs include costs related to the power take-off (PTO) system [34–36], the supply of all its constituting components and its delivery to the port. Mooring [24,25,33,37,38] and foundation [30,39] costs are also included in this cost centre together with the electrical installation necessary to render the farm grid-connected [30,33,40–44]. All in all, this cost centre has the second most significant impact on the CAPEX. From the costs gathered in [26], this cost centre can be estimated at 33% of the CAPEX.

Installation and Commissioning

Installation costs [21,24,33] include the installation of the WECs on site and the commissioning of these to a fully operational state, up to the point of issue of any takeover certificate. Those costs are to a great extent driven by vessel-chartering costs. Installation methods that require small and, thus, cheaper vessels lead to a lower cost, and the installation in port followed by towing can provide significant cost-reduction opportunities. According to the review of costs presented in [26], installation and commissioning costs include the costs of installing the foundation or moorings, the offshore substation, the WEC and the cables; they typically fall in the range of 8% to 17% of the CAPEX. In this particular exercise, they are estimated at 13% of the CAPEX.

Decommissioning

Decommissioning costs [24,41] include all costs related to the removal of the WECs, the foundation or mooring system and the electrical cables according to the legally binding contract. The decommissioning of an offshore project is often seen at the reverse operation of the installation and commissioning process. Experience from the wind sector [45–47] helps estimating costs for decommissioning of a wave energy farm project, and according to [26], 10% of the CAPEX is a reasonable estimate for this cost centre.

2.3.2. OPEX

The OPEX relates to all expenditures associated with the operation of a WEC farm from the moment a takeover certificate is issued, including the cost of all operation and maintenance (O&M) activities as well as the cost associated to site leasing and insurance. Table 1 summarises the cost breakdown for the OPEX obtained from the literature review presented in [26]. The main two cost centres are the site lease and insurance, accounting for 6% of the OPEX, and the O&M, accounting for the remaining part of OPEX.

Table 1. OPEX breakdown of costs.

Cost Center	Fraction of OPEX
Operation and Maintenance (O&M)	94%
Site lease and Insurance	6%

For a wave energy farm for which data is scarce, the annual OPEX can be estimated as a percentage of the CAPEX. As shown in the literature, estimates of the total OPEX per year roughly range from 1.5% to 9% of the CAPEX [29,33,48,49]. This is due to different factors (e.g., single prototype or utility-scale project, distance to shore, floating or submerged WEC, innovative or traditional O&M techniques applied, etc). For example, the OPERA project [25] has shown that when specific innovative O&M techniques are applied, the

OPEX can be as low as 1.8% to 2.2% of the CAPEX, depending on the deployment location and size of the array. If a utility-scale project is considered, then the experience from the offshore wind energy sector (with the annual OPEX as 4.5% of the CAPEX [6] and the annual OPEX as 3% of the CAPEX [50]) can be used.

2.4. Estimates for CAPEX, OPEX and Other Relevant Economic Indicators

It is very pertinent to provide different references that may assist in the application of the proposed guidance to a particular WEC. The purpose of this subsection is to present estimated relations between the CAPEX and OPEX as well as targeted values of the CAPEX, OPEX, capacity factors, availability and interest rates provided by different technology developers in the wave energy sector. The references have been selected according to their relevance. All of them aim to take into account the views and state of the art of several wave energy developments.

Ocean Energy System [4] reviewed current and projected costs (CAPEX, OPEX and LCoE) for wave energy converters at a TRL 6 and above by engaging with a large number of international stakeholders globally involved in wave energy developments. Three different development stages were considered: i) first array deployed, ii) second array deployed and iii) first commercial-scale project (first project that is constructed with a view to generate commercial return without the need for capital or public sector support outside of an authorised feed-in-tariff). The costs of a generic WEC were derived by considering the different TRLs of the concepts being consulted and the uncertainty behind the data. A summary of the main findings is provided in Table 2, and the reader is referred to [4] for the detailed methodology employed.

Table 2. An example of estimated CAPEX and OPEX values for different deployment stages [4]. The maximum value is either that from the responses of consulted developers or from any of the reference studies analysed. This is particularly true for the OPEX, where developers were presenting costs significantly more optimistic than past studies have suggested. An exchange rate of USD 1.11 to EUR has been applied. Data adapted from [4].

Deployment Stage		Minimum Value	Maximum Value
First array	CAPEX (EUR/kW)	3600	16,300
	OPEX (EUR/kW/year)	125	1350
Second array	CAPEX (EUR/kW)	3240	13,800
	OPEX (EUR/kW/year)	90	450
	Availability (%)	85%	98%
	Capacity factor (%)	30%	35%
First commercial scale project	CAPEX (EUR/kW)	2400	8200
	OPEX (EUR/kW/year)	65	340
	Discount rate (%)	10%	10%
	Availability (%)	95%	98%
	Capacity factor (%)	35%	40%

In 2018–2019, a second study [51] had the main goal of updating the previous findings. With the aim of targeting a higher number of respondents, all active wave energy developers around the world were invited to participate in the study. Based on the respondents, the typical features and costs of a generic, utility-scale floating wave energy farm were obtained and are provided in Table 3. These costs represent only the start of the learning curve, and the values are expected to decrease as more farms are deployed.

Table 3. Typical features and costs of a generic, utility-scale floating wave energy farm gathered from a survey sent to many developers around the world [51]. Data adapted from [51].

Project Characteristics	
Project capacity (MW)	160 MW
Project lifetime	22 years
Discount rate	7%
Overall CAPEX	3100 EUR/kW
Overall annual OPEX	4% of CAPEX
Capacity factor	36%

To complement the previous figures, the estimates of future costs for wave power included in the Technology Catalogue of Denmark [15] are shown in Table 4. The costs presented aim to provide an estimate for what capital and operational costs of wave power converters might be in the future assuming most of the research and development challenges have been overcome, economics of scale have been realised and efficiencies in production and operation due to the learning curve effect have been achieved.

Table 4. Wave power data sheet from the Technology Data Catalogue [15]. Data adapted from [15].

Technical and Financial Data	Units	2030	2050
Generating capacity for one power plant	MW	10–100	50–500
Length of installation of one power plant	km	1–20	5–100
Annual generated electricity production	MWh/MW	3500	4500
Availability	%	97	98
Technical lifetime	years	25	30
Capital Investment	MEUR/MW	2.2–4.5	1.6
OPEX	EUR/kW/year	60	47

3. Example and Discussion

The methodology presented in the previous section is applied in this section to the LiftWEC concept. The LiftWEC project [52] aims to develop a new type of wave energy converter (the LiftWEC concept) that couples with the waves through lift forces generated by one or more hydrofoils that rotate in a single direction. LiftWEC is currently at TRL 2, and it is expected to reach TRL 4 [52] by the end of the project (late 2022). The concept is ultimately designed to work in wave energy farms and supply electricity at grid-scale. Unlike other projects, the concept is yet to be defined during the LiftWEC project. No developer is involved, and the resulting concept will be the fruit of the unique development process detailed in [52].

The target deployment location for the LiftWEC concept is off the North Atlantic coast of France, close to Quimper, where the water depth at the deployment location is 50 m and the wave resource is estimated at 40 kW/m. The targeted rated power is still an unknown for the final concept but it should be in the range [0.75, 2] MW, while the lifetime of the project is set at 25 years.

The project has set two reasonable economic goals to its technology development. The first goal is to prove an LCoE of 200 EUR/MWh by mid-project, coinciding with TRL 2. The second goal is proving an LCoE of 120 EUR/MWh by project end, i.e., when reaching TRL 4. The latter LCoE is aligned with target values to be achieved for a utility-scale project, and both target values will be used in the following section to perform the reverse calculation and obtain ranges for the CAPEX and OPEX for the concept using the technology agnostic breakdown of costs presented in Section 2.3.

3.1. LCoE Calculation

The goal behind the LCoE calculation is to get an indication of the values that would allow achieving the LiftWEC project mid-term target (TRL 2) LCoE of 200 EUR/MWh and end-of-project (TRL 4) target LCoE of 120 EUR/MWh. In Section 2.2, the assumptions and equations leading to the calculation of the LCoE were introduced. The values for the parameters affecting the LCoE calculation, i.e., the capacity factor, the discount rate, the project lifetime and the availability, are presented in the top four rows of Table 5.

From the discount rate and the lifetime, according to Equation (1), the annualisation factor becomes $A_f(25, 0.05) = 0.0696$ for both cases. The normalised annual energy production is then obtained from Equation (2), providing the value of NAEP = 2.50 MWh/kW/year for the mid-term target and NAEP = 2.98 MWh/kW/year for the end-of-project target.

As also presented in Section 2, a relation can be established between the CAPEX and OPEX. An estimate of the OPEX accounting for 5% the CAPEX seems reasonable for the mid-term project (TRL 2), where no O&M optimisation has been done. For the end-of-project (TRL 4), where O&M optimisation techniques shall be considered, a reduction of the OPEX is expected, and hence, it seems reasonable to estimate an OPEX value of about 2.5% CAPEX.

Isolating the CAPEX and OPEX in Equation (3) and considering the OPEX as x% of the CAPEX gives:

$$\text{CAPEX} = \frac{\text{LCoE} \cdot \text{NAEP}}{A_f + x} \tag{4}$$

resulting in a CAPEX value of 4181 EUR/kW for the mid-term project, assuming the OPEX accounts for 5% the CAPEX; and a CAPEX value of 3780 EUR/kW for the end-project, assuming the OPEX accounts for 2.5% the CAPEX. For clarity purposes, the two estimates of the CAPEX are rounded to 4200 EUR/kW and 3800 EUR/kW, respectively. OPEX values are consecutively calculated and obtained at 210 EUR/kW/year and 95 EUR/kW/year, respectively, providing the two target LCoEs to be achieved by mid-project and end-of-project of 200 EUR/MWh and 120 EUR/MWh, respectively.

Table 5 summarises the values for the different parameters and the results to be achieved in order to reach the two target LCoEs of 200 EUR/MWh and 120 EUR/Mwh.

Table 5. A summary of the parameters of the two LCoE targets along the project.

Parameter	Symbol	Value in Mid-Term Project	Value in End-of-Project	Unit
Capacity factor	C_f	30	35	%
Availability	$a_{\%}$	95	98	%
Discount rate	r	5	5	%
Lifetime	n	25	25	year
Annualisation factor	A_f	0.0696	0.06968	-
Annual energy production	AEP	2.50	2.98	MWh/kW/year
Capital expenditures	CAPEX	4200	3800	EUR/kW
Operational expenditures	OPEX	210	95	EUR/kW/year
Levelised cost of energy	LCoE	200	120	EUR/MWh

By comparing the numbers shown in Table 2 to the estimates presented in this work, it can be seen that the present estimates of CAPEX (4200 EUR/kW and 3800 EUR/kW) and OPEX (210 EUR/kW/year and 95 EUR/kW/year) are in the same order of magnitude that the minimum values expected in the three deployment stages addressed by the OES study (first array, second array and first-commercial scale project). From this, we could argue that the calculations and targets presented in this paper—and also worked throughout the LiftWEC project—are aligned to the sector's targets. However, it is also important to notice that there are still considerable R&D efforts to go from the LCoE of 200 EUR/MWh to the 120 EUR/MWh, especially in terms of increasing the ability of the LiftWEC concept to capture more energy (and hence, increasing the AEP without compromising costs) and to lower the OPEX by applying innovative O&M techniques. Two aspects that have proven to be of the utmost importance to the sector [53]. It is also important to note that the presented values (i.e., LCoE, CAPEX, OPEX, capacity factors and availability) are estimates with a relevant degree of uncertainty (the LifWTEC Concept is currently in TRL 2), as indicated in Section 3.3.

3.2. Estimates of the CAPEX Breakdown for the Mid-Term Project (TRL 2)

As presented in Section 2 and shown in Figure 1, the expected breakdown of the CAPEX for a generic wave energy converter has been suggested, which gives a reasonable starting point for a low-TRL project where detailed information of all costs is not yet available. Based on the assumption at the mid-term project and TRL 2 of CAPEX at 4200 EUR/kW, the estimates for the different CAPEX cost centres can be inferred too. They are presented in Table 6.

Table 6. Breakdown of costs for the mid-term project (TRL 2), assuming a CAPEX value of 4200 EUR/kW targetting the LCOE of 200 EUR/MWh.

CAPEX Cost Centres	% of CAPEX	Cost in EUR/kW
Development costs	6	250
WEC structure & prime mover	33	1340
Balance of plant	38	1600
Installation & commissioning	13	590
Decommissioning	10	420
Total	100	4200

To exemplify the meaning of these numbers, Table 7 and Figure 2 provides a deeper insight into the cost centre labeled *Balance of plant*. With an allocated percentage contribution to CAPEX of 38%, the following estimates can be inferred for the different parts that compose it. These are the power take-off system, the foundation or support structure, the offshore electrical cables, the offshore substation and the onshore transmission and connection [26].

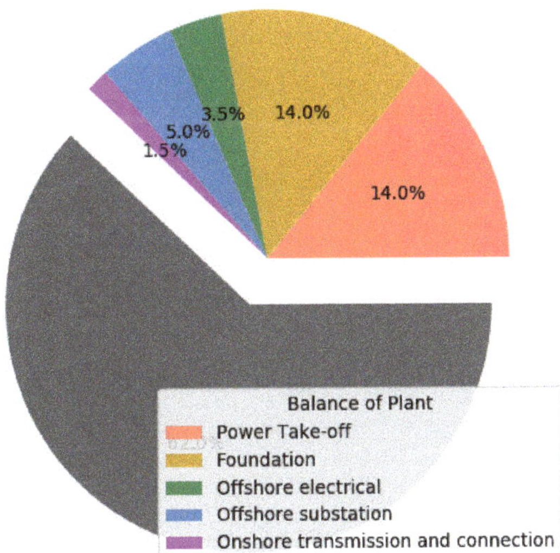

Figure 2. A breakdown of costs for the Balance of plant cost centre. The dark grey part represents the rest of the CAPEX.

Table 7. A breakdown of costs for the Balance of plant costs based on [26] at the mid-term project (TRL 2).

Cost Centres	% of CAPEX	Cost in EUR/kW
Balance of plant	38	1600
Power take-off	14	590
Foundation	14	590
Offshore electrical cables	3.5	150
Offshore substation	5	210
Onshore transmission and connection	1.5	60

By looking into each of the categories depicted in Table 7, it is relevant to compare how the suggested numbers coming from the reverse calculation fall into the costs experienced by the wave energy sector. Ricci et al. [36] suggest that 600 EUR/kW for a linear generator PTO-type or 800 EUR/kW for a hydraulic PTO-type are reasonable estimates. Other studies [34,35] suggest that a unit cost of 340 EUR/kW can be used for the different PTO systems (mechanical, air, water and hydraulic) if series production is considered. Therefore, assuming a value around 600 EUR/kW seems to be reasonable.

The estimated costs for the foundation (590 EUR/kW) are also comparable with the costs presented in [30] for a monopile structure at 30 m water depth.

With regards to the costs of the electrical connections, they are in the same range as the ones presented in [32] for the inter-array electric cable, in [30,42] for the offshore substation and in [40] for the onshore transmission and connection.

It is important to note that the costs of an offshore electrical connection are very much project-specific and site-dependent. Denmark has traditionally financed the electrical connections for offshore wind energy projects. This fact has had an important impact on the final LCoE for offshore wind energy in Denmark compared to the one obtained in other countries, i.e., Great Britain, where developers shall self-finance the export infrastructure, and the difference of these on the final LCoE is estimated at 25% [54].

Overall, the example presented in this section has allowed drawing some estimates of the values and costs that will allow the LiftWEC concept to be competitive in the energy market. It is the primary goal that this economic exercise is done in combination and in collaboration with technological development, so every advancement in the concept's design decision is considering all the technological and economic aspects together. It is also important to note that the presented values may be read as orders of magnitudes rather than absolute figures, and therefore, the overall exercise is also found to be useful in helping to identify expected costs ranges for the different categories, elements or cost-centres that compose a WEC. Those values and the breakdown of costs are likely to change as the project evolves to higher TRL.

3.3. *Uncertainties and Improvement Possibilities*

It is important to be aware that there are uncertainties in all the data handled throughout the calculations. Therefore, there will be uncertainties in the output results.

A quantification of the uncertainties that accompanied the cost assessment of wave energy technologies was proposed in [55]. The quantification depends on the technology's development stage (TRL) and the type of assessment (preliminary, baseline, detailed, etc.) that is carried out. This quantification has been used in [4,29,51], among others, and has allowed us to provide a sensitive evaluation of uncertainties. Table 8 summarises the uncertainties associated to a preliminary cost evaluation.

Table 8. The uncertainty of the LCoE estimates as a function of the development stage [55]. Data adapted from [55].

Deployment Stage	Uncertainty Range
Phase 1/TRL 1, 2 and 3	[−30, 50]%
Phase 2/TRL 4	[−25, 30]%
Phase 3/TRL 5 and 6	[−20, 20]%
Phase 4/TRL 7 and 8	[−15, 15]%
Phase 5/TRL 9	[−10, 10]%

Accordingly, the associated uncertainty to the estimated values at the present development stage of the LiftWEC concept (at TRL 2) is about (−30/%, 50)%. This is an important fact to take into account, as it reflects that there are also lot of uncertainties on the technological side. The end-of-project target LCoE of 120 EUR/MWh (expected at TRL 4) has an associated uncertainty of (−25, 30). This reduction in uncertainty is associated to the advancements of technology development. The first results form tank testing, and refined numerical modeling will be available. Furthermore, annual energy production estimates will include the contribution of a specific-designed control strategy. The inclusion of such a control strategy will also affect the OPEX (to a large extent) and CAPEX, which will be evaluated and examined in detail. At this stage, a more specific description of costs will be drawn.

4. Conclusions

The economic assessment of wave energy projects is highly recommended at any stage of development. Setting target LCoE goals for a specific market is paramount to understand the key values that shall be achieved in order to be competitive in that specific market. The concepts to be evaluated in the LiftWEC project are at a very early stage of development. Nevertheless, as the target is to deliver electricity to the grid at a competitive price, some key economic indicators can be used to facilitate achieving the end goal. The present work has defined estimate values for overall project interest rate, capacity factor, WEC availability, CAPEX and OPEX when a specific LCoE is set as a target. The estimates have shown to be aligned with the targets of the wave energy sector and provide reasonable orders of magnitude for the different elements that constitute a wave energy converter. The uncertainty associated with these estimates has also been discussed. With an extensive database of costs, the estimated values enable to define ranges of costs for all cost centres of a WEC project. By keeping these in mind throughout the design process, it is the aim to ensure that the development of a WEC concept is following a competitive pathway to commercialisation.

Author Contributions: Conceptualization, A.T. and J.F.C.; methodology, A.T. and J.F.C.; formal analysis, A.T. and J.F.C.; writing—original draft preparation, A.T. and J.F.C.; writing—review and editing, A.T. and J.F.C. All authors have read and agreed to the published version of the manuscript.

Funding: This research was funded by European Union's Horizon 2020 research and innovation programme under grant agreement No 851885.

Acknowledgments: This paper is supported by the European Union's Horizon 2020 research and innovation programme under grant agreement No 851885, project LiftWEC (the development of a novel wave energy converter based on hydrodynamic lift forces). The authors would also like to acknowledge the fruitful discussions and comments from the LiftWEC project participants, especially Matt Foley and Paul Lamont-Kane (Queens University Belfast).

Conflicts of Interest: The authors declare no conflict of interest.

Abbreviations

The following abbreviations are used in this manuscript:

ACE	Ratio of the Average climate capture width to the Characteristic capital Expenditure
NAEP	Normalised Annual Energy Production
EU	European Union
WEC	Wave energy converter
LCoE	Levelised cost of energy
CAPEX	Capital expenditure
OPEX	Operating expenditure
O&M	Operation and maintenance
TRL	Technology readiness level
LiftWEC	lift-based WEC
PTO	Power Take-Off

References

1. European Commission. *An EU Strategy to Harness the Potential of Offshore Renewable Energy for a Climate Neutral Future*; Technical Report; European Commission: Brussels, Belgium, 2020. Available online: https://ec.europa.eu/energy/sites/ener/files/offshore_renewable_energy_strategy.pdf (accessed on 8 February 2021).
2. European Commission—SET Plan Secretariat. *SET Plan–Declaration of Intent on Strategic Targets in the Context of an Initiative for Global Leadership in Ocean Energy*; Technical Report; European Commission: Brussels, Belgium, 2016. Available online: https://setis.ec.europa.eu/system/files/integrated_set-plan/declaration_of_intent_ocean_0.pdf (accessed on 8 February 2021).
3. Magagna, D.; Monfardini, R.; Uihlein, A. *JRC Ocean Energy Status Report*, 2016th ed.; Technical Report; Joint Research Centre, Directorate for Energy, Transport and Climate Change; Publications Office of the European Union: Luxembourg, 2016. Available online: https://publications.jrc.ec.europa.eu/repository/bitstream/JRC104799/kj1a28407enn.pdf (accessed on 8 February 2021).
4. Fernandez Chozas, J.; MacGillivray, A.; Raventos, A.; Jeffrey, H.; Nielsen, K.; Aderibigbe, D. *International Levelised Cost of Energy for Ocean Energy Technologies*; Technical Report; Ocean Energy Systems (OES): Lisbon, Portugal, 2015.
5. Falcão, A. Wave energy utilization: A review of the technologies. *Renew. Sustain. Energy Rev.* **2010**, *14*, 899–918. [CrossRef]
6. Pecher, A.; Kofoed, J.P. *Handbook of Ocean Wave Energy*; Springer, Ocean Engineering & Oceanography: Cham, Switzerland, 2017.
7. Barbarit, A. A database of capture width ratio of wave energy converters. *Renew. Energy* **2015**, *80*, 610–628. [CrossRef]
8. Jenne, S.; Weber, J.; Thresher, R.; Bull, D.; Driscoll, F.; Dallman, A.; Newborn, D.; Quintero, M.; LaBonte, A.; Karwat, D. Methodology to Determining the ACE Metric Used in the Wave Energy Prize. In Proceedings of the 12th European Wave and Tidal Energy Conference, Cork, Ireland, 27 August–1 September 2017.
9. EIA. *Levelised Costs of New Generation Resources in the Annual Energy Outlook 2021*; Technical Report; U.S. Energy Information Administration: Washington, DC, USA, 2021. Available online: https://www.eia.gov/outlooks/aeo/pdf/electricity_generation.pdf (accessed on 8 February 2021).
10. Correia da Fonseca, F.; Amaral, L.; Armayor, A.; Cândido, J.; Arede, F.; Henderson, J.; Hudson, B.; Nava, V.; Tunga, I.; Petrov, A. System Lifetime Costs Tools–Alpha Version; Technical Report; H2020 DTOceanPlus Project Deliverable. 2019. Available online: https://ec.europa.eu/research/participants/documents/downloadPublic?documentIds=080166e5ca810ef1&appId=PPGMS (accessed on 8 February 2021).
11. Weber, J. WEC Technology Readiness and Performance Matrix – finding the best research technology development trajectory. In Proceedings of the 4th International Conference on Ocean Energy (ICOE), Dublin, Ireland, 17–19 October 2012.
12. European Commission. *Horizon 2020 Work Programme 2014–2015. 19. General Annexes Revised*; Technical Report; European Commission: Brussels, Belgium, 2014. Available online: http://ec.europa.eu/research/participants/portal/doc/call/h2020/common/1617621-part_19_general_annexes_v.2.0_en.pdf (accessed on 8 February 2021).
13. WindEurope. *Unleashing Europe's Offshore Wind Potential*; Technical Report; WindEurope: Brussels, Belgium, 2017.
14. SPC; GIZ. A Least Cost Analysis of Electricity Generation Options for Kiritimati Island (Draft). Available online: http://prdrse4all.spc.int/system/files/first_draft_kiritimati_lca_report.pdf (accessed on 8 February 2021).
15. Danish Energy Agency and Energinet. *Technology Data–Energy Plants for Electricity and District Heating Generation*; Technical Report; Danish Energy Agency and Energinet: Copenhagen, Denmark, 2016
16. 75 % Cost Savings Compared to Diesel Generator-Christmas Island, Australia. Available online: https://www.tesvolt.com/en/projects/75-cost-savings-compared-to-diesel-generator.html (accessed on 8 February 2021).
17. Ramos, V.; Giannini, G.; Calheiros-Cabral, T.; Rosa-Santos, P.; Taveira-Pinto, F. Legal framework of marine renewable energy: A review for the Atlantic region of Europe. *Renew. Sustain. Energy Rev.* **2021**, *137*, 110608. [CrossRef]
18. Short, W.; Packey, D.; Holt, T. *A Manual for the Economic Evaluation of Energy Efficiency and Renewable Energy Technologies*; Technical Report; National Renewable Energy Laboratory: Golden, CO, USA, 1995. Available online: https://digital.library.unt.edu/ark:/67531/metadc678552/m2/1/high_res_d/35391.pdf (accessed on 8 February 2021).
19. Paredes, M.; Padilla-Rivera, A.; Güereca, L. Life Cycle Assessment of Ocean Energy Technologies: A Systematic Review. *J. Mar. Sci. Eng.* **2019**, *7*, 322. [CrossRef]

20. Capacity Factors at Danish Offshore Wind Farms. Available online: https://energynumbers.info/capacity-factors-at-danish-offshore-wind-farms (accessed on 8 February 2021).
21. Carbon Trust. *Future Marine Energy, Results of the Marine Energy Challenge: Cost Competitiveness and Growth of Wave and Tidal Stream Energy*; Technical Report; Carbon Trust: London, UK, 2006.
22. Carbon Trust. *Accelerating Marine Energy: The Potential for Cost Reduction–Insights from the Carbon Trust Marine Energy Accelerator*; Technical Report; Carbon Trust: London, UK, 2011.
23. SI Ocean. Ocean Energy: Cost of Energy and Cost Reduction Opportunities; Technical Report; SI Ocean Project. 2013. Available online: https://oceanenergy-sweden.se/wp-content/uploads/2018/03/130501-si-ocean-cost-of-energy-report.pdf (accessed on 8 February 2021).
24. OPERA Project. *Tracking Metrics for Wave Energy Technology Performance*; Technical Report; Deliverable D7.3; University of Edinburgh for the OPERA Project: Edinburgh, UK, 2019.
25. OPERA Project. Final Assessment and Recommendations; Technical Report; Deliverable D7.5; Tecnalia for the OPERA Project. 2019. Available online: http://opera-h2020.eu/wp-content/uploads/2019/09/OPERA_D7.5-Final-assessment-and-recommendations_TECNALIA_20190724_v1.0.pdf (accessed on 8 February 2021).
26. Têtu, A.; Fernandez Chozas, J. Deliverable D8.1-Cost Database. Technical Report, LiftWEC–Development of a New Class of Wave Energy Converter Based on Hydrodynamic Lift Forces. 2020. Available online: https://liftwec.com/wp-content/uploads/2020/06/LW-D08-01-1x3-Cost-database.pdf (accessed on 8 February 2021).
27. Têtu, A.; Fernandez Chozas, J. Database of Costs for Wave Energy Projects. Available online: https://zenodo.org/record/4442080#.YQj69EARVhE (accessed on 8 February 2021).
28. Nielsen, K. *Point Absorber Feasibility and Development Requirements*; Technical Report; Danish Energy Agency: Copenhagen, Denmark, 2001.
29. Fernandez Chozas, J.; Kofoed, J.P.; Helstrup Jensen, N.E. *User Guide—The COE Calculation Tool for Wave Energy Converters*; Technical Report; DCE Technical Report No. 161; Aalborg University: Aalborg, Denmark, 2014.
30. Siegel, S. Final Scientific Report. In *Cycloidal Wave Energy Converter*; Technical Report; DE-EE0003635; Atargis Energy Corporation: Pueblo, CO, USA, 2012.
31. Nielsen, K.; Bingham, H.; Bjerg Thomsen, J. On the Absorption of Wave Power Using Ship Like Structures. In Proceedings of the Twenty-eighth International Ocean and Polar Engineering Conference, Sapporo, Japan, 10–15 June 2018.
32. Vikkelsø, A.; Larsen, J.H.M.; Sørensen, H.C. *The Middelgrunden Offshore Wind Farm*; Technical Report; Copenhagen Environment and Energy Office CEEO: Copenhagen, Denmark, 2003; ISBN 87-986690-3-6. Available online: https://base.socioeco.org/docs/a118_doc1.pdf (accessed on 8 February 2021).
33. Previsic, M. *System Level Design, Performance, and Costs of California Pelamis Wave Power Plant*; Technical Report; EPRI: Washington, DC, USA, 2004.
34. Nielsen, K. *Development of Recommended Practices for Testing and Evaluating Ocean Energy Systems*; Technical Report; OES (Ocean Energy Systems), Annex II: Lisbon, Portugal, 2003.
35. Meyer, N. *Rambøll: Bølgekraftprogram-Afsluttende Rapport fra Energistyrelsens Rådgivende Bølgekraftudvalg*; Technical Report; Rambøll: Copenhagen, Denmark, 2002.
36. Ricci, P.; Lopez, J.; Touzón, I.; Duperray, O.; Villate, J. A methodology for the global evaluation of wave energy array performance. In Proceedings of the Fourth International Conference on Ocean Energy, Dublin, Ireland, 17–19 October 2012.
37. Nielsen, K.; Friis-Madsen, E. *Deliverable 6 on the LCOE*; Technical Report; Danish Beton til Bølgekraft Project, b2b Project: Copenhagen, Denmark, 2020.
38. Harris, R.; Johanning, L.; Wolfram, J. Mooring systems for wave energy converters: A review of design issues and choices. *Proc. Inst. Mech. Eng. Part J. Eng. Manuf.* **2004**, *220*, 159–168.
39. Serrano González, J.; Burgos Payán, M.; Riquelme Santos, J.M. An improved evolutive algorithm for large offshore wind farm optimum turbines layout. In Proceedings of the 2011 IEEE Trondheim PowerTech, Trondheim, Norway, 19–23 June 2011.
40. O'Connor, M.; Lewis, T.; Dalton, G. Techno-economic performance of the Pelamis P1 and Wavestar at different ratings and various locations in Europe. *Renew. Energy* **2013**, *50*, 889–900. [CrossRef]
41. Bimep. *Best Practice for Developers When Preparing Sea Trials*; Technical Report; Bimep, Biscay Marine Energy Platform: Armintza, Spain, 2018.
42. Catapult, O. *Wind Farm Costs–Guide to an Offshore Wind Farm BVG Associates*; Technical Report; ORE Catapul: Scotland, UK, 2019. Available online: https://guidetoanoffshorewindfarm.com/wind-farm-costs (accessed on 14 May 2020).
43. Iglesias, G.; Astariz, S.; Vazquez, A. The Economics of Wave and Tidal Energy. In *Wave and Tidal Energy*; Greaves, D., Iglesias, G., Eds.; John Wiley & Sons, Ltd: Hoboken, NJ, USA, 2018; Chapter 11, pp. 513–532.
44. Beels, C.; Troch, P.; Kofoed, J.; Frigaard, P.; Vindahl Kringelum, J.; Carsten Kromann, P.; Heyman Donovan, M.; De Rouck, J.; De Backer, G. A methodology for production and cost assessment of a farm of wave energy converters. *Renew. Energy* **2011**, *36*, 3402–3416. [CrossRef]
45. Myhr, A.; Bjerkseter, C.; Ågotnes, A.; Nygaard, T.A. Levelised cost of energy for offshore floating wind turbines in a life cycle perspective. *Renew. Energy* **2014**, *66*, 714–728. [CrossRef]
46. Maslov, N.; Charpentier, J.F.; Claramunt, C. A modelling approach for a cost-based evaluation of the energy produced by a marine energy farm. *Int. J. Mar. Energy* **2015**, *9*, 1–19. [CrossRef]

47. Kaiser, M.J.; Snyder, B. Modeling the decommissioning cost of offshore wind development on the U.S. Outer Continental Shelf. *Mar. Policy* **2012**, *36*, 153–164. [CrossRef]
48. Carbon Trust. *Oscillating Water Column Wave Energy Converter Evaluation Report*; Technical Report; Carbon Trust: London, UK, 2005.
49. Nielsen, K.; Thomsen, J.B.; Ferri, F.; Friis-Madsen, E.; Rasmussen, K.D.; Thomas, S.; Kofoed, J. *Impact of Cost of Selected Mooring Solutions on CoE of Partner WECs*; Technical Report; DCE Contract Report No.197; Aalborg University: Aalborg, Denmark, 2018.
50. IRENA. *Renewable Power Generation Costs in 2017*; Technical Report; International Renewable Energy Agency: Abu Dhabi, UAE, 2017.
51. OES. *Cost of Ocean Energy: Energy Cost Analysis and Forecasts for Ocean Energy Converters*; Technical Report; Ocean Energy Systems (OES): Lisbon, Portugal, 2019.
52. A Horizon 2020 Project Developing a Lift-Based Wave Energy Converter. Available online: http://https://liftwec.com/ (accessed on 8 March 2021).
53. Hodges, J.; Henderson, J.; Ruedy, L.; Soede, M.; Weber, J.; Ruiz-Minguela, P.; Jeffrey, H.; Bannon, E.; Holland, M.; Maciver, R.; et al. *An International Evaluation and Guidance Framework for Ocean Energy Technology*; Technical Report; IEA_OES: Paris, France, 2021.
54. R. Wiser, J.S.; Paulos, B. *Fact Sheet: Cost Reductions for Offshore Wind*; Technical Report; Lawrence Berkeley National Laboratory: Berkeley, CA, USA, 2016. Available online: https://eta-publications.lbl.gov/sites/default/files/offshore-wind-fact-sheet.pdf (accessed on 3 May 2021).
55. Previsic, M. Cost Reduction Pathways for Wave Energy. *Ocean Energy Systems Annual Report 2012*; Ocean Energy Systems (OES): Lisboa, Portugal, 2012; pp. 108–115

energies

MDPI

Article

Reliability and Maintenance Management Analysis on OffShore Wind Turbines (OWTs)

Jose V. Taboada [1], Vicente Diaz-Casas [1] and Xi Yu [2,*]

[1] Integrated Group for Engineering Research, Centro de Innovacions Tecnolóxicas, Campus de Esteiro, Universidade da Coruña, 15403 Ferrol, Spain; jose.taboada1@udc.es (J.V.T.); vicente.diaz.casas@udc.es (V.D.-C.)

[2] FEMOTECH Ltd., 43 Berkeley Square, London W1J 5AP, UK

* Correspondence: femotech@126.com

Abstract: Due to the extreme marine operating environment, the remoteness from the maintenance base, and the expensive specialized accessibility and overhaul equipment needed (e.g., barges, boats, ships, and vessels), offshore O&M costs are greater than those for onshore-based installations. In the operation of wind farms, the main challenges are related to sudden and unexpected failures and downtimes. This paper has three main objectives. The first is to compare and optimize implementation techniques for maintenance strategies. The second is to analyze the cost-benefit of each maintenance strategy model. The third objective is to demonstrate the optimization and effectiveness of maintenance procedures and strategies recreated with stochastic and probabilistic life cycle cost (LCC) models, depending upon the degree of reliability and the maintenance process for offshore wind farms. The cost of operation and maintenance is directly dependent on failure rates, spare parts costs, and the time required by technicians to perform each task in the maintenance program. Calculations for each case study, with either light vessel/transfer boats (Alternative 1) or oilfield support vessels (Alternative 2), focused on the operational costs for transportation. In addition, each case study demonstrated which maintenance conditions and strategies are operational and optimal, and their corresponding cost–risk impacts. Results from this paper suggest that O&M costs are highly correlated with maintenance round frequency (offshore trips) and the operating costs for transportation by light vessel/transfer boat (CTV) and oil-field support vessel (FSV). The paper analyzes cumulative lifecycle costs and finds that for long-term life cycles (25 years), the implement of light vessels (Alternative 1) is more suitable and cost-effective. In contrast, oilfield support vessels (Alternative 2) are more expensive to operate, but they guarantee major capabilities, as well as the advantage of achieving the access levels need to efficiently operate. According to the results obtained by the outcome analysis, it can be concluded that the implementation of light vessels (Alternative 1) shows a lower overall LCC (<million $), which is mainly due to the fact that corrective maintenance and minor repairs are less costly. It should be noted that the cost of major repair operations with light vessels in Alternative 1 is still less than the high costs for minor repairs in Alternative 2 (with FSV).

Keywords: accessibility; O&M plan; O&M strategies/alternatives; condition-monitoring; LCC; condition base monitoring (CBM); integrity maintenance reliability (IMR); levelized cost of energy (LCOE); cumulative lifecycle costs; risk-based inspection and maintenance analysis (RIMAP)

Citation: V. Taboada, J.; Diaz-Casas, V.; Yu, X. Reliability and Maintenance Management Analysis on OffShore Wind Turbines (OWTs). *Energies* **2021**, *14*, 7662. https://doi.org/10.3390/en14227662

Academic Editors: Davide Astolfi, Eugen Rusu, Kostas Belibassakis and George Lavidas

Received: 29 July 2021
Accepted: 11 November 2021
Published: 16 November 2021

Publisher's Note: MDPI stays neutral with regard to jurisdictional claims in published maps and institutional affiliations.

Copyright: © 2021 by the authors. Licensee MDPI, Basel, Switzerland. This article is an open access article distributed under the terms and conditions of the Creative Commons Attribution (CC BY) license (https://creativecommons.org/licenses/by/4.0/).

1. Introduction

The continued growth in the size and complexity of offshore wind turbines means that more profitable O&M actions will be needed to optimize the upper ranges of robustness for RAMS, in order to fulfill the size increase [1].

Previous research has indicated that O&M constitutes up to 20–30% of the overall cost of OWTs during their lifetimes. However, lowering the O&M cost per unit power will rely on larger OWTs, due to the greater cost per failure of smaller OWTs, their high demand for palliative actions (e.g., corrective maintenance), and their loss of production during downtimes [1]. Therefore, increasing turbine size implies decreasing O&M costs. Larger OWTs provide a lower number of individual machines that need to be conserved and could therefore provide lower O&M costs [2]. The design and modeling of O&M costs is essential to the screening of cost-effective maintenance strategies and decision-making, as well as the development of specific methodologies for O&M. In addition, design and modeling increase trust for wind energy investors financing OWTs. Therefore, this analysis is a significant step for the growth of wind power [3]. The O&M costs quantified and measured in this paper are the cost for personnel, spare parts, and vessels required for the accomplishment of maintenance requirements of the wind farms. Normally, maintenance is understood as a general concept that includes all interventions (inspections, repairs, replacement of components/elements, etc.). The analysis of current and previous O&M strategies for OWTs takes into account industrial achievements made in the oil and gas industry and the manufacturing industry in order to identify the most important functional drivers for O&M planning, and management for OWTs. Thus, previous trials and achievements in other industries act as an input driver for O&M in the offshore wind industry.

To gain insight into current advances in O&M knowledgebase standardization, offshore wind farm models are based on today's state-of-the-art OWTs, approximately 25 years after the first generation of conventional OWTs was designed, manufactured, and installed.

On the other hand, the use of larger wind turbines generates much greater uncertainty. Operation and maintenance costs represent a large part of the total life cycle cost (LCC), with operation and maintenance costs being approximately 22 to 40% of the overall total cost of an offshore wind farm [4,5]. Those costs are related to the risk cost incurred by the profit lost due to downtimes of OWTs.

O&M activities account for around $\frac{1}{4}$ of the life-time costs of a regular offshore wind farm. Over the next twenty years, offshore wind O&M will turn into a significant industrial sector in its own right [3]. For instance, in the UK government's forecasts for the deployment of offshore wind, O&M activities for more than 5500 OWT's could be worth almost £2bn/year by 2025. The graphs are shown below in Figure 1.

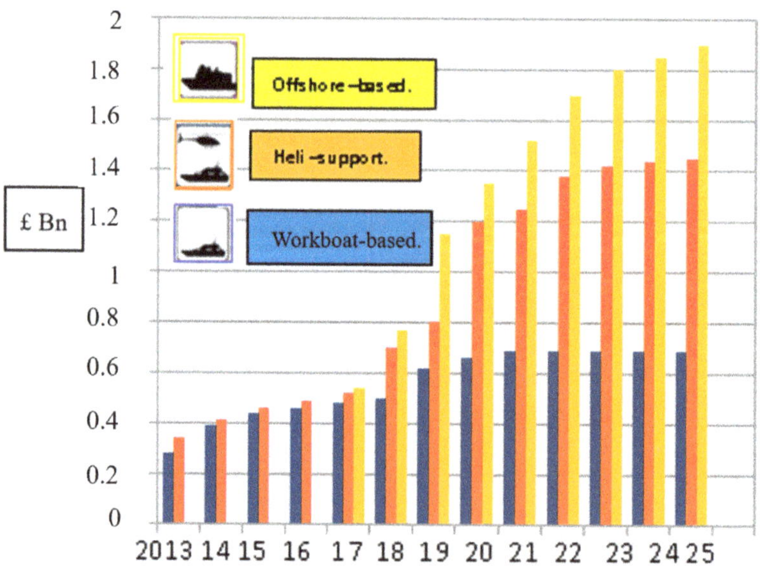

Figure 1. UK O&M spending by strategy class during years 2013–2025, data from [6].

Figure 2 represents a simple understanding of O&M research results for common offshore wind projects at different distances from the nearby O&M harbor. From the analysis, the junction points are at around 12 nautical miles (NM) (to have helicopter support) and at 40 NM (to trigger offshore wind-based strategies). However, it is vital to remember that there also many site-specific external aspects (environmental conditions, aviation regulations, safety considerations, and suitability) of existing ports that affect decisions about the exact positions of these junction points [3].

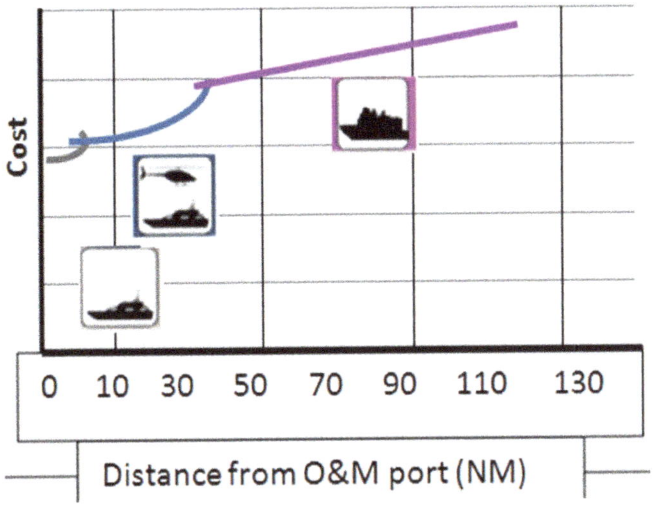

Figure 2. O&M cost as a function of distance from the O&M port, data from [5].

On the other hand, the prominence and challenges of O&M for OWTs are recognized in both academia and industry. The availability of OWTs is much less favorable and their costs can be more than 1.5 times higher than onshore wind. Furthermore, onshore wind turbines are capable of achieving 95–99% availability and producing electricity at a reasonable price in the market. There is clear cost reduction potential for O&M, which contributes around 30% of the total cost of offshore wind.

The emphasis of this document is to research and develop methods to improve and optimize the efficiency of operation and maintenance in offshore wind farms. Efficiency is related to the optimization of maintenance organization in offshore wind farms. The decrease of O&M costs is directly addressed in this document and the research results are supportive.

The research presented throughout this document analyzes the existing approaches and methods used for access, design, operation, maintenance planning, and life cycle engineering in offshore wind farms.

1.1. Challenges and Solutions for OWT Maintenance Activities

1.1.1. Weather Conditions

The meteorological window is represented in the model by a time series accounting for significant wave height and wind speed when determining the hourly time. The weather forecast notes when a given set of offshore or marine activities (operations, construction, etc.) can be carried out within their maximum limits for wave height, wind speeds, etc. Specifically, marine operations are planned based on a reference period; the operation reference period is (TR) = planned operation period (TPOP) + estimated maximum contingency time (TC) [6]. Incorporating wave height and wind speed into a weather window is crucial to ensuring the accessibility of offshore wind farms. For operations to be considered not limited by meteorological factors, it is necessary that the planned operating time (TPOP) be less than 72 h and the reference period (TR) be less than 96 h.

The meteorological time series are created using a Markov chain model based on historical meteorological record input data from the specific site of an offshore wind farm. The Markov chain model reproduces and recreates random time based on models and estimated stochastic probabilities [4].

Failure occurrence can fit an exponential probability distribution dependent on failure rates. Given the failure rates λ (e.g., $\lambda^{year\ 04} = 1$, $\lambda^{year\ 510} = 0.75$, etc.) for a component/element in an OWT, the distribution probability function for the time duration Δt until a failure happens on that explicit component/element, is set as:

$$p\ (\Delta t) = \lambda e^{\lambda \Delta t} \tag{1}$$

where (Δt) is the time interval until the next fault. Two cases are defined:

- At the beginning of the simulation, the OWT components at the time that the "first failure" occurs are extracted independently of the exponential probability distribution, considering the relevant failure rate as an input parameter;
- After a corrective maintenance process, when the next failure occurs in the maintained element, the distribution is extracted based on the failure rate relevant for this task and the current time. Therefore, feedback is provided for a corrective maintenance entry.

The maintenance model, therefore, is able to repeat simulations. Each one takes weather scenarios as diverse and random, and uses arbitrary times for failures to account for doubt in the times for failure rates and weather effects [7].

1.1.2. Weather Delays and Repair Timing

The total downtime per failure is the sum of the downtime originating because of:

- Waiting for appropriate weather window conditions;
- Queuing resulting from a lack of maintenance technicians;
- Repairs in the OWTs.

Safety weather window and work shift constraints create expected maintenance delays, which are statistically determined for the given time duration (r_m & r_M) based on the environmental time series sum for the offshore location, with the vessels considered limited by wave height and wind speed [7].

Downtime repair comprising of waiting for weather (without the effect of queuing) is referred to as d_m^s (minor repairs) and d_M^s (major repairs). The average failure rate (λ^S) and repair time (d_{CM}^s) per failure and per season is calculated as:

$$\lambda^S = \lambda_m^S + \lambda_M^S \tag{2}$$

$$d_{CM}^S = \frac{\lambda_m^S d_m^S + \lambda_M^S d_M^S}{\lambda^S} = \frac{1}{\mu^S} \tag{3}$$

where μ^S is the resulting repair rate [8].

1.1.3. Accessibility

As stated above, both wave height and wind speed are essential to guaranteeing the safety and accessibility of an offshore wind farm. Accessibility itself is particularly essential for offshore wind power systems, to guarantee reduction of the great financial risks due to doubts to the accessibility and reliability of OWT [9].

Maintenance technicians' transportation to the OWTs shall be carried out by workboats, which are limited by wave height [8].

1.1.4. Operation and Maintenance Plan

Maintenance planning is the prioritization of maintenance tasks ahead of available resources (for example, personnel, maintenance equipment, and spare parts). Maintenance planning involves all maintenance tasks, and the optimization process can achieve great savings. Mainly, cost savings are correlated with current assets (fuel, mobilization costs, production losses, and logistics costs) [9].

Managing operation and maintenance activities to reduce OPEX (operating expenses) costs is one of the most decisive challenges of offshore wind farms, due to the distribution of maintenance varying with time depending on the performance of OWTs and their sub-assemblies, as well as the weather window. Thus, to determine operation and maintenance activities, project managers need to have a clear understanding of sub-assembly history, background, performance, and weather [9,10]

Maintenance program activity triggers are usually failures of a component/element or a time interval based on operational service principles.

1.1.5. Objectives

The O&M programs and models rely on condition monitoring (CM)-based technologies such as dynamic load characteristics, oil analysis, strain measurements, physical condition of the materials, acoustic monitoring, performance monitoring, etc., which are helpful for monitoring wind turbines. The primary research goal is oriented around the condition monitoring of wind turbines and CM data is used to decide on maintenance planning and strategies/alternatives to be implemented, as well as to define deterioration models and develop mathematical models. The second objective is operational and maintenance (O&M) cost reduction coupled with less downtime. Due to offshore wind farm locations being much further from shore, new challenges will emerge which may interfere with reducing O&M costs.

The third objective shall be to overcome such challenges to minimize O&M expenditure.

1.1.6. Condition Assessment and Condition Indicators

The degradation speed curve of the technical condition of a component/element is an on-going process from an "as new" condition until failure happens, as illustrated in Figure 3.

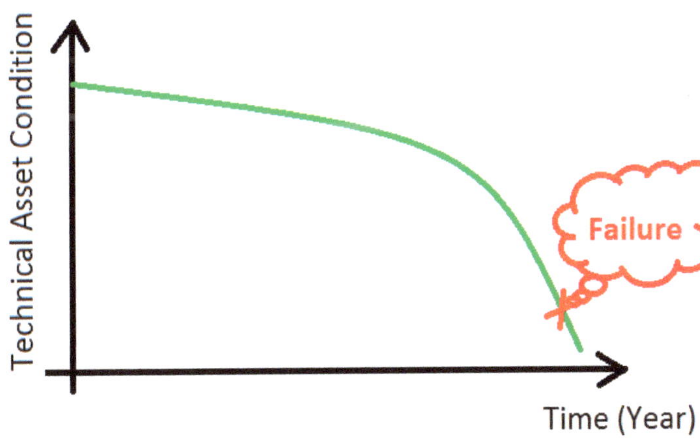

Figure 3. Generic continuous degradation curve of a component.

Very few, if any, condition monitoring methods give a direct and accurate description of the actual technical condition of the component/element. Methods used for the condition monitoring normally result in an indication of the technical condition of the components/elements. Energy companies have also carried out condition monitoring using visual inspections. As we know, visual inspections have higher uncertainty when giving precise knowledge of the point in time and momentum in space on an on-going deterioration curve. Besides O&M activities by power companies, visual inspection in modern industrial manufacturing plants has applied condition monitoring, based upon specific software solutions installed in each piece of equipment (for their respective production machines), which incorporate a tracking system for their technical condition.

Above all, O&M demands four key principles [11]:

- Maximize the level of turbine availability;
- Enable regular service and quick troubleshooting intervention;
- Enable component change, ensuring compatibility with the component exchanged;
- Ensure the cost effectiveness of the O&M concept.

Most publications have focused on quantifying the limitations of the three key O&M variables [10]:

I. The distance of the service station;
   - Service personnel stationed at an onshore site to service offshore platforms;

II. Logistics to and from the offshore site;
   - Service needs (e.g., vessels and helicopters);

III. The availability of cranes or jack-ups;
   - Adequate safe access to vessels for operational needs (e.g., replacing or transferring large components) [10].

1.2. Scope Work

This document reviews O&M management research on OWT operations and maintenance, including strategies, critical challenges and proposed solutions, on-site operations, and endpoints. Capable solutions are recognized with regard to the future development of O&M strategies. In addition, the negative effects of weather conditions, weather delays, repair times, and accessibility on offshore maintenance are presented. This analytical review presents a comprehensive overview of the OWT maintenance literature and provides a basis for improving O&M strategies and alternatives (1 vs. 2) in the future for offshore wind power installation facilities. To solve the information gaps, the comparison of scientific publications, technical reports and projects, and open databases has been used. The analysis is organized as follows. In Section 2, the research methodology, vessel data, personnel data, maintenance data, and online health monitoring are introduced and discussed, as well as the case studies (O&M Strategy 1 and O&M Strategy 2). Based on the designated maintenance methodology adopted, optimal maintenance direction-finding and scheduling are analyzed in Section 3. Several characteristics of the associated cost optimization problem are analyzed, including their advances, challenges, and targets. O&M strategies and alternatives, namely, O&M Alternative/Strategy 1 and O&M Alternative/Strategy 2 and their respective assumptions, are highlighted. A life cycle cost (LCC) analysis is conducted to evaluate both O&M alternatives/strategies (1 vs. 2) and determine which one is better. In Section 4, conclusions are drawn and discussed regarding operational and maintenance related issues from the outcomes obtained from the O&M alternatives-strategies analyzed, such as that a long-term life cycle (25 years) is more suitable for implementing Alternative 1, as it is more cost-effective. In contrast, it is more suitable to switch to Alternative 2 in order to guarantee major capabilities, as well as to have the advantage of achieving the access levels need to efficiently operate.

2. Methodology for Detailed Maintenance—Parameters Analyzed

2.1. Vessel Data

The O&M tasks to be carried out involve a fleet of diverse vessels. A standardized vessel consists of a vessel with a pre-established access system; therefore, maintenance technicians can easily access the OWTs. Some boats have additional capabilities (for example, cranes for lifting elements). There are questions about "high climate dependency" to access the OWT and "specific functional ship climate requirements" for the operational restrictions, in terms of the maximum possible, to access an OWT. Offshore vessels will not be able to participate in the maintenance tasks if the height of the waves or the speed of the wind exceeds their own meteorological limits, so they are not capable as such.

2.2. Personnel Data

The associated parameters associated with maintenance personnel are based on the availability of human resources. These resources focus on the number of maintenance technicians at different offshore site locations, as well as their own scheduled work shifts. Maintenance technicians are stationed at land or marine bases, while ships remain at sea for several days. Motherships have their own staff dedicated to the maintenance crew, who can operate the ships in their entirety for maintenance work purposes. The scheduled work preparation time for maintenance personnel is preset and identified by the scheduled work time (per shift combined + the n° of shifts/day).

2.3. Maintenance Data

On each component/element for the OWTs, the maintenance model relays one or more maintenance activities and rounds. To accomplish each maintenance activity, the model takes into account three kinds of assets:

- Vessel(s);
- Consumables and spare parts required;
- Maintenance work force.

Spare parts and consumables are included in the model by assigning them a delivery time and a cost linked to it. It is also necessary to detach all the maintenance tasks involving traveling to the offshore wind farm, which requires an offshore vessel to transport the maintenance technicians. However, some maintenance tasks require specific capabilities, such as high load capacity. For these maintenance tasks, vessels with additional capacity are required, which creates an additional cost for the LCC chain. All these factors have to be considered within the developed model [5].

2.4. Online Health Monitoring

OWTs demand appropriate online monitoring, in order to measure the industrial assets in real-time. Therefore, online data management for maintaining OWT is needed, to be exported to monitoring systems (i.e., SCADA, CMS, etc.). This online data measures reliability, availability, and maintenance from the control monitoring room of the OWTs' OEM.

As we can see from the figure above, online asset management data gives robust health monitoring, allowing continuous monitoring of OWTs as well as ensuring that the OEM controls and operates in a cost-efficient and reliable manner, in order to guarantee the lowest LCC of the OWTs.

Description of the Case Studies.

We assume two different maintenance contracts, both lasting 20 years, including transport systems. Each contract carries out a hypothetical O&M strategy.

O&M Strategy 1:

The transport of maintenance crews offshore uses a light vessel (CTV) without access systems (MCA class 2), with 20 knots of cruising speed, a catamaran hull design, 12 personnel and 2 crews needed to operate, and suitable for 10–20 km offshore travels. This vessel has a limitation of 1.5 m in significant wave height, since availability cannot be over 98%.

O&M Strategy 2:

The transport of maintenance crews offshore uses an oilfield support vessel (FSV) with 12 knots of cruising speed and 18–68 personnel and crews needed to operate, suitable for long stays offshore up to 5–7 weeks. FSVs have dynamic positioning and access systems suitable for transferring heavier equipment to the OWT, so they can do heavier repair operations than Alternative 1. This vessel has a limitation of 4 m in significant wave height, so it is suitable for year-round maintenance. Hourly operation costs can be summarized as follows [10].

3. Analysis Review

3.1. Life Cycle Cost Analysis

Life cycle costs regarding O&M activities related to a general configuration can be calculated considering the following terms:

LCC = Capital Costs (C_{cap}) + Operating Costs (C_{op}) + Cost of Deferred Production (C_{pr}).

These terms have to be calculated yearly and corrected with a discount rate that accounts for inflation, interest rate, and investor risk, as is usual in economic analyses. A more general approach can be formulated as:

$$\mathrm{LCC} = \sum_{\mathrm{i}}^{\mathrm{N}} \frac{\left(C_{cap} + C_{op} + C_{pr}\right)_{\mathrm{i}}}{(1+r)^{\mathrm{i}}} \quad (4)$$

where "N" is the life of the project in years (20).

This equation also complies with NORSOK O-CR-001 (for systems and equipment) and O-CR-002 (for production facilities). However, since this is an example comparing two different strategies for O&M in offshore wind power, not for equipment or production facilities, an optimum alternative solution will be used.

Now we compute the LCC for two alternatives, that is, for two different O&M strategies and two different transport concepts for maintenance crews:

Alternatives:

The two different maintenance contracts each last 25 years (the minimum life cycle of the OWT). Both alternatives are for an offshore wind location at a distance to the shore of 20 km (10,7238 NM) from where the wind farm is placed (i.e., WindFloat). Each O&M strategy will include different transport systems [11–16]:

- ○ O&M Strategy 1 (Alternative 1): Using a light vessel (CTV) without access systems. Parameters of Alternative 1 are summarized in Table 1.

Table 1. Input data for Alternative 1 using a light vessel (CTV) without access systems.

Power of WT	8.4	MW
Number of OWT	3	
Distance from shore	20	km
Water depth	85–100	m
N0 of Trips/round	30	Trips/Round
Cost per maintenance trip	2218.5	$/TRIP
Failure rate (minor repairs) %	75.00%	
Failure rate (major repairs) %	25.00%	
Cost of man-hours offshore	93.07	$/h
Number of crew members	4	people
Cost of electricity	50	$/Mwh
Offshore trips for minor repair	1	trips
Hours/WT preventive maintenance	36	h
Hours/WT corrective maintenance (minor repairs)	36	h

- ○ O&M Strategy 2 (Alternative 2) Using an oilfield support vessel (FSV). Parameters of Alternative 2 are summarized in Table 2.

Table 2. O&M Strategy 2 (Alternative 2) using an oilfield support vessel (FSV).

Power of WT	8.4	MW
Number of OWT	3	
Distance from shore	20	km
Water depth	85–100	m
N^0 of trips/round	1	Trips/Round
Cost per maintenance trip (12 days)	216,500.1	$/trip
Cost per corrective maintenance trip (per days)	21,650.01	$
Failure rate (minor repairs) %	75.00%	
Failure rate (major repairs) %	25.00%	
Number of crew members (major repairs)	8	people
Number of crew members (minor repairs)	4	people
Cost of man-hours offshore	501.07	$/h
Number of crew members (preventive)	28	people
Cost of electricity	50	$/Mwh
Offshore days for minor repair	3	days
Hours/WT preventive maintenance	32	h
Hours/WT corrective maintenance (minor repairs)	36	h
Offshore days for major repair	3	days
Hours/WT corrective maintenance (major repairs) (2 shifts 12 h x 2)	72	h

3.2. Assumptions

The upcoming analysis requires a list of assumptions. The two different strategies will be compared based on the following assumptions. The preventive maintenance program shall be done every 3500 h (2 times/year), taking 2–3 days/WT per year. In this case study, we have N = 20 years of duration of the transport contract and, since this transport alternative is externally hired, capital costs are 0, so:

$$\text{LCC} = \sum_{\text{i}=1}^{20} \frac{(C_{op} + C_{pr})_{\text{i}}}{(1+r)^{\text{i}}} \tag{5}$$

Operating costs will be divided between preventive and corrective, since both are mandatory and the LCC of each needs different treatment:

$$\text{LCC} = \sum_{\text{i}=1}^{20} \frac{\left(C_{op}^{prev} + C_{op}^{corr} C_{pr}\right)_{\text{i}}}{(1+r)^{\text{i}}} \tag{6}$$

Relevant operating costs for comparing both alternatives are due to transportation strategies (including energy/fuel consumption) and man-labor hours. Spare parts, insurance, and other operating costs are considered constant for both alternatives.

We will consider a failure rate that changes with time to be more realistic, since WTs are more likely to fail the older that they get, following the bathtub curve approach:

$\lambda^{year\ 04} = 1$; $\lambda^{year\ 510} = 0.75$; $\lambda^{year\ 1116} = 0.5$; $\lambda^{year\ 1720} = 0.75$
$\lambda^{year\ 2125} = 1$ in failures/year.

This equation comprises minor and major failures (needing minor and major repair). In this context, we define failure as an event that prevents the WT from producing energy at all.

Preventive maintenance will be based on planned maintenance rounds, which are also assumed to change with time, and according to the feedback from each settled and applied maintenance program, in order to better optimize O&M strategies with the failure rates:

- From year 1–8: 2 Maintenance rounds/year.
- From year 9–20: 4 Maintenance rounds/year.

The cost of man-labor in offshore conditions is considered to be 250 $/h.

During corrective maintenance, minor failures on each WT will take 1 day to repair (9 h of offshore labor by 1–4-man crews); major failures will take 3 days offshore with accommodation, in 4 shifts (4 × 4-man crews, 8 h each) [12].

3.3. Operational Cost Results

The hourly costs (costs of operation) of each alternative are shown in the Tables 3 and 4 below:

Table 3. Hourly cost for Alternative 1.

Transport Alternative 1: CTV MCA CLASS C HIRE AND FUEL			
	Fuel costs $/ton	381	$/ton
	12 h trip rental	1905	$
	Fuel costs 2 × 20 km trip (2 × 10.71 NM; 0.12 tons of fuel used)	45.72	$
	Fuel costs on location 8 h, no heavy seas and light sailing (0.4 MT fuel used)	152.4	$
	TOTAL CTV INCLUDING FUEL	2103.12	$/trip
WORKING HOURS PER WT	3 × 4-man teams working 9 h	36	h/WT

Alternative 1 cost of transportation per hour of O&M work is equal to 58.42 $/h.

Table 4. Hourly cost for Alternative 2.

Alternative Transport Alternative 2: FSV HIRE AND FUEL	DAYS	12	
	Fuel costs $/ton	381	$/ton
	12 day trip rental	152,400	$
	Fuel costs 2 × 20 km trip (2 × 10.71 NM at 12 knots, 29 tons/24 h fuel used)	952.5	$
	Fuel consumption at port (1 day for picking and leaving shifts)	571.5	$
	Fuel costs on location, no heavy seas and light sailing (6.5 MT/24 h fuel used) during 12 days	29,718	$
	TOTAL FSV INCLUDING FUEL	183,642	$/trip
WORKING HOURS PER WT	4 × 7-man maintenance team working 8 h in 3 shifts (4× morning, 4× afternoon, and 3× night) working 12 days	32	h/WT

Alternative 2 cost of transportation per hour of O&M work is equal to 5738.8 $/h.

Both case studies (Alternatives 1 and 2), are calculated using the same distance to the shore, at 20 km, from where the wind farm (i.e., WindFloat) is placed.

3.4. Cost of Deferred Production

According to NORSOK O-CR-001 and O-CR-002, the costs of deferred production can be calculated, in general form, as:

$$C_{pr} = \lambda * p * D * L \tag{7}$$

where λ is the failure rate per year (which is assumed to be varying with time, as stated above), p is the probability of interrupted production reduction, D is the duration of production reduction (downtime), and L is the production loss per time unit.

- λ is assumed to be: $\lambda^{year\ 04} = 1$; $\lambda^{year\ 58} = 0.75$; $\lambda^{year\ 912} = 0.5$; $\lambda^{year\ 1316} = 0.75$; $\lambda^{year\ 1720} = 1$.
- P is taken as 0.01, so a 1 $\times$ 100% train configuration is assumed.
- L is taken as 8.4 MWh, which is the power of a WT wind farm (for example, WindFloat) every hour, so all production is assumed to stop at every failure. The price of electricity is taken as 50 $/MWh [13].

The downtime (D) is the main difference between the two alternatives. Alternative 2 can have a much higher availability and lower downtime. For this, we follow some of the concepts and procedures indicated by [11].

In general, the failure rate during a season (year) can be divided into failure needing major repair (change of rotor blades) and minor repair (change of lubricating boxes):

$$\lambda^s = \lambda^s_m + \lambda^S_M = \frac{1}{MTBF} \tag{8}$$

We will assume $\lambda = \lambda_m + \lambda_M = 0.75\lambda + 0.25\lambda$ failures/year, so 75% of failures are solved with minor repair operations, while 25% need major repair. When considering both major and minor repairs, the repair time per failure MTTR can be calculated as (this downtime includes waiting for the weather window, but does not include queuing, when maintenance crews are not available to repair the failures, or logistics, such as waiting time for spares; these are supposed to be constant in both alternatives):

$$d^s_{CM} = \frac{\lambda^S_m * d^s_m + \lambda^s_M * d^s_M}{\lambda^S} = \frac{1}{\mu^S} = MTTR \tag{9}$$

Where d^s_m is the mean downtime due to failure needing minor repairs, d^s_M is the mean downtime due to failures needing major repairs, and μ^S is the average repair rate.

For Alternative 1, we will assume that d^s_m is around 3 days/turbine and d^s_M is large, in the order of 20 days/turbine, since no major repairs can be done with these vessels. Notice that in this case, we would need another vessel for that purpose (major repairs), which is outside of the scopes of the contract. So, considering the time varying failure rate per year:

$$d^{alt1}_{CM} = \frac{0.75 * 3 + 0.25 * 20}{1} = 7.25\frac{days}{failure} = \frac{1}{\mu^{alt1}} \tag{10}$$

For Alternative 2, we will assume that d^s_m is around 1.5 days/turbine, since 24 h shifts can be considered, and d^s_M is in the order of 10 days/turbine, since major repairs can be done with the FSV vessel.

$$d^{alt2}_{CM} = \frac{0.75 * 1.5 + 0.25 * 10}{1} = 3.625\frac{days}{failure} = \frac{1}{\mu^{alt2}} \tag{11}$$

With these assumptions, we can finally obtain an estimate for the costs of deferred production. A more detailed calculation on downtimes, including queuing issues, is discussed in [10], by means of Markov chain models.

The expressive summary for the whole life cycle of the project, comparing the given O&M options, is showed in Table 5 and Figure 4:

Table 5. Comparison between Alternatives 1 and 2.

	Corrective					
	Minor Repairs			**Major Repairs**		
	Transport	Man-labor	Total	Transport	Man-labor	Total
1	$51.14477	$77.24208	$128.38685	$1.99645656	$998.05372	$2.99451028
2	$499.11414	$415.85572	$914.96986	$1.66371380	$831.71143	$2.49542523
	Overall Life Cycle Costs (Discounted)					
1	$13.44641325					
2	$24.03934295					
	Preventive					
	Transport		**Man-labor**			**Total**
1	$174.25252		$1.32804120			$1.1537868
2	$833.90240		$4.58711952			$3.75321712
	Deferred Production Costs					
1	$93.59827					
2	$46.79913					

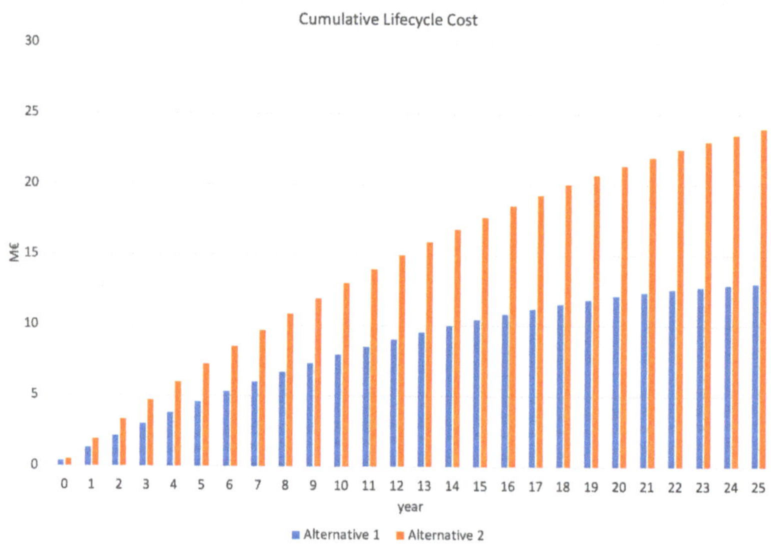

Figure 4. Cumulative lifecycle costs.

Alternative 1 shows lower overall LCC (less than a million USD); this is mainly because corrective maintenance due to minor repairs is less costly due to the characteristics of the chosen transportation (CTV). The penalization in the costs of major repair operations (120%) is not enough to compensate for the high costs for minor repair of Alternative 2 (FSV).

Deferred production costs are not high enough to be decisive in the selection between alternatives. If this were an oil and gas project, this may have been different.

This way of obtaining LCC leads us to average values. In order to assess the variability of these assumptions and costs, a Monte Carlo simulation can be carried out on the decisive parameters (cost of man-labor, cost of fuel, costs due to major repairs, downtimes, failure rates), assuming a variance of those, with a certain distribution (usually a triangular one, with the mode at the center). After this simulation, we can obtain an estimate of the uncertainty and sensitivity of some assumptions, such as quantities for the obtained LCC or the probability that these are in a certain range, confidence intervals, or any other quantification of uncertainty. This though, is beyond the scope of this article.

4. Conclusions

The potential impact from maintenance at the operating and logistical level (flexibility, throughput time, quality management, etc.) is considerable, and, therefore, the financial impact of maintenance can be substantial.

This work analyzes decreasing the O&M cost depending upon failure rates, downtimes, the timing needed for each maintenance schedule work activity, and the associated spare part costs.

The O&M cost results proved a great variability in cost of transportation between each alternative. In Alternative 1, the cost of transportation per hour of O&M work is 58.42 $/h, but for Alternative 2, it goes up to 5738.8 $/h. In summary, the total O&M cost of transportation per hour of O&M work differs from Alternative 1 to 2 by 5680.38 $/h, showing that a reachable decrease in O&M cost is highly dependent upon the technical assumptions set into the initial alternative/strategy and on the development of O&M requirement values (parameters and variables), which are key to recreating and covering the full spectrum of each case study.

Availability rises with a higher degree of accessibility and faster transportation times from support organizations. In contrast, the availability itself depends upon the O&M principles (effective working hours scheduled and number of technicians) set in each O&M strategy (Alternative 1 vs. Alternative 2).

In addition, as the cumulative lifecycle cost proves, for almost half of the life cycle (25 years), the costs-discounted are higher for Alternative 2 (using FSV) than for Alternative 1. Therefore, the long-term life cycle (25 years) is more suitable for implementing Alternative 1, as it is more cost-effective. In contrast, it is more suitable to switch to Alternative 2 in order to guarantee major capabilities, as well as the advantage of achieving the access levels needed to efficiently operate.

Increasing the size of OWTs demands a higher robustness of the O&M implementation, in comparison with traditional and conventional offshore wind farms.

Finally, the optimal O&M strategy maximizes availability at the lowest cost by ensuring safety and the best access to offshore wind farms, minimizing unscheduled maintenance activities, and carrying out scheduled maintenance tasks as efficiently as possible, ultimately resulting in the lowest possible LCOE.

Author Contributions: Writing—original draft preparation, J.V.T.; reviewing and editing, V.D.-C.; supervision, X.Y. All authors have read and agreed to the published version of the manuscript.

Funding: This paper has been partially funded by the Xunta de Galicia ED431C 2021/39 program, and for the open access charge, the Universidade da Coruña/CIUG.

Institutional Review Board Statement: Not applicable.

Informed Consent Statement: Not applicable.

Conflicts of Interest: The authors declare no conflict of interest.

References

1. Questa, J.E.C.; Vanhonacker, T.; Karyotakis, A.; Lang, Z.-Q.; de la Cuadra, C.R.; Oseas, I.E.; Marquez, F.P.G.; Lee, D.; Hillmansen, S.; Hajiabady, S. Optimisation of operational reliability of large-scale industrial wind turbines. In Proceedings of the 1st International Conference on Renewable Energies Offshore, Lisbon, Portugal, 24–26 November 2014.
2. Hofmann, M.; Sperstad, I.B. Will 10 MW wind turbines bring down the operation and maintenance cost of offshore wind farms? In Proceedings of the 11th Deep Sea Offshore Wind R & D Conference (EERA DeepWind'2014), Trondheim, Norway, 22–24 January 2014.

3. Phillips, J.; Fitch-Roy, O.; Reynolds, P.; Gardner, P. A guide to UK offshore wind operations and maintenance. *Scottish Enterprise and the Crown Estate.* 2013. Available online: http://csmres.co.uk/cs.public.upd/article-downloads/Offshore-wind-guide-June-2013-updated.pdf (accessed on 13 November 2021).
4. Westwood, D. *Offshore Wind Assessment for Norway*; The Research Council of Norway: Oslo, Norway, 2010.
5. Krohn, S.; Morthorst, P.E.; Awerbuch, S.; Blanco, M.I.; van Hulle, F.; Kjaer, C. (Eds.) *The Economics of Wind Energy*; EWEA: Brussels, Belgium, 2009.
6. *DNV-OS-H101: Marine Operations, General*, 31 October 2011. Available online: https://nanopdf.com/download/dnv-os-h101-marine-operations-general_pdf (accessed on 13 November 2021).
7. Kerres, B.; Fischer, K.; Madlener, R. Economic Evalutaion of Maintenance Strategies for Wind Turbines: A Stochastic Analysis. FCN Working Paper No. 3/2014. *SSRN Electron. J.* **2015**, *9*, 766–774.
8. Hagen, B.; Simonsen, I.; Hofmann, M.; Muskulus, M. A multivariate Markov Weather Model for O & M Simulation of Offshore Wind Parks. In Proceedings of the DeepWind'2013—10th Deep Sea Offshore Wind R & D Conference, Trondheim, Norway, 24–25 January 2013.
9. Hagen, B.A.L. Sensitivity Analysis of O & M Costs for Offshore Wind Farms. Master's Thesis, Norwegian University of Science and Technology, Trondheim, Norway, June 2013.
10. Peter, T. *Offshore Wind Turbines: Reliability, Availability and Maintenance*; The Institution of Engineering and Technology: London, UK, 2012.
11. Besnard, F.; Fischer, K.; Tjernberg, L.B. A Model for the Optimization of the Maintenance Support Organization for Offshore Wind Farms. *IEEE Trans. Sustain. Energy* **2013**, *4*, 443–450. [CrossRef]
12. Besnard, F. On Maintenance Optimization for Offshore Wind Farms. Ph.D. Thesis, Chalmers University of Technology, Gothenburg, Sweden, 2013.
13. Diz-Lois, G.; Poungkrajorn, T. Design for operation and maintenance of offshore wind farms. In *Technical Report at the OFF 510 (Operation and Maintenance Management)*; University of Stavanger (UiS): Stavanger, Norway, 2014.
14. Tranfield, D.; Denyer, D.; Smart, P. The modern approach to industrial maintenance management. *Br. J. Manag.* **2003**, *14*, 207–222. [CrossRef]
15. Eggen, A.O.; Rommetveit, O.; Reitlo, A.; Midtbø, E.O. *Handbook on Condition Monitoring of Wind Turbines*; Sintef: Trondheim, Norway, 2009.
16. Roland Berger Strategy Consultants. *Offshore Wind Toward 2020—On the Pathway to Cost Competitiveness*; Roland Berger Strategy Consultants: London, UK, 2013.

processes

Article

Tech-Economic Assessment of Power Transmission Options for Large-Scale Offshore Wind Farms in China

Qin Jiang, Baohong Li * and Tianqi Liu

College of Electrical Engineering and Information Technology, Sichuan University, Chengdu 610000, China; jiangqin_jq@126.com (Q.J.); tqliu@scu.edu.cn (T.L.)
* Correspondence: scu_lbh@163.com

Abstract: China is taking initiative in energy transition to cope with the long-term controversy of its enormous energy consumption, aiming to use less carbon. Wind power, especially offshore wind energy, has become a prevailing alternative due to its low carbon emissions, renewability, competitiveness, and operation security. The layout of a transmission channel is a key consideration in marine project implementation. This paper investigates the technical characteristics, application status, and viable advantages of a conventional AC transmission, voltage source converter-based high-voltage direct current (VSC-HVDC) transmission, gas-insulated line (GIL) transmission, and hybrid HVDC transmission. A component-resolved evaluation model was proposed to estimate the costs to be incurred of four electrical transmission options for offshore wind power along the coast of Eastern China, with technical feasibility and economical considerations. Cost comparisons and component sensitivity analyses were developed with different transmission distances and capacities. Results suggest HVAC transmission and VSC-HVDC are the preferable solutions for present offshore wind farm development in Eastern China, and the economic potential of the hybrid HVDC makes it feasible for future deployment. Some conclusions can be applied in disparate regions across the globe.

Keywords: offshore wind; China; HVDC; opex; economic evaluation

Citation: Jiang, Q.; Li, B.; Liu, T. Tech-Economic Assessment of Power Transmission Options for Large-Scale Offshore Wind Farms in China. *Processes* **2022**, *10*, 979. https://doi.org/10.3390/pr10050979

Academic Editors: Eugen Rusu, Kostas Belibassakis and George Lavidas

Received: 11 April 2022
Accepted: 11 May 2022
Published: 13 May 2022

Publisher's Note: MDPI stays neutral with regard to jurisdictional claims in published maps and institutional affiliations.

Copyright: © 2022 by the authors. Licensee MDPI, Basel, Switzerland. This article is an open access article distributed under the terms and conditions of the Creative Commons Attribution (CC BY) license (https://creativecommons.org/licenses/by/4.0/).

1. Introduction

Deployment of variable renewable energy resources are technical solutions driving global climate change. In order to sharply decrease the carbon emission and accelerate the global energy transition [1], wind power has experienced a rapid development in the last 20 years, which has become the mainstream renewable energy around the world now [2]. In 2019, China maintains the first place in terms of cumulative installed capacity of wind power and is vigorously promoting wind power on a priority basis [3]. Compared with onshore wind power, offshore wind farms have much less negative impacts on humankind as no land resource is needed, which also makes them usually have a larger scale and the offshore turbines have a larger capacity, which means a fall in the capital costs [4].

Because of the above advantages, plenty of studies have been conducted in the cost assessment area of the offshore wind farms (OWFs), which concentrates on the cost evaluation methodologies, potential economical technologies, and cost reduction. The infrastructure costs of OWFs are strongly related to the spatial condition [5,6]. Myhr et al. presented a cost sensitivity analysis and pointed out that the results suffer significant spatial bias and may differ in various countries [7], such as spatially-explicit assessment for the United Kingdom (UK) [8,9], Australia [10], Thailand [11], India [12], and Nigeria [13]. Thus, a Geographic Information System (GIS) makes costs and energy potential estimations possible based on spatially clustered data [14,15]. To obtain the cost reduction potential, the GIS-based levelized production cost (LPC) methodology is a common analysis model [16–18]. Furthermore, some assessments take the impacts of marine ecosystem and weather or climate variance into consideration [19,20].

The costs of the OWFs are more expensive than the onshore farms due to its complex foundation, installation, and submarine cabling; with the construction of the marine economy, the transmission vehicle becomes an important part [21]. Furthermore, the costs of different transmission methods are distance- and capacity-dependent functions [22], because the required diameter and number of cables are capacity-resolved, especially for projects with GWs capacity, and there exists a "breakeven" distance [23]. Two prevailing approaches are conventional: alternating current (AC) transmission, which is effective for near shore farms [24], and extensive voltage source converter-based high-voltage direct current (VSC-HVDC) transmission, which is the preferred solution for long-distance transmission. The HVDC transmission technology has many advantages, such as a fast power control speed [25] and oscillation damping control [26], and can be used in ultra-high-voltage occasions [27–29]. However, the HVDC implementation also has disadvantages. The first disadvantage compared to AC transmission is the cost, as the VSC is based on so many IGBT components that finally lead a relatively high investment. The second disadvantage is the stability problem, where the VSC often suffers from oscillation risk, especially when the power fluctuates. The third disadvantage is that the IGBT is very sensitive to the fault current, and it requires a fast protection scheme. However, the advantage of VSC makes it still be suitable for offshore wind farm integration. Offshore wind power is often located in the far sea area and the transmission cable also decreases the fault possibilities. The same as with HVDC transmission, a gas-insulated transmission Line (GIL) provides another way due to its advantage of considerably larger capacity, but its exaggerated expense makes it less competitive. Consequently, VSC-HVDC becomes more eye-catching for investors with predominant capability and desirable loss, which is suitable for crossing long-distance water transmission, such as in the North Sea of Germany [30]; but, the terminal converter stations are more expensive. However, the choice of electrical transmission ultimately depends on both technological potential and economic potential [31]. For future technical development of OWFs, there is another competitive option—hybrid VSC/LCC-HVDC technology—which is a novel form of HVDC transmission not widely applied, but it greatly decreases the costs and is planned for use China, possessing huge technological potential.

In China, there are many large-scale blocks with a capacity of more than hundreds of MW planned for OWFs [32]. The existing research has mostly focused on a single offshore wind farm project [33], lacking the overall research on regional offshore transmission systems [34]. There is a need to explore the optimal technical transmission method of the regional offshore transmission network for wind farms, which is conducive to wind energy utilization. This paper conducted regional cost analysis and economic feasibility comparisons of four electricity transmission options for offshore wind power in Guangdong Province using component-resolved evaluation models. Economic costs and sensitivity have been derived using the Discounted Cashflow Model (DCF). This contributes to determining the reasonable scope of technical and economic application of various transmission modes, giving perceptible information for stakeholders for offshore wind transmission infrastructure under indigenous development, economic perspective of relevant technologies, and possible potential to future deployment and implementation of marine projects.

To make clear the characteristics of the different wind power transmission technologies, this paper compares various offshore wind farms with the HVAC, HVDC, GIL, and hybrid HVDC output channels. The novelties of the paper are as follow.

(1) The evaluation models for different wind power integration technologies are investigated, and the techno-economic costs can be calculated according to the proposed method for different technologies.
(2) The economic characteristics of each technology are clarified based on the proposed analysis model, and the compositions of these various transmission solutions are studied and compared.
(3) The influence factors for the investment of different technologies are also investigated, and the suitable application situations are proposed for different wind power output solutions.

This paper is structured as follows: The study area and technical potentials of the four transmission methods are introduced in Section 2. Section 3 proposes specific evaluation models of the various transmission solutions. Then, the techno-economic costs and sensitivity to transmission distance and capacity analysis results are investigated using the DCF approach in Section 4. The conclusions and areas of further work are discussed in Section 5.

2. Study Area and Methodology

2.1. Study Area

Guangdong is located on the eastern coast of China, which is rich in offshore wind energy; it is also the core area of economic development in southern China. To reach the new goal of deployment of the Guangdong–Hong Kong—Macao Greater Bay area (GBA). It has made great efforts to develop offshore wind power, which is effective to adjust coastal resources in line with a prosperous economy. By 2030, more than 1000 km of transmission lines will be built for grid connections for offshore wind power. For this trend of future planning of OWFs in Guangdong, policymakers are concerned with the cost assessments of efficient electrical transmission options to transport large quantities of offshore energy across great distances.

Offshore wind resources of Guangdong Province are in western and eastern Guangdong. Based on the Notice of Guangdong Development and Reform Commission on Guangdong offshore wind power development plan (2017–2030), 15 offshore wind farm sites are located in the offshore shallow water area, and 8 sites are in the offshore deep-water area. Yangjiang city is the closest with a stable wind power supply base in the west to the GBA. There are three regions for offshore wind farms in the plan: Nanpeng Island OWFs, Hailing Island OWFs, and Shapa OWFs. The total planned installed capacity of the renewable energy is about 36 GW, as indicated in Figure 1.

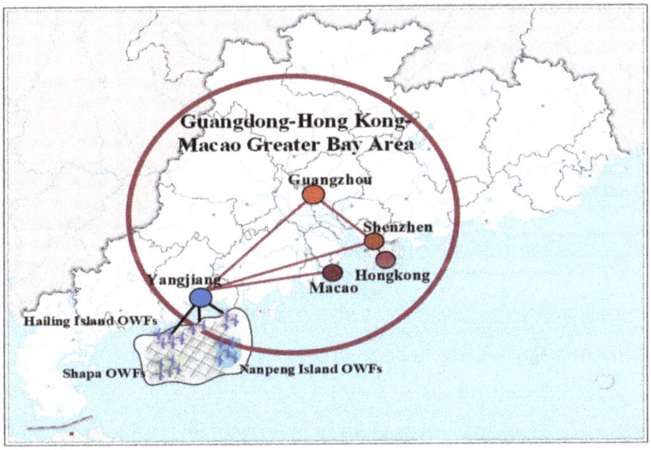

Figure 1. Overview of the study area.

2.2. Technical Evaluation of the Transmission Solutions

Four transmission solutions are studied in this paper, as shown in Figure 2. The offshore wind power from each farm is collected and transmitted to the offshore step-up transformer station. Then the voltage will be raised and the electrical power will be delivered to the onshore step-up transformer station via a submarine high-voltage transmission line (AC/DC cable or GIL line) and delivered to the onshore booster station.

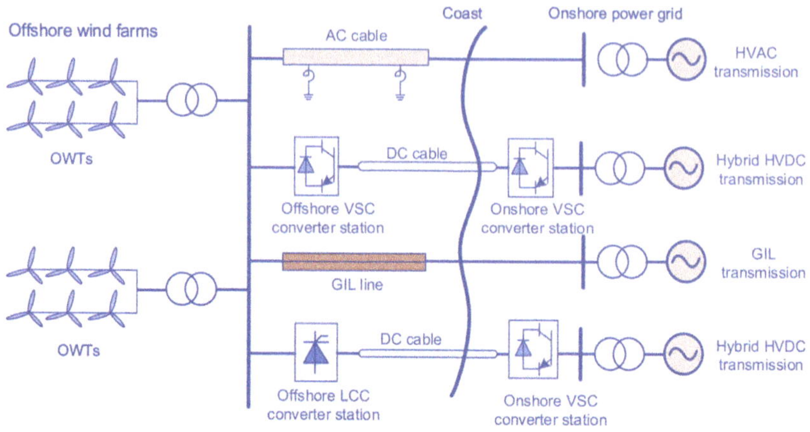

Figure 2. Structure of the four transmission methods for OWFs.

As mentioned above, AC transmission is widely used in near sea OWFs, compared with others. The distributed capacitance of the AC cable will become larger and larger, and the ampacity will decrease with the increase in length. This significantly reduces the transmission ability; also, multi lines are needed to transfer the large amount of wind power, which means more investment cots in the AC cable. In addition, due to the close electrical connection between the wind farm and the onshore power grid, the fault of either side will quickly spread to the other side; this will cause voltage oscillation and power instability, which reduces the power quality. It is necessary to install dynamic reactive power compensation devices to improve the stability and available transfer capability. A DC cable is cheaper and able to transfer more capacity with lower loss, which is popular in OWF transmission, but an offshore converter station needs to be assembled and a large DC platform should be built for it, which makes the economic investment of VSC-HVDC higher in the early infrastructure. However, it is convenient to build and expand by stages, and the asynchronous connection to onshore grid can suppress the synchronous transmission of faults.

With regard to GIL lines, as derived from GIS, GIL only needs to have basic electrical performance, such as insulation and dynamic thermal stability, and there is no switchgear; it thus has obvious reliability advantages over either AC/DC cable or overhead lines in long-distance and large-capacity power transmission. However, the high costs and high technical requirements of the construction design and the long project period are difficult problems for the actual project. In China, a new hybrid DC transmission mode combines the superior performance of LCC-HVDC and VSC-HVDC technologies and has a lower cost than current VSC-HVDC transmission. Yet, the available transmission power is determined by the VSC-HVDC side, and it is hard to realize power flow reversal due to the voltage polarity that needs to be changed in the LCC converter station. Still, it is a new trend of innovation and becomes an alternative for offshore wind power transmission though it has not been applied due to a lack of research, except in China. In summary, the technical potential of the four methods is in Table 1.

Table 1. Technical comparisons of the four transmission options.

Methods	Advantage	Restrictions	Potential
HVAC	Easy layout, High reliability, Rich experience	Large distributed capacitance, Additional reactive power compensation, Multi lines for larger capacity, Synchronous fault propagation	Popular for near sea OWFs
VSC-HVDC	Better stability, Low line cost and loss, Restrain fault propagation, Easy for construction and capacity expansion	Layout of converter station, Additional offshore platform	Developing rapidly Large-capacity transmission, and long-distance transmission
GIL	Best operation reliability, High ampacity, Large transmission capacity of single line and less loss	High cost, High technical requirements, Long project period	Limited application for large-capacity transmission
Hybrid HVDC	Better performance than VSC-HVDC or LCC-HVDC, Lower cost than VSC-HVDC	The available transmission power is determined by VSC side, Hard to power flow reversal, Lack research	New trend of transmission, Worth developing

3. Methodology

The costs evaluation can be broken down into multiple components, such as site-dependent variables, fixed water depth, the distance to grid connection point, and fixed costs [35]. Total investment cost equals the summation of the capital cost components, calculated as suggested by Dicorato et al. [35] and Hong and Möller [14,33].

The methodology establishes an empirical component-resolved evaluation model from an industry standard or outline to evaluate four electrical transmission concepts. The economic costs under each concept are intricate, so the main resolved components, including capital costs, OPEX, and loss costs, are considered and calculated in this paper.

3.1. Costs Calculation of HVAC Cables Transmission

The HVAC cable transmission concept is a popular way for offshore wind farms, and the principal cost drivers include capital costs $C_{cap.AC}$, operation and maintenance costs $C_{opex.AC}$, and loss costs $C_{loss.AC}$. The calculation is given by

$$C_{AC} = C_{cap.AC} + C_{opex.AC} + C_{loss.AC} \tag{1}$$

3.1.1. Capital Costs

In the concept of HVAC cable transmission, $C_{cap.AC}$ covers the relative substation foundation costs $C_{station.AC}$, underwater cable foundation and installation costs $C_{cable.AC}$, and reactive power compensation foundation costs $C_{reacpc.AC}$, estimated by

$$C_{cap.AC} = C_{station.AC} + C_{cable.AC} + C_{reacpc.AC} \tag{2}$$

1. Substation foundation cost

Substation foundation cost in the HVAC transmission system is the total costs of each transformer substation capital expenditure, which is dependent on the infrastructure investment of the substation, expense of the transformer and the investment cost of auxiliary electrical equipment, including the reactive compensation capacitor and switchgear. Then the calculation of $C_{station}$ is based on cost C_{perMVA} and determined by the capacity of substation S.

$$C_{station.AC} = C_{perMVA} \cdot S \tag{3}$$

2. Cable foundation and installation cost

The underwater cable for offshore wind energy is utilized for the link between the transformer substation and offshore substation; hence, $C_{cable.AC}$ is estimated as a proportion of distance to station L.

$$C_{cable.AC} = 3(P_1 + P_2)L \tag{4}$$

where P_1 and P_2 are the expense and installation cost of one unit (length, km) of cable, respectively.

3. Reactive power compensation foundation cost

In the HVAC transmission system, the distributed capacitance of the cable is generally much larger than the overhead line, so a large capacitance current will be generated in the AC line, which significantly reduces the available transfer capability. Therefore, reactive power compensation devices should be installed on sides of the cable according to the actual operation. Thus, compared with the VSC-HVDC transmission method, the foundation cost of the reactive power compensation should be considered additionally, which mainly includes the cost of the shunt reactors. To calculate it, the reactive power Q_{reac} (MVAR) of the line capacitance is calculated.

$$Q_{reac} = 2\pi \times f \times c \times l \times U_{cable}{}^2 \tag{5}$$

where f is the operational frequency of system, c is the capacitance value per km of the cable, and U_{cable} is the voltage of AC cable.

The capacity of reactors $C_{reacpc.AC}$ can be determined by

$$C_{reacpc.AC} = P_3 \times Q_{reac} \tag{6}$$

where P_3 is the expense of the reactors.

3.1.2. Operation and Maintenance Costs

$C_{opex.AC}$ is usually estimated in the form of percentage A of the annual maintenance cost to total investment cost (excluding land occupation cost and offshore platform costs) or percentage A_1 of lifetime maintenance costs to total investment cost. The relation between A and A_1 is

$$A = A_1 \times \frac{i(1+i)^n}{(1+i)^n - 1} \tag{7}$$

where i is the annual interest rate; n is the lifetime; Van Eeckhout gives the specific data of A equals to 1.2%, n is 20 years, i is 5% [36]. Then, $C_{opex.AC}$ is estimated:

$$C_{opex.AC} = C_{cap.AC} \cdot A \tag{8}$$

3.1.3. Costs of Loss

The loss costs $C_{loss.AC}$ comprise of substation loss $C_{sub.loss}$ and transmission line loss $C_{line.loss}$. $C_{sub.loss}$ is dependent on the substation loss rate $P_{sub.loss}$, as referred to in the literature [36]. The $P_{sub.loss}$ of two substations is 0.8%, that means the loss rate of each substation is 0.4%. $C_{line.loss}$ includes conductor losses $C_{con.loss}$ and losses of sheath and armor $C_{shar.loss}$. $C_{con.loss}$ can be formulated by the current I_{cable} of the copper conductor, which can be approximately calculated by

$$I_{cable} = \frac{P}{\sqrt{3}U_{cable}\cos\varphi} \tag{9}$$

where P is the active power; the power factor $\cos\varphi$ is 0.95.

Therefore, with the resistance of conductor R_{cu}, $C_{con.loss}$ is given by

$$C_{con.loss} = 3I_{cable}{}^2 \cdot R_{Cu} \tag{10}$$

The losses of sheath and armor $C_{shar.loss}$ is estimated.

$$\begin{bmatrix} \Delta U_C \\ \Delta U_S \\ \Delta U_A \end{bmatrix} = \begin{bmatrix} Z_1 & Z_2 & Z_3 \\ Z_4 & Z_5 & Z_6 \\ Z_7 & Z_8 & Z_9 \end{bmatrix} \begin{bmatrix} I_C \\ I_S \\ I_A \end{bmatrix} \tag{11}$$

where ΔU_c, ΔU_s, ΔU_A, I_c, I_s, I_A are the voltage and current of the copper core, sheath, and armor, respectively; Z_1–Z_9 are the matrix of parameters of the cable.

Moreover, since both ends of the sheath are grounded, the armor layer is linked with the sea, with the assumption of $U_s = U_A = 0$ and $I_c = I_{cable}$; so, $C_{shar.loss}$ can be given by the power loss $P_{ar} = 3{I_A}^2 \times R$, $P_{sh} = 3{I_S}^2 \times R$, and I_S and I_A are

$$\begin{cases} I_A = \left(Z_9 - Z_8 Z_5^{-1} Z_6\right)^{-1} \left(Z_8 Z_5^{-1} Z_4 - Z_7\right) I_C \\ I_S = -Z_5^{-1}(Z_4 I_C + Z_6 I_A) \end{cases} \tag{12}$$

The costs of C_{arsh} is dependent on the operation time of full generation per year T_f and the on-grid price of electricity $P_{on\text{-}grid}$, which are

$$C_{shar} = (P_{sh} + P_{ar}) \times T_f \times P_{on-grid} \tag{13}$$

The evaluation of $C_{loss.ac}$ is obtained by total C_{sh} and C_{ar}.

$$C_{loss.AC} = C_{sub.loss} + C_{shar} + C_{con.loss} \tag{14}$$

3.2. Costs Calculation of VSC-HVDC Transmission

As for the VSC-HVDC transmission concept, the total costs of C_{VSC} compose of capital costs $C_{cap.VSC}$, operation and maintenance costs $C_{opex.VSC}$, and loss costs $C_{loss.VSC}$.

$$C_{VSC} = C_{cap.VSC} + C_{opex.VSC} + C_{loss.VSC} \tag{15}$$

3.2.1. Capital Costs

$C_{cap.VSC}$ consists of the converter station foundation cost $C_{station.VSC}$, and the cable foundation and installation costs $C_{cable.VSC}$.

$$C_{cap.VSC} = C_{station.VSC} + C_{cable.VSC} \tag{16}$$

1. Converter station foundation cost

$C_{station.VSC}$ is the total infrastructure investment of each converter station. Furthermore, the additional costs of IGBT, converter controller and reactor, DC capacitor and AC filter, as well as the cost of civil construction of the offshore platform for converter station layout are estimated. Then $C_{station.VSC}$ is computed as a proportion of the capacity of per converter station P.

$$C_{station.VSC} = C_{perMW} \cdot 2P \tag{17}$$

2. Cable foundation and installation cost

Similar to HVAC cable, $C_{cable.VSC}$ of DC cable is calculated by the transmission distance.

$$C_{cable.VSC} = 2(P_1 + P_2)L \tag{18}$$

where P_1 and P_2 are the expense and installation costs of per km DC cable.

3.2.2. Operation and Maintenance Costs

$C_{opex.VSC}$ is obtained in Equation (8), and the A of the DC submarine cable equals to 0.5%, n is 20 years, and i is 5%, which were applied to this study.

3.2.3. Costs of Loss

The loss costs $C_{loss.VSC}$ consist of converter station loss $C_{sub.loss}$ and line loss $C_{line.loss}$. Converter station loss rate $P_{sub.loss}$ is the percentage of station power loss to the transmitted power. $P_{sub.loss}$ of two converter stations is 1.6–2.4%, and Zhen points out that $P_{sub.loss}$ is 1–2% [36]. The levelized loss rate of each substation is

$$P_{sub.loss} = (\frac{1.6\% + 2.4\%}{2} + \frac{1\% + 2\%}{2})/2 = 1.75\% \quad (19)$$

Meanwhile, the line losses $C_{line.loss}$ can be evaluated as

$$C_{line.loss} = (P/U_{DC})^2 \cdot R \cdot 2L \cdot T_f \cdot P_{on-gird} \quad (20)$$

where U_{DC} is the DC voltage; R is resistance.

The evaluation of $C_{loss.VSC}$ is described as

$$C_{loss.VSC} = P_{sub.loss} \cdot P \cdot P_{on-gird} + C_{line.loss} \quad (21)$$

3.3. *Costs Calculation of GIL Transmission Concept*

The GIL transmission concept is similar to the AC transmission concept, but there is no reactive power compensation costs.

$$C_{GIL} = C_{cap.GIL} + C_{opex.GIL} + C_{loss.GIL} \quad (22)$$

3.3.1. Capital Costs

The capital expenditure $C_{cap.GIL}$ is dependent on the transformer substation foundation cost $C_{station.GIL}$ and cable foundation and installation cost $C_{cable.GIL}$.

$$C_{cap.GIL} = C_{subfound.GIL} + C_{cable.GIL} \quad (23)$$

The foundation cost $C_{station.GIL}$ is similar to $C_{station.AC}$ in Equation (3); similarly, the calculation of $C_{cable.GIL}$ is as Equation (4).

3.3.2. Operation and Maintenance Costs

Based on the OPEX in HVAC transmission system, $C_{opex.GIL}$ is expressed by the percentage A as in Equation (7).

3.3.3. Costs of Loss

The loss costs $C_{loss.GIL}$ in the GIL transmission system comprises of $C_{sub.loss}$ and line loss $C_{line.loss}$ as well.

$$C_{loss.GIL} = C_{sub.loss} + C_{edcir} + C_{con.loss} \quad (24)$$

where $C_{sub.loss}$ is dependent on the $P_{sub.loss}$, $C_{line.loss}$ comprises of conductor losses $C_{con.loss}$, as computed by Equation (10), and the eddy current and circulating current loss $C_{edcir.loss}$ of the shell.

3.4. *Costs Calculation of Hybrid HVDC Transmission*

3.4.1. Capital Costs

Capital costs in the hybrid HVDC transmission $C_{cap.HybDC}$ covers the converter station foundation cost $C_{station.HybDC}$, cable foundation cost, and installation cost $C_{cable.HybDC}$.

$$C_{cap.HybDC} = C_{station.HybDC} + C_{cable.HybDC} \quad (25)$$

where $C_{station.HybDC}$ is dependent on the sum of the investment costs of the different types of converter stations on both sides.

3.4.2. Operation and Maintenance Costs

Since the OPEX in the hybrid HVDC transmission system is in the same way in the VSC-HVDC transmission system, $C_{opex.HybD}$ can be calculated by Equation (8).

3.4.3. Costs of Loss

Similarly, $C_{loss.HybDC}$ consists of substation loss $C_{sub.loss}$ and line loss $C_{line.loss}$. $C_{sub.loss}$ is dependent on the converter station loss rate $P_{sub.loss}$, which is 1.75% for VSC-HVDC and 0.8% for LCC-HVDC. The average value of $P_{sub.loss}$ is 1.275%. The transmission line loss $C_{line.loss}$ is the same as in Equation (20).

4. Results and Discussion

The empirical component-resolved evaluation models give a crucial message to stakeholders that the economic costs are sensitive to transmission distance and capacity. The cost comparisons of the four electrical transmission options for wind farms with different distances and transmitted power were carried out. The rated voltage is 220 kV, and the frequency is 50 Hz, and the operation hour of full capacity per year is 2500 h. If the capacity is 300 MW, 600 MW, and 900 MW, respectively, the economic evaluations from 25 km to 75 km were conducted.

4.1. Essential Evaluation Data

Based on the DCF model, the costs evaluation results can be converted to cash value. Unlike an onshore power grid, the specific environment and operational conditions of the offshore substation are more complicated; it is necessary to adopt more strict standards for long-term stability. For the AC cable, one line is needed for 300 MW, two lines for 600 MW, and three lines for 900 MW wind power. However, it is important to point out that in the GIL transmission concept, the rated current is 3.15 kA, so the transmission capacity of a single line is 1200 MVA, and there is no need to install additional lines with different capacity. The data are given in Table 2.

Table 2. Overview of basic data related to the economic evaluation of offshore wind farm transmission.

Components	**Cost**	**HVAC**	**VSC-HVDC**	**GIL**	**Hybrid HVDC**
		220 kV, 1200 mm^2 Singlecore Underwater AC Cable	**±200 kV, 500 mm^2 Core Optical DC Cable**	**220 kV GIL Line**	**±200 kV, 500 mm^2 Core Optical DC Cable**
Capital costs	Foundation costs of substation (converter station)	CNY 0.45 million/MVA [a]	CNY 1.1 million/MW [36]	CNY 0.45 million/MVA [a]	CNY 0.9621 million/MW [b]
	Expense for P_1	CNY 3.732 million/km [c]	CNY 1.077 million/km [b]	CNY 20 million/km [37]	CNY 1.077 million/km [b]
	Installation cost $P_2(P_3)$	CNY 0.30533 million/km [a,d]	CNY 0.3 million/km [a]	CNY 0.3 million/km [a]	CNY 0.3 million/km [b]
OPEX	Annual percentage A	1.2% [36]	0.5% [36]	0.5% [36]	0.5% [b]
Loss costs	Power loss of substation (converter station)	0.4% [36]	1.75% [36]	0.4% [36]	1.275% [b]
	Loss costs of cables	CNY 0.6145 million/km	CNY 0.0876 million/km [38]	CNY 0.077 million/km [37]	CNY 0.0876 million/km [b]

[a] Design Control Index of Power Grid Project in China (2014). [b] Presented in this paper considering both LCC-HVDC and VSC-HVDC cost. [c] Materials provided by Dongfang Cable Factory in Ningbo city, China. [d] Materials of the project of 66kV Xin-Guang underwater cable in Dalian city, China.

4.1.1. Capital Costs Evaluation

The $C_{station}$ of the AC cable varies among different projects. For example, according to the materials in Design Control Index of Power Grid Project (2014 standard) provided by the Electric Power Planning and Engineering Institute of China [39], the investment of the 220 kV Yucai substation project (indoor) is 0.303 million RMB per megavolt-ampere (MVA), and the cost of the 220 kV Pingli substation project (Laizhou, Shandong) is CNY 0.435 million/MVA. In this study, for 35 kV wind farms with a 220 kV single-core underwater cable, the foundation costs of substation C_{perMVA} is CNY 0.45 million/MVA. Procurement materials provided by Dongfang Cable Factory in Ningbo city indicates that the expense basis of a 220 kV single-core underwater cable with the 1200 mm^2 cross-sectional area of copper core is CNY 3.732 million/km. The installation costs refer to cable crossing barge, sea sweeping, and trench laying. The project of the 66 kV Xin-Guang submarine cable in Dalian city gives the cost P_2 around CNY 0.3 million/km. As for the reactive power compensation, the rated power of the AC cable is 427 MVA and the capacitance for the 1200 mm^2 cable is 0.179 μF/km. The maximum DC resistances of the 20 °C and AC resistance of 90 °C are 0.0151 Ω/km and 0.02 Ω/km, respectively.

Several studies provide various foundation costs for the converter station for reference. The costs of the traditional ±500 kV and ±800 kV LCC-HVDC converter stations are around CNY 0.52428 million/MW and CNY 0.56228 million/MW, respectively. There is a lack of reports on the cost of a VSC-HVDC converter station in China, which varies widely across the globe. Reference [36] applied the technical materials of ABB Ltd. to evaluate the costs of a VSC-HVDC station as CNY 1.155 to 1.343 million/MW, and the costs of the ±300 kV converter station are CNY 1.2 million/MW. Taking the development of offshore wind power technology into account, the standard of C_{perMW} is CNY 1.1 million/MW. According to the industry date provided by Dongfang Cable Factory, the expense the P_1 of XLPE-insulated DC submarine cable (Model: DC200 kV YJQ411 500 + 2 × 12 (core optical cable)) with a cross-sectional area of 500 mm^2 is CNY 1.077 million/km. Moreover, considering the difficulty of hybrid HVDC transmission technology, then C_{perMW} is CNY 0.9621 million/MW.

The expense of GIL P_1 is CNY 20 million/km. P_2 of the four transmission methods equals CNY 0.3 million/km. Thus, the capital costs under different capacity can be obtained.

4.1.2. Operation and Maintenance Costs Evaluation

The annual percentage A of the operation and maintenance cost of the AC submarine cable accounts for 1.2%, and 0.5% is adopted in the other three transmission methods.

4.1.3. Costs of Loss Evaluation

It is assumed that the operation hours are 2500 h per year, referred to in [37], and the on-grid price of offshore wind power is CNY 0.0085 million/MW·h [40]. The substation loss rate $P_{sub.loss}$ in AC cable and GIL transmission system is 0.4%. The apparent power is 427 MVA, based on Equation (9), and the current of copper core with I_{cable} is 0.8287 kA. R_{cu} is 0.006 Ω/km. Some industry gives I_s is 502.4 A and I_A is 313.2 A. The resistances of the sheath and armor are 0.21 Ω/km and 0.301 Ω/km. The eddy current loss $P_{ed.loss}$ and circulating current loss $P_{cir.loss}$ in the GIL lines are 0.0177 MW/km and 0.0062 MW/km, respectively.

For the XLPE-insulated DC submarine cable, the rated power is 324 MW and the DC resistance is 0.0366 Ω/km. Based on Equation (20), the conductor loss of 300 MW is 0.0412 MW/km, 0.0824 MW/km, and 0.1648 MW/km. The substation loss rate $P_{sub.loss}$ in the hybrid HVDC system is 1.275%.

4.2. Evaluation Results

It can be seen from the above analysis that offshore distance and capacity have an important impact on the capital costs of the four types of transmission. Based on the DCF model, the comparisons with different transmission distances and capacities were calculated, and the results shown in Table 3. Components analysis of the total costs was carried out to acquire an importance view for stakeholders, shown in Table 4. The gradual change in color from green to red represents an increase in costs.

Table 3. Economic costs comparisons of different *P* and different *L*.

P (MW)	*L* (km)	HVAC (CNY Million)	VSC-HVDC (CNY Million)	GIL (CNY Million)	Hybrid HVDC (CNY Million)
300	25	471.624	745.839	1686.1	659.657
	50	796.791	817.222	3218.2	731.04
	75	1121.96	888.605	4750.2	802.423
600	25	943.247	1491.68	1840.2	1250.12
	50	1593.58	1634.45	3372.2	1323.69
	75	2243.93	1777.21	4904.3	1397.26
900	25	1414.47	2237.52	1993.8	1840.58
	50	2389.99	2451.67	3525.9	1916.34
	75	3365.48	2665.82	5057.9	1992.1

Table 4. Economic costs comparisons of different *P* and different *L*.

Options	*L* (km)	C_{cap} (CNY Million)	C_{opex} (CNY Million)	C_{loss} (CNY Million)	C_{total} (CNY Million)
HVAC	25	448.33	5.38	17.914	471.624
	50	754.46	9.054	33.278	796.791
	75	1060.59	12.727	48.641	1121.958
VSC	25	728.85	3.644	13.345	745.839
	50	797.7	3.989	15.534	817.222
	75	866.55	4.332	17.723	888.605
GIL	25	1664.7	8.323	13.083	1686.11
	50	3187.2	15.936	15.009	3218.15
	75	4709.7	23.549	16.935	4750.18
Hybrid	25	646.11	3.231	10.317	659.657
	50	714.96	3.575	12.506	731.04
	75	783.81	3.919	14.694	802.423

4.3. Comparisons of Economic Evaluation

To obtain the best transmission method for Guangdong offshore wind power, the costs comparisons were calculated.

4.3.1. Total Costs Comparisons with Different *L*

According to the data in Table 2, the relationships between the total costs of transmission distance *L* from 25 km to 75 km are shown in Figure 3.

It is clear that the economic costs of the GIL electrical transmission concept are considerably much more than either the HVAC or HVDC transmission concept. When the transmission distance is not so long, such as 25 km, the costs of the GIL system is more than twice of that in other systems. In addition, the costs of the GIL changes great when *L* increases, which means it is most sensitive to transmission distance; it can even be four times that of the others at a distance of 75 km. On the other hand, the hybrid HVDC transmission concept has economic advantages to the VSC-HVDC system, and when the installed capacity increases, the preferred distance range under the hybrid HVDC technology becomes longer from 50 km (300 MW), 38.6 km (600 MW), to 36.4 km (900 Mw). It is feasible if the technology is developed widely in the future.

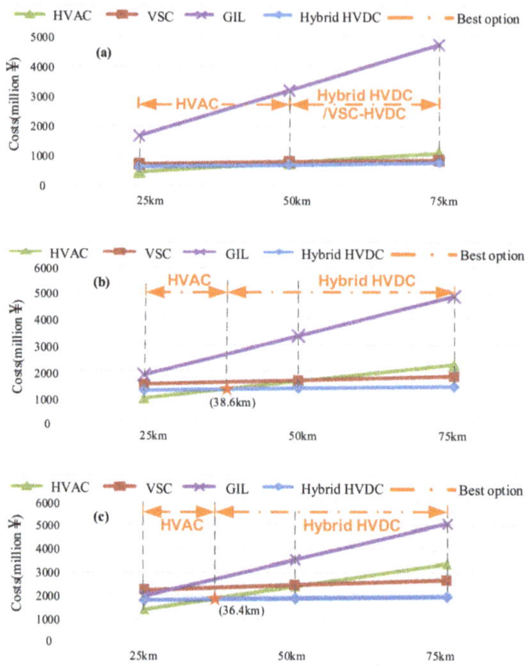

Figure 3. Total costs of different L of 300 MW, 600 MW, and 900 MW wind farms: (**a**) shows the 300 MW offshore wind farm total costs with different distances; (**b**) shows the 600 MW offshore wind farm total costs with different distances; (**c**) shows the 900 MW offshore wind farms total costs with different distances.

If the hybrid HVDC is not taken into present planning consideration due to its limited development, compared with VSC-HVDC electrical transmission, the HVAC concept is more superior when L is less than around 50 km in both 300 MW, 600 MW, and 900 MW wind farms. Otherwise, the HVDC transmission concept is preferable with a longer distance. HVDC is also less sensitive to transmission distance than the HVAC system.

4.3.2. Costs Components Comparisons with Different L

Taking 300 MW wind farms as an example, as Figure 4 shows, in the GIL and AC transmission system, the cable costs account for a large proportion of the total cost, especially the extravagant cable costs of the GIL transmission concept. That means the capital costs are the most important component to be considered, and the HVDC system has huge technological potential for offshore wind power transmission.

For the VSC-HVDC and hybrid HVDC transmission systems, cable costs are cheaper than for the AC transmission system. In turn, he costs of the converter station are much higher than that of the substation, as well as the costs of the converter station loss. It is important to notice that the capital costs in the AC system and GIL system increase greater than in HVDC systems.

4.3.3. Total Costs Comparisons with Different P

The economic costs of various transmission concepts are sensitive to transmission capacity, as shown in Figure 5. In near sea wind energy transmission, the economic costs of GIL transmission for 900 MW wind farms are even lower than that of VSC-HVDC transmission near sea wind farm transmission due to its advantages of large-capacity transmission. Thus, the GIL concept may be a better choice in the scenario under short distances with large-capacity transmission.

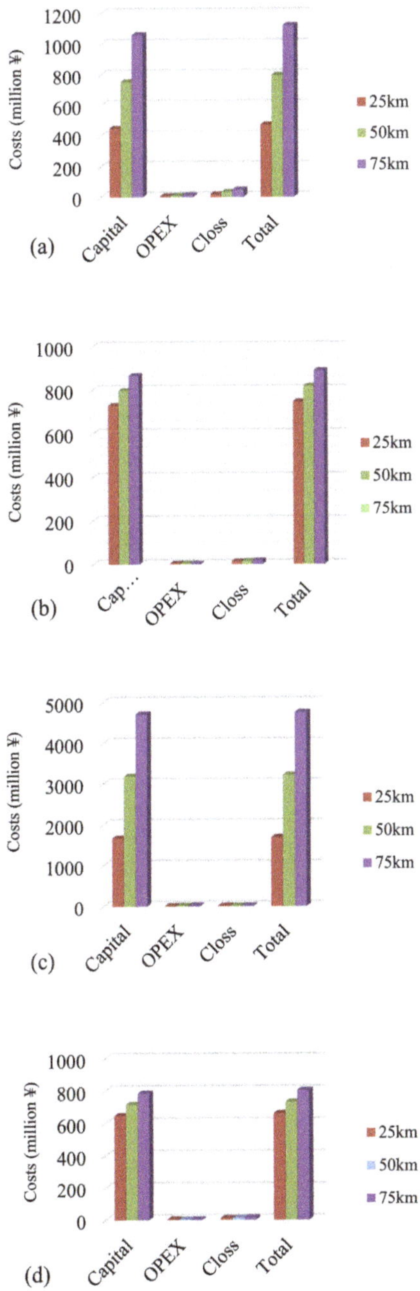

Figure 4. Cost component comparisons with different L (P is 300 MW): (**a**) shows the HVAC system; (**b**) shows the VSC-HVDC system; (**c**) shows the GIL system; (**d**) shows the Hybrid HVDC system.

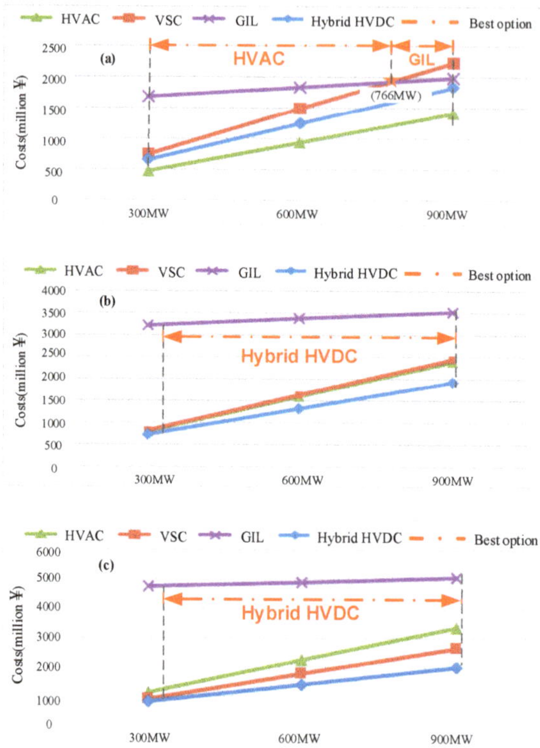

Figure 5. Total costs of different P of 25 km, 50 km, and 75 km wind farms: (**a**) shows the 25 km offshore wind farms total costs with different capacity; (**b**) shows the 50 km offshore wind farms total costs with different capacity; (**c**) shows the 75 km offshore wind farms total costs with different capacity.

Still, the lowest economic costs with different capacities are seen in the hybrid HVDC transmission concept, followed by HVAC electrical transmission and VSC-HVDC transmission. The GIL system increases significantly with a longer transmission distance; so, GIL is not recommended. Considering both economic and technical feasibility, for the offshore wind farms with different distances and transmission capacities, at present, HVAC transmission and VSC-HVDC transmission are selected according to the actual situation. The total costs are more sensitive to the distance than capacity. What is more, there is a need to notice that the hybrid HVDC transmission system is a preferred choice with the economic potential for either large-capacity or long-distance wind farms.

4.3.4. Costs Components Comparisons with Different P

Components comparisons of different P are carried out to acquire a view for investors. Taking the 25 km wind farms as an example, a sensitivity analysis was conducted, and the results shown in Figure 6.

It can be seen that there is an obvious advantage of the GIL system, as the capital costs of the GIL transmission system changed the least compared to the other systems, even when the capacity reached 900 MW. That means it is less sensitive to transmission capacity. However, the HVAC system and HVDC systems are so sensitive to capacity, which should be considered in OWF planning.

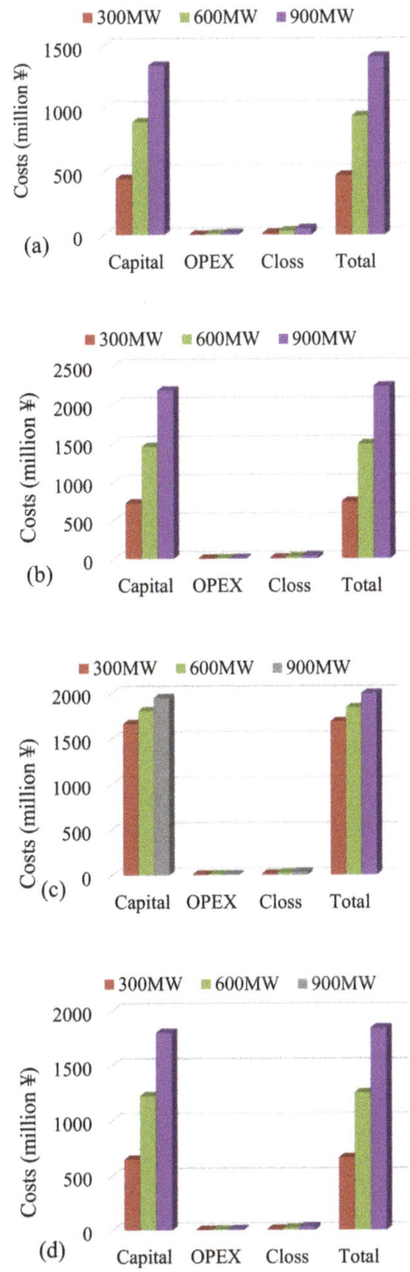

Figure 6. Costs components comparisons with different P (L is 25 km): (**a**) shows the HVAC system; (**b**) shows the VSC-HVDC system; (**c**) shows the GIL system; (**d**) shows the Hybrid HVDC system.

5. Conclusions

This paper investigates the four electrical transmission options for current and future Western Guangdong offshore wind farm implementation, including their technical characteristics, application status, and economic costs. Based on the component-resolved evaluation model, the capital costs, OPEX, and loss costs of four concepts of electrical transmission were studied, with the results showing that capital costs are the major component. The capital cost of the DC-type transmission technology is mainly related to the converter investment, while the cost of the AC type and GIL transmission technology is mainly related to the line and compensation cost. Meanwhile, the analysis also indicates that the offshore distance and capacity have an important impact on the capital costs of the four types of transmission. The sensitivity analysis of the four transmission solutions regarding transmission distance and capacity recommends the powerful competitive alternative of the HVAC transmission concept if the transmission system is less than 50.48 km for Yangjiang offshore wind farms, and VSC-HVDC and hybrid HVDC transmission for longer distances and larger capacities. For future planning, the GIL transmission system should be the preferred option in near sea and large-capacity wind farms, and the hybrid HVDC transmission possesses significant economic potential with a wide range of transmission distances and capacities.

Author Contributions: Q.J. performed the conceptualization, formal analysis and original draft preparation; B.L. provided methodology and reviewed the manuscript; T.L. supported the funding acquisition. All authors have read and agreed to the published version of the manuscript.

Funding: This research was funded by Science and Technology project of State Grid Corporation of China "Cloud energy storage framework-based AI dispatching strategy of renewable energy integration and contingency response", grant number 5100-202199274A-0-0-00.

Data Availability Statement: Data available on request due to privacy restrictions.

Conflicts of Interest: The authors declare no conflict of interest.

References

1. Snyder, B.; Kaiser, M.J. Ecological and economic cost-benefit analysis of offshore wind energy. *Renew. Energy* **2009**, *34*, 1567–1578. [CrossRef]
2. IRENA. *Global Renewables Outlook: Energy Transformation 2050*; International Renewable Energy Agency: Abu Dhabi, United Arab Emirates, 2020.
3. GWEC. *Global Wind Report 2019*; Global Wind Energy Council: Brussels, Belgium, 2020.
4. IEA. *Offshore Wind Outlook 2019*; International Energy Agency: Paris, France, 2019.
5. BNEF. *Offshore Wind Energy Market Outlook 2017*; Bloomberg New Energy Finance: Shanghai, China, 2017.
6. Sachs, J.; Moya, D.; Giarola, S.; Hawkes, A. Clustered spatially and temporally resolved global heat and cooling energy demand in the residential sector. *Appl. Energy* **2019**, *250*, 48–62. [CrossRef]
7. Staffell, I.; Pfenninger, S. Using bias-corrected reanalysis to simulate current and future wind power output. *Energy* **2016**, *114*, 1224–1239. [CrossRef]
8. Myhr, A.; Bjerkseter, C.; Ågotnes, A.; Nygaard, T.A. Levelised cost of energy for offshore floating wind turbines in a life cycle perspective. *Renew. Energy* **2014**, *66*, 714–728. [CrossRef]
9. Cavazzi, S.; Dutton, A.G. An Offshore Wind Energy Geographic Information System (OWE-GIS) for assessment of the UK's offshore wind energy potential. *Renew. Energy* **2016**, *87*, 212–228. [CrossRef]
10. Hdidouan, D.; Staffell, I. The impact of climate change on the levelised cost of wind energy. *Renew. Energy* **2017**, *101*, 575–592. [CrossRef]
11. Hearps, P.; McConnell, D. *Renewable Energy Technology Cost Review*; Melbourne Energy Institute: Melbourne, Australia, 2011.
12. Waewsak, J.; Landry, M.; Gagnon, Y. Offshore wind power potential of the Gulf of Thailand. *Renew. Energy* **2015**, *81*, 609–626. [CrossRef]
13. Nagababu, G.; Kachhwaha, S.S.; Savsani, V. Estimation of technical and economic potential of offshore wind along the coast of India. *Energy* **2017**, *138*, 79–91. [CrossRef]
14. Effiom, S.O.; Nwankwojike, B.N.; Abam, F.I. Economic cost evaluation on the viability of offshore wind turbine farms in Nigeria. *Energy Rep.* **2016**, *2*, 48–53. [CrossRef]
15. Möller, B.; Hong, L.; Lonsing, R.; Hvelplund, F. Evaluation of offshore wind resources scale of development. *Energy* **2012**, *48*, 314–322. [CrossRef]
16. Yamaguchi, A.; Ishihara, T. Assessment of offshore wind energy potential using mesoscale model and geographic information system. *Renew. Energy* **2014**, *69*, 506–515. [CrossRef]
17. Abdelhady, S.; Borello, D.; Shaban, A. Assessment of levelized cost of electricity of offshore wind energy in Egypt. *Wind. Eng.* **2017**, *41*, 160–173. [CrossRef]
18. Bosch, J.; Staffell, I.; Hawkes, A.D. Global levelised cost of electricity from offshore wind. *Energy* **2019**, *189*, 116357. [CrossRef]

19. Ebenhoch, R.; Matha, D.; Marathe, S.; Muñoz, P.C.; Molins, C. Comparative Levelized Cost of Energy Analysis. *Energy Procedia* **2015**, *80*, 108–122. [CrossRef]
20. Nagababu, G.; Kachhwaha, S.S.; Naidu, N.K.; Savsani, V. Application of reanalysis data to estimate offshore wind potential in EEZ of India based on marine ecosystem considerations. *Energy* **2017**, *118*, 622–631. [CrossRef]
21. Papathanasopoulou, E.; Beaumont, N.; Hooper, T.; Nunes, J.; Queirós, A.M. Energy systems and their impacts on marine ecosystem services. *Renew. Sustain. Energy Rev.* **2015**, *52*, 917–926. [CrossRef]
22. Kim, J.Y.; Oh, K.Y.; Kang, K.S.; Lee, J.S. Site selection of offshore wind farms around the Korean Peninsula through economic evaluation. *Renew. Energy* **2012**, *54*, 189–195. [CrossRef]
23. Pacheco, A.; Gorbeña, E.; Sequeira, C.; Jerez, S. An evaluation of offshore wind power production by floatable systems: A case study from SW Portugal. *Energy* **2017**, *131*, 239–250. [CrossRef]
24. Elliott, D.; Bell, K.R.; Finney, S.J.; Adapa, R.; Brozio, C.; Yu, J.; Hussain, K. A comparison of AC and HVDC options for the connection of offshore wind generation in Great Britain. *IEEE Trans. Power Deliv.* **2016**, *31*, 798–809. [CrossRef]
25. Cui, Y.; Song, X.; Zhao, L.; Yuan, H.; Wu, G.; Wang, C. WSN-Based Measurement of Ion-Current Density Under High-Voltage Direct Current Transmission Lines. *IEEE Access* **2019**, *7*, 10947–10955. [CrossRef]
26. Kimura, N.; Funaki, T.; Matsu-ura, K. Damping of current oscillation in superconductive line applied for high voltage direct current transmission system. *IEEE Trans. Appl. Supercond.* **1993**, *3*, 223–225. [CrossRef]
27. Lu, T.; Feng, H.; Zhao, Z.; Cui, X. Analysis of the Electric Field and Ion Current Density Under Ultra High-Voltage Direct-Current Transmission Lines Based on Finite Element Method. *IEEE Trans. Magn.* **2007**, *43*, 1221–1224. [CrossRef]
28. Jacobs, K.; Heinig, S.; Johannesson, D.; Norrga, S.; Nee, H.-P. Comparative Evaluation of Voltage Source Converters with Silicon Carbide Semiconductor Devices for High-Voltage Direct Current Transmission. *IEEE Trans. Power Electron.* **2021**, *36*, 8887–8906. [CrossRef]
29. Yuan, H.; Yang, Q.; Liu, Y.; Lu, J.; Akhtar, S.A. Development and Application of High-Frequency Sensor for Corona Current Measurement under Ultra High-Voltage Direct-Current Environment. *IEEE Trans. Instrum. Meas.* **2012**, *61*, 1064–1071. [CrossRef]
30. Blanco, M.I. The economics of wind energy. *Renew. Sustain. Energy Rev.* **2009**, *13*, 1372–1382. [CrossRef]
31. ABB. *World's First Power-from-Shore Dynamic ac Cable—Gjoa Floating Oil and Gas Platform*; North Sea: Zurich, Switzerland, 2015.
32. Schell, K.R.; Claro, J.; Guikema, S.D. Probabilistic cost prediction for submarine power cable projects. *Int. J. Electr. Power Energy Syst.* **2017**, *90*, 1–9. [CrossRef]
33. Wu, J.; Wang, J.; Chi, D. Wind energy potential assessment for the site of Inner Mongolia in China. *Renew. Sustain. Energy Rev.* **2013**, *21*, 215–228. [CrossRef]
34. Hong, L.; Möller, B. Offshore wind energy potential in China: Under technical, spatial and economic constraints. *Energy* **2011**, *36*, 4482–4491. [CrossRef]
35. Da, Z.; Xiliang, Z.; Jiankun, H.; Qimin, C. Offshore wind energy development in China: Current status and future perspective. *Renew. Sustain. Energy Rev.* **2011**, *15*, 4673–4684. [CrossRef]
36. *2014 Standard*; Design Control Index of Power Grid Project. Electric Power Planning and Engineering Institute of China: Beijin, China, 2014.
37. Dicorato, M.; Forte, G.; Pisani, M.; Trovato, M. Guidelines for assessment of investment cost for offshore wind generation. *Renew. Energy* **2011**, *36*, 2043–2051. [CrossRef]
38. Eeckhout, B.V. *The Economic Value of VSC-HVDC Compared to HVAC for Offshore Wind Farms*; Katholieke Universiteit Leuven: Leuven, Belgian, 2008.
39. Guozhen, X.; Yonggang, G. Analysis and estimation of circulation and loss of GIS and Gil shells. *High Volt. Technol.* **2009**, *35*, 247–249.
40. National Development and Reform Commission of the People's Republic of China. *Yi Kang, Notice of the National Development and Reform Commission on the on Grid Tariff Policy of Offshore Wind Power*; National Development and Reform Commission of the People's Republic of China: Beijing, China, 2014.

Journal of
Marine Science and Engineering

MDPI

Article

Economic Viability Analysis for an OTEC Power Plant at San Andrés Island

Jorge Herrera [1,*], Santiago Sierra [2], Hernando Hernández-Hamón [3], Néstor Ardila [4,5], Andrés Franco-Herrera [1] and Asier Ibeas [6]

1 Faculty of Natural Sciences and Engineering, Universidad Jorge Tadeo Lozano, Bogotá 110311, Colombia; andres.franco@utadeo.edu.co
2 Department of Mechatronics Engineering, Universidad ECCI, Bogotá 111311, Colombia; ssierraa@ecci.edu.co
3 Ecomapp Geospatial Solutions, Bogotá 110231, Colombia; hernando.hernandez@ecomapp.co
4 División de Biología Marina, ECOMAR Consultoría Ambiental, Bogotá 110311, Colombia; nestorardila@ecomarconsultoria.co
5 Programa de Biología Aplicada, Facultad de Ciencias Básicas y Aplicadas, Campus Nueva Granada, Universidad Militar Nueva Granada, Cajicá 250247, Colombia
6 Department of Telecommunications and Systems Engineering, Universidad Autónoma de Barcelona, 08193 Barcelona, Spain; asier.ibeas@uab.cat
* Correspondence: jorgea.herrerac@utadeo.edu.co; Tel.: +57-1-242-7030

Abstract: This paper presents the economic feasibility analysis of a 2 MW Ocean Thermal Energy Conversion (OTEC) power plant in the open cycle. The plant can supply 6.35% of the average annual consumption of the electricity demand located at San Andrés Island (Colombia). On the one hand, the work presents the selection of the place to locate an offshore facility considering the technical viability while, on the other hand, the economic viability analysis is performed. The latter considers two scenarios: one without desalinated water production and another one with desalinated water. In this way, it is intended to first determine its construction's technical requirements to analyse its economic performance. This approach allows us to have a general idea of the implementation costs and the benefits obtained with this type of plant, for the particular case of San Andrés, an island in the Colombian Caribbean with sustained stress on electricity production and freshwater generation. The results obtained show that the technology is viable and that the investment can be recovered in an adequate time horizon.

Keywords: floating OTEC plant; marine energy; power energy; economic feasibility analysis

Citation: Herrera, J.; Sierra, S.; Hernandez-Hamón, H.; Ardila, N.; Franco-Herrera, A.; Ibeas, A. Economic Viability Analysis for an OTEC Power Plant at San Andrés Island. *J. Mar. Sci. Eng.* **2022**, *10*, 713. https://doi.org/10.3390/jmse10060713

Academic Editor: Eugen Rusu

Received: 30 April 2022
Accepted: 18 May 2022
Published: 24 May 2022

Publisher's Note: MDPI stays neutral with regard to jurisdictional claims in published maps and institutional affiliations.

Copyright: © 2022 by the authors. Licensee MDPI, Basel, Switzerland. This article is an open access article distributed under the terms and conditions of the Creative Commons Attribution (CC BY) license (https://creativecommons.org/licenses/by/4.0/).

1. Introduction

Islands emit fewer greenhouse gases (GHG) globally than the emissions generated on the continents; however, they have high per capita emissions. For example, the Caribbean Islands generated 0.4% of the world's total GHG emissions in 2011, but per capita emissions exceeded 120 tons while the world average was 5 tons per person [1]. These emissions are derived mainly from the generation of energy through fossil fuels.

Due to its location far from the rest of the country, the San Andrés Island is part of the non-interconnected areas of Colombia. Currently, electricity at San Andrés Island is generated by diesel power plants, with an approximate cost of USD 0.3 per kWh, which is relatively high compared to an interconnected city whose electricity price fluctuates around USD 0.08 per kWh. This difference is due to the costs of the thermoelectric operation plant and because diesel fuel must be brought by boat from Cartagena. It should also be noted that thermoelectric plants produce greenhouse gases that reach the atmosphere and contribute to global warming. Furthermore, most developing countries have insufficient financial and legislative resources to meet the challenges of climate change [2]. In addition, sustainable energy supplies are needed to reduce greenhouse gas emissions, thus mitigating the devastating effects of climate change [2]. Furthermore, the seventh Sustainable

Development Goals of the United Nations Organization emphasise energy affordability and clean energy use. In this way, Colombia and, in particular, the Colombian Caribbean is consequently in need of reliant sources of energy capable of guaranteeing a continuous supply of energy sustainably.

Furthermore, the well-being of communities can be improved by increasing the supply of electricity. However, there are historical reasons why this has not happened in some regions of the Colombian Caribbean. The lack of enough energy is added to other problems concerned with access to potable water, adequate housing, quality food, and high infant mortality rate, among others [3,4]. The reports [5,6] reveal that it will be necessary to increase political commitment and investment in energy at San Andrés Island, or else energy poverty will increase.

If a small part of the energy stored in the oceans could be recovered, the world's energy demand could be satisfied. However, the technology to recover energy from the oceans is at a very incipient level of development, with the production of marine power being only residual compared to other sources of renewable energy [7,8]. Among the different technologies to harvest energy from the sea, OTEC is one of the most incipient and promising ones. According to [9], the installed capacity of ocean energy in 2050 could reach 337 GW [7]. Along the same lines, a recent analysis by [10] estimates that the industrial-scale potential of an OTEC system is around 13 terawatts worldwide [7]. Moreover, the development of OTEC technologies is an opportunity to generate an industry around this type of energy generation, generating new jobs, both for the construction and maintenance of OTEC plants.

Countries located in tropical areas have the ideal conditions to develop OTEC technology; due to their proximity to the equator, the necessary temperature gradients are generated for the OTEC system operation [11,12]. Additionally, OTEC systems can be built to generate power on a small scale [13]. On the other hand, OTEC systems have greater economic viability in tropical islands not connected to the mainland electricity grid, with potable water deficiencies and air conditioning needs, since these needs could be mitigated simultaneously by an OTEC system [7]. Some developing countries that are investigating the feasibility of OTEC technology are Colombia [14], Indonesia [15–17], Panama [18] and Pakistan [7], among others.

OTEC systems use only the temperature gradient (ΔT) generated between the sea surface and deep water as an energy source, converting it into renewable energy. The ΔT is directly related to the performance of the OTEC power cycles; $\Delta T \approx 20$ °C is generally needed for an OTEC system to be viable [8]. Therefore, OTEC systems have a low thermal efficiency compared to other renewable energy sources. For example, an OTEC system based on the Rankine cycle generally has an efficiency of no more than 5% [19,20]. However, an OTEC system has the following two advantages. First of all, the power generation system can constantly work 24 h a day, something that is not possible with photovoltaic or wind systems; the temperature of the sea in tropical areas does not change considerably throughout the year, presenting small variations in the order of degrees due to the seasons and climate changes. Additionally, the variation of ΔT between day and night is around 1 °C [15,21]. Second, the OTEC system can generate freshwater as an indirect product [22–25].

Research on the OTEC system has focused on evaporators, turbines, generators, condensers, pumps, pipes to transport water, and moorings. For example, some investigations focus on the design of turbines to obtain the highest efficiency and net power [15]. Other investigations work on cycles; the most used are the Rankine, and double-stage Rankine [26] cycles; OTEC systems have also been implemented using various cycles such as Kalina [27], Uehara [28], which is an improvement of the Kalina cycle [19,20,27]. At the pipeline level, research is being carried out on the material and coating of the pipeline and the control that can be implemented to keep it stable. The pipe that transports deep seawater is a fundamental component of the OTEC system; this pipe, generally greater than 1000 m, must be designed to withstand the vibrations generated by deep water. Research on flexible structures can be found in [25]. Finally, studies on the economic viability of OTEC systems can be found in [15]. Various studies [16,29] have shown that the economic viability of an OTEC system requires plants with a maximum power of 100 MW, since the cost and complexity increase considerably from systems with the higher power.

Although the technical aspects are essential for developing this technology, it is also true that it is necessary to know the economic viability due to its low efficiency and high implementation costs. In this line, different works with multiple approaches have been developed. For example, in [30], a thermo-economic analysis of a 20 kW OTEC system is conducted by calculating the unit cost of electricity generated. They concluded that OTEC systems are economically viable in regions where the surface seawater temperature is greater than 25 °C. On the other hand, Ref. [31] presents the design of an OTEC Ecopark consisting of a 60 MW OTEC system coupled to a marine aquaculture farm located near the Island of Cozumel. The proposed system meets part of the needs of coastal communities for energy production, desalinated water and food production. The work was based on the technical-economic evaluation of the OTEC Ecopark, and the financial evaluation showed that the OTEC Ecopark is economically viable, having a CAPEX of USD 655.38 M, an OPEX of USD 69.66 M and an annual income of USD 348 M. Studies that carry out economic viabilities can be found in [32–34], and an interesting review on economic feasibility studies is presented in [35].

This paper conducts an economic feasibility study of installing an OTEC plant at San Andrés Island. In this sense, it is assumed that the economic viability is closely related to the installation location and the plant's technical design. The economic viability is based on two analyses, (i) a cash flow analysis of the project, and (ii) a levelized cost of energy (LCOE) are carried out [36]. The latter way of measuring is usually used to assess the cost of employing different methods for generating renewable energy [35]. It should be noted that the LCOE in this work is calculated individually for the proposed OTEC system. Ideally, if it wants to have a total generation from renewable energy, the LCOE should be calculated for the entire system [37].

This study is a starting point to develop this technology on the Island, developing an OTEC system that improves electricity and potable water supply needs. Furthermore, the environmental conditions at San Andrés Island sea (optimal surface temperature, ΔT around 20 °C all year, low frequency of hurricanes, ideal depth at a short distance from the coast) favour the implementation of an OTEC plant, which could generate electricity for several homes on the island without producing polluting waste and operating costs lower than those of the diesel plants.

The paper is organised as follows. Section 2 explains the current energy and water situation at San Andrés Island and briefly describes an OTEC system. Section 3 summarises the methodology for performing the economic analysis. Section 4 exposes an OTEC system location, the technical conditions of the OTEC system, and the economic viability of two possible scenarios presented, without potable water and with potable water. Finally, Section 5 summarises the main conclusions.

2. Problem Formulation

Next, the energy and water needs of San Andrés Island are presented to complete the description of the OTEC system.

2.1. Energy and Water Needs of San Andrés Island

San Andrés Island is located to the west of the Caribbean Sea in the Atlantic Ocean. The island has a warm climate, between 26 °C and 29 °C all year round. Throughout the year, there are two seasons (i) the dry season, usually between January and April, but it can last a maximum of five months, (ii) the other months are part of the rainy season with strong winds, usually between May and December [38].

The economy of the department is based mainly on tourism and commerce. Its main export product is coconut, but it also produces avocado, sugar cane, mango, orange and yucca [39].

It is the least extensive department in the country and has the highest population density, which places the islands in a delicate resource management situation. It is estimated that around 75,000 people live on San Andrés Island, and every year, one million tourists arrive [40]. A considerable figure for a territory of 26 square kilometres with no rivers.

The particular geography of the San Andrés Island, its condition of insularity, having only two aquifers, the amount of population that inhabits the islands, as well as the floating population that arrives throughout the year and depends on imported food are just some of the characteristics that make this territory highly vulnerable to climate change and shortage of potable water.

Currently, the San Andrés Island electricity demand is approximately 160–187 GWh/year [41], which is supplied by diesel-powered thermoelectric plants, consuming around 40 million L of diesel each year, which are brought by boat from Cartagena. In addition, there is also a thermoelectric power plant that works with the burning of garbage, which has an installed capacity of 1.6 MW and a helpful power of 1 MW [42]. This situation, combined with the variable costs of fuel and the transport high prices, serves as an incentive for companies and communities to seek new energy alternatives.

On 17 July 2018, the Colombian government delivered the new desalination system and treated water line for the neighbourhoods of La Loma, El Cove and San Luis, at San Andrés Island, which facilitates the supply of treated water in the communities of this sector and avoids that emergencies occur due to the shortage of water in times of drought. The work had a total initial investment of 4.2 million dollars, and the plant can treat 25 L of water per second, enough to benefit more than 23,000 inhabitants [43].

It is convenient to study more thoroughly the possibilities offered by an OTEC plant for the San Andrés Island since not only can the natural resources available be used to generate electricity, but in addition, OTEC plants can convert seawater into potable water at a cost similar to that of a conventional desalination plant [44]. This last one would be quite beneficial for the island population because it would help to supply this indispensable resource.

2.2. Ocean Thermal Energy

OTEC is a type of renewable energy that uses the ΔT between the surface and deep layers of the sea to move a thermal machine and produce valuable work, usually in the form of electricity [12]. On the other hand, the oceans cover more than 70% of the earth's surface, and by absorbing heat, it can store a large part of the energy emitted by the sun. In this sense, using a small portion of this stored energy could meet the energy needs of a country [45]. The water column temperature in the Colombian Caribbean depends on the origin of the water masses from different latitudes, such as the North or Central Atlantic, and the contribution of the great rivers of the southern Orinoco and Amazon. Each mass of water has a characteristic temperature, salinity, and density [46]. In the Colombian Caribbean, three thermal layers can be differentiated in the water:

1. The surface is between 0 and 100 m deep and has a temperature between 25 °C and 30 °C all year round.
2. The intermediate one is between 100 and 1000 m deep and is a thermal barrier between the upper and lower.
3. The depth is more than 1000 m deep, where the temperature gradually drops to 4 °C.

In an OTEC system, the cycle efficiency is directly related by the ΔT [47]. The larger the ΔT is, the higher efficiency obtained. OTEC systems should be located close to shore to reduce the transmission costs of the electricity generated [14].

2.2.1. OTEC Cycles

Surface water heats a liquid using a heat exchanger, transforming it into steam, which drives a turbine that generates electricity. The cycle cools the steam with another heat exchanger in deep water, restarting the generation cycle. Currently, the primary cycles are open, closed or hybrid. In an open cycle OTEC system, the hot water found on the surface is taken to a vacuum chamber using a vacuum pump, which operates at a maximum of 3% of atmospheric pressure [48]. The water evaporates rapidly through this pressure drop, and the expanding steam drives a low-pressure turbine connected to an electric turbine. An advantage of this cycle is that the steam leaves the minerals in the vacuum chamber, producing desalinated water, which can be used depending on its physical-chemical characteristics for water for human consumption or irrigation [49]. Closed cycle OTEC systems, on the other hand, use refrigerant, which is heated directly by heat from surface water; the evaporated refrigerant drives an electric turbine and is cooled in deep water. The main advantage of this cycle is lower construction and operation costs; however, it requires more outstanding care in the handling of the refrigerant; it should be noted that the closed cycle does not produce desalinated water [50]. The hybrid cycle contains characteristics of both the open-loop and the closed-loop. In this case, the water is taken through a pump to a vacuum chamber to be evaporated. The water vapour heats a refrigerant that activates an electric turbine; the water vapour is condensed in a heat exchanger located in deep water. The hybrid cycle has desalinated water as an indirect product [51].

2.2.2. Desalinated Water as a Derived Product

An advantage of OTEC systems is the possibility of generating desalinated water. For example, a 1 MW hybrid OTEC system can produce around 4500 m^3/day of desalinated water [52]. However, it should be noted that the cost of producing desalinated water by this method is comparable to standard desalination plants [48].

OTEC systems can be a solution to both the water and energy needs of San Andrés Island. However, the implementation costs are high. Considering that the Island has limited financial resources, it is necessary to carry out a financial assessment that serves as a starting point when implementing a size project to ensure its long-term viability and profitability. This work is aimed at contributing to this point.

3. Methodology to Perform the Economic Analysis

The proposed methodology is divided into three parts: first, selecting a location for the system; second, a power plant's technological description and third, an economic analysis, which are described below.

3.1. Methodology Used in the System Location

The OTEC system's location selection aims to determine the place to settle the plant and select if an onshore or offshore installation is carried out. Bathymetry and temperature differences at San Andrés Island, along with the best location selection are presented. To this end, environmentally data collected from NASA are used. Then, the differential temperature between the sea surface and the 1000 m depth around San Andrés Island is obtained. Next, bathymetry is performed at 7 points on the island, and the point with the most appropriate profile is selected according to [14].

3.2. Power Plant's Technological Description

The power plant's technological description is introduced to perform its economic analysis. To this end, various software will be used to simulate the operation of an OTEC power plant at San Andrés Island, understand its benefits and obtain its technical characteristics. Google Earth for the positioning of the plant, Autodesk Inventor for the 3D sketch and Cadesimu for the electrical circuit.

The technical conditions of the OTEC system are defined, and it should be clarified that simplifications are made to obtain an overview of the implementation. The energy model, the sketch of the power plant, the electrical transfer scheme, and the emissions analysis are provided in this step.

3.3. Economic Analysis

Finally, the economic analysis is carried out for two scenarios, namely, without and with potable water production, based on the particular characteristics designed in the previous items. The economic viability is based on two analyses.

3.3.1. Cash Flow Analysis of the Project

The project's cash flow analysis is made with RETScreen, which calculates and analyses variables associated with the project, being these either technological, economic or environmental. The information that must be supplied to the software is as follows:

- It is taken from its database based on location, parameters such as temperature, relative humidity, precipitation, atmospheric pressure, wind speed, and daily solar radiation.
- Type of installation (type of generation)
- Turbine Specifications: steam flow, operating pressure, temperature, efficiency, generation capacity.
- Initial and annual costs
- Transmission and distribution losses
- Financial parameters: 3% of the inflation rate, project lifetime of 30 years, a debt ratio of 50%, debt interest rate of 3%, debt duration of 15 years.
- Meteorological parameters correspond to the particular location, such as temperature, relative humidity, precipitation, atmospheric pressure, wind speed, and daily solar radiation. These parameters are taken from the RETScreen database.

The parameters specifying the above characteristic are provided in the technical description of the system in Section 4.2.

3.3.2. Levelized Cost of Energy (LCOE)

The LCOE can be used to assess the economic viability of energy projects. The LCOE can be interpreted as the minimum average price at which the generated electricity should be sold (throughout its useful life) to reach a similar cost to other energy projects. All expenses associated with the project must be considered [35,53]. It can be calculated using Equations (1) and (2).

$$LCOE = \frac{CRF \cdot CAPEX + OPEX}{E_t} \tag{1}$$

with

$$CRF = \frac{i \cdot (1+i)^N}{(1+i)^N - 1} \tag{2}$$

and:
$CAPEX$: capital expenses.
$OPEX$: operational expenses.
E_t: produced electricity in year t.
CRF: capital recovery factor.
N: project lifetime
i: interest rate.

4. Results

4.1. OTEC System Location

Initially, selecting an appropriate place to locate the OTEC facility is discussed. According to NASA data presented in Table 1, San Andrés Island has the appropriate sea and weather conditions to host an OTEC facility. Table 1 presents the air temperature

(AT), relative humidity (RH), daily solar radiation (DSR), wind speed (WS) and earth temperature (ET).

Table 1. Climatological data of San Andrés Island, average of the last ten years. Source NASA.

Month	AT (°C)	RH (%)	DSR (kWh/m^2/d)	WS (m/s)	ET (°C)
January	26.6	75.0	5.4	7.6	27.3
February	26.3	75.0	6.1	7.2	27.0
March	26.1	75.8	6.9	6.7	27.1
April	26.4	78.0	7.0	6.0	27.6
May	26.7	82.8	6.1	4.9	28.1
June	27.0	83.9	5.4	5.1	28.3
July	2.0	82.5	5.5	5.9	28.0
August	26.9	83.5	5.6	5.1	28.4
September	26.8	84.5	5.4	4.1	28.8
October	27.0	82.7	4.9	4.3	28.9
November	27.2	79.0	4.6	5.6	28.5
December	27.1	76.0	4.7	7.4	27.8
Anual	26.8	79.9	5.6	5.8	28.0

The main requirement to meet is the condition $\Delta T \geq 20$ °C between the surface and deep layers of the ocean. Conveniently, the ΔT is achieved as superficially as possible to minimise the water pumping cost.

It can be seen in Figure 1 that the ΔT at San Andrés Island is greater than 20 °C throughout the year, between the surface and 1000 m depth. In the period of highest temperature, which occurs in September, it is observed that the ΔT is around 22.5 °C. In the season of lowest temperature, which occurs in March, the ΔT is around 20 °C. It should be noted that the blank spaces in Figure 1 mean that the depth is less than 1000 m in those places. Consequently, it is convenient to select locations for the OTEC facility where the access to 1000 m depth is easy. To this end, the bathymetry of the San Andrés Island is presented in [14].

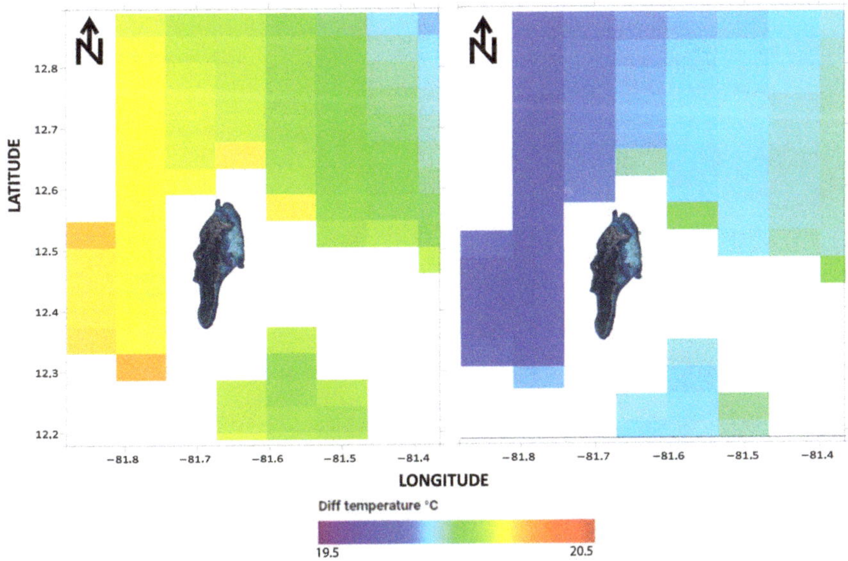

Figure 1. Differential on water temperature between surface and 1000 m depth around San Andrés Island on March (**left**) and September (**right**).

Figure 2 shows the bathymetric profile for seven points at San Andrés Island. The seven points selected are Airport, Barrio Obrero, Hans Dive Shop, La Loma, Playa Rocky Cay, San José, and Tana. The specific location of these points can be found in [14]. The bathymetric and temperature conditions of the seven points are presented in Table 2. Where $size_{mp}$ is the distance from the coast to where the platform break is generated, $depth_{pb}$ is the depth where the platform break starts, $distance_{fc}$ is the distance from the coast to reach 1000 m depth, ΔT_{min} is the minimum temperature differential and ΔT_{max} is the maximum temperature differential. It can be seen that the maximum width is around 4 km.

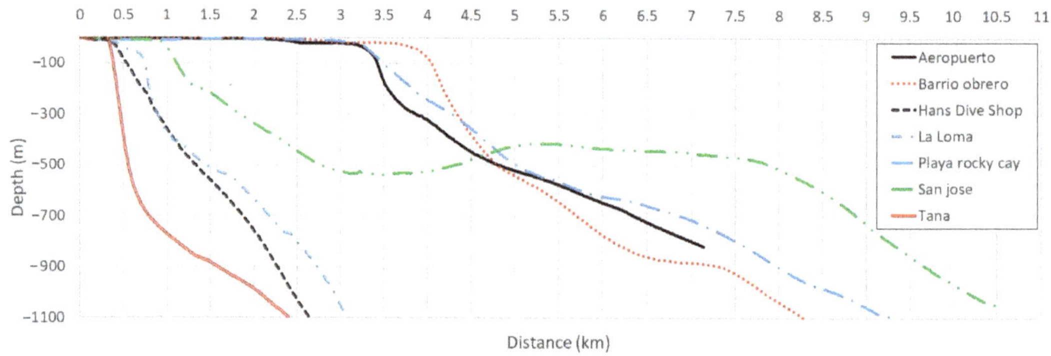

Figure 2. Bathymetric profiles around San Andrés Island, showing how high depths are reached near the coast.

Table 2. Continental platform break in 7 points of San Andrés Island.

Cities	$size_{mp}$ (km)	$depth_{pb}$ (m)	$distance_{fc}$ (km)	Slope	ΔT_{min} (°C)	ΔT_{max} (°C)
Aeropuerto	3.27	−39.52	7.85	165.46	19.88	22.38
Barrio Obrero	3.94	−58.74	8.00	191.21	19.97	22.49
Playa Rocky Cay	3.29	−36.39	8.75	160.34	19.94	22.40
San José	1.01	−41.37	10.32	59.48	20.03	22.40
Hans Dive Shop	0.42	−38.64	2.68	403.53	19.91	22.25
La Loma	0.75	−113.16	3.18	298.73	19.91	22.26
Tana	0.35	−43.86	2.49	324.77	19.91	22.24

Based on the climatological data, the San Andrés Island has an ideal temperature to implement an OTEC plant, in addition to the other necessary characteristics, such as a short distance between 1000 m deep and the coast, with a steep slope and a seabed without many reliefs, the location must have low waves (height less than 3.7 m), and ocean surface currents less than 1.5 m/s, and a low incidence of natural phenomena such as storms, earthquakes, hurricanes, among others [54].

It is concluded that the best location for the OTEC facility is at Tana Point, at coordinates 81°44′ W and 12°29′ N. Tana presents the 1000 m depth at 2.49 km; another advantage is that the slope decreases rapidly from the break in the platform at 0.35 km until the first kilometre. After that, it descends gently, being an advantage when establishing the mooring system. On the other hand, the ΔT_{min} is 19.91 °C, and the ΔT_{max} is 22.24 °C, being an ideal temperature for the optimal efficiency of an OTEC system. Therefore, this will be the location chosen to analyse the economic feasibility of the OTEC plant.

4.2. Technical Conditions of the OTEC System

The technical conditions of the proposed system are described below.

4.2.1. Development of the Energy Model

According to the calculations made in [55], they determined that about 4 m^3/s of surface water and 2 m^3/s of deep water are required with $\Delta T \approx 20$ °C for each MW of net electricity generated. An average speed of around 2 m/s is required for the seawater to circulate through the pipes from deep water to the surface; in this way, pumping losses more significant than 30% of the gross power are avoided [56]. Taking as a reference [57], which states that 10% of the steam that enters the turbine can be converted into desalinated water, this value is taken as a reference point in the calculations presented here.

An open-cycle OTEC of 2MW is chosen for the design, representing approximately 9.4% of the yearly electricity demand at San Andrés Island (160–187 GWh/year). Table 3 shows the variables used in the energy model.

Table 3. Summary of the variables with their respective values calculated for a 2 MW OTEC.

Symbol	Quantity	Values	Units
HWF	Hot water flow	8	m^3/s
CWF	Cold water flow	4	m^3/s
WS	Water speed		m/s
IDHWP	Internal diameter of hot water pipe	2.25	m
IDCWP	Internal diameter of cold water pipe	1.59	m
SFT	Steam flow in the turbine		kg/s
MFDW	Mass flow of desalinated water	9.37	kg/s
DWF	Desalinated water flow		L/s

Equations (3)–(10) present the calculations for the energy model; it should be noted that these calculations are based on recommendations made by [55]. The values were adjusted to meet the 2 MW objective of the proposed plant.

$$HWF = 4\ m^3/s \cdot 2 = 8\ m^3/s \tag{3}$$

$$CWF = 2\ m^3/s \cdot 2 = 4\ m^3/s \tag{4}$$

$$WS = 2\ m/s \tag{5}$$

$$IDHWP = \sqrt{\frac{4 \cdot 8\ m^3/s}{2\ m/s \cdot ß}} = 2.25\ m \tag{6}$$

$$IDCWP = \sqrt{\frac{4 \cdot\ 4m^3/s}{2\ m/s \cdot ß}} = 1.59\ m \tag{7}$$

$$SFT = 8\ m^3/s \cdot 1000\ kg/m^3 \cdot 1.17\% = 93.75\ kg/s \tag{8}$$

$$MFDW = 93.75\ kg/s \cdot 10\% = 9.375\ kg/s \tag{9}$$

$$DWF = 9.37\ L/s \tag{10}$$

In Figure 3, it can be seen the system sketch, along with the amounts of water in both liquid and gaseous states, which are calculated for each processing part. Table 4 presents the values calculated in the RetScreen software, taking into account the thermodynamics of a steam turbine [58]. A 2 MW plant can produce up to 15.8 GWh/year.

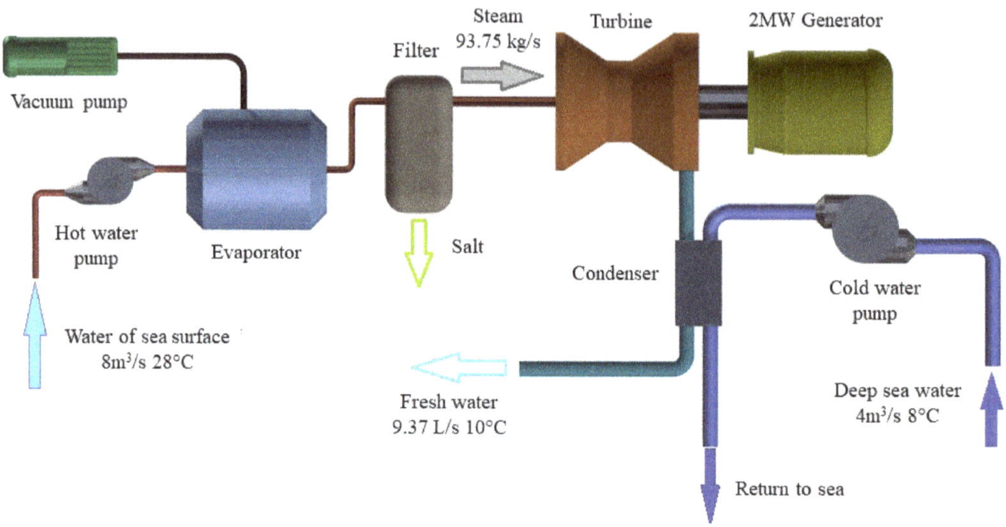

Figure 3. A schematic diagram showing values for an open-cycle OTEC system of 2 MW.

Table 4. Operating Specifications for Steam Turbine.

Item	Unity	Value
Avaliability	90.0%	7884 h
Steam flow	kg/h	337.50
Operating pressure	kPa	3.00
Saturation temperature	°C	23.00
Steam temperature	°C	28.00
Back pressure	kPa	1.00
Steam turbine efficiency	%	22.00
Actual steam rate	kg/kWh	166.88
Power capacity	kW	2.02
Electricity exported to grid	MWh	15.95
Electricity exported rate	MWh	166.67

4.2.2. Sketch of the Power Plant

Due to the island's conditions, an offshore floating OTEC is chosen. An offshore installation is selected instead of an onshore one since San Andrés Island has a high density and limited onshore space. On the other hand, the 1000 m depth is obtained near the coast, reducing transmission costs to the coast of both the electrical energy and the water produced. By the calculations shown above, the pipes' sizes and the plant itself are dimensioned. A preliminary design is created and shown in Figure 4.

4.2.3. Electrical Transfer Scheme

Since many companies and businesses have diesel plants, these could continue to be used if the OTEC plant needs to stop for maintenance or it cannot provide the required energy. For these cases, the electrical design is presented in Figure 5.

This transfer circuit would allow switching between the OTEC and the Diesel generation to the grid. The change occurs automatically when the power of the OTEC network is not available, in which case it will take eight seconds, and the backup system will be activated. When the OTEC power returns, the backup system will be automatically deactivated, eight seconds will pass, and the central system will re-enter.

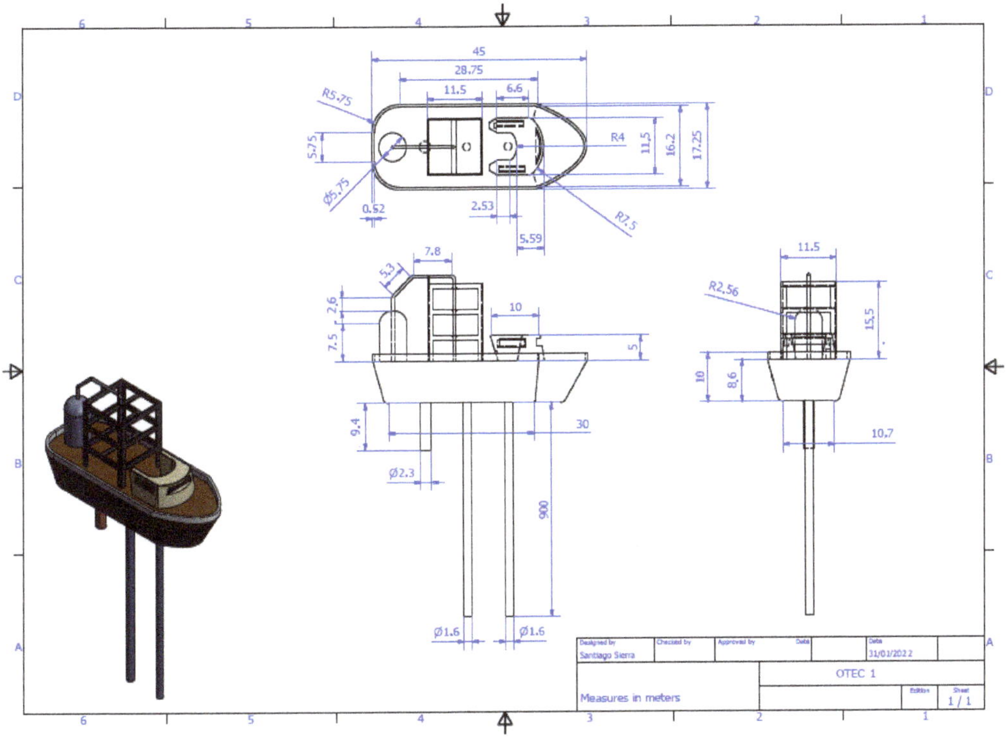

Figure 4. A layout for 2 MW OTEC floating device.

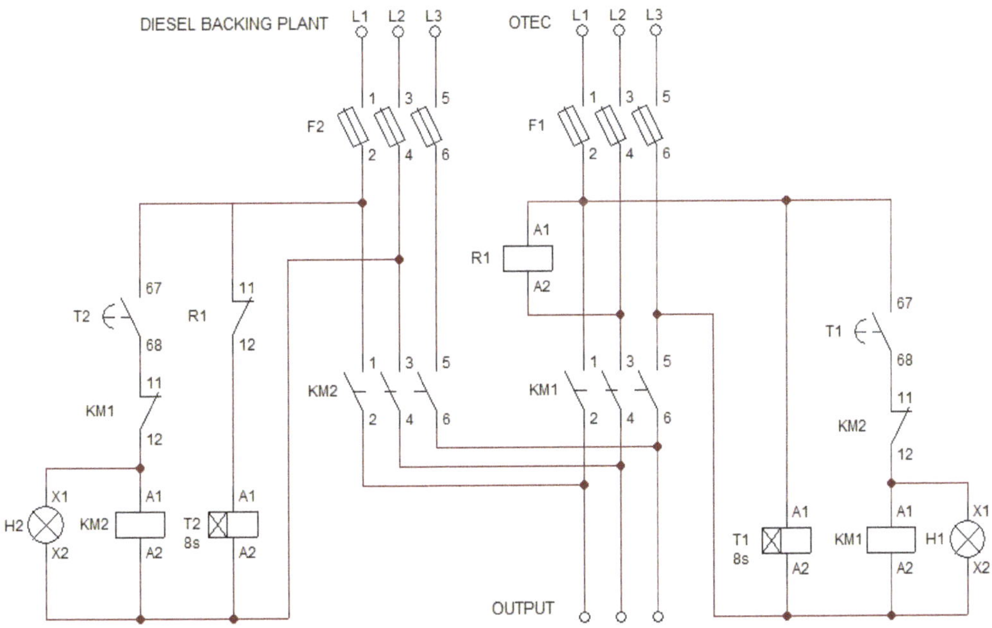

Figure 5. Electrical transfer circuit. In case of a failure, the electric transfer circuit is in charge of switching between energy coming from the OTEC to a backup diesel plant.

4.2.4. Analysis of Emissions

The emissions analysis gives a favourable result for the OTEC plant since, if it compares the same generation of energy made with the existing diesel plants, an annual reduction of 2404 tons of CO_2 can be obtained, equivalent to 1,032,932 L of gasoline not consumed.

In Table 5, it can be seen the broken down of this data, all of them calculated using the RetScreen software by using statistical data on the production of greenhouse gases in power generation plants.

Table 5. Emissions analysis. Net annual GHG emission reduction 2404 tCO_2 is equivalent to 1,032,932 L of gasoline not consumed.

Base Case Electricity System (Baseline)				
Country-region	Fuell type	GHG emission factor (excl. T&D) tCO_2/MWh	T&D losses %	GHG emission factor tCO_2/MWh
Colombia	All types	0.15 L	12.3	0.17
GHG emision reduction summary				
	Base case GHG emission tCO_2	Proposed case GHG emission tCO_2	Gross annual GHG emission reduction tCO_2	Net annual GHG emission reduction tCO_2
Power project	2741.7	337.2	2404.5	2404.50

4.3. *Result of Economic Analysis*

Apart from generating electricity, the OTEC power plant offers another essential advantage: potable water production. In this paper, two possible scenarios for the OTEC plant at San Andrés Island are examined, the first is without potable water production, and the second is if potable water is produced and sold as a public service.

In both cases, an initial investment of 50% of the total project value is assumed by the company that builds the OTEC system, while the other 50% is requested from external financing with a 15-year loan. According to the current OTEC plants, it is estimated that the cost of construction of one of these plants is, on average, USD 15,000 per kW, without counting the transmission lines or the adjustments that must be made to the terrain. It should be noted that the economic data for the construction of the plant and the energy production itself were taken based on the OTEC built by the Natural Energy Laboratory of Hawaii Authority (NELHA) [59], also taking into account importation costs and Colombian labour.

4.3.1. Scenario 1: Without Production of Potable Water

Table 6 presents the information from the financial analysis. In this case, the total initial cost is USD 39,245,257 and has an annual cost of USD 1,528,675 for salaries, maintenance and payments for the external financing. The only income is the sale of electric energy, approximately 15,945 MWh per year, which generates an average annual income of USD 2,657,472. Figure 6 shows the behaviour of the cash flow of scenario 1. The project starts to be profitable after year 13; then, the profit exceeds the annual costs.

Table 6. Initial and annual costs, and annual income, for Scenario 1.

Project Costs and Saving/Income Summary		
Initial costs		
Feasibility study	2.5%	$1,000,000
Engineering	12.7%	$5,000,000
Power system	79.9%	$31,376,435
Balance of system & misc.	4.8%	$1,868,822
Total initial costs	100%	$39,245,257
Annual costs and debt payments		
O&M		$220,500
Fuel cost-proposed case		0
Debt payments-15 years		$1,308,175
Total annual costs		$1,528,675
Annual saving and income		
Electricity export income		$2,657,472
Total annual saving and income		$2,657,472

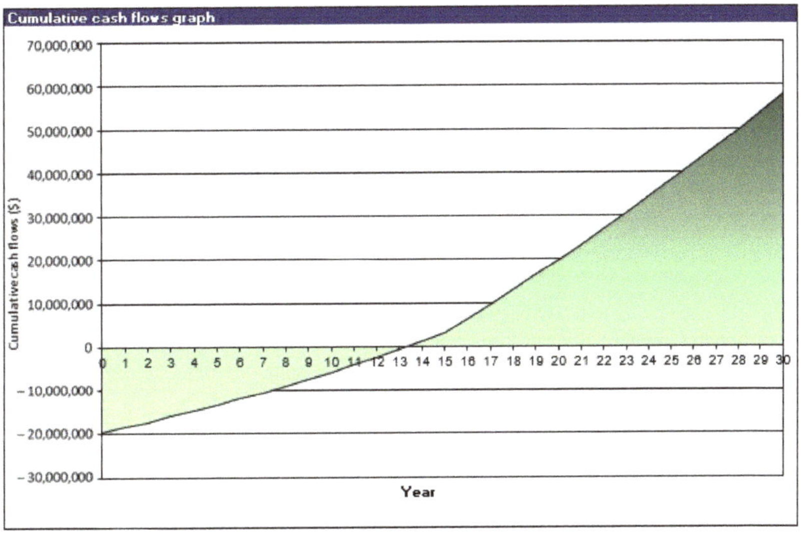

Figure 6. Return on investment, and long-term cash-flow, for Scenario 1.

Using Equations (1) and (2) a LCOE of 0.22 USD/kWh is obtained for scenario 1. Despite this value may be slightly high compared to [35], whose LCOE is around 0.2 USD/kWh, it is observed that the proposed system has a cost of 36% lower than diesel production, which at this time is around 0.3 USD/kWh. Consequently, installing an OTEC facility without desalination is economically viable on San Andrés Island.

4.3.2. Scenario 2: Potable Water as Public Service

In this case, the total initial cost is USD 42,395,257, as can be seen in Table 7. It is more expensive than the first scenario because equipment must be purchased, and rooms must be conditioned to convert the desalinated water into potable water (it is necessary to adjust the pH to 7.7 and eliminate all types of pathogenic organisms by chlorination, perchlorination or ozonation).

Table 7. Initial and annual costs, and annual income, for Scenario 2.

Project Costs and Saving/Income Summary		
Initial costs		
Feasibility study	2.4%	$1,000,000
Engineering	18.9%	$8,000,000
Power system	74.0%	$31,376,435
Balance of system & misc.	4.8%	$2,018,822
Total initial costs	100%	$42,395,257
Annual costs and debt payments		
O&M		$517.465
Fuel cost-proposed case		0
Debt payments-15 years		$1,413,175
Total annual costs		$1,930,641
Annual saving and income		
Electricity export income		$2,657,472
Other income (cost) - 30 year		$295,492
Total annual saving and income		$2,952,964

The annual cost is about USD 1,930,641 for salaries, maintenance, payments to the bank, and potable water production. It has two incomes:

The sale of electric energy, approximately 15,945 MWh per year, which generates an average annual income of USD 2,657,472.

The sale of potable water as public service, approximately 295,492 m^3 per year, with a standard price of 1 USD/m^3. It represents an additional annual income of USD 295,492.

The two incomes add up to an annual total of USD 2,952,964. The project starts to be profitable after year 15. After that, some profits greatly exceed the annual costs, as shown in Figure 7.

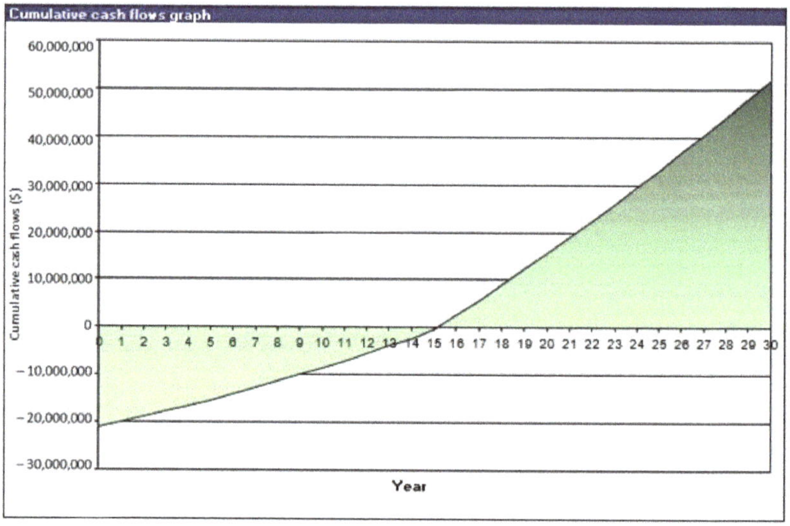

Figure 7. Return on investment, and long-term cash-flow, for Scenario 2.

For scenario 2, the LCOE obtained is 0.26 USD/kWh, which remains competitive concerning diesel production and makes installing an OTEC facility with desalination economically viable at San Andrés Island.

5. Conclusions

San Andrés Island has the ideal condition for the location of an OTEC power plant since it meets all the technical and environmental conditions required, and a power plant of this type would bring significant benefits to the island, such as an electrical system more ecological, economic and stable, together with a more continuous potable water service.

The results show that the operation of an OTEC plant at San Andrés can be viable. Of course, it requires a high initial investment, but given that it is a clean technology that does not consume fuels and that can cogenerate associated products, in the long term, the investment can be recovered and eventually give benefits.

However, it must be taken into account that the values obtained in this paper are based on a theoretical analysis; these values would change when implementation is carried out since it is possible that environmental factors and the type of soil, among others, impact the costs of the plant. In this sense, it is recommended to carry out more detailed studies regarding the installation and moorings costs in different parts of the island, which may be lines of future research. However, the analysis performed in this work provides the positive overall conclusion that it is worthy to seriously explore the installation of such a facility in San Andrés since the system is economically feasible.

Author Contributions: Conceptualization, J.H., S.S. and A.F.-H.; investigation, S.S. and J.H.; writing—original draft preparation, J.H., H.H.-H., S.S., N.A., A.F.-H. and A.I.; writing—review and editing, J.H., H.H.-H., S.S., N.A., A.F.-H. and A.I.; funding acquisition, J.H., A.F.-H. and A.I. All authors have read and agreed to the published version of the manuscript.

Funding: This work was partially supported by the ERANET-LAC project, which has received funding from the European Union Seventh Framework Programme, and has been funded by the Ministry of Science, Innovation, Universities through the Spanish Research Agency (PCI2019-103376, ERANet17/ERY 0168) and by the Faculty of Natural Sciences and Engineering of the Universidad Jorge Tadeo Lozano.

Institutional Review Board Statement: Not applicable.

Informed Consent Statement: Not applicable.

Data Availability Statement: Not applicable.

Conflicts of Interest: The authors declare no conflict of interest.

Abbreviations

The following abbreviations are used in this manuscript:

AT	Air temperature
CWF	Cold water flow
ΔT	Temperature Gradient
DSR	Daily solar radiation
DWF	Desalinated water flow
ET	Earth temperature
GHG	Greenhouse Gases
HWF	Hot water flow
IDCWP	Internal diameter of cold water pipe
IDHWP	Internal diameter of hot water pipe
LCOE	levelized cost of energy
MFDW	Mass flow of desalinated water
OTEC	Ocean Thermal Energy Conversion
RH	Relative humidity
SFT	Steam flow in the turbine
T&D	Transmission and distribution
WS	Wind speed

References

1. Group, W.B. *World Development Indicators 2014*; World Bank Publications: Washington, DC, USA, 2014.
2. Menyah, K.; Wolde-Rufael, Y. Energy consumption, pollutant emissions and economic growth in South Africa. *Energy Econ.* **2010**, *32*, 1374–1382. [CrossRef]
3. Mazur, A. Does increasing energy or electricity consumption improve quality of life in industrial nations? *Energy Policy* **2011**, *39*, 2568–2572. [CrossRef]
4. Aristizábal, A.J.; Herrera, J.; Casta neda, M.; Zapata, S.; Ospina, D.; Banguero, E. A new methodology to model and simulate microgrids operating in low latitude countries. *Energy Procedia* **2019**, *157*, 825–836. [CrossRef]
5. Opoku, E.E.O.; Kufuor, N.K.; Manu, S.A. Gender, electricity access, renewable energy consumption and energy efficiency. *Technol. Forecast. Soc. Chang.* **2021**, *173*, 121121. [CrossRef]
6. Danielsen, K. *Gender Equality, Women's Rights and Access to Energy Services*; Ministry of Foreign Affairs of Denmark: Copenhagen, Denmark, 2012.
7. Farhan, M.; Qureshi, S.R.; Tayyab, S.M.; Shahid, M. Ocean Thermal Energy Conversion(OTEC)—A Techno-economic Analysis for Coastal Area of Pakistan. In Proceedings of the 2018 International Conference on Power Generation Systems and Renewable Energy Technologies (PGSRET), Islamabad, Pakistan, 10–12 September 2018; pp. 1–8. [CrossRef]
8. Mansour, M.Z.; Shehata, A.S.; Shehata, A.I.; ElSafty, A.F. Techno Selection Approach of Working Fluid for Enhancing the OTEC System Performance. In Proceedings of the 2020 3rd International Conference on Power and Energy Applications (ICPEA), Busan, Korea, 9–11 October 2020; pp. 154–158. [CrossRef]
9. Ocean Energy Europe. *Ocean Energy Project Spotlight-Investing in Tidal and Wave Energy*; Technical Report; Ocean Energy Europe: Brussels, Belgium, 2017.
10. Rajagopalan, K.; Nihous, G.C. An assessment of global Ocean Thermal Energy Conversion resources under broad geographical constraints. *J. Renew. Sustain. Energy* **2013**, *5*, 063124. [CrossRef]
11. Semmari, H.; Stitou, D.; Mauran, S. A novel Carnot-based cycle for ocean thermal energy conversion. *Energy* **2012**, *43*, 361–375. [CrossRef]
12. Herrera, J.; Sierra, S.; Ibeas, A. Ocean Thermal Energy Conversion and Other Uses of Deep Sea Water: A Review. *J. Mar. Sci. Eng.* **2021**, *9*, 356. [CrossRef]
13. Moriarty, P.; Honnery, D. What is the global potential for renewable energy? *Renew. Sustain. Energy Rev.* **2012**, *16*, 244–252. [CrossRef]
14. Herrera, J.; Hernández-Hamón, H.; Fajardo, L.; Ardila, N.; Franco, A.; Ibeas, A. Colombian Caribbean Bathymetry for an OTEC System Location. *J. Mar. Sci. Eng.* **2022**, *10*, 519. [CrossRef]
15. Arief, I.S.; Aldara, D.R. Preliminary Design of Ocean Thermal Energy Conversion (OTEC) Axial Turbine for Laboratory Scale. In Proceedings of the 2018 Asian Conference on Energy, Power and Transportation Electrification (ACEPT), Singapore, 30 October–2 November 2018; pp. 1–8. [CrossRef]
16. Bachtiar, I.K.; Putra, R.D. OTEC Potential Studies For Energy Sustainability In Riau Islands. In Proceedings of the 2019 6th International Conference on Electrical Engineering, Computer Science and Informatics (EECSI), Bandung, Indonesia, 18–20 September 2019; pp. 385–391. [CrossRef]
17. Prakasa, L.O.R.N.; Sholichah, H.; Wikaningrum, T. WOTEC Technology as the Potential Renewable Energy: Literature Study in East Nusa Tenggara. In Proceedings of the 2019 International Conference on Sustainable Engineering and Creative Computing (ICSECC), Bandung, Indonesia, 20–22 August 2019; pp. 83–88. [CrossRef]
18. Tenorio, G.L.; James, A.; Ortega, M.D.L.A.; Jurado, F. OTEC Alternative for the Electric Power Generation in Panama. In Proceedings of the 2019 7th International Engineering, Sciences and Technology Conference (IESTEC), Panama City, Panama, 9–11 October 2019; pp. 197–202. [CrossRef]
19. Matsuda, Y.; Oouchida, R.; Sugi, T.; Goto, S.; Morisaki, T.; Yasunaga, T.; Ikegami, Y. Simultaneous Regulation of Multiple Flow Rates for Power Generation Control of OTEC Plant Using Double-Stage Rankine Cycle. In Proceedings of the 2018 57th Annual Conference of the Society of Instrument and Control Engineers of Japan (SICE), Nara, Japan, 11–14 September 2018; pp. 983–988. [CrossRef]
20. Aosaki, Y.; Matsuda, Y.; Sugi, T.; Goto, S.; Yasunaga, T.; Ikegami, Y. Model Construction of OTEC Plant Using Double-stage Rankine Cycle with Time Delay by Considering Separator and Working Fluid Tank. In Proceedings of the 2019 12th Asian Control Conference (ASCC), Kitakyushu, Japan, 9–12 June 2019; pp. 358–363.
21. Soesilo, K.K. Simulasi Sistem Pembangkit OTEC Siklus Tertutup Dengan Variasi Fluida Kerja Ammonia (NH3) dan Refrigerant (R-12, R-22, R-23, R-32, R134a) Menggunakan Cycle Tempo. Ph.D. Thesis, Institut Teknologi Sepuluh Nopember, Surabaya, Indonesia, 2017.
22. Buitrago, J.; Rada, M.; Hernández, H.; Buitrago, E. A single-use site selection technique, using GIS, for aquaculture planning: Choosing locations for mangrove oyster raft culture in Margarita Island, Venezuela. *Environ. Manag.* **2005**, *35*, 544–556. [CrossRef]
23. Rajagopalan, K.; Nihous, G.C. Estimates of global Ocean Thermal Energy Conversion (OTEC) resources using an ocean general circulation model. *Renew. Energy* **2013**, *50*, 532–540. [CrossRef]
24. Nihous, G.C. An estimate of Atlantic Ocean thermal energy conversion (OTEC) resources. *Ocean. Eng.* **2007**, *34*, 2210–2221. [CrossRef]

25. He, X.; He, W.; Liu, Y.; Wang, Y.; Li, G.; Wang, Y. Robust Adaptive Control of an Offshore Ocean Thermal Energy Conversion System. *IEEE Trans. Syst. Man Cybern. Syst.* **2020**, *50*, 5285–5295. [CrossRef]
26. Ikegami, Y.; Yasunaga, T.; Morisaki, T. Ocean thermal energy conversion using double-stage Rankine cycle. *J. Mar. Sci. Eng.* **2018**, *6*, 21. [CrossRef]
27. Antonelli, M.; Baccioli, A.; Francesconi, M.; Desideri, U.; Martorano, L. Operating maps of a rotary engine used as an expander for micro-generation with various working fluids. *Appl. Energy* **2014**, *113*, 742–750. [CrossRef]
28. Ikegami, Y. Performance experiments on ocean thermal energy conversion system using the Uehara cycle. *Bull. Soc. Sea Water Sci. Jpn.* **2006**, *60*, 32–38.
29. Malik, A. Assessment of the potential of renewables for Brunei Darussalam. *Renew. Sustain. Energy Rev.* **2011**, *15*, 427–437. [CrossRef]
30. Jung, J.Y.; Lee, H.S.; Kim, H.J.; Yoo, Y.; Choi, W.Y.; Kwak, H.Y. Thermoeconomic analysis of an ocean thermal energy conversion plant. *Renew. Energy* **2016**, *86*, 1086–1094. [CrossRef]
31. Tobal-Cupul, J.G.; Gardu no-Ruiz, E.P.; Gorr-Pozzi, E.; Olmedo-González, J.; Martínez, E.D.; Rosales, A.; Navarro-Moreno, D.D.; Benítez-Gallardo, J.E.; García-Vega, F.; Wang, M.; et al. An Assessment of the Financial Feasibility of an OTEC Ecopark: A Case Study at Cozumel Island. *Sustainability* **2022**, *14*, 4654. [CrossRef]
32. Martel, L.; Smith, P.; Rizea, S.; Van Ryzin, J.; Morgan, C.; Noland, G.; Pavlosky, R.; Thomas, M.; Halkyard, J. *Ocean Thermal Energy Conversion Life Cycle Cost Assessment, Final Technical Report, 30 May 2012*; Technical Report; Lockheed Martin: Manassas, VA, USA, 2012.
33. Langer, J.; Cahyaningwidi, A.A.; Chalkiadakis, C.; Quist, J.; Hoes, O.; Blok, K. Plant siting and economic potential of ocean thermal energy conversion in Indonesia a novel GIS-based methodology. *Energy* **2021**, *224*, 120121. [CrossRef]
34. Langer, J.; Ferreira, C.I.; Quist, J. Is bigger always better? Designing economically feasible ocean thermal energy conversion systems using spatiotemporal resource data. *Appl. Energy* **2022**, *309*, 118414. [CrossRef]
35. Langer, J.; Quist, J.; Blok, K. Recent progress in the economics of ocean thermal energy conversion: critical review and research agenda. *Renew. Sustain. Energy Rev.* **2020**, *130*, 109960. [CrossRef]
36. Gardu no-Ruiz, E.P.; Silva, R.; Rodríguez-Cueto, Y.; García-Huante, A.; Olmedo-González, J.; Martínez, M.L.; Wojtarowski, A.; Martell-Dubois, R.; Cerdeira-Estrada, S. Criteria for optimal site selection for ocean thermal energy conversion (OTEC) plants in Mexico. *Energies* **2021**, *14*, 2121. [CrossRef]
37. Brecha, R.J.; Schoenenberger, K.; Ashtine, M.; Koon Koon, R. Ocean Thermal Energy Conversion—Flexible Enabling Technology for Variable Renewable Energy Integration in the Caribbean. *Energies* **2021**, *14*, 2192. [CrossRef]
38. Zea, S.; Geister, J.; Garzón-Ferreira, J.; Díaz, J. Biotic changes in the reef complex of San Andres Island (Southeastern Caribbean Sea, Columbia) occuring over three decades. *Atoll Res. Bull.* **1998**, *456*, 1–30. [CrossRef]
39. Baine, M.; Howard, M.; Kerr, S.; Edgar, G.; Toral, V. Coastal and marine resource management in the Galapagos Islands and the Archipelago of San Andres: Issues, problems and opportunities. *Ocean. Coast. Manag.* **2007**, *50*, 148–173. [CrossRef]
40. Fabian, C.L.; Iba nez, J.W.; Prieto, F.S.; Camargo, C.C. Groundwater Sustainability Assessment in Small 2 Islands: The Case Study of San Andres in the 3 Caribbean Sea 4. *Environ. Sci.* **2018**. [CrossRef]
41. Osorio, A.F.; Arias-Gaviria, J.; Devis-Morales, A.; Acevedo, D.; Velasquez, H.I.; Arango-Aramburo, S. Beyond electricity: The potential of ocean thermal energy and ocean technology ecoparks in small tropical islands. *Energy Policy* **2016**, *98*, 713–724. [CrossRef]
42. Payan, L.F.B.; Lopez, D.C. Small-scale steam turbine and electric generation from municipal solid waste. *Int. J. Environ. Waste Manag.* **2018**, *22*, 74–86. [CrossRef]
43. Grueso López, N. *Gestión Predial y Geo-Especialización de Soluciones Hídricas Independientes para el Plan Director del Recurso Hídrico de San Andrés Isla*; Repositorio Institucional Universidad Distrital (RIUD): Bogotá, Colombia, 2018.
44. Devis-Morales, A.; Montoya-Sánchez, R.A.; Osorio, A.F.; Otero-Díaz, L.J. Ocean thermal energy resources in Colombia. *Renew. Energy* **2014**, *66*, 759–769. [CrossRef]
45. Bruch, V.L. *An Assessment of Research and Development Leadership in Ocean Energy Technologies*; Technical Report; Sandia National Labs.: Livermore, CA, USA, 1994.
46. Correa-Ramirez, M.; Rodriguez-Santana, Á.; Ricaurte-Villota, C.; Paramo, J. The Southern Caribbean upwelling system off Colombia: Water masses and mixing processes. *Deep. Sea Res. Part I Oceanogr. Res. Pap.* **2020**, *155*, 103145. [CrossRef]
47. Vega, L.A. Ocean thermal energy conversion primer. *Mar. Technol. Soc. J.* **2002**, *36*, 25–35. [CrossRef]
48. Kim, A.S.; Kim, H.J.; Lee, H.S.; Cha, S. Dual-use open cycle ocean thermal energy conversion (OC-OTEC) using multiple condensers for adjustable power generation and seawater desalination. *Renew. Energy* **2016**, *85*, 344–358. [CrossRef]
49. Habib, W.B.; Hassam, M.; Rupan, T.H. Study On Electrification of Remote and Isolated Tropical Islands Using OTEC. In Proceedings of the International Exchange and Innovation Conference on Engineering & Sciences, Fukuoka, Japan, 14–15 October 2016; Volume 2, p. 1.
50. Bernardoni, C.; Binotti, M.; Giostri, A. Techno-economic analysis of closed OTEC cycles for power generation. *Renew. Energy* **2019**, *132*, 1018–1033. [CrossRef]
51. Bechtel, M.; Netz, E. OTEC-Ocean Thermal Energy Conversion. *Comp. Anal. Power Gener. Ocean.* **2016**, *1*, 14–18.
52. Edwards, D. OTEC Cold Water Retrieval and Desalination Systems. U.S. Patent 9,181,932, 10 November 2015.

53. Visser, E.D.; Held, A.; Jager, D.D. *Methodologies for Estimating Levelised Cost of Electricity (LCOE)*; European Commission: Brussels, Belgium, 2014.
54. Avery, W.H.; Wu, C. *Renewable Energy from the Ocean: A Guide to OTEC*; Oxford University Press: Oxford, UK, 1994.
55. Nihous, G.; Vega, L. A review of some semi-empirical OTEC effluent discharge models. In Proceedings of the Ocean Technologies and Opportunities in the Pacific for the 90's Proceedings (OCEANS'91), Honolulu, HI, USA, 1–3 October 1991; Volume 1, pp. 25–29.
56. Vega, L.A. Ocean thermal energy conversion. In *Renewable Energy Systems*; Springer: Berlin/Heidelberg, Germany, 2013; pp. 1273–1305.
57. Bose, B.K. *Power Electronics in Renewable Energy Systems and Smart Grid: Technology and Applications*; Wiley: New York, NY, USA, 2019.
58. Starzmann, J.r.; Schatz, M.; Casey, M.; Mayer, J.; Sieverding, F. Modelling and validation of wet steam flow in a low pressure steam turbine. In Proceedings of the Turbo Expo: Power for Land, Sea, and Air, Vancouver, BC, Canada, 6–10 June 2011; Volume 54679, pp. 2335–2346.
59. Vega, L.A. Economics of ocean thermal energy conversion (OTEC): An update. In Proceedings of the Offshore Technology Conference, Houston, TX, USA, 3–6 May 2010.

MDPI
St. Alban-Anlage 66
4052 Basel
Switzerland
www.mdpi.com

MDPI Books Editorial Office
E-mail: books@mdpi.com
www.mdpi.com/books

Disclaimer/Publisher's Note: The statements, opinions and data contained in all publications are solely those of the individual author(s) and contributor(s) and not of MDPI and/or the editor(s). MDPI and/or the editor(s) disclaim responsibility for any injury to people or property resulting from any ideas, methods, instructions or products referred to in the content.

www.ingramcontent.com/pod-product-compliance
Lightning Source LLC
Chambersburg PA
CBHW061456220926
51850CB00004B/102
* 9 7 8 3 0 3 6 5 9 1 3 4 6 *